AF327273

Mechanics and Reliability of Flexible Magnetic Media

Bharat Bhushan

Mechanics and Reliability of Flexible Magnetic Media

With 287 Illustrations

Springer-Verlag

New York Berlin Heidelberg London Paris
Tokyo Hong Kong Barcelona Budapest

Bharat Bhushan, Ph.D., D.Sc., P.E.
Ohio Eminent Scholar Professor
Director, Computer Microtribology and
  Contamination Laboratory
Department of Mechanical Engineering
Ohio State University
Columbus, OH 43210
USA

On the cover is the tape path in an IBM 3480/3490 data-processing tape drive.

Library of Congress Cataloging-in-Publication Data
Bhushan, Bharat, 1949–
    Mechanics and Reliability of Flexible Magnetic Media / Bharat
Bhushan.
        p.   cm.
    Includes bibliographical references and index.
    ISBN 0-387-97708-2
    1. Magnetic tapes—Reliability.   2. Magnetic disks—Reliability.
3. Magnetic recorders and recording.   I. Title.
TK5984.B48   1992
621.34—dc20   91-34833

Printed on acid-free paper.

Production coordinated by Brian Howe and managed by Francine Sikorski; manufacturing
supervised by Jacqui Ashri.
Typeset by Asco Trade Typesetting Ltd., Hong Kong.
Printed and bound by Hamilton Printing Company, Rennselaer, NY.
Printed in the United States of America.

9 8 7 6 5 4 3 2 1

ISBN 0-387-97708-2 Springer-Verlag New York Berlin Heidelberg
ISBN 3-540-97708-2 Springer-Verlag Berlin Heidelberg New York

To

my wife (Sudha), my son (Ankur), and my daughter (Noopur)

Preface

Magnetic recording is presently a $80 billion industry. It spans audio, video, and data-processing applications in the form of tapes and disks. Annual sales for tape drives are about 2.5 million units at a cost of about $2.5 billion. According to some estimates, 95% of the information today is stored on paper, 3% on microfiche, and only 2% on magnetic/optical and semiconductor storage devices. Semiconductor storage devices are almost exclusively used for dynamic random-access-memory (D-RAM) in computers and constitute a very small fraction of the total storage capacity. Thus, theoretically, magnetic/ optical storage can grow up to fifty times its current volume. The industry is expected to grow by a factor of five or more in the next decade. The growth will be accompanied by dramatic improvements in technology. Magnetic storage densities as high as ten times (more than one gigabit per square inch) the current densities have been demonstrated in the laboratory by IBM.

The magnetic-recording process is accomplished by relative motion between a magnetic head and a magnetic medium. Types of magnetic media for magnetic recording are flexible media (tapes and flexible disks) and rigid disks. Physical contact between head and medium occurs during starts and stops, and hydrodynamic air-film develops at operating speeds. Flying heights (mean separation between head and media) are of the order of 0.1 μm comparable to surface roughnesses of the mating members. The need for higher and higher recording densities requires that surfaces be as smooth as possible and flying heights be as low as possible. Smoother surfaces lead to increased wear. In the case of magnetic tapes, in order to have high bit density for a given size of spool, we like to use as thin a tape substrate as possible. Thinner tapes are prone to local or bulk viscoelastic deformation during storage. This may lead to variations in head–tape separations resulting in problems in data reliability. Anisotropic mechanical properties and viscoelastic deformations of the flexible disks may result in data-reliability problems. All magnetic media have to be sufficiently lubricated to minimize head and magnetic-medium wear. The lubrication is carried out either topically or in bulk. A fundamental understanding of the tribology, mechanics, and reliability of the head–magnetic-medium interface is, therefore, very crucial for the future of the fast-growing magnetic-recording industry.

The tribology and mechanics of magnetic storage devices have been covered in a separate book published by Springer-Verlag in 1990. The present book is a systematic compilation of the current knowledge of mechanics and the reliability of flexible magnetic media and is the first of its kind. The organization of the book is straightforward. Chapter 1 presents brief descriptions of the physics of magnetic recording, magnetic storage systems, and the manufacturing processes of magnetic media. Chapter 2 presents a brief description of the manufacturing processes of poly(ethylene terephthalate) films commonly used as flexible media substrate, and the physical and chemical properties of the PET films and coated magnetic media. Chapter 3 then presents the viscoelastic properties of the PET films and coated magnetic media. Chapter 4 presents analytical predictions of stresses in wound magnetic tapes and flexible disks. In Chapter 5, long-term reliability problems of magnetic tapes encountered during storage and use are discussed. Their descriptions, mechanisms, and methods of preventing them are presented. In Chapter 6, the long-term reliability problems of flexible disks encountered during storage and use are discussed.

I have tried, wherever possible, to discuss theories and types of experimental measurements that can usefully be made in corroborating theories and/or in developing our understanding. Emphasis has been on the fundamental understanding of the subject matter before proceeding to a diversity of practical applications. I have presented ample experimental data, relevant properties of materials and surfaces, to make this book useful to engineers working in the industry. The book is intended for three types of reader: the graduate student of magnetic recording, the research worker who is active or intends to become active in this field, and the practicing engineer who has encountered a reliability problem and hopes to solve it as expeditiously as possible. Viscoelasticity theories presented in this book are very general and are applicable in other than magnetic-storage systems.

I wish to thank my wife, Sudha, my son, Ankur, and my daughter, Noopur, who have been very forbearing during the years when I spent long days and nights in conducting the research and preparing this book. I would also like to thank many colleagues at the IBM Corporation who contributed to some of the research reported in this book. I have immensely benefited from my association with F.W. Hahn, D. Connolly, J.C. Heinrich, D.J. Plazek (University of Pittsburgh), E.A. Bartkus, D.J. Winarski, B.S. Sharma, R.L. Bradshaw, C.J. Heffelfinger (DuPont), B.S. Berry, T.L. Smith, J.J. Gniewek, S.M. Vogel, W.B. Phillips, M.A. Marchese, and A.A. Gaudet.

Columbus, Ohio   BHARAT BHUSHAN

Contents

Preface vii

1. Introduction 1
  1.1. Physics of Magnetic Recording 1
    1.1.1. Basic Principle 1
      1.1.1.1. Magnetism 1
      1.1.1.2. Electromagnetic Induction 8
      1.1.1.3. Magnetic Recording 8
    1.1.2. Vertical Recording 10
    1.1.3. Signal-Processing Methods 12
    1.1.4. Design Considerations 15
      1.1.4.1. Recording Density 15
      1.1.4.2. Reproduced Signal Amplitude 16
      1.1.4.3. Signal-to-Noise Ratio 18
  1.2. Magnetic Storage Systems 19
    1.2.1. History of Magnetic Recording 19
      1.2.1.1. Storage Hierarchy 21
    1.2.2. Examples of Modern Storage Systems Using Flexible Media ... 22
      1.2.2.1. Tape Drives 22
      1.2.2.2. Flexible-Disk Drives 43
    1.2.3. Head Materials 51
      1.2.3.1. Permalloys 53
      1.2.3.2. Mu-Metal and Hy-Mu 800B 53
      1.2.3.3. Sendust Alloys 53
      1.2.3.4. Alfenol Alloys 53
      1.2.3.5. Amorphous Magnetic Alloys 53
      1.2.3.6. Ferrites 54
      1.2.3.7. Some Examples of Head Constructions 56
    1.2.4. Flexible Media Materials 58
      1.2.4.1. Base Film 59
      1.2.4.2. Magnetic Medium 59
      1.2.4.3. Particulate Magnetic Coatings 63
      1.2.4.4. Magnetic Thin Films 64
    1.2.5. Functional Requirements 67

x   Contents

1.3. Manufacturing Processes of Flexible Magnetic Media 70
    1.3.1. Particulate Media 70
        1.3.1.1. Tapes 70
        1.3.1.2. Flexible Disks 74
    1.3.2. Thin-Film Media 74
        1.3.2.1. Metal-Evaporated Media 76
        1.3.2.2. Sputtered Media 78
        1.3.2.3. Electro/Electroless Plated Media 79
References 79

2. Physical and Chemical Properties of PET Substrate and Coated Magnetic Media 85
  2.1. Manufacturing Process of PET Films 88
  2.2. Structure of PET Films 93
    2.2.1. One-Way Stretching 95
    2.2.2. Two-Way Stretching 97
    2.2.3. Heat Setting (or Crystallization) 99
    2.2.4. Post-Stretching 100
    2.2.5. Strain Relaxation (or Annealing) 100
    2.2.6. Commercial Biaxially-Oriented PET (Mylar A) 100
    2.2.7. Summary 102
  2.3. Physical and Chemical Properties 104
    2.3.1. Density 105
    2.3.2. Refractive Index, Birefringence, and Infrared Dichroism 105
    2.3.3. In-Plane Mechanical Properties 111
        2.3.3.1. Effect of Temperature and Strain Rate 118
        2.3.3.2. Effect of Annealing 119
        2.3.3.3. Effect of Solvents 119
    2.3.4. Elastic Modulus in the Thickness Direction 121
    2.3.5. Radial Elastic Modulus of the Wound Reels 123
        2.3.5.1. Effect of Winding Parameters and Magnetic Coating ... 125
        2.3.5.2. Effect of Storage 125
        2.3.5.3. Radial Relaxation Modulus 126
    2.3.6. Thermal Expansion Properties 128
        2.3.6.1. Heating Rate Effects 131
        2.3.6.2. Effect of Annealing 132
        2.3.6.3. Mechanisms of Thermal Expansion 134
    2.3.7. Hygroscopic Expansion Properties 140
        2.3.7.1. Effect of Annealing 143
    2.3.8. Long-Term Dimensional Stability (Shrinkage) 145
        2.3.8.1. Effect of Annealing 147
    2.3.9. Hydrolytic Stability 147
    2.3.10. Summary 148
  2.4. Outlook for Improved Substrates 149
    2.4.1. Mechanical Properties 150
    2.4.2. Coefficient of Thermal Expansion 154
        2.4.2.1. Isotropic Polymer Films 154
        2.4.2.2. Oriented PET Films 155

      2.4.2.3. Laminates 156
      2.4.2.4. Incorporation of Fibers and Filaments 156
    2.4.3. Coefficient of Hygroscopic Expansion 157
    2.4.4. Long-Term Dimensional Stability (Shrinkage) 158
    2.4.5. High-Temperature Polyimide Substrate 158
    2.4.6. Summary 160
  References 160

3. Viscoelastic Properties of PET Substrate and Coated Magnetic Media 164
  3.1. Introduction to Viscoelasticity 164
    3.1.1. Elasticity 166
      3.1.1.1. Generalized Hooke's Law 167
      3.1.1.2. Material Constants for Orthotropic Material 174
      3.1.1.3. Material Constants for Isotropic Material 178
    3.1.2. Viscous Liquids 181
    3.1.3. Viscoelasticity 181
      3.1.3.1. Constitutive Equation 182
      3.1.3.2. Description of Time-Dependent Deformation Experiments 185
      3.1.3.3. Mechanical Model Analogies of Linear Viscoelastic Behavior 195
      3.1.3.4. Time (or Frequency) and Temperature Effects 202
  3.2. Dynamic Modulus Data 204
    3.2.1. Measurement Techniques 204
      3.2.1.1. DMTA 204
      3.2.1.2. DMA 206
    3.2.2. Experimental Results 207
  3.3. Tensile Relaxation and Creep Data 209
    3.3.1. Descriptions of Creep and Relaxation Test Apparatuses 212
      3.3.1.1. Creep 212
      3.3.1.2. Relaxation 218
    3.3.2. Constitutive Relations for Analysis of Isothermal Experimental Data 219
    3.3.3. Experimental Results 224
      3.3.3.1. Linearity of PET Material Response 224
      3.3.3.2. Effect of PET Film Thickness 225
      3.3.3.3. Effects of Temperature and Humidity in PET Films ... 227
      3.3.3.4. Thermoviscoelastic Behavior of PET Films 246
      3.3.3.5. Viscoelastic Behavior of Coated Tapes and Magnetic Coatings 250
      3.3.3.6. Effects of Thermal Treatment of PET Films 260
  3.4. Compressive Creep Data 276
    3.4.1. Description of Creep Test Apparatus 276
    3.4.2. Experimental Results 278
      3.4.2.1. PET Films and Cast Films of Magnetic Coatings 278
      3.4.2.2. Calendered Versus Uncalendered Magnetic Coatings .. 278
      3.4.2.3. Recovery Experiments 279
      3.4.2.4. Summary 282

References 282
Appendix 3.A. Analysis of Flexural and Tensile Stress Relaxation of
     a Multilayered Tape 287
     3.A.1. Flexural Relaxation at Constant Curvature 287
     3.A.2. Tensile Relaxation at Constant Elongation 289
Appendix 3.B. Analysis of Thermal Curling of a Multilayered Magnetic
     Tape in the Elastic Regime 291

4. Stress Analysis of Flexible Media 295
  4.1. Wound Magnetic Tape Reels 295
    4.1.1. Initial Stress Field 296
      4.1.1.1. Analytical Techniques 298
      4.1.1.2. Measurement of Radial Stresses in a Wound Tape Reel . 305
      4.1.1.3. Role of Reel Geometry and Winding Parameters 305
      4.1.1.4. Environmental Stresses 310
      4.1.1.5. Summary 314
    4.1.2. Stress Relaxation 314
      4.1.2.1. Constitutive Relationships 315
      4.1.2.2. Relaxation Matrix from Experimental Measurements .. 317
      4.1.2.3. Axisymmetric Finite-Element Model 318
      4.1.2.4. Numerical Results 323
      4.1.2.5. Summary 328
  4.2. Flexible Disks 331
    4.2.1. Elasticity Solution 332
      4.2.1.1. Orthotropic Solid Disk 333
      4.2.1.2. Annular Disk: Fixed Inner Boundary 335
      4.2.1.3. Approximate Scheme for Annular Disk: $a/b \ll 1$ 339
      4.2.1.4. Estimates Based on Incomplete Knowledge of
      Elastic Compliances 341
    4.2.2. Viscoelasticity Solution 345
      4.2.2.1. Solution 345
      4.2.2.2. Estimates for Large Times 347
    4.2.3. Summary 349
  References 350

5. Long-Term Reliability of Magnetic Tapes 353
  5.1. Interlayer Slip (ILS) 355
    5.1.1. Various Forms of Tape Distortions 357
      5.1.1.1. Cinching 357
      5.1.1.2. Spoking 357
      5.1.1.3. Windowing 357
      5.1.1.4. Telescoping 357
    5.1.2. Model for Reel Slippage 358
    5.1.3. Methods to Contain ILS 358
    5.1.4. Estimate of Periods Between Rewinds 359
      5.1.4.1. Environment for Archival Storage 359
  5.2. Instantaneous Speed Variations (ISV) 360
    5.2.1. Loose Wraps 362
    5.2.2. Tape Vibrations 363

5.3. Uneven Tape-Stack Profile (Hardband) 365
   5.3.1. Sources and Methods of Preventing Distortions 367
      5.3.1.1. Substrate and Coating-Thickness Variations 368
      5.3.1.2. Entrapped Air Pockets, Tension Ridges, and Scratches . 368
   5.3.2. How Does an Uneven Tape Stack Affect Data Reliability? 382
   5.3.3. Summary 390
5.4. Mechanical Print-Through 390
5.5. Staggered Wraps 393
   5.5.1. Sources of Tape Stagger 397
   5.5.2. Effect of Winding Parameters and Storage Conditions 398
      5.5.2.1. Analysis of Elastic Droop of a Staggered Wrap 398
      5.5.2.2. Experimental Study 403
   5.5.3. How Does Staggered Tape Cause Errors? 406
   5.5.4. Methods of Preventing Tape Stagger 411
   5.5.5. Summary 416
5.6. Design of Tape Reels 417
   5.6.1. Hub 417
   5.6.2. Flanges 422
   5.6.3. Reel Materials 423
References 423
Appendix 5.A. Tension-to-Flatten Analysis for a Tape Reel with
           a Circumferential Bump 425
      5.A.1. Tape Substrate 427
      5.A.2. Composite Tape 428
Appendix 5.B. Tension-Gradient Measurement Technique 429
Appendix 5.C. Instantaneous Failure Rate Model to Assess Failures
           Due to Viscoelastic Deformation-Related Defects 431
      5.C.1. Effect of "Burn-In" 435

6. Long-Term Reliability of Flexible Disks 438
6.1. Analysis of Disk Deformation 439
   6.1.1. Thermal Expansion 439
   6.1.2. Hygroscopic Expansion 439
   6.1.3. Shrinkage 440
   6.1.4. Centrifugal Stresses and Displacements 440
   6.1.5. Summary 441
6.2. Measurements of Disk Deformation 441
   6.2.1. Description of Disk Deformation Measurement Apparatuses .. 442
      6.2.1.1. Stroboscopic-Disk Deformation Measurement
            Apparatus 442
      6.2.1.2. Scanning-Laser-Disk Deformation Measurement
            Apparatus 443
   6.2.2. Experimental Results 445
      6.2.2.1. Thermal and Hygroscopic Deformations 446
      6.2.2.2. Shrinkage 448
      6.2.2.3. Creep 449
      6.2.2.4. Summary 450
References 450

Appendix A. Requirements and Supporting Test Methods for
Magnetic Tapes and Tape Reels 452

Appendix B. Analysis of Life Data 476

Author Index 547
Subject Index 555

CHAPTER 1

Introduction

In this chapter, we present an overview of the physics of magnetic recording. This emphasizes the need for closer and constant head-to-medium spacing during the read/write operation. We then describe various magnetic storage systems which use flexible media, namely, tapes and flexible disks (floppy disks or diskettes). Examples of various audio, video, and data (computer) storage systems and materials used in the construction of the magnetic head and medium are given. Finally, we describe the manufacturing processes used to produce magnetic media. The primary purpose is to provide the reader, who is relatively new to magnetic recording, with the background required to assimilate the chapters that follow. For additional reviews and general reading in magnetic recording, one should consult books on magnetic recording such as: Haynes (1957); Hoagland (1963); Chikazumi and Charap (1964); Mee (1964); Athey (1966); Lowman (1972); Pear (1967); Kalil and Buschman (1985); White (1985); Mallinson (1987); Camras (1988); Jorgensen (1988); and Mee and Daniel (1990).

1.1. Physics of Magnetic Recording

1.1.1. Basic Principle

Magnetic recording technology is founded on magnetism and on electromagnetic induction.

1.1.1.1. Magnetism

It is well known that the currents in electric wires produce a magnetic field. Consider an infinitely long, straight conductor carrying an electric current of I amperes as shown in Fig. 1.1(a). The current induces a magnetic field which circles the conductor. The magnetic field is everywhere tangential, that is, normal to the radius. The direction of the magnetic field is that given by the

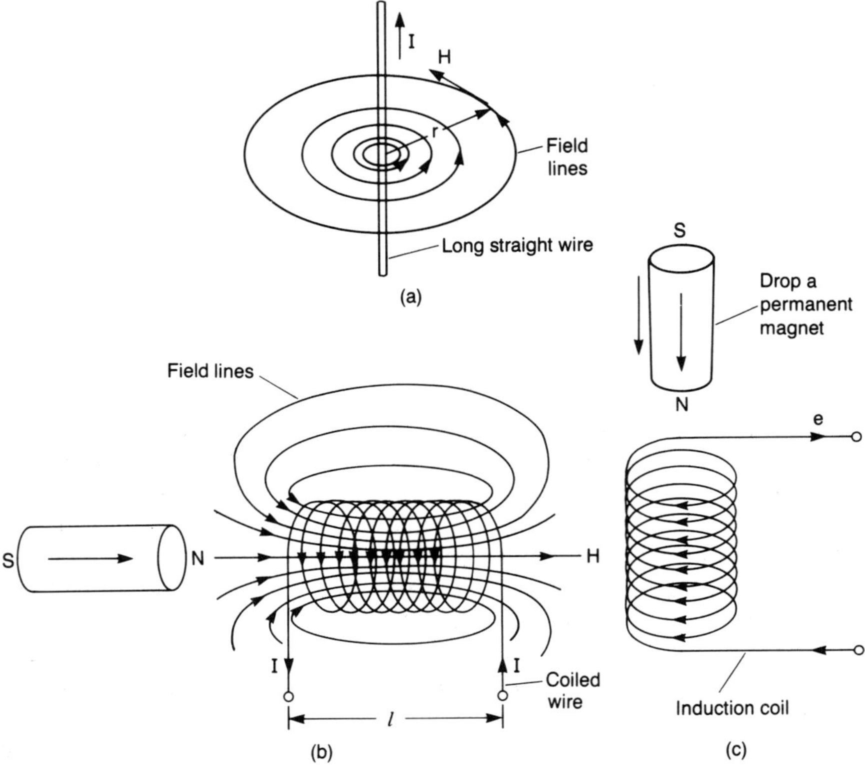

Fig. 1.1. (a) Magnetic field lines from a current in a long straight wire, (b) magnetic field lines from a current in a coiled wire, and (c) electric current in a coiled wire generated by dropping a permanent magnet through a coil.

right-hand rule; point the thumb of the right hand in the direction of current flow, and the magnetic field direction is given by the way the curved fingers point. The intensity of the magnetic field is given by (in SI units)

$$H = \frac{I}{2\pi r},\tag{1.1}$$

where H is the field intensity in A[1]/m (or Oe[1]; 1 A/m = $4\pi \times 10^{-3}$ Oe), I is the electric current in amperes, and r is the radial distance in meters. Now suppose that the conductor is coiled to form the solenoid shown in Fig. 1.1(b). It carries an electric current of I amperes which induces a magnetic field. The

---

[1] A—Ampere, Oe—Oersted, G—Gauss, Wb—Weber (=1 volt.s), Henry (Hy) = Wb/A.

field lines run through the length of the solenoid and close to themselves. The field intensity inside a long solenoid is uniform (parallel and of the same magnitude) and is (in SI units)

$$H = \frac{NI}{l},$$ (1.2)

where H is the magnetic field in A/m, N is the number of turns, I is the current in amperes, and l is the coil length in meters. Note that the field does not depend upon the cross-sectional area of a long coil.

If we now insert a piece of magnetic material in the coil, the material would be magnetized. The magnetic moment per unit volume (1 m^3) of a magnetic substance is called the intensity of magnetization M. The M vector points from the S pole to the N pole, which would appear if the portion in which the magnetization is being specified, were isolated from the rest of the specimen. The right-hand rule shows the direction of the field lines or where the N pole is located; take the solenoid or electromagnet in the right-hand while wrapping the fingers in the direction of the current, and the thumb will then point toward the N pole. The unit of M is the Wb^1/m^2 (tesla or G^1; 1 Wb/m^2 = 1 tesla; 1 Wb/m^2 = $(1/4\pi) \times 10^4$ G). The field lines associated with H are called flux lines; a stronger field has more flux lines running through a given area. The magnetic induction or magnetic flux density B is also commonly used in engineering applications to describe the magnetization. It is expressed in Wb/m^2 (or tesla or G). The relationship between B and M is given as (in SI units) (Chikazumi and Charap, 1964)

$$B = \mu_0 H + M,$$ (1.3a)

where μ_0 is the permeability of vacuum, the value of which is

$$\mu_0 = 4\pi \times 10^{-7} \text{ Wb/Am } (= \text{Hy}^1/\text{m}).$$ (1.3b)

$\mu_0 H$ is the B field in vacuum and M is from the material's spontaneous magnetization. We note that the right-hand term in Eq. (1.3a) is $(H + 4\pi M)$ in the cgs system, so that the conversion factor between SI and cgs units is different for B and M (1 Wb/m^2 of $B = 10^4$ G in cgs units).

The relation between the intensity of magnetization M and the magnetic field H can be expressed by

$$M = \chi H,$$ (1.4)

where χ is the magnetic susceptibility. The unit of χ is the Wb/Am (or Hy/m; 1 Wb/Am = 1 Hy/m) which is the same as that of μ_0; hence it is possible to measure χ in units of μ_0. The susceptibility thus measured is called a relative susceptibility χ_r, which is

$$\chi_r = \chi/\mu_0.$$ (1.5)

χ_r is a dimensionless quantity, and its value is 4π times larger than the susceptibility χ_r measured in the cgs system. Substituting for M [of Eq. (1.3a)] in the expression (1.4), we have

$$B = (\chi + \mu_0)H = \mu H, \tag{1.6}$$

where μ is the permeability, the unit of which is also the Wb/Am (Hy/m).

We define a relative permeability μ_r,

$$\mu_r = \mu/\mu_0 = \chi_r + 1. \tag{1.7}$$

The value of μ_r is the same as that of the permeability measured in the cgs system. The μ_r is a material property. When a magnetic material is placed in a magnetic field H, then the number of flux lines inside the material is increased by its relative permeability. The observed value of the relative susceptibility ranges from 10^{-5} for very weak magnetism to 10^6 for very strong magnetism. μ_r for vacuum is equal to 1.

Magnetization Curve

A magnetization curve for a material which exhibits spontaneous magnetization is shown in Fig. 1.2(a). Starting from a demagnetized state ($H = M = 0$) the magnetization increases with an increase in the field along the curve OABC, and finally reaches the saturation magnetization which is normally denoted by M_s. In the region OA the process of magnetization is reversible, that is, the magnetization comes back to zero upon removal of the field. The inclination of the curve OA (at $H = 0$) is called the initial susceptibility χ_0. Beyond this region, the processes of magnetization are no longer reversible. If the field is decreased from its initial value at point B, the magnetization comes back, not along BAO, but along the minor loop BB'. The inclination of BB' is called the reversible susceptibility χ_{rev} or the incremental susceptibility. The slope of each portion of the initial magnetization curve OABC is called the differential susceptibility χ_{diff}, and the slope of the line which connects the origin O and each point on the initial magnetization curve

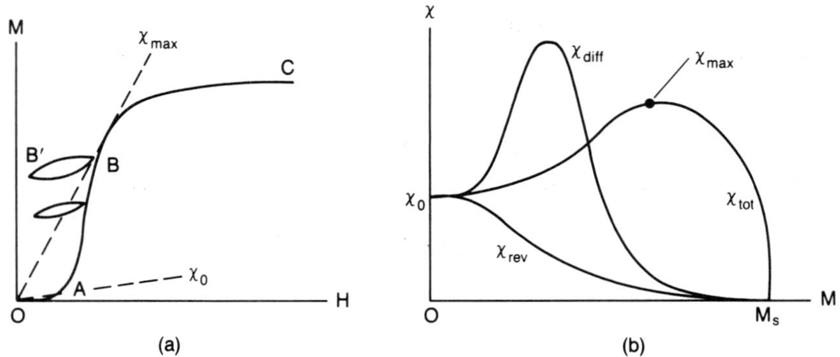

Fig. 1.2. (a) Initial magnetization curve and minor loops, and (b) various kinds of magnetic susceptibilities as a function of the intensity of magnetization.

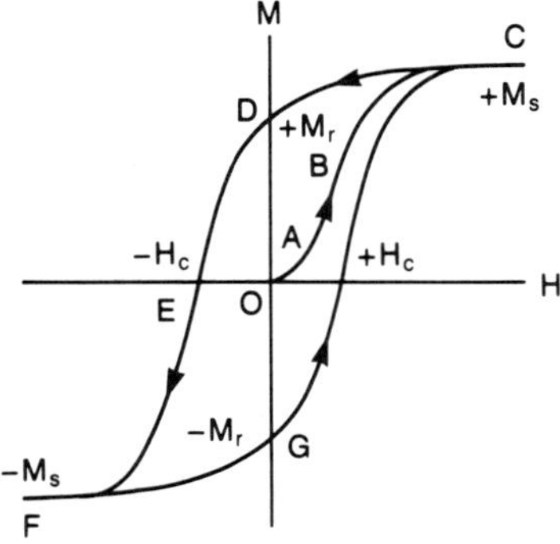

Fig. 1.3. $M-H$ hysteresis loop.

OABC is called the total susceptibility χ_{tot}. The maximum value of the total susceptibility, that is, the slope of the tangent line drawn from the origin to the initial magnetization curve, is called the maximum susceptibility χ_{max}; it is a good measure of the average inclination of the initial magnetization curve. Changes in χ_{rev}, χ_{diff}, and χ_{tot} along the initial magnetization curve are shown in Fig. 1.2(b). Starting from the value of χ_0, χ_{rev} decreases monotonically, while χ_{diff} has a sharp maximum, and χ_{tot} goes through its maximum value χ_{max} and drops off at $M = M_s$. The difference between χ_{diff} and χ_{rev} represents the susceptibility due to irreversible magnetization; it is called the irreversible susceptibility χ_{irr}, that is,

$$\chi_{diff} = \chi_{rev} + \chi_{irr}. \tag{1.8}$$

If the magnetic field is decreased from the saturated state C (Fig. 1.3), the magnetization M is gradually decreased along CD, not along CBAO, and at $H = 0$ it reaches the finite value M_r ($=$OD), which is called the residual magnetization or remanent magnetization or the remanence. A further increase in the magnetic field in a negative sense results in a continued decrease in the intensity of magnetization, which finally falls to zero. The field at this point is called the coercive force or coercivity H_c ($=$OE). This portion, DE, of the magnetization curve is often referred to as a demagnetizing curve. A further increase in H in a negative sense results in an increase in the intensity of magnetization in a negative sense and finally leads to a negative saturation magnetization. If the field is then reversed to the positive sense, the magnetization will change along FGC. The closed loop CDEFGC is called the *hysteresis loop* (Chikazumi and Charap, 1964).

The true magnetization loop ($M-H$ curve) for the material alone is obtained by subtracting $B = \mu_0 H$ ($B-H$ loop for vacuum) from the $B-H$ loop [Eq. (1.6)] normally measured, e.g., using a toroidal geometry. The resulting $M-H$ loop from a $B-H$ loop is shown in Fig. 1.4. We note that the $M-H$ loop flattens at a lower value of H compared to the $B-H$ loop. There are three basic types of equipment used to characterize the magnetic properties of the

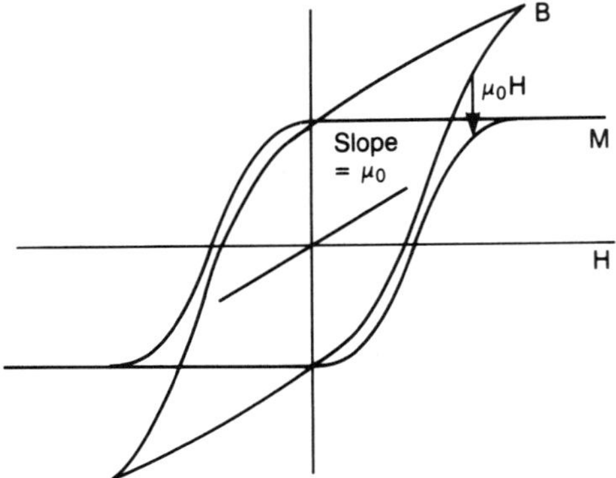

Fig. 1.4. Construction of the $M-H$ loop from the $B-H$ loop; subtract $\mu_0 H$.

materials: the 60-Hz $M-H$ looper, the toroidal $B-H$ looper, and the vibrating sample magnetometer (VSM) which measures $M-H$ and other loops (Mallinson, 1987; Jorgensen, 1988).

Now we discuss the work necessary to magnetize a material. Suppose that the magnetization is increased from M to $M + \delta M$ under the action of a magnetic field H which is parallel to M. The work necessary to magnetize a unit volume of the magnetic material from M_1 to M_2 is expressed by

$$W = \int_{M_1}^{M_2} H \, dM. \tag{1.9a}$$

This is equal to the area surrounded on the ordinate axis of Fig. 1.2(a). The energy supplied by this work is partially stored as potential energy, and also partly dissipated as heat which is generated in the substance. During one cycle of the hysteresis loop, the potential energy should return to its original value, so that the resultant work must be consumed as heat. This heat is called the hysteresis loss and is given by

$$W_h = \oint H \, dM, \tag{1.9b}$$

which is equal to the area surrounded by the hysteresis loop.

Substances which are magnetized, more or less, by a magnetic field are called magnetic substances. Electromagnets are those substances where magnetism is created by currents in electric wires. For engineering applications, magnetic substances can be classified into soft and hard magnetic materials, Fig. 1.5 (Cullity, 1972). Soft magnetic materials generally exhibit high permeability (easily magnetized), low remanence, low coercivity ($\lesssim 1$ Oe), and small hysteresis loss, yet of high saturation magnetization. The slope of the $M-H$ curve (or the susceptibility χ) is fairly constant until saturation occurs. On the other hand, hard magnetic materials are used as permanent magnets composed of a ferromagnetic or ferrimagnetic material in which the material holds its magnetization after it has been moved away from the magnetic field.

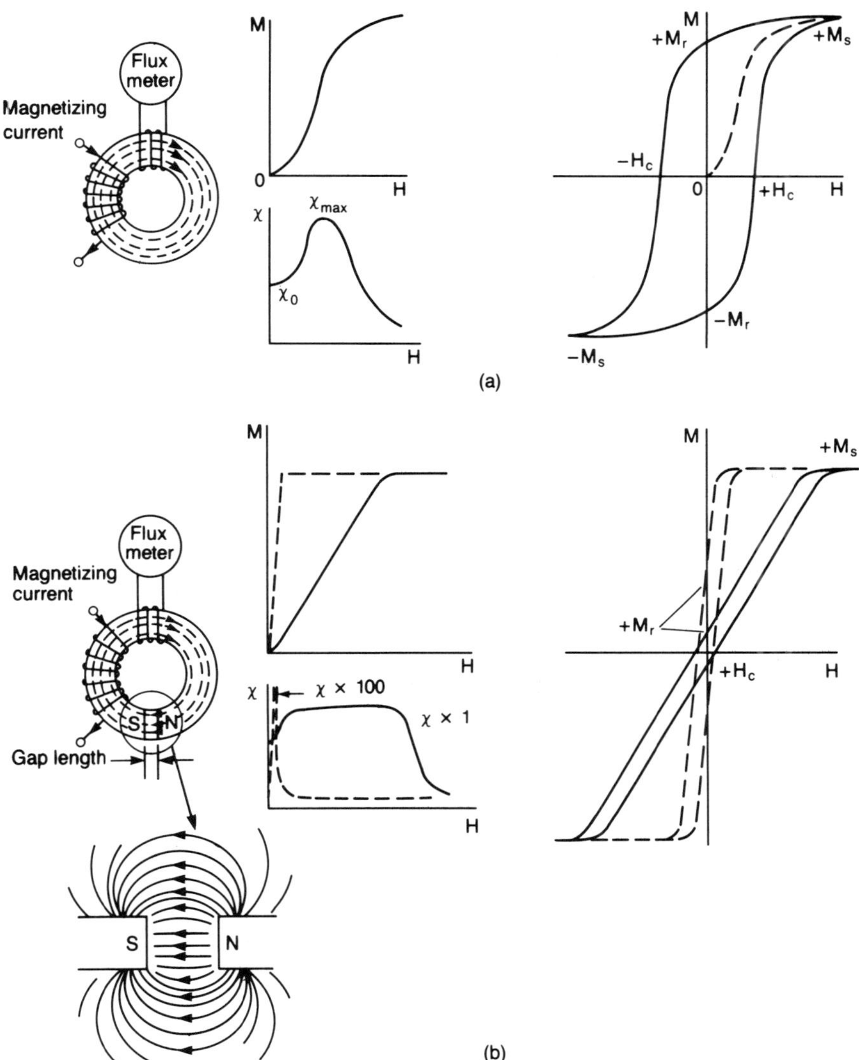

Fig. 1.5. Initial magnetization curve, magnetic susceptibility, and hysteresis loop of (a) a hard magnetic material and (b) a soft magnetic material with an air gap (broken lines for material without an air gap). (Toroidal geometry measures $B-H$ loop, $\mu_0 H$ has been subtracted from the $B-H$ loop to obtain the $M-H$ loop, see Fig. 1.4.)

These generally exhibit high coercivity ($\gtrsim 300$ Oe), high remanence, and large hysteresis losses. Magnetic heads are made of soft magnetic materials such as Permalloy (Ni–Fe), Mu-metal, Sendust, Alfenol, ferrites, and thin-films of nickel–iron composites. The head material should exhibit high M_s because the transducer must provide an intense magnetic field at the surface of the medium to write the transition. The hard magnetic materials are found in the powders used for media coatings: gamma–ferric oxide, cobalt–iron oxide, chromium dioxide, barium ferrite, and iron. Also used are deposited films composed of nickel, cobalt, chromium, etc.

1.1.1.2. Electromagnetic Induction

A permanent magnet will generate an electric voltage when it is dropped through a coil (bringing about a change in the number of flux lines), Fig. 1.1(c). The magnitude of the generated voltage is

$$e(t) = -N\frac{d\phi}{dt}, \tag{1.10}$$

where $e(t)$ is in volts, N is the number of turns in the coil, and ϕ is the number of flux lines in Webers. (The unit Weber is equal to volts.s.) Thus $d\phi/dt$ is the rate of change of magnetization or flux.

1.1.1.3. Magnetic Recording

Magnetic recording and playback are dependent on the phenomenon of magnetic hysteresis. Most magnetic materials exhibit this phenomenon. The hysteresis behavior for a magnetic material is represented in Fig. 1.3, where M is the intensity of magnetization induced in the material (an intrinsic material property) in the presence of a field of intensity H. The intensity of magnetization M is expressed as magnetic moment per unit volume (Wb/m^2 or G) or, more conveniently in the case of porous material or powders, as magnetic moment per unit mass or called as specific magnetization σ (emu/g; 1 emu $=$ $4\pi \times 10^{-10}$ Wb m or 1 emu/cc $= 1$ G). Important attributes of hysteresis behavior are the remanent magnetization or remanence (M_r), the magnetization remaining after saturation when H is reduced to zero, and the coercivity $-H_c$ (A/m or Oe), the field required to reduce magnetization to zero.

Figure 1.5(a) shows an example of the M–H loop for the hard magnetic material (typical for magnetic medium) in which the material holds its magnetization after it has been moved away from the magnetic field. Figure 1.5(b) shows the M–H loop for a soft magnetic material used for head core. Coercivity of a soft magnetic material is very low and its permeability is very high. The ring core is now provided with a narrow air gap that we find in read, write, and erase heads. The introduction of a small air gap results in the formation of N and S free poles on the two gap surfaces, and it drastically reduces the M_r to a very small value, lower permeability but unchanged coercivity; if not reduced, M_r causes serious noise and distortion problems in recording. (It is good practice to demagnetize a recorder head periodically to remove any remanence.) Note that only a small field is required to saturate the ring core without an air gap. This clearly shows the material's ability to produce a large flux density for small fields, which is extremely desirable in magnetic heads to write the transitions.

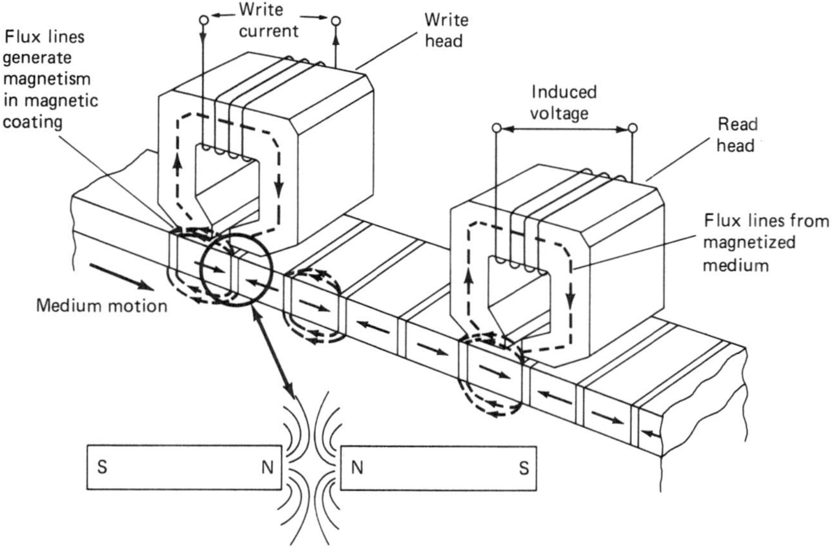

Fig. 1.6. Principle of horizontal magnetic recording and playback. Arrows indicate the direction of magnetization.

We have seen that a current can generate a magnetic field that can magnetize a hard magnetic material permanently. This permanent magnet can, in turn, generate an electric voltage: if it is dropped through a coil, it will generate a voltage pulse. We have just described the basic principle of magnetic recording (writing) and playback (reading), and this is shown schematically in Fig. 1.6, which consists of the relative motion between a magnetic medium and a read/write ring head (Jorgensen, 1988; Mee and Daniel, 1990). Read and write heads (inductive type) consist of a ring of high-permeability magnetic material with an electrical winding and a gap in the magnetic material at or near the surface of the storage medium. Writing is accomplished by passing a current through the coil. The flux is confined to the magnetic core, except in the region of the small nonmagnetic gap. The fringe field in the vicinity of the gap, when sufficiently strong, magnetizes the medium, moving past the write head. The magnetic medium consists of high-coercivity magnetic material that retains its magnetization after it has passed through the field from the write head gap. The medium passes over the read head, which, like the write head, is a ring core with an air gap. Each particle in the medium is a miniature magnet, and its flux lines will join up with those of the other particles to provide an external medium flux, proportional in magnitude to the medium magnetization. The flux lines from the medium permeate the core and induce a voltage in the head winding. This voltage, after suitable amplification, reproduces the original signal. A single head can be used for both read and write functions.

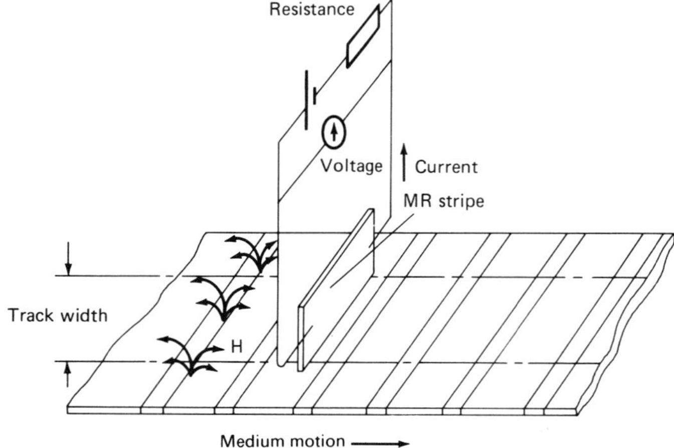

Fig. 1.7. Principle of operation of a magnetoresistive (MR) type read head.

More recently, some read heads are of the magnetoresistive (MR) type in which a strip of a ferromagnetic alloy (e.g., $Ni_{80}Fe_{20}$) is mounted vertically. The variation of the magnetic-field component in the magnetic medium (perpendicular to the plane of medium, H) causes the variation in the electrical resistance of the MR stripe (Fig. 1.7) which can be readily measured (Potter, 1974; Shelledy and Brock, 1975; Van Gestel et al., 1977; Tsang, 1984; Cannon et al., 1986; Tsang et al., 1990). MR-type read heads are attractive because they can be miniaturized and exhibit high signal amplitude, high signal-to-noise ratio, high resolution, and low distortion. This is particularly of interest for systems where more than one track has to be read out. Furthermore, as a flux (ϕ) sensing device (the inductive heads are a $d\phi/dt$ sensor), the MR outputs are independent of sensor-medium velocity. This feature renders the MR elements especially attractive in low-velocity recording environments such as tape drives, where inductive head outputs might be inadequate. One disadvantage of the MR-type head over the inductive-type head is that the pole tip of a MR head is of the order of 30 nm (compared to several microns for the inductive head) so it does not have enough magnetic flux to be able to write effectively, therefore, the MR head is only used as a *read* head and both read and write functions cannot be combined in one head.

1.1.2. Vertical Recording

So far, we have described longitudinal (horizontal) recording. In 1977, perpendicular (vertical) recording was proposed, for ultrahigh density magnetic recording, by Iwasaki and Nakamura (1977), see Fig. 1.8. In vertical recording, magnetization is oriented perpendicular to the plane of medium rather than in its plane. Single-pole heads (Fig. 1.8) are generally used for recording

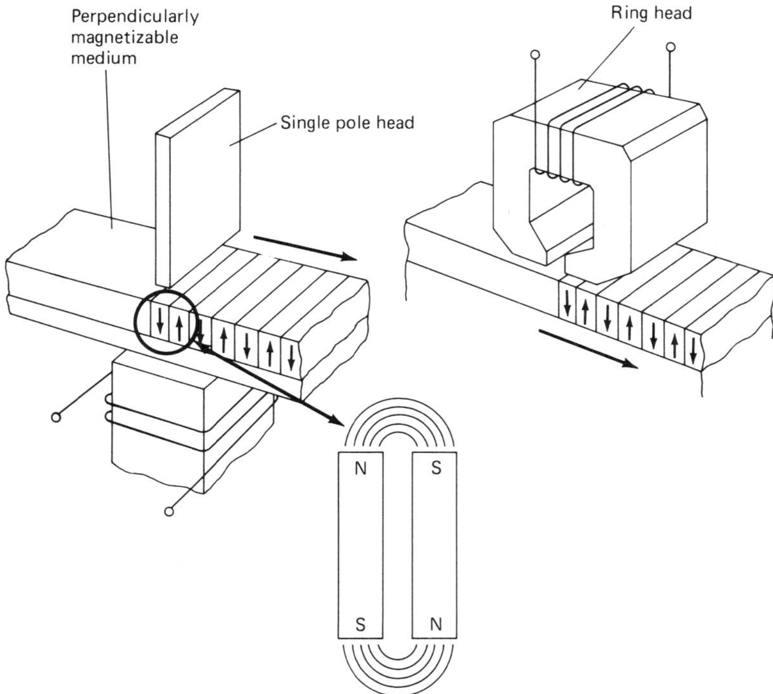

Fig. 1.8. Principle of vertical recording using either a single-pole head or ring head. Demagnetization is reduced because the flux lines of neighboring opposing domains reinforce rather than conflict with one another (as in horizontal recording).

on vertical media. Ring-shaped heads, previously discussed, have also been used with some design changes (Iwasaki, 1984; Yamamori et al., 1983).

Vertical recording has the advantage of reduced self-demagnetization. In horizontal recording, when a transition is made from one magnetic orientation to another, the N or S poles of the neighboring cells are adjacent. Just as the poles of two magnets laid end to end in this manner weaken each other, so the cells tend to demagnetize each other, and the tendency gets worse with smaller cells (Fig. 1.6). In vertical recording, however, the opposing poles of the two cells lie next to each other, so the magnetic fields reinforce each other (Fig. 1.8). This produces an extremely sharp cell boundary and thus the *possibility of smaller cells.*

Magnetic media for vertical recording should have only vertical anisotropy and no intrinsic in-plane orientation and, therefore, should allow themselves to be vertically oriented by an external field. Several alternatives for magnetic media for vertical recording are available. One such anisotropic medium is based on a particulate γ-Fe_2O_3 and is used in tapes marketed for instrumentation recorders. A more widely pursued approach is to develop a medium that can be magnetized only vertically (Ouchi, 1990), such as particulate media based on barium–ferrite particles (Fujiwara, 1985) and thin-film media. Iwasaki (1984) has shown that when a cobalt–chromium alloy is sputtered onto a substrate, a vertical magnetic anisotropy can be induced. Electroplated cobalt has also been deposited in the form of needlelike crystals, with both the axis of the particle and the hexagonal crystal axis normal to the film (Chen and Cavallotti, 1982).

1.1.3. Signal-Processing Methods

A magnetic recording system stores information and then permits accurate retrieval of the information at a later date (sometimes years after it is re-corded). The data integrity is of paramount importance in computer applications. The illustration for the transmission of a signal from source to destination is shown in Fig. 1.9. It shows the application of analog-to-digital conversion, which is exclusively used for computer data and high definition audio and video signal processing. The fundamental advantage of the digital signal over an analog signal is that the digital form can be processed, transmitted, and stored almost without error with the help of error-correction codes. Small perturbations of digital signals caused by the transmission channel can be recognized and corrected, whereas analog signals are left at their perturbed values, with no possibility of correction. The superiority of digital signals is that they can be regenerated.

In a typical magnetic-recording channel, the incoming data are first encoded with an error-correction code with a well-defined data format. The data format is designed to fit the requirements imposed by the device architecture, and the error-correction code is designed to protect the data against commonly encountered error modes. A second encoder then converts data

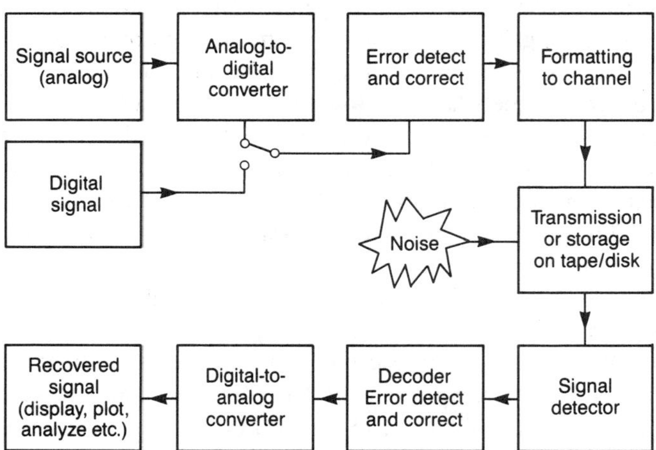

Fig. 1.9. Block diagram of a communication channel.

into an analog signal in accordance with a waveform encoding technique or modulation code.

Computers, in general, use the binary code. The existence of two stable magnetization states M_r and $-H_c$ [in Fig. 1.5(a)] is particularly well suited to the storage of binary information. In a binary code, all characters are represented by combinations of 1's and 0's. These 1's and 0's are called binary digits, abbreviated "bits." (Each letter in the alphabet is represented by a byte that consists of eight bits, where the last bit is an added "0," and the remaining seven are the ASCII code.) Magnetic flux reversals on a storage medium, in either one direction (S → N) or the opposite (N ← S), lend themselves readily to the representation of bits. The direction of magnetization of the medium depends on the direction of current flow in the coil. A pattern of reversals in magnetization can thus be produced in the medium by a pattern of reversals in the direction of current. Reading is accomplished by sensing magnetization reversals in the medium by the voltage induced in the winding of the read head.

Various methods of encoding binary information have been used, e.g., nonreturn-to-zero (NRZ), nonreturn-to-zero-inverted (NRZI), phase encoding (PE), frequency modulation (FM), modified (delay) frequency modulation (MFM), (2, 7) RLL code, and (1, 7) RLL code (Mee and Daniel, 1990). One of the common methods of writing is NRZI. One (1) is represented by a change in the recorded magnetization direction and zero (0) by the absence of change in magnetization in the medium. Thus, a 1 and a 0 of the binary data are represented by a presence and an absence of a transition in the binary waveform, respectively. We define the principles of NRZI encoding by an example. The logic and timing circuits turn on and off the write amplifiers. The magnetic medium is moved under the write head with a constant velocity V. The timing clock is running with a spacing between pulses of time interval Δt (bit length $= V \Delta t$), Fig. 1.10(a). The logic is set as shown in Fig. 1.10(b). The current waveform in the recording head for this logic is as shown in Fig. 1.10(c). The current in the positive direction is defined to give a positive direction to the magnetization (to the right) and vice versa. The magnetization direction and magnetization magnitude in the medium produced by the recording-head current are shown in Figs. 1.10(d) and (e), respectively. A simple reading system resembles the writing system. Any discontinuity in magnetization produces a fringe field. When the fringe field is moved past the read head, a change in flux per unit time induces a voltage in the coil; the polarity of the voltage is reversed for every NRZI 1, Fig. 1.10(f). The signal in the head is amplified, and at each clock pulse the system examines the output of the head to determine whether a signal exists or not. If no signal can be found, the bit is a 0; if a signal exists, it is a 1. The recovered code is shown in Fig. 1.10(h).

The read head output may further be processed by integration, Fig. 1.10(g). The signal degradation by passage through a magnetic drive signal channel can be seen by comparing Figs. 1.10(e) and (g).

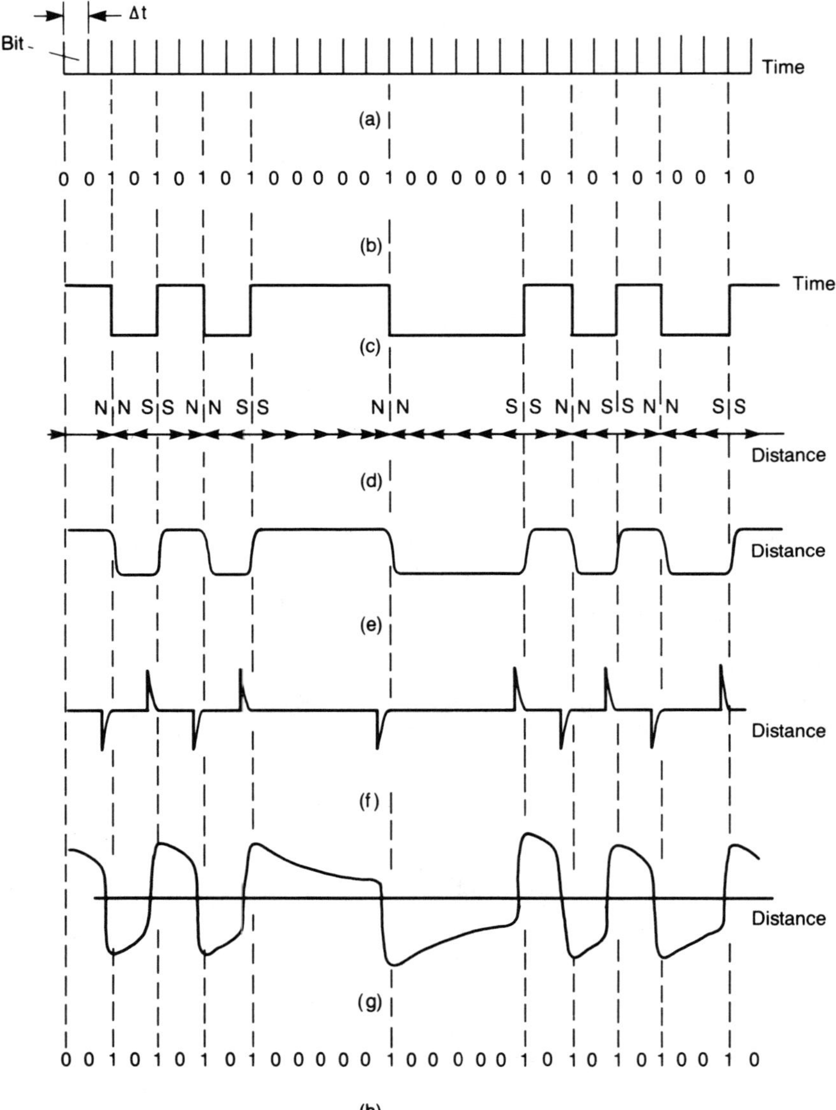

Fig. 1.10. Demonstration of the principle of NRZI encoding (a) clock for timing (bit length = $V\,\Delta t$), (b) binary code to be written using NRZI code, (c) current waveform in the write head, (d) magnetization direction in the medium, (e) magnetization magnitude in the medium, (f) voltage induced in the read head, (g) integrated read voltage, and (h) recovered code.

1.1.4. Design Considerations

Requirements for high-quality systems are virtually error-free recording and faithful reproduction of the input signal of the recorder (number of missing data bits to be a minimum; e.g., one in a million or even in a trillion). It is extremely desirable to have the highest areal recording density [track density (or number of tracks per unit width) times the linear density (or number of bits per unit length)] and fastest data transfer rate (maximum number of data bits that can be read per unit time).

1.1.4.1. Recording Density

Magnetic tapes use a multitrack head that reads and writes on parallel tracks on the tape with some blank space in between. A track is written wider than what is read. Flexible disks use circular tracks with some blank space in between. For the highest track densities, servo techniques are used to locate tracks during recording and playback in the disks. The three most commonly used servo techniques are sector, profile, and embedded (or buried). In sector servo, a small sector of the disk contains servo information, and in profile servo, few tracks (either at the outer or inner diameter) consist of servo information. Disadvantages of sector servo are that these tie up part of usable disk space. In an embedded servo system read/write heads alternately read data and servo sectors embedded within each track. In a multiple-frequency embedded or buried servo, the embedded servo information occupies the same position as data tracks without reducing track capacity. A servo track is written on the media at a frequency much lower than the data recording frequency, then filtering is used to separate the readback signal into a data component and a servo tracking component. Flexible disks normally use profile, sector, or embedded servo, the latter being most common for high-density flexible disks (track density greater than about 5.31 tracks/mm). The rotating-head tape drives for data-processing applications also use embedded servo.

The achievable track density depends primarily on the precision of the head position. In order to compensate for imprecision in track location, either a track is written wider than the read gap width or to tunnel erase a track zone wider than the read/write gap. [The erase heads are generally similar in construction to read–write heads, but have a gap (or double gaps) of greater gap length which permits higher magnitude erase fields at the far side of thick media.] In both instances, the objective is to avoid reading extraneous signals from adjacent tracks due to misregistration of the head and the desired track. As the track becomes narrow, the playback signal amplitude decreases more rapidly than the noise decreases. In order to maintain good signal-to-noise ratio and resolution at higher track densities, the spacing between head and medium is reduced.

The linear density that can be stored in a medium is dependent upon the minimum recordable wavelength. The wavelength λ of a recorded bit is given as

$$\lambda = V/f = 2/k, \tag{1.11}$$

where V is the medium speed, f is the recorded frequency, and k is the linear flux density (number of flux reversals per unit distance). For a high linear flux density, it is general practice to run storage drive at slow speed and to record at high frequency. However, speed should be fast enough for a high data rate. For example, in a computer tape drive, if the tape speed is 2 m/s and the recorded frequency is 1 MHz, then the wavelength is 2 μm or a *bit length* (equal to half of a wavelength) of 1 μm for binary recording. Small wavelengths put stringent requirements on the medium surface defects (caused by coating porosity and coating and substrate roughnesses). Incidentally, penetration of magnetic flux in the coating is typically $\lambda/3$.

The smallest wavelength of recorded amplitude is governed by the extent to which the transition zone between opposed magnetized regions can be made narrow and sharply defined. The mutual demagnetizing effect of adjacent, oppositely magnetized regions tends to broaden the transition region. It has been shown that the minimum transition length can be related to the 2π times remanence-thickness product ($M_r\delta$) divided by the coercivity (H_c) (Potter, 1970; Chen, 1981). Thus, the lower transition length can be obtained by increasing the H_c of the magnetic particles[2] and reducing the M_r or δ.

1.1.4.2. Reproduced Signal Amplitude

Reproduced signal amplitude is very sensitive to the recorded wavelength and the track width. As we reduce the wavelength and/or track width, head-to-medium separation, gap length, and magnetic properties of the medium and its thickness need to be adjusted to get maximum amplitude. A general expression for the reproduced signal amplitude for a sinusoidal recording is given as (Wallace, 1951)

$$e(t) \propto \alpha N w \frac{\mu_r}{\mu_r + 1} M_r V (2\pi\delta/\lambda) T(\lambda) S(\lambda) G(\lambda) \cos(2\pi x_0/\lambda), \qquad (1.12)$$

where

$$
\begin{aligned}
M_r &= \text{remanence (remanent magnetization) of the medium (Wb/m}^2 \text{ or G)} \\
    &= Sp\sigma_s\rho, \\
T(\lambda) &= \text{thickness loss} \\
           &= \frac{1 - \exp(-2\pi\delta/\lambda)}{2\pi\delta/\lambda}, \\
S(\lambda) &= \text{separation loss} \\
           &= \exp(-2\pi d/\lambda), \\
G(\lambda) &= \text{read gap length loss} \\
           &= \frac{\sin(\pi g/\lambda)}{(\pi g/\lambda)},
\end{aligned}
$$

---

[2] Coercivity depends on the particle size and shape. For example, for a constant acicularity, as the particle size is reduced, it is typically found that the coercivity increases, goes through a maximum, and then tends to go toward zero.

α = head efficiency factor,
N = number of turns in the head,
w = track width (m),
μ_r = relative permeability of the core,
V = sliding speed (m/s),
x_0 = Vt, the longitudinal position of the head with respect to an arbitrary reference in the medium (μm),
t = time (μs),
λ = recorded wavelength (μm),
δ = thickness of the medium (μm),
d = effective magnetic spacing between the head and the surface of the medium (μm) (which can be higher than the mechanical spacing),
g = gap length of the head measured from one pole face to another (μm),
S = remanence squareness (reduced saturation remanence) of the hysteresis loop of the medium = M_r/M_s,
M_s = saturation magnetization of the medium (Wb/m^2 or G),
p = packing fraction of particles in the coating (0–1.0),
σ_s = specific saturation magnetization of particles (emu/g where 1 emu = $4\pi \times 10^{-10}$ Wb m), and
ρ = density of particles (kg/m^3).

In order to compare the computed effect of thickness, separation, or read gap length loss with the experimentally observed data, it is useful to put the loss function in decibel (dB) form. This is done by computing twenty times the $\log_{10}$ of the loss function. Thus, we get (Wallace, 1951)

$$T(\lambda) = -20 \log_{10}\left[\frac{2\pi\delta/\lambda}{1 - \exp(-2\pi\delta/\lambda)}\right], \tag{1.13a}$$

$$S(\lambda) = -54.6\, d/\lambda, \tag{1.13b}$$

and

$$G(\lambda) = -20 \log_{10}\left[\frac{\pi g/\lambda}{\sin(\pi g/\lambda)}\right]. \tag{1.13c}$$

Computed loss functions are plotted in Fig. 1.11.

The head efficiency factor in Eq. (1.12) depends on many things, including the orientation ratio (or factor) also called the squareness ratio, which is a ratio of the remanence squareness parallel $[S(\|) = M_r(\|)/M_s]$ and transverse to the direction of the recording $[S(\perp) = M_r(\perp)/M_s]$. Ideally, the orientation ratio should be infinite for maximum magnetization. It primarily depends on the particle shape and size and the coating process. As a rule of thumb, the orientation ratio for a perfectly aligned particle is roughly the ratio of its length to width. Anisotropic (ellipsoidal) particles are selected to provide magnetization along the long axis of the ellipse.

We note that the signal amplitude decreases as we increase either track

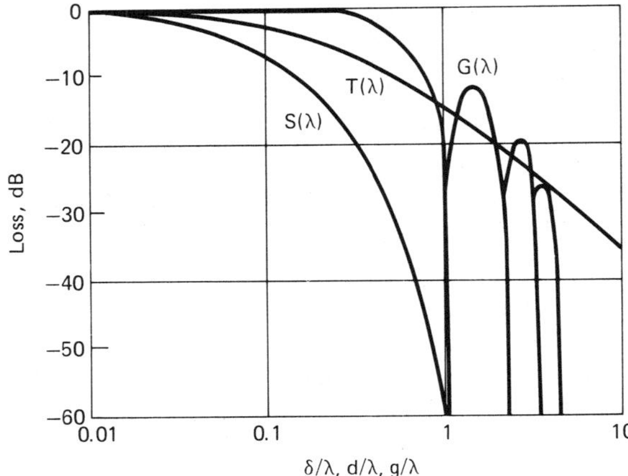

Fig. 1.11. Thickness loss $[T(\lambda)]$, separation loss $[S(\lambda)]$, and gap loss $[G(\lambda)]$ in recording.

density or linear density. Therefore, for a high density recording, the amplitude has to be increased by increasing N and $M_r\delta$ and by decreasing d and g. We have noted previously that a thinner coating and a high coercivity are desirable to reduce the effect of demagnetization. In addition, a thinner coating has a smaller thickness loss. Therefore, a high $M_r\delta$ should be achieved by selecting a magnetic medium with a high M_r $(= Sp\sigma_s\rho)$ and small δ. The medium should have a high H_c compatible for a given head.

Next, we study the separation loss. We note that the separation should be as little as possible. Typical values for d in modern storage systems are less than 0.1–0.3 μm. From Eq. (1.13b), we note that for d/λ of 0.2, the signal loss is 11 dB (Fig. 1.11). Smaller spacings require the surface roughness of the head and medium to be very small, which affects the tribological properties of the interface. Last, we discuss the read gap length loss that is caused by the finite length of the gap. A limit is reached when the recorded wavelength is equal to the gap length: the flux contributions from the two oppositely magnetized half-wavelengths cancel, and the induced voltage is zero (Fig. 1.6). Therefore, gap length should be less than the wavelength.

1.1.4.3. Signal-to-Noise Ratio

The noise in the reproduced signal needs to be minimized. Thus, the signal-to-noise ratio (SNR) must be as high as possible. This is characterized by the decibel (dB), which is the logarithmic ratio of the power output of a given signal to the noise power in a given bandwidth

$$\text{SNR} = 20\log_{10}(V_{\text{signal}}/V_{\text{noise}}). \tag{1.14}$$

Current technology can provide a SNR to be in excess of 30 dB at a bandwidth of 25 MHz and at a transition density of 1000 flux reversals/mm. The bandwidth within which a satisfactory SNR is obtained is called *dynamic range*.

The narrow-band SNR for a slot of flux density width Δk is given as (Mallinson, 1969, 1987)

$$\text{SNR}_{\text{narrow}} = \frac{2\pi nwF^2[1 - \exp(-|k|\pi\delta')]^2}{|k|[1 - \exp(-2|k|\pi\delta)]\Delta k}, \tag{1.15}$$

where

n = the number of magnetic particles/volume ($1/\text{mm}^3$),
w = track width of medium (μm),
F = fraction of total magnetic moment of the medium,
δ' = depth of recording (μm),
k = the number of flux reversals/distance ($1/\text{mm}$) = $2/\lambda$, and
δ = thickness of the medium.

The narrow-band SNR is, of course, independent of head-to-medium spacing. The adverse effects of nonsaturation (1-F) and partial penetration recording δ' are evident; both reduce the SNR because, while only a limited number of particles contribute to the signal, all particles contribute to the noise.

The wide-band SNR for a slot of flux density width ($k_{\text{max}} - k_{\text{min}}$) can be approximated as (Mallinson, 1969, 1987)

$$\text{SNR}_{\text{wide}} \sim 4\pi nwF^2/(k_{\text{max}}^2 - k_{\text{min}}^2). \tag{1.16}$$

The wide-band SNR is dependent upon head-to-medium spacing. From Eqs. (1.15) and (1.16) we note that the SNR is dependent on the track width and n; n can be increased by using smaller particles (or particles with large surface area).

1.2. Magnetic Storage Systems

1.2.1. History of Magnetic Recording

The first demonstration that a magnetic medium could be used to record information was provided by the Danish engineer Valdemar Poulsen in 1898. In particular, Poulsen recorded and reproduced sound with an invention called a "telegraphone," which used a strung-out steel piano wire and an electromagnet connected to a microphone. Poulsen moved the electromagnet along the wire as he spoke into the microphone, and by later connecting the wires from the magnet to the telephone receiver he heard his voice reproduced. In the 1920s, the first recorders with steel tapes were made by Germans, and in 1928, Fritz Pfleumer filed a patent for coating iron particles onto a strip of paper as a recording medium. A machine using such a tape, the German magnetophone, was exhibited in Berlin in 1935. This development used a plastic base with a magnetic coating. The 3M Company finished their first oxide tapes in 1947. Ampex started delivering the first commercial audio tape recorders in 1948. Since that time magnetic tapes developed rapidly.

Video recording was demonstrated in 1951 by the 3M Company followed by RCA in 1953. In 1956, the Ampex Corporation announced a rotating-head video recorder using 50.8-mm-wide tape. Rotating-head technology afforded a high head-to-tape speed of about 38 m/s, which made it possible to record sufficient bandwidth of the video signal by FM. The rotary-head recorder employed a transverse format in which recording was done by four heads (quadruplex head) mounted on a rotating drum. In the quadruplex (or quad) system, each transverse track on the magnetic tape could record only one-sixteenth of one field and complicated switching was required to reassemble a complete field. In order to record one field continuously on the tape, it was necessary to lengthen the video track; the one way to do this was to record the video track on the tape diagonally. To achieve this, the tape had to be physically wrapped around the rotating-head drum in a helical-shaped tape path. This helical scan recording method was introduced in 1960. Instrumentation recording was introduced in the early 1950s when national security efforts required ever-increasing surveillance bandwidths. The earliest instrumentation recorders were adapted from professional audio machines to accommodate frequency modulation (FM) and other higher-than-audio frequency electronics.

After World War II, the digital magnetic recording of data on a tape grew synergistically with the electronic computer. In 1953, IBM marketed the first magnetic tape drive, the IBM 726, with a data density of 4 bytes/mm. For the evolution of tape drives at IBM for mainframe (high-end) computers which use 12.7-mm-wide tape, see Table 1.1 (Harris et al., 1981).

There was a need for a data-storage system that offered random access, that is, a device that would allow information to be recorded and retrieved in any order, rather than serially, as required by magnetic tape. In 1957, IBM announced the first magnetic rigid disk drive, the IBM 305 or RAMAC (random access memory accounting machine) with a storage capacity of

Table 1.1. Tape-drive (12.7-mm-wide tape) evolution for mainframe computers in IBM

IBM product number	726	3420	3480	3490E
First customer shipment	1953	1973	1985	1991
Data density (kB/mm)	0.004	0.245	1.5	3
Number of tracks	7	9	18	36
Reel capacity (MB)	2.2	156	200	400
Data transfer rate (MB/s)	0.075	1.25	3	3
Modulation code	NRZI	(0, 2)	(0, 3)	
Tape transport	Vacuum column	Vacuum column	Single reel cartridge	Single reel cartridge

3 bits/mm^2, which utilized an externally pressurized (hydrostatic) slider air bearing. IBM products have changed in order to provide ever-increasing recording densities (Harker et al., 1981; Bajorek, 1991; Bhushan, 1990). In 1970, IBM introduced a cheaper alternative, the flexible disk or diskette or floppy disk, a 200-mm, round, flexible disk in a rectangular shell [with a shell width of 203 mm (8 in.)] with a storage capacity of a few hundred kilobytes. The initial 200-mm-diameter disks were followed in 1976 by 130.2-mm-diameter disks [with a shell width of 133 mm (5.25 in.)] and in 1980 by microfloppy disks having 85.8-mm diameter [with a shell width of 90 mm (3.5 in.)] and in 1987 by 75-mm (3 in.) diameter and 47-mm or 50.8-mm (2 in.) diameter [with a shell width of 60 mm (2.4 in.)]. The principal applications for flexible-disk drives are found in a number of small business and home computers. Further historical backgrounds on audio, instrumentation, video, and data processing tape drives and flexible disk drives are presented by Engh (1981), Harris et al. (1981), Sugaya (1985), Teramura (1985), Bashe et al. (1986), Mee and Daniel (1990), and Pugh et al. (1991).

1.2.1.1. Storage Hierarchy

In a data storage system, tapes and disks often appear in complementary roles that utilize their unique features. Thus disk drives with their random access have access times much shorter than tape drives. The tapes have extremely high volumetric density (up to several hundred megabytes or even gigabytes per tape reel), have high data rates, and are much cheaper than the rigid disks in dollars per megabyte but are not random access. These are primarily used for off-loading the data from rigid disks for archival storage in mainframe computers. The tape drives with rotating-head configuration have extremely high data rates (up to 400 Mbits/s) and are used for storing voluminous data such as in instrumentation and video recorders. Incidentally, data rates can be increased to as much as 1 Gbits/s by parallel accessing (drive arrays).

Large mainframe computer systems require large amounts of on-line storage. Such storage requirements are satisfied by having large, rigid disks (up to 275 mm in diameter) on one spindle with a total capacity of a few gigabytes. On the other hand, a personal computer or work station may only require hundreds of megabytes, which can be provided by a (typically 65-, 95-, or 130-mm) rigid-disk drive. Flexible disks have a low capacity (a fraction of a megabyte to a few tens of megabytes) and are a cheaper alternative for inexpensive portable models. The interchangeability of data among drives is accomplished by removable (flexible) media. Optical disk drives use removable media and have the potential of providing on-line storage density much higher than the magnetic storage devices, and optical drives do not require the read/write optical head to be close to the disk surface. However, their access times (50–80 ms) and data rates (0.1–1 MB/s) are slower than that of the rigid disk drives. Semiconductor memory devices (D-RAM and flash memory cards) have very high data rates and have very high volumetric density, but these are extremely expensive and are not expected to become popular for information storage.

1.2.2. Examples of Modern Storage Systems Using Flexible Media

Typical magnetic characteristics of modern, high-end, magnetic data-storage systems for computer applications are presented in Table 1.2. The heads in modern storage systems are designed so that they develop hydrodynamic (self-acting) air bearings during operation. Formation of air bearing minimizes the head–medium contact. Physical contact between head and medium occurs during starts and stops. Head–medium separation (or flying height) is minimized for optimum magnetic performance. Flying height in the tape drives is generally smaller than in the rigid-disk drives, and in the flexible-disk drives it is generally very small. We also note that data-processing tapes are much more demanding in smoothness and low-error rates than the instrumentation and video tapes, which in turn are more demanding than audio tapes.

In most audio, video, and instrumentation systems, an analog signal is recorded on, and reproduced from, a tape. Digital recording is used for broadcast audio, broadcast video, and modern instrumentation systems and all data-storage systems (tapes, and flexible and rigid disks). Data rate requirements of most instrumentation, broadcast video, and many high-end computer storage systems are very high. Typical operating conditions of modern data-storage systems will be presented later. Examples of tape drives and flexible-disk drives follow (Anonymous, 1990, 1990/1991, 1991a–f; Peterson, 1989, 1990; Waid et al., 1990; Abraham and Freeman, 1991a, b).

1.2.2.1. Tape Drives

Tape drives typically use a 25.4-mm (1-in.), 19-mm (0.75-in.), 12.7-mm (0.5-in.), 8-mm (0.315-in.), 6.35-mm (0.25-in.), or 4-mm nominal [actual 3.81 mm (0.15 in.)] wide tapes in open-reel and single-reel cartridge, or double-reel cassette formats. Tape drives in both stationary and rotating-head configurations are used. Rotating-head configuration is more complex and is generally used for applications requiring very high data rates and very high volumetric densities.

Audio Tape Drives

The linear analog technique is most commonly used for most domestic audio recorders. A schematic of a tape path in an audio recorder is shown in Fig. 1.12. A 4-mm-wide, about 6 (C120)- to 12 (C60)-μm-thick, and 1080–2160-m-long magnetic tape wound on a cassette reel is transported at about 0.5 m/s over a fixed read–write head. The head has a cylindrical surface contour and the wrap angle is 0.6–1 radian. The tape is driven by a capstan and the tension (0.2–0.5 N) is applied by spring-loaded rollers. There is always physical

Table 1.2. Typical magnetic characteristics of high-end data-storage systems

Drive	Linear density, kb/mm	Linear flux density,[a] kfr/mm	Track density, 1/mm	Areal density,[b] kb/mm^2	Total formatted capacity, MB	Data transfer rate,[c] kB/s	Access time between any track, ms	Typical medium form factors
Tape (12.7-mm wide and 165-m long—IBM 3490E)	0.86	0.97	2.84	2.44	400	3000		3.81, 6.35, 8, and 12.7-mm wide, several hundred meters long
Flexible disk (90-mm diameter)	0.52		5.31	2.78	1.6 (unformatted)	62.5	3	50, 90, 130, and 200-mm nominal diameter
Rigid disk (275-mm diameter—IBM 3390)	1.08	0.81	88.2	95.5	504^d	4200	16	65, 95, 130, and 275-mm diameter

[a] Linear density = linear flux density × recording ratio (information bits per flux reversal).

[b] Areal density = linear density × track density.

[c] Data transfer rate = data density × linear speed, and data density = (1 − redundancy) × areal density × medium width/number of bits per byte, and 1B (byte) = 8b (bits).

[d] 3.784 GB per spindle having a total of nine dual-sided 275-mm disks with 15 data surfaces; 6 spindles per box.

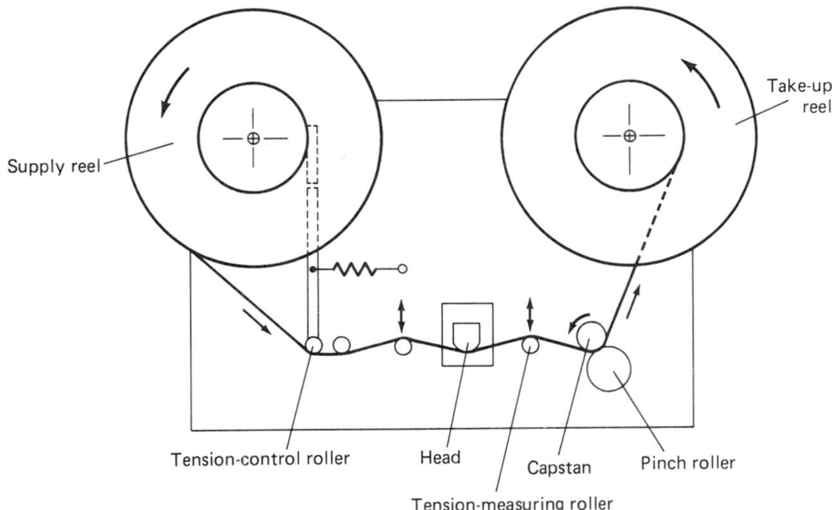

Fig. 1.12. Tape path in an audio tape recorder.

contact between the head and the tape. In a broadcast audio (professional) recorder, a 25.4-mm-wide tape is moved over a multitrack head. Commonly-used magnetic particles in the construction of the magnetic tapes are γ-Fe_2O_3 and CrO_2.

To alleviate the high cost of multitrack heads, helical-scanning rotary single-track heads developed for video recording are also used for high data rate and high recording density capabilities. For professional audio recording, dedicated digital audio technologies have emerged based, again, on either a multiple-track stationary-head (S-DAT) approach or a helical scanning rotating-head (R-DAT) approach. The tape width is 3.81 mm. The metal particles or metal-evaporated tapes are generally used.

Video Tape Drives

A video recorder uses a helical-scanning rotating-head configuration as shown in Fig. 1.13. The term helical scan derives its name from the shape of the recorded track on the tape. The read and write heads are positioned on a rapidly rotating drum called a scanner. Because a rotating head can be moved at a greater speed than a heavy roll of tape, much higher data rates can be achieved in a rotating head drive than in a linear tape drive. Relative head-to-tape velocity in a linear tape drive is typically from a fraction of a meter to several meters per second, while in a rotating head system, it can be as high as 40 m/s. Cassettes with a 12.7-mm- or 8-mm-wide particulate tape (with magnetic particles—γ-Fe_2O_3, Co–γ-Fe_2O_3, CrO_2, or metal) are wound helically around the rotating-head drum, and a magnetic head supported on the

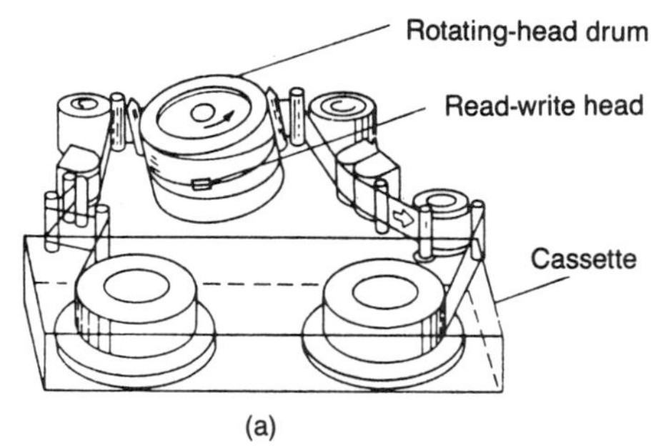

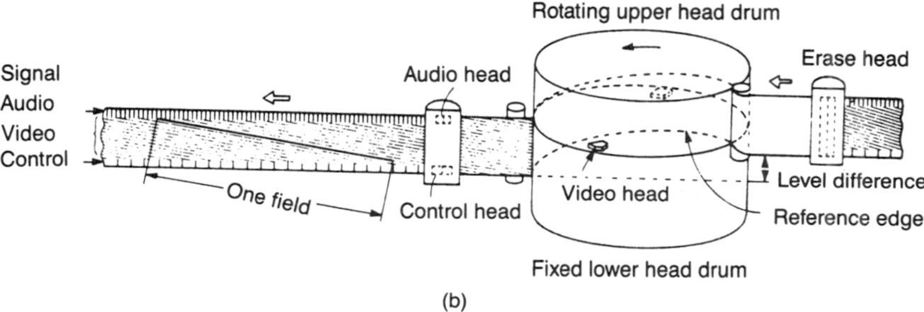

Fig. 1.13. (a) Tape path in a helical-scanning rotating-head video recorder, and (b) schematic of a head-tape configuration in a two-head helical-scan video recorder and recording format.

rotor rotates at high speed with respect to the tape. This forms long tracks at an acute angle (on the order of 5°) to the tape length. This results in recorded tracks much longer than the tape width (an aspect ratio of on the order of 10). Head-scanning mechanisms with two or four heads, mounted on a rotating wheel, are commercially used, which scan successively across the width of the tape at high speed (Kirk, 1981; Ginsburg, 1986). The head has a spherical or ellipsoid contour with or without tapered edges. The tape is wrapped through 360° around the pressurized mandrels, which support the tape hydrostatically in the mandrel area. Hydrodynamic air film between the rotor and the tape is developed to a thickness of up to about 10 μm when the rotor is running at its operating speed, therefore the head is protruded about 50 μm to obtain ultra-low head-tape spacing typically of 25 nm. Frequency modulation (FM) video recorders are most commonly used.

For broadcast video (professional) recording, digital video recorders with metal-particle tapes or metal-evaporated (Co–Ni) tapes of 25.4-mm or 19-mm width are used for high signal-to-noise ratio. The 12.7-mm-wide tapes are also expected to become popular for broadcast video recording. Broadcast video recorders (D1 and D2) have the data rate requirements of 100–400 Mbits/s and the storage capacity of cassettes is typically of the order of 10–100 GB.

Data-Processing Tape Drives

(a) Mainframe Computers

Digital tape drives are used for data-processing (computer) applications (Harris et al., 1981; Mee and Daniel, 1990). For evolutions for tape drives in 12.7-mm (0.5-in.) reel tape format for mainframe computers, see Table 1.1. A schematic of the tape path in an IBM 3420 data-processing tape drive is shown in Fig. 1.14. A 12.7-mm-wide[3] (0.50 in.) and about 30-μm-thick γ-Fe_2O_3 particulate tape [Fig. 1.15(a)] wound on a 267-mm (10.5-in.)-diameter reel (Fig. 1.16) is transported over a fixed read–write head (write and read heads placed side by side) and the tape is accurately guided using frictionless rollers. The read and write heads have a cylindrical surface contour. There is very little (if any) physical contact when the tape is running at its operating speed, however, physical contact occurs during starts and stops. The read and write heads are an inductive-coil type, nine-track head. Intermittent tape transport requires the supply and take-up reel motors and a separate capstan

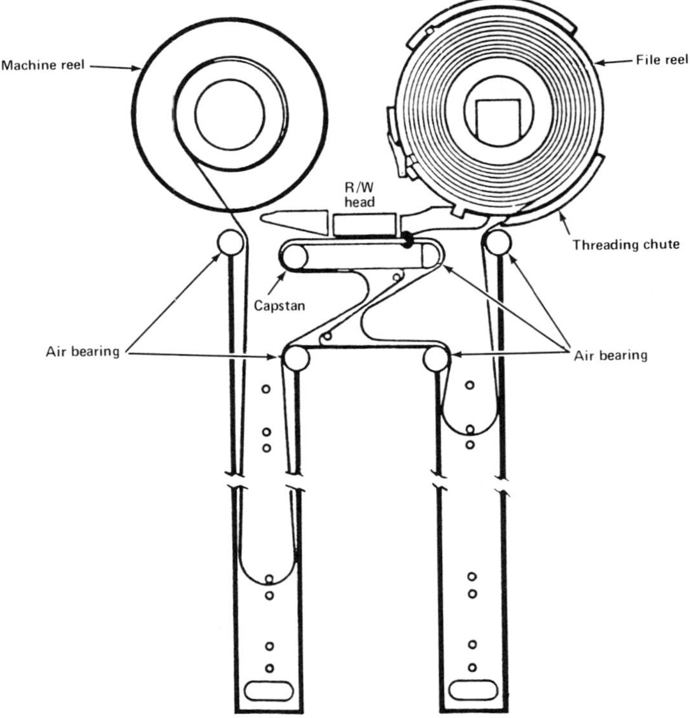

Fig. 1.14. Tape path in an IBM 3420 data-processing tape drive.

---

[3] We note that the IBM 3850 mass storage system used a 133-mm-wide, 202-m-long tape. The mass storage used a rotary head configuration (Harris et al., 1981).

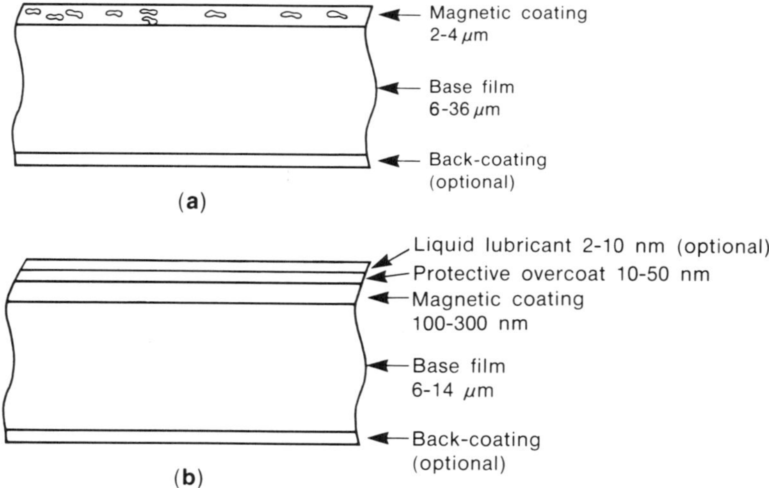

Fig. 1.15. Sectional views of (a) a particulate magnetic tape and (b) a thin-film magnetic tape.

motor to control the tape velocity. The tape is buffered in a vacuum column which provides the tape tension. Tension is controlled by the vacuum pressure, and the tape position is controlled by the capacitive sensing units. Figure 1.17 shows the top view (a) and the cross section (b) of the inductive-coil type head consisting of a magnetic core, electrical windings, and a housing. In recent years, various thin-film designs have miniatured internal electrical circuits, but basic external characteristics of the standard inductive head have been preserved.

Figure 1.18(a) shows a schematic of the high-density, high-data-rate, data-processing IBM 3480 tape drive introduced in 1985. For this drive, a 165-m-long, 12.7-mm-wide, and roughly 26-μm-thick particulate tape (having a 23.4-μm-thick substrate and about a 3-μm-thick magnetic coating containing CrO_2 magnetic particles with a coercivity of about 500 Oe) wound on a reel is housed inside a rectangular cartridge ($100 \times 125 \times 25$ mm), Fig. 1.19. The internal brake mechanism of the cartridge, a spring-loaded brake button prevents reel rotation when the cartridge is not in the drive, Fig. 1.19(a). The brake button is released automatically by the magnetic clutch of the drive as the cartridge is loaded into the drive and is automatically re-engaged upon the removal of the cartridge from the drive. After loading the cartridge in the drive, the tape leader is threaded in the drive to the take-up reel by a pentagon threading mechanism. A decoupler column placed near the cartridge entrance decouples any tape vibrations that may occur inside the cartridge. It isolates the shock and absorbs the lateral motion and absorbs the change in radius of the supply reel, from a full reel to an empty one. The decoupler [Fig 1.18(b)] is designed to support the tape by means of two hydrostatic bearings at the

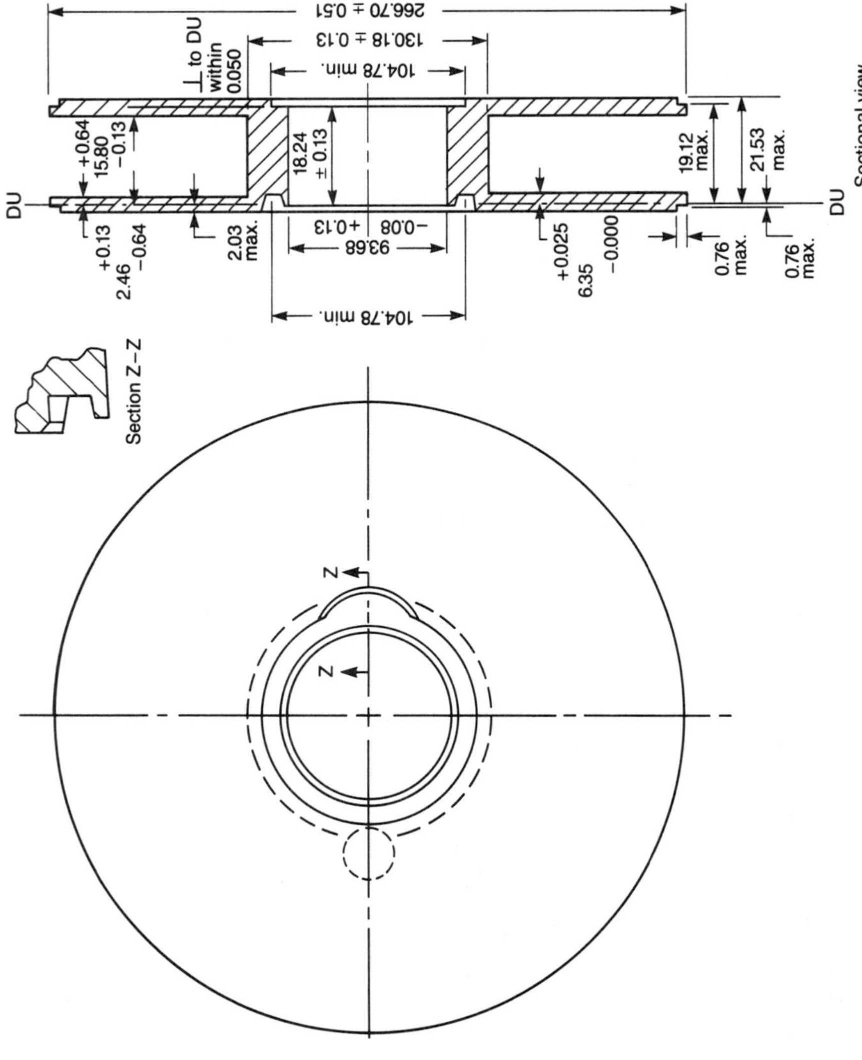

Fig. 1.16. Schematic of the IBM 3420 reel assembly (all dimensions are in millimeters) for a 12.65 ± 0.05-mm-wide tape.

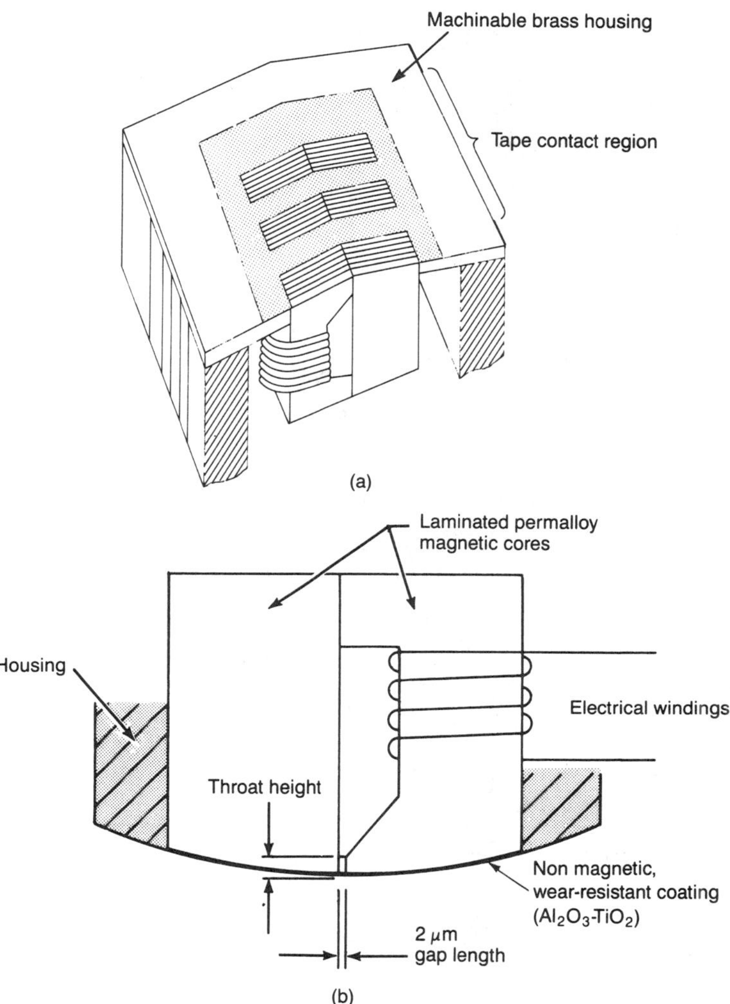

Fig. 1.17. Schematic of an inductive-coil-type magnetic head for an **IBM** 3420 tape drive (a) top view (only three tracks are shown schematically although it is a nine-track head) and (b) cross section.

inlet and the exit. At the midpoint, the tape is pulled in with a vacuum pocket to form an omega-shaped loop. This miniature vacuum column softens the tape path and absorbs transients in tape stretch and tape slack. Before the tape reaches the head, it passes over a cleaner-blade assembly to scrape debris from the tape surface and to vacuum it away.

The tape is wrapped around a cylindrical guide surface (a 90° segment of a cylinder) or D-bearing with the magnetic head mounted in the center, to increase its buckling strength. A hydrostatic air bearing causes it to float on

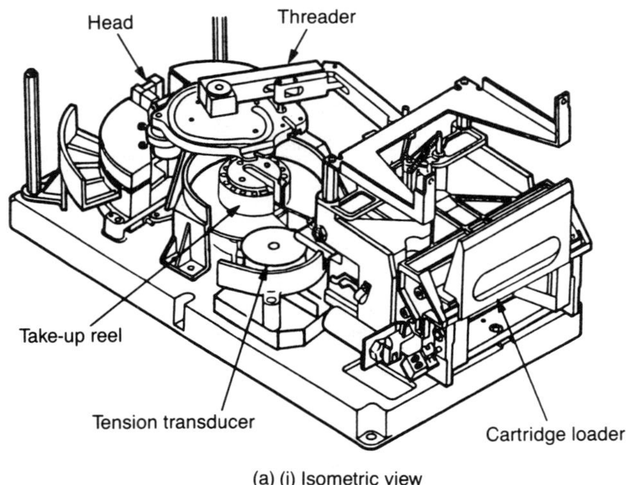

Fig. 1.18. (a) Tape path in an IBM 3480/3490 data-processing tape drive, (b) schematic of decoupler elements, (c) schematic of tape-guiding mechanism at the head assembly with a leaf-spring compliant tape guide, (d) ceramic compliant tape guide, and (e) schematic of a tension transducer (Winarski et al., 1986). © 1986 International Business Machines Corporation; reprinted with permission.

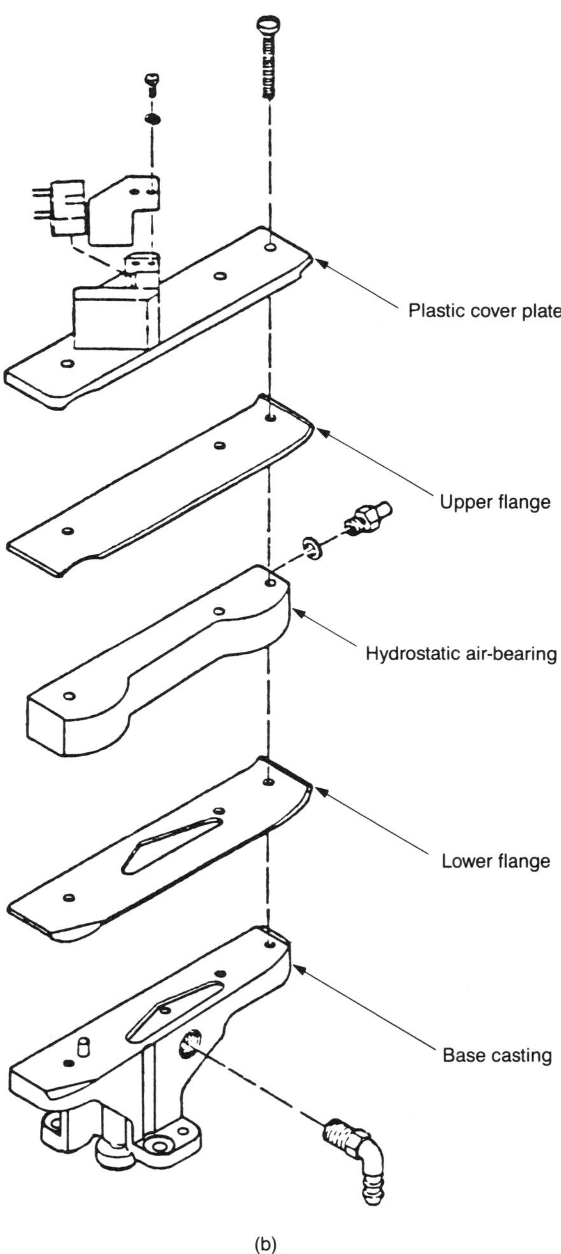

Fig. 1.18 (*continued*)

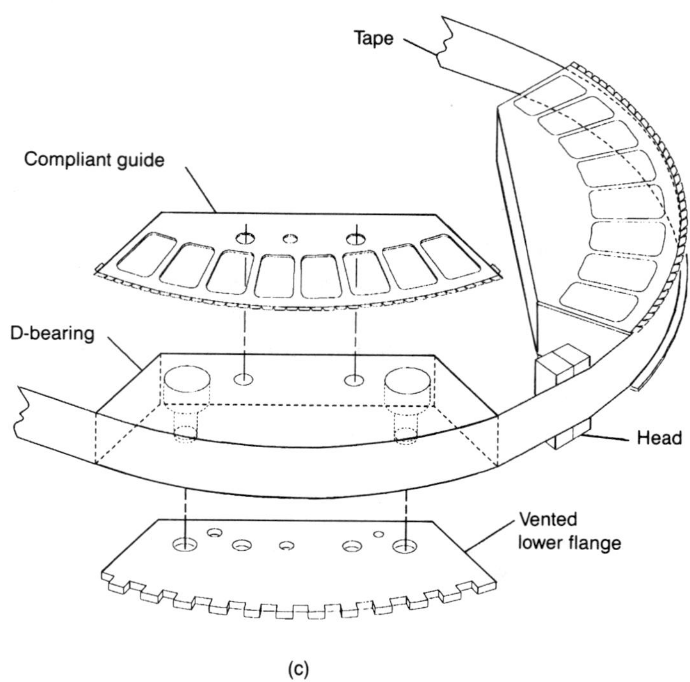

(c)

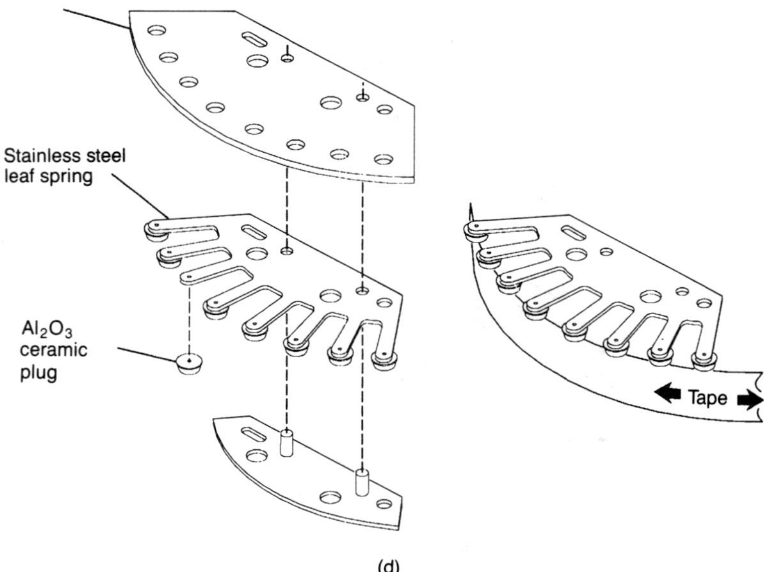

(d)

Fig. 1.18 (*continued*)

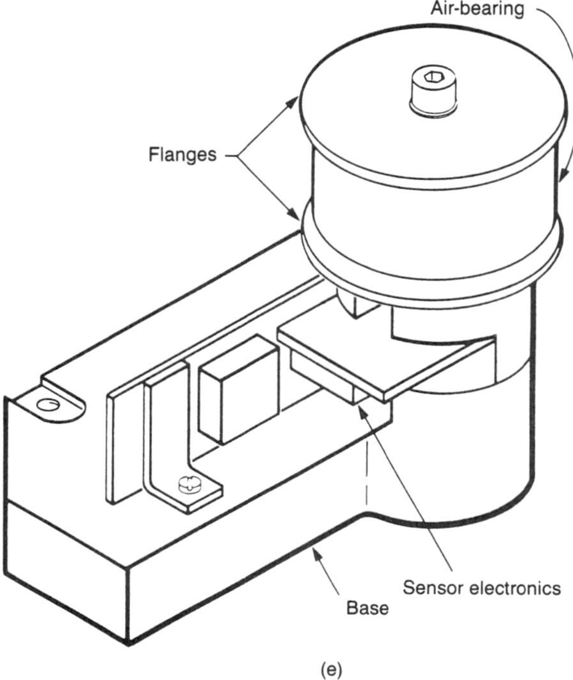

Fig. 1.18 (*continued*)

a thin film of air. Guides, one on each side of the magnetic head, are also provided to move the magnetic tape repeatedly across the read and write elements of the heads [Fig. 1.18(c)]. The top guides are compliant (made of a leaf spring which has been photoetched from thin, nonmagnetic stainless steel) and exert a distributed load on one edge of the tape (~ 50 mN/guide). The compliant flange was later modified to improve its wear resistance and minimize tape vibrations (Bhushan, 1987). The new ceramic compliant tape guide now has alumina pads at the end of the stainless-steel leaf spring, Fig. 1.18(d). This causes seating of the opposite edge of the tape against a reference (rigid) lower flange, initially made of tungsten carbide and later of bulk alumina. The lower flange is vented (material removed in the area in which the tape actually contacts the flange) to provide a uniform pressure distribution in the hydrostatic air bearing and is used to prevent tape vibrations that would degrade the read–write performance. A pneumatic tape lifter between the read and write heads is used to minimize head–tape contact during starting and stopping.

During the use of a cartridge, the tape is then wound onto the machine reel through a tension transducer which senses and controls the tension and which also acts as a buffer, Fig. 1.18(e). The tension transducer is a hydrostatic

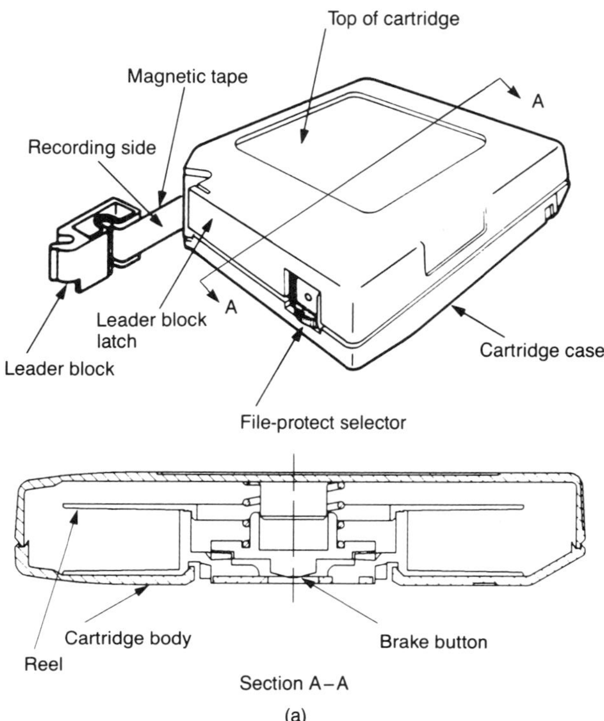

Fig. 1.19. Schematic of an IBM 3480 tape cartridge (a) cartridge assembly and (b) reel assembly (all dimensions are in millimeters) for a 12.675 ± 0.025-mm-wide tape.

air bearing of 15-mm radius with a tape wrap of approximately 180° on the bearing. A solid-state pressure transducer is used to sense tape tension by sensing the air pressure between the moving tape and the fixed surface of the air bearing through a sensor hole in the center of its face. The output of the pressure transducer is applied differentially to the reel motors through the motor drivers to control the desired tape tension. The tape velocity is controlled by supply and take-up reel servo motors; this is a substantial simplification since it eliminates the capstan and its motor. Reading and writing is done at a tension of 2.2 N and at a sliding velocity of 2 m/s and rewinding is done at 4 m/s. The tape wraps over the head about 16° (Winarski et al., 1986).

A schematic of the 18-track read–write head made using thin-film technology for the IBM 3480 drive is shown in Fig. 1.20. The write head is an inductive type and the read head is a magnetoresistive (MR) type (Cannon et al., 1986). The radii of the cylindrical heads are approximately 20 mm. The (blind) bleed slots are provided to reduce the flying height for maximum reproduced amplitude. Edge slots are provided for flying uniformity. The write and read head track widths are 540 and 410 μm, respectively, and the track-to-track centerline distance is 630 μm.

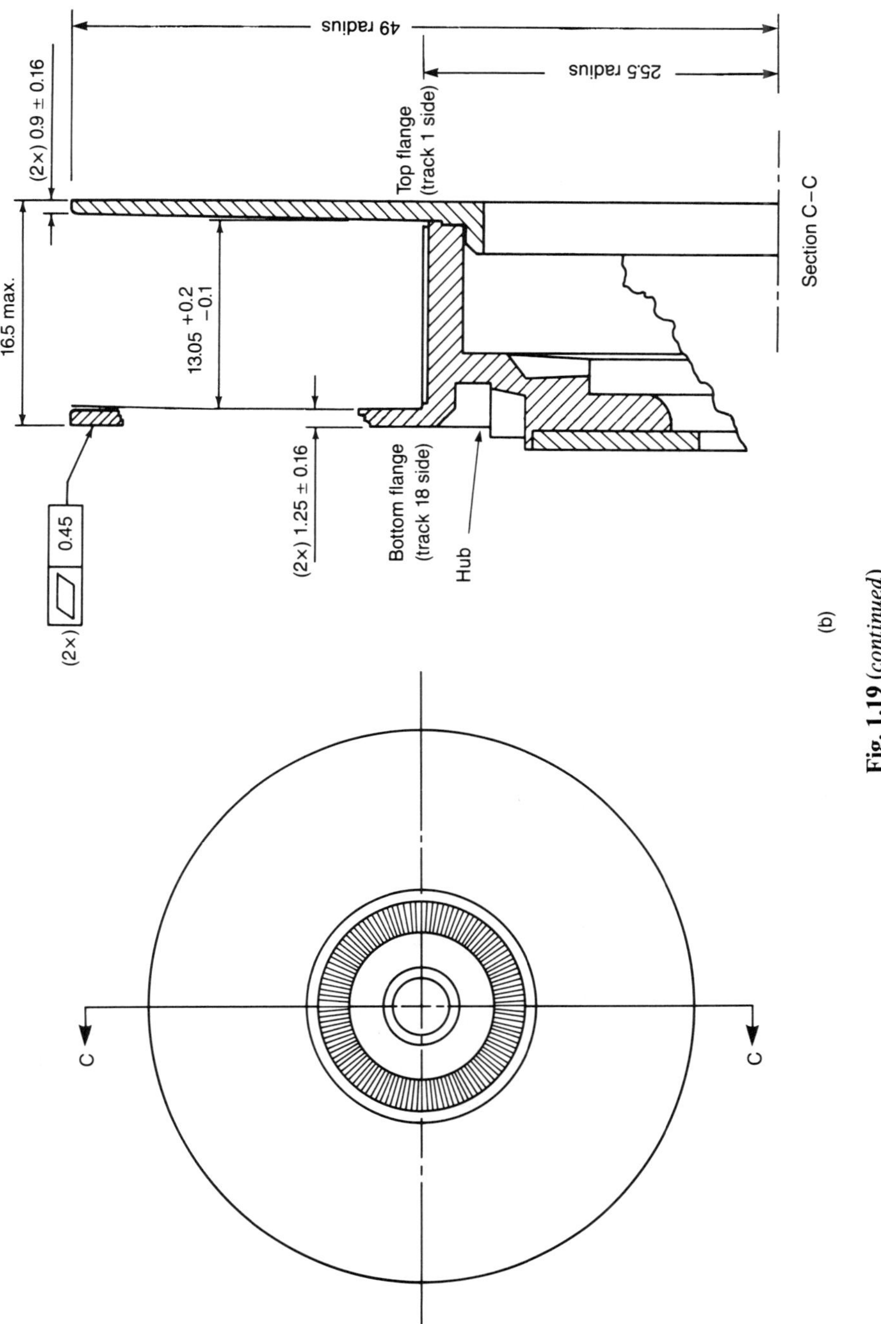

(b)

Fig. 1.19 (*continued*)

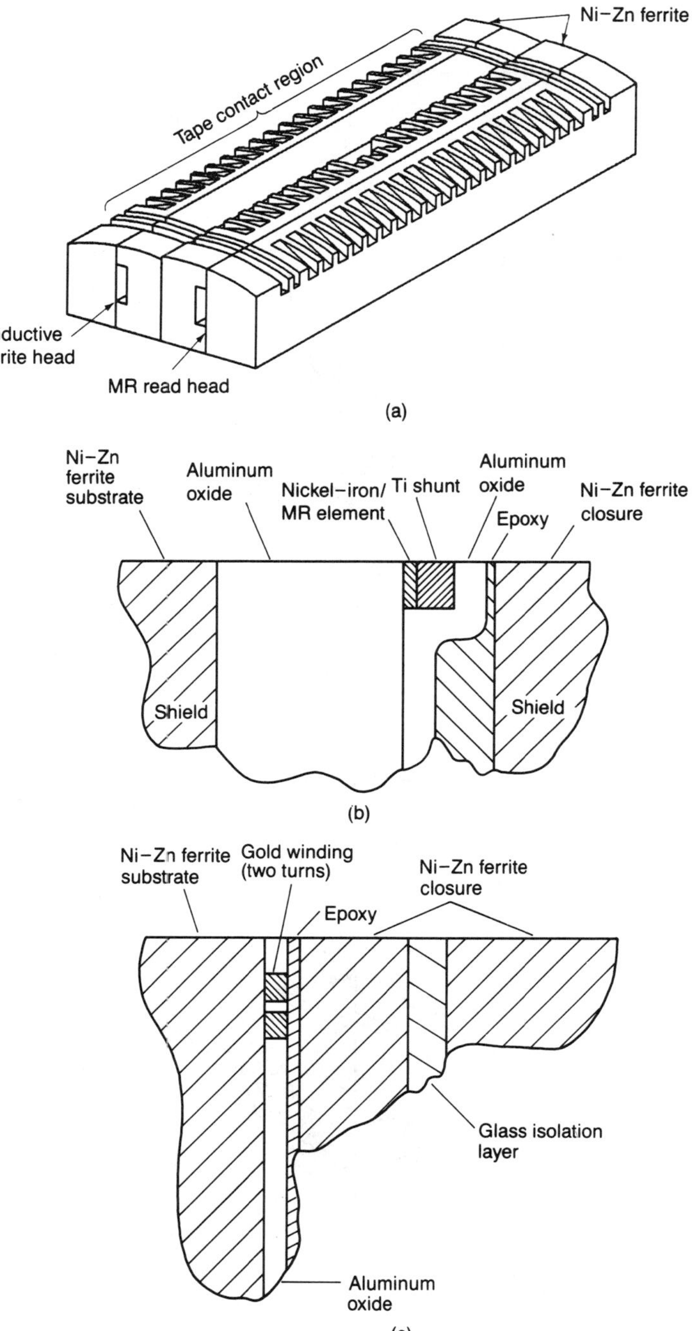

Fig. 1.20. Schematic of a magnetic thin-film head (with a radius of cylindrical contour of about 20 mm) for an **IBM 3480** tape drive: (a) top view, (b) cross section of the **MR** read module, and (c) cross section of the inductive write module (Cannon et al., 1986). © 1986 International Business Machines Corporation; reprinted with permission.

In 1990, IBM introduced an IBM 3490 drive which is repackaged as 3480 with two drives on top of each other and an automatic loader for 32 cartridges. In 1991, IBM introduced an enhanced tape drive IBM 3490E. The cartridge capacity was doubled over the earlier IBM 3480 (1985) and IBM 3490 (1990) base models by means of a new 36-track bidirectional enhanced capability format. Rewind time is virtually eliminated when reading or writing a full cartridge. The 3490E drive uses a 36-track read–write head with two modules having alternating read and write elements. These modules are geometrically similar to that in a 18-track read–write head. When the head is assembled, the write element on module 1 lines up with a read element on module 2. Similarly, the read elements on module 1 line up with the write elements on module 2. Now, with the tape moving in the "forward" direction, write elements are active on module 1, with reading taking place on module 2. When the tape is reversed, it is being written by elements on module 2, and read by module 1. "Odd-numbered" tracks—18 of them—are written in the "forward" tape direction; "even-numbered" tracks are written in the "backward" tape direction. Each module therefore has 18 write and 18 read elements. The tape velocities in the forward and reverse directions are 2 m/s.

Evolution of the designs for IBM data-processing tape-head contours is shown in Fig. 1.21. The figure shows six basic contours and indicates the first implementation (Hahn, 1989).

(b) Mid-Range Computers

The 8-mm and 4-mm helical-scanning data cartridge, 6.35-mm (0.25-in.) (longitudinal) data cartridge and 4-mm (0.15-in.) (longitudinal) data cassette tape subsystems, combine removable, rewritable magnetic media with a compact, integrated drive/controller to provide high storage capacity (on the order of five times the capacity of a linear tape drive) at low cost. The 8-mm and 4-mm helical-scanning data cartridges share a common base of recording characteristics, similarly, 6.35-mm (longitudinal) data cartridges and 4-mm (longitudinal) cassettes share a common base of recording characteristics.

Helical-scanning (dual) rotatory-head configuration in an 8-mm tape format similar to video recorders is used in tape drives requiring very high volumetric density (on the order of 6.4 kB/mm^3) for work stations (e.g., IBM RS-6000). A drive using a 130-mm full height form factor (height—82.6 mm, width—146.1 mm, and depth—203.2 mm) can have a capacity of up to 2.5–5 GB (EXB 8200 and EXB 8500, manufactured by Exabyte Corp., Boulder, Colorado). The magnetic particles generally used for high density tapes are metal particles with a coercivity of about 1450 Oe. An inductive read-after-write with separate full-width erase head (made of Mn–Zn ferrite) is used. A dedicated servo head is used for (embedded) servo control, Fig. 1.22(a). Typical characteristics of the highest capacity data cartridge and heads are presented in Table 1.3.

Helical-scanning, rotating-head configurations in a 4-mm tape format, the same as in R-DAT recorders, were introduced in 1989 for personal computers and work stations. The tape width is 3.81 mm and the tape lengths are 60 m

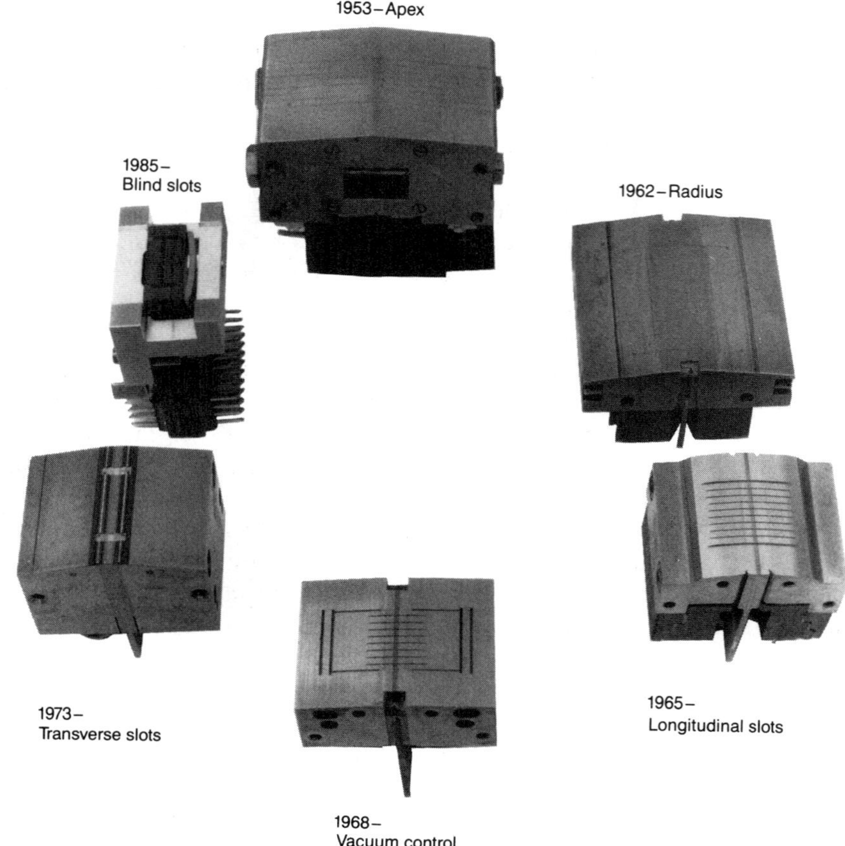

Fig. 1.21. Photographs of six basic contours for tape heads used by IBM since 1953 (Hahn, 1989).

or 90 m. The tape substrate and coated tape thicknesses are typically 9.4 μm and 13 μm, respectively for 60-m tape length and 6.4 μm and 9 μm, respectively, for 90-m tape length. Typical cartridge dimensions are as follows: $73.7 \times 53.3 \times 10.1$ mm. The metal particles with a coercivity of about 1500 Oe are generally used. Two sets of read-after-write heads are used, Fig. 1.22(b). Storage capacity ranging from 0.6–1.3 GB in a 95-mm half-height form factor can be obtained.

Belt-driven data cartridges using a 6.35-mm (0.25-in.)-wide tape (von Behren, 1971; Anonymous, 1977; Simpson, 1984; von Behren and Smith, 1991) and with a storage capacity of up to several gigabytes (made by the 3M Corp., St. Paul, Minnesota) are commonly used for relatively high-capacity mid-range computers (e.g., DC 2080 and DC 2120 for IBM PS/2 and DC 6150 and DC 6250 for IBM AS/400 and DC 6525 and Magnus 1.35 for IBM RS-6000 work stations). A data cartridge tape is a self-contained reel-to-reel

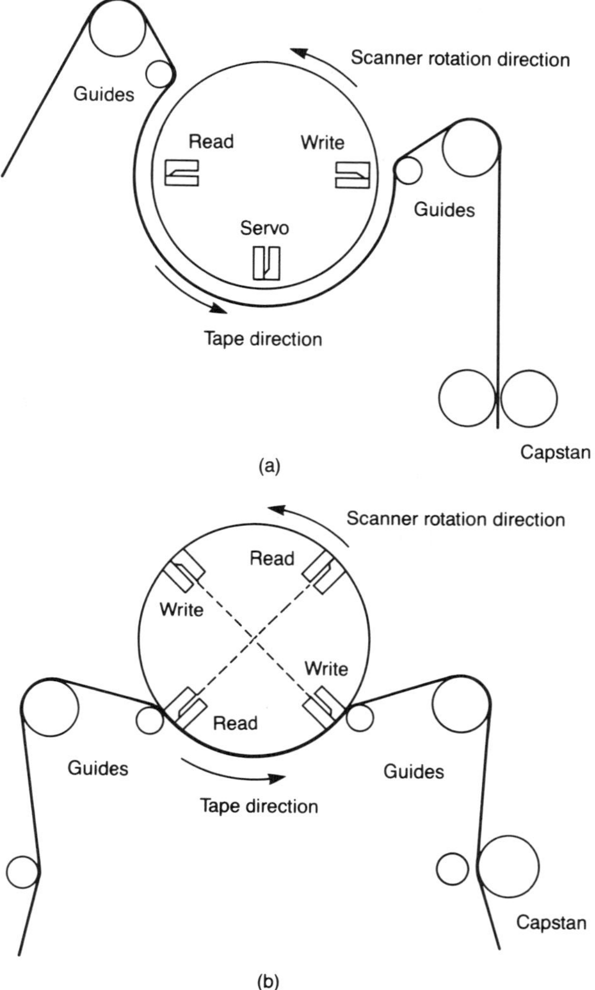

Fig. 1.22. Tape path in a helical-scanning rotating-head data-processing tape drive in (a) an 8-mm tape format and (b) a 4-mm tape format.

tape deck without a motor or read–write head and is belt driven via a capstan. The drive is much smaller than that required by a conventional 12.7-mm computer tape drive and costs much less. A schematic of the tape path of a belt-driven data cartridge is shown in Fig. 1.23. A drive roller, tape guides, a head window (with a hinged protective door), and a light sensing window are disposed along one long edge of the cartridge. Inside are the supply and take-up spools of magnetic tape. The belt-drive roller and two additional idler rollers for the belt complete the drive train in the cartridge. The drive motor puck turns the cartridge drive roller (made of polyurethane)

Table 1.3. Typical characteristics of 8-mm data cartridges and heads for work stations

Exabyte product number	EXB 8200	EXB 8500
First customer shipment	1987	1990
Overall cartridge dimensions	$95 \times 62.5 \times 15$ mm	
Drive size ($H \times W \times D$)	$82.6 \times 146.1 \times 203.2$ mm	
	(130-mm full height)	
Magnetic particles	Metal	
coercivity, Oe	1450	
Tape length, m	112	
Tape thickness, μm		
total	10	
substrate	6	
Tape velocity, mm/s	10.9	
Effective head-to-tape speed, m/s	3.81	
Linear flux density, kfr/mm	1.7	
Data transfer rate, kB/s	246	500
Track density, /mm	32.2	64.5
Formatted capacity, GB	2.5	5
Head	Inductive coil R/W with separate erase head and dedicated servo head	

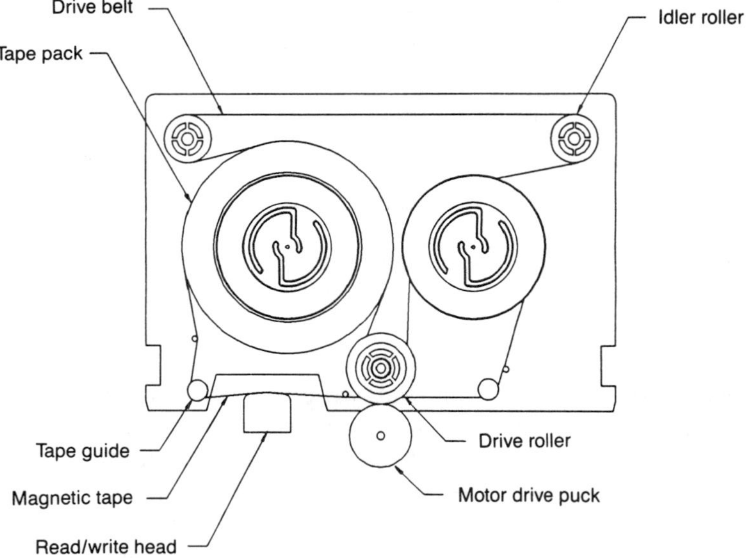

Fig. 1.23. Schematic of a 3M belt-driven 6.35-mm tape cartridge used for data processing in mid-range computers (von Behren and Smith, 1991).

which in turn moves the elastomeric drive belt. The drive belt transfers the movement to tape motion. For economy and reliability, designs embodying a self-tensioning drive belt without any movable tensioning rollers are used. In such designs, the belt tension is maintained by the elastic properties of the textured drive belt (typically made of polyurethane). The textured surface (matte finish) on the drive belt allows air to escape, which minimizes belt slippage even at high speeds and thus improves the grip for better tape tension. The drive belt simultaneously controls the rotation of both the winding and unwinding tape packs, without the need for any complicated servo mechanisms or capstans. The attractive feature of such a cartridge is that the single-point drive and the management of tape tension within the cartridge greatly simplify the design of the cartridge drives utilizing such cartridges. The tape cartridge is capable of operating bidirectionally in either a streaming (or continuous) mode or a start–stop mode.

The tape is wound with its magnetic side facing out and the textured and compressible backcoat facing toward the tape pack center. The textured and compressible backcoat provides enough friction to hold each layer of tape in place and diminishes interlayer slippage during long-term storage. The magnetic particles generally used are $Co-\gamma-Fe_2O_3$ with a coercivity of about 550–900 Oe. The tape length ranges from 63–311 m. Typical tape speeds range from 0.64–3.05 m/s and the tape tensions at the magnetic head positions range from 0.28–1.25 N throughout the full recorded tape length. The overall dimensions of the cartridges are 152 mm $\times$ 102 mm $\times$ 17 mm and 81 mm $\times$ 61 mm $\times$ 15 mm for 130-mm (41.3 mm) half-height and 95-mm (25.4 mm) half-height form factors, respectively. The smaller cartridges are generally referred to as data minicartridges.

An inductive read-after-write single-track head (made of Mn–Zn ferrite with a track width of about 90 μm or less), with a separate erase bar or read-after-write thin-film three-track MR heads, is used to write multiple tracks in serpentine longitudinal form (by vertically displacing the head). Typical track densities of 26–32 for a 6.35-mm-wide tape are used. Typically, linear recording flux densities of 492–1526 flux reversals/mm (or bits/mm) are utilized to provide 80–1350 MB of on-line data storage per cartridge. Typical characteristics of selected high capacity data cartridges and heads are presented in Table 1.4. 3M cartridges DC 2120 and DC 6250 are extended-length versions (93.7 m and 310.9 m) of DC 2080 and DC 6150, respectively.

The 4-mm reel-to-reel data cassettes similar to audio cassette recorders are used for relatively small computers. The tape is enclosed in a protective package and wound onto two spools, one for supply and one for takeup, Fig. 1.24. The tape width is 3.81 mm, the tape length is 183 m, and coated-tape thicknesses are typically 9–11 μm. Typical cassette dimensions are 99.1 $\times$ 63.5 $\times$ 12.2 mm. The $Co-\gamma-Fe_2O_3$ particles with a coercivity of about 310–900 Oe are generally used. The data cassettes also employ serpentine recording with stepper heads and are also packaged in 95 mm and 130 mm half-

Table 1.4. Typical characteristics of high-density 6.35-mm data cartridges and heads for mid-range computers and work stations

3M Product number	DC 2080	DC 6150	DC 6525	Magnus 1.35
First customer shipment	1988	1988	1988	1991
Overall cartridge dimensions	$81 \times 61 \times 15$ mm	$152 \times 102 \times 17$ mm	$152 \times 102 \times 17$ mm	$152 \times 102 \times 17$ mm
Form factor	95 mm HH[a]	130 mm HH	130 mm HH	130-mm HH
Magnetic particles	$Co-\gamma-Fe_2O_3$	$Co-\gamma-Fe_2O_3$	$Co-\gamma-Fe_2O_3$	$Co-\gamma-Fe_2O_3$
coercivity, Oe	550	550	550	900
Tape length, m	62.5	189.0	310.9	231.6
Tape thickness, μm	15.2	15.2	10.2	14.2
Tape velocity, m/s	0.64–2.29	1.52–3.05	1.52–3.05	1.52–3.05
Tape tension, N	0.14–0.98	0.28–0.98	0.28–0.83	0.34–1.25
Linear flux density, kfr/mm	0.579	0.492	0.788	1.526
Number of data tracks	32	32	26	32
Number of servo tracks	—	—	—	16
Formatted capacity, GB	0.080	0.150	0.525	1.35
Data transfer rate, kB/s	100	—	100	385
Head		Single track inductive coil read after write head (with 89-μm track width) with separate erase bar		Read after write thin film/3-track MR head and dedicated thin-film servo head, track following servo-16 tracks

[a] Half-height.

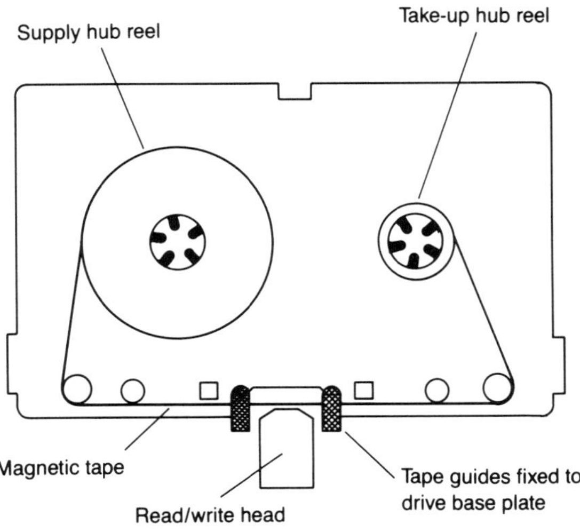

Fig. 1.24. Schematic of a 4-mm data cassette.

height form factors similar to the 6.35-mm data cartridges. Storage capacity ranging from 20 MB to 600 MB can be obtained.

(c) Instrumentation Recorders

Instrumentation recording covers the broad field of data recording in which analog or digital information is stored in other than the standard audio, video, or computer formats. In a linear instrumentation recorder, a 12.7-mm or 25.4-mm (more common) wide instrumentation tape is moved over a multitrack head (with 14, 28, or 42 tracks). Because of an extremely high data rate (100–400 Mbits/s) and high storage-density requirements (several GB), helical-scanning rotating-head configuration with an 8-mm- or 12.7-mm-wide tape (similar to broadcast video) is more commonly used today. The 19-mm-wide tapes in helical-scanning recorders are also expected to become popular for instrumentation recorders. Digital recording is now generally used to obtain low error rates. We note that there is lot of redundancy in the data recorded by the instrumentation recorders, therefore, their error rate requirements (bit error rate of the order of 10^{-6}) are not as stringent as for computer storage applications.

1.2.2.2. Flexible-Disk Drives

A flexible disk, also called a floppy disk or diskette, is a magnetic recording medium which is physically a thin and pliable disk and functionally a removable, random-access cartridge. The flexible-disk drive is a mechanism in electronics which writes and reads data from the disk. Figure 1.25 shows a schematic of the key elements of a flexible-disk drive (Engh, 1981). The drive is

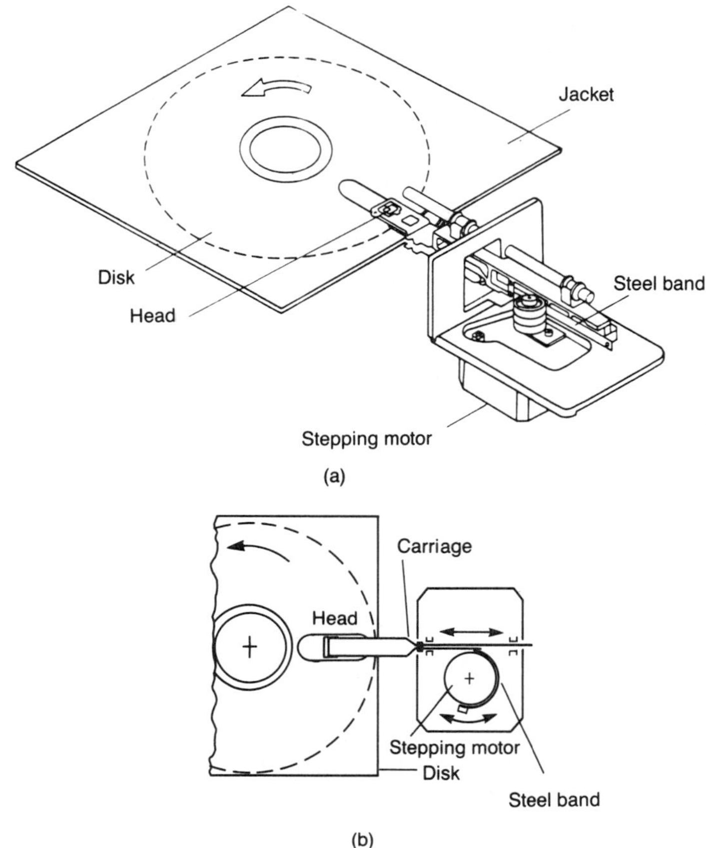

Fig. 1.25. Schematic of the key elements of a flexible disk drive with head positioning by a linear actuator using a stepping motor and capstan-band system (a) isometric view and (b) top view.

designed for easy insertion and removal of the single flexible disk (diskette) in its enclosing jacket. When mounted in the drive, the disk is clamped at its center and rotated at a relatively low speed while the read–write head accesses the disk through a slot in the jacket. In most designs the accessing arms traverse above and below the disk, with the read–write head elements mounted on spring suspensions. When reading and writing take place, the suspensions push the heads together with the double-coated disk interposed. In this way, writing and reading can be achieved on both sides of the disk. Since the heads are in contact with the recording surface, high linear densities can be achieved. On the other hand, track densities are relatively low compared with rigid-disk files, because of the poor dimensional stability of the plastic disk substrate, the limited positioning accuracy of the disk clamp, and the lack of a track servo system in a low-cost drive. Track position inaccuracy is allowed for by providing a tunnel-erasing head. The new track is written wide and trimmed narrow with tunnel-erase gaps which erase either side of the track to clean out unwanted information.

Disk speed stability is critical for a flexible-disk drive. Most drives today use a brushless dc servo motor to drive the disk spindle directly. Some disks use an index hole or sector hole located inside the innermost track. When the index hole is sensed with a detector, it indicates the beginning of a track. Individual sectors are located at uniform angular positions from the index position and are identified by markers recorded in the same way as were data on the track. High track-density systems use a sector or embedded servo.

The magnetic head must be positioned in contact with the correct track before writing or reading. This is accomplished by mounting the head on an arm that moves radially to the section of the disk allocated to the read–write operation. These types of actuators are known as linear actuators. Current trends are toward smaller, more compact disk storage devices and the low-mass, low-cost, rotary actuators are used to save space in the drive in which the head is mounted on a rotary arm. With a rotary actuator, the head orientation (with respect to disk radius) changes continuously (Bhushan, 1990). The actuator is operated either by a stepping motor or voice-coil motor (VCM). The VCM is very much like a loudspeaker coil/magnet mechanism. The VCM has the advantage of providing the desired linear or rotary motion directly, whereas circular motion provided by the stepping motor needs to be converted to the desired linear or rotary motion, generally by a lead screw or a capstan-band method. Furthermore, the VCM provides faster and smaller stepping than that by the stepping motor. However, VCMs are more expensive than stepping motors. Figure 1.23 shows a schematic of a linear actuator in which the head positioning is accomplished by a stepping motor, and whose rotational motion is translated to a linear motion of the actuator via a steel band. The steel band is actually a spring that is fastened to the capstan on the stepping motor shaft. The compliance of the band preloads the actuator such that backlash is minimized, and repeat positioning of the head improved.

The 200-mm (8-in.) form factor disk drives which use 200-mm-diameter disks (with a shell width of 203 mm (8 in.)) were introduced in 1970 by IBM and their use is very limited; some of the very high density (10–21.4 MB) Bernoulli-type disk drives made by Iomega (to be discussed later) are also 200-mm disk drives. The initial 200-mm form factor drives were followed in 1976 by 130-mm (5.25-in.) form factor drives which use 130.2-mm-diameter disks [with a shell width of 133 mm (5.25 in.)] introduced by IBM, and in 1980 by 90-mm (3.5-in.) form factor microfloppy disk drives which use 85.8-mm-diameter disks [with a shell width of 90 mm (3.5 in.)] introduced by Sony. In 1987, Matsushita Electronic Components introduced 75-mm (3-in.)-diameter disks (which are rare) and Matsushita Communications Industrial (followed by Chinon, Mitsumi Electric, and Sony) introduced 50-mm (2-in.) form factor drives which use 47-mm or 50.8-mm [nominal 50 mm (2 in.)]

diameter disks (with a shell width of 60 mm) for "notebook" computers. Vertical heights of the drive have come down significantly from full height [116.8 mm (4.6 in.), 82.6 mm (3.25 in.), and 50.4 mm (2 in.) for drives with 200-mm, 130-mm, and 90-mm diameter drives, respectively] to half-height [58.4 mm (2.3 in.), 41.3 mm (1.625 in.), 25.4–27.9 mm (1–1.1 in.) for drives with 200-mm, 130-mm, and 90-mm form factor, respectively]. The newest 130-mm-diameter drives have one-third height of 25.4–27.9 mm (or 1–1.1 in.). The newest drives with 90-mm and 50-mm form factor have only 17–19-mm (0.75 in.) of vertical front panel space for notebook and laptop computers. This is significant because it permits two drives to be mounted in the space of a standard 41.3-mm drive bay. Some manufacturers expect to be able to eventually manufacture flexible drives with 12.7-mm (0.5-in.) height form factors.

The most commonly-used disk diameters today are 85.8 mm and 130.2 mm. The 85.8-mm disks (with 25-mm inner diameter) use a metal hub and are encased in a hard plastic jacket which does not bend like the soft jackets of a 130.2-mm-diameter disk (with 40-mm inner diameter) and incorporates a shutter to protect the disk surface, Fig. 1.26 (Katoh et al., 1981, 1983; Tsukahara and Yanaga, 1984). Disk clamping, accuracy, and security are important factors in achieving the disk interchangeability. Drives for 130.2-mm-diameter disks clamp the center hole of a disk between a metal spindle hub and a plastic collet, as shown in Fig. 1.27(a). Most 85.8-mm disks are incorporated with a metal hub at the center which is captured by the spindle with a magnet. Drives for these disks have a thin spindle pin and a self-indexing pin, to chuck and rotate the disk hub, instead of a wide center hole, as shown in Fig. 1.27(b) (Takahashi, 1984).

The disk heads are inductive-coil-type composite heads. In a single-sided flexible-disk drive the head button has a spherical surface contour, and the spinning disk is locally sandwiched between the spherical head and a spring-loaded pressure pad. In a dual-sided flexible-disk drive (almost exclusively used today), the pressure pad is replaced by another read–write head. The two heads are either spherically contoured and slightly offset (to reduce normal pressure) or are flat and loaded against each other. In most disk heads, a single core is used for both read and write functions as opposed to separate sets of read cores and write cores used on tape drives. The head structure contains a separate tunnel-erase head to erase previously recorded information (Baasch and Luecke, 1981). These components are assembled in the magnetically-inert structure shown in Fig. 1.28 (Engh, 1981).

The flexible disk typically about 82-μm thick consists of a 76.2-μm flexible-base film coated on both sides with a magnetic coating. Most data-processing drives for personal computers today use either 85.8-mm- or 130.2-mm-diameter disks, and laptop computers use 85.8-mm-diameter disks with γ-Fe_2O_3 particles. Typical unformatted capacities and track densities for the high-density 85.8-mm- and 130.2-mm-diameter disks are 1.6 MB and 0.8 MB and 5.31 tracks/mm and 1.89 tracks/mm, respectively, Table 1.2. The 47-mm-

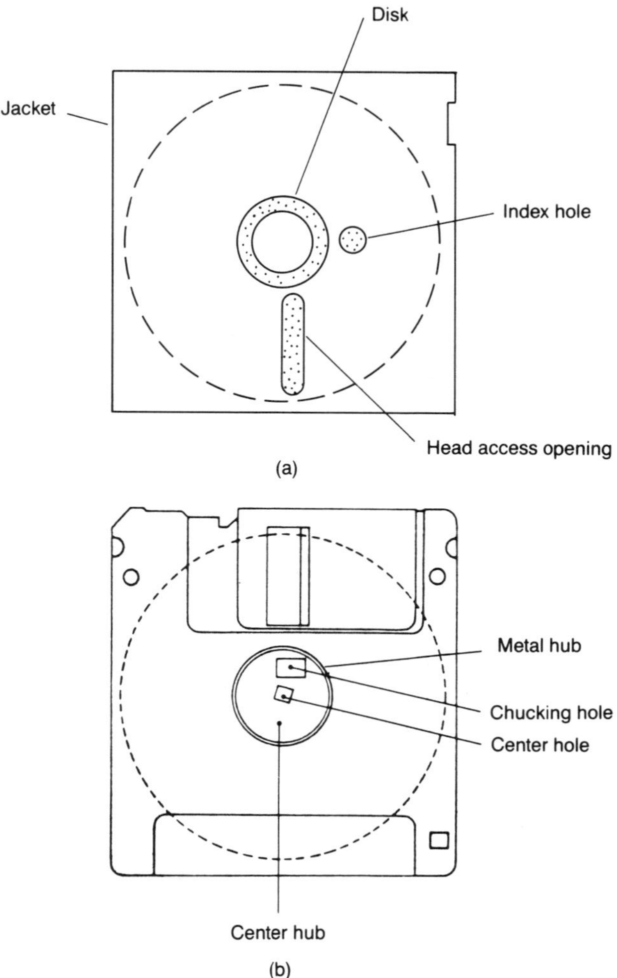

Fig. 1.26. Schematics of (a) a conventional 130-mm (5.25-in.) form factor flexible disk and (b) a hard plastic 90-mm (3.5-in.) form factor disk cartridge showing shutter that closes over the head-accessible window to protect the disk surface.

or 50.8-mm-diameter disks with $Co-\gamma-Fe_2O_3$ particles with an unformatted capacity of 1 MB and a track density of 10 tracks/mm are common for "notebook" computers. The 85.8-mm and 130.2-mm disks containing barium ferrite or metal particles (primarily barium ferrite) with a capacity of 4–21.4 MB are available for high-density applications, Table 1.5 (e.g., manufactured by Brier Technology Inc., San Jose, California; Citizen Watch Co., Tokyo; Insite Peripherals Inc., San Jose, California; Matsushita Communication Industrial Co., Yokohama, Japan; NEC Corp., Tokyo; Teac Corp., Tokyo; Toshiba Corp., Tokyo; Y-E Data, Tokyo, and Verbatim Corp.). Be-

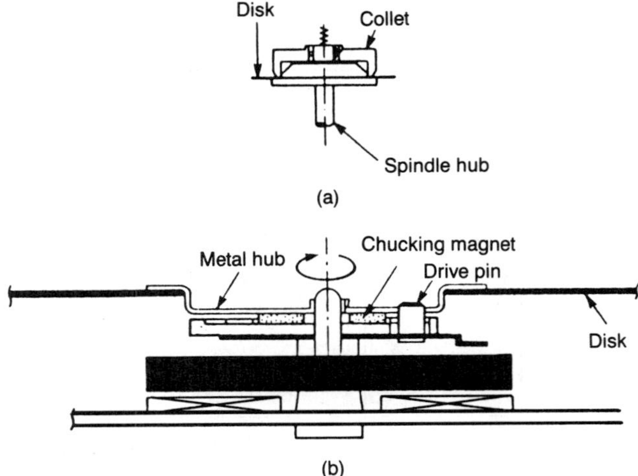

Fig. 1.27. (a) Direct clamping method used to connect 130-mm form factor flexible disks to the drive shaft, and (b) a magnet chucking method used to connect 90-mm form factor disks (in a hard plastic case) with a metal hub, to the drive shaft.

cause of requirements of high track densities for high density drives, these use embedded servo techniques.

A so-called "Bernoulli" disk has been developed by Iomega Corp., Roy, Utah (the Bernoulli box, Alpha and Beta products) to obtain a stable head–disk interface at higher operating speeds to achieve high data transfer rates. In a design proposed by Norton and Wright (1978), an 200-mm-diameter flexible disk is rotated at high speed (1500 rpm) in close proximity to a fixed flat (polished aluminum) plate called the Bernoulli plate (Fig. 1.29). A foundation stiffness results from hydrodynamic loading (air bearing) created by

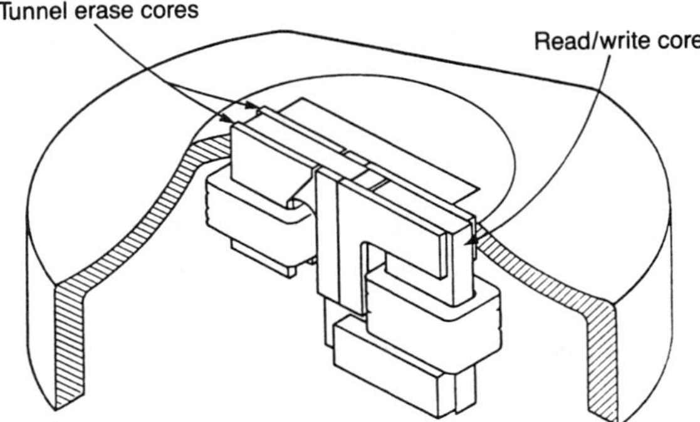

Fig. 1.28. Schematic of an inductive-coil-type magnetic head with tunnel erase and read/write cores for a flexible disk drive (Engh, 1981). © 1981 International Business Machines Corporation; reprinted with permission.

Table 1.5. Typical characteristics of 90-mm and 130-mm form factor flexible-disk drives for high-density data storage applications

Product number	Toshiba PD-211	Toshiba PD-401	Brier Technology BR-3225	Verbatim
First customer shipment	1988	—	1987	1989
Disk diameter, mm	85.8	85.8	85.8	130.2
Drive size ($H \times W \times D$)	25.4 × 101.6 × 149.9 mm (half-height)	25.4 × 101.6 × 149.9 mm (half-height)	25.4 × 101.6 × 146.1 mm (half-height)	41.3 × 146.1 × 203.2 mm (half-height)
Magnetic particles		Barium ferrite		
Rotational speed, rpm	300	1500	720	720
Maximum linear density, kb/mm	0.34/1.37	1.38	1.02	0.95
Track density, /mm	5.31	21.34	30.59	26.22
Unformatted capacity, MB	1.0/2.0/4.0	16.0	21.4	20.2
Data transfer rate, kB/s	31.25/62.5/125	625	1250	330
Track-to-track access time, ms	3	3	15	25
Actuator type	Lead screw, stepping motor	Linear, voice coil	Linear, voice coil	Voice coil

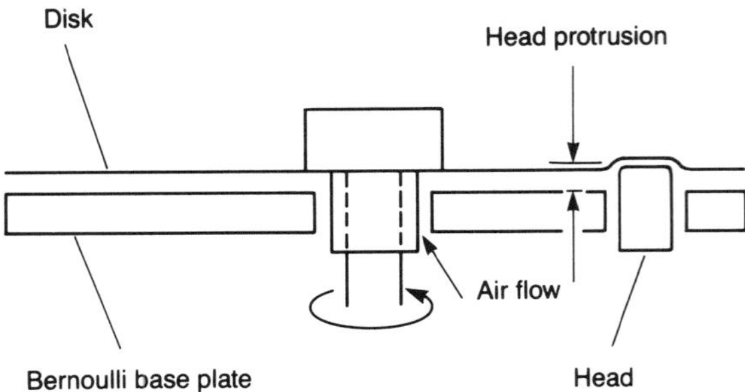

Fig. 1.29. Schematic of head–flexible-disk interface in a Bernoulli disk drive.

spinning the disk in close proximity to the base plate. This is an aerodynamic (Reynolds) effect and it is misnomered as the Bernoulli effect. The high disk stiffness pushes up the first critical speed. Therefore, these disks can be operated at high speeds without disk flutter. A slot in the Bernoulli plate allows the magnetic head to access the disk. In the more recent design of smaller drives with 130-mm-diameter disks (Beta drives), the Bernoulli plate (a heavy aluminum plate) has been eliminated. In these drives, the inner surface of the cartridge (used to house the disk) acts as a rigid Bernoulli plate. Two disks are used to provide stability in the case of opposed heads. Bernoulli drives in the form factor of 130.2-mm diameter, with a formatted capacity of 21.4 MB (Beta 20) and 44.5 MB (Beta 44, introduced in 1989), are commercially available, Table 1.6 (Tripathi, 1990). The capacity, performance, and cost of these drives have traditionally placed them in competition with small rigid-disk

Table 1.6. Typical characteristics of 200-mm and 130-mm form factor flexible-disk drives for high-density data storage applications

Iomega product number	A120H/A220H Bernoulli box	Beta 20
First customer shipment	1985	1986
Disk diameter, mm	200	130.2
Drive size ($H \times W \times D$)	$160.5 \times 317.5 \times 387.4$ mm (full height)	$41.3 \times 146.1 \times 203.2$ mm (half-height)
Magnetic particles	$Co-\gamma\text{-}Fe_2O_3$	
Rotational speed, rpm	1500	1845
Maximum linear density	0.94 kb/mm	0.94 kb/mm
	0.71 kfr/mm	0.69 kfr/mm
Track density, /mm	25.2	22.4
Formatted capacity, MB	21.4	21.4
Data transfer rate, kB/s	1130	666
Track-to-track access time, ms	10	6.2
Actuator type	Rotary, voice coil	Linear, voice coil

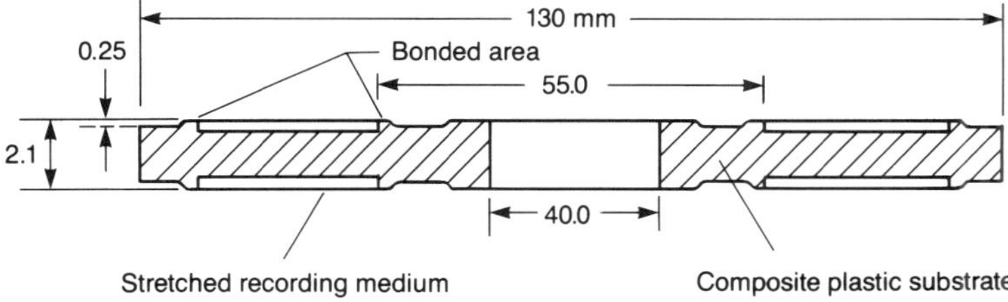

Fig. 1.30. Schematic of a cross section of a stretched surface recording (SSR) medium.

drives and removable rigid-disk cartridge drives, rather than with the most of the fixed flexible-disk drives available in the past. These drives are relatively expensive and are used for specialized applications, e.g., secure data applications.

A rigid disk in combination with a stretched, web-coated flexible medium, called stretched surface recording (SSR), has been developed by 3M, St. Paul, Minnesota, for its better dimensional stability than the unsupported flexible disk, giving the potential for higher track densities and data rate (Knudsen, 1985). The ability of SSR disks to operate at 3600 rpm allows the same data rate and access speed as rigid-disk products. It consists of an injection-molded composite plastic (typically polyimide-based) substrate with raised edges at the inner and outer diameters of the disk. A web-coated, flexible medium (typically PET substrate) is isotropically stretched and bonded to the raised edges at the inner and outer diameters of the disk, Fig. 1.30. This results in a compliant surface with a 250-μm gap between the substrate and the media. The SSR disk was developed to provide the performance of a rigid disk at a much lower cost, owing to the use of the more economical polymer film substrate to attain the required surface smoothness. However, SSR disks have not yet been introduced commercially.

1.2.3. Head Materials

There are several classes of soft magnetic materials that have been evolved for use in the construction of the magnetic core of magnetic heads. Table 1.7 lists some commercial materials used in modern recording technology, their magnetic properties, and their microhardnesses.[4] Details on these materials follow (Cullity, 1972; White, 1985; Jorgensen, 1988; Mee and Daniel, 1990).

---

[4] Materials under development for ultrahigh density (~ 1 Gb/in.^{2}) recording include Co–Zr–Nb and FeN with saturation magnetization of 14,000 and 20,000 G, respectively; these can be used to write and read on a magnetic medium of coercivity as high as 1800 Oe and 2500 Oe, respectively.

Table 1.7. Properties of soft magnetic materials for heads

Material	Composition, wt.%	Permeabilities		Coercivity, Oe	Saturation magnetization, G	Curie temperature, K	Electrical resistivity, μ ohm·cm	Vickers microhardness, GPa (kg/mm^2)
		μ_0	μ_{max}					
Molybdenum permalloy	79 Ni 17 Fe 4 Mo	20,000	100,000	0.05	8,700	730	100	1.2 (120)
Mu-metal	77 Ni 14 Fe 5 Cu 4 Mo	30,000	100,000	0.02	7,500		100	
Hy-Mu 800B	70 Ni 17 Fe 4 Mo Nb[a] Ti[a]	70,000	250,000	0.002	8,000		110	2.5 (250)
Sendust	85 Fe 9 Si 6 Al	8,000		0.03	10,000	770	85	5.4 (550)
Alfenol	84 Fe 16 Al	4,000		0.04	8,000	670	150	2.8 (290)
Hot-pressed ferrites	Ni–Zn	1,500	2,000	0.30	4,300		10^{11}–10^{13}	6.9 (700)
	Mn–Zn	10,000	15,000	0.15	5,000	420	5×10^4–5×10^5	5.9 (600)
	Single–crystal Mn–Zn	1,000		0.05	4,000	440	5×10^3–5×10^4	5.9 (600)

[a] Small additions.

1.2.3.1. Permalloys

The general term permalloys refers to alloys of Ni and Fe. A particular composition of 79% Ni, 17% Fe, and 4% Mo (molybdenum permalloy) has very small anisotropy and magnetostriction (a measure of change in magnetic properties such as permeability as a function of strain applied during original formation, machining, and lapping or through wear and use). This leads to very low coercivity and high permeability. The low resistivity of permalloys, however, requires a laminated structure in a conventional ring head in order to minimize eddy current losses. They are mechanically soft, which limits their use to noncontact applications. Permalloys can be made with near-zero magnetostriction, rendering them relatively insensitive to strain and high-saturation magnetization and good writing performance.

1.2.3.2. Mu-Metal and Hy-Mu 800B

Mu-metal has been successfully used in heads and efforts continue to develop better alloys. For example, Hy-Mu 800B has been shown to have an enhanced hardness while still retaining very high permeability and low magnetostriction.

1.2.3.3. Sendust Alloys

Sendust alloys (Fe–Si–Al) have high permeability, low coercivity, and near-zero magnetostriction comparable to molybdenum permalloy. The principal advantage of Sendust is its high hardness; it is often used in pole tips.

1.2.3.4. Alfenol Alloys

Alfenol is an alloy with properties intermediate between those of molybdenum permalloy and Sendust. Alfenol is somewhat easier to form than Sendust and has relatively high hardness and resistivity in comparison to molybdenum permalloy. The permeability of Alfenol is somewhat lower than either of the other materials, but its magnetic performance is still adequate for many applications.

1.2.3.5. Amorphous Magnetic Alloys

Ferromagnetic amorphous alloys are a relatively new class of materials typically formed of 75–85 at. % transition elements (Fe, Co, Ni), with the balance being a glass-forming metalloid (B, C, Si, P, or Al) (Luborsky, 1980). Like glasses, these materials have no structural order that can be detected, except that associated with the near-neighbor interaction and, consequently, are free of some of the characteristics of crystalline materials, such as crystalline anisotropy, porosity associated with crystal grains, and stresses associated with grain growth. Amorphous magnetic materials generally have smaller permeability losses due to eddy currents than crystalline materials, since the

Table 1.8. Microhardness and wear rates for various amorphous magnetic alloys

Composition	Vickers microhardness, GPa (kg/mm^2)	Relative wear factor
Fe–Si–Al (Sendust, for comparisons)	5.4 (550)	1
$(Co_{85.5}Nb_{14.5})_{98}B_2$	8.3 (850)	0.3
$(Co_{85.5}Nb_{14.5})_{95}B_5$	7.8 (800)	0.4
$(Co_{85.5}Nb_{14.5})_{90}B_{10}$	8.8 (900)	0.6
$Fe_{2.5}Co_{71.5}Mn_3Si_8B_{15}$	8.8 (900)	2
$Fe_{80}P_{13}C_7$	7.5 (760)	3
$Fe_{40}Ni_{40}P_{14}B_6$	7.4 (750)	4

resistivity of these alloys is typically two to four times larger than the corresponding transition metal crystalline alloys without the metalloid glass-forming elements. Amorphous magnetic alloys have shown desirable permeabilities and saturation magnetic inductions, low magnetostriction, and suitable hardness for head applications.

Table 1.8 shows Vickers microhardness and wear rates for a number of amorphous alloys in comparison with Fe–Si–Al (Sendust) (Kohmoto et al., 1979; Sakakima et al., 1981). The Co–Fe–Si–B alloy has a higher saturation magnetic induction than the Mn–Zn ferrite head and a higher permeability than Sendust heads. Thin ribbons (10–25-μm thick) of Co–Fe–Si–B amorphous alloys have been used in single-lamination designs for narrow-track video recording heads. Amorphous Co–Zr (90–10 at. %) alloys produced by magnetron sputtering have been used to fabricate film heads. A serious concern regarding amorphous alloy heads is their stability in the presence of frictional heat and poor wear performance.

1.2.3.6. Ferrites

The crystal structure of Mn–Zn ferrite (17% MnO, 11% ZnO, 72% Fe_2O_3) and Ni–Zn ferrite (11% NiO, 22% ZnO, 67% Fe_2O_3) is that of the spinels, as illustrated for Mn–Zn ferrite in Fig. 1.31. Oxygen ions are in a nearly close-packed cubic array. However, the distribution of the cations, such as Fe^{3+}, Mn^{2+}, and Zn^{2+}, in the available sites must be determined experimentally. This distribution influences not only the magnetic properties of Mn–Zn ferrite but also its mechanical properties and slip systems. Polycrystalline ferrites are produced by cold- or hot-pressing powders obtained by ball milling or precipitation from solution. The heat can be supplied by rf induction or resistance furnaces. The pressure can be applied in any of three basic modes: unidirectional, bidirectional, and isostatically (via a pressurized gas to an encapsulated sample). High-density ferrites are generally produced by hot

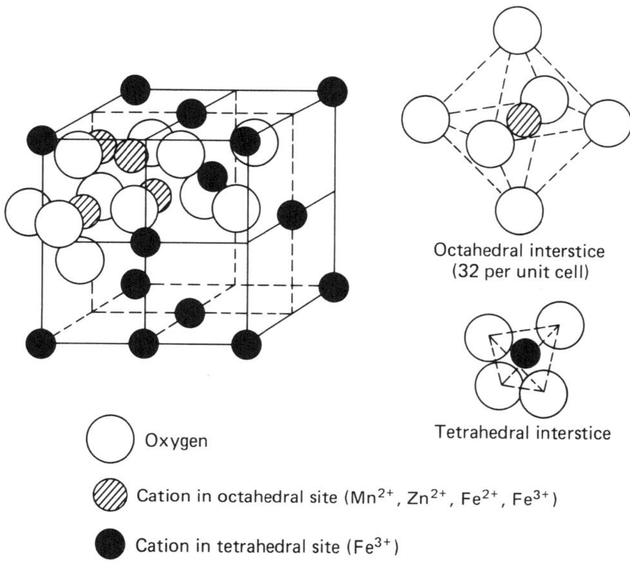

Fig. 1.31. Spinel structure of a Mn–Zn ferrite crystal (Kingery et al., 1976).

isostatic pressing, commonly referred to as the HIP process (Monforte et al., 1971). These have a porosity of less than 0.1% with a grain size in the 5–7-μm range. Single crystals of Mn–Zn ferrite are usually grown by the Bridgman method (Mizushima, 1971). A carbon heater in an inert gas and silicon–carbide heat elements in an atmosphere of CO gas are commonly used to grow large, single crystals.

The resistivity of ferrite materials is at least three orders of magnitude greater than that of most metallic magnetic materials, so that eddy currents and associated permeability losses are relatively small. As a consequence, these materials have dominated the field of high-frequency head applications for the last 20 years. Ferrites are also hard (Table 1.7) and are wear resistant. We note that in the case of single-crystal ferrites, the hardness depends on the crystallographic orientation. The most closely packed plane of single-crystal Mn–Zn ferrites (i.e., {110}) exhibits the greatest hardness (Table 1.9). With regard to magnetic properties, the most serious problem with ferrites is their

Table 1.9. Microhardness of single-crystal Mn–Zn ferrites

Surface	{110}	{100}	{111}	{211}
Direction	$\langle 001 \rangle \langle 110 \rangle$	$\langle 001 \rangle \langle 011 \rangle$	$\langle 112 \rangle \langle 110 \rangle$	$\langle 111 \rangle \langle 011 \rangle$
Knoop microhardness, GPa (kg/mm^2)	6.38 (650)	6.18 (630)	6.08 (620)	5.98 (610)

inherently low value of saturation magnetization, generally less than 5000 G, which is significantly lower than that for metallic magnetic materials.

Ni–Zn is preferred over Mn–Zn ferrite for very high frequency operations because its relatively high resistivity suppresses permeability losses due to eddy currents. On the other hand, Mn–Zn is preferred at frequencies below a few megahertz because it has a lower coercivity, higher permeability, higher saturation magnetization, and lower magnetostriction (Mizushima, 1971; Hirota et al., 1971). Although the Ni–Zn ferrite is harder and is desirable for low wear, it tends to have a thicker "magnetic-dead layer" (or nonmagnetic layer) due to lapping or wear damage. The dead layer increases the magnetic spacing between the head and the medium on the order of tens of nanometers.

Single-crystal Mn–Zn ferrites were developed to improve porosity and wear resistance, and in several ways, to represent superior magnetic (low coercivity) and tribological material. The microcracking and grain pullout problems of polycrystalline ferrite during processing (chipping) and use are largely avoided. If specified crystal orientations are used, single-crystal heads also exhibit excellent wear properties. However, the permeability of single-crystal ferrites is low and magnetostriction-induced noise due to media contact can be a problem. Single-crystal ferrites are also very expensive. With certain polycrystalline ferrites, the existence of many grains and grain boundaries suppresses magnetostriction noise. The choice of single-crystal orientation is a compromise among wear rate, ease of processing, permeability, magnetostriction noise, and cost. Improved polycrystalline ferrites are available from hot-pressing sintering techniques, which seem to give fairly comparable wear performance to single-crystal material (Mizushima, 1971; Fisher and Blades, 1971).

1.2.3.7. Some Examples of Head Constructions

Most drives use inductive read–write heads. The conventional heads are combinations of a body forming the air-bearing surface and a magnetic ring core carrying the wound coil with a read–write gap. For tape heads, the core materials that have been typically used are permalloy and Sendust. However, since these alloys are good conductors, it is sometimes necessary to laminate the core structure to minimize losses due to eddy currents. Laminations with a thickness as low as 25 μm have been used to reduce high-frequency eddy-current losses. The laminations are epoxied together and coils are wound onto the bonded stacks. The stacks (each stack represents a track) with a nonmagnetic spacer (e.g., Cu–Be) in the gap are then assembled in a machinable brass housing and subsequently the top surface is lapped and polished to the desired contours. As shorter gap lengths (< 1 μm) are required for short-wavelength recording and playback, other gap-spacer materials, such as glass or silica, are used and are deposited by evaporation or sputtering. The copper wire is generally used for the coil. In many computer-tape drive applications, the air-bearing surfaces are coated with wear resistant coatings, such as electroplated Cr, plasma sprayed Al_2O_3–TiO_2 (87–13 wt. %), or yttria-stabilized zirconia (8–92 wt. %). (See, for example, Fig. 1.17.)

Trends to film-head design have been driven by the desire to capitalize on semiconductorlike processing technology to reduce fabrication costs. In addition, thin-film technology allows the production of high-track density heads with accurate positioning control of the tracks and high reading sensitivity. Read and write heads in modern tape drives are miniaturized using thin-film technology. Separate read- and write-head assemblies can be obtained by using two coil structures on their ferrite substrates, with a common ferrite yoke to form the closure paths (Brock and Shelledy, 1975). With a few exceptions, film heads have used a magnetic film of permalloy (Ni_{80}– Fe_{20}) deposited by evaporation, sputtering, or plating (Howard, 1986; Mee and Daniel, 1990). High-conductivity metals, such as copper deposited by evaporation, sputtering, or plating and gold deposited by plating, have been used for conductors (coil). The most frequently used insulation and gap materials in film heads are SiO_2, SiO, and Al_2O_3, which are deposited by evaporation or sputtering and are reasonably durable and hard (Berghof and Gatzen, 1980). In some cases, organic insulators, such as cured photoresist polyimide, can be used, although they are much softer and less thermally stable materials.

Film heads must be deposited on substrates that must exhibit good wear characteristics, high thermal conductivity, good machining characteristics, and in some instances, soft magnetic characteristics. Several designs have been built that have a hybrid thin-film bulk magnetic material construction. In these cases, ferrite is chosen to serve as a pole or shield, and the balance of the structure is in film form. Sometimes Ni–Zn ferrite is preferred over Mn–Zn ferrite because its high resistivity does not lead to a shorting out of deposited-film conductors (e.g., IBM 3480 tape drives).

Modern tape drives (e.g., the IBM 3480 tape drive) have introduced thin-film magnetoresistive (MR) *read* heads and planar thin-film write turns (inductive write head), Fig. 1.20 (Cannon et al., 1986). The MR read head includes a Ni–Zn ferrite substrate on which thin films have been deposited and patterned, and a Ni–Zn ferrite closure which is held in place with epoxy. The films consist of a MR element of 81% Ni–19% Fe, a biasing conductor of Ti, and two spacing layers of Al_2O_3. The MR element is center-tapped and is provided with bias current at the two outside taps, where the signal is sensed differentially. The ferrite shields that surround the MR element provide the necessary resolution for high-linear-density signals. The MR element amplitude is independent of tape speed and has a much simpler topology than an inductive film head to an equivalent amplitude. However, in order to contain the signal degradation, the total wear of the MR element should be less than that of the inductive element. The inductive write head consists of a Ni–Zn ferrite substrate on which is deposited a film of gold patterned into a two-turn spiral winding. Al_2O_3 is used as indicated. A Ni–Zn ferrite closure

Table 1.10. Selected physical properties of hard head materials

Material	Density, kg/m^3	Young's modulus, GPa	Knoop microhardness, GPa (kg/mm^2)	Flexural strength, MPa	Electrical resistivity, μ ohm · cm
Ni–Zn ferrite	4570	122	6.9 (700)	150	10^{11}–10^{13}
Mn–Zn ferrite	4570	122	5.9 (600)	120	5×10^4–5×10^5
ZrO$_2$–Y$_2$O$_3$ (94–6)	6360	210	12.8 (1300)	500–700	10^{16}
BaTiO$_3$	4320	110	10.3 (1050)		

is held in place with epoxy. The closure includes a glass isolation layer and forms a second magnetic pole-piece. The closure is more complex than in the read module, because it must perform the function of defining the width of the written track. To do this, it is slotted so that for each track there is a pole-piece magnetically isolated from all other tracks. The selection of Ni–Zn ferrite for the substrate and pole-tip material results in a wear-resistant structure which, because of negligible electrical conductivity, does not require any insulating layers.

Flexible-disk heads are inductive-coil-type composite heads. Multigap structures have evolved due to the requirement to erase previously recorded information in adjacent track regions. Media interchangeability could be degraded by head–track misalignment caused by tolerances from parts, adjustments, and environmental effects. A tunnel-erase head was adopted to absorb these variations. The new track is written wide and trimmed narrow with tunnel-erase gaps which erase either side of the track to clean out unwanted information. Figure 1.26 illustrates the structure of the flexible-disk composite head. Mn–Zn ferrite and Ni–Zn ferrite are used for flexible-disk head cores in the read/write and tunnel erase sections and barium titanate is selected for the magnetically inert structures which support the cores. The ferrite materials have higher permeability at high frequencies and higher resistance to wear than permalloy. Mn–Zn ferrite has replaced Ni–Zn ferrite, which is easy to process but inferior in hardness and electrical resistivity. For high density drives, because of narrow track width, sintered single-crystal Mn–Zn ferrite, in place of polycrystalline ferrite, has the advantage of easier mechanical processing. Some physical properties of hard head materials are presented in Table 1.10.

The expected life of modern heads is at least 3000–4000 h for a tape drive and at least 7500 h for a flexible-disk drive, equivalent to about 3 years of drive use. The drive use includes about 7500 contact starts/stops and remaining streaming (flying) under required operating conditions.

1.2.4. Flexible Media Materials

Magnetic media fall into two categories: particulate media where magnetic particles are dispersed in a polymeric matrix and coated onto the polymeric substrate, and thin-film media where continuous films of magnetic material are deposited onto the polymeric substrate by vacuum techniques (primarily evaporation). The overwhelming preponderance of media made to date have used particulate coatings. However, requirements of higher recording densities with low error rates have resulted in an increased use of thin films, which are smoother and considerably thinner than the particulate media; they have begun to be used for high-density audio/video and data processing tape applications. Recently, a few thin-film flexible disks have also become available on a prototype basis for high performance applications.

The chief advantages of thin films are the higher remanent magnetization (M_r) compared to a particulate disk of comparable thickness (which allows the use of thinner recording films while maintaining signal amplitude) and high coercivity. In thin films, the packing fraction of magnets is 1. Thin films lead to a better defined magnetization reversal and consequently to higher recording densities. Film technology also makes possible the tailoring of magnetic properties to meet specific design requirements. Very smooth thin films can be manufactured with fewer isolated high asperities, which allow lower flying heights resulting in higher recording densities. Although thin-films provide a recording density and performance advantage over particulate media, thin-film manufacturing processes have low productivity resulting in much higher cost compared to the particulate media.

1.2.4.1. Base Film

The base film for flexible media is almost exclusively biaxially-oriented film of poly(ethylene terephthalate) (PET). Other substrate materials such as polyimides and polyamide–polyimide copolymers have been explored for isotropy and better dimensional stability of flexible disks. In the production of thin-film metal-alloy flexible media, a tremendous amount of heat is generated. PET film can be barely used for the production of tapes with evaporated films of Co–Ni. In the deposition of some magnetic films, the high substrate temperature has to be maintained (e.g., $>200°$ C for Co–Cr films) to obtain high coercivity and perpendicular orientation of the easy axis. A film base considered suitable is the polyimide which can withstand temperatures of at least $200°$ C. However, these films are frightfully expensive. Typically, a 6.35–36.07-μm-thick (25, 35, 57, 88, 92, and 142 gage) PET substrate (rms roughness ~ 1.5–2.5 nm for particulate media and even smoother ~ 1.5–2 nm rms for thin-film media) is used for tapes and 76.2-μm thick (300 gage) for flexible disks.

1.2.4.2. Magnetic Medium

The hard permanent magnetic materials used in magnetic media with their magnetic properties are listed in Table 1.11 (Mori, 1985; Sharrock, 1989). Electron micrographs of various particles are shown in Fig. 1.32. The most commonly-used magnetic particles are γ-Fe_2O_3 (low coercivity particles, used

Table 1.11. Properties of hard magnetic materials for flexible media

Material		Particle length, μm	Remanence squareness	Coercivity, Oe	Remanence, G	Specific saturation magnetization, emu/g	Curie temperature, K	Density, kg/m^3	SSA,[a] m^2/g	Mean particle volume, 10^{-4} μm^3
Gamma-ferric oxide[b]	γ-Fe$_2$O$_3$	0.3–1	0.75	200–350	1100–1300	70–75	870	4800	12–35	1–25
Cobalt—modified[b]	Co–γ-Fe$_2$O$_3$	0.3–0.6	0.75	600–900	1200–1700	70–75	793	4800	16–33	1–10
Chromium dioxide[b]	CrO$_2$	0.3–1	0.9	400–900	1400–1700	75–85	387	4800	18–36	1–10
Barium ferrite[c]	BaO·6Fe$_2$O$_3$	0.06–0.15	0.5	500–1500	1100–1500	55–70	728	5300	20–35	<2
Metal particles[d]	Fe	0.1–0.2	0.8	900–2000	3500	125–170	1043	5800	35–50	<1
Co–Ni (evaporated)	82 Co 18 Ni		0.9	400–2000	13,000	200		8000		
Co–Cr[e] (Sputtered/evap.)	83 Co 17 Cr		0.9	400–2000	10,000	160	1073	8000		

[a] SSA, specific surface area of particles.

[b] Acicular particles with an aspect ratio of about 5–10.

[c] Hexagonal platelets with long dimension to thickness (easy axis) ratio of 3–30; axis of magnetization perpendicular to the substrate.

[d] Acicular particles with an aspect ratio of about 3–4.

[e] Hexagonal structure with axis of magnetization perpendicular to the substrate.

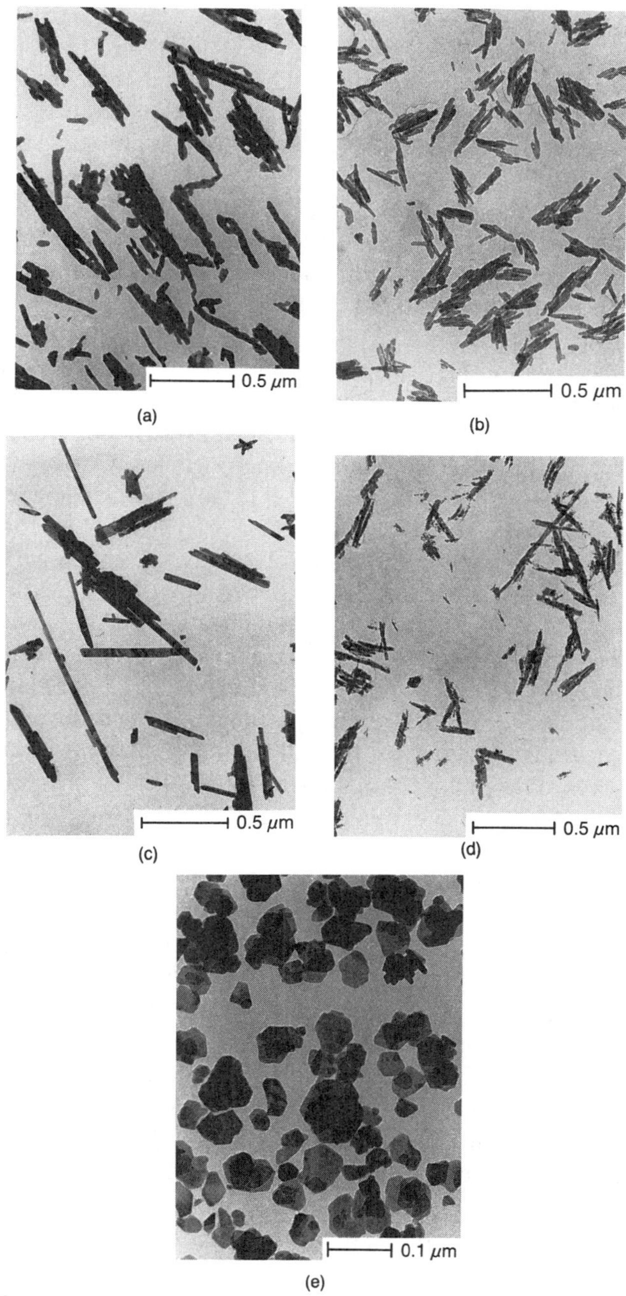

Fig. 1.32. Transmission electron micrographs of (a) γ-Fe$_2$O$_3$ particles, (b) Co-doped γ-Fe$_2$O$_3$ particles, (c) CrO$_2$ particles, (d) metal (Fe) particles, and (e) BaO·6Fe$_2$O$_3$ (barium ferrite) particles (Sharrock, 1989). © IEEE, 1989.

for low recording density), Co-modified γ-Fe$_2$O$_3$ (γ-Fe$_2$O$_3$ particles to which 3–6 wt.% cobalt has been added), and CrO$_2$ for tapes and γ-Fe$_2$O$_3$ and Co-modified γ-Fe$_2$O$_3$ for flexible disks. These particles are acicular in shape with an aspect ratio of about 5–10. These particles are magnetically uniaxial and derive their magnetic anisotropy partly from their acicular shape and partly from magnetocrystalline sources.

Barium ferrite (BaO · 6Fe$_2$O$_3$) and metal (iron usually alloyed with cobalt or nickel) particles with very high coercivity are used commercially for high-recording density/performance audio/video and data-processing applications (Katoh et al., 1983; Chubachi and Tamagawa, 1984; Imamura et al., 1986; Fujiwara et al., 1986; Speliotis, 1987; Suzuki et al., 1989; Yamamori and Tanaka, 1989; Noda et al., 1990). In secure credit-card applications, tapes with magnetic particles of high coercivity particles such as barium ferrite are primarily used. Barium ferrite particles are extensively used in flexible disks. High-density applications require very fine particles with high coercivity. The size of barium ferrite particles (0.06–0.15 μm) is much smaller and their coercivity is higher than that of γ-Fe$_2$O$_3$ and CrO$_2$ (Table 1.9), and thus, they are very desirable for high-density recording. Furthermore, since these particles are hexagonal platelets with long dimensions to a thickness ratio of 3–30, they have their easy axis (axis of magnetization) easily oriented perpendicular to the substrate. When such platelets are coated onto a substrate, they tend to lie flat and a vertically-oriented medium is obtained (Fujawara, 1985). Further orientation of the particles, by a magnetic field during the manufacturing process, is necessary for vertical recording. However, barium ferrite has poorer recording performance characteristics (signal amplitude, signal-to-noise ratio, overwrite) than metal particles.

The Curie temperature (at which magnetization becomes zero) of the metal particles is the highest of any of other particles. The disadvantage of the metal particles is the need for protection against the tendency to rapid oxidation (or other reactions) with the atmosphere that could result from the very high specific surface areas (~ 50 m^2/g). Metal particles are also more catalytic to hydrolytic degradation of the coating material and susceptible to chemical or electrochemical corrosion. Therefore, their use is limited for archival storage of data, and they are primarily used for high performance audio/video tape applications. Metal particles for use in recording media are stabilized against corrosion by the use of alloying elements and additives, by the protective action of the ceramic coatings Al$_2$O$_3$ and SiO$_2$, and by a controlled oxidation of surfaces.

The most commonly-used magnetic thin films (100–300 nm in thickness) are evaporated Co–Ni (85/15 to 70/30) and evaporated/sputtered Co–Cr (85/15 to 70/30). (Sputtering is avoided because of low deposition rates compared to that of evaporation.) The columnar microstructure with tilted columns results in high magnetic anisotropy (Feuerstein and Mayr, 1984; Nishikawa et al., 1985). Vacuum-deposited Co–Cr films have a hexagonal-closed-packed (HCP) structure and these can be produced with magnetization easy axis of the film oriented perpendicular to the film plane thus it can be used for vertical recording (Iwasaki, 1984; Ouchi, 1990). Electroplated Co has also been deposited in the form of needlelike crystals, with the hexagonal crystal axis normal to the film (Chen and Cavallotti, 1982). The materials which have been electroless plated include Co–P, Co–Ni–P, and Co–Ni–Re–P (Judge et al., 1965; Matsuda et al., 1985). Electro/electroless plated coatings are not in commercial use for flexible media probably because of inherent disadvantages of plated media such as poor magnetic noise characteristics and high film porosity, which leads to poor corrosion resistance compared to that of vacuum-deposited films. We note that the corrosion resistance of Co–Cr alloys is significantly superior to that of Co or Co–Ni alloys without Cr (Dubin et al., 1982; Kuhn and Wellwood, 1986; Bhushan, 1990). High-performance magnetic films are sometimes applied in several (typically 2 or 3) discrete layers, possibly separated by nonmagnetic insulation layers which result in high coercivity and low noise.

Barium ferrite has also been applied by using dc diode sputtering for the purpose of vertical recording rigid-disk applications. They have not become popular for flexible media since the substrate temperature requirement during the deposition is over 500° C in order to obtain the coating with high coercivity and perpendicular orientation of the easy axis (Matsuoka and Naoe, 1985; Morisako et al., 1986).

The thin-film magnetic media (primarily evaporated Co–Ni) are commercially used for high-performance audio/video tape applications. Flexible media require relatively flexible coatings in order to avoid crazing and cracking of the coating during use. The metal films have elastic properties much different to those of its plastic substrate. Another disadvantage of thin-film media include, in some cases, the lack of the necessary chemical stability (corrosion susceptibility) and durability unless overcoated with a protective magnetic layer (Bhushan, 1990; Ouchi, 1990).

1.2.4.3. Particulate Magnetic Coatings

The base film is coated with a magnetic coating, typically 2–4-μm thick (coated on both sides with a magnetic coating in the case of flexible disks), composed of acicular magnetic particles (such as γ-Fe_2O_3, Co-modified γ-Fe_2O_3, CrO_2, and metal particles for horizontal recording or hexagonal platelets of barium ferrite for both horizontal and vertical recording), polymeric binders [such as polyester–polyurethane, polyether–polyurethane, nitrocellulose, poly(vinyl chloride), poly(vinyl alcohol–vinyl acetate), poly (vinylidene chloride), VAGH, phenoxy, and epoxy], lubricants (mostly fatty acid esters, e.g., tridecyl stearate, butyl stearate, butyl palmitate, butyl myristate, stearic acid, and myrstic acid), a cross linker or curing agent (such as functional isocyanates), a dispersant or wetting agent (such as lecithin), and solvents (such as tetrahydrofuran and methyl isobutylketone). In some media, carbon black is added for antistatic protection if the magnetic particles are highly insulating and abrasive particles (such as Al_2O_3) are added as a head cleaning agent and to improve the media's wear resistance. Magnetic particles are typically 70–80% by weight (or 43–50% by volume), lubricants are typically 1–3% by weight, carbon black is 0–2.5% by weight, and abrasive particles are 0–6% by weight of the total solid coating (Tochihara, 1982). The coating is calendered to a surface roughness of 8–15 nm rms.

Most flexible media-binder materials have their glass-transition temperature below the operating temperature (i.e., they are in the rubbery state) to provide the desired flexibility. Some block copolymer binders, such as polyester–polyurethanes, have one segment in the rubbery state and another in the glass state. Most magnetic tapes have a 1–3-μm-thick backcoating for antistatic protection and for improved tracking [Fig. 1.15(a)]. The backcoat is generally a polyester–polyurethane coating containing conductive carbon black and TiO_2. The TiO_2 and carbon contents are typically 50% and 10% by weight, respectively. For a cross section of a particulate medium, see Fig. 1.33(a).

Flexible disks are packaged inside a soft polyvinyl chloride (PVC) jacket or an acrylonitrile–butadiene–styrene (ABS) hard jacket (for a 90-mm form factor). Inside the jacket, a soft liner, a protective fabric is used to minimize wear or abrasion of the media. The wiping action of the liner on the medium coating removes and entraps particulate contaminants which may originate from the diskette manufacturing process, the jacket, the head–disk contact (wear debris), and the external environment. The liner is made of nonwoven fibers of polyester [poly(ethylene terephthalate)], rayon, polypropylene, or nylon. The liner fibers are thermally or fusion bonded to the plastic jacket at spots (Ostrowski, 1983; Tse and Lewis, 1986). The thermoplastic staple fibers, carded into a homogeneous web, are heated and consolidated under pressure between embossed calender rolls to form an array of diamond-shaped bond spots throughout the web. These bond-spot areas constitute about 25% of the total area, Fig. 1.34. The soft jacket near the data window is pressed with a sponge pad to create a slight friction and hence stabilize the disk motion under the heads. The hard cartridge is provided with an internal plastic leaf spring for the same purpose.

1.2.4.4. Magnetic Thin Films

Thin-film (also called metal-film) flexible media consist of polymer substrate (PET or polyimide) with an evaporated film of Co–Ni (with about 18% Ni) and less commonly evaporated/sputtered Co–Cr (with about 17% Cr) for vertical recording, which is typically 100–300 nm thick. For a cross section of a thin-film video tape, see Fig. 1.33(b). Electroplated Co and electroless plated Co–P, Co–Ni–P, and Co–Ni–Re–P have also been explored but are not commercially used. Since the thickness of the magnetic layer is only 100–300 nm, the surface of the thin-film medium is greatly influenced by the surface of the substrate film. Therefore, an ultra-smooth PET substrate film (rms rough-

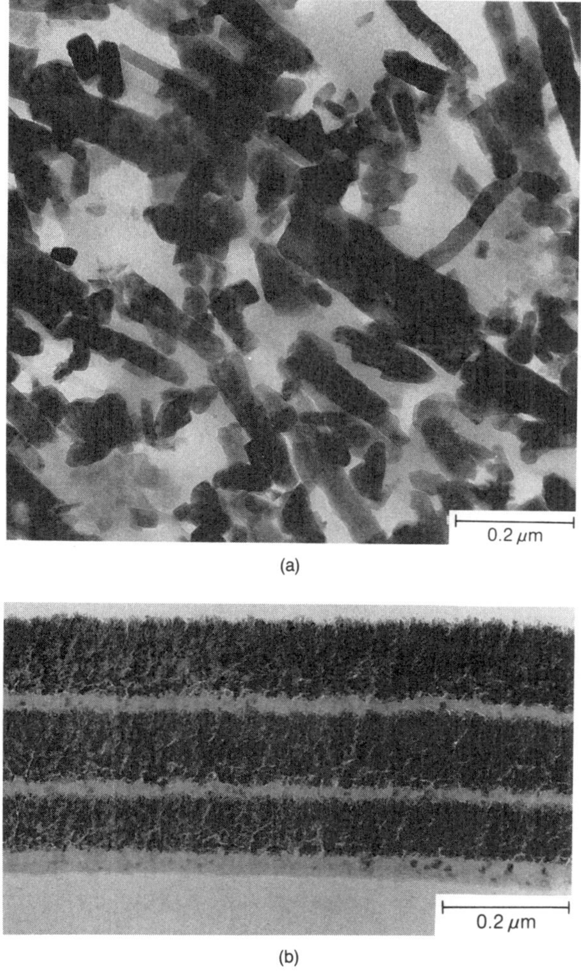

Fig. 1.33. Transmission electron micrograph of the cross section of (a) a particulate Co–γ-Fe$_2$O$_3$ tape and (b) a thin-film video tape (with a primer layer and three active evaporated Co–Ni magnetic layers separated by nonmagnetic spacer layers, the uppermost active layer is coated with a very thin lubricant/anticorrosion layer which is not very visible in the photograph).

ness ~ 1.5–2 nm) is used to obtain a smooth tape surface (rms roughness ~ 5–6 nm) for high-density recording.

Several undercoatings, overcoatings, and oxidation treatments (by adding oxygen into the vacuum chamber during evaporation) are used to increase corrosion resistance and durability. Tomago et al. (1988) and Chiba et al. (1989) employed an undercoating layer consisting of very fine particles, each protruding a few nanometers. By setting the diameter and density of these

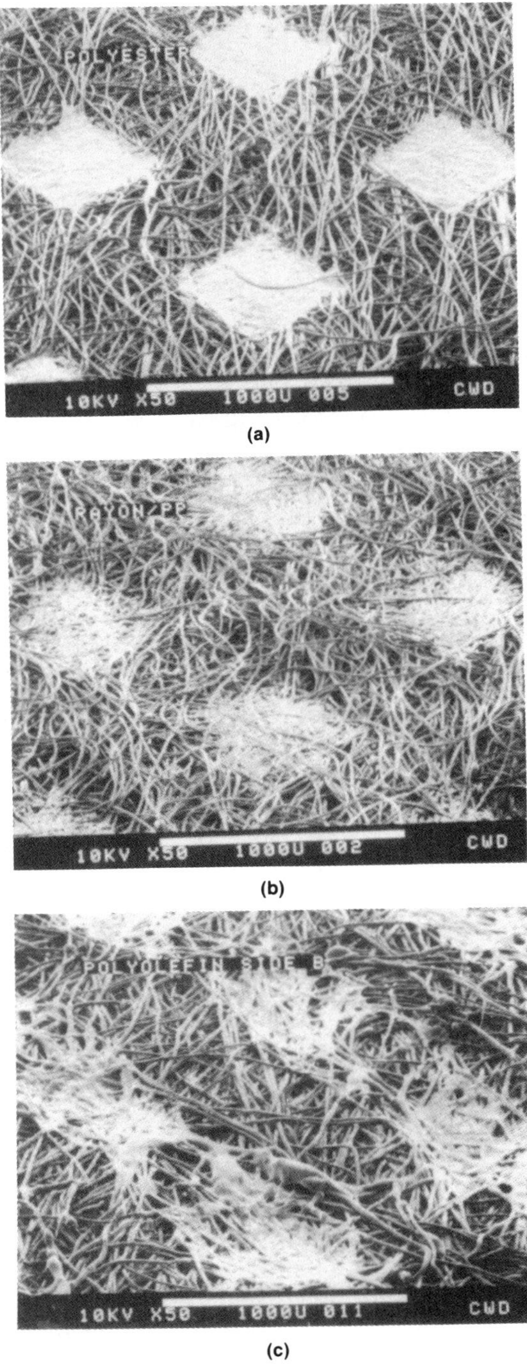

Fig. 1.34. Scanning electron micrographs of various thermally-bonded nonwoven liner fabrics (samples tilted at 60°) (a) polyester, (b) Rayon/polypropylene, and (c) polypropylene (Tse and Lewis, 1986).

Table 1.12. Selected physical properties of magnetic media and its components

Material	Density, kg/m^3	Young's modulus, GPa	Knoop microhardness, GPa (kg/mm^2)
γ-Fe$_2$O$_3$ particles	4800		11.8 (1200) Est.
Magnetic flexible medium substrate	1520	2.75–4.50	0.20 (20)
Particulate flexible medium coating	1660	1.25–2.25	0.25 (25)
Sputtered Co–Cr alloy metal film	8030	170–210	6.9–8.8 (700–900)
Diamondlike carbon overcoat for metal film	2100	160–180	14.7–19.6 (1500–2000)

particles at the optimum level, they attained good durability and excellent magnetic performance. Various organic (such as acrylic polymer in 50–80 nm thickness) and inorganic overcoats (such as diamondlike carbon in about 10–20 nm thickness, SiO$_2$ and ZrO$_2$) have been used to protect against corrosion and wear (Kadokura et al., 1987; Nagao et al., 1987; Mitani et al., 1989; Bhushan and Gupta, 1991). Alumina ($\sim 0.2\ \mu$m) particles are also added in some polymeric overcoats. Feuerstein and Mayr (1984), Kunieda et al. (1985), and Skorjanec and Moe (1989) have used another approach in which oxygen is leaked into the vacuum chamber at a controlled rate during the Co–Ni evaporation. This has the effect of forming an oxide top layer of 10–30 nm, and probably introduces oxidized grain boundaries. This oxidation improves durability and corrosion resistance and decreases magnetic noise. (It also increases coercivity at the expense of saturated magnetization and hysteresis loop squareness.) Surface oxidation of evaporated Co–Cr coating by annealing in air has also been reported to improve the durability of the coating (Honda et al., 1988; Skorjanec and Moe, 1988).

In addition to a solid overcoat, generally a thin layer (2–10 nm in thickness or 4–20 mg/m^2 in weight) liquid lubricant is applied on the medium surface to further reduce friction, wear, and corrosion. Fatty acid esters or fluorine-based compounds (such as perfluoropolyethers) are most commonly used (Buttafava et al., 1985; Burguette and Foss, 1987; Kondo et al., 1988; Tagami et al., 1988; Tomago et al., 1988; Chiba et al., 1989; Bhushan, 1990).

Typical physical properties of the components of magnetic media are presented in Table 1.12. A summary of materials used in different magnetic media and typical operating conditions is presented in Table 1.13.

1.2.5. Functional Requirements

Since different media are used for different applications, the environmental requirements of different media (which affect the choice of media materials) are different. Magnetic tapes and flexible disks can be shipped, stored, and operated at various environmental conditions. Examples of the environmental requirements for a tape cartridge to be used in the IBM 3480/3490 tape

Table 1.13. Typical operating conditions and typical materials used in different magnetic flexible media for computer use

Magnetic medium	Normal pressure, kPa	Sliding speed, m/s	Flying height, μm	Substrate	Magnetic coating binder	Magnetic medium/thickness	Solid/liquid lubricant	Lubricant application method/quantity
Particulate tape	7–28[a]	2–4	0.1–0.2	PET[b] 6.4–23.4 μm thick	Polyester–polyurethane[c]	γ-Fe$_2$O$_3$, Co–γ-Fe$_2$O$_3$, CrO$_2$, Fe, or BaO.6Fe$_2$O$_3$/2–4 μm	Fatty-acid ester	Internal/1–3% by weight
Metal-evaporated tape				PET[b] 6.4–14.5 μm thick	None	Evaporated Co–Ni or Co-Cr/ 100–300 nm	Polymer or inorganic and perfluoropolyether	Solution or vacuum/ 10–50 nm, Topical/2–10 nm thick
Particulate flexible disk	14–70 (10–20g)	1–12 (300–1800 rpm)	Partial contact (< 0.1)	PET[b] 76.2 μm thick	Polyester–polyurethane and epoxy[c]	γ-Fe$_2$O$_3$ Co–γ-Fe$_2$O$_3$, BaO.6 Fe$_2$O$_3$, or Fe/2–4 μm	Fatty-acid ester	Internal/1–3% by weight

[a] Interlayer pressure on a tape surface near the hub of a wound reel (end of tape) can be as high as 1.38 MPa.

[b] PET—Poly(ethylene terephthalate).

[c] Load-bearing alumina particles are added to increase the wear resistance of the medium.

Table 1.14. Environmental requirements
of a particulate tape for the IBM 3480/
3490 tape drive for high-end applications

Tape operating environment
   Temperature: 15.6° to 32.2° C
   Relative humidity: 20 to 80%
   Maximum wet bulb temperature: 25.6° C

Tape storage environment
   Temperature: 4.4° to 32.2° C
   Relative humidity: 5 to 80%
   Maximum wet bulb temperature: 26.7° C
   Maximum duration: 4 weeks

Tape shipping environment
   Temperature: $-23°$ to 49° C
   Relative humidity: 5 to 100%
   Maximum wet bulb temperature: 26.7° C
    (with no condensation formation)
   Maximum duration: 10 days

---

drive (for high-end application) and a flexible disk drive are presented in Tables 1.14 and 1.15, respectively. Computer tape reels are used on the average of 200–400 file passes and very few reels are used for 5000–7500 file passes. The tape reel may be subjected to up to several thousand contact starts/stops. The head–tape interface is expected to have acceptable friction and magnetic reliability (or flyability) after storage in the extreme environments for at least 1 week and under extreme operating conditions for about 7 years of life with less than 1% failure rate or an average life of 30 years. Flexible disks may be subjected to up to several thousand contact starts/ stops. The head–flexible-disk interface is expected to have acceptable friction/

Table 1.15. Typical environmental
requirements of a flexible disk

Disk operating environment
   Temperature: 10° to 32.2° C
   Relative humidity: 8 to 80% RH
   Maximum wet bulb temperature: 29.4° C

Disk storage environment
   Temperature: 10° to 51.7° C
   Relative humidity: 8 to 80%
   Maximum wet bulb temperature: 29.4° C

Disk shipping environment
   Temperature: $-40°$ to 51.7° C
   Relative humidity: 8 to 80%
   Maximum wet bulb temperature: 29.4° C
   Maximum duration: 10 days

---

flyability after storage in the extreme environments for 1–4 weeks and acceptable flyability (or magnetic performance) for 10^7 or more revolutions (equivalent life of 7 years) under extreme operating conditions, with a failure rate of less than 1%.

1.3. Manufacturing Processes of Flexible Magnetic Media

Particulate contamination severely impacts the life of a head–medium interface, therefore, all media are manufactured and the final file assembly is done in a clean room, typically class 100 (not more than 100 particles of size larger than 0.3 μm per cubic foot).

1.3.1. Particulate Media

1.3.1.1. Tapes

Figure 1.35(a) is a block diagram representing the manufacturing process of particulate tapes and Fig. 1.36 shows the equipment layout for the coat-

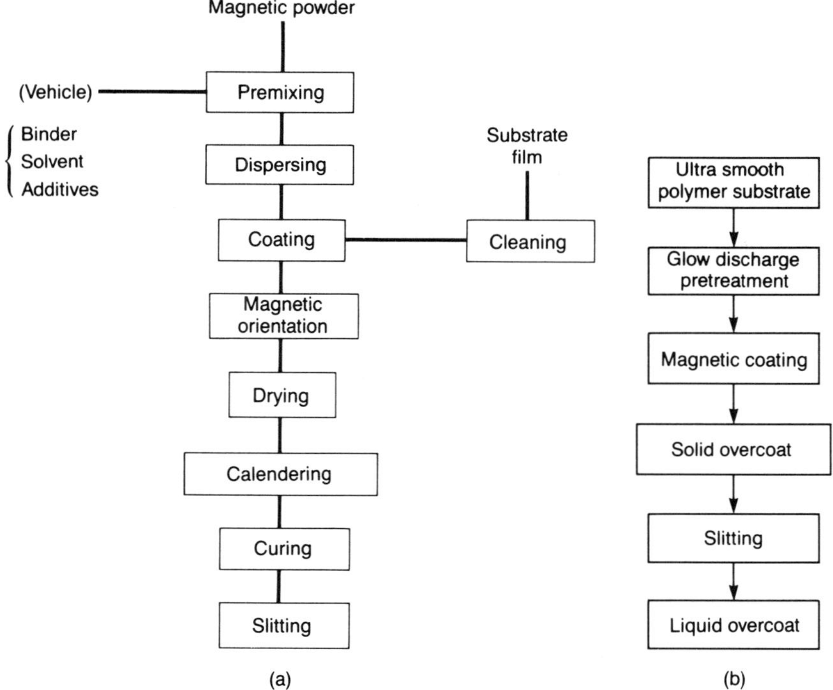

Fig. 1.35. Block diagrams representing the manufacturing process of (a) particulate tapes and (b) vacuum deposited thin-film tapes.

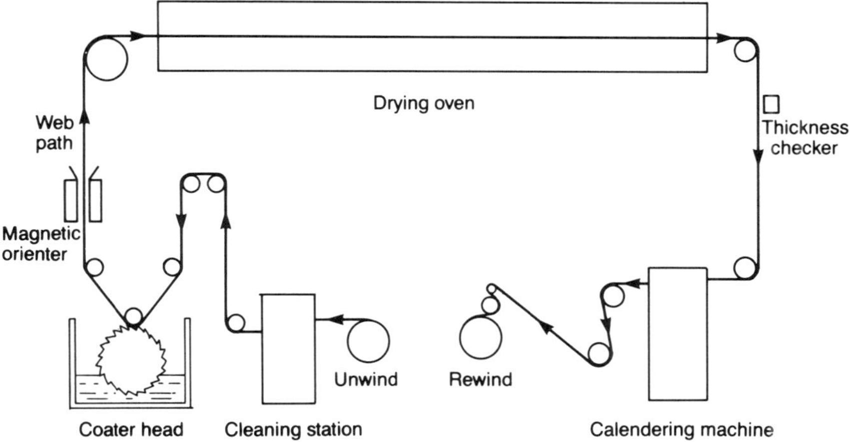

Fig. 1.36. Schematic arrangement for deposition of particulate magnetic coatings onto a continuous web.

ing deposition. The coating formulation (binder, magnetic particles, solvents, cross-linker, dispersant, and lubricant) is thoroughly mixed by means of a mixing disperser. Ball mills and sand mills are commonly used. The mix is then filtered in order to eliminate agglomerates. The time of milling may last from a few hours to several days, and it is dependent on the chemical composition of the binder and particles used. The particles are liable to agglomerate in the mix because of their magnetic attraction. Coating slurry next passes to the coater head for application onto a clean and UV-treated polymer substrate, poly(ethylene terephthalate) webs typically 3500-m long, 0.6-m wide, and 6.4–36.1 μm thick for data-processing applications. Several types of coater heads are used, such as the knife coater, the reverse roll coater, and the gravure coater (Fig. 1.37). Coating speeds are typically 2.5 m/s at a web tension of about 175 N/m. Immediately after coating to a thickness of 20–30 μm (four to ten times thicker than the finished coating of about 2–4 μm in thickness), while the binder is still wet, the coated web undergoes orientation in a magnetic field of 1000–5000 Oe strength, whereby the magnetic particles are aligned. The direction of acicular particles is here made to coincide with the magnetization direction of magnetic recording. The effectiveness of orientation is generally referred to as the squareness ratio. For randomly oriented particles, the orientation (squareness) ratio is 0.5, and for ideally and perfectly oriented particles, the ratio is 1 (Jorgensen, 1988; Bhushan, 1990).

After orientation, the coated web enters a drying oven, which is of the order of 30 m long at 50°–80° C, where all the solvents are evaporated leaving a dry web. The web is supported on Bernoulli or air-cushion shoes in the drying oven, as shown in Fig. 1.38. The spacing between the shoe and the web is about 12.7–25 mm. After drying, the coating thickness is measured using an in-line measuring instrument such as a specular gloss meter (Bhushan and

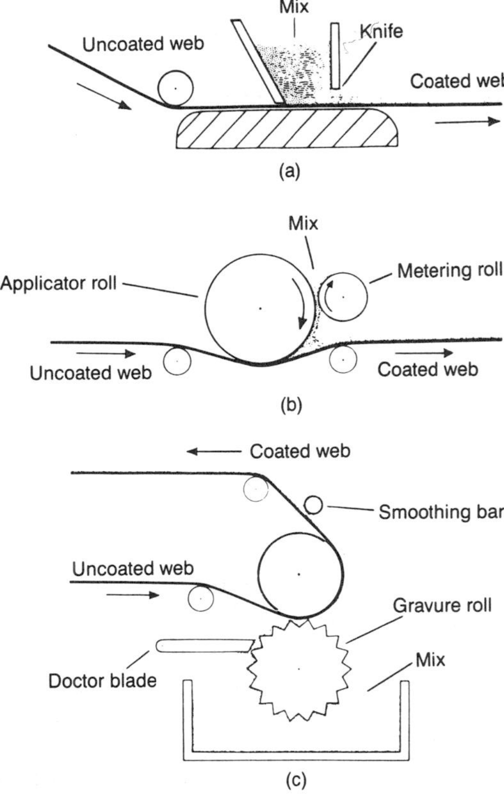

Fig. 1.37. Schematics of coater heads (a) knife coater, (b) reverse roll coater, and (c) gravure coater (Jorgensen, 1988).

Gupta, 1991). The coated surface must be made smoother, and this is done by calendering to a surface roughness of about 8–15 nm rms. In calendering, the coated web is passed between a set of hard (steel) and soft (highly compressed paper) rolls under high pressure (of the order of 300–500 N/mm or 60–100 MPa) and high temperature (60°–80° C) which compacts the binder and smooths the surface, Fig. 1.39. To provide high smoothness, the web is sometimes calendered using a set of hard (steel) rolls. The tapes are generally backcoated with a binder containing TiO_2 and C to a thickness of 2–4-μm for improved tracking and antistatic protection. In the case of backcoated tapes, the web may be calendered before a backcoat is applied or more commonly after both sides have been coated. In the manufacturing of high performance tapes, the web may be coated on both sides in a single pass and then calendered; formulation for the backcoat is such that it remains textured after calendering. After coating/calendering, the web is wound onto large rolls (known as jumbos) at a tapered web tension of about 90–60 N/m (lower web tension at larger roll diameter). Then the thermosetting resin binder in the coating is cured at either ambient or elevated temperatures to obtain

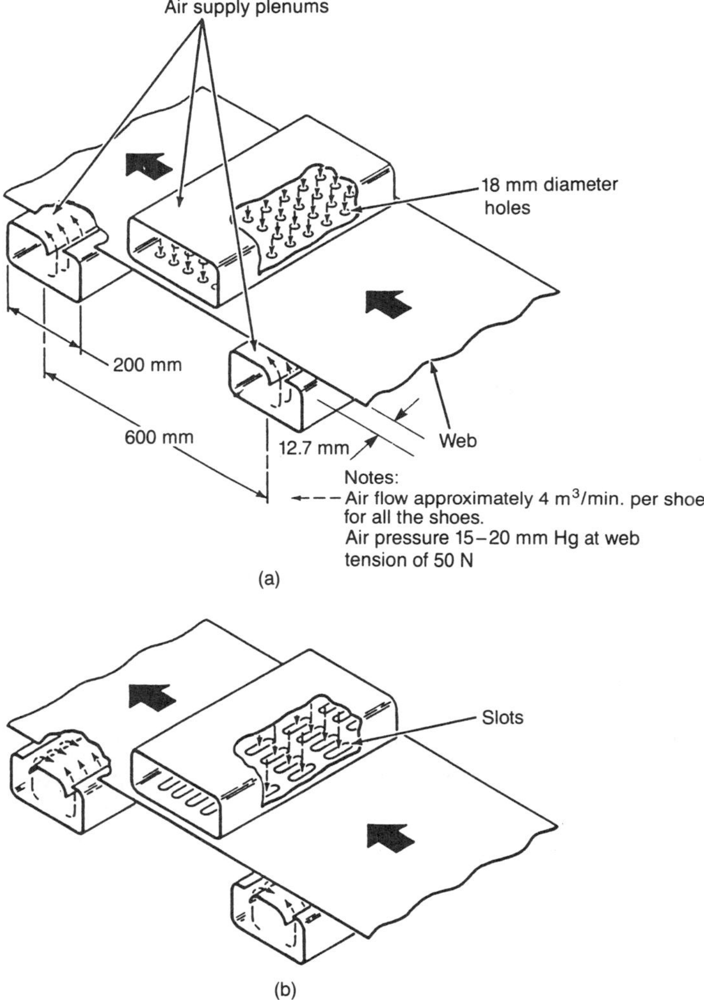

Fig. 1.38. Schematics of shoes used in the drying oven for web support (a) Bernoulli shoe and (b) air-cushion shoe.

optimum mechanical and chemical properties. The curing or cross-linking has also been reported by electron beam curing, which is claimed to have a thorough, consistent cure (Rand, 1983).

After curing, the web is slit to the desired tape widths (typically 12.675 mm $\pm$ 25 μm wide) at fairly high speeds (about 2.5 m/s). The slitting of the web is normally done by shear slitting. Shear cut slitting uses pairs of WC–Co blades ($\sim$ 100-mm diameter, 0.5-mm thickness) rotating in opposite directions (Anonymous; Rienau, 1980). The engagement of the blades is typically 100–200 μm and the spacing between the blade pairs is set to give the desired tape width (Fig. 1.40). After slitting, the total length of tape is then wound

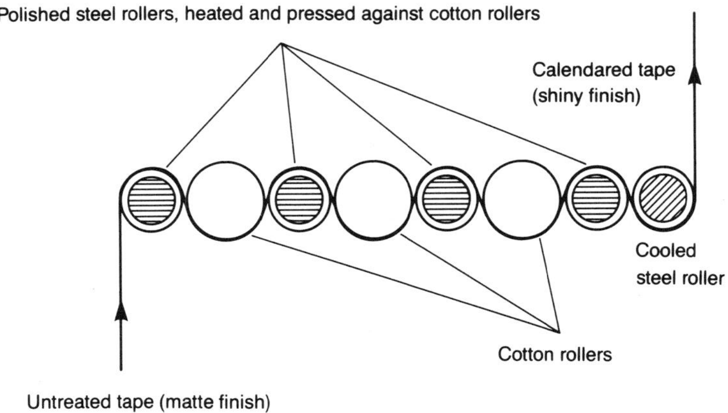

Fig. 1.39. Schematic of tape calendering process.

onto intermediate reels (known as pancakes), prior to being wound onto data reels and cartridges and audio and video cassettes (Anonymous).

1.3.1.2. Flexible Disks

Flexible disks are manufactured after the same principles as the tapes. Almost all flexible-disk substrates are 76.2-μm-thick PET. There are two exceptions: the particles are not oriented but left as random as possible; and the calendered web is first punched into disk form and is buffed by an abrasive tape to remove high asperities. In the tape-buffing process, an abrasive tape (e.g., alumina abrasive with a grit size of 3–5 μm on a 50-μm-thick polyester substrate) supported by a compliant roller (typically of polyurethane) is pressed against the rotating-disk surface, Fig. 1.41. The worn tape is continually advanced.

1.3.2. Thin-Film Media

Figure 1.35(b) is a block diagram representing the manufacturing process of thin-film tapes. Three coating methods can be used to produce thin-film metal-alloy flexible-media—vacuum evaporation, sputtering, and electro/electroless plating. (For a comparison of features of various coating processes, see Table 1.16). The first method employs high-vacuum evaporation of the metallic alloy, using an electron gun to melt and vaporize alloy ingots (at $\sim 2000°$ C) in the source crucible. Evaporation methods have the advantage of producing rather larger quantities of metal vapor in unit time, permitting fairly rapid web speeds. They also generate a tremendous amount of heat, which must be dissipated rapidly and efficiently to avoid any adverse effects to the moving plastic web. To minimize the heating problem, the film is moved past the source rapidly, and very thin coatings of the order of 15–20 nm are laid down. Cooling can therefore be very efficient. If the final active

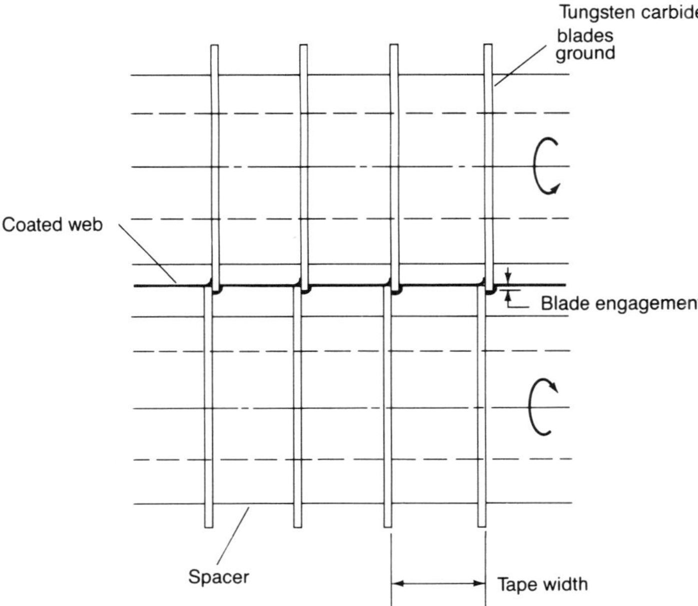

Fig. 1.40. Schematic of tape slitting process.

coating thickness desired is, for example, 200 nm, it is possible to deposit a plurality of active layers, one over the other, until the final desired thickness is achieved. Moreover, these active layers can be of differing composition, and can be separated by very thin insulating layers, which, if properly chosen, improve the flexibility of the final composite coating to a considerable degree, and can increase coercivity due to coupling effects between the active layers. The penalty here is the need for a giant machine, with many deposition stations. Metal evaporated tapes are used for high performance audio, video, and data-processing applications.

The second method of producing a metal-film flexible medium is to use a multichamber plasma sputtering system which is commercially used for the

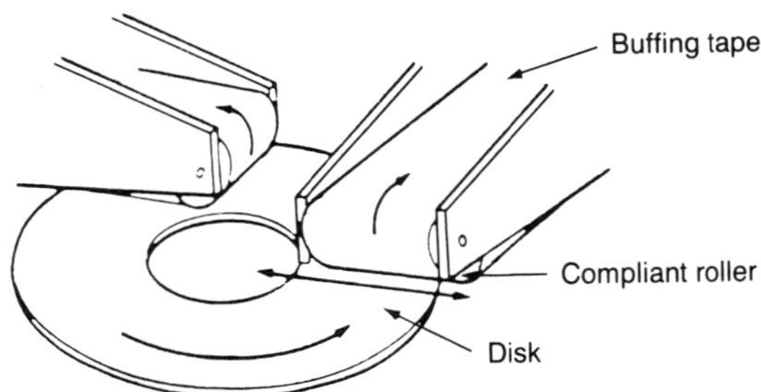

Fig. 1.41. Schematic of tape buffing process used to remove high spots from the flexible-disk surface.

Table 1.16. Comparison of the features of various coating processes used in the production of flexible media

Processes	Deposition rate, μm/min	Features
Particulate coating	High	High productivity
		Flexibility
		Low cost
Vacuum process		
Evaporation	60	High-rate deposition
		Lower cost than sputtering
Sputtering	0.005–2	High energy
		High thermal stability
		Strong adhesion
		Excellent crystal orientation
Chemical process		Low temperature
		High productivity
		Low cost
Electroplating[a]	0.5	
Electroless plating	0.5	

[a] Cannot be used on insulating substrate.

mass production of longitudinal hard-disk media (Mee and Daniel, 1990). Sputtering provides a strong adhesion, excellent crystal orientation, and high thermal stability. Sputtering as a method provides much less deposition rate than evaporation methods do. This is a limitation as coating thicknesses required for flexible media are fairly large (100–300 nm). With high-speed sputtering systems, web speeds of about 50 m/min can be obtained. For continuous fabrication of a web of thin-film medium, the deposition area should be considered as well to estimate the throughput. Although cooling is not the horrendous problem as in an evaporator method, sufficient heat is generated in the sputtering process to require special attention to matters of controlling the temperature of web during coating. The sputtering process has been explored to deposit Co–Cr media for vertical recording. Primarily because of high production costs, sputter-coated media has not become popular.

The third method includes the application of the magnetic coating by chemical–electroplating or electroless plating methods. The advantages of plating process are low substrate temperature during coating, high productivity, and low cost. However, electro or electroless plating has not yet found use because of the inherent disadvantages of plated media, such as poor magnetic noise characteristics and high film porosity which lead to poor corrosion resistance compared to that of vacuum-deposited media.

1.3.2.1. Metal-Evaporated Media

Because of high deposition rates obtained in the evaporation method, this is primarily used in the production of thin-film media, usually referred to as metal-evaporated (ME) media (Nakamura et al., 1982; Feuerstein and Mayr,

1984; Feuerstein et al., 1985; Kunieda et al., 1985). Magnetic anisotropy for horizontal recording can be induced in evaporated ferromagnetic films by causing the vapor flux to strike the substrate at an oblique angle instead of normally (Smith et al., 1960; Schuele, 1964). The anisotropy increases with the angle of incidence for angles from 0° (normal) to about 55°. The magnetic easy axis is perpendicular to the plane of incidence of the vapor. At about 55°, the anisotropy decreases rapidly and is zero near the angle of incidence at about 65°. Beyond about 65°, the easy axis rotates parallel to the plane of incidence and the anisotropy again becomes large with an angle of 85°. When evaporating Co–Cr alloy for vertical recording, the perpendicular orientation for the easy axis of magnetization can be promoted by normally incident/sputtered atoms.

The metal-film media for horizontal recording is fabricated by evaporation of ferromagnetic (primarily Co–Ni or Co–Cr for vertical recording) alloy onto an ultrasmooth PET web at oblique incidence, such that the incident vapor stream impinging on the substrate forms a large angle (approximately 80°) relative to the substrate normal, see Fig. 1.42. The alloy is evaporated by an electron beam gun from water-cooled copper or ceramic crucibles in a vacuum of 10^{-4}–10^{-5} Torr. Beam powers of about 300 kW are necessary to achieve the deposition rate. The web is transported via rollers and rotating cooled drums past one or more deposition stations. (The drum is water cooled in order to avoid heat damage to the polymer film.) As the web moves around the drum, it passes by an aperture mask which controls the range of vapor-stream incident angles seen by the web. If the web is moved in the direction of decreasing incident angles, higher coercivities and squareness ratios result than if the tape is run in the opposite direction. The oblique evaporation

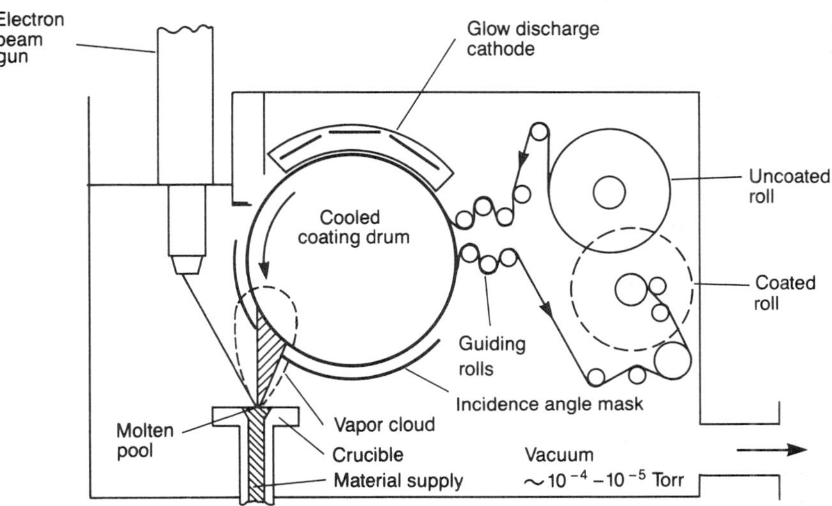

Fig. 1.42. Schematic arrangement for deposition of obliquely evaporated metal films onto a continuous web (Feuerstein and Mayr, 1984). © IEEE, 1984.

process is capable of high throughputs of the order of 100 m of web per minute and deposition rates of up to about 1 μm/s. Magnetic films are sometimes applied in several discrete layers, possibly separated by nonmagnetic insulation layers. For some applications, oxygen is introduced into the vacuum chamber during deposition, in order for the film to be a mixture of ferromagnetic Co–Ni alloy and their nonmagnetic alloys (protective overcoats). Magnetic layer thicknesses of 100–300 nm are used for Co–Ni alloys (75/25 to 70/30), with coercivities of 750–1000 Oe or greater.

An unusual problem is caused by the volume of metal deposited. Only some 10% of this is collected by the substrate; the rest must be cleaned from the chamber at the end of a coating run. Because of this, the rolls of base film are coated one at a time, the vacuum chamber being opened at the end of each run for reloading and cleaning.

The different internal processing stations are used. Glow discharge pretreatment of the uncoated web is generally used for cleaning (degassing) and for improved adhesion. In some cases, the web is precoated by, for example, 20–30 nm Al by evaporation or Cr–Ni by magnetron sputtering for improved adhesion and the precoat also acts as a diffusion barrier against outgassing during the latter coating process. Glow discharge post-treatment of the web is also conducted to remove residual stresses.

After metallizing, the roll of film is coated with an overcoat (10–50 nm in thickness) and slit into individual tapes or disks, and then these are generally lubricated with a 2–10 nm (or 4–20 mg/m^2) film of generally fatty acid ester or a perfluoropolyether type of oil from a dilute solution to give a low coefficient of friction and low wear.

1.3.2.2. Sputtered Media

Experimental flexible media with magnetic coatings applied by sputtering have been produced by several investigators such as Iwasaki (1984), Buttafava et al. (1985), Nishikawa et al. (1985), Kadokura et al. (1981), Nagao et al. (1987), Nakamura et al. (1982), and Tagami et al. (1988). Primarily, Co–Cr has been sputtered for vertical recording applications. Typically, the media are prepared by rf diode sputtering on PET or polyimide (for depositions requiring high substrate temperatures) substrates using Co–Ni or Co–Cr alloy targets to a thickness ranging from 100–300 nm. When perpendicular media are prepared by sputtering of Co–Cr alloy, it is usually necessary to raise the substrate temperature ($> 200°$ C) in order to establish high coercivity and perpendicular orientation of the easy axis (c-axis of HCP crystal for Co–Cr media) of magnetization. For this purpose, a heat-resistant polymer base, such as polyimide, is widely used. However, since the required temperature is not very high for Co–Cr media, even a PET film can be used with an appropriate water cooling system.

After metallizing, the subsequent processing procedure is similar to that of evaporated media.

1.3.2.3. Electro/Electroless Plated Media

The chemical processes including both electro and electroless plating, have been applied to Co and Co-based (e.g., Co–Ni, Co–P, Co–Ni–P, Co–Ni–Re–P) *experimental* flexible media (Judge et al., 1965; Chen and Cavallotti, 1982; Matsuda et al., 1985). The benefit of electroless plating over electroplating is that a nonconducting substrate can be coated.

The cobalt films were electroplated by Chen and Cavallotti (1982) from solutions of cobalt sulfamate or sulfate at a bath temperature ranging from $20°$–$60°$ C, and the current density varied between 1 and 100 mA/cm^2. A conductive substrate precoated with electroless Ni–P was used. The deposited coating had a HCP Co structure with the *c*-axis of all the crystallites (near the film surface) oriented normal to the film plane.

The Co–Ni–P films were electroless plated by Judge et al. (1965) using a bath of $CoSO_4$, $NiSO_4$, NaH_2PO_2, $(NH_4)_2 SO_4$, 0.12M-Na_3 (citrate) at $80°$ C. The substrate surface of PET film was treated with a hot chromic sulfuric acid solution, followed by a hot NaOH solution. It was then sensitized with a stannous chloride solution and finally with a palladium chloride solution.

The plated films are also believed to require overcoats for wear and corrosion protection.

References

Abraham, R. C. and Freeman, R. C. (1991a). "Mass Storage Outlook." Freeman Associates Inc., 311 E. Carrillo Street, Santa Barbara, California 93101.

Abraham, R. C. and Freeman, R. C. (1991b). "Computer Tape Outlook," Vol. 1—Helical Scan Products, Vol. 2—Data Cassette and Data Cartridge Products, Vol. 3—Half-Inch Products. Freeman Associates Inc., 311 E. Carrillo Street, Santa Barbara, California 93101.

Anonymous. "Techniques for Slitting and Winding." John Dusenbery Co., Inc., Randolph, New Jersey.

Anonymous (1977). "Unrecorded Magnetic Tape Cartridge for Information Exchange, 0.250 in., 1600 bpi, Phase Encoded." ANSI X3.55–1977.

Anonymous (1990). "1990 Disk/Trend Report—Flexible Disk Drives." Disk Trend Inc., 1925 Landings Drive, Mountain View, California 94043.

Anonymous (1990/1991). "Rigid Disk Drive—Magnetic Head/Media Market and Technology Report." Peripheral Research Corporation, 351 South Hitchcock Way B-200, Santa Barbara, California 93105.

Anonymous (1991a). "Computer Storage—Source: Dataquest." Dataquest Inc., 1290 Ridder Park Drive, San Jose, California 95131.

Anonymous (1991b). "Computer Storage—Tape Drives." Dataquest Inc., 1290 Ridder Park Drive, San Jose, California 95131.

Anonymous (1991c). "Computer Storage—Flexible Disk Drives." Dataquest Inc., 1290 Ridder Park Drive, San Jose, California 95131.

Anonymous (1991d). "Computer Storage—Rigid Disk Drives." Dataquest Inc., 1290 Ridder Park Drive, San Jose, California 95131.

Anonymous (1991e). "Computer Storage—Dataquest Perspective." Dataquest Inc., 1290 Ridder Park Drive, San Jose, California 95131.

Anonymous (1991f). "Storage—An IDC Systems Market Planning Service." International Data Corp., 5 Speen Street, Framingham, Massachusetts 01701.

Athey, S. W. (1966). "Magnetic Tape Recording." SP-5038, NASA, Washington, D.C.

Baasch, H. J. and Luecke, F. S. (1981). "Read–Write and Tunnel Erase Magnetic Head Assembly." U.S. Patent 4,276,574.

Bajorek, C. H. (1991). Trends in recording and control technologies and evolution of subsystem architectures for data storage. *Adv. Info. Storage Syst.* **1**, 1–14.

Bashe, C. J., Johnson, L. R., Palmer, J. H., and Pugh, E. W. (1986). "IBM's Early Computers." MIT Press, Cambridge, Massachusetts.

Berghof, W. and Gatzen, H. H. (1980). Sputter deposited thin-film multilayer head. *IEEE Trans. Magn.* **MAG-16**, 782–784.

Bhushan, B. (1987). Development of a wear test apparatus for screening bearing-flange materials in computer tape drives. *ASLE Trans.* **30**, 187–195.

Bhushan, B. (1990). "Tribology and Mechanics of Magnetic Storage Devices." Springer-Verlag, New York.

Bhushan, B. and Gupta, B. K. (1991). "Handbook of Tribology: Materials, Coatings and Surface Treatments." McGraw-Hill, New York.

Brock, G. W. and Shelledy, F. B. (1975). Batch-fabricated heads from an operational standpoint. *IEEE Trans. Magn.* **MAG-11**, 1218–1220.

Burguette, M. D. and Foss, G. D. (1987). "Magnetic Recording Media Having Perfluoropolyether Coating." U.S. Patent 4,671,999, June 9.

Buttafava, P., Bretti, V., Ciardiello, G., Piano, M., Caporiccio, G., and Scarati, A. M. (1985). Lubrication and wear problems of perpendicular recording thin film flexible media. *IEEE Trans. Magn.* **MAG-21**, 1533–1535.

Camras, M. (1988). "Magnetic Recording Handbook." Van Nostrand Reinhold, New York.

Cannon, D. M., Dempwolf, W. R., Schmalhorst, J. M., Shelledy, F. B., and Silkensen, R. D. (1986). Design and performance of a magnetic head for a high-density tape drive. *IBM J. Res. Develop.* **30**, 270–277.

Chen, T. (1981). The micromagnetic properties of high-coercivity metallic thin films and their effects on the limit of packing density in digital recording. *IEEE Trans. Magn.* **MAG-17**, 1181–1191.

Chen, T. and Cavallotti, T. (1982). Electroplated cobalt film for perpendicular recording medium. *Appl. Phys. Lett.* **41**, 205–207.

Chiba, K., Sato, K., Ebine, Y., and Sasaki, T. (1989). Metal evaporated tape for high band 8 mm video system. *IEEE Trans. Consumer Elec.* **35**, 421–427.

Chikazumi, S. and Charap, S. H. (1964). "Physics of Magnetism." R. E. Krieger, Malabar, Florida.

Chubachi, R. and Tamagawa, N. (1984). Characteristics and applications of metal tape. *IEEE Trans. Magn.* **MAG-20**, 45–47.

Cullity, B. D. (1972). "Introduction to Magnetic Materials." Addison-Wesley, Reading, Massachusetts.

Dubin, R. R., Winn, K. D., Davis, L. P., and Cutler, R. A. (1982). Degradation of Co-based thin-film recording materials in selective corrosive environments. *J. Appl. Phys.* **53**, 2579–2581.

Engh, J. T. (1981). The IBM diskette and diskette drive. *IBM J. Res. Dev.* **25**, 701–710.

Feuerstein, A. and Mayr, M. (1984). High vacuum evaporation of ferromagnetic materials—a new production technology for magnetic tapes. *IEEE Trans. Magn.* **MAG-20**, 51–56.

Feuerstein, A., Lammermann, H., Mayr, M., and Ranke, H. (1985). Video tape manufacture by high vacuum evaporation—aspects of machine technology. *Vacuum* **35** (7), 277–281.

Fisher, R. D. and Blades, J. D. (1971). Single crystal manganese zinc ferrite recording heads. *IEEE Trans. Magn.* **MAG-7**, 350–351.

Fujiwara, T. (1985). Barium ferrite media for perpendicular recording. *IEEE Trans. Magn.* **MAG-21**, 1480–1485.

Fujiwara, H., Mizushima, K., Sakemoto, A., Miyake, A., Naganawa, N., and Doi, T. (1986). Recording characteristics of high density metal powder floppy disks. *J. Magnetism and Mag. Mat.* **54–57**, 1567–1570.

Ginsburg, C. P. (1986). Development of the videotape recorder. In "Videotape Recording." Ampex Corporation, Redwood City, California.

Hahn, F. W. (1989). Historical perspective of tape head contours. In "Tribology and Mechanics of Magnetic Storage Systems," Vol. 6 (B. Bhushan and N. S. Eiss, eds.), pp. 21–27. SP-26, STLE, Park Ridge, Illinois.

Harker, J. M., Brede, D. W., Pattison, R. E., Santana, G. R., and Taft, L. G. (1981). A quarter century of disk file innovation. *IBM J. Res. Develop.* **25**, 677–689.

Harris, J. P., Phillips, W. B., Wells, J. F., and Winger, W. D. (1981). Innovations in the design of magnetic tape subsystems. *IBM J. Res. Develop.* **25**, 691–699.

Haynes, N. M. (1957). "Elements of Magnetic Tape Recording." Prentice-Hall, Englewood Cliffs, New Jersey.

Hirota, E., Mihara, T., Ikeda, A., and Chiba, H. (1971). Hot-pressed Mn–Zn ferrite for magnetic recording heads. *IEEE Trans. Magn.* **MAG-7**, 337–341.

Hoagland, A. S. (1963). "Digital Magnetic Recording." Wiley, New York.

Honda, K., Sugita, R., Echigo, N., Sakamoto, Y., and Murakami, Y. (1988). Surface oxidation of CoCr evaporated films. *IEEE Trans. Magn.* **24**, 2664–2666.

Howard, J. K. (1986). Thin-films for magnetic recording technology: A review. *J. Vac. Sci. Technol.* **A4**, 1–13.

Imamura, M., Ito, Y., Fujiki, M., Hasegawa, T., Kubota, H. and Fujiwara, T. (1986). Barium ferrite perpendicular recording flexible disk drive. *IEEE Trans. Magn.* **MAG-22**, 1185–1187.

Iwasaki, S. (1984). Perpendicular magnetic recording—evaluation and future. *IEEE Trans. Magn.* **MAG-20**, 657–668.

Iwasaki, S. and Nakamura, Y. (1977). An analysis for the magnetization mode for high density magnetic recording. *IEEE Trans. Magn.* **MAG-13**, 1272–1277.

Jorgensen, F. (1988). "The Complete Handbook of Magnetic Recording," 3rd ed. Tab Books, Blue Ridge Summit, Pennsylvania.

Judge, J. S., Morrison, J. R., Speliotis, D. E. and Bate, G. (1965). Magnetic properties and corrosion behavior of thin electroless Co–P deposits. *J. Electrochem. Soc.* **112**, 681–684.

Kadokura, S., Tomie, T., and Naoe, M. (1981). Deposition of Co–Cr films for perpendicular magnetic recording by improved opposing target sputtering. *IEEE Trans. Magn.* **MAG-17**, 3175–3177.

Kadokura, S., Kamei, K., Teranishi, K., and Sobajima, S. (1987). On improved durability characteristics for a Co–Cr flexible disk system. *IEEE Trans. Magn.* **MAG-23**, 2404–2406.

Kalil, F. and Buschman, A. (1985). "High-Density Digital Recording." Reference Publication 1111, NASA, Washington, D.C.

Katoh, Y., Nakayama, M., Taneka, Y., and Takahashi, K. (1981). Development of a new compact floppy disk drive system. *IEEE Trans. Magn.* **MAG-17**, 2742–2744.

Katoh, Y., Nakayama, M., Chubachi, R., and Okamoto, N. (1983). High density magnetic recording on a 3.5 inch micro floppy disk drive system. *IEEE Trans. Magn.* **MAG-19**, 1707–1709.

Kingery, W. D., Bowen, H. K., and Uhlmann, D. R. (1976). "Introduction to Ceramics," 2nd ed. Wiley, New York.

Kirk, D., ed. (1981). "25 Years of Video Tape Recording," 3M U.K. Ltd., Bracknell, England.

Knudsen, J. K. (1985). Stretched surface recording disk for use with a flying head. *IEEE Trans. Magn.* **MAG-21**, 2588–2591.

Kohmoto, O., Ohya, K. and Ojima, T. (1979). Wear-resistant audio head using amorphous alloy material. *Trans. IEICE* **E72** (9), 993–997.

Kondo, H., Kawana, T. and Yatagai, H. (1988). "Magnetic Recording Medium." U.S. Patent No. 4,735,848, April 5, 1988.

Kuhn, A. T. and Wellwood, J. (1986). Aqueous and atmospheric corrosion of cobalt–nickel alloys. *Br. Corros. J.* **21** (4), 228–234.

Kunieda, T., Shinohara, K., and Tomago, A. (1985). Metal evaporated video tape. *J. IERE* **55** (6), 217–222.

Lowman, C. E. (1972). "Magnetic Recording." McGraw-Hill, New York.

Luborsky, F. E. (1980). Amorphous ferromagnets. In "Ferromagnetic Materials," Vol. 1. North-Holland, Amsterdam.

Mallinson, J. C. (1969). Maximum signal-to-noise ratio of a tape recorder. *IEEE Trans. Magn.* **MAG-5**, 182–186.

Mallinson, J. C. (1987). "The Foundations of Magnetic Recording." Academic Press, San Diego, California.

Matsuda, H., Nishiguchi, T. and Takano, O. (1985). Preparation of magnetic recording tape by electroless Co–P plating. *J. Met. Finish. Soc. (Japan)* **36** (3), 124–130.

Matsuoka, M. and Naoe, M. (1985). Sputter deposition and read/write characteristics of Ba–ferrite thin film disk. *IEEE Trans. Magn.* **MAG-21**, 1474–1476.

Mee, C. D. (1964). "The Physics of Magnetic Recording." North-Holland, Amsterdam.

Mee, C. D. and Daniel, E. D., eds. (1990). "Magnetic Recording Handbook." McGraw-Hill, New York.

Mitani, T., Kurokawa, H., and Yonezawa, T. (1989). Application to the metal-evaporated videotape of diamond-like carbon film deposited by PI–CVD method. *J. Japan Soc. Prec. Eng.* **55**, 299–304.

Mizushima, M. (1971). Mn–Zn single crystal ferrite as a video-head material. *IEEE Trans. Magn.* **MAG-7**, 342–344.

Monforte, F. R., Chen, R., and Baba, P. D. (1971). Pressure sintering of MnZn and NiZn ferrites. *IEEE Trans. Magn.* **MAG-7**, 345–350.

Mori, T. (1985). Progress of magnetic recording materials. *The BKSTS Journal*, September, 526–529.

Morisako, A., Matsumoto, M., and Naoe, M. (1986). Ba–ferrite thin film rigid disk for high density perpendicular magnetic recording. *IEEE Trans. Magn.* **MAG-22**, 1146–1148.

Nagao, M., Sano, K., Kojima, M., Iwasaki, H., Nahara, A., and Kitamoto, T. (1987). Improvement in durability of CoCr films with carbon protective layer. *IEEE Trans. Magn.* **MAG-23**, 2395–2397.

Nakamura, K., Ohta, Y., Itoh, A., and Hayashi, C. (1982). Magnetic properties of thin films prepared by continuous vapor deposition. *IEEE Trans. Magn.* **MAG-18**, 1077–1079.

Nishikawa, R., Suzuki, T., and Sonoda, T. (1985). Co–Cr sputtered films for perpendicular magnetic recording. *Toshiba Rev.* **154**, 32–36.

Noda, M., Okazaki, Y., Hara, K., and Ogisu, K. (1990). Characteristics of barium ferrite tape for magnetic contact duplication. *IEEE Trans. Magn.* **26**, 81–86.

Norton, D. G. and Wright, G. T. (1978). "Flexible Disk Storage Apparatus." U.S. Patent 4,074,330, February 14.

Ostrowski, H. S. (1983). Nonwoven liners for floppy disks and related jacket manufacturing methods. Presented at Symp. on Mag. Media Manuf. Methods, held at Honolulu, Hawaii in May, sponsored by Mag. Media Information Service, Chicago, Illinois.

Ouchi, K. (1990). Properties, preparation, advances, and limitations of materials and media for perpendicular recording. *IEEE Trans. Magn.* **26**, 24–29.

Pear, C. B. (ed.) (1967). "Magnetic Recording in Science and Industry." Reinhold, New York.

Peterson, M. L. (1989). "Data Cartridge and Data Cassette Market Strategy Report." Peripheral Strategies Inc., 351 So. Hitchcock Way, Suite B-200, Santa Barbara, California 93105.

Peterson, M. L. (1990). "1990 Helical Data Storage Market Strategy Report." Peripheral Strategies Inc., 351 So. Hitchcock Way, Suite B-200, Santa Barbara, California 93105.

Potter, R. I. (1970). Analysis of saturation magnetic recording based on arctangent magnetization transitions. *J. Appl. Phys.* **41**, 1647–1651.

Potter, R. I. (1974). Digital magnetic recording theory. *IEEE Trans. Magn.* **MAG-10**, 502–508.

Pugh, E. W., Johnson, L. R., and Palmer, J. H. (1991). "IBM's 360 and Early 370 Systems." MIT Press, Cambridge, Massachusetts.

Rand, W. M. (1983). Electron curing of magnetic coatings. *Radiation Curing*, February, 26–30.

Rienau, J. H. (1980). Achieving higher productivity in slitting/rewinding. *Paper, Film & Foil Converter*, **54** (8), 62–64, **54** (9), 82–86.

Sakakima, H., Yanaguchi, Y., Satomi, M., Senno, H., and Hirota, E. (1981). Improvement in amorphous magnetic alloys for magnetic head core. *Proc. Fourth Int. Conf. Rapidly Quenched Metals* (T. Masumoto and K. Suzuki, eds.), p. 941. Jpn Soc. Metals, Sendai, Japan.

Schuele, W. J. (1964). Coercive force of angle of incidence films. *J. Appl. Phys.* **35**, 2558–2559.

Sharrock, M. P. (1989). Particulate magnetic recording media: A review. *IEEE Trans. Magn.* **25**, 4374–4389.

Shelledy, F. B. and Brock, G. W. (1975). A linear self-biased magnetoresistive head. *IEEE Trans. Magn.* **MAG-11**, 1218–1220.

Simpson, D. (1984). Quarter-inch cartridges command tape drive market. *Mini-Micro Systems*, November 19, 117.

Skorjanec, J. and Moe, C. D. (1988). "Metallic Thin Film Magnetic Recording Medium Having a Hard Protective Layer." U.S. Patent 4,729,924, March 8.

Skorjanec, J. and Moe, C. D. (1989). "Reactive Sputtering Process for Recording Media." U.S. Patent 4,803,130, February 7.

Smith, D. O., Cohen, M. S., and Weiss, G. P. (1960). Oblique incidence anisotropy in evaporated permalloy films. *J. Appl. Phys.* **31**, 1755–1762.

Speliotis, D. E. (1987). Barium ferrite magnetic recording media. *IEEE Trans. Magn.* **MAG-23**, 25–28.

Sugaya, H. (1985). Mechatronics and the development of the video tape recorder. In "Tribology and Mechanics of Magnetic Storage Systems," Vol. 2 (B. Bhushan and N. S. Eiss, eds.), pp. 64–71. SP-19, ASLE, Park Ridge, Illinois.

Suzuki, T., Ito, T., Isshiki, M., and Saito, N. (1989). Barium ferrite tape for DAT magnetic contact duplication. *IEEE Trans. Magn.* **25**, 4060–4062.

Tagami, K., Tamai, H., Hoyashida, H. and Arai, T. (1988). Pass wear resistance for perpendicular recording media. *IEEE Trans. Magn.* **24**, 2655–2657.

Takahashi, S. (1984). "The Latest Floppy Disk Drives and Their Application Know-How (Japanese)." C Q Publishing, Tokyo.

Teramura, N. (1985). Recent progress in floppy disk recording technology (review paper). In "Tribology and Mechanics of Magnetic Storage Systems," Vol. 2 (B. Bhushan and N. S. Eiss, eds.), pp. 27–35. SP-19, ASLE, Park Ridge, Illinois.

Tochihara, S. (1982). Magnetic coatings and their applications in Japan. *Progress in Organic Coatings* **10**, 195–204.

Tomago, A., Murai, M., and Enomoto, S. (1988). Tribology of vacuum deposited magnetic thin film recording tape. *IECE Japan.* **87**, 365.

Tripathi, K. C. (1990). Development of high speed head-disk interface for flexible storage system through Bernoulli principle. In *Proc. Japan Int. Trib. Conf.* Nagoya, 1923–1927.

Tsang, C. (1984). Magnetics of small magnetoresistive sensors. *J. Appl. Phys.* **55**, 2226–2231.

Tsang, C., Chen, M. M., Yogi, T., and Ju, K. (1990). Gigabit density recording using dual-element MR/inductive heads on thin-film disks. *IEEE Trans. Mag.* **26**, 1689–1693.

Tse, M. K. and Lewis, A. F. (1986). Triboacoustics of nonwoven fabric/floppy disk dynamic contact. In "Tribology and Mechanics of Magnetic Storage Systems," Vol. 3 (B. Bhushan and N. S. Eiss, eds.), pp. 63–71. SP-21, ASLE, Park Ridge, Illinois.

Tsukahara, N. and Yanaga, M. (1984). The 3.5-inch floppy and its merits for use with computers at home. *J. Electron. Eng.* **21**, July, 74–77.

Van Gestel, W. J., Gorter, F. W., and Kuijk, K. E. (1977). Read-out of a magnetic tape by the magnetoresistance effect. *Philips. Tech. Rev.* **37** (2/3), 42–50.

von Behren, R. A. (1971). "Belt Driven Tape Cartridge." U.S. Patent 3,692,255.

von Behren, R. A. and Smith, D. P. (1991). Mechanical design of a belt-driven data cartridge. *Adv. Info. Storage Syst.* **1**, 49–60.

Waid, D. D., Peterson, M. L., and Casey, D.C. (1990). "Tape Head/Media Market and Technology Report." Peripheral Strategies Inc. & Peripheral Research Corp., 351 So. Hitchcock Way, Suite B-200, Santa Barbara, California 93105.

Wallace, R. L. (1951). The reproduction of magnetically recorded signal. *Bell Syst. Tech. J.* **30**, 1145–1173.

White, R. M. (1985). "Introduction to Magnetic Recording." IEEE Press, New York.

Winarski, D. J., Chow, W. W., Froehlich, F. B., Bullock, J. G., and Osterday, T. G. (1986). Mechanical design of the cartridge and transport for the IBM 3480 subsystem. *IBM J. Res. Develop.* **30**, 635–644.

Yamamori, K., Nishikawa, R., Muraoka, T., and Suzuki, T. (1983). Perpendicular magnetic recording floppy disk drive. *IEEE Trans. Magn.* **MAG-19**, 1701–1703.

Yamamori, K. and Tanaka, T. (1989). Recording characteristics for Ba–ferrite floppy disk. *IEEE Translation J. Magn. (Japan)* **4**, 631–640.

CHAPTER 2

Physical and Chemical Properties of PET Substrate and Coated Magnetic Media

Figures 2.1 and 2.2 compare the mechanical and thermal characteristics of several different film-forming polymers. Poly(ethylene terephthalate) or PET films have very desirable stress–strain characteristics (high modulus of elasticity, high elongation or strain, high yield strength, and high breaking strength). The "upper limit service temperature" (which is the maximum temperature that should be considered for extended periods) and the "zero strength temperature" (which is the temperature where the film can no longer support its own weight) of PET film are the highest of all films considered. Commercial advantages of these PET films can be summarized as follows: highly resistant to most organic solvents and mineral acids; no plasticizers and very low moisture retention [less than 0.5% at room temperature and 50% relative humidity (RH)]; high strength, toughness, and durability; excellent flex durability; high melting point; high upper limit service temperature; and excellent electrical properties (Heffelfinger and Knox, 1971; Sweeting, 1971). PET films are manufactured in large quantities by various companies, Table 2.1. By means of suitable chemical, structural, and other modifications, PET films are made with a broad range of properties that serve many diverse types of applications; for example, Du Pont offers eight basic types of Mylar (Du Pont Registered Trademark) in more than thirty varieties. The various types are A, D, S, HS, M, W, T, and C.

PET films are used as a substrate for magnetic recording media (tapes and flexible disks) and photographic film, as a capacitor dielectric and shrinkable packaging material. PET films such as Mylar A,[1] DB (data base) are commonly used as substrates for magnetic tapes and flexible disks. These films are biaxially oriented and heat stabilized and are made of a semicrystalline (about 45–50% crystallinity) material that is strong, tough, and flexible, and exhibit low-dimensional change with humidity, compared with other polymer sub-

---

[1] A general-purpose balanced film made in all thicknesses; transparent to 23.4-μm thickness and translucent in thicker films. Mylar D, a highly transparent film available in thicknesses greater than 76.2 μm, has only been experimented with for flexible disks; it has not been used only because of looser manufacturing specifications.

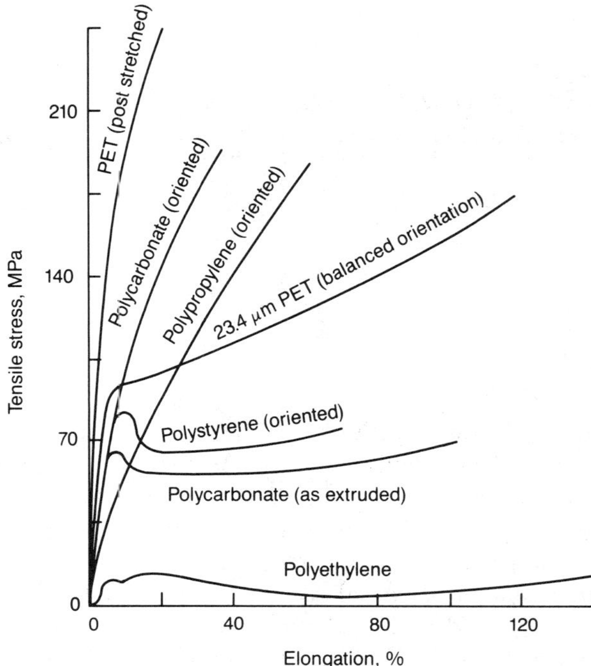

Fig. 2.1. Stress–strain curves for several polymer films (Heffelfinger and Knox, 1971).

strates. PET film is produced commercially by the extrusion of a polymer melt that is quenched, oriented, and heat set. The orientation involves stretching of the film in the machine direction (MD) and then in the transverse direction (TD) to produce balanced mechanical properties. The film is pulled more in the TD than in the MD for better thickness control in the TD and balanced mechanical properties. The film is then restrained at its stretched dimension and heated briefly to increase its crystallinity and reduce its tendency to shrink. Finally, the film is cooled and wound onto rolls.

The degree of crystallinity and biaxial orientation in the crystalline and amorphous regions of PET film determine its physical properties. Because of variations in the degree of crystallinity and in the orientation of the crystalline and amorphous regions, the properties of the biaxially-oriented PET film have in-plane anisotropy and vary across the TD of the film. The film also exhibits anisotropy in the thickness direction. In addition, commercial film is metastable in two respects. First, the percent crystallinity of the film is much lower than the equilibrium crystalline content. Second, noncrystalline regions of the film contain frozen-in strains which tend to relax and allow the film to contract.

Such physical properties as Young's modulus of elasticity, viscoelastic deformation, the long-term shrinkage, and the coefficients of linear thermal

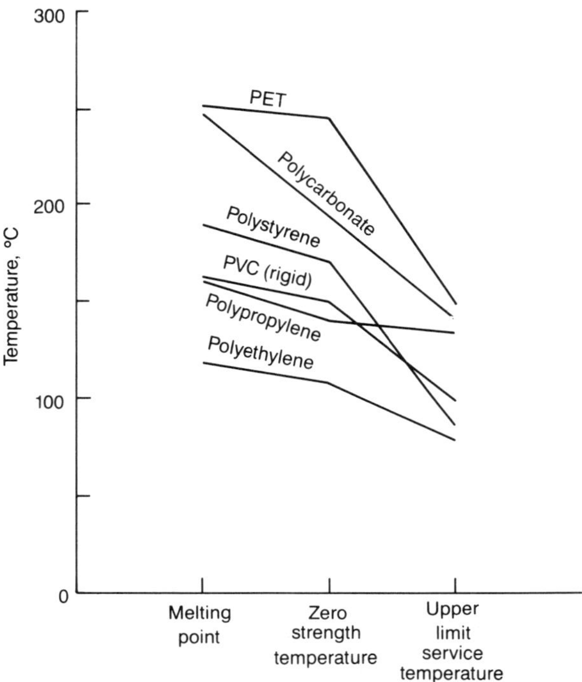

Fig. 2.2. Thermal properties of several polymer films (Heffelfinger and Knox, 1971).

Table 2.1. Some poly(ethylene terephthalate) film
manufacturers for magnetic recording industry

Manufacturer	Product trademark
E. I. duPont de Nemours & Co., Circleville, Ohio; Florence, South Carolina; and Luxembourg	Mylar
Imperial Chemical Industries Ltd., Hopewell, Virginia, and Welwyn Garden City, England	Melinex
Minnesota Mining & Mfg. (3M) Co., St. Paul, Minnesota	Scotchpar
American Hoechst, Greenville, South Carolina	Celanar
Toray Industries, Tokyo, Japan	Lumirror
Teijin–Konishiroku Film Co., Japan	Teteron
Fuji Photo Film Co., Japan	Fuji Film

and hygroscopic expansions depend on the in-plane direction at any point, as well as on the position across a web, they are also anisotropic across the thickness direction (Bhushan, 1985). Along a line parallel to the machine direction, the properties are essentially constant but not necessarily over a long distance. One of the principal long-term dimensional changes occurs from shrinkage due to residual stresses.

The anisotropic mechanical properties affect the tracking of moving films; the anisotropic mechanical (including viscoelastic), thermal, and hygroscopic properties also cause anisotropic deformation of flexible disks resulting in the radial runout of an initially circular track. To accommodate high track densities on a flexible disk without resorting to a prohibitively expensive servo system requires that the disk be dimensionally stable. For high areal densities on a tape, the tape must be dimensionally stable. Anisotropy and dimensional instabilities are limiting factors of PET films. The annealing of PET films under relatively mild conditions provides a more dimensionally stable film with less long-term shrinkage. Although other films (such as polyimide films—Kapton grade by Du Pont) provide isotropic properties, PET films continue to be used for low cost. An injection-molded composite plastic disk in combination with a stretched, web-coated flexible media (typically PET substrate), called stretched surface recording (SSR), has been developed for better dimensional stability than the unsupported flexible disk. In the production of metal-evaporated flexible media, a tremendous amount of heat is generated. A film base which can withstand temperatures of at least 200° C is the polyimide film. (Several companies, most notably Ube Kosan and Mitsubishi Jushi, offer suitable polyimide films made on pilot plant equipment.) However, these films are frightfully expensive.

In this chapter, we describe the manufacturing process of PET films and present the various physical and chemical properties relevant to the performance of magnetic media. Viscoelastic (dynamic modulus, creep, and relaxation) properties will be presented in Chapter 3.

2.1. Manufacturing Process of PET Films

The production of a biaxially-oriented PET film involves melt extrusion of the polymer, rapid quench from the extrusion temperature, a subsequent heating and stretching of the film, and finally, some heat setting or stabilization process, Fig. 2.3 (Heffelfinger and Knox, 1971; Heffelfinger, 1978). The films are produced at a rate of 150–300 m/s from a PET resin whose number-average molecular weight is typically between 15 and 50×10^3. [The intrinsic viscosity of the polymer in a phenol-o-dichlorobenzene mixture (60/40) at 25° C is 0.6 for a molecular weight of 37,000.] In essence, the PET resin is extruded at about 280° C ($T_m \sim 255°$ C) through a flat die onto a cold quenching drum to give an amorphous glassy film. The amorphous film is then preheated above the glass transition temperature ($T_g \sim 70°$ C) typically at 70°–90° C and passed through two sets of nip rolls. The second set of nip

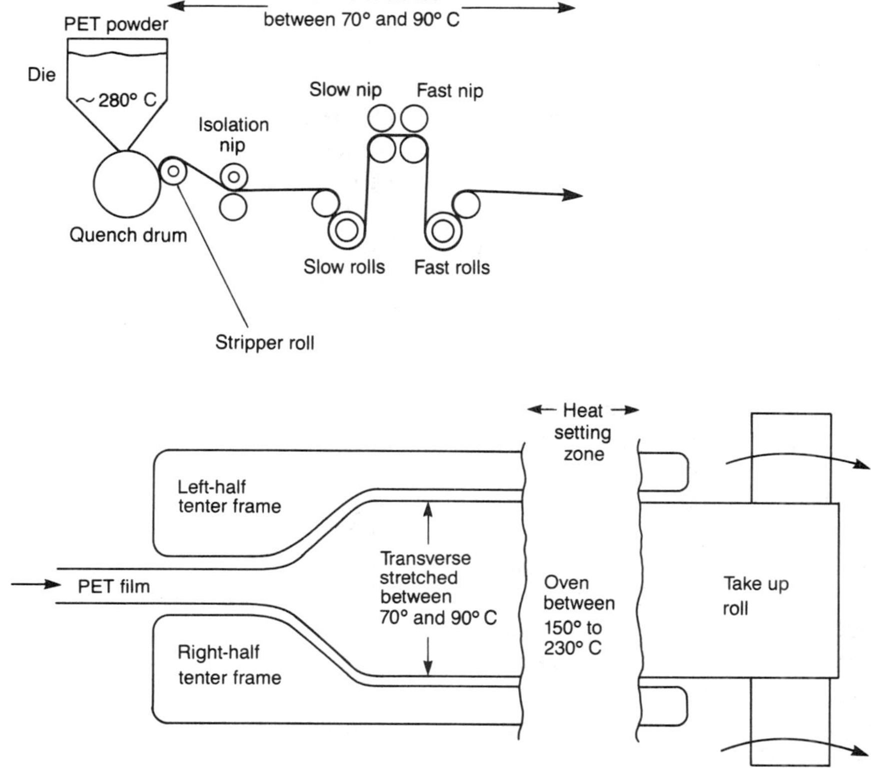

Fig. 2.3. Schematic of the PET manufacturing process.

rolls operate at two to six times faster than the first set such that the film is stretched longitudinally, i.e., in the machine direction (MD) by about 250–275% (stretch or draw or deformation ratio ∼ 3.5–3.75). During this operation, the film is oriented uniaxially and some crystallinity (about 20%) develops.

The uniaxially drawn and partially crystalline film is then stretched in the transverse direction (TD) through stressing the outer edges of the film by increasing the width of the so-called tenter frame which holds them. This equipment consists of film-gripping devices attached to a moving endless chain. Film is transported by the chain through the heating zone to increase ductility and crystallinity. The chain path diverges and the film is stretched in the transverse direction by about 325–350% (stretch ratio ∼ 4.25–4.5), and concomitantly the crystallinity increases a few percent. The extruded film web is thicker in the center than at the edges (convex-shaped web), therefore, the film is pulled more in the TD than in the MD for better thickness control in the TD. Typically, the amount of transverse stretch is controlled to provide balanced mechanical properties in the film.

At the end of the stretch zone, the doubly drawn film sheet exhibits high mechanical properties in both directions; it is tough and flexible but will shrink if heated above the stretching temperature. To impart dimensional stability, the film next passes through a chamber at temperatures in the range of 150°–230° C, while still restrained. During this "heat-set" operation (at a temperature significantly above T_g and under tension), the crystallinity increases to about 45–55%. Typically, a heat-set film may still shrink slightly (a fraction of a percent) especially when heated above its glass transition temperature, generally between 70°–90° C, because of residual strains. This shrinkage can be reduced further by heating the film under very low tension (to minimize induced stresses) at a temperature lower than the heat-set temperature (but above T_g) to relax frozen-in strains (strain-relaxation or annealing process), for further details, see Chapter 3. Crystallinity also increases during the annealing process. However, commercial films are not stabilized by the annealing process. Finally, the film is cooled to ambient temperature, the film edges that were on the gripping devices are cut off and are wound onto compliant (e.g., cardboard) hubs.

The films produced using a normal tenter process (such as Mylar A produced by Du Pont) have nonuniform physical properties across the web width. Specialized films are available (e.g., from ICI, U.K.) which are produced using a modified process in which MD and TD orientation is carried out simultaneously. Such films tend to be more isotropic, as in the blown film process, where orientation is again simultaneous. A "tensilized" film (i.e., one more highly oriented in one direction than the other) can be made by additional processing (post-stretching) of the biaxially oriented film (such as Mylar T). This is done by stretching again, usually in the machine direction by a stretch ratio of about 1.5–3. Young's modulus of such a film may be about two times that of the balanced film (as high as 8.5 GPa) in the post-stretched direction. The 9.4 μm (37 gage) or thinner PET films used for video and audio tape applications are tensilized to reduce deformation and to increase the breaking strength. We note that coefficients of linear thermal expansion and hygroscopic expansion decrease in the stretch direction, however, the shrinkage increases proportional to modulus in the stretch direction.

Throughout manufacture, conversion, and use, the film must possess a property called "handling" (Heffelfinger, 1978). Good handling infers that the film can be wound at high speeds without stagger in the transverse direction. It is essential that films be wound into large rolls at high speeds and have a pleasing appearance. Among the most important parameters which influence handling are surface topography, hydrodynamic and boundary friction, and the stiffness (proportional to the cube of thickness) of the web. The properties called "slip" and "blocking" are important to the way films can be wound. "Slip" infers the ability of films to slide over themselves. As a roll winds, it is necessary for each layer to tighten and to clinch upon itself. It must stay in registry with the layer underneath, and also accommodate larger perturbations in winding speeds, vibrations, and tension without forming wrinkles or damaging the surfaces. Alternatively, the film cannot be so slippery that the roll telescopes. "Blocking" is the tendency of films to stick together. Often the surfaces are so smooth that they tend to fuse together and part only when the surfaces tear. Because of slip and blocking, it becomes obvious that optimization of interlayer friction and surface topography become important.

The coefficients of static and kinetic friction of polymer films (interlayer or boundary friction of film surfaces against each other) is generally reduced by increasing the surface roughness of the films and/or by surface lubrication. The hydrodynamic air film generated during winding gets trapped between the layers of film and acts like a lubricant. If a large amount of air is included, film-to-film contact becomes negligible; the film rides on an air bearing and the winding system becomes either essentially uncontrollable or the air layer will limit production speeds. Film surface roughness, winding tensions, lay-on rolls, and many other methods are utilized to control these air-bearing effects.

For magnetic recording applications, a coextruded film (such as made by ICI) is desirable in which one surface is extremely smooth for magnetic coating application, while the reverse surface has sufficient roughness for film handling. Several available methods for controlling the amount and kind of roughness include dispersing submicron/micron inorganic/organic particulates in the polymer before extrusion, embossing the film (e.g., knurling the edges of the web), sandblasting the surface with an abrasive, coating, sizing, and chemical etching (Heffelfinger, 1978; Bhushan et al., 1987).

Dispersion of particulates in the polymer before extrusion is most common. Two size distributions of particulates are generally used. Submicron particulates with a height of approximately 0.5 μm are used to reduce interlayer friction, and larger particles with a height of approximately 2–3 μm are used to control the air film (to provide leakage path for trapped air). The concentration of larger particles is almost an order of magnitude lower than that of small particles. Inorganic particles used include silica, titania, bentonite, or clays of different kinds. [The use of additives may affect the abrasion resistance of the PET film (unbackcoated tape surface) when used against a rough guide post in a tape drive.] After extrusion, subsurface voids are formed in the vicinity of additives. The shape and orientation of the surface asperities, as a result of additives, depends on the shape of the additives as well as mechanical properties of the web. Since the mechanical properties of the web vary across the width of the web, the shape and the orientation of the asperities vary across the width of the web which affects the air entrapment (Bhushan, 1990). Typical changes that occur in the surface roughness[2] as a result of polymer additives are shown in Fig. 2.4. Generally, PET substrate for magnetic media applications should not have more than about 2 asperities/ 100 cm^2 with heights larger than 1.5 μm. It is known that log (number of asperities/area) decrease linearly with an increase in asperity height. PET substrate roughness for particulate media typically ranges from 1.5–2.5 nm

---

[2] Increase in surface roughness by adding particulates reduces transparency desirable in many applications such as solar films.

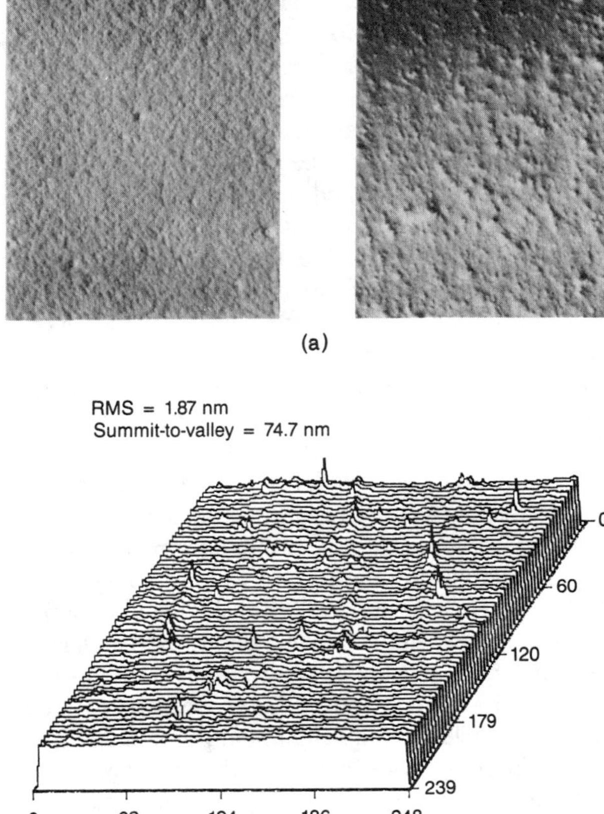

Fig. 2.4. Surface roughness of the PET film obtained from polymer additives (a) photomicrograph and (b) surface plot taken using an optical profiler.

rms and for thin-film media it is even smoother. However, peak-to-valley distance is generally high on the order of 50–100 nm or higher because of particulates added to the PET (Oden et al., 1992). We can also lubricate the PET surface, if required to reduce friction. Often, a slip agent can be added to the polymer melt, which exudes to the surface of the quenched film (Oswin, 1975). This technique has had great success with some polyolefins where the amides of fatty acids are often used.

To give an example of web dimensions in many Du Pont manufacturing processes for data-processing applications, the oriented web is 6.1-m wide (or 3.05-m wide in older manufacturing lines) and 1370-m long. The web is slit into four sections, each of which is rolled onto a master roll 1.4-m wide. The

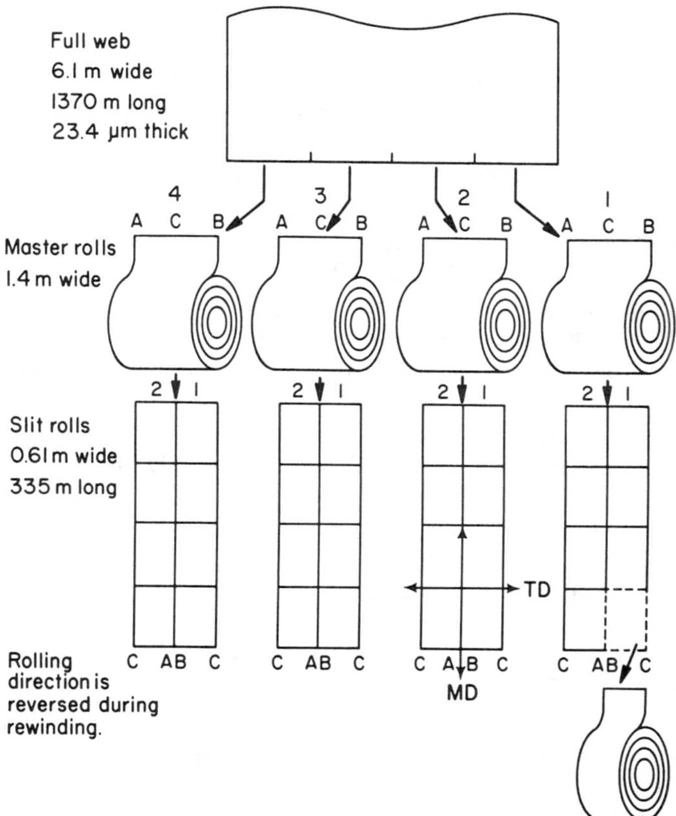

Fig. 2.5. Manufacturing sequence of cut PET film rolls.

master rolls are then slit into two sections, each 0.61-m wide. These sections are slit into four 335-m-long sections and wound onto slit rolls, Fig. 2.5 (Bhushan, 1985). Slit rolls are then coated for manufacturing of magnetic media. Typical thicknesses of PET films for magnetic tapes include 6.35 µm (25 gauge), 8.89 µm (35 gauge), 14.5 µm (57 gauge), 22.4 µm (88 gauge), 23.4 µm (92 gauge), 36.1 µm (142 gauge), and 76.2 µm (300 gauge) thick films are used for flexible disks. The thickness uniformity of better than 3% across TD and along MD has been reported by Ishai et al. (1968).

2.2. Structure of PET Films

Molten PET is highly viscous and can be extruded and quenched into a glassy state in film form. When rapid cooling occurs to a temperature at which the crystalline state is expected to be more stable, molecular movement is too sluggish to take up a crystalline conformation. Therefore the random arrangement characteristic of the liquid persists down to temperatures at which the viscosity is so high that the material is considered to be a solid. The formation of an amorphous or glassy film is essential to the development of the desired ordered PET structure. An amorphous PET film is in a metastable state at room temperature.

Amorphous PET film is of little commercial importance, since it has low strength and a tendency to be brittle. However, by appropriate combinations of deformation (stretching and/or rolling) and thermal processes, films with high degrees of orientation and crystallinity (ordered structure) can be made which exhibit exceptional strength and toughness. Temperature has a major role in the deformation processes of amorphous films. At low temperatures, molecular flexibility is suppressed, but at a sufficiently high temperature, referred to as glass transition temperature (T_g), the material passes from a relatively rigid glass to a flexible, ductile viscous phase. [We note that T_g is the temperature at which a rapid decrease in the rate of change of volume with temperature occurs, Choy and Plazek (1968).] For PET, T_g is about 70° C and the melting temperature (T_m) is about 255° C. The film is generally processed at temperatures in the range 70°–90° C. At these temperatures, the film can be drawn or rolled easily. During the deformation of the heated film, the stresses developed cause the uncoiling and translational motion of the molecular chain segments. Concurrently, nucleation and strain-induced crystallization occur. Therefore, the net result of the deformation process is the partial alignment of molecular segments in the direction of deformation, coupled with the formation of a new phase. Highly oriented films provide an effective molecular network where crystallites formed in the oriented state restrain amorphous polymer chain segments from diffusing to random conformations. The structure introduced by these orientation steps is immobilized (for increased thermal stability) through crystallization by a heat setting or crystallization process (annealing at temperatures significantly above T_g and under tension). The thermal stability can be further improved by annealing at a temperature slightly above T_g and under very low tension.

The structure, degree of crystallinity, and the glass transition temperature depend strongly on the stretching rate, amount of stretch, crystallization (heat setting) temperature, and the prior and post-thermal treatments—annealing and physical aging (Uematsu and Uematsu, 1960; Stein and Misra, 1973; Heffelfinger and Schmidt, 1965; Biangardi and Zachmann, 1977; Groeninckx et al., 1976, 1980a, b; Choy et al., 1983; Gupta et al., 1984a, b; Vallat et al., 1986; Vallat and Plazek, 1988). The degree of crystallinity generally increases with an increase in the crystallization temperature (above T_g). The maximum degree of crystallinity normally attained is between 50% and 60%. The glass transition temperature increases with an increase in the degree of crystallinity, and also the morphology of the PET has a strong influence on the size of the increase. Glass transition temperature also increases with annealing (above T_g). In semicrystalline polymers (PET) a range of order exists and both the crystalline and amorphous regions contribute to the properties.

The structure, relative crystallinity, and orientation distributions of PET films have been studied with x-ray diffraction (Cobbs and Burton, 1953; Daubeny et al., 1954; Heffelfinger and Schmidt, 1965; Groeninckx et al., 1976; Viswanathan et al., 1976; Gupta et al., 1984a, b), light scattering and other optical techniques (Dulmage and Geddes, 1958; Heffelfinger and Knox, 1971; Chu and Smith, 1973), infrared spectroscopy (Dulmage and Geddes, 1958; Schmidt, 1963; Heffelfinger and Schmidt, 1965; Gupta et al., 1984b), and nuclear magnetic resonance (Biangardi and Zachmann, 1977). One way to obtain the crystalline content of PET is by use of the fractional change in density relative to the density difference between the PET crystal and the amorphous polymer. (The amorphous and crystalline phases have densities of 1331 kg/m^3 and 1470 kg/m^3, respectively.)

Configuration of PET molecules is shown in Fig. 2.6(a). Crystalline PET has a triclinic unit cell with the dimensions shown in Fig. 2.6(b). In addition to the crystalline–amorphous division for PET films, a more detailed division of the structure is based on the molecular conformations of the ethylene glycol linkage, which in PET may exist in either of two rotational isomeric forms (*trans* or *gauche*) and can be observed using infrared techniques, Fig. 2.7. The conformation in the crystalline regions is *trans*, whereas the molecules in the amorphous region assume both the *trans* and *gauche* conformations, with an abundance of *trans* (Schmidt, 1963).

The structural changes that take place in an initially amorphous unoriented film, as it is taken through several stretching procedures, are described below. The effects of film orientation on the structure and mechanical properties are demonstrated by the experimental data presented in Table 2.2 (Heffelfinger and Schmidt, 1965; Heffelfinger, 1978). Young's modulus is a measure of the stiffness of the film. Breaking strength is indicative of the perfection of orientation. Strain at break (elongation) can be considered as a measure of the mobility of chains under stress, while the stress required to elongate the sample 5% is a measure of the tautness of the polymer chains. Each of these characteristics provides a qualitative idea of the sample's structure.

2.2.1. One-Way Stretching

Quenched amorphous PET film is a mixture of molecular chains of varying lengths having no unique directionality. In this state, the chains are most likely coiled, folded, or both and extremely tangled. The orientation of this structure is called random. When this amorphous film is stretched in the MD, the molecular chains become partially aligned along the stretch direction and some crystallinity occurs. The crystallites formed, in general, have their *c*-axis, i.e., the long axis (chain axis) of the PET crystallite, parallel to the direction of stretch, and some fraction of the (100) crystalline planes are aligned parallel to the surface of the film, Table 2.2. The resulting structure is classified as uniplanar–axial orientation. Since the molecular chains were most likely

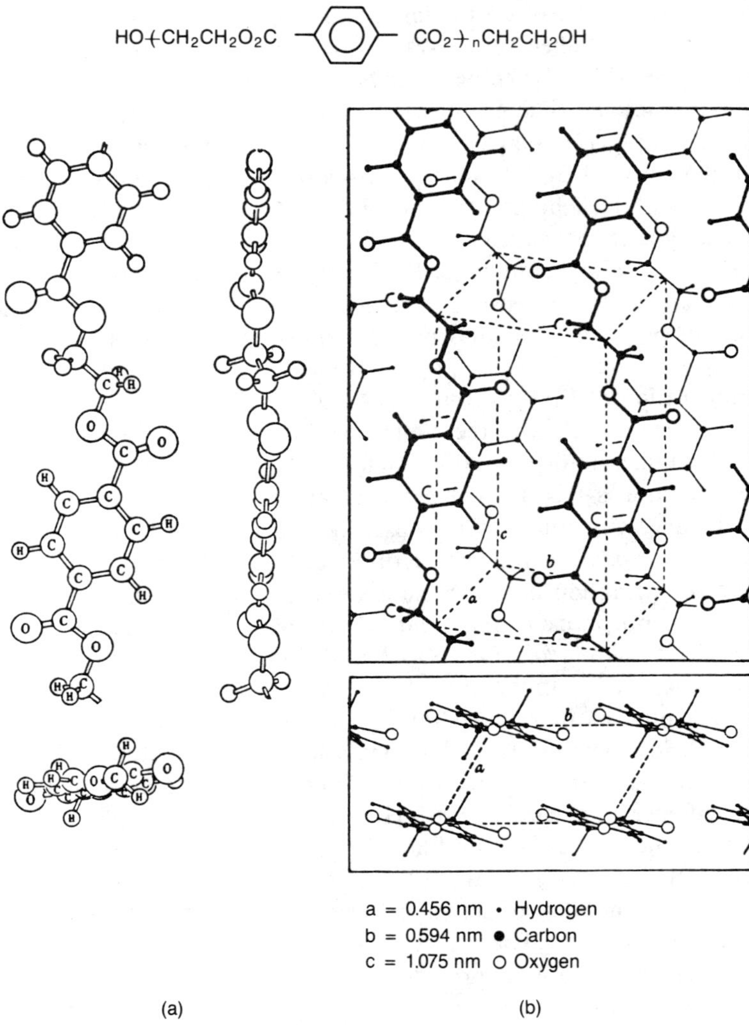

Fig. 2.6. (a) Configuration of a PET molecule, and (b) arrangement of molecules in PET crystal: above, projection nominal to (010) plane; below, projection along c-axis (Daubeny et al., 1954).

initially entangled and randomly oriented, many chain segments probably remain entangled after the MD stretch. Under high stretching conditions (stretch ratio > 3.5), the chains are believed to be so strained that there are many "taut" oriented amorphous regions between the crystallites. In other words, many chain segments are extended, and rather limited, by entanglements with other taut chains or crystallites in how far they can be further stretched. As a result, Young's modulus is high and high yield strengths are produced. The transverse direction properties remain essentially unchanged.

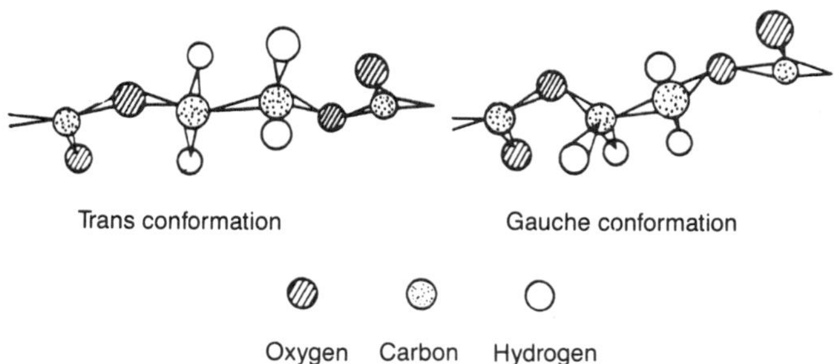

Fig. 2.7. *Trans* and *gauche* conformations of the glycol sections in PET.

X-ray measurements show that the distance between crystalline centers (long period spacing) is about 11.3 nm, and the distance between crystalline edges (the amorphous length) decreases with increased orientation. The crystallite length increases with increased stretch ratio, since the crystalline regions can grow more easily, as the alignment of chains in this direction is improved (Table 2.2).

Infrared studies show the cast PET films to be randomly oriented and to contain a fixed amount of *trans–gauche* structure. In these films, there is approximately 13% *trans* structure and 87% *gauche* structure. Since the *trans* structure is an extended form, this indicates that the majority of molecules are in a relaxed, unextended state. When the cast film is stretched in the MD direction and the molecules become aligned along the stretch direction, the individual chain segments become extended; that is, some *gauche* PET is converted into *trans* PET. The *trans* isomer is the only one that exists in the crystalline regions, and some crystallization will occur as the *trans* isomers are brought into proximity while being aligned along the stretch axis. As stretching is continued, the *trans* content of the film increases and the *gauche* content decreases, the amount of amorphous *trans* increasing at a rate more rapid than the amount of *trans* from crystallization. The alignment of the c- axis relative to the direction of stretch and the alignment of the (100) planes relative to the film surface is also improved. Since the *trans* isomer of PET is extended relative to the *gauche* isomer, the amount of *trans* structure at a given level of crystallinity is a measure of the "tautness" of the amorphous region of the polymer and is directly related to the mechanical properties of the film (Table 2.2).

2.2.2. Two-Way Stretching

Subsequent stretching of the one-way MD stretched film in the TD causes the chain segments (in the amorphous regions) and the crystallites to fan out in the second stretching direction, the amount of this movement being dependent upon stretch ratio, temperature, and other process conditions. From this

Table 2.2. Structure and typical mechanical properties of PET films stretched in various directions and at various stretch ratios (SR)

Property		Amorphous and unoriented	One-way stretched (SR = 2 in MD)	One-way stretched (SR = 3.5 in MD)	Two-way stretched (SR = 2.5 in MD then SR = 2.5 in TD), heat set	Two-way stretched (SR = 3.5 in MD then SR = 3.5 in TD), heat set	Two-way stretched + post stretched (SR = 1.5 in MD)
				(a) *Structure*			
$\sigma_{(a,a)}$,[a] degrees		—	45.0	18.0	47.3	36.1	36.5
$\sigma_{(p,p)}$,[a] degrees		—	45.0	25.9	16.7	11.9	15.9
Long period spacing, nm		—	10.7	11.3	11.9	12.0	13.2
Crystallite length, nm		—	4.4	5.8	4.8	6.4	6.4
Amorphous length, nm		—	6.3	5.5	6.4	5.6	6.8
Crystalline *trans*, %		0	7	17 (17)[c]	48	48 (17)[b]	48
Amorphous *trans*, %		13	14	24 (11)[c]	11	26 (35)[b]	40
Total *trans*, %		13	21	41 (28)[c]	59	74 (52)[b]	88
				(b) *Mechanical properties*			
Young's modulus, GPa	MD	2.4	3.2	7.0	4.3	5.2	7.0
	TD	2.4	2.3	1.8	4.8	4.7	3.2
Stress to produce	MD	—	60	130	90	105	180
5% elongation, MPa	TD	—	50	40	95	100	75
Breaking strength, MPa	MD	50	120	220	135	210	260
	TD	50	45	50	160	170	120
Strain at break	MD	>5	2.0	0.7	1.3	0.8	0.5
	TD	>5	5.0	4.0	1.4	1.0	2.5

[a] $\sigma_{(a,a)}$—Average position of the c-axis of the cystallite relative to the direction of stretching; $\sigma_{(p,p)}$—Average position of a major crystalline plane, (100) relative to the film surface. The 0° denotes the perfect parallelism and 45° denotes complete randomization of the operator to the reference system.
[b] Not heat set.
[c] Heat relaxed at 100° C without the application of tension.

treatment, some of the crystallites c-axes move from the MD toward the TD, the amount of this movement measured by the crystallite orientation distribution parameter $\sigma_{(a,a)}$. In addition, the (100) planes of the crystallites are forced into better parallelism with the surface of the film, as measured by the crystallite orientation parameter $\sigma_{(p,p)}$ (Table 2.2). It is believed that during the second direction stretch some chain disentanglements occur, since the long period (the distance between crystallite centers in the MD) increases to 11.9 nm (Table 2.2). The crystallite length, however, increases slightly over that produced by the first direction stretch and, therefore, the average amorphous length is essentially unchanged. The structural changes that take place during the second direction stretch are primarily associated with the reapportionment of the structure that has been produced by the first direction stretch. The amorphous regions of this film are not as highly strained in either MD or TD, and as a consequence we have lower modulus and lower yield strength than one-way stretched films. The shifting of some of the orientation from MD toward the TD has now generated comparable properties in the two directions.

2.2.3. Heat Setting (or Crystallization)

The production of dimensionally stable films requires that the structure introduced by these orientation steps be immobilized (for better dimensional stability) through crystallization by a process known as "heat setting" [annealing at high temperatures (significantly above T_g) under tension]. Only the *trans* isomers are capable of crystallizing. As noted before, *gauche* PET is transformed into *trans* PET during the stretching process. Here many *trans* isomers are brought into proximity with one another, hence crystallization occurs. Crystallization may be induced in either of two ways: by polymer stretching or by heat setting at temperatures substantially higher than T_g. Crystallization induced by the heat-setting process may come about by either the movement of amorphous *trans* molecules into the necessary proximity or by the thermal isomerization of *gauche* near neighbors with a subsequent crystallization of the newly-formed *trans* material. The latter explanation is more plausible, since the amorphous *trans* content changes only about one-third that of the *trans* crystalline change during the crystallization of two-way stretched films (Table 2.2).

Tension on the film sample during crystallization plays an important role in the development of structure. Generally, if a one-way or two-way stretched film is heat set under tension, to try to maintain the orientation produced by stretching, the total *trans* content of the film increases as the crystallinity increases (Table 2.2). The degree of molecular orientation of this, however, will decrease. Evidently, the orientation stresses are so great that they cannot be completely restrained by tension on the film during crystallization, so that stress decay, localized melting, and recrystallization occur.

2.2.4. Post-Stretching

If the essentially balanced, crystallized, two-way stretched film is subsequently given an additional MD stretch, i.e., by "post-stretching," the molecular chain segments are once again shifted toward the MD but are now more constrained in their motion by the additional crystallinity present. As a result, the molecular chain segments are drawn taut. Infrared measurements show that the amount of *trans* crystalline structure remains essentially unchanged, but that the amount of *trans* amorphous structure has increased. X-ray measurements show that the crystallite lengths remain essentially unchanged, but that the long-period spacings have increased (to beyond 12.5 nm), so that the distance between crystallites (the amorphous length) has increased (Table 2.2). This very taut structure is difficult to deform further, since the chain segments in the amorphous regions are already extended in a variety of directions, thereby limiting the freedom that can occur during additional stretching, e.g., during tension testing. Low elongation, high modulus, and high yield strength are, therefore, produced.

2.2.5. Strain Relaxation (or Annealing)

One of the most significant features exhibited by these taut extended amorphous structures is the ability to relax easily and generate new structural and property features. This results in an overall contraction (shrinkage) of the film in the direction of orientation. For example, if a one-way stretched film is heated to a temperature of 100° C for a short time without the application of tension, there is a large decrease in the amorphous *trans* content, little or no change in the crystalline *trans* content, along with a decrease in the amount of molecular alignment (Table 2.2). The annealed films (by heating at a temperature above T_g and under very low tension) exhibit good thermal stability (low shrinkage). However, relaxation results in degeneration of yield strength caused by a change in the tautness of the segments in the amorphous regions. Density and elongations increase as a result of relaxation.

2.2.6. Commercial Biaxially-Oriented PET (Mylar A)

In the manufacture of biaxially-oriented PET films, according to the process described in Section 2.1, fundamental differences exist between the MD and TD orientation processes. The nip rolls used in MD stretching exert a constant stress across the complete web which results in fairly uniform stretching across the TD and along the MD. Consequently, during the stretch in the MD, crystallites form; the crystallites and amorphous chains tend to align with MD. On the other hand, during the stretch in the TD, stretching is greatest near the edges and lowest in the center across the width and essential-

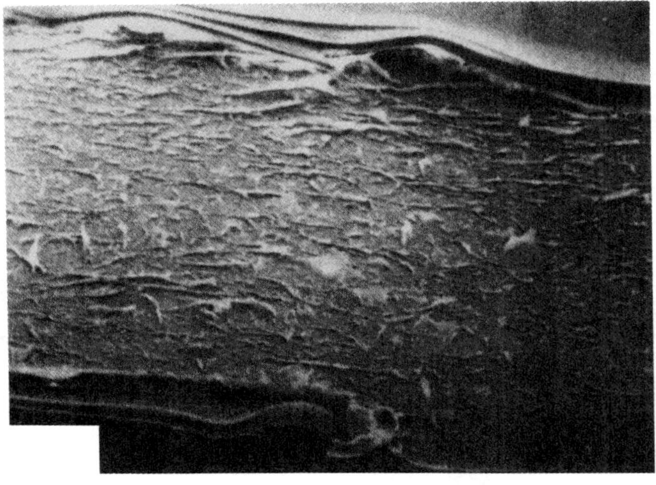

Fig. 2.8. Electron micrograph of biaxially-oriented PET film.

ly uniform stretching along the roll.[3] Consequently, the amount of reorientation of crystallites and amorphous chains during the stretch in the TD is greatest near the edges and least in the center; thus, an "unbalanced" film whose physical properties vary across the web (with highest modulus near the edges) is obtained, i.e., the film has in-plane anisotropy. In addition, the film has anisotropy in the thickness directions with the highest anisotropy in the center of the film.

Morphology of a biaxially-oriented PET film is shown in Fig. 2.8. In the electron micrograph, branched lamellae with small lateral dimensions are observed. Amorphous regions exist between the ordered crystal lamellae. Crystalline thicknesses range from 10 nm to 50 nm and may be several microns in breadth (Heffelfinger, 1978; Groeninckx et al., 1976, 1980a, b).

The crystalline phase of the biaxially-oriented PET does not exist in the spherulite macrostructure, normally obtained from melt-crystallized PET (Yeh and Geil, 1967). Studies have shown that the melt-extrusion process complicates the morphology. Because the molecules have been pre-oriented by stretching along two perpendicular directions and then crystallized, the crystallites tend to orient and distribute themselves in a specific way. Chu and Smith (1973) have shown, by using a small-angle, light-scattering technique, that the folded-chain elements are gathered together in rodlike superstruc-

---

[3] During TD stretching, there is some relaxation in the MD especially near the center of the film since there is nothing to hold the film in the MD.

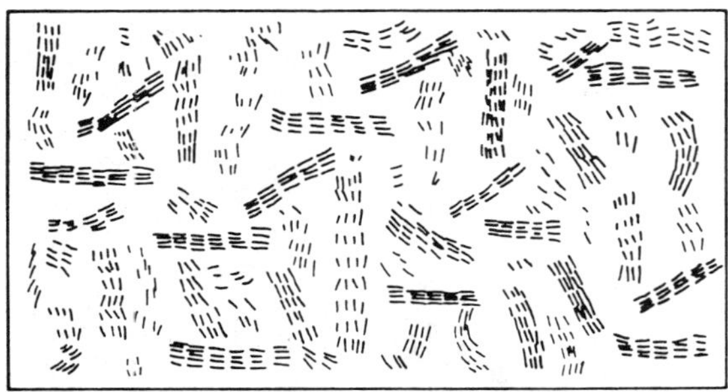

Fig. 2.9. Schematic diagram to illustrate the possible orientation distribution of rod-like superstructures in biaxially-oriented PET films based on light-scattering patterns (Chu and Smith, 1973).

tures, and tend to be oriented in the various directions depending upon the location in the film sheet for biaxially-oriented PET. The possible orientation distribution of rodlike structures is shown schematically in Fig. 2.9. Barrall and Logan (1973) have observed that the crystalline phase structure is intermediate between a true regularly folded polymer crystal and an extended chain crystal structure.

The amorphous phase of the biaxially-oriented PET is in the glass state at room temperature. Orientation occurs in the same general direction as the rodlike crystal macrounits. Miyagi and Wunderlich (1972) have demonstrated that the amorphous phase in oriented PET is far from equilibrium with respect to the crystalline phase. That is, the crystallinity of an oriented PET film is far lower than that of the same film after heating to near the melting point. Numerous frozen-in strains exist in the coiled and deformed coil regions of the oriented amorphous phase. The word *metastable* most completely defines biaxially-oriented PET. An attempt to illustrate the macrostructure of biaxially-oriented PET film is shown in Fig. 2.10.

2.2.7. Summary

The PET is an initially amorphous unoriented polymer. The deformation of the heated film results in the partial alignment of molecular segments in the direction of deformation, coupled with strain-induced crystallinity. Subsequent thermal treatment results in a secondary crystallization. Thus, the biaxially-oriented film consists of an oriented, strained amorphous phase in which is embedded a crystalline phase. Crystalline PET has a triclinic unit cell. The molecular conformation in the crystalline region is *trans*, whereas the molecules in the amorphous region assume both the *trans* and *gauche* conformations with an abundance of *trans*. The crystalline phase exists in the folded-chain configuration oriented in macrounits depending upon the loca-

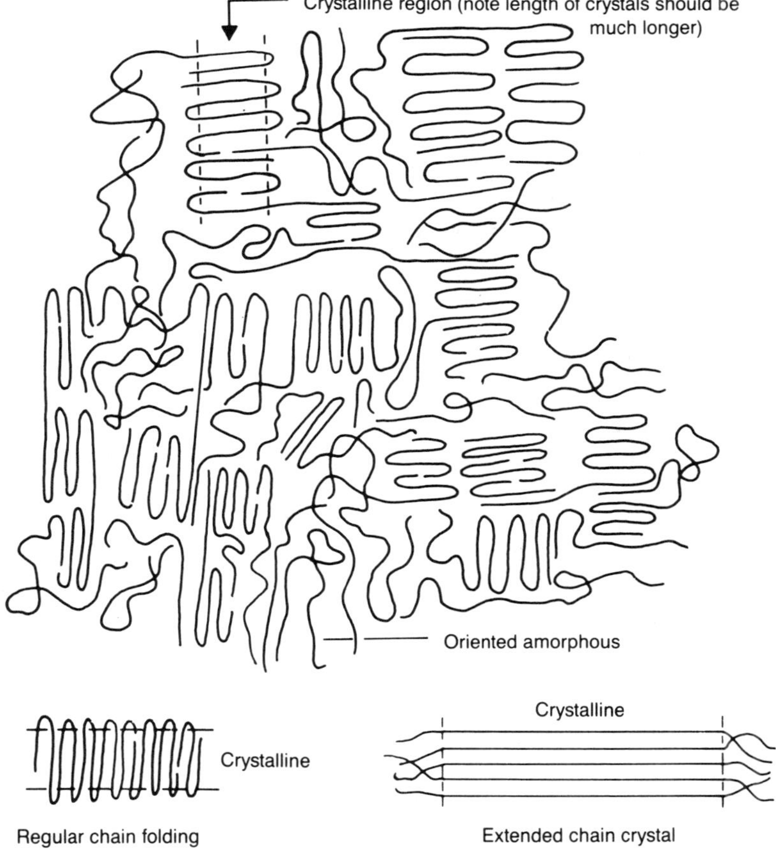

Fig. 2.10. A schematic representation of the oriented amorphous and crystalline structures found in biaxially-oriented PET films (Barrall and Logan, 1973).

tion in the film sheet. Neither the crystalline nor the amorphous phase is at equilibrium. The film is metastable in two respects. First, the percent crystallinity of the film is much lower than the equilibrium crystalline content. Second, amorphous regions of the film contain frozen-in strains which tend to relax and allow the film to contract.

The structural factors which influence resultant film properties are:

(1) the kind and perfection of crystallite orientation;
(2) the amount and direction of the *trans–gauche* isomerization in the amorphous regions; and
(3) the amount of crystallinity.

The degree of interaction of these factors results in substantially different film properties and relaxation behavior.

2.3. Physical and Chemical Properties

Physical and chemical properties and their in-plane anisotropy of PET are directly related to the relative proportion of the molecules aligned or oriented in the same direction at which the physical properties are being measured. Think of the molecules in PET as pencils, Fig. 2.11. If all the pencils are lined up in the same direction, then the very high mechanical properties (e.g., Young's modulus of elasticity) will be in the longitudinal direction, compared to values measured at right angles. If the pencils are now uniformly oriented as a function of angle, and since the proportions are the same in any direction contributing to strength, the mechanical and other physical properties would be found to be uniform and invariant in any direction. Thus, molecular orientation of the PET molecules control the orientation characteristics of physical properties.

The "extraordinary" or "slow" or "strong" optical axis is a *principal optical axis* direction in which the net orientation of crystals and amorphous material is a maximum. We will see that this orientation for biaxially-oriented PET film used for magnetic recording applications exhibits maximum in-plane refractive index, maximum density, maximum in-plane Young's modulus of elasticity, maximum breaking strength but minimum in-plane thermal and hygroscopic expansion coefficients and strain at break. The "ordinary" or "fast" or "weak" axis is a *transverse optical axis*, orthogonal to the extraordi-

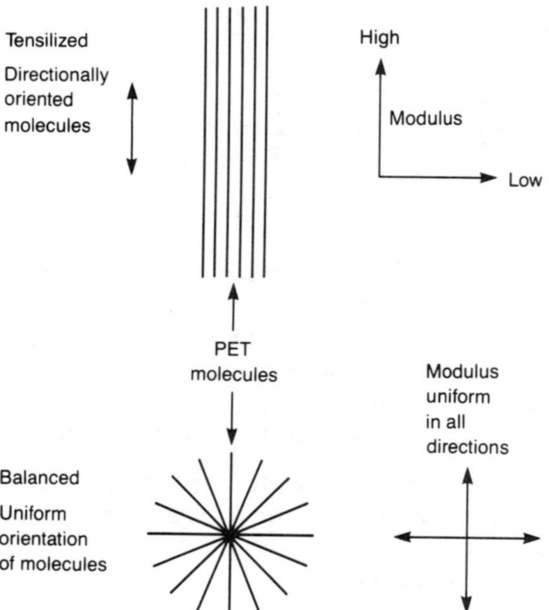

Fig. 2.11. Schematics of the relationship between molecular orientation and Young's modulus.

nary or slow or strong optical axis and exhibits minimum in-plane refractive index, etc. The yield strength exhibits in-plane near isotropy. The shrinkage at high temperatures and high humidities is also anisotropic with maximum shrinkage along the MD.

In this section, we present anisotropic, physical, and chemical properties— density, optical properties, mechanical properties, thermal expansion properties, hygroscopic expansion properties, and shrinkage properties. Visco-elastic (dynamic modulus, creep, and relaxation) properties will be presented in Chapter 3. A summary of the typical properties of a 23.4-μm-thick biaxially-oriented PET (Mylar A, 92 DB) film is presented in Table 2.3.

2.3.1. Density

The density of biaxially-oriented PET films (Mylar A) was measured by Barrall and Logan (1973) and Chu and Smith (1973) using a water–zinc chloride density-gradient column, which had been calibrated with glass spheres of known densities. The density of 36.1-μm-thick PET film was found to be 1392.5 kg/m^3. If the amorphous and crystalline phases have densities of 1331 kg/m^3 and 1470 kg/m^3 (Cobbs and Burton, 1953), the density data correspond to crystallinity of 46.7%.

Chu and Smith (1973) also studied the effect of annealing on the density and degree of crystallinity of the PET film. (Also see Fakirov et al., 1977 and Biangardi and Zachmann, 1977.) The data presented in Table 2.4 show that the density, as well as crystallinity, increase progressively with the annealing temperature. The increased thickness because of annealing results from a lateral contraction associated with the partial release of frozen-in tensile strain. Because each annealed specimen has nearly the same thickness, the amount of frozen-in strain released at each annealing temperature is essentially the same.

2.3.2. Refractive Index, Birefringence, and Infrared Dichroism

One of the easiest and nondestructive methods used to determine in-plane molecular orientation in transparent films is the refractive index. The only equipment needed is a suitable refractometer or a polarizing microscope with compensators (Wahlstrom, 1969; Samuels, 1981). The orientation of the film results in an increased density leading to an increased refractive index, in-creased Young's modulus of elasticity, and decreased thermal and hygroscop-ic expansion properties. Therefore, by measuring the refractive indices along different directions or by measuring the birefringence (difference in refractive index in the slow and fast optical axes[4]) we can determine whether the film

---

[4] The slow and fast optical axes can be determined by measuring the extinction angle under a cross polarizer (made of two polymer films). The axes that show extinction of light are axes with maximum and minimum refractive indices. Maximum transmission identifies the direction 45° to the slow and fast axes. In an isotropic material (having no birefringence), there will be extinction at all angles.

Table 2.3. Typical properties of a 23.4-μm-thick biaxially-oriented PET (Mylar A)[a] film

Property	Typical value(s)	Condition	Method[b]
(a) *Physical properties*			
Density, kg/m^3	1392–1395	22° C	D1505-63T
Poisson ratio	0.25–0.3		
Young's modulus (tensile), GPa		22° C	D882-64T
(SOA—45°–90° to MD)			
along SOA[c] (a)	4.4–5.8		
along FOA[c] (b)	3.5–3.8		
along MD[c]	3.7–3.9		
along TD[c]	4.4–5.0		
anisotropy ratio (a/b)	1.2–1.6		
Yield strength (tensile), MPa		22° C	D882-64T
along SOA	90		
along FOA	90		
along MD	90		
Breaking strength (tensile), MPa		22° C	D882-64T
along SOA	220		
along FOA	130		
along MD	170		
Strain at break (tensile)		22° C	D882-64T
along SOA	0.4		
along FOA	0.5		
along MD	0.45		
Compressive modulus (325-μm thick), GPa			Thermomechanical
along Th D[c]	2.7		analysis system
Shear strength (127-μm thick), MPa	150		D732-46
Impact strength, kg-cm	6.0	22° C	Pneumatic impact tester
Folding endurance, kcycles	100	22° C	D2176-63T (1 kg loading)
Tear strength-initial (Graves), g/25 μm	800	22° C	D1004-61
Bursting strength (Mullen), kPa	450	22° C	D774-63T
Refractive index (Abbe)		22° C	D542-50
along SOA	1.66–1.68		
along FOA	1.62–1.64		
along Th D	1.49		
Birefringence	0.02–0.04	22° C	
Dichroic ratio	1.3–1.6	22° C	
Coefficient of friction (kinetic, film-to-film)	0.45	22° C	D1894-63
(b) *Thermal properties*			
Glass transition temp., °C	68–70		
Melting point, °C	250–265		
Service temp., °C	−60–150		
Coefficient of thermal expansion, × 10^{-5}/°C		20°–50° C	D696-44
along SOA	1.0		
along FOA	2.0		
along MD	1.4		
along TD	1.3		
along ThD	16		
volumetric	19		

Table 2.3 (*continued*)

Property	Typical value(s)	Condition	Method[b]
Thermal conductivity, W/m K	0.15	25°–75° C	
Specific heat, J/g K	1.18	25° C	
Shrinkage, %	0.08	24 h at 65° C	D-1204
	0.2–0.7	30 min at 105° C	
	2–3	30 min at 150° C	
(c) *Electrical properties*			
Dielectric strength, V/μm	300	22 °C, 60 Hz	D149-64
	200	150° C, 60 Hz	
Dielectric constant	3.3	22° C, 60 Hz	D150-65T
	3.25	22° C, 1 kHz	
	3.0	22° C, 1 MHz	
	3.7	150° C, 60 Hz	
Dissipation factor	0.0025	22° C, 60 Hz	D150-65T
	0.0050	22° C, 1 kHz	
	0.0160	22° C, 1 MHz	
	0.0040	150° C, 60 Hz	
Volume resistivity, ohm-cm	10^{18}	22° C	D257-66
	10^{14}	150° C	
Surface resistivity, ohm/cm^2	10^{15}	22° C, 30% RH	D257-66
	10^{11}	22° C, 80% RH	
(d) *Chemical properties*			
Water absorption, %	<0.8		Immersion for 24 h at 22° C D570-63
Coefficient of hygroscopic expansion, $\times 10^{-5}$/% RH			8–80% RH at 22° C
along SOA	0.5		
along FOA	0.6		
along MD	0.6		
along TD	0.5		
along ThD	7.4		
volumetric	8.5		
Permeability, Gas, cc/100 cm^2-24 h-atm		22° C	D1434-66
carbon dioxide	2.6		
hydrogen	1.6		
nitrogen	0.16		
oxygen	0.96		
Vapor,[d] g/100 cm^2-24 h		40° C	Modified E96-66
acetone	0.35		
benzene	0.58		
carbon tetrachloride	0.013		
ethyl acetate	0.013		
hexane	0.019		
water	0.029		
Fungus resistance	Inert	12 months' soil burial	E96-66T

Table 2.3 (*continued*)

	Tensile strength retained	Elongation retained	Tear strength retained	Days immersed at room temperature
Chemical resistance				
acetic acid, glacial	100	100	100	31
hydrochloric acid, 10%	100	100	100	31
sodium hydroxide, 2%	100	100	70	31
ammonium hydroxide, conc.	0	0	0	3
trichloroethylene	100	100	100	31
hydrocarbon oil	92	88	87	500 h at 100° C baked 168 h at 150°C

[a] Mylar is a Du Pont registered trademark.
[b] American Society for Testing and Materials, Philadelphia, PA., 1968 Standards, except as noted.
[c] SOA—slow optical axis; FOA—fast optical axis, MD—machine (longitudinal) direction; TD—transverse direction, ThD—thickness direction.
[d] Vapor permeabilities are determined at the partial pressure of the vapor at the temperature of the test.

is anisotropic, because an isotropic film will have zero birefringence. The birefringence is determined using a Babinet compensator and a monochromatic light (e.g., of wavelength 546.1 nm).

Typical values of the refractive index or of the birefringence ($n_\alpha - n_\beta$) versus the amount of stretch of unidirectionally stretched PET film, are shown in Figs. 2.12(a) and (b) (Dulmage and Geddes, 1958; Heffelfinger and Knox, 1971; Ward, 1967). Accompanying changes in Young's modulus of elasticity as a function of the amount of stretch are also shown in Fig. 2.12(c). Curve B of Fig. 2.12(b) illustrates that when amorphous films are stretched, they develop strong birefringence due to the gradual alignment of polymer chains parallel to the direction of stretch. At high elongation, the curve flattens out, owing to an enhancement of the refractive index in the transverse direction, arising from higher orientation of the plane of the molecules in the plane of

Table 2.4. Density, crystallinity, and thickness of untreated and annealed PET (Mylar A) films (Chu and Smith, 1973)

Annealing temp., °C	Density, kg/m^3	Crystallinity, wt.%[a]	Thickness, μm
Untreated	1392.5	46.7	36.1
95	1394.1	47.9	40.9
120	1395.9	49.2	40.6
180	1401.9	53.5	40.4

[a] Calculated from density assuming that densities of amorphous and crystalline phases are 1331 kg/m^3 and 1470 kg/m^3, respectively.

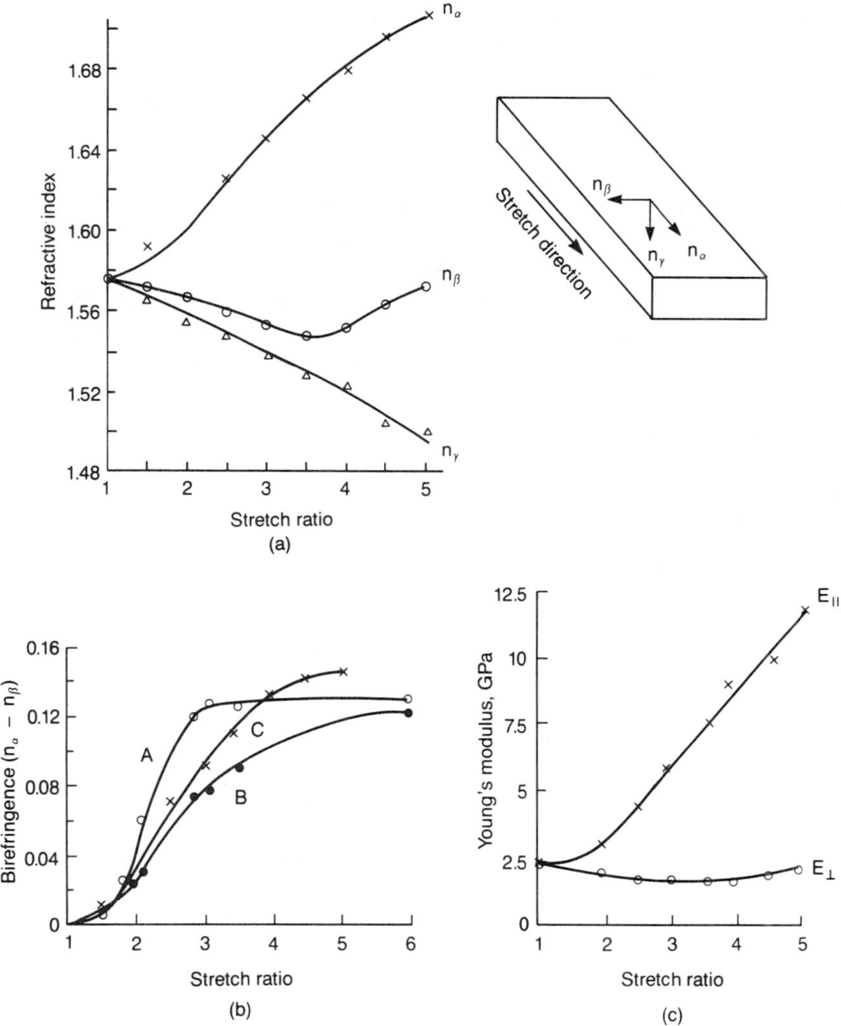

Fig. 2.12. (a) Refractive indices as a function of the stretch ratio of unidirectionally stretched PET films in the stretch direction (n_α), normal to the stretch direction (n_β), and the thickness direction (n_γ). (b) Birefringence of unidirectionally stretched PET film: A, stretched at 90° C and thermally crystallized; B, stretched at 90° C and not thermally crystallized; C, stretched at 85° C and not thermally crystallized (birefringence is defined here as the difference between the refractive indices in the stretch direction and normal to the stretch direction). (c) Young's modulus of elasticity as a function of the stretch ratio of unidirectionally stretched PET films in the stretch ($E_{||}$) and normal to the stretch direction ($E_\perp$) (Dulmage and Geddes, 1958).

the sheet. Curve A of Fig. 2.12(b) illustrates the effect of a post-stretching thermal crystallization on the birefringence ($n_\alpha - n_\beta$). Curve C of Fig. 2.12(b) illustrates the effect on the birefringence of a decrease in the stretching temperature. To observe the changes in the refractive index for various amounts of biaxial stretching requires sequential analysis. A particular point on the refractive index curve for a one-way stretched film is chosen (e.g., 200% extension). Then the changes in the refractive indices in two (in-plane) orthogonal directions (n_α, n_β) are determined, as stretching is performed in the transverse direction. Refractive index versus the amount of stretch (like those illustrated in Fig. 2.13) are produced. Normally, the film is considered to be balanced when the birefringence approaches zero.

Infrared dichroism, a technique employing polarized infrared light, also detects directly the molecular orientation of the PET molecules and, therefore, also the orientational characteristics of the elastic modulus (Dulmage and Geddes, 1958; Cuddihy, 1972; Heffelfinger, 1984). It works as follows. A chemical grouping exists in the PET molecule which strongly absorbs infrared radiation at a frequency of exactly 973 cm^{-1}. Chemical groups absorb infrared radiation by having internal motions whose energy levels are related to a fixed frequency (Planck's Law. energy, $E = hv$ where h is Planck's constant

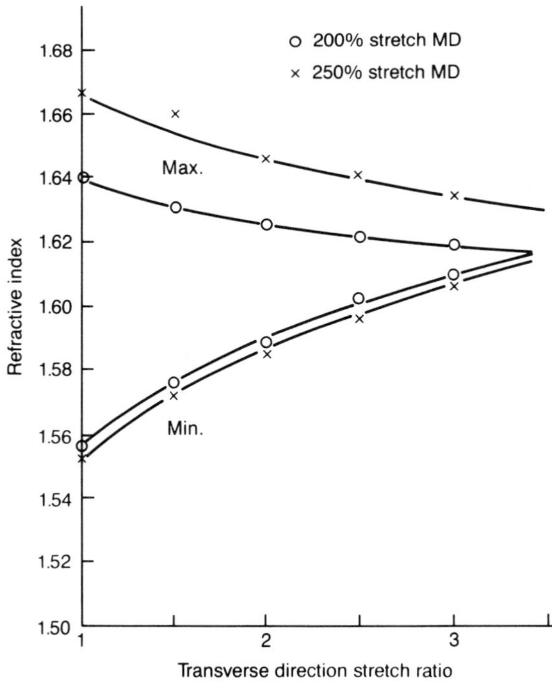

Fig. 2.13. Refractive indices as a function of the stretch ratio in the transverse direction of biaxially-drawn PET film (Heffelfinger and Knox, 1971).

and v is the frequency) in the infrared spectrum. When that frequency is employed, a resonance occurs with the group absorbing energy, causing an attenuation of the initial beam intensity. This attenuation can be detected, which provides testimony to the presence of the group as well as indicating the concentration of the group by the extent of attenuation. The internal motion of the chemical group in PET films which causes the resonance at 973 cm^{-1} is uniquely aligned along the long axis of the PET molecules. The polarized source will interact and resonate with only those chemical groupings which are also aligned in the direction of the plane of polarization of the incident infrared radiation. The extent of attenuation of the beam is now directly proportional to the proportion of PET molecules that are aligned in the direction of the plane of polarization. Hence, directional variations in the level of attenuation of the incident polarized infrared radiation are directly relatable to the variation in the physical properties such as Young's modulus of the material; the optical density of attenuated radiation is inversely related to Young's modulus.

The ratio of attenuated infrared radiation along two orthogonal directions (along fast divided by along slow optical axes) is called the *dichroic ratio*. The dichroic ratio represents the material anisotropy, which equals 1 for an isotropic material and is greater than 1 for an anisotropic material. The infrared dichroic ratio method is only useful for thin (< 25 μm) films because of the lack of light transmission in thicker films. The dichroic ratios of stretched PET films vary as a function of stretch ratio and crystallinity (Dulmage and Geddes, 1958).

2.3.3. In-Plane Mechanical Properties

Slow and fast axes of refractive indices correspond to the major and minor axes (principal material directions) of Young's modulus of elasticity in tension, respectively, see Fig. 2.12 (Heffelfinger and Knox, 1971; Chu and Smith, 1973; Blumentritt, 1979a; Bhushan, 1985). In fact, the refractive indices at any web location follow an elliptical distribution similar to that in Young's modulus (measured by Instron, Ishai et al., 1968; Chu and Smith, 1973; Viswanathan et al., 1976; Bhushan, 1985). A typical refractive index of PET film is about 1.65 with a birefringence of about 0.02–0.04. The mean of the refractive index in the MD changes very little (0.005) along the web. The optical density of the attenuated polarized infrared radiation at a frequency of 973 cm^{-1} in infrared dichroism measurements also follows an elliptical distribution, in which the axis with lowest optical density corresponds to the axis with the highest Young's modulus and vice versa, Fig. 2.14 (Heffelfinger, 1984).

Bhushan (1985) measured Young's modulus, yield strength, breaking strength, and strain at break of PET films (Mylar A) cut from a 6.1-m-wide web. Tensile properties were measured on die-cut dogbone-shaped samples (ASTM D-1708) with a 17.5-mm gauge length, using a constant strain-rate

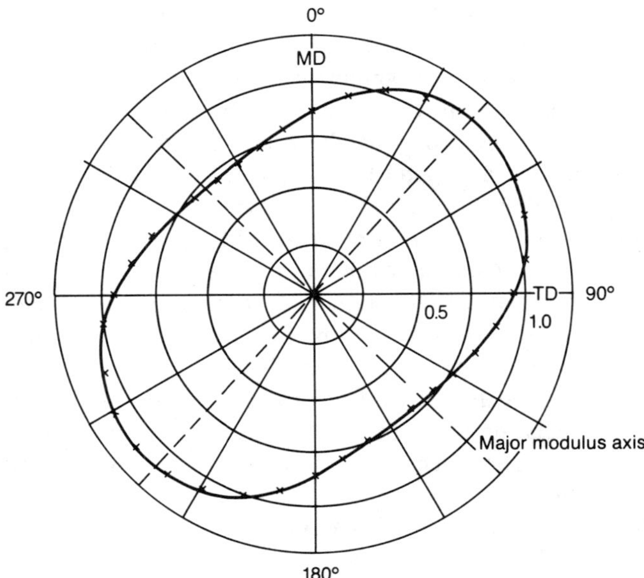

Fig. 2.14. Optical density of the attenuated polarized infrared radiation at a frequency of 973 cm^{-1}, as a function of the in-plane direction of a 23.4-μm-thick biaxially-oriented PET (Mylar A) film near the outer edge of the web. Dotted lines are the major and minor axes of Young's modulus.

test machine (Instron) at room temperature and with a crosshead speed of 12.7 mm/min, which translates to a strain rate of about 1.2×10^{-2}/s. An example of the polar plots of various mechanical properties for a PET sample is given in Fig. 2.15 which shows that Young's modulus, breaking strength, and strain at break follow an elliptical pattern, and that the yield strength remains constant with orientation. The major axes of modulus and breaking strength and the minor axis of strain-at-break were found to coincide. Strain-at-break data had a high degree of scatter.

Figure 2.16 shows typical stress–strain curves along the major and minor axes. Region A up to the yield point of the stress–strain curves can be thought of as the work required to move the molecular segments from their equilibrium positions, that is, simple elastic extension. Region B is indicative of massive viscoelastic and plastic deformation of the film in which the molecular chain segments rotate, translate, and unfold. Once the system is set into motion, little additional force is required. Finally, the stress will exceed the strength of the film and fracture occurs. The shape and magnitude of the stress–strain curve depend on many parameters such as the temperatures used. (For more details, also see Chapter 3.)

The orientations of the major and minor axes of Young's modulus across the film web are plotted in Fig. 2.17(a). We note that the minor axis is along

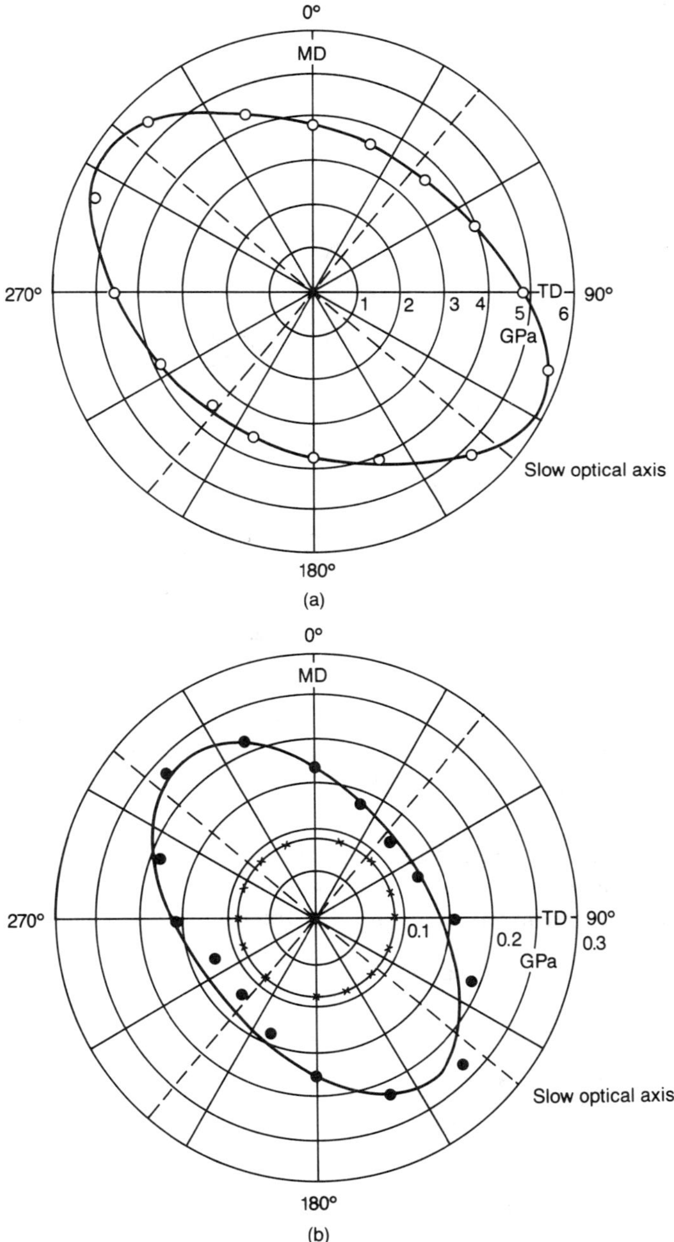

Fig. 2.15. Polar plot for a 23.4-μm-thick PET (Mylar A) sample (taken near the outer edge of the web) measured at 22° C and at a strain rate of 1.2×10^{-2} s^{-1}. (a) Young's modulus, (b) yield strength ($\times$), breaking strength ($\bullet$), and (c) strain-at-break. Dotted lines are the slow and fast optical axes that show the extinction of light under a cross polarizer.

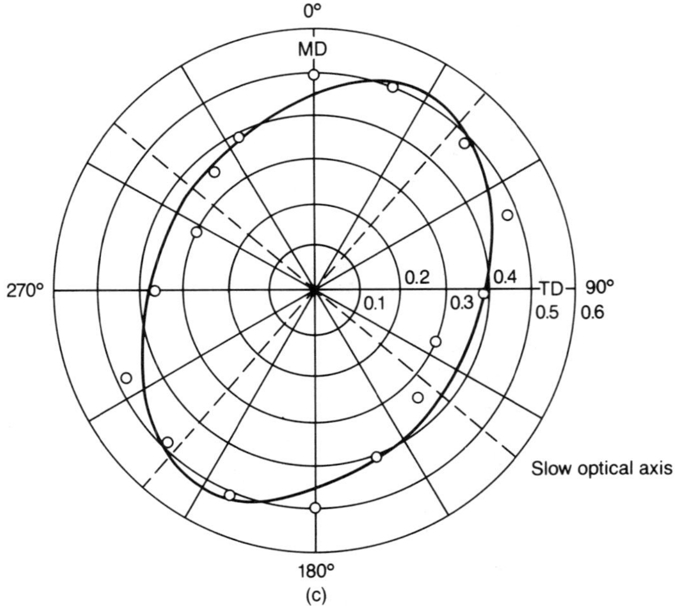

Fig. 2.15 (*continued*)

the MD at the center of the web and it rotates to about 45° (typically ranging from 40 to 55°) with respect to the MD at the edges. If we could get samples closer to the clamps (distorted, so slit off and recycled), we should see close to 90° rotation.

Young's modulus along the major axes (*a*) and minor axes (*b*) and the anisotropy ratio *a/b* across the web are plotted in Fig. 2.17(b). The magnitude of the modulus is symmetrical about the web center. The value of *a* and *a/b*

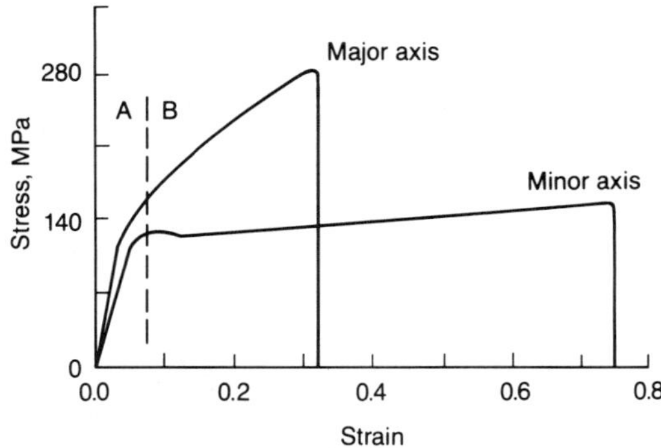

Fig. 2.16. Typical stress–strain curves for a PET film along the major and minor axes. See text for an explanation of A and B regions.

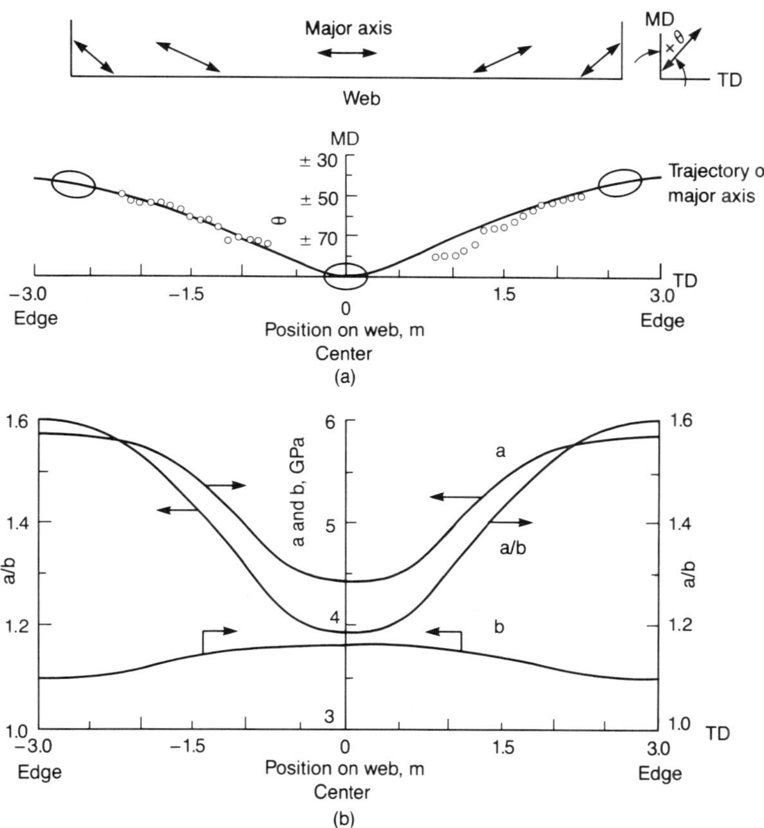

Fig. 2.17. (a) Orientation of the major axis of Young's moduli with respect to the MD across the 6.1-m-wide PET (23.4-μm-thick Mylar A) web (clockwise positive). (b) Variation of Young's moduli in the major axis a and minor axis b and the anisotropy ratio a/b across the 23.4-μm-thick and 6.1-m-wide PET (Mylar A) web.

increases from the web center to either edge (because of increased orientation toward edges during TD stretch), while b decreases slightly. The anisotropy ratio is typically 1.1–1.2 at the web center, increasing to as much as 1.6–1.7 at the web edges. The large angular dependence of the modulus ellipse across the web causes variations in the actual MD and TD moduli, Fig. 2.18. We note that modulus in the MD remains almost constant being slightly lower at the edges ($< 10\%$) and modulus in the TD (higher than that in MD) increases from center to edges.

Anisotropy of the mechanical properties can also be obtained by non-destructive test techniques—birefringence and infrared dichroic ratios. The dichroic ratios can only be used for thin ($< 25\ \mu$m) films. Birefringence is more commonly used. Figure 2.19 shows birefringence and dichroic ratios for one-half of the tenter frame width (3.05 m) of a web (Heffelfinger, 1984). The

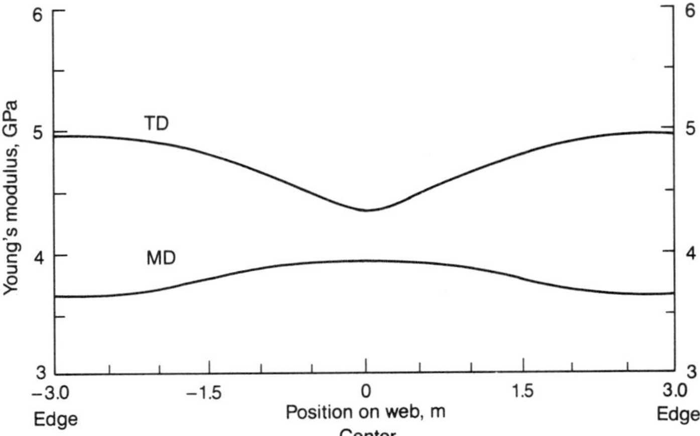

Fig. 2.18. Variation of Young's moduli in the MD and TD across the 6.1-m-wide PET (23.4-μm-thick Mylar A).

data suggest that the anisotropy ratio of Young's modulus is low at the center and maximum at the edges of the web.

Blumentritt (1979a, b) measured the mechanical properties of biaxially-oriented PET films ranging in thickness from 23.4 μm to 335.6 μm. He found that the thinner film had the higher values of Young's modulus along slow optical axis and greater anisotropy (a difference of about 10% from thinnest to thickest film), which can be explained by this film having the higher degree of in-plane orientation in the thin film. We have seen in Fig. 2.12(c) that Young's modulus in the stretch direction increases with the stretch ratio.

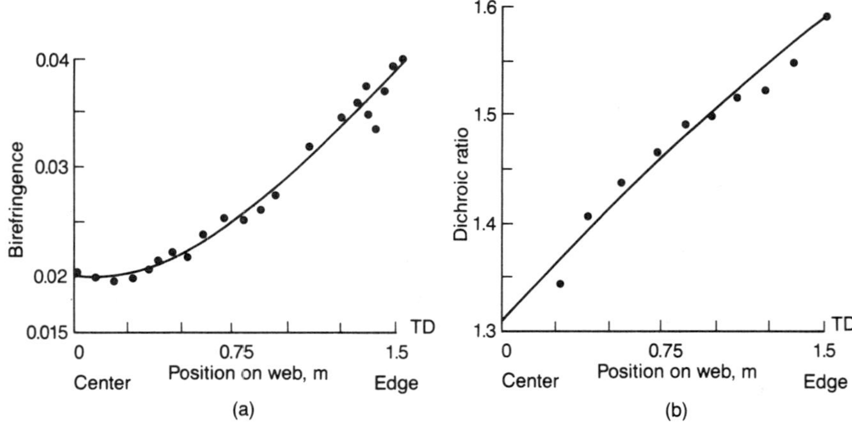

Fig. 2.19. Variation of (a) the birefringence and (b) the dichroic ratio across the 36.1-μm-thick and 3.05-m-wide PET (Mylar A) web.

Bogy et al. (1979) measured the Poisson ratio of 38.1-μm-thick biaxially-oriented PET (Mylar A) and a CrO_2-coated tape. Measurements were made by uniaxial loading on the PET strips in increments of about 5 N, and measuring the longitudinal and lateral strain changes from an initial slightly loaded (2.5 N) reference configuration. (For details of the experimental apparatus, see Chapter 3.) The ratio of the slope of the lateral strain versus stress to the slope of the longitudinal strain versus stress gives the Poisson ratio. They reported a value of about 0.25–0.3 and, to a first approximation, it was time-independent.

PET drive belts have been used on flight tape recorders. The failure mode of these belts is flexural fatigue initiating in mechanically-weak locations, which are introduced into the belt during fabrication as well as by exposure to some solvents used for cleaning during fabrication. Cuddihy (1972) measured the flexural fatigue strength of 23.4-μm-thick biaxially-oriented PET (Mylar A) films. He found that with increasing flexing angle at constant stress, the cycles to failure decreased rapidly until a critical angle was surpassed, where a greatly reduced dependence was encountered, Fig. 2.20. The fatigue endurance limit for PET film, cut in the low modulus duration at a flexing angle *within the critical zone*, was reported to be about 65 MPa with cycles to failure $>10^5$, Fig. 2.21. We note that the fatigue endurance limit is much

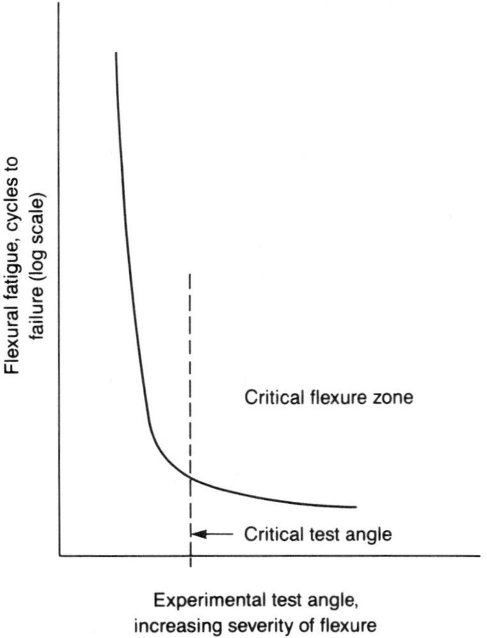

Fig. 2.20. General flexural fatigue characteristic of a polymer as a function of a test angle at constant stress.

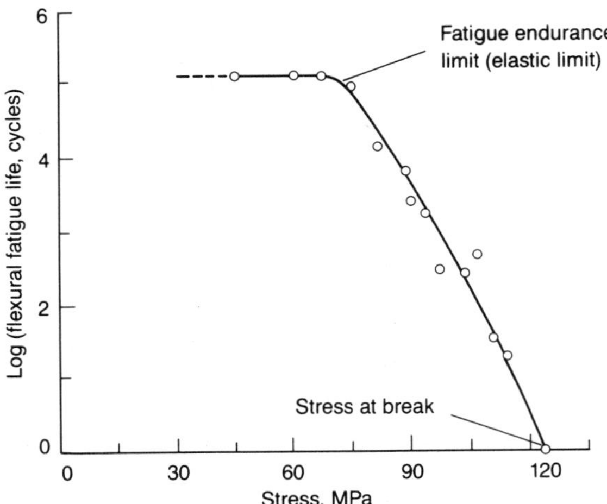

Fig. 2.21. Flexural fatigue curve for the 23.4-μm-thick biaxially-oriented PET (Mylar A) film (cut in the weak modulus direction) tested at a flexing angle within the critical zone (Cuddihy, 1972).

larger at smaller flex angles. Cuddihy further reported that the low modulus direction was significantly weaker in resistance to flexural failure than the corresponding high modulus direction. (Also see Biangardi and Zachmann, 1977.)

2.3.3.1. Effect of Temperature and Strain Rate

In-plane mechanical properties as a function of strain rate and temperature were measured by Bhushan (1985) using an Instron machine. Properties were measured of 23.4-μm-thick PET, 0.25-mm-thick free film of magnetic coating A (polyester polyurethane with a 32% hard polyurethane segment and a 50% pigment volume concentration of CrO_2 magnetic particles) as cast and calendered using a two-roll calender at 74° C and a load of 385 N/mm ($\sim$ 72 MPa), and a 31.8-μm-thick coated web (coated with magnetic coating A on both sides and with one side calendered) (Bhushan, 1990). The results are presented in Table 2.5. We note that Young's modulus increases with an increase in strain rate, as expected. The percent change in Young's modulus of the free film for an increase in the strain rate by a factor of 10 is more than that for the PET film or the coated web. Young's modulus, the yield strength, and breaking strength decrease and the strain-at-break increases with temperature. We also observe that Young's modulus of both calendered and un-calendered free film is lower than that for the PET film, and the modulus of the coated web is slightly lower than that of the PET film. Additional data of the effect of strain rate and temperature on the dynamic modulus will be presented in Chapter 3.

Table 2.5. Mechanical properties data at different
temperatures and strain rates

Temp., °C	Strain rate, $\times 10^{-2}$/s	Young's modulus, GPa	Yield strength, MPa	Breaking strength, MPa	Strain at break
1. PET 23.4-μm thick (Mylar A) (MD)					
22	0.012	3.59	76.1	115	0.47
22	0.12	3.66	81.2	122	0.48
22	1.2	3.69	86.8	126	0.44
38	1.2	2.84	74.7	100	0.26
52	1.2	2.52	72.4	101	0.45
60	1.2	2.12	72.3	97.4	0.30
2. 0.25-mm-thick free film of magnetic coating A					
Uncalendered					
22	0.12	0.97	6.6	7.4	0.06
22	1.2	1.10	8.1	9.3	0.06
38	1.2	0.60	4.6	7.0	0.06
52	1.2	0.48	4.0	4.5	0.05
60	1.2	0.48	3.5	4.6	0.07
Calendered					
22	1.2	1.75	9.9	11.7	0.10
3. Tape web A					
Coated on both sides with one side calendered					
22	0.12	3.25	76.4	95.6	0.49
22	1.2	3.31	80.5	95.5	0.42
38	1.2	2.91	70.6	92.1	0.42
52	1.2	2.49	68.0	90.8	0.52
60	1.2	2.18	66.1	83.1	0.52

2.3.3.2. Effect of Annealing

Chu and Smith (1973) and Blumentritt (1979b) studied the effect of annealing (thermal treatment above T_g) on Young's modulus, the yield strength, breaking strength, and strain-at-break. Blumentritt found that annealing for 10 min at 120° C and 2 h at 107° C resulted in essentially no change in the dependence of orientation on Young's modulus, the yield strength, and breaking strength. Annealing tends to reduce slightly the modulus (by less than 10% for the annealing conditions used in the study) and yield strength (by less than 5%) of oriented PET films by allowing relaxation of ordered noncrystalline regions in the films (Fig. 2.22). The annealing treatments tend to increase very slightly (by less than 5%) the breaking strength and strain-at-break, possibly due to higher crystallinity.

2.3.3.3. Effect of Solvents

PET film is sensitive to solvent-stress cracking and susceptible to premature failure when stressed in some solvent environments (Cuddihy, 1972, 1975).

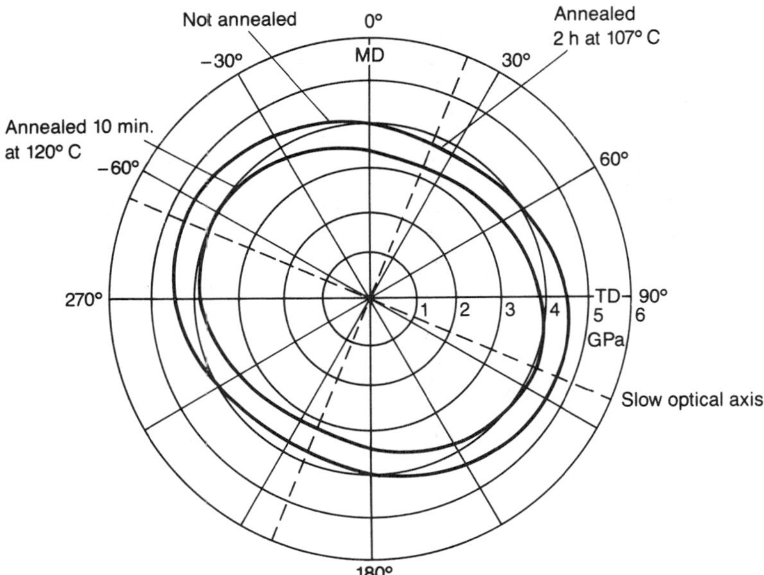

Fig. 2.22. Young's modulus as a function of the in-plane direction of a 36.1-μm-thick PET film (Mylar A) after annealing. Dotted lines are the axes that show the extinction of light under a cross polarizer (Blumentritt, 1979b). © 1979 International Business Machines Corporation; reprinted with permission.

Cuddihy (1972) measured the sensitivity of PET film in the high and low modulus directions (major and minor axes) by measuring the stress–strain curve while completely immersed in the methyl ethyl ketone (MEK) solvent, Fig. 2.23. We note that in the presence of MEK, the stress-at-break of PET in the low modulus duration is reduced from 11.4 MPa to 5.9 MPa. However, very little effect, only in the plastic draw region, is observed along the PET's high modulus direction. Cuddihy exposed PET films to MEK for periods ranging from 1 min to 96 h, after which he removed the samples from the solvent, air dried them, and then subjected them to stress–strain measurements in the low-modulus direction. The results, given in Fig. 2.24, clearly demonstrate the occurrence of a partial, but permanent, loss of mechanical properties. Even after only 1 min of exposure, there was a significant decrease in the yield stress and stress-at-break. The 5 min to 4 h of MEK exposure resulted in a reduction in modulus and yield stress, respectively, from 2.76 GPa and 82.7 MPa to near 2.24 GPa and 68.9 MPa.

Cuddihy (1972) measured the sensitivity of a series of solvents. The results shown in Table 2.6 reveal that solvent effects on PET films can be broken down into three general classes. The first, called Class I solvents, cause serious damage with loss of yield stress and breaking strength and their use must be avoided. The second, called Class II solvents, affect the plastic draw region (breaking strength) to much lesser degree and they may be used for brief

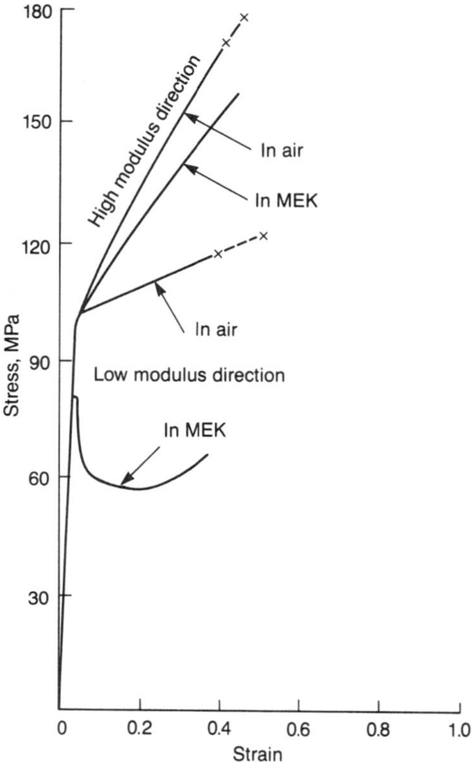

Fig. 2.23. Effect of MEK on the stress-strain behavior of a 23.4-μm-thick PET (Mylar A) film, measured at a strain rate of 25.4 mm/min (Cuddihy, 1972).

exposure periods. The third, called Class III solvents, may be used safely for long periods.

The coating process of PET films for magnetic media exposes the film's surface to various solvents for a few minutes. Appropriate selection of solvents may minimize the degradation of mechanical properties.

2.3.4. Elastic Modulus in the Thickness Direction

Young's modulus of PET film in the thickness direction (compressive modulus) was measured by Bhushan (1985) using a thermomechanical analysis system (Bhushan and Smith, 1985; Bhushan, 1990; for further details of the experimental apparatus and procedure, see Chapter 3) which involves measuring the deformation of a sample by a quartz probe having a radius of 0.457 mm. It is difficult to make measurements on a thin (such as 92 gauge) film because: (1) it is difficult to hold the thinner film flat; (2) the deflection of thinner films is too small to measure; and (3) the analysis to calculate the

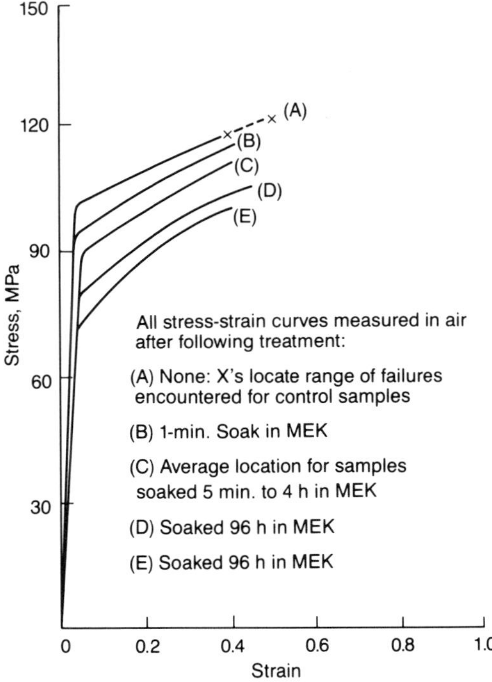

Fig. 2.24. Change in the stress-strain behavior of a 23.4-μm-thick PET (Mylar A) film after MEK exposure, measured at a strain rate of 25.4 mm/min (Cuddihy, 1972).

Table 2.6. Ranking of solvents relative to compatibility with PET film (Cuddihy, 1972)

Class I:	Solvent stress cracking
(1) MEK	agents: use to be avoided.
(2) Acetone	
(3) MEK/Freon	
(4) Toluene	
(5) Ethanol	
(6) Isopropyl alcohol	
Class II:	Selectively compatible:
(1) Hexane	exercise care in use.
(2) Freon TA-11	
(3) M-50 (1, 1, 1-trichloroethane)	
(4) Freon-TE-35	
(5) Freon-TP-35	
Class III:	Compatible for all
(1) Freon-TF	applications
(2) Freon-TE	

Table 2.7. Young's modulus in the thickness direction of 0.325-mm-thick PET (Mylar A) film

Temperature, °C	Young's modulus, GPa
22	2.70
38	2.04
52	1.65
60	1.36

Note: The in-plane modulus measured by tensile test on this film varied from 5.4 GPa to 3.4 GPa.

modulus of a thin film placed on a substrate is complex and only approximate. Therefore, a thicker biaxially-oriented film of 0.325 mm was used.

Measurements of Young's modulus in the thickness direction were made at four temperatures between 22° and 60° C, and the results are reported in Table 2.7. The modulus in the thickness direction was about 2.7 GPa at room temperature; that measured in the longitudinal direction varied from 3.4 GPa to 5.4 GPa (comparable to the values reported earlier on the 23.4-μm-thick PET). Therefore, we find that the modulus in the thickness direction is lower than that in the in-plane direction. Because the measurements in the thickness direction are conducted at a much lower strain rate, part of the decrease in modulus in the thickness direction might be caused by the strain-rate effect; however, this effect is believed to be on the order of 10% (see Chapter 3) which is a small fraction of the total difference.

2.3.5. Radial Elastic Modulus of the Wound Reels

The properties of a wound tape reel would be quite different from those of a single layer. The value of the radial modulus is dependent to a large extent on the amount of air entrapped during winding (Chapter 5). Because of stress relaxation, the mechanical properties of a wound reel also change with time, especially at higher temperatures and higher humidities (Bhushan et al., 1984b; Eshel et al., 1984). A 12.7-mm-wide and 23.4-μm-thick PET film and a 31.8-μm-thick tape A (23.4-μm-thick PET film coated with magnetic coating A on both sides and with one side calendered) each of about 165 m in length were wound by Bhushan (1985) onto spools at selected tensions and winding speeds. The reel was held on a specially made fixture, as shown in Fig. 2.25. A flat-ended probe 1.25 mm in diameter was brought into contact with the reel at a crosshead speed of 1.27 mm/min or at a strain rate of 1.2×10^{-3}/s. [Initially, the modulus of a tape reel was measured at various crosshead speeds (12.7, 1.27, and 0.127 mm/min), but we found that they do not have any effect on the modulus.] A dial indicator having an accuracy of

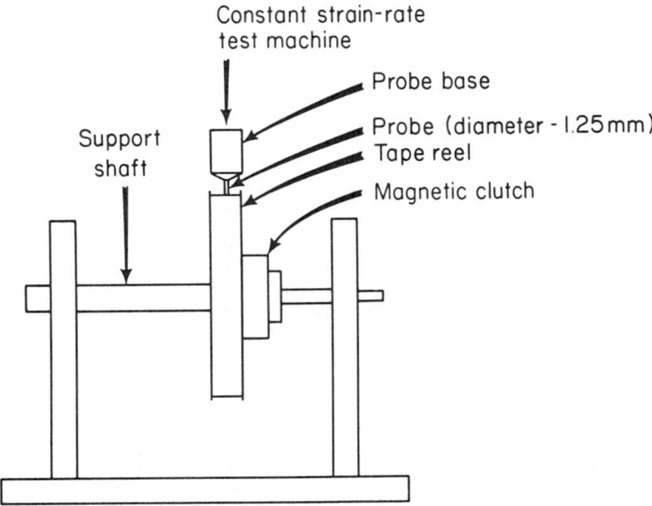

Fig. 2.25. Schematic of a test fixture for measuring the radial elastic modulus in the compression of a tape reel.

2.5 μm was used at the bottom of the reel to ensure that there was no deflection of the support shaft. A minimum of three measurements of load versus penetration were made at three locations: at the center of the reel and on either side (roughly 1.5 mm in from the edges). The variations in the three data points at each location and from one location to another were less than $\pm 10\%$.

The load versus penetration was plotted and the relationship was linear in the load range of 0–44.5 N. To calculate the modulus from the load–penetration data, the probe loading on the tape reel was modeled as a circular load on a semi-infinite body. For a circular load of radius a, a vertical deflection h at the outer radius ($r = a$) is related to the normal load per unit area σ as follows (Timoshenko and Goodier, 1970):

$$E_r = 4(1 - v_{r\theta}^2)\sigma a/\pi(h_{r=a}),\qquad(2.1)$$

where $v_{r\theta}$ is the radial Poisson ratio, that is, the ratio of the circumferential strain to the radial strain for an element under pure radial stress, and E_r is the radial elastic modulus. Even though $v_{r\theta}$ is unknown, the value of E_r is not very sensitive to a variation in a physically reasonable value of $v_{r\theta}$. $v_{r\theta}$ is generally related to $v_{\theta r}$ (circumferential Poisson ratio, an easily measured quantity ~ 0.25–0.3 for PET) by the following relationship (Lekhnitskii, 1968)

$$v_{r\theta} = v_{\theta r}(E_r/E_\theta),\qquad(2.2)$$

where E_θ is Young's modulus of the tape in the circumferential direction in tension. Based on the values of E_r and E_θ in this chapter, $v_{r\theta}$ is roughly equal to 0.02. A measured value of $v_{r\theta} \sim 0.05$ is presented by Umanskii et al. (1978).

2.3.5.1. Effect of Winding Parameters and Magnetic Coating

Experiments were conducted to study the effect of winding speed and tension and magnetic coating on the elastic modulus. The tests were done right after the tape was wound, generally within 1 h, and the results are presented in Table 2.8. Bhushan (1985) observed that a change in the winding speed from 1 m/s to 4 m/s had no influence on the modulus, but that changes in the tension from 1.1 N to 3.3 N increased the modulus. There is less air entrapped at a higher tension, which results in a higher modulus. A rough backcoat on tape A was found to have a higher modulus than the PET film or a tape coated only on one side and calendered. We will see in Chapter 5 that the interlayer friction in the case of a backcoated tape is higher than that in a tape that is not backcoated or in PET layers. The higher interlayer friction results in increased pinning of the layers and in less air entrapment with a consequent higher modulus.

2.3.5.2. Effect of Storage

PET film and tape A were wound at 2.2 N and 2 m/s, and the reels periodically taken out of storage (at 22°, 43°, and 52° C); the radial modulus was measured at ambient temperature. From the results presented in Fig. 2.26 we find that the radial modulus in the PET film and tape A reels reached a maximum after about 6 h. Air is entrapped during winding and *escapes in about* 6 h, which increases the modulus. After about 6 h, the modulus of the PET film reel remained constant during 1 week of testing, but the modulus of tape A decreased logarithmically. It is known that the magnetic coating goes through stress relaxation at a much more rapid rate than the PET film (see

Table 2.8. Radial elastic modulus data of wound reels at a strain rate of 1.2×10^{-3}/s

Tape type	Winding tension, N	Winding speed, m/s	Radial modulus, MPa
1. Effect of winding speed			
Tape A	2.2	4	238
		2	235
		1	234
2. Effect of winding tension			
Tape A	1.1	2	206
	1.7		241
	2.2		235
	2.8		245
	3.3		251
3. Effect of coating			
Tape A	2.2	2	235
Tape A, not back-coated			145
PET film substrate (23.4-μm thick)			148

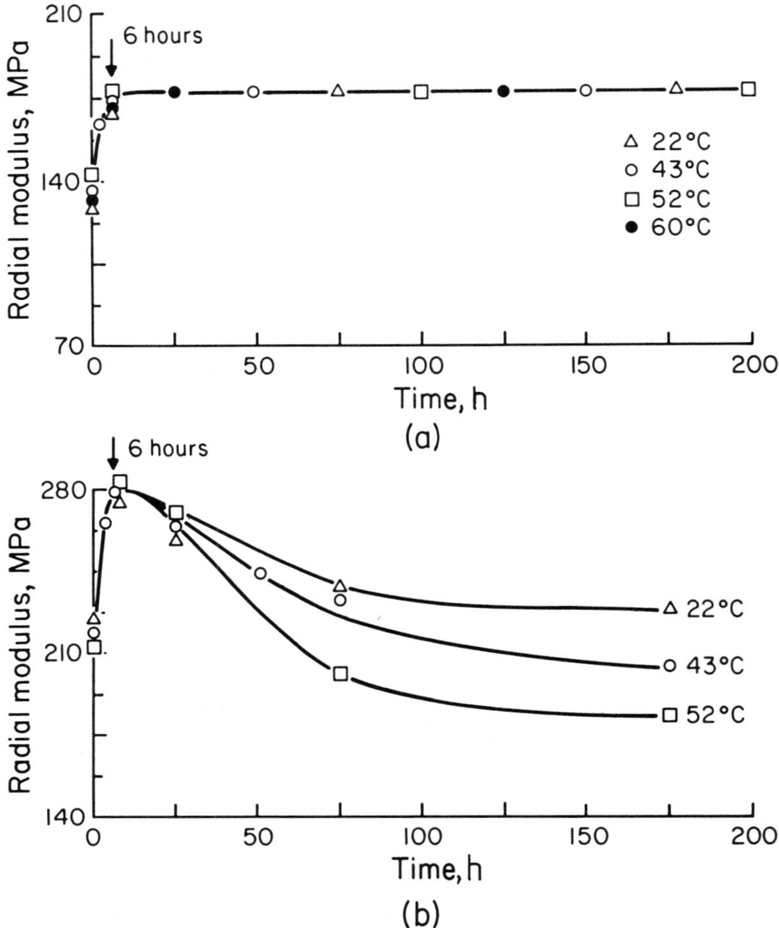

Fig. 2.26. Radial elastic modulus measured at 22° C as a function of the storage time at various temperatures of (a) a PET-film reel and (b) a tape A reel wound at 2.2 N and 2 m/s. The arrow indicates the time at which the peak of radial modulus occurs.

Chapter 3) and that the stress relaxation of the coating subjected to an interlayer pressure of about 1 MPa results in a decrease of the radial modulus of the tape reel after storage (Bertram and Eshel, 1980; Bhushan et al., 1984a; Eshel et al., 1984; Bhushan, 1985).

2.3.5.3. Radial Relaxation Modulus

A probe was left in contact with the reel surfaces of PET film and tape A at various ambient temperatures, and the decay in the normal load was continuously monitored for up to about 8 h. From these data, radial relaxation moduli were calculated and are presented in Fig. 2.27, which show a rapid

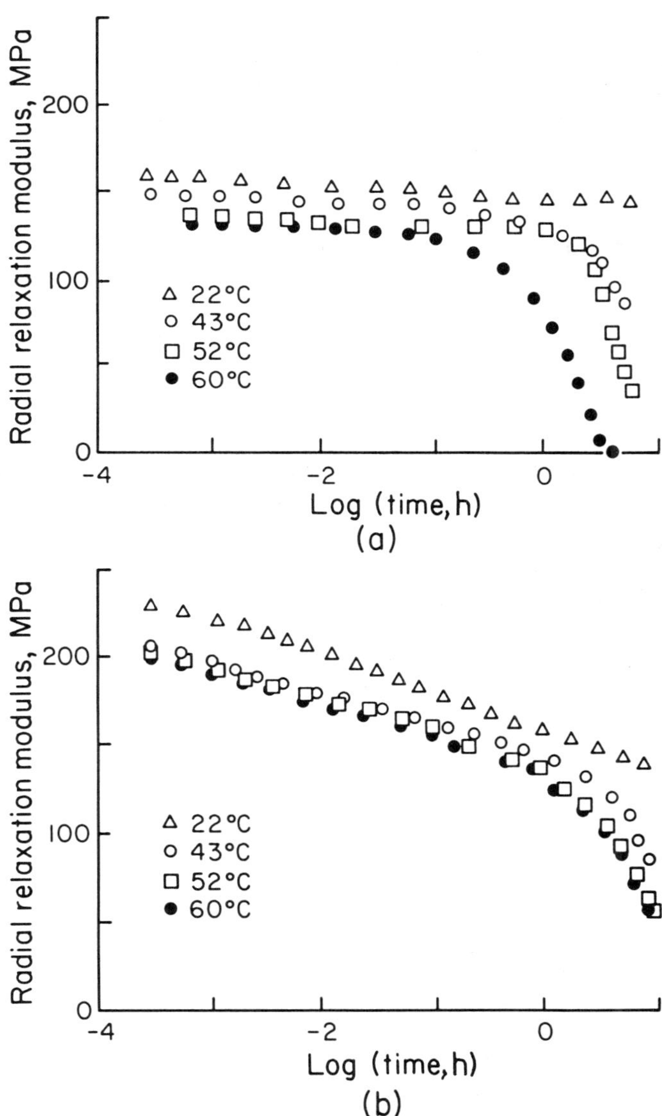

Fig. 2.27. Radial relaxation elastic modulus as a function of the test time at various test temperatures of (a) a PET-film reel and (b) a tape A reel wound at 2.2 N and 2 m/s.

drop in the modulus at higher temperatures, especially after about 1 h of testing (Bhushan et al., 1984b; Bhushan, 1985).

2.3.6. Thermal Expansion Properties

Coefficients of linear thermal expansion (α) of the PET films were measured by Barrall and Logan (1973), Blumentritt (1979a, b), Chan and Smith (1979), and Choy et al. (1983) using a Du Pont 942 thermomechanical analyzer (TMA) system. The specimens used were 3-mm wide with 7-mm long between the Invar chucks which were then attached to quartz stirrups; one stirrup was fixed and the other was coupled to the core of the LVDT probe. To eliminate any oscillations of the core-spring assembly, a small weight was placed on the weight tray of the TMA which applied a tensile stress of about 64 kPa to the sample. Tests were run at a heating rate of 5° C per minute. The coefficient of linear thermal expansion was calculated from the slope of the best straight line drawn through the temperature range of 25°–50° C, on a plot of probe displacement versus temperature (Barrall and Logan, 1973).

Choy et al. (1983) reported that the coefficient of linear thermal expansion of the oriented PET film along the stretch direction ($\alpha_{\parallel}$) decreases sharply with increasing orientation and may become negative at sufficiently high stretching, however, in the transverse direction, ($\alpha_{\perp}$) shows a slight increase. The large drop in $\alpha_{\parallel}$ and the accompanying increase in the axial Young's modulus $E_{\parallel}$ and the axial refractive index $n_{\parallel}$ can be ascribed to chain alignment in the crystalline regions. At a very high stretch ratio (>4 or $>300\%$ extension), the crystalline orientation becomes saturated so that the effect is diminished (Fig. 2.28) and approaches a limit—the coefficient of thermal

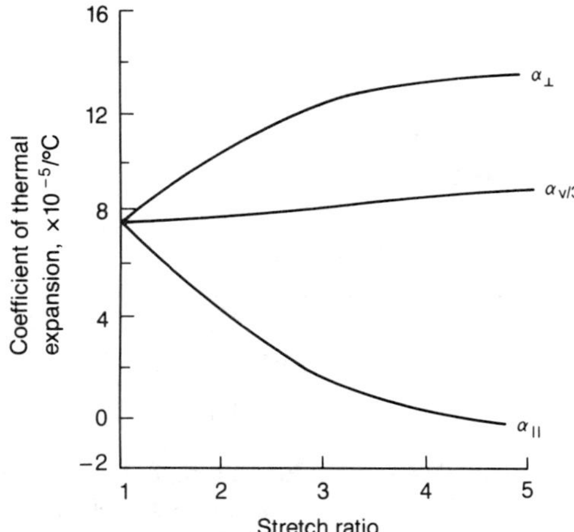

Fig. 2.28. Coefficient of the thermal expansion (at a temperature of 27° C and at a heating rate of 10° C/min) as a function of the stretch ratio of PET films unidirectionally stretched at 90° C (Choy et al., 1983).

expansion of the perfect polymer crystal in the chain direction. The coefficient of volumetric thermal expansion (α_v) increases by less than 10% as stretch ratio increases from 1 to 4.8. Uematsu and Uematsu (1960) reported that the coefficient of the linear thermal expansion of PET films decreases with an increase in the amount of crystallinity. The decrease in the coefficient of linear thermal expansion with an increase in the amount of crystallinity is higher for the films crystallized at lower temperatures ($\sim 120°$ C) than those crystallized at higher temperatures ($\sim 225°$ C).

Figure 2.29 is a polar plot of the coefficient of thermal expansion versus in-plane direction in the 36.1-μm-thick biaxially-oriented film and shows the dependence of thermal expansion on the orientation direction. The coefficient of thermal expansion is strongly anisotropic and is lowest along the orientation direction with maximum Young's modulus (slow optics axis at 70° to MD).

Chan and Smith (1979) found that the angular dependence of the coefficient of linear thermal expansion (α) could be represented closely by the following equation:

$$\alpha(\theta) = (\alpha_\parallel - \alpha_\perp) \cos^2 \theta + \alpha_\perp, \tag{2.3}$$

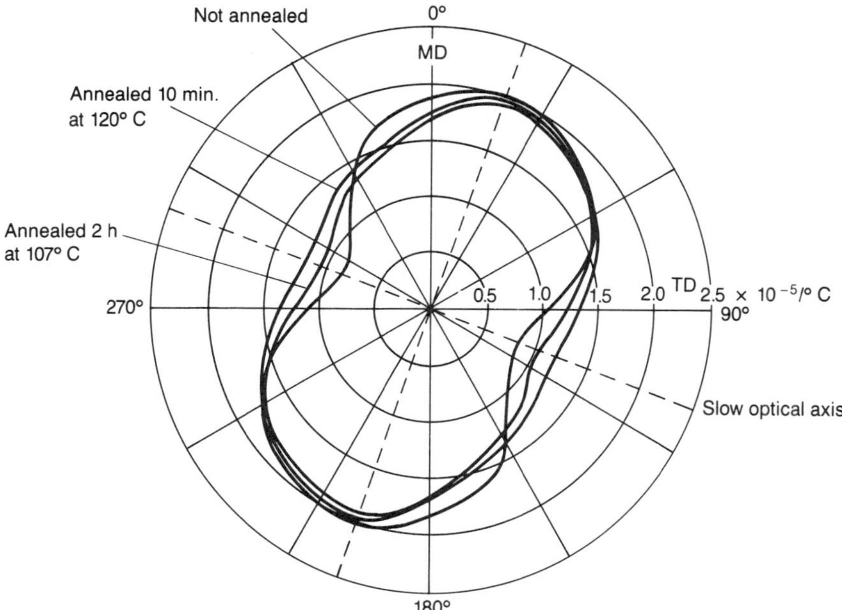

Fig. 2.29. Coefficient of the linear thermal expansion (in the temperature range of 25°–50° C and at a heating rate of 5° C/min) as a function of the in-plane direction of a 36.1-μm-thick biaxially-oriented PET film (Mylar A). Dotted lines are the axes that show the extinction of light under a cross polarizer. The minimum thermal expansion coefficient is along the major axis of Young's modulus (Blumentritt, 1979b). © 1979 International Business Machines Corporation; reprinted with permission.

where $\alpha_\parallel$ and $\alpha_\perp$ are the coefficients of linear thermal expansion parallel and perpendicular, respectively, to the slow optical axis of an anisotropic film. $\alpha(\theta)$ is the coefficient along the line that makes an angle θ with the slow optical axis. Although θ changes with temperature, the change is negligible. The rate of change is given by

$$d\theta/dT = \tfrac{1}{2}(\alpha_\perp - \alpha_\parallel)\sin 2\theta; \tag{2.4}$$

this derivative is maximum when $\theta = \pi/4$. For the biaxially-oriented PET (Mylar A) with $\sigma_\perp - \sigma_\parallel$ is about $1 \times 10^{-5}/°C$ and $\Delta T = 30°$ C, the maximum change in θ is about 2×10^{-4} radians or $0.01°$.

Chan and Smith (1979) also found that $\alpha_\parallel/\alpha_\perp$ is related semiquantitatively to the birefringence

$$(1 - \alpha_\parallel/\alpha_\perp) \sim 6.8 \times 10^{-2}\,\Delta n, \tag{2.5a}$$

where Δn is the birefringence given as

$$\Delta n = n_\parallel - n_\perp, \tag{2.5b}$$

where $n_\parallel$ and $n_\perp$ are the refractive indices parallel and perpendicular, respectively, to the slow optical axis. Thus, the anisotropy of the coefficient of thermal expansion can be related roughly to the birefringence.

For the PET films, it is instructive to consider the coefficients of thermal expansion in the three mutually perpendicular directions. Toward this end, it is necessary to invoke the well-established fact that the coefficient of volumetric thermal expansion α_v is independent of the degree of orientation, i.e., the coefficient for an isotropic polymer is the same as that for the polymer in an anisotropic state. (If the polymer is semicrystalline, this state is true only if the degree of crystallinity for the isotropic and anisotropic materials is the same.) For isotropic PET, whose crystallinity is approximately 50%, the reported value of α_v below and above glass temperature are 19×10^{-5} and 36×10^{-5} mm^3/mm^3/°C (Uematsu and Uematsu, 1960). For the biaxially-oriented PET films, as well as other anisotropic materials,

$$\alpha_v = \alpha_\parallel + \alpha_\perp + \alpha_t, \tag{2.6}$$

where α_t is the coefficient of linear thermal expansion in the thickness direction. ($\alpha_\parallel$, $\alpha_\perp$, and α_t can be the coefficients in any three mutually perpendicular directions.) For an isotropic material, each of the three coefficients equals $\alpha_v/3$. We further note that α_v is always positive, therefore, no more than two of the three linear expansion coefficients can be negative.

Table 2.9 shows measured values of $\alpha_\parallel$ and $\alpha_\perp$ and the calculated values of α_t for 23.4-μm, 36.1-μm, and 76.2-μm-thick oriented PET (Mylar A) films. We note that α_t is considerably larger than $\alpha_v/3$, the value for isotropic semicrystalline PET. Chan and Smith (1979) reported that the fractional increase in the coefficient of areal expansion in going from $25°$ C to $130°$ C in the second heating (after annealing at $190°$ C), using a heating rate of $5°$ C/min, is larger than that for the expansion coefficient in the thickness direction; in other words, α_t/α_v is smaller at $130°$ C than at $25°$ C.

Table 2.9. Coefficients of the thermal expansion of PET (Mylar A) film at 25–50° C , using a heating rate of 5° C/min

Film	$\alpha_{\parallel}$, $\times 10^{-5}/°C$	$\alpha_{\perp}$, $\times 10^{-5}/°C$	$\alpha_{t},$[a] $\times 10^{-5}/°C$	$\alpha_{\perp}/\alpha_{\parallel}$	α_{t}/α_{v}
23.4-μm-thick biaxially-oriented thick PET (Mylar A)[b]	0.96	2.01	16.03	2.09	0.84
36.1-μm-thick biaxially-oriented PET (Mylar A)[b]	1.18	2.17	15.65	1.84	0.76
76.2-μm-thick biaxially-oriented PET (Mylar A)[b]	1.50	3.07	14.43	2.05	0.76
Isotropic PET	6.33	6.33	6.33	1	0.33

[a] The value of α_v for isotropic PET film with about 50% crystallinity (about the same as that of biaxially-oriented PET film) is $19 \times 10^{-5}/°C$ (Uematsu and Uematsu, 1960).
[b] MD is 49°–45° to slow optical axis (Blumentritt, 1979a).

From the data presented in Table 2.9 we further note that the thinnest film had the lowest coefficient of linear thermal expansion, which can be explained by this film having the highest degree of in-plane orientation (Blumentritt, 1979a). All films showed similar anisotropy in the coefficient. Blumentritt (1979a, b) reported the coefficients of thermal expansion for 76.2-μm-thick uncoated PET substrate and for the coated magnetic media (with 76.2-μm-thick PET substrate coated on both sides with a 2.5-μm coating of γ-Fe_2O_3 particles dispersed in a polymeric binder). They found that coefficients of the uncoated substrate and coated magnetic media were comparable (within 5%). The coefficient for the magnetic coating with CrO_2 particles is estimated to be about $1.5 \times 10^{-5}/°C$, based on some thermal curling experiments (see Chapter 3), somewhat lower than PET in some orientations (Berry and Pritchet, 1988).

2.3.6.1. Heating Rate Effects

Barrall and Logan (1973) investigated the expansion and shrinkage properties of 36.1-μm-thick PET films cut parallel, perpendicular, and 45° to the MD. The samples were heated at three heating rates, 2, 5, and 10° C/min, and at temperatures up to about 230° C. The samples were under a tension of 64 kPa. During heating, the PET sample was found to expand up to some temperature, between about 73° and 97° C, that depended on the heating rate and the angle between the MD and the longitudinal axis of the specimen. At higher temperatures, shrinkage occurred. In the representative curve for the MD sample shown in Fig. 2.30(a), the sample expanded uniformly up to 77° C where contraction proceeds in three relatively well-defined stages up to 223° C where the sample begins to elongate under load. The stages of linear expansion and contraction are summarized in Table 2.10. The heating rate did not affect the coefficients calculated from expansion portions of the curves. However, the rates of contraction and total shrinkage of the films up to the yield points ($\sim 220°$ C) show a large heating rate-dependence. The

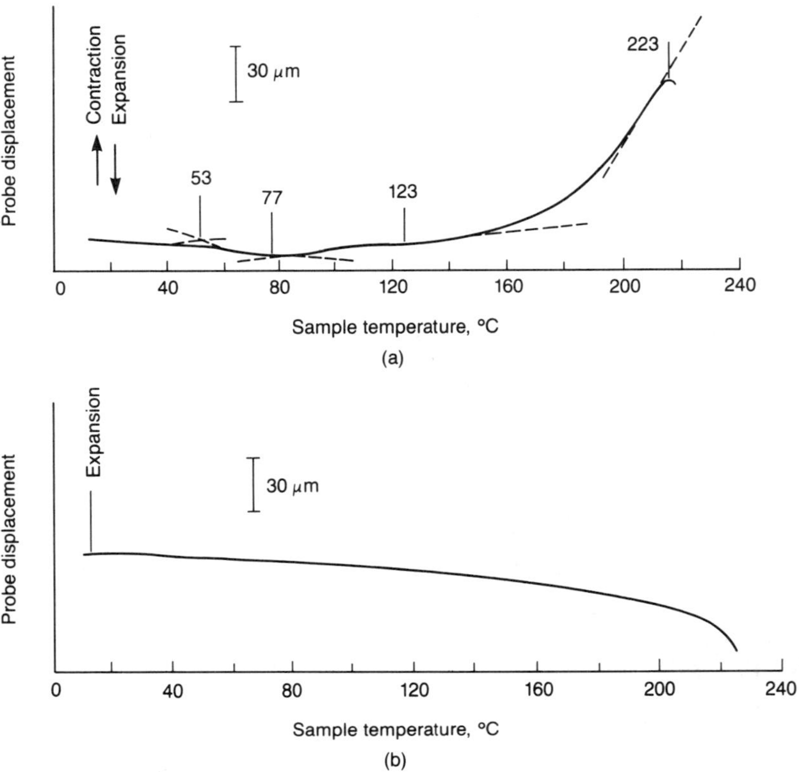

Fig. 2.30. TMA probe displacement as a function of temperature for a 36.1-μm-thick as-received PET (Mylar A) cut in the machine direction. The sample was under a 64 kPa tension and the heating rate was 2° C/min (a) as-received sample and (b) reheated after annealing at 230° C for 3 h (Barrall and Logan, 1973).

faster the rate of heating, the greater the observed shrinkage. In addition, the constants are very temperature sensitive. (We note that the expansion coefficients of an amorphous polymer above its T_g are on the order of threefold greater than below.) The samples cut 45° to the MD and at the MD follow much the same pattern of expansion and contraction. The total shrinkage was maximum along the MD. More shrinkage data are presented in Chapter 3.

2.3.6.2. Effect of Annealing

Blumentritt (1979b) studied the effect of annealing (thermal treatment above T_g) on the coefficient of thermal expansion (near room temperature) of 36.1-μm-thick PET films cut in various orientations, Fig. 2.29. The two annealing treatments appeared to affect slightly the values of the coefficients of thermal expansion but it remained anisotropic.

Barrall and Logan (1973) carried out a detailed study. The samples were

Table 2.10. Thermal expansion properties of the "as-received" 36.1-μm-thick PET (Mylar A) film at various heating rates (Barrall and Logan, 1973)

Heating rate, °C/min	Temperature range, °C	Coefficient of linear thermal expansion, $\times 10^{-5}$/°C	Shrinkage, % from 25° to 230° C
Parallel to Machine Direction (MD)			
2	25 to 54	1.84	5.0
2	54 to 77 (80)[a]	2.25	
2	88 to 132	−5.44[b]	
2	207 to 220	−91.0	
5	25 to 84 (90)[a]	1.80	8.3
5	93 to 145	−6.60	
5	222 to 230	−199	
10	25 to 97 (97)[a]	1.84	11.2
10	97 to 143	−7.82	
10	211 to 226	−292	
45° Angle to Machine Direction (MD)			
2	25 to 53	1.80	4.4
2	53 to 71 (82)[a]	2.72	
2	88 to 142	−6.25	
2	210 to 221	−98.3	
5	25 to 81 (83)[a]	1.85	8.3
5	81 to 146	−6.94	
5	208 to 222	−210	
10	25 to 88 (88)[a]	1.86	11.2
10	88 to 150	−6.70	
10	212 to 224	−266	
Perpendicular to Machine Direction (MD)			
2	25 to 41	1.19	3.2
2	41 to 73 (78)[a]	2.04	
2	79 to 103	−11.5	
2	110 to 154	−5.09	
2	207 to 222	−75.1	
5	25 to 71 (78)[a]	1.19	7.3
5	82 to 112	−10.7	
5	112 to 160	−2.36	
5	210 to 222	−176	
10	25 to 77 (83)[a]	1.20	8.2
10	86 to 111	−14.1	
10	111 to 160	−4.04	
10	212 to 227	−248	

[a] Temperature at which shrinkage is first noted.
[b] Negative coefficients indicate contraction.

annealed by placing them on glass plates in ovens at 95°, 120°, 180°, and 230° C for 5–8 h. The samples were removed and cooled quickly to room temperature. Samples were then reheated at various heating rates for thermal expansion measurements. They found that when a specimen had been annealed at some temperature above that at which a nonannealed sample begins to shrink significantly, the sample no longer undergoes shrinkage on heating but expands continuously at an ever-increasing rate, until the temperature exceeded the annealing temperature, Fig. 2.30(b) and Table 2.11. The temperature at which contraction starts (inversion temperature) is very near the annealing temperature, Fig. 2.31. The expansion coefficient in the room temperature range is very dependent upon the annealing temperature, see Table 2.11. At 95° C and below, heat treatment causes the coefficient to increase over the unheat-treated sample (Table 2.10). At 120° C and above, a marked reduction in the coefficient is noted. In addition, the total shrinkage of the film at 230° C is also affected by annealing at low temperatures. The effect is very temperature range specific.

The effects of stress during annealing (thermal treatment above T_g) on the coefficient of linear thermal expansion, inversion (onset of contraction) temperature on second heating, and the shrinkage properties are shown in Figs. 2.32 and 2.33. The application of stress during annealing results in a decreasing room temperature linear expansion coefficient (as well as inversion temperature and shrinkage during annealing) with an increase in stress. However, the sample does not relax uniformly. The rate of contraction is reduced significantly with increasing load, Fig. 2.34. In general, frozen-in strains are preserved during annealing at high stresses. Barrall and Logan (1973) reported that the above is not true of the properties in the film direction normal to the loading direction during annealing.

2.3.6.3. Mechanisms of Thermal Expansion

In the simplest form, the expansion of a solid is due to two separate functions: the increase in average interatomic distance with an increase in temperature (vibrational), and the sweeping out of an ever-increasing volume by movement of larger segments of the molecule with every increasing temperature. The various vibrational modes of the bonded atoms account for only a very small part of the net expansion observed in a polymer solid. By far, the largest detectable portion is due to in-place chain rotation and minor translations. Due to crystal lattice constraints, molecular motion is somewhat reduced. Thus, for most polymers, the crystal expansion coefficient is less than the amorphous expansion coefficient. Also, the amorphous phase can exhibit many more degrees of local freedom than the crystal. Orienting the amorphous phase serves to reduce this inherent freedom by imposing a quasi-crystal extension of long polymer-chain sequences.

Now we try to understand the mechanisms responsible for the unique thermal expansion (and contraction) properties of oriented PET films. These

Table 2.11. Thermal expansion properties of 36.1-μm-thick PET (Mylar A) film annealed at various temperatures for 5–8 h and reheated at a heating rate of 2° C/min (Barrall and Logan, 1973)

Sample	Temperature range, °C	Coefficient of linear thermal expansion, $\times 10^{-5}$/°C	Total shrinkage, % from 25° to 230° C
Annealed at 95° C			
MD	25 to 48	2.12	4.4
	48 to 81	3.14	
	81 to 98 (101)[a]	8.02	
	101 to 130	−12.1	
	198 to 220	−69.0	
45° to MD	25 to 46	2.57	5.7
	56 to 73	4.34	
	85 to 93 (100)[a]	7.18	
	109 to 133	−6.64	
	210 to 224	−148	
TD	25 to 40	0.61	3.6
	58 to 79 (100)[a]	2.73	
	90 to 97	7.04	
	104 to 147	−6.82	
	210 to 221	−91.3	
Annealed at 120° C			
MD	25 to 50	1.56	5.7
	84 to 92	4.43	
	92 to 119	5.44	
	121 to 133 (137)[a]	7.42	
	151 to 162	−28.0	
	209 to 223	−177	
45° to MD	25 to 49	0.72	2.9
	69 to 83	2.98	
	96 to 115	8.68	
	115 to 128 (145)[a]	12.4	
	208 to 220	−110	
TD	25 to 48	0.80	2.2
	58 to 76	2.35	
	93 to 110	6.14	
	122 to 130 (140)[a]	9.82	
	201 to 218	−76.4	
Annealed at 180° C			
MD	25 to 48	1.67	0.7
	60 to 157	5.86	
	170 to 181 (183)[a]	8.47	
	206 to 221	−93.4	
45° to MD	25 to 45	1.05	1.0
	60 to 77	2.62	
	89 to 96	5.78	
	101 to 181 (181)[a]	7.66	
	214 to 226	−106	

Table 2.11 (*continued*)

Sample	Temperature range, °C	Coefficient of linear thermal expansion, $\times 10^{-5}/°C$	Total shrinkage, % from 25° to 230° C
TD	25 to 50	0.76	0.9 expand
	61 to 74	2.46	
	92 to 100	5.14	
	105 to 168	7.09	
	171 to 184 (191)[a]	12.6	
	208 to 218	−33.2	
Annealed at 230° C			
MD	20 to 88	3.55	
	118 to 160	5.16	
	175 to 200	11.1	
45° to MD	20 to 88	3.13	
	118 to 160	7.75	
	175 to 200	12.2	
TD	20 to 88	2.91	
	118 to 160	11.4	
	175 to 200	17.2	

[a] Temperature at which shrinkage is first noted.

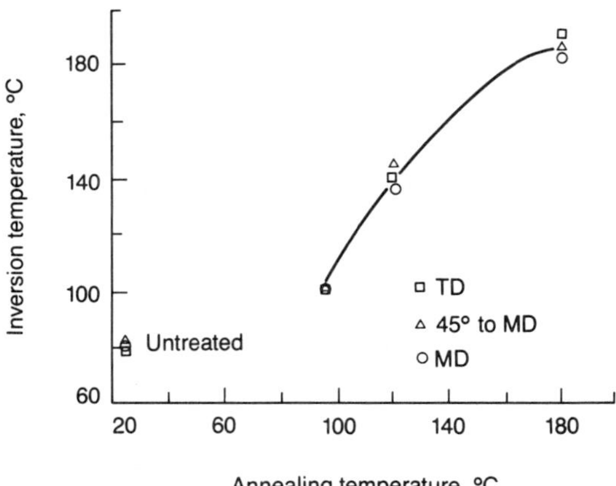

Fig. 2.31. Relationship between the annealing temperature and the inversion (onset of contraction) temperature for a 36.1-μm-thick PET (Mylar A). The sample was under a 64 kPa tension and the heating rate during reheating was 2° C/min (Barrall and Logan, 1973).

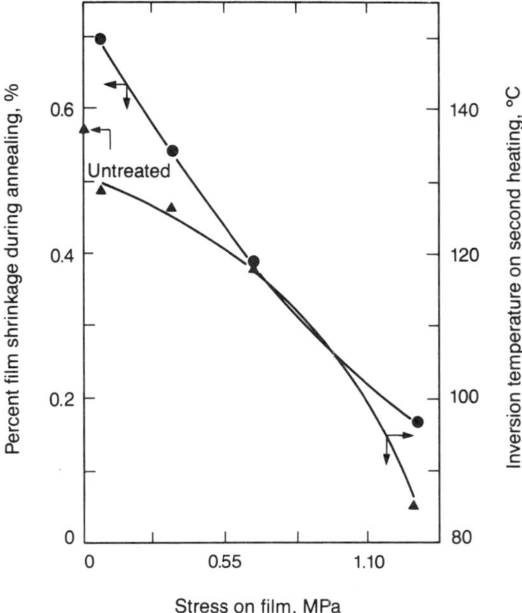

Fig. 2.32. Effect of annealing under stress at 120° C for 3 h on the film shrinkage and the inversion (onset of contraction) temperature for a 36.1-μm-thick PET (Mylar A) cut in the machine direction, heating rate during reheating was 2° C/min (Barrall and Logan, 1973).

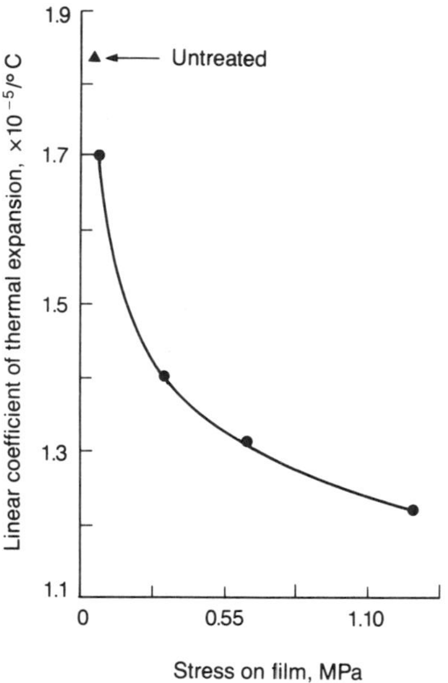

Fig. 2.33. Effect of annealing under stress at 120° C for 3 h on the room temperature linear coefficient of thermal expansion for a 36.1-μm-thick PET (Mylar A) cut in the machine direction, heating rate during reheating was 2° C/min (Barrall and Logan, 1973).

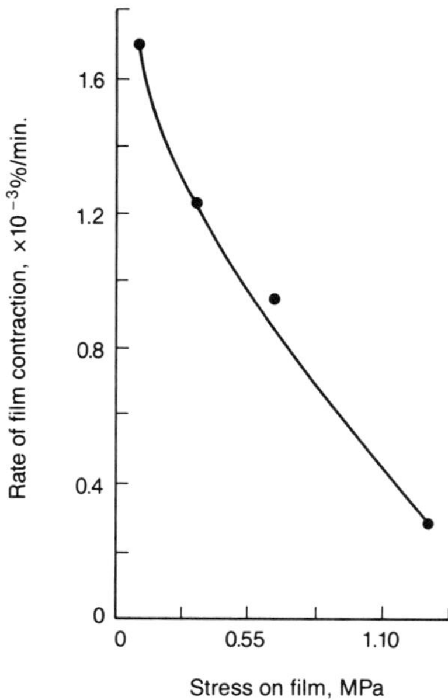

Fig. 2.34. Effect of annealing under stress at 120° C for 3 h on the rate of film contraction at 120° C for a 36.1-μm-thick PET (Mylar A) cut in the machine direction, heating rate during reheating was 2° C/min (Barrall and Logan, 1973).

properties are not singly the result of the existence of crystalline and amorphous phases. The unique properties of oriented PET reside in the oriented nature of both the crystalline and the amorphous phases. The thermal expansion of oriented PET film up to the onset of contraction has been found to follow two distinct slopes, Fig. 2.30(a). The slope established in the room temperature range increases near 50° C to another linear range. The sample will expand a characteristic amount upon temperature increment due to rotational and vibrational increases of a certain kind in the sample. Should a new, yet still constricted, mode become possible, the expansion coefficient must increase. Expansion will continue by the previous process as well as by any volume contribution due to the new process. That is, expansion in a given state is the sum of the available modes. Thus, PET in the glass range must have at least two types of molecular rotation and vibration in large chain segments. When the glass transition is reached, the amorphous fraction is no longer constrained, and totally different processes become possible.

The expansion properties of oriented PET in the glass range appear to be governed to a great extent by the oriented amorphous chains. So long as orientation is preserved, the change in coefficient at 50° C is observable in the TMA curves. Large-segment rotation in the oriented phase is unusually quantized in PET films. The sharp slope increases near 50° C and becomes progressively less distinct as amorphous orientation is lost by annealing at temperatures at which rapid crystallization is possible. In these experiments, this was found to be near 210° C (Barrall and Logan, 1973).

The contraction phenomena observed in oriented PET below 120° C is almost certainly due to the release of frozen-in stresses present in the oriented amorphous phase. Below 120° C, contraction due to crystallization is unimportant in the time scale of these experiments, in fact, it has been suggested that crystallization is very slow below ~ 130° C (Zachmann and Stuart, 1960). The release of frozen-in stresses is in no way correlated with the glass transition temperature [which is imperfectly defined for an oriented phase (Petrie, 1972)]. Contraction occurs when the stresses in the sample exceed the modulus of the material at a given temperature. On a short time scale, these rapid stress releases result in a fairly sharp contraction. This is not to imply that release does not occur at a lower temperature, i.e., amorphous creep. However, on the time scale of less than 1 h, these slow effects are negligible (Barrall and Logan, 1973).

The effect of imposing an external stress on the film gives a qualitative measure of the size of the internal stresses. These must be greater than 0.64 MPa. Stresses greater than this are able to hold amorphous chains in the oriented position in the direction of the applied force. This greatly reduces the amount of shrinkage in the stress direction. The expansion coefficient in the room temperature range is also reduced significantly in stressed samples, as opposed to samples heat treated with no applied stress (Barrall and Logan, 1973).

In Table 2.10 the total shrinkage of the film in the TMA is demonstrated to be directly related to the heating rate. The faster the rate of heating, the greater the observed shrinkage. The increase is nearly linear, with the equilibrium shrinkage (infinitely low heating rate) being 3–4%. The heating rate effect is due to the several crystal forms into which the amorphous phase may enter. At very low heating rates, small folded chain crystals form at low temperatures. These are relatively stable up to 220° C (the yield point of this measurement). These small crystals, being imperfect and perhaps epitaxial on larger crystals, have a larger volume than larger more perfect crystals. At higher rates of heating, the crystallization rate of the amorphous phase is too slow to form small crystals rapidly. Thus, the sample is in the higher temperature region before significant small crystals form. At higher temperatures, it is well known that larger, more perfect, and denser crystals form.

With the relaxation of the oriented amorphous phase and the removal of frozen-in strains, the major process up to 95° C, some change in the expansion coefficients of the sample could be expected after annealing, Table 2.11. In the MD and oblique direction, the release of strains has resulted in a loss of order, as indicated by an increase in the room temperature expansion coefficient. However, loss of order in these directions has increased the net total order in the TD, note the sharp decrease in the 25°–40° C expansion coefficient in Table 2.11.

At annealing temperatures of 120° C and above, crystallization of the amorphous phase becomes important. The end-product of stress release is a crystalline phase whose origins are the amorphous phase. The new crystalline phase preserves somewhat the order induced by orientation, since the expansion coefficients in the TD and MD are decreased rather than increased. The volume of the system is not conserved, for it becomes smaller—a good indication for the formation of a crystalline phase. The sample in the thickness direction remains essentially the same as that attained by the relaxation of the oriented amorphous phase at 95° C. Given a somewhat disoriented amorphous phase, the question arises as to *how orientation of the crystal phase is preserved as this amorphous phase is crystallized*. This question can be satisfactorily answered if it is assumed that the amorphous phase crystallizes near to or on the existing oriented crystalline phase. This assumption is logical, since no new nuclei for crystallization should be formed at a temperature so far removed from the melting point. The existing crystals act as a nuclei and favor-oriented growth.

2.3.7. Hygroscopic Expansion Properties

The absorption of water or other liquids by polymers is commonly not a simple process (Barrie, 1968). The resulting increase in volume does not always equal the volume of the liquid absorbed. In some cases, this phenomena has been attributed to sorption of the liquid into pre-existent microcavities, giving little or no change in volume. The absorption of water by PET has been studied by several investigators. Ravens and Ward (1961) showed that water is more soluble in amorphous than in semicrystalline unoriented PET. Cuddihy (1972) observed ready reversibility of the water content of the PET with changes in RH.

The coefficients of the linear hygroscopic expansion of biaxially-oriented PET (Mylar A) were measured by Blumentritt (1979a, b) using 210-mm square specimens. Targets were made by lightly scribing crosses on ink spots placed at various points on a circle of approximately 150 mm in diameter. Dimensional changes along the various diameters were measured with a coordinate measuring microscope. Specimens were conditioned to $8 \pm 1\%$ RH at 22° C for at least 24 h. Then the specimens were pressed between glass plates and the distances between targets were measured. The specimens were then conditioned to $80 \pm 1\%$ RH at 22° C for at least 24 h. After pressing between glass plates, the specimen dimensions were remeasured. The specimens were then returned to 8% RH and measured again. Coefficients of hygroscopic expansion were calculated from the average of the two measurements made by increasing, and then by decreasing, the relative humidity ranging from 8% to 80% RH. A plot of the coefficient of hygroscopic expansion versus the in-plane direction of the 36.1-μm PET film is shown in Fig. 2.35. We note that the anisotropy in hygroscopic expansion is much less than the anisotropy in the coefficient of thermal expansion or mechanical properties. The coefficient of hygroscopic expansion is lowest along the orientation

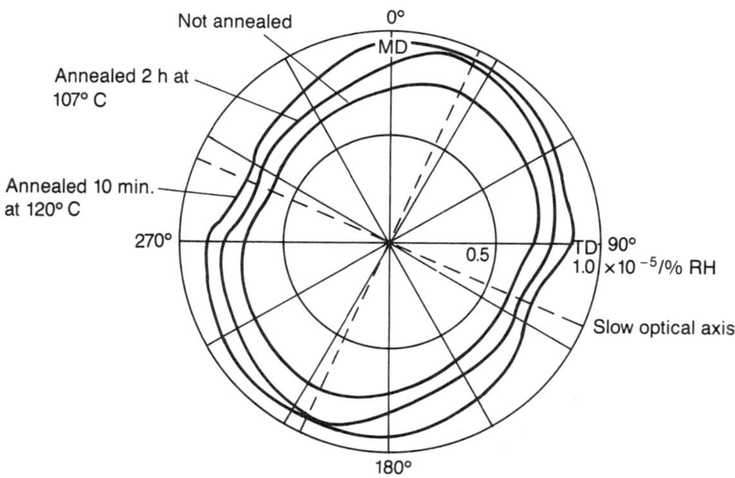

Fig. 2.35. Coefficient of the linear hygroscopic expansion (in the humidity range of 8–80% RH at 22° C, exposure time ≥ 24 h) as a function of the in-plane direction of a 36.1-μm-thick biaxially-oriented PET film (Mylar A). Dotted lines are the axes that show the extinction of light under a cross polarizer. The minimum hygroscopic expansion coefficient is along the major axis of Young's modulus (Blumentritt, 1979b). © 1979 International Business Machines Corporation; reprinted with permission.

direction with maximum Young's moduli (slow optics axis 70° to the MD). We note that Greenberg et al. (1977, 1978) have reported that the coefficient of linear hygroscopic expansion for 76.2-μm-thick Mylar A film is essentially isotropic.

For an anisotropic material, we relate the coefficient of the volumetric hygroscopic (β_v) to the linear coefficients in the three mutually perpendicular directions

$$\beta_v = \beta_\parallel + \beta_\perp + \beta_t, \tag{2.7}$$

a relation analogous to that for thermal expansion [Eq. (2.5)]. Ravens and Ward (1961) and Yasuda and Stannett (1962) report that β_v for amorphous PET and biaxially-oriented semicrystalline PET (Mylar A) are equal to 17.4×10^{-5} mm³/mm³/% RH and to 8.54×10^{-5} mm³/mm³/% RH, respectively. Table 2.12 presents the measured values of $\beta_\parallel$ and $\beta_\perp$ (Blumentritt, 1979a) and the calculated values of β_t for 23.4-μm, 36.1-μm, and 76.2-μm-thick oriented PET (Mylar A) film. We note that β_t is considerably larger than $\beta_v/3$, the value for isotropic semicrystalline PET.

From the data of Table 2.12 we further note that the thinner film had the lower coefficient of linear hygroscopic expansion, and higher anisotropy similar to the thermal expansion. Blumentritt (1979a, b) also reported the coefficients for 76.2-μm-thick uncoated PET substrate and coated magnetic media (76.2-μm-thick PET substrate coated on both sides with a 2.5-μm coating of γ-Fe₂O₃ particles dispersed in a polymeric binder). They found that

Table 2.12. Coefficients of the hygroscopic expansion of PET films, 8–80% RH at 22° C

Film	$\beta_\parallel$, $\times 10^{-5}/\%$ RH	$\beta_\perp$, $\times 10^{-5}/\%$ RH	β_t,[a] $\times 10^{-5}/\%$ RH	$\beta_\perp/\beta_\parallel$	β_t/β_v
23.4-μm-thick biaxially-oriented thick PET (Mylar A)[b]	0.52	0.63	7.39	1.21	0.87
36.1-μm-thick biaxially-oriented PET (Mylar A)[b]	0.63	0.80	7.11	1.27	0.83
76.2-μm-thick biaxially-oriented PET (Mylar A)[b]	0.81	0.86	6.87	1.06	0.80
Isotropic PET	2.85	2.85	2.84	1	0.33

[a] The value of β_v for biaxially-oriented PET (Mylar A) is 8.54×10^{-5} mm^3/mm^3/% RH (Yasuda and Stannett, 1962).
[b] MD is 49°–45° to slow optical axis (Blumentritt, 1979a).

coefficients of the uncoated substrate and coated media were comparable (within 10%).

Ravens and Ward (1961) reported that the solubility coefficient of water for unoriented PET is roughly independent of temperature from 20° to 150° C, especially at relative humidities less than about 20%. Hence, except at elevated temperatures, at which the degree of crystallinity and the crystal morphology may change, the coefficient of hygroscopic expansion should be temperature-independent. Yasuda and Stannett (1962) found that the solubility coefficient for biaxially-oriented PET (Mylar A) is constant between 0% and 100% RH. Hence, to calculate the apparent hygroscopic coefficient resulting from changes in both temperature and water content, it is quite reasonable to assume that

$$W_{H_2O} = k\frac{p}{p^0}, \tag{2.8}$$

where W_{H_2O} is the weight of water in grams absorbed by 1g of PET, k is the (temperature-independent and humidity-independent) solubility coefficient ($\sim 6.15 \times 10^{-5}$, weight of water in grams in 1 g of Mylar A at 1% RH), p is the partial pressure of water vapor in the specimen vicinity, p^0 is the vapor pressure of water at the test temperature, and RH in percent is equal to $100 \times p/p^0$. This equation shows that if p is held constant, the solubility of water will decrease with a temperature increase because of the increase in p^0. Hygroscopicity is a reversible property, and thus, hygroscopic materials will also readily desorb water as relative humidity is lowered.

Cuddihy (1976) measured the hygroscopic properties of a γ-Fe_2O_3 tape and its components using a commercial 12.7-mm in width backcoated wide-band instrumentation tape. The hygroscopic property of the samples was determined by measuring their weight changes by a Cahn microbalance when exposed to changing levels of RH. The equilibrium curves at 25° for a finished tape, tape with its magnetic coating removed, and for its PET substrate stripped of both coatings are shown in Fig. 2.36. From this data, the individual equilibrium curves for the γ-Fe_2O_3 magnetic coatings and the carbon backcoat can be calculated, Fig. 2.37. We note that the magnetic coating has the maximum solubility coefficient. Cuddihy also measured the time-dependence for the tape pack to change its water content resulting from changes in the RH of the environment. Unwound tapes required approximately 0.6 min to achieve a water content in equilibrium for each percent change in RH and a tape pack required several (~ 3–4) days to achieve a constant weight condition.

If we assume the additivity of volumes, the variations in dimensions affected by combined temperature and humidity changes can be calculated directly from the coefficients of thermal and hygroscopic expansion.

2.3.7.1. Effect of Annealing

The effect of annealing on the coefficient of linear hygroscopic expansion is shown in Fig. 2.35. Annealing increased the coefficient of hygroscopic expan-

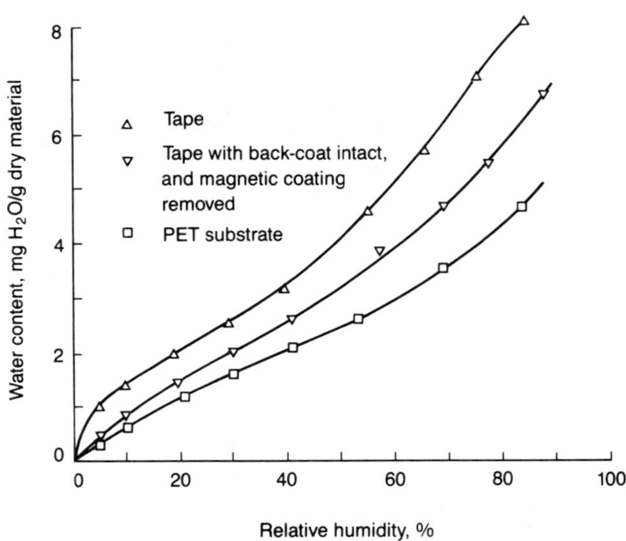

Fig. 2.36. Equilibrium curves of the absorbed water content as a function of the relative humidity at 25° C for a tape and its components.

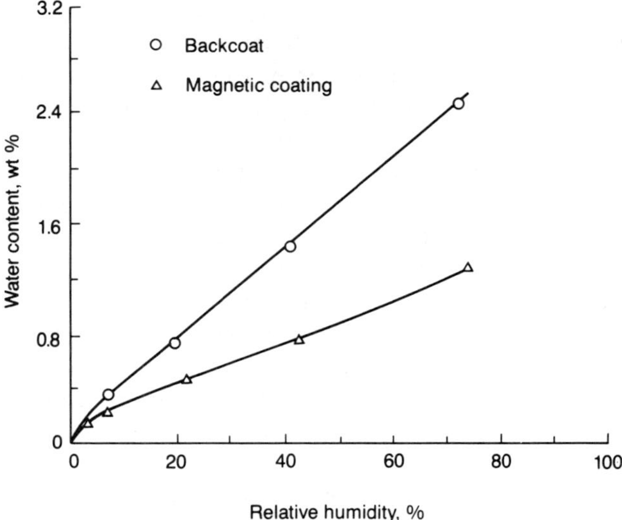

Fig. 2.37. Equilibrium absorbed water contents as a function of the relative humidity at 25° C of the γ-Fe_2O_3 magnetic coating and carbon backcoat material.

sion but it remained anisotropic. The annealing conditions used allowed relaxation of the films, and this relaxation may allow the materials to absorb more moisture.

2.3.8. Long-Term Dimensional Stability (Shrinkage)

One of the principal long-term dimensional instabilities of these materials is shrinkage (nonrecoverable deformation) due to residual stresses. The amount of shrinkage measured after a particular time interval depends upon the time-temperature-stress history of the film (Greenberg et al., 1977, 1978; Blumentritt, 1979a, b; Chan and Smith, 1979; Vallat and Plazek, 1988). High temperatures and humidities will greatly accelerate the shrinkage. Shrinkage is highly anisotropic. In some cases, there may even be shrinkage along one axis and expansion along the other. Shrinkage versus temperature at constant exposure times for the 76.2-μm-thick PET film (slow optical axis at 45° to the MD) is shown in Fig. 2.38. Shrinkage of the film at temperatures below 60° C was negligible with the times used. At temperatures at and above 60° C the shrinkage of the PET was approximately linear with temperature when the exposure time was held constant. When the temperature was held constant and the exposure time was varied, the 76.2-μm film shrank, as shown in Fig. 2.39. The shrinkage of the PET was linear with the log (time); supposedly this rate continued for several years. At temperatures higher than 60° C, shrinkage accelerated very rapidly. Thus any temperature above 60° C is not a realistic

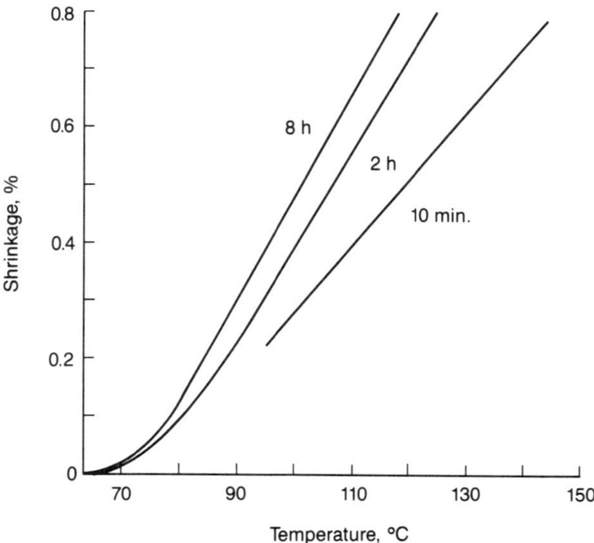

Fig. 2.38. Shrinkage in a 76.2-μm-thick biaxially-oriented PET film (Mylar A) in the machine direction after exposure at a given temperature for constant times (Blumentritt, 1979b). © 1979 International Business Machines Corporation; reprinted with permission.

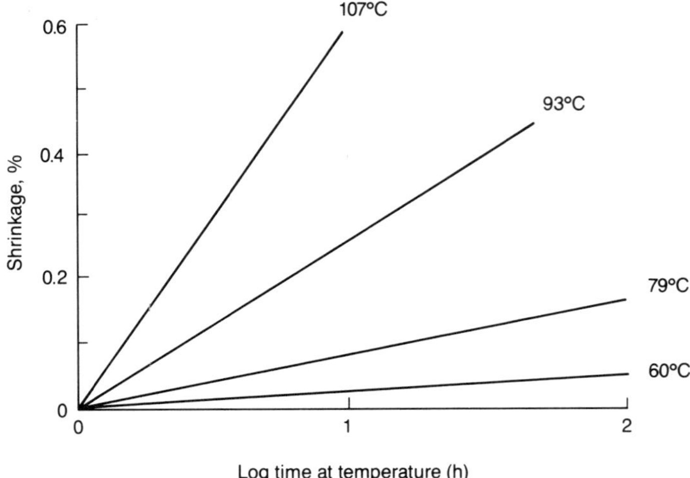

Fig. 2.39. Shrinkage in a 76.2-μm-thick biaxially-oriented PET (Mylar A) film in the machine direction at constant temperature (Blumentritt, 1979b). © 1979 International Business Machines Corporation; reprinted with permission.

operating temperature. Maximum shrinkage tends to occur along the machine direction of the film, and minimum shrinkage tends to occur along the transverse direction, Table 2.13. The shrinkage of the thinner film is higher than that of the thicker film, possibly because of increased straining required to manufacture the thin material.

The shrinkage phenomena is due to the relaxation of oriented molecules in

Table 2.13. Shrinkage of PET film (Mylar A) held at 60° C (Blumentritt, 1979b)

Material	Hours at 60° C	Shrinkage, %			
		0°[a]	45°	90°	135°
36.1-μm-thick PET	168	0.035	0.029	0.021	0.030
not annealed	2016	0.061	0.057	0.039	0.062
annealed 2 h	168	0.010	0.010	0.007	0.009
at 107° C	2016	0.009	0.012	0.011	0.013
annealed 10 min	168	0.007	0.015	0.016	0.015
at 120° C	2016	0.019	0.020	0.022	0.021
76.2-μm-thick PET	168	0.029	0.024	0.014	0.026
not annealed	2016	0.074	0.044	0.027	0.065
annealed 2 h	168	0.004	0.012	0.011	0.018
at 107° C	2016	0.010	0.008	0.016	0.017
annealed 10 min	168	0.008	0.011	0.013	0.018
at 120° C	2016	0.010	0.009	0.018	0.017

[a] With respect to the MD of the substrate. The slow optical axis was about 45° to the MD.

the amorphous regions of the film, resulting in an overall contraction of the film in the direction of orientation. For orthotropic, biaxially-oriented PET films, the observed contraction in each direction of orientation is the sum of contraction in that direction plus the expansion due to the orthogonal contraction. Crystalline regions in which molecular motion is severely restricted would not contribute appreciably. The shrinkage rate depends on the fraction of the oriented amorphous phase, the degree of orientation of the amorphous phase, and the speed or frequency of molecular motion. More data on shrinkage are presented in Chapters 3 and 6.

2.3.8.1. Effect of Annealing

Annealing (thermal treatment above T_g) reduces internal stresses in oriented PET films (Groeninckx et al., 1976, 1980a, b). After annealing, PET films shrank a small amount at temperatures below the annealing temperatures (Blumentritt, 1979b; Chan and Smith, 1979). The shrinkage occurs due to continued relaxation of the strained amorphous regions in the film. The amount of shrinkage observed in annealed materials depends upon the time and temperature, and upon the residual stresses in the film after annealing. Blumentritt (1979b) conducted long-term dimensional stability tests at 60° C and zero stress. The data for shrinkage of the unannealed and annealed films with two thicknesses are given in Table 2.13. The annealed film shrank much less than the unannealed film during long-term exposure to 60° C, as would be expected. After annealing, shrinkage does not appear to relate to film machine direction or to material. Similar results, showing the reduced shrinkage of annealed PET film, have been reported by Greenberg et al. (1977).

The observed changes, extrapolated to long times, are great enough to be of concern for future systems in which data are stored at high density on flexible media with such PET substrates. However, annealing PET substrate films and allowing them to shrink freely at high temperatures substantially reduces the rate of long-term shrinkage. The annealing of PET under relatively mild conditions (such as 10 min at 120° C and 2 h at 107° C) did not significantly change the mechanical properties or the coefficients of thermal expansion and hygroscopic expansion at ambient conditions, as reported earlier.

2.3.9. Hydrolytic Stability

PET film will hydrolyze and become brittle under conditions of high temperature and humidity, as shown by the effect of steam on the tensile properties of PET, Fig. 2.40. Therefore, care should be taken to ensure that there is a minimum of water in any hermetically sealed unit. Adequate removal of water from PET is usually obtained by heating for 4 h at 160° C. Drying at these conditions should reduce the water content of the film to less than 0.1%; more than this amount of water must be present in a system before the film can become embrittled due to hydrolysis.

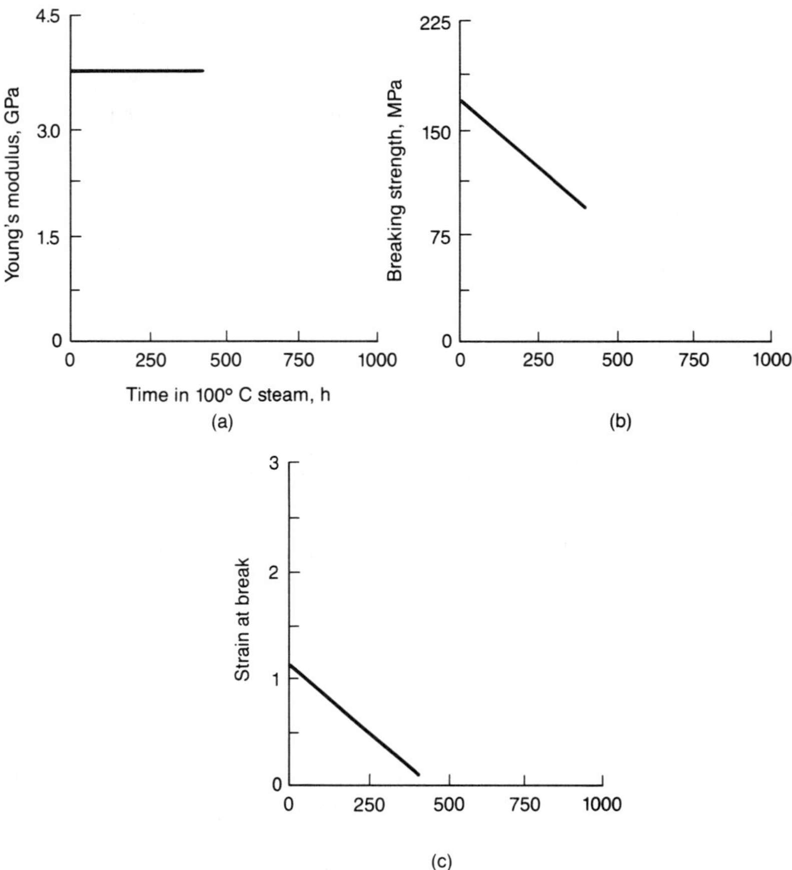

Fig. 2.40. Effect of the 100° C steam exposure on the mechanical properties of biaxially-oriented PET (Mylar A) film.

2.3.10. Summary

The degree of crystallinity and the degree of orientation with crystalline and amorphous regions of the PET film determine its resultant properties. The mechanical properties of PET films are a strong function of temperature- and strain-rate used for measurements. PET film is sensitive to solvent-stress cracking and susceptible to premature failure when stressed in some solvent environments. Biaxially-oriented PET films exhibit significant in-plane anisotropy in the refractive index, Young's modulus of elasticity, breaking strength, strain-at-break, and the coefficient of linear thermal expansion, and follow an elliptical distribution. The coefficient of linear hygroscopic expansion and yield strength are more nearly isotropic. Maximum Young's modulus, maximum breaking strength, minimum strain-at-break, minimum coefficient of linear thermal expansion, and the minimum coefficient of linear hygroscopic expansion are found to be parallel to the slow optical axis (in which the net orientation of crystals and the amorphous material is a maximum and exhibits maximum refractive index). The long-term shrinkage (nonrecoverable deformation) is maximum along the machine direction. The thinner film has the higher value of Young's modulus, lower values of the coefficients of thermal and hygroscopic expansion, and a higher value of long-term shrinkage along the slow optical axis and greater anisotropy than a thicker film.

The orientation of the major and minor axes of Young's modulus, as well as the anisotropy ratio, changes across the web. We note that the minor axis is along the MD at the center of the web and it rotates to about 45° with respect to the MD at the edges. The magnitude of the modulus is symmetrical about the web center. The anisotropy ratio is minimum (typically 1.1–1.2) at the web center and maximum (typically 1.6–1.7) at the web edges. The modulus in the MD remains almost constant, being slightly lower at the edges and the modulus in the TD (higher than that in MD) increases from the center to the edges.

Young's modulus in the thickness direction is less than that in the in-plane. The radial elastic modulus of the wound reel is dependent on the winding parameters, the magnetic coating, and storage time.

Upon heating, a PET specimen is found to expand up to some temperature, between about 73° and 97° C, that depends on the heating rate and the angle between the longitudinal axis of the specimen and the MD. At higher temperatures, shrinkage occurs. When a sample has been annealed, at some temperature above that at which a nonannealed sample begins to shrink significantly, the sample no longer undergoes shrinkage on heating but expands continuously at an ever-increasing rate, until the temperature exceeds the annealing temperature.

Annealing the biaxially-oriented PET under relatively mild conditions, such as 107° C for 2 h or 120° C for 10 min, does not significantly change the mechanical properties or the coefficient of linear thermal expansion at ambient conditions. Young's modulus values are reduced slightly by annealing, and the coefficients of linear hygroscopic expansion are increased by annealing. Annealing greatly reduces the long-term shrinkage. Thus, annealing a film under low tension at an elevated temperature for some appropriate period improves the long-term dimensional stability of the film at low temperatures (for more details, see Chapter 3).

2.4. Outlook for Improved Substrates

Anisotropy and dimensional instabilities are limiting factors of the PET films for flexible-media applications which require dimensions and properties that are as precise as possible. The anisotropic mechanical properties affect the tracking of moving films; the anisotropic mechanical (including viscoelastic), thermal, and hygroscopic properties also cause anisotropic deformation of flexible disks, resulting in a radial run-out of an initially circular track (Greenberg et al., 1977, 1978; Bhushan et al., 1984a; Eshel et al., 1984). (We note that hygroscopic extension is much less anisotropic than the thermal expansion and mechanical properties.) For flexible disks, it is acceptable, to an extent, for the disk to change its dimensions due to changes in the environment, the forces of rotation, and even to undergo time-dependent changes (creep), but it is crucial that these dimensional changes be the same or nearly so in all directions (isotropic). The most important property required for future applications is the dimensional stability, especially very low coefficients of thermal and hygroscopic expansion (about half or less of that of biaxially-oriented PET such as Mylar A) with in-plane isotropy; some improvement in stability under mechanical stress and elevated temperatures is also desired.

Several approaches have been considered for producing flexible films with improved dimensional stability and low in-plane anisotropy. A properly annealed PET film should shrink less than an unannealed film over long periods of time at room temperature. But the thermal and hygroscopic expansions of annealed films do not change, and the annealed films remain anisotropic in mechanical properties and coefficients of thermal and hygroscopic expansions.

We ask whether a polymer can be found whose coefficients of volumetric thermal and hygroscopic expansions are appreciably less than that for isotropic PET. Biaxially-oriented PET (Mylar) has adequate mechanical properties, especially toughness. Hence, the question is whether PET can be oriented more highly than Mylar to obtain a reduction in the coefficients of linear thermal and hygroscopic expansions (although the material remains anisotropic). Cross-ply lamination is an approach to produce isotropic in-plane properties in films (Park, 1969). Other approaches include the incorporation of fibers and filaments with improved dimensional stability and low in-plane anisotropy (Ashton et al., 1969). Ideally, we should combine more than one approach, e.g., a laminate prepared from films, that are highly oriented in one direction, can provide isotropic properties with high elastic modulus and low thermal and hygroscopic expansion properties.

In the production of thin-film metal-alloy flexible media, a tremendous amount of heat is generated. In the deposition of some magnetic films, high substrate temperature has to be maintained (e.g., $> 200°$ C for Co–Cr films) to obtain high coercivity and perpendicular orientation of the easy axis. A film base considered suitable is the polyimide which can withstand temperatures of at least 200° C. However, these films are frightfully expensive.

2.4.1. Mechanical Properties

Mechanical properties (such as Young's modulus) of biaxially-oriented PET are adequate, although it will be desirable to have a somewhat higher Young's modulus. Certain molecular motions are inhibited, or frozen out, in an oriented material; hence, the modulus of an oriented material in the direction of orientation is greater than when it is unoriented (Heffelfinger and Schmidt, 1965). Hence, if PET is more highly drawn than Mylar A, or if reinforcing fibers or filaments are added to reduce the coefficient of thermal expansion, the modulus of the resulting material will undoubtedly exceed the value desired for an improved substrate. However, it will be accomplished at the expense of the low toughness.

It is desirable to have a substrate with in-plane isotropic mechanical properties. Cross-ply PET film laminates appear to be suitable in providing isotropic in-plane properties in films. Blumentritt (1979c) produced PET-film and composite-film laminates with two- and three-plies. Laminates of more than three plies offer very little improvement over three-ply films, and they probably are not cost effective. PET-film laminates were produced by bonding plies of unannealed or annealed PET film so as to form more nearly isotropic films. In two-ply laminates, the plies of 36.1-μm-thick Mylar A were oriented at 0° and 90° with respect to one another. In three-ply laminates, the orientation of 23.4-μm-thick Mylar A plies was $-60°$, 0, $+60°$. The adhesive used was 13-μm-thick thermoplastic polyester film (GT-100, manufactured by Sheldahl Corp., Northfield, Minn.) and the bonding took place at 150° C and 0.7 GPa for 1–2 min. The laminates were then cooled under pressure.

While the lamination of polymer films can be used to produce isotropic properties, films with maximum dimensional stability require some form of reinforcement. If the reinforcement is precisely oriented and uniformly distributed, the films will also have isotropic in-plane properties. Woven fabric reinforcements allow good control of fiber orientation and distribution. Sheldahl Corp. (Northfield, Minn.) produces a 114.3-μm-thick three-ply laminate with two plies of 8.9-μm PET film and one ply of a plain weave PET fabric (85 by 85 count) bonded together with a thermosetting polyester adhesive.

The values for Young's modulus of PET-laminated films are given in Table 2.14(a). The single-ply films showed significant anisotropy, especially in the case of the 23.4-μm film. The two-ply laminate was less anisotropic in modulus and the three-ply film had the least anisotropy. Figure 2.41 is a plot using polar coordinates of modulus versus direction for single-, two-, and three-ply films. (The elastic modulus value of single-ply PET film is somewhat higher than that reported earlier; this difference is due to a variation in the samples and as well as because of the effect of the strain rate.) It is apparent that as the number of plies in a film increased, the modulus became more nearly isotropic. Also, as the number of plies increased, the magnitude of the modulus decreased, since the modulus of the adhesive is much lower than the modulus of the oriented film. The other mechanical properties of the laminates were also lower than the properties of the oriented PET films due to the presence of the adhesive. In the PET film laminates, the volume fraction of adhesive was 0.22 in the two-ply film and 0.27 in the three-ply material. The

Table 2.14. Properties of the PET film and laminates (Blumentritt, 1979c).

(a) Young's modulus at 22° C and 50% RH and at a strain rate of 10^{-3}/s

Film	Young's modulus, GPa			
	0°[a]	45°	90°	135°
23.4-μm-thick Mylar A	4.61	3.54	4.46	5.42
36.1-μm-thick Mylar A	4.73	4.01	4.51	4.96
Two-ply (0°, 90°) laminate	3.92	3.74	4.18	3.84
Three-ply ($-60°, 0°, +60°$) laminate	3.43	3.72	3.56	3.66
Woven PET fabric laminate	1.14	0.52	0.74	0.57

(b) Coefficient of the linear thermal expansion at
25°–50° C using a heating rate of 5° C/min.

Film	Coefficient of linear thermal expansion, $\times 10^{-5}$/°C			
	0°[a]	45°	90°	135°
23.4-μm-thick Mylar A	1.40	2.01	1.26	0.96
36.1-μm-thick Mylar A	1.62	1.73	1.00	0.95
Two-ply (0°, 90°) PET film laminate	1.55	1.70	1.86	1.42
Three-ply ($-60°, 0°, +60°$) PET film laminate	2.00	2.08	2.15	2.05
Woven PET fabric laminate	1.76			1.87

(c) Coefficient of the linear hygroscopic expansion at 8–80% RH and 22° C.

Film	Coefficient of linear hygroscopic expansion, $\times 10^{-5}$/% RH			
	0°[a]	45°	90°	135°
23.4-μm-thick Mylar A	0.62	0.96	0.61	0.54
36.1-μm-thick Mylar A	0.73	0.77	0.71	0.66
Three-ply ($-60°, 0°, +60°$) PET film laminate	0.63	0.63	0.64	0.60
Woven PET fabric laminate	0.58	0.56	0.60	0.58

(d) Shrinkage due to stress relaxation after 1344 h at 60° C.

Film	Shrinkage, %			
	0°[a]	45°	90°	135°
23.4-μm-thick Mylar A, not annealed	0.113		0.100	
36.1-μm-thick Mylar A, not annealed	0.066	0.067	0.052	0.059
annealed	0.013	0.013	0.018	0.012
Two-ply PET film laminate, not annealed	0.050	0.057	0.060	0.052
Three-ply PET film laminate, not annealed	0.048	0.068	0.053	0.060
annealed	0.014	0.015	0.008	0.004

[a] Orientation with respect to the MD of film or of one-ply in laminate.

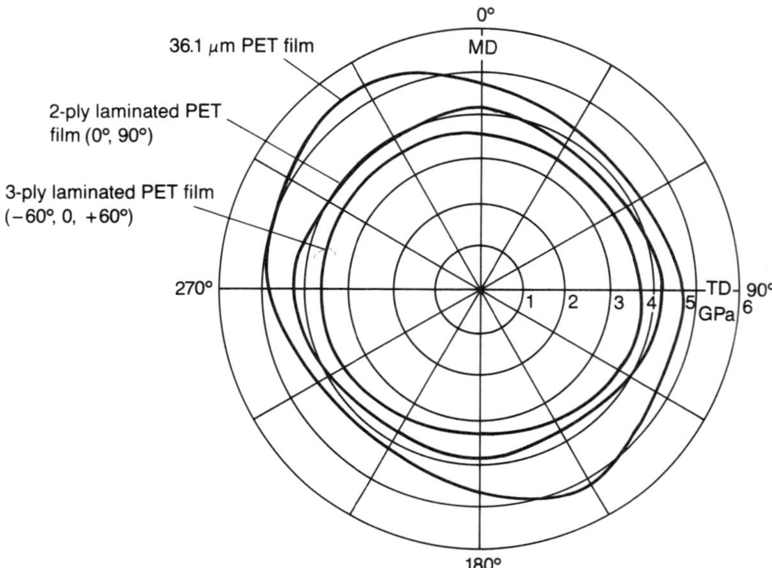

Fig. 2.41. Young's modulus measured at 22° C and a strain rate of 10^{-3}/s as a function of the in-plane direction, of the PET film, and of the laminated PET films (Blumentritt, 1979c). © 1979 International Business Machines Corporation; reprinted with permission.

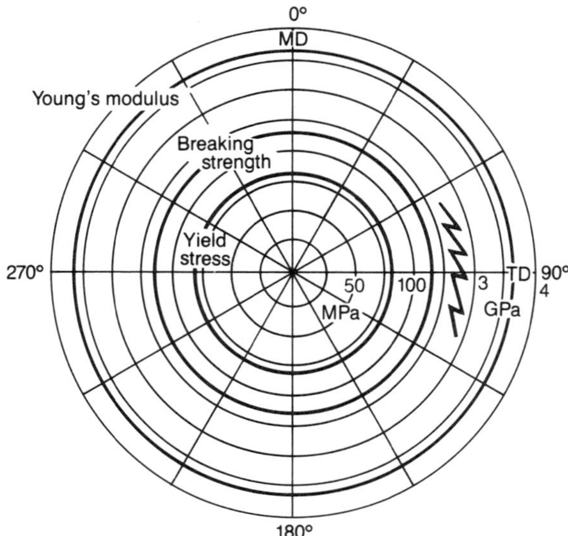

Fig. 2.42. Mechanical properties measured at 22° C and a strain rate of 10^{-3}/s as a function of the in-plane direction of a three-ply PET-film laminate (Blumentritt, 1979c). © 1979 International Business Machines Corporation; reprinted with permission.

laminate, reinforced with woven PET fabric, showed considerable anisotropy in elastic modulus.

Figure 2.42 shows the mechanical properties of a three-ply PET film laminate with the plies oriented at $-60, 0$, and $+60$ degrees. Modulus values ranged from 3.43 GPa to 3.72 GPa. The yield strength varied from 76.5 MPa to 86.2 MPa in the three-ply laminate. The breaking strength values ranged from 118 MPa to 134 MPa. The ultimate elongation ranged from 33% to 46% in the laminate. In general, it can be stated that the three-ply PET film was nearly isotropic in mechanical properties.

2.4.2. Coefficient of Thermal Expansion

2.4.2.1. Isotropic Polymer Films

The coefficient of volumetric thermal expansion, α_v, decreases as the temperature is lowered. For crystalline solids and, apparently, for glasses (including organic glasses), the coefficient approaches zero as absolute zero is approached. A schematic sketch of the temperature dependence of the α_v for a polymer is shown in Fig. 2.43. At sufficiently low temperatures, only molecular, or atomic, vibrations contribute to α_v; these diminish progressively in intensity as absolute zero is approached.

For a polymer to have low thermal expansion in the vicinity of room temperature, no thermally activated motions of chain segments or side groups, except molecular or atomic vibrations, can occur; that is, the structure must be very rigid. A polymer whose chains presumably are somewhat rigid is polypyromellitimide, sold by the Du Pont Co. in film form under the

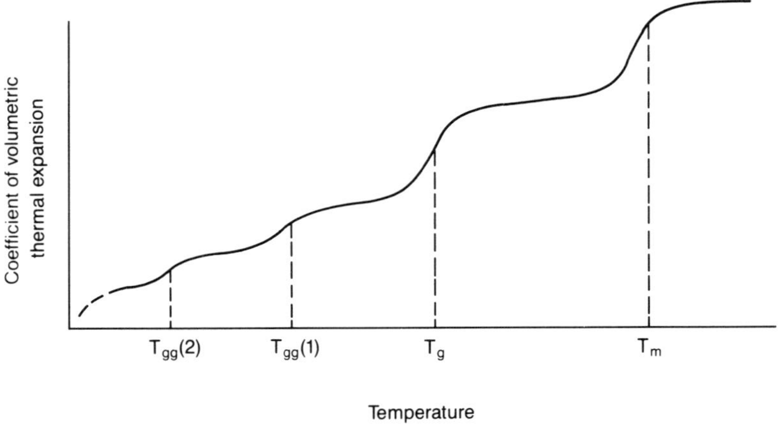

Fig. 2.43. Schematic of the coefficient of the volumetric thermal expansion (α_v) as a function of the temperature for a polymer. T_m and T_g are the melting and glass temperatures, respectively. Many polymers show several secondary transitions below T_g [e.g., $T_{gg}(1)$ and $T_{gg}(2)$ shown].

trade name Kapton. The glass-transition temperature of Kapton films is between 360° and 410° C with no melting point. Thus the film can be used to much higher operating temperatures, compared to PET. However, the coefficient of linear thermal expansion is high, reported to be about 2×10^{-5} mm/mm/°C, comparable to the biaxially-oriented PET currently used in magnetic media. The film shrinks about 0.05% at 200° C for 60 min and 0.3% at 250° C for 30 min. In view of the semirigid chains in Kapton, its thermal expansion coefficient seems high. Be that as it may, evidence that molecular motions occur at room temperature is provided by the observation that the ultimate elongation is about 70% at 25° C; Kapton remains flexible (Sroog et al., 1965). These observations indicate that to obtain a very low coefficient of volumetric expansion at room temperature, the polymer must have a structure more rigid than that of Kapton. Such a material quite probably would be essentially as brittle as inorganic glasses. (Inorganic glasses typically have expansion coefficients in the range $0.3-1 \times 10^{-5}$ mm/mm/°C.)

2.4.2.2. Oriented PET Films

The coefficient of volumetric thermal expansion (α_v) is independent of the degree of orientation, i.e., the amount of anisotropy, especially for amorphous polymers. For semicrystalline polymers, any change in α_v affected by orientation is a direct consequence of a change in the degree of crystallinity. That is, α_v depends on the degree of crystallinity (below T_g the dependence is slight), and only to a minor extent on crystal morphology, and not on the degree of orientation. Choy et al. (1983) have reported that a change is effected in the coefficient of the linear thermal expansion of PET films in the stretch direction, by stretching a specimen at an elevated temperature and then quenching it to obtain a permanently oriented material (Fig. 2.28). The coefficient in the direction parallel to the stretch is reduced, and that in the direction perpendicular to the stretch is increased.

For specimens of polymethyl methacrylate (PMMA)[5] that had been stretched uniaxially by different amounts up to 162% (stretch ratio = 2.62), the expansion coefficients have been measured from $-40°$ to $40°$ C both parallel and perpendicular to the stretch direction (Tjader and Protzman, 1956). The data at 20° C, for example, show that the linear expansion coefficient in the direction parallel to the stretch is given by

$$\alpha_{\parallel} \times 10^5 = 7.2 - 1.3(\lambda - 1) \qquad \text{for} \quad 1 \le \lambda \le 2.6, \qquad (2.9a)$$

or

$$\frac{\alpha_{\parallel}}{\alpha_0} = 1 - 0.18(\lambda - 1), \qquad (2.9b)$$

---

[5] The smallest coefficient of thermal expansion observed for an isotropic polymer at cryogenic temperatures is probably that for PMMA $\alpha \simeq 1 \times 10^{-5}$ mm/mm/°C at temperatures between 20 and 60 K.

where λ is the stretch ratio, i.e., the length of the stretched specimen divided by the length in the isotropic state, and α_0 is the coefficient of linear thermal expansion of the isotropic material $(= 7.2 \times 10^{-5})$. Smith (1984) has advanced the following relationship for the coefficient of the linear thermal expansion of material with uniform biaxial orientation as a function of the stretch λ

$$\frac{\alpha_{or}}{\alpha_0} \simeq 1 - 0.4(\lambda - 1), \tag{2.10}$$

where λ_{or} is the coefficient of the linear thermal expansion parallel to either of the stretch directions. Although it seems unlikely that the predictions by this equation are correct, the trends are correct. It suggests that $\alpha_{or} = 0$ can be attained for highly oriented PET.

2.4.2.3. Laminates

Coefficients of the linear thermal expansion of PET films and laminated films are given in Table 2.14(b). The single-ply PET films enforcement showed pronounced anisotropy in thermal expansion coefficients, with a factor of 2 difference between high and low values in a given film. The PET film laminates were significantly more uniform in their coefficient of thermal expansion than the single-ply films, and the three-ply film was nearly isotropic. The PET film laminates have coefficients of expansion which are higher than the coefficients for the single-ply material. This is to be expected, since the coefficient of thermal expansion of the adhesive is several times greater than the coefficient for oriented PET film.

2.4.2.4. Incorporation of Fibers and Filaments

The coefficient of the linear thermal expansion of a polymer can be reduced considerably by incorporating fibers and filaments. Consider a composite containing continuous filaments oriented parallel to each other in a film. The coefficient of thermal expansion in the direction parallel to the filaments can be estimated using the following equation (Ashton et al., 1969):

$$\frac{\alpha_{\parallel}}{\alpha_m} = \frac{(1 - v_f) + (E_f\alpha_f/E_m\alpha_m)v_f}{(1 - v_f) + (E_f v_f/E_m)}, \tag{2.11}$$

where v_f is the volume fraction of filaments, E_m and E_f are the moduli of the matrix material and filaments, respectively, and α_m and α_f are similarly the coefficients of linear thermal expansion. We note that to obtain a minimum value of $\alpha_{\parallel}$, when the volume fraction of the filament is specified, E_f/E_m should be maximized, and α_f/α_m should be minimized. For example, α_f for glass, graphite, and tungsten carbide are 0.5, 0.27, and 0.4×10^{-5} mm/mm/°C, respectively.

The expansion coefficient perpendicular to the filament direction is not strongly affected by the filaments unless their volume fraction is very high

$$\frac{\alpha_\perp}{\alpha_m} = (1 + v_m)v_m + (1 + v_f)\frac{\alpha_f}{\alpha_m}v_f - \alpha_\parallel(v_m v_m + v_f v_f) \qquad (2.12a)$$

and

$$v_m = 1 - v_f, \qquad (2.12b)$$

where v_m and v_f are the Poisson ratios for the matrix and filament materials, respectively.

Oriented noncontinuous fibers also effect a substantial reduction in the expansion coefficient, the amount dependent on the length-to-diameter ratio, L/D, of the fibers. An approach for obtaining the isotropic expansion in one plane consists of incorporating a random array of fibers into a sheet.

In considering laminates, or composites containing a random array of fibers, whose expansion is isotropic in the plane of the film, it is worth noting that the in-plane coefficient of linear expansion is reduced at the expense of an increase in the coefficient of linear expansion in the thickness direction. This arises because the filler decreases the coefficient of volumetric expansion by a smaller amount than the in-plane coefficient of linear expansion.

2.4.3. Coefficient of Hygroscopic Expansion

Among the numerous polymers listed by Barrie (1968), PET dissolves the least water, except for a few polymers—polyolefins, polydimethyl siloxane, polystyrene, polytetrafluoroethylene, polyvinyl chloride, and polyvinyliso-butyl ether. For Kapton film, the hygroscopic coefficient is 2.2×10^{-5} mm/mm/% RH, a value two- to threefold greater than for Mylar A. Although certain polymeric films quite probably have a coefficient of hygroscopic expansion less than that for PET, it is also probable that certain of their other physical properties would not satisfy the requirements for magnetic medium substrate.

However, the steps that may decrease the coefficient of the thermal expansion of PET or other polymers, e.g., orienting film to high degree, PET-film, and composite-film laminates, or incorporating reinforcing fibers and filaments, should also reduce the coefficient of hygroscopic expansion in the plane of the film.

Table 2.14(c) shows coefficients of the linear hygroscopic expansion of single-ply PET films and laminated films. The coefficient of linear hygroscopic expansion was anisotropic in the single-ply films, especially in the thinner film. In the three-ply PET film laminate, the variation in the coefficient of linear hygroscopic expansion with direction was much less. The laminate reinforced with woven PET fabric has slightly smaller coefficient of linear hygroscopic expansion than the three-ply laminate and is also nearly isotropic in this property.

2.4.4. Long-Term Dimensional Stability (Shrinkage)

We have discussed, at length earlier, the effect of annealing on the long-term dimensional stability (shrinkage) of PET films. Various annealing treatments can be used to obtain desirable long-term dimensional stability.

Values for the shrinkage of various single-ply and laminated films (whose thermal and hygroscopic expansion data have been presented earlier) are presented in Table 2.14(d). The thinner PET film had the most shrinkage, presumably due to higher induced stress during manufacture of the film. Annealing significantly reduces shrinkage of both single-ply and laminated films. The unannealed PET film laminates had about the same shrinkage as the 36.1-μm single-ply PET film. Because the plies cannot shrink freely during the bonding operation, the lamination process does not allow stress relaxation to take place. A separate annealing process, either before or after lamination, appears necessary if laminated.

2.4.5. High-Temperature Polyimide Substrate

A base film which can withstand temperatures of at least 200° C is the polyimide film. Another advantage of polyimide films is that they provide isotropic properties. However, these films are frightfully expensive. Several companies offer suitable polyimide films, most notable are: du Pont (Kapton grade), Ube Kosan, and Mitsubishi Jushi. Kapton is produced in three forms, Type H, Type V, and Type F. Type H is the basic uncoated polyimide film. Type V is similar to Type H but has superior dimensional stability. Type F is coated on one or both sides with FEP fluorocarbon resin which imparts heat sealability, provides a moisture barrier, and enhances chemical resistance. Kapton films are commercially used in a variety of high-temperature electrical and electronic insulation applications, e.g., wire and cable tapes, formed coil insulation, and substrates for flexible printed circuits.

Kapton is manufactured by casting a 10–20% solid solution of polyamic acid [synthesized by a polycondensation reaction between an aromatic diamine (ODA) and an aromatic diahydride (PMDA)] on a belt or drum. Water is removed by a thermal conversion or a chemical conversion process to form polyimide. The polymer is dried and taken off the belt or the drum. It is then restrained in a tenter frame and stretched, followed by curing at a temperature higher than 400° C. Kapton has the structure:

The Kapton films are very expensive because of low throughput due to slow drying and expensive monomers. Kapton film manufacturing is more susceptible to particulate contamination from the environment than PET film manufacturing, because Kapton starts with a liquid solution and PET starts with a melt. There is no known organic solvent for the film and it is infusible and flame resistant. The glass-transition temperature of Kapton films is between 360°–410° C with no melting point or softening point, Table 2.15. A $25 \cdot 4$-μm-thick film has a zero-strength temperature of 815° C, the

Table 2.15. Typical properties of a 25.4-μm-thick polyimide (Kapton H) film

Property	Typical value(s)	Condition	ASTM method
(a) *Physical properties*			
Density, kg/m^3	1420	22° C	D1505-63T
Young's modulus (tensile), GPa	3.0	22° C	D882-64T
Yield strength (tensile), MPa	70	22° C	D882-64T
Breaking strength (tensile), MPa	170	22° C	D882-64T
Strain at break (tensile)	0.7	22° C	D882-64T
Impact strength, kg-cm/25 μm	6.0	22° C	Pneumatic impact test
Folding endurance, kcycles	10	22° C	D643-43
Tear strength—initial (Graves), g/25 μm	510	22° C	D1004-61
Bursting strength (Mullen), kPa	515	22° C	D774-63T
Refractive index	1.78	22° C	
(b) *Thermal properties*			
Glass transition temp., °C	360–410		
Melting point, °C	none		
Zero strength temperature, °C	815° C	140 kPa load for 5 s	Hot bar (Du Point test)
Coefficient of thermal expansion, $\times 10^{-5}$/°C	2	-14–38° C	D696-44
Thermal conductivity, W/m K	0.62	25–75° C	
Specific heat, J/g K	0.98	40° C	
Shrinkage, %	0.05	60 min at 200° C	D-1204
	0.3	30 min at 250° C	
	0.5	30 min at 300° C	
(c) *Electrical properties*			
Dielectric strength, V/μm	280	22° C, 60 Hz	D149-64
Dielectric constant	3.5	22° C, 1 kHz	D150-64T
Dissipation factor	0.003	22° C, 1 kHz	D150-64T
Volume resistivity, ohm · cm	10^{18}	22° C	D257-61
Surface resistivity, ohm/cm^2	10^{15}	22° C, 50% RH	D257-61
(d) *Chemical properties*			
Coefficient of hygroscopic expansion, $\times 10^{-5}$/% RH	2.2	22° C, 20–80% RH	
Water absorption, %	1.3	Immersion for 24 h at 22° C	D570-63

temperature at which Kapton begins to char. (Zero-strength is measured as the maximum temperature at which the film will sustain a load of 140 kPa for 5 s. Thus the film can be used to much higher operating temperatures compared to PET. This is much less than that of PET.) The film shrinkage is a direct indication of frozen-in stresses (Tatum et al., 1963). The crystallinity of Kapton is low (about 15–20%) compared to that of a biaxially-oriented PET (about 50–60%), as a result, the viscoelastic properties (creep and relaxation behavior) of Kapton film is poorer than that of biaxially-oriented PET; however, at higher temperatures ($> T_g$ of PET), Kapton exhibit superior viscoelastic behavior. Other mechanical properties and coefficient of thermal expansion are comparable to that of biaxially-oriented PET, however, hygroscopic expansion is two- to three-fold greater than for biaxially-oriented PET.

2.4.6. Summary

Polyimide films (such as Kapton) provide isotropic properties and can withstand temperatures of at least 200° C. However, it appears that no isotropic polymer is likely to have low coefficients of linear and hygroscopic expansion (significantly lower than that of biaxially-oriented PET), as well as the other required physical properties. It appears that a sufficiently high degree of orientation in the PET films can provide low expansions and high elastic modulus in the orientation direction. A two- or three-ply laminate film prepared from films that are highly oriented in one direction can provide in-plane isotropy. Another promising approach involves the incorporation of a random array of fibers that have a low expansion coefficient and also a very high modulus. Annealing significantly reduces the shrinkage of oriented-polymer films.

References

Ashton, J. E., Halpin, J. C., and Petit, P. H. (1969). Primer on composite materials. Technomic Publishing, Stamford, Connecticut.

Barrall, E. M. and Logan, J. A. (1973). Some physical properties of biaxially oriented poly(ethylene terephthalate). IBM Research Report No. RJ 1223, May 21.

Barrie, J. A. (1968). Water in polymers. In "Diffusion in Polymers" (J. Crank and G. S. Park, eds.), Chapter 8. Academic Press, New York.

Berry, B. S. and Pritchet, W. C. (1988). Elastic and viscoelastic behavior of a magnetic recording tape. *IBM J. Res. Develop.* **32**, 682–694.

Bertram, N. and Eshel, A. (1980). "Recording Media Archival Attributes (Magnetic)," Report No. RADC-TR-80-123. Ampex Corp., Redwood City, California.

Bhushan, B. (1985). Anisotropic mechanical properties of biaxially oriented poly-(ethylene terephthalate) films. In "Tribology and Mechanics of Magnetic Storage Systems," Vol. 2 (B. Bhushan and N. S. Eiss, eds.), pp. 119–126. SP-19, ASLE, Park Ridge, Illinois.

Bhushan, B. (1990). "Tribology and Mechanics of Magnetic Storage Devices." Springer-Verlag, New York.

Bhushan, B., Hahn, F. W., Sharma, B. S., and Connolly, D. (1984a). Long-term reliability of magnetic tapes for digital recording. In "Tribology and Mechanics of Magnetic Storage Systems" (B. Bhushan et al., eds.), pp. 132–147. SP-16, ASLE, Park Ridge, Illinois.

Bhushan, B., Heinrich, J. C., and Connolly, D. (1984b). Orthotropic viscoelastic behavior of polyethylene terephthalate film under plane-stress conditions. In "Tribology and Mechanics of Magnetic Storage Systems" (B. Bhushan et al., eds.), pp. 158–171. SP-16, ASLE, Park Ridge, Illinois.

Bhushan, B., Bradshaw, R. L., and Sharma, B. S. (1984c). Friction in magnetic tapes II: Role of physical properties. *ASLE Trans.* **27**, 89–100.

Bhushan, B. and Smith, D. R. (1985). Measurement of creep properties in compression and their influence on the friction of magnetic tapes. *ASLE Trans.* **28**, 325–335.

Bhushan, B., Sharma, B. S., and Srinivasan, R. (1987). Texturing a magnetic tape surface. *Res. Discl.* February, No. 274.

Biangardi, H. J. and Zachmann, H. G. (1977). Influence of the arrangement of the crystals and the structure of the noncrystalline regions of the mechanical properties of polyethylene terephthalate. *J. Poly. Sci., Poly. Symp.* **58**, 169–183.

Blumentritt, B. F. (1979a). Anisotropy and dimensional stability of biaxially oriented poly(ethylene terephthalate) films. *J. Appl. Poly. Sci.* **23**, 3205–3217.

Blumentritt, B. F. (1979b). Annealing of poly(ethylene terephthalate)-film-based magnetic recording media for improved dimensional stability. *IBM J. Res. Develop.* **23**, 56–65.

Blumentritt, B. F. (1979c). Laminated films with isotropic in-plane properties. *IBM J. Res. Develop.* **23**, 66–74.

Bogy, D. B., Bugdayci, N., and Talke, F. E. (1979). Experimental determination of creep functions for thin orthotropic films. *IBM J. Res. Develop.* **23**, 450–458.

Chan, A. and Smith, T. L. (1979). Anisotropy of the coefficients of linear thermal expansion and other properties of six biaxially oriented films of poly(ethylene terephthalate). IBM Research Report No. RJ 2669 (34184), San Jose, Califomia, October.

Choy, C. L., Masayoshi, I., and Porter, R. S. (1983). Thermal expansivity of oriented poly(ethylene terephthalate). *J. Poly. Sci.* **21**, 1427–1438.

Choy, I. C. and Plazek, D. J. (1968). The physical properties of bisphenol-*a*-based epoxy resins during and after curing. *J. Poly. Sci., Part B: Poly. Phys.* **24**, 1303–1320.

Chu, W. H. and Smith, T. L. (1973). Rodlike superstructures in and mechanical properties of biaxially oriented polyethylene terephthalate film. In "Structure and Properties of Polymer Films" (R. W. Lenz and R. S. Stein, eds.), pp. 67–69. Plenum Press, New York.

Cobbs, W. H. and Burton, R. L. (1953). Crystallization of polyethylene terephthalate. *J. Poly. Sci.* **10**, 275–290.

Cuddihy, E. F. (1972). Analysis of the failure of a polyester peripheral drive belt on the mariner mars 1971 flight tape recorder. *JPL Quarterly Tech. Rev.* **2**, 82–99.

Cuddihy, E. F. (1975). Solvent-sensitivity of biaxially oriented poly(ethylene terephthalate) film. *Poly. Lett. Edition* **13**, 595–601.

Cuddihy, E. F. (1976). Hygroscopic properties of magnetic recording tape. *IEEE Trans. Magn.* **MAG-12**, 126–135.

Daubeny, R. deP., Bunn, C. W., and Brown, C. J. (1954). The crystal structure of polyethylene terephthalate. *Proc. Roy. Soc. (London)* **A226**, 531–542.

Dulmage, W. J. and Geddes, A. L. (1958). Structure of drawn polyethylene terephthalate. *J. Poly. Sci.* **31**, 499–512.

Eshel, A., Baker, S., Hartman, A., and Orcutt, F. K. (1984). Mechanical aspects of archival storage of magnetic tape. In "Tribology and Mechanics of Magnetic Storage Systems" (B. Bhushan et al., eds.), pp. 148–157. SP-16, ASLE, Park Ridge, Illinois.

Fakirov, S., Fischer, E. W., Hoffman, R., and Schmidt, G. F. (1977). Structure and properties of poly(ethylene terephthalate) crystallized by annealing in the highly oriented state: 2. Melting behavior and the mosaic block structure of the crystalline layers. *Polymer* **18**, 1121–1129.

Greenberg, H. J., Stephens, R. L., and Talke, F. E. (1977). Dimensional stability of floppy disks. *IEEE Trans. Magn.* **MAG-13**, 1397–1399.

Greenberg, H. J., Stephens, R. L., and Talke, F. E. (1978). Measurement of dimensional changes in thin polymer films. *Experimental Mechanics* **18**, 115–120.

Groeninckx, G., Berghmans, H., and Smets, G. (1976). Morphology and modulus-temperature behavior of semicrystalline poly(ethylene terephthalate) (PET). *J. Poly. Sci., Poly. Phys. Ed.* **14**, 591–602.

Groeninckx, G., Reynaers, H., Berghmans, H., and Smets, G. (1980a). Morphology and melting behavior of semicrystalline poly(ethylene terephthalate) (PET). I. Isothermally crystallized PET. *J. Poly. Sci., Poly. Phys. Ed.* **18**, 1311–1324.

Groeninckx, G. and Reynaers, H. (1980b). Morphology and melting behavior of semicrystalline poly(ethylene terephthalate). II. Annealed PET. *J. Poly. Sci., Poly. Phys. Ed.* **18**, 1325–1341.

Gupta, V. B., Ramesh, C., and Gupta, A. K. (1984a). Structure-property relationship in heat-set poly(ethylene terephthalate) fibers. I. Structure and morphology. *J. Appl. Poly. Sci.* **29**, 3115–3129.

Gupta, V. B., Ramesh, C., and Gupta, A. K. (1984b). Structure-property relationship in heat-set poly(ethylene terephthalate) fibers. II. Thermal behavior and morphology. *J. Appl. Poly. Sci.* **29**, 3727–3739.

Heffelfinger, C. J. (1978). A survey of film processing illustrated with poly(ethylene terephthalate). *Poly. Eng. Sci.* **18**, 1163–1173.

Heffelfinger, C. J. (1984). Personal communications.

Heffelfinger, C. J. and Schmidt, P. G. (1965). Structure and properties of oriented poly(ethylene terephthalate) films. *J. Appl. Poly. Sci.* **9**, 2661–2680.

Heffelfinger, C. J. and Knox, K. L. (1971). Polyester films. In "The Science and Technology of Polymer Films" (O. J. Sweeting, ed.), pp. 587–639. Wiley-Interscience, New York.

Ishai, O., Weller, T., and Singer, J. (1968). Anisotropy of Mylar A sheets. *J. Mat. JMLSA* **3**, 337–351.

Lekhnitskii, S. G. (1968). "Anisotropic Plates." Gordon and Breach, New York.

Miyagi, A. and Wunderlich, B. (1972). Superheating and reorganization on melting of poly(ethylene terephthalate). *J. Poly. Sci.* **A-2**, **10**, 1401–1405.

Oden, P. I., Majumdar, A., Bhushan, B., Padmanabhan, A., and Graham, J. J. (1992). AFM imaging, roughness analysis and contact mechanics of magnetic tape and head surfaces. *J. Trib., Trans. ASME* (in press).

Oswin, C. R. (1975). "Plastic Films and Packaging." Wiley, New York.

Park, W. R. R. (1969). "Plastic Film Technology." Van Nostrand Reinhold, New York.

Petrie, S. E. B. (1972). Thermal behavior of annealed organic glasses. *J. Poly. Sci.* **A-2**, **10**, 1255–1272.

Ravens, D. A. S. and Ward, I. M. (1961). Chemical reactivity of polyethylene terephthalate. *Trans. Faraday Soc.* **57**, 150–159.

Samuels, R. J. (1981). Application of refractive index measurements to polymer analysis. *J. Appl. Poly. Sci.* **26**, 1383–1412.

Schmidt, P. G. (1963). Polyethylene terephthalate structural studies. *J. Poly. Sci.* **A1**, 1271–1292.

Smith, T. L. (1984). Personal communications.

Sroog, C. E., Endrey, A. L., Abramo, S. V., Berr, C. E., Edwards, W. M., and Olivier, K. L. (1965). Aromatic polypyromellitimides from aromatic polyamic acids. *J. Poly. Sci.* **A3**, 1373–1390.

Stein, R. S. and Misra, A. (1973). Kinetics of growth of developing spherulites. *J. Poly. Sci.* **A-2**, **11**, 109–116.

Sweeting, 0. J. (1971). "The Science and Technology of Polymer Films." Wiley, New York.

Tatum, W. E., Amborski, L. E., Gerow, C. W., Heacock, J. F., and Mallouk, R. S. (1963). H film—DuPont's new polyimide film. Electrical Insulation Conf., Materials and Applications, Chicago.

Timoshenko, S. P. and Goodier, J. N. (1970). "Theory of Elasticity," 3rd ed., McGraw-Hill, New York.

Tjader, T. C. and Protzman, T. F. (1956). The thermal expansion of stretched methyl methacrylate polymer. *J. Poly. Sci.* **20**, 591–592.

Uematsu, I. and Uematsu, Y. (1960). The effect of crystallinity on glass transition temperature. *Report on Progress in Poly. Physics (Japan)* **3**, 102–104.

Umanskii, E. S., Kryuchov, V. V., and Rakovskii, V. A. (1978). Determination of the stressed state of a coil of magnetic tape. *Problemy Prochnosti* 3, March, 83–85.

Vallat, M. F., Plazek, D. F., and Bhushan, B. (1986). Effects of thermal treatment of biaxially oriented poly(ethylene terephthalate). *J. Poly. Sci., Part B: Poly. Phys.* **24**, 2123–2134.

Vallat, M. F. and Plazek, D. F. (1988) Effects of thermal treatment of biaxially oriented poly(ethylene terephthalate), II. The anisotropic glass temperature. *J. Poly. Sci., Part B: Poly Phys.* **26**, 545–554.

Viswanathan, A., Wiff, D. R., and Adams, W. W. (1976). "An Investigation of Structure–Property Correlations in Polyethylene Terephthalate Films." Report No. AFML-TR-75-207, Air Force Materials Laboratory, April.

Wahlstrom, E. E. (1969). "Optical Crystallography." 4th ed. Wiley, New york.

Ward, I. M. (1967). The mechanical behavior of poly(ethylene terephthalate). *J. Macromol. Sci.-Phys.* **B1**(4), 667–694.

Yasuda, H. and Stannett, V. (1962). Permeation, solution, and diffusion of water in some high polymers. *J. Poly. Sci.* **57**, 907–923.

Yeh, G. S. Y. and Geil, P. H. (1967). Crystallization of polyethylene terephthalate from the glassy amorphous state. *J. Macromol. Sci. (Phys.)* **B1**, 235–249.

Zachmann, H. G. and Stuart, H. A. (1960). Partial melting and new crystallizing of terylene. *Makromol. Chem.* **41**, 148–173.

Viscoelastic Properties of PET Substrate and Coated Magnetic Media

Flexible polymer films, primarily biaxially oriented, semicrystalline poly (ethylene terephthalate) (PET) films, are used as a base film for flexible, magnetic-recording media. When high recording densities are used, strict requirements are imposed on the dimensional stability of the films. Time-dependent dimensional changes, from variations of temperature and humidity under some stress, can cause shift and distortion in the data tracks severe enough to cause data loss (Greenberg et al., 1977, 1978; Bhushan et al., 1984b; Eshel et al., 1984). These time-dependent deformations can be directly related to the viscoelastic (creep/relaxation) properties of the polymer.

In many cases, flexible media are stored at one temperature and used at another. For example, tape can be wound over a contaminant or some perturbation that causes an associated localized stress field and then stored at one temperature; during unwinding and rewinding, the contaminant or perturbation is removed, and then the tape is stored at a different stress field and, maybe, at a different temperature and humidity. Therefore, it is of interest to know the viscoelastic behavior of magnetic media when stress and temperature fields are changed during storage (nonisothermal or thermoviscoelastic behavior).

In this chapter, we present an overview of the basics of viscoelasticity. Then we present the viscoelastic and thermoviscoelastic properties of PET and coated media.

3.1. Introduction to Viscoelasticity

The elastic materials are those materials which have a capacity to store mechanical energy with no dissipation of energy. On the other hand, a New-tonian viscous fluid in a nonhydrostatic stress state has a capacity for dissipating energy, but none for storing it. But, then, materials which must be outside the scope of these two theories are those for which some, but not all, of the work done to deform them can be recovered. Such materials possess a capacity both to store and dissipate mechanical energy.

A different way of characterizing these materials is through the nature of their response to a suddenly applied uniform distribution of load on a specimen. The term "suddenly applied" loading state does not imply rates sufficiently great to cause the excitation of a dynamic response in the specimen. An elastic material, when subjected to such a suddenly applied loading state held constant thereafter (a step load), responds instantaneously with a state of deformation which remains constant (a step change of deformation). A Newtonian viscous fluid responds to a suddenly applied state of uniform shear stress by a steady flow process. There are, however, materials for which a suddenly applied and maintained state of uniform stress (a step change in stress) induces an instantaneous deformation followed by a flow process which may or may not be limited in magnitude as time grows. A material which responds in this manner is said to exhibit both an instantaneous elasticity effect and creep characteristics. This behavior is clearly not described by either an elasticity or a viscosity theory but combines features of each and is referred to as *viscoelastic behavior*. If the material goes through permanent (or nonrecoverable) deformation, the materials are known to exhibit plastic or viscoplastic behavior.

The prominence of viscoelasticity in polymers is not unexpected when one considers the complicated molecular adjustments which must underlie any macroscopic mechanical deformation. In a polymer, each flexible threadlike molecule pervades an average volume much greater than atomic dimensions, and is continually changing the shape of its contour as it wriggles and writhes with its thermal energy. Under stress, a new assortment of configurations is obtained; the response to the local aspects of the new distribution is rapid, the response to the long-range aspects is slow, and there is a very wide and continuous range of time scale covering the response of such a system to external stress.

Polymers exist in four physical states: the crystalline and the three amorphous states (glassy, rubbery, and viscous flow). Glassy material is liquidlike with a nonequilibrium density. Its glass transition temperature (T_g) is that temperature which marks its departure from an equilibrium density at a given rate of cooling. Equilibrium is lost because the coordinated molecular motions, involved in the molecular rearrangements needed to reach the average liquid structure that is characteristic at a particular temperature, do not have time to occur. Polymers that exist in glassy or crystalline states are called rigid. At temperatures lower than the glass transition, the writhing thermal motions essentially cease, and polymers deform in the manner of a rigid glass (or elastic solid). A significant increase in reversible strain occurs at temperatures above this, entering the rubbery state. In this range, the elastic modulus changes little with temperatures up to the flow or melting temperature, T_m. The mode of deformation of rubbery polymers is viscoelastic. Above this flow temperature, a polymer behaves as a liquid of high viscosity (see Fig. 3.1). The flexible magnetic media are in the rubbery regime which exhibit significant viscoelastic behavior.

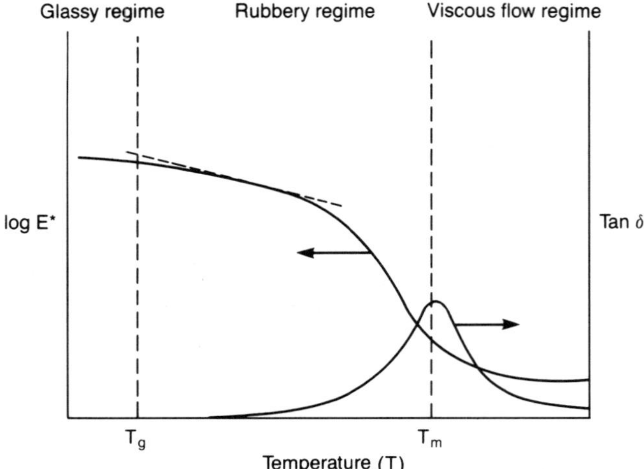

Fig. 3.1. Temperature dependence on the complex modulus (E^*) and the tangent modulus (tan δ) of polymers (T_g glass transition temperature and T_m melting or flow temperature).

3.1.1. Elasticity

Consider a thin rod which is subjected to a variable tensile load. If the stress is assumed to be distributed uniformly over the area of cross section, then the *nominal stress* σ can be calculated by dividing the load with the original cross-sectional area. The *actual stress* σ_a is obtained, under the assumption of a uniform stress distribution, by dividing the load at any stage of the test by the actual area of the cross section at that stage. The difference between the nominal and the actual stress is negligible, however, throughout the elastic range of the material. If the nominal stress is plotted as a function of strain ε (change in length per unit length of the specimen), then for some ductile metals a graph like that in Fig. 3.2(a) is obtained. The graph is very nearly a straight line until the stress reaches the proportional limit [point P in Fig. 3.2(a)]. The slope of the straight line ($\sigma/\varepsilon = E$) is known as Young's modulus (Hooke's Law). The greatest stress that can be applied without producing a plastic (permanent) deformation is called the *elastic limit* of the material.

When the stress increases beyond the elastic limit, a point is reached (Y on the graph) at which the rod suddenly stretches with little or no increase in the load. The stress at point Y is called the *yield-point stress*.

The nominal stress may be increased beyond the yield point until the ultimate tensile stress or UTS (point U) is reached. The corresponding force is the greatest load that the rod will bear. When the ultimate stress is reached, a brittle material breaks suddenly, while a rod of some ductile metal begins to "neck," i.e., its cross-sectional area is greatly reduced over a small portion of

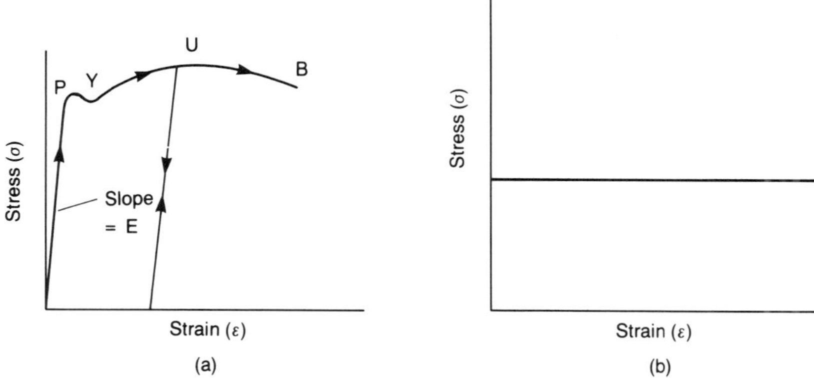

Fig. 3.2. Typical stress–strain curve for (a) a ductile metal and (b) a linear viscous liquid.

the length of the rod. Further elongation is accompanied by an increase in actual stress but by a decrease in total load, in cross-sectional area, and in nominal stress until the rod breaks (point *B*). The stress at break is called the *breaking stress*.

Now if the rod is unloaded after it was loaded beyond *P*, only the elastic strains are recovered and the plastic strains are permanent. We shall consider only the behavior of elastic materials subjected to stresses below the proportional limit; that is, we shall be concerned only with those materials and situations in which Hooke's Law is valid.

3.1.1.1. Generalized Hooke's Law

The classical theory of elasticity deals with the mechanical properties of *linear elastic solids*, for which, in accordance with Hooke's Law, stress is always directly proportional to strain in small deformations but independent of the rate of strain or past history (Love, 1944; Sokolnikoff, 1956; Hearmon, 1961; Lekhnitskii, 1968, 1981).

Cartesian Coordinates

From the classical theory of elasticity, any arbitrary load field may be composed into at most six independent stress components σ_{ij} (Fig. 3.3) with

$$\sigma_{ij} = \sigma_{ji}, \qquad i, j = 1, 2, 3. \tag{3.1}$$

The arabic indices 1, 2, 3 correspond to x_1, x_2, x_3 or x, y, z Cartesian coordinate axes. (By a simple consideration of the equilibrium of the body, we obtain the symmetry of the stress tensor.) Similarly, an arbitrary deformation can be constructed as a linear combination of at most six independent

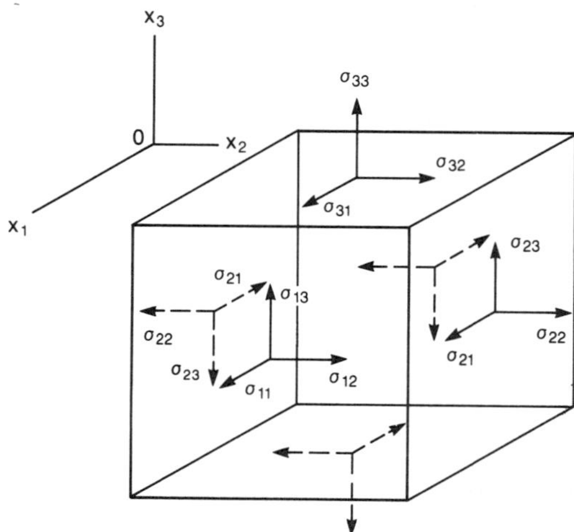

Fig. 3.3. Schematic representation of normal and shear stresses on the sides of a cubic element in Cartesian coordinates.

relative deformations or strains ε_{ij} with

$$\varepsilon_{ij} = \varepsilon_{ji}, \qquad i, j = 1, 2, 3. \tag{3.2}$$

If u_i are the components of a displacement vector (Fig. 3.4), then the infinitesimal strain tensor ε_{ij} is defined by

$$\varepsilon_{ij} = \tfrac{1}{2}(u_{i,j} + u_{j,i}), \tag{3.3}$$

where $u_{i,j} = \partial u_i / \partial u_j$.

The strain-energy density function (the energy per unit volume of the

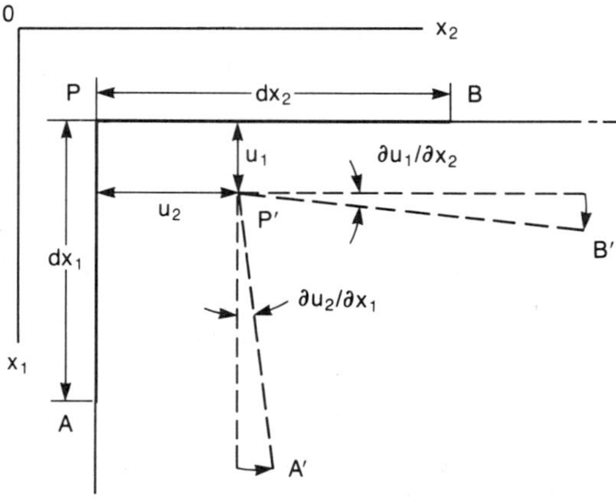

Fig. 3.4. Schematic representation of displacements of the elements PA and PB to $P'A'$ and $P'B'$.

material) associated with an arbitrary relative deformation may be written as

$$U = C_{ijkl}\varepsilon_{ij}\varepsilon_{kl}, \qquad i, j, k, l = 1, 2, 3. \tag{3.4}$$

The repeated indices imply the summation convention. The stress is given by

$$\sigma_{ij} = \frac{\partial U}{\partial \varepsilon_{ij}} \tag{3.5a}$$

and the material constants are given by

$$C_{ijkl} = (\partial^2 U / \partial \varepsilon_{ij} \, \partial \varepsilon_{kl}), \tag{3.6}$$

so that

$$\sigma_{ij} = C_{ijkl}\varepsilon_{kl} \tag{3.5b}$$

or

$$\boldsymbol{\sigma} = \mathbf{C}\boldsymbol{\varepsilon}. \tag{3.7}$$

If the inverse of **C** exists then

$$\boldsymbol{\varepsilon} = \mathbf{C}^{-1}\boldsymbol{\sigma} \tag{3.8a}$$

or

$$\boldsymbol{\varepsilon} = \mathbf{S}\boldsymbol{\sigma}, \tag{3.8b}$$

where

$$\mathbf{S} = \mathbf{C}^{-1} \tag{3.8c}$$

and

$$\mathbf{C} = \mathbf{S}^{-1}. \tag{3.8d}$$

$\boldsymbol{\sigma}$ is the stress tensor and $\boldsymbol{\varepsilon}$ is the strain tensor. σ_{ij} and ε_{ij} $(i = j)$ are the normal stresses and normal strains, respectively, and σ_{ij} and ε_{ij} $(i \neq j)$ are the shear stresses and shear strains, respectively.[1] **C** and **S** are the fourth-order tensors, **C** being the elastic tensor and **S** being the compliance tensor. The coefficients C_{ijkl}, in general, will vary from point to point of the medium. If the material is *elastically homogeneous*, the C_{ijkl} are independent of the position of the point.

This important result connects stress and strain and is referred to as generalized Hooke's Law. The coefficients C_{ijkl} (or S_{ijkl}) may be considered to be generalized material descriptors since they describe the ability of the material to transfer strain into stress. The symmetry of stress and strain tensors implies

---

[1] σ_{ij} and ε_{ij} are also written as follows:

$$\sigma_{ij} = \begin{bmatrix} \sigma_x & \tau_{xy} & \tau_{xz} \\ \tau_{yx} & \sigma_y & \tau_{yz} \\ \tau_{zx} & \tau_{zy} & \sigma_z \end{bmatrix} \quad \text{and} \quad \varepsilon_{ij} \equiv \begin{bmatrix} \varepsilon_x & \frac{1}{2}\gamma_{xy} & \frac{1}{2}\gamma_{xz} \\ \frac{1}{2}\gamma_{yx} & \varepsilon_y & \frac{1}{2}\gamma_{yz} \\ \frac{1}{2}\gamma_{zx} & \frac{1}{2}\gamma_{zy} & \varepsilon_z \end{bmatrix},$$

σ and ε with a single suffix are normal stresses and strains, respectively; here the suffix indicates the direction of the normal to the side of the element on which the normal component acts. (Sometimes these are written with two suffices, e.g., $\sigma_x \rightarrow \sigma_{xx}$.) τ and γ with two suffices are the shear stress and shear strains, respectively; here the first suffix indicates the direction of the normal to the side of the element on which the component acts and the second suffix indicates the axis to which the stress or strain component arrow is parallel. Note that γ_{ij} $(= 2\varepsilon_{ij}, i \neq j)$ is the reduction of the original right angle between line elements dx_i, dx_j at x_1, x_2, x_3 (Fig. 3.4).

that $\mathbf{C}$ and $\mathbf{S}$ are symmetric tensors, i.e., they are symmetric with respect to the first two and the last two indices,

$$C_{ijkl} = C_{jikl} = C_{ijlk}, \tag{3.9a}$$

so that there are at most 36 independent material constants. Furthermore, one assumes that the quadratic form (3.4) is symmetric, and it then follows from (3.5a) that

$$C_{ijkl} = C_{klij} \tag{3.9b}$$

and the number of constants is reduced to 21. It is now generally accepted that, for the most general case of an anisotropic elastic body, the number of independent elastic constants in the generalized Hooke's Law is 21. If the medium is elastically symmetric in certain directions, then the number of independent constants C_{ij} is reduced from 21.

Material symmetry can drastically reduce the number of required descriptors. Consequently, it is advantageous to establish the intrinsic material descriptors with respect to unique material directions in order to exploit fully the simplifications introduced by material symmetry. There is no guarantee, however, that the applied load (or deformation) field will correspond to the unique directions of the material. Nonetheless, the stress and strain tensors referenced with respect to axes associated with the applied load (or deformation) field remain connected through a relationship of the form

$$\sigma'_{ij} = C'_{ijkl}\varepsilon'_{kl}, \tag{3.10}$$

where the C'_{ijkl} are apparent material descriptors for the current choice of reference axes.

Clearly, since the material has not been modified by the choice of coordinate axes, the C'_{ijkl} must be related to the C_{ijkl} defined to exploit fully the symmetry of the material. It can be readily demonstrated that the apparent descriptors C'_{ijkl}, referenced with respect to an arbitrary orthogonal axes system (x'_1, x'_2, x'_3) are related to the C_{ijkl} referenced with respect to a base orthogonal axes system (x_1, x_2, x_3) by the transformation (Lekhnitskii, 1981)

$$C'_{ijst} = \omega_{ki}\omega_{lj}\omega_{ps}\omega_{qt}C_{klpq}, \tag{3.11}$$

where the $\omega_{\alpha\beta}$ are the elements of the rotation matrix associated with the coordinate transformation. The components $\omega_{\alpha\beta}$ may be identified as the direction cosine between x'_α and x_β; however, it is convenient to express the $\omega_{\alpha\beta}$'s in terms of three Eulerian angles, in order to emphasize the fact that only three of the nine components $\omega_{\alpha\beta}$ are independent. The elements of the rotation matrix corresponding to successive rotations of ϕ (around the x_1 axis), θ (around the x_2 axis), and ψ (around the x_3 axis) (Fig. 3.5) are (Goldstein, 1953)

$$\Omega = \begin{bmatrix} \omega_{11} & \omega_{12} & \omega_{13} \\ \omega_{21} & \omega_{22} & \omega_{23} \\ \omega_{31} & \omega_{32} & \omega_{33} \end{bmatrix}, \tag{3.12}$$

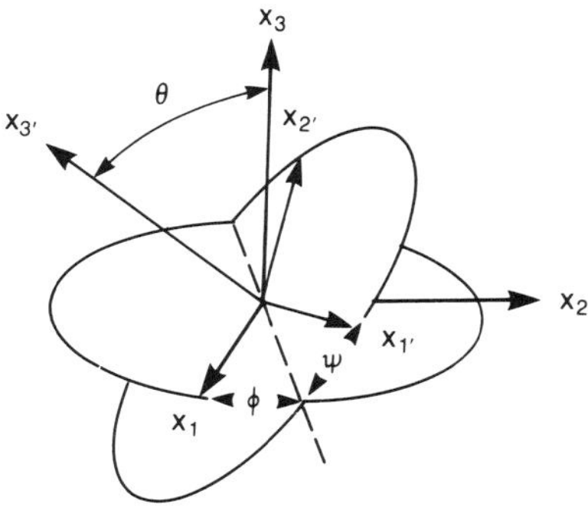

Fig. 3.5. Rotation of the coordinate axes by three Eulerian angles.

where

$$\omega_{11} = \cos\psi\,\cos\phi - \cos\theta\,\sin\phi\,\sin\psi,$$

$$\omega_{12} = \cos\psi\,\sin\phi - \cos\theta\,\cos\phi\,\sin\psi,$$

$$\omega_{13} = \sin\psi\,\sin\theta,$$

$$\omega_{21} = -\sin\psi\,\cos\phi - \cos\theta\,\sin\phi\,\cos\psi,$$

$$\omega_{22} = -\sin\psi\,\sin\phi + \cos\theta\,\cos\phi\,\cos\psi,$$

$$\omega_{23} = \cos\psi\,\sin\theta,$$

$$\omega_{31} = \sin\theta\,\sin\phi,$$

$$\omega_{32} = -\sin\theta\,\cos\phi,$$

$$\omega_{33} = \cos\theta.$$

To avoid the inconvenience of the quadruple indices on the C_{ijkl}, it has become customary to contract the stress and strain tensors to a six-component vector. Under this convention, the stress and strain components are redefined as

$$\sigma_{11} = \sigma_1, \quad \sigma_{22} = \sigma_2, \quad \sigma_{33} = \sigma_3, \quad \sigma_{23} = \sigma_4, \quad \sigma_{31} = \sigma_5, \quad \sigma_{12} = \sigma_6, \quad (3.13)$$

$$\varepsilon_{11} = \varepsilon_1, \quad \varepsilon_{22} = \varepsilon_2, \quad \varepsilon_{33} = \varepsilon_3, \quad \varepsilon_{23} = \varepsilon_4, \quad 2\varepsilon_{31} = \varepsilon_5, \quad 2\varepsilon_{12} = \varepsilon_6. \quad (3.14)$$

The elements of the fourth-order tensor are similarly contracted according to the convention $C_{[ij][kl]} \to C_{[p][q]}$ where $[ii] \to [i]$ for $i = 1$, 2, or 3; $([23], [32]) \to ([4])$; $([31], [13]) \to [5]$; and $([12], [21]) \to [6]$. For example, $C_{1111} \to C_{11}$ and $C_{1112} \to C_{16}$. Following this convention, the generalized Hooke's Law relationship between the elements of the stress and strain tensor (represented as a six-element column vector) can be compactly written as

$$\boldsymbol{\sigma} = \mathbf{C}\boldsymbol{\varepsilon}$$

or

$$\sigma_i = C_{ij}\varepsilon_j, \qquad i, j = 1, 2, \ldots, 6, \qquad (3.15a)$$

and

$$C_{ij} = C_{ji}, \qquad (3.15b)$$

where **C** is a 6×6 symmetric matrix array of the elements with 21 independent constants. Similarly Eq. (3.8) can be compactly written as

$$\varepsilon = \mathbf{S}\sigma$$

or

$$\varepsilon_i = S_{ij}\sigma_j, \qquad i, j = 1, 2, \ldots, 6, \qquad (3.16a)$$

and

$$S_{ij} = S_{ji}. \qquad (3.16b)$$

Traditionally, material response has been characterized by engineering "constants" rather than by the elements of the elastic constant or compliance array. Young's modulus E is used to describe the ability of a material to transfer a pure extensional strain into a pure extensional stress; the Poisson ratio v is used to indicate the extent to which the lateral dimensions of a body decrease (or increase) in response to a pure extensional (or compressional) strain, and finally, the shear modulus G is used to describe the ability of a material to transfer a pure shear strain into a pure shear stress. For anisotropic materials, the engineering constants are reported for each unique materials direction.

Curvilinear Coordinates

The possibility of obtaining a simple solution to a given boundary-value problem often crucially depends upon the choice of that coordinate system in which the boundary conditions assume a simple form. In addition to the rectilinear anisotropy, there is another kind of anisotropy, the curvilinear. The latter is characterized by the fact that in such a body equivalence is not found in parallel directions but follows some other direction. In dealing with axially symmetrical (axisymmetric) bodies, for example, it is usually advisable to use cylindrical or spherical coordinates. The cylindrical anisotropy is frequently encountered in practice. The variables involved in cylindrical coordinates are r, θ, z (Fig. 3.6). The indices 1, 2, and 3 assume r, θ, and z, respectively. The stress and strain components are represented as

$$\sigma = \begin{bmatrix} \sigma_{rr} \\ \sigma_{\theta\theta} \\ \sigma_{zz} \\ \sigma_{\theta z} \\ \sigma_{rz} \\ \sigma_{r\theta} \end{bmatrix} \quad \text{and} \quad \varepsilon = \begin{bmatrix} \varepsilon_{rr} \\ \varepsilon_{\theta\theta} \\ \varepsilon_{zz} \\ 2\varepsilon_{\theta z} \\ 2\varepsilon_{rz} \\ 2\varepsilon_{r\theta} \end{bmatrix}, \qquad (3.17)$$

and stress–strain relationships are given by Eqs. (3.15) and (3.16). σ_{rr}, $\sigma_{\theta\theta}$, and σ_{zz} (also written as σ_r, σ_θ, and θ_z) are normal stresses and $\sigma_{\theta z}$, σ_{rz}, and $\sigma_{r\theta}$ are shear stresses. Similarly, ε_{rr}, $\varepsilon_{\theta\theta}$, ε_{zz} (also written as ε_r, ε_θ, and ε_z) are normal

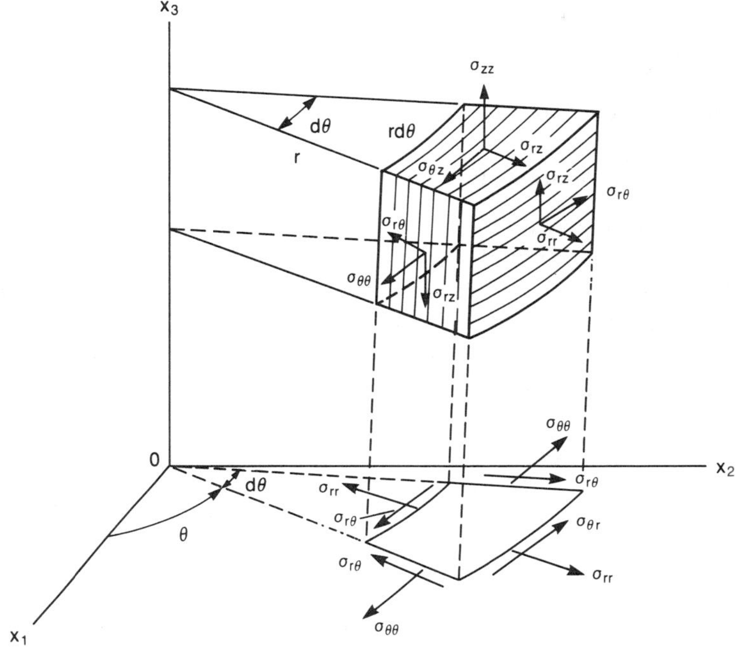

Fig. 3.6. Schematic representation of normal and shear stresses on the sides of a cubic element in cylindrical coordinates.

strains and $\varepsilon_{\theta z}$, ε_{rz}, and $\varepsilon_{r\theta}$ are shear strains. Strain-displacement relationships are given as

$$\varepsilon_{rr} = \frac{\partial u_r}{\partial r},$$

$$\varepsilon_{\theta\theta} = \frac{1}{r}\frac{\partial u_\theta}{\partial \theta} + \frac{u_r}{r},$$

$$\varepsilon_{zz} = \frac{\partial u_z}{\partial z},$$

$$\varepsilon_{\theta z} = \frac{1}{2}\left(\frac{\partial u_\theta}{\partial z} + \frac{1}{r}\frac{\partial u_z}{\partial \theta}\right),$$

$$\varepsilon_{rz} = \frac{1}{2}\left(\frac{\partial u_z}{\partial r} + \frac{\partial u_r}{\partial z}\right),$$

$$\varepsilon_{r\theta} = \frac{1}{2}\left(\frac{1}{r}\frac{\partial u_r}{\partial \theta} + \frac{\partial u_\theta}{\partial r} - \frac{u_\theta}{r}\right).$$

$$(3.18)$$

The number of independent elastic constants C_{ij} for the most general case are 21. When each point of a body has a plane of elastic symmetry normal to the axis of anisotropy x_3 (axisymmetric), then

$$C_{14} = C_{15} = C_{24} = C_{25} = C_{34} = C_{35} = C_{46} = C_{56} = 0. \tag{3.19}$$

The number of independent elastic constants thus decreases to 13.

3.1.1.2. Material Constants for Orthotropic Material

Orthotropic symmetry is characteristic of materials having elastic symmetry with respect to three mutually orthogonal planes. Stretched films may exhibit orthotropic symmetry. For this symmetry, nine independent material constants are required to completely specify the mechanical response characteristics in a Cartesian or curvilinear coordinate system. The matrix of elastic constants (or the compliance) for the orthotropic material takes the following form:

$$\begin{bmatrix} C_{11} & C_{12} & C_{31} & 0 & 0 & 0 \\ C_{12} & C_{22} & C_{23} & 0 & 0 & 0 \\ C_{31} & C_{23} & C_{33} & 0 & 0 & 0 \\ 0 & 0 & 0 & C_{44} & 0 & 0 \\ 0 & 0 & 0 & 0 & C_{55} & 0 \\ 0 & 0 & 0 & 0 & 0 & C_{66} \end{bmatrix}. \tag{3.20}$$

The relationship between the fundamental descriptors S_{ij} (or C_{ij}) and the experimental engineering descriptors (E, v, and G) for an orthotropic material is obtained as follows. Suppose that a pure tensile load is applied in the 1-direction of an orthotropic material and the resulting deformations in the 1-, 2-, and 3-directions determined. For this situation $\sigma_{11} = f_1$, $\sigma_{22} = \sigma_{33} = \sigma_{23} = \sigma_{31} = \sigma_{12} = 0$. The following strain response will be observed for orthotropic material: $\varepsilon_{11} = e_1, \varepsilon_{22} = e_2, \varepsilon_{33} = e_3, 2\varepsilon_{23} = 2\varepsilon_{31} = 2\varepsilon_{12} = 0$. Accordingly, $S_{11} = e_1/f_1$, $S_{12} = e_2/f_1$, and $S_{13} = e_3/f_1$. But $e_1/f_1 = 1/E_1$, Young's modulus of the material in the 1-direction; also $e_2/e_1 = v_{12}$ and $e_3/e_1 = v_{13}$, the Poisson ratios. Accordingly, $S_{11} = 1/E_1$, $S_{12} = -v_{12}/E_1$, and $S_{13} = -v_{13}/E_1$. Repetition of the tensile loading experiment in the 2- and then in the 3-direction gives the following results:

$$S_{11} = E_1^{-1}, \qquad S_{21} = -v_{21}E_2^{-1}, \qquad S_{31} = -v_{31}E_3^{-1},$$

$$S_{12} = -v_{12}E_1^{-1}, \qquad S_{22} = E_2^{-1}, \qquad S_{32} = -v_{32}E_3^{-1}, \tag{3.21}$$

$$S_{13} = -v_{13}E_1^{-1}, \qquad S_{23} = -v_{23}E_2^{-1}, \qquad S_{33} = E_3^{-1}.$$

However, since $S_{12} = S_{21}$, $S_{13} = S_{31}$, and $S_{23} = S_{32}$, the nine measured engineering constants cannot all be independent. Thus, $v_{21} = v_{12}(E_2/E_1)$, $v_{31} = v_{13}(E_3/E_1)$, and $v_{32} = v_{23}(E_3/E_2)$. Consequently, three more experiments must be performed to obtain the nine independent engineering constants. This data may be extracted from three shear experiments, viz., $S_{44} = 1/G_{23}$, $S_{55} = 1/G_{13}$, and $S_{66} = 1/G_{12}$. The elastic constants may be related to the engineering constants through the relationship $\mathbf{C} = \mathbf{S}^{-1}$. The results for this transformation are summarized in Table 3.1.

Table 3.1. Relationships between elastic constants and engineering constants

Orthotropic materials

$$S_{11} = E_1^{-1}, \qquad S_{12} = -v_{12}E_1^{-1}, \qquad S_{13} = -v_{13}E_1^{-1}, \qquad S_{44} = G_{23}^{-1},$$

$$S_{21} = S_{12}, \qquad S_{22} = E_2^{-1}, \qquad S_{23} = -v_{23}E_2^{-1}, \qquad S_{55} = G_{13}^{-1},$$

$$S_{31} = S_{13}, \qquad S_{32} = S_{23}, \qquad S_{33} = E_3^{-1}, \qquad S_{66} = G_{12}^{-1},$$

$$v_{ij} = v_{ji}(E_i/E_j)$$

$$C_{11} = E_1[1 - (E_3/E_2)v_{23}^2]A, \qquad C_{22} = E_2[1 - (E_3/E_1)v_{13}^2]A, \qquad C_{44} = G_{23},$$

$$C_{12} = C_{21} = [E_2 v_{12} + E_3 v_{13} v_{23}]A, \qquad C_{23} = C_{32} = (E_3/E_1)[E_1 v_{23} + E_2 v_{12} v_{13}]A, \qquad C_{55} = G_{13},$$

$$C_{13} = C_{31} = E_3[v_{12}v_{23} + v_{13}]A, \qquad C_{33} = E_3[1 - (E_2/E_1)v_{12}^2]A, \qquad C_{66} = G_{12},$$

$$A^{-1} = 1 - 2(E_3/E_1)v_{12}v_{23}v_{13} - v_{13}^2(E_3/E_1) - v_{23}^2(E_3/E_2) - v_{12}^2(E_2/E_1).$$

Isotropic materials

$$C_{11} = C_{22} = C_{33},$$

$$C_{12} = C_{13} = C_{23},$$

$$C_{44} = C_{55} = C_{66} = \tfrac{1}{2}(C_{11} - C_{12}).$$

$$E = E_1 = E_2 = E_3,$$

$$v = v_{12} = v_{21} = v_{13} = v_{31} = v_{23} = v_{32},$$

$$G = G_{12} = G_{13} = G_{23} = E/[2(1 + v)].$$

Further simplifications occur if the loading is *plane stress* or *plane strain* (Timoshenko and Goodier, 1970). This situation may be encountered, e.g., by a thin film subjected to the simultaneous action of loads (or deformation) confined to the film plane. For these special cases, only *four* independent material constants are required to specify the in-plane material response.

If a thin plate is loaded by forces applied at the boundary, parallel to the plane of the plate, and distributed uniformly over the thickness, the stress components σ_3 $(=\sigma_{33})$, σ_4 $(=\sigma_{23})$, σ_5 $(=\sigma_{31})$ are zero on both faces of the plate, and it may be assumed that they are zero within the plate. The state of stress is then specified by σ_1 $(=\sigma_{11})$, σ_2 $(=\sigma_{22})$, σ_6 $(=\sigma_{12})$ only, and is called plane stress. It is assumed that these components are independent of x_3, i.e., they don't vary through the thickness. The zero material constants in Eq. (3.20) are

$$C_{31} = C_{23} = C_{33} = C_{44} = C_{55} = 0. \tag{3.22}$$

A similar simplification is possible at the other extreme when the dimension of the body in the x_3 direction is very large. If a long cylindrical or prismatical body is loaded by forces that are perpendicular to the longitudinal elements and do not vary along the length, it may be assumed that all cross sections are in the same condition. It is simplest to suppose that the end conditions are confined between fixed smooth rigid planes, so that displacement in the axial direction is prevented. Since there is no axial displacement at the ends

Table 3.2. Elastic and engineering constants of orthotropic material

Plane stress	Plane strain
$\sigma_3 = 0,\ \sigma_4 = 0,\ \sigma_5 = 0$	$\varepsilon_3 = 0,\ \varepsilon_4 = 0,\ \varepsilon_5 = 0$
$\sigma_1 = C_{11}\varepsilon_1 + C_{12}\varepsilon_2,$	$\varepsilon_1 = S_{11}\sigma_1 + S_{12}\sigma_2,$
$\sigma_2 = C_{12}\varepsilon_1 + C_{22}\varepsilon_2,$	$\varepsilon_2 = S_{12}\sigma_1 + S_{22}\sigma_2,$
$\sigma_6 = C_{66}\varepsilon_6,$	$\varepsilon_6 = S_{66}\sigma_6,$
$C_{11} = E_1/(1 - v_{12}v_{21}),$	$S_{11} = (1 - v_{13}v_{31})/E_1,$
$C_{12} = E_1[v_{21}/(1 - v_{12}v_{21})],$	$S_{12} = -(v_{12} + v_{23}v_{32})/E_2,$
$C_{22} = E_2/(1 - v_{12}v_{21}),$	$S_{22} = (1 - v_{23}v_{32})/E_2,$
$C_{66} = G_{12},$	$S_{66} = G_{12}^{-1},$
$v_{ij} = v_{ji}(E_i/E_j)$	

and, by symmetry, at the mid-section, it may be assumed that the same holds at every cross section. Since the axial displacement u_3 is zero, $\varepsilon_4\ (=2\varepsilon_{23})$, $\varepsilon_5\ (=2\varepsilon_{31})$, and $\varepsilon_3\ (=\varepsilon_{33})$ are equal to zero which implies that $\sigma_4\ (=\sigma_{23})$ and $\sigma_5\ (=\sigma_{31})$ are equal to zero and $\sigma_3\ (=\sigma_{33})$ is related to $\sigma_1\ (=\sigma_{11})$ and $\sigma_2\ (=\sigma_{22})$ such that $\varepsilon_3 = 0$. Thus the plane strain problem, like the plane stress problem, reduces to the determination of σ_{11}, σ_{22}, and σ_{12} as functions of x_1 and x_2 only.

The relationships between engineering constants and elastic constants and the associated stress–strain relations for an orthotropic material subjected to the plane stress or plane strain are summarized in Table 3.2.

We now consider an orthotropic material subjected to plane stress or plane strain with loads (or deformations) imposed at an angle ϕ with respect to the material direction, such that the rotation transformation between the material axes and the arbitrary load (or deformation) axes is a simple rotation around the common x_3 axis. Since the C_{ij} and S_{ij} arrays undergo the same transformations, the analysis can be generalized in terms of apparent material descriptors $\bar{A}_{ij}(\phi)$ and base descriptors A_{ij}. The expressions for A_{ij} are (Lekhnitskii, 1981)

$$
\begin{aligned}
\bar{A}_{11} &= B_1 + B_7 \cos 2\phi + B_8 \cos 4\phi, \\
\bar{A}_{12} &= \bar{A}_{21} = B_2 - B_8 \cos 4\phi, \\
\bar{A}_{16} &= \bar{A}_{61} = \tfrac{1}{2}(B_7 \sin 2\phi) + B_8 \sin 4\phi, \\
\bar{A}_{22} &= B_1 - B_7 \cos 2\phi + B_8 \cos 4\phi, \\
\bar{A}_{26} &= \bar{A}_{62} = \tfrac{1}{2}(B_7 \sin 2\phi) - B_8 \sin 4\phi, \\
\bar{A}_{66} &= B_6 - B_8 \cos 4\phi,
\end{aligned}
\tag{3.23}
$$

where

$$B_1 = \tfrac{1}{8}(3A_{11} + 3A_{22} + 2A_{12} + 4A_{66}),$$

$$B_7 = \tfrac{1}{2}(A_{11} - A_{22}),$$

$$B_8 = \tfrac{1}{8}(A_{11} + A_{22} - 2A_{12} - 4A_{66}),$$

$$B_2 = \tfrac{1}{8}(A_{11} + A_{22} + 6A_{12} - 4A_{66}),$$

$$B_6 = \tfrac{1}{8}(A_{11} + A_{22} - 2A_{12} + 4A_{66}).$$

Replacement of the A_{ij} by elastic constants C_{ij} yields expressions for the transformed elastic constants $(\bar{A}_{ij} = \bar{C}_{ij})$: replacement of the A_{ij} by the compliance constants S_{ij} yields expressions for the transformed compliance constants $(\bar{A}_{ij} = \bar{S}_{ij}/m_i m_j)$; $m_k = 1$ for $k = 1, 2,$ or 3; and $m_k = 2$ for $k = 4, 5,$ or 6. It should be noted that while zeros appear in the C_{ij} [or S_{ij}] array, the $\bar{C}_{ij}(\phi)$ [or $\bar{S}_{ij}(\phi)$] array is fully populated with nonzero elements. This is to be expected, since both shear and extension are involved in off-axes $(\phi \neq 0)$ loads and deformations. As would be expected, if the material is isotropic then $B_7 = B_8 = 0$ so that the material response is independent of direction.

To illustrate the application of these transformations, we obtain an expression for Young's modulus at $45°$ (E_{45}) in terms of the material constants (C_{ij}) for an orthotropic material under plane stress conditions. We first need to rotate by $45°$, the $\mathbf{C}$ tensor about the x_3 axis. If we denote $\bar{\mathbf{C}}$ as the transformed matrix representing tensor $\mathbf{C}$ in the rotated system, the components of $\bar{C}$ will be given by [Eq. (3.23)]

$$\bar{C}_{11} = \tfrac{1}{4}[C_{11} + C_{22} + 2C_{12}],$$

$$\bar{C}_{12} = \tfrac{1}{4}[C_{11} + C_{22} + 2C_{12} - 4C_{66}],$$

$$\bar{C}_{16} = \tfrac{1}{4}[C_{11} - C_{22}],$$

$$\bar{C}_{22} = \bar{C}_{11}, \tag{3.24}$$

$$\bar{C}_{26} = \bar{C}_{31},$$

$$\bar{C}_{66} = \tfrac{1}{4}[C_{11} + C_{22} - 2C_{12}],$$

$$\bar{C}_{ij} = \bar{C}_{ji}.$$

The rotated tensor $\bar{\mathbf{C}}$ is now a fully populated matrix and all its components are nonzero.

The modulus $E_{45}(t)$ is then obtained from the following expression:

$$E_{45} = [(\bar{C}_{11}\bar{C}_{22}\bar{C}_{66} - \bar{C}_{11}\bar{C}_{26}^2 - \bar{C}_{12}^2\bar{C}_{66} - \bar{C}_{16}^2\bar{C}_{22} + 2\bar{C}_{12}\bar{C}_{16}\bar{C}_{66})/$$

$$(\bar{C}_{22}\bar{C}_{66} - \bar{C}_{26}^2)]. \tag{3.25}$$

We now consider another example in which we obtain an expression for Young's modulus as a function of the angle $E(\phi)$ in terms of the familiar engineering constants E_1, E_2, v_{12}, and G_{12}. We first get an expression for

$\bar{S}_{11}(\phi)$ from Eq. (3.23) then get the relationship for $E(\phi)$

$$E(\phi)^{-1} = E_1^{-1}[3 + 4\cos(2\phi) + \cos(4\phi)]/8$$
$$+ E_2^{-1}[3 - 4\cos(2\phi) + \cos(4\phi)]/8 + \alpha[1 - \cos(4\phi)]/4, \qquad (3.26)$$

with

$$\alpha = (\tfrac{1}{2}G_{12}^{-1} - v_{12}E_1^{-1}).$$

Clearly, the expression for $E(\phi)$ contains only three terms which are dependent upon the nature of the material: E_1, E_2, and α. Consequently, the evaluation of $E(\phi)$ for three distinct directions is sufficient to characterize the directional dependence of Young's modulus. Hence

$$E_1 = E(0°), \qquad (3.27a)$$

$$E_2 = E(90°), \qquad (3.27b)$$

$$\alpha = 2E(45°)^{-1} - [E(0°)^{-1} + E(90°)^{-1}]/2. \qquad (3.27c)$$

3.1.1.3. Material Constants for Isotropic Material

When the elastic properties of a body are identical in all directions, i.e., if the body is elastically isotropic, the number of essential material components reduce to 2. The relationships between engineering constants and material constants are summarized in Table 3.1. Hooke's Law [Eqs. (3.7) or (3.15)] for an isotropic material reduces to

$$\sigma_{ij} = 2\mu\varepsilon_{ij} + \lambda\delta_{ij}\varepsilon_{kk}, \qquad i,j = 1, 2, 3, \qquad (3.28)$$

or

$$\varepsilon_{ij} = \frac{1}{2\mu}\sigma_{ij} - \frac{\lambda\delta_{ij}}{2\mu(3\lambda + 2\mu)}\sigma_{kk}, \qquad (3.29)$$

where λ and μ are called the Lame constants and δ_{ij} is referred to as the Kronecker delta. The symbol means zero when $i \neq j$ and it means unity when $i = j = 1, 2,$ or 3. λ and μ are related to E, v, and G (only two of the three are independent constants) by

$$\lambda = \frac{Ev}{(1 + v)(1 - 2v)} \qquad (3.30a)$$

and

$$\mu = G = \frac{E}{2(1 + v)}. \qquad (3.30b)$$

$\sigma_{kk}/3$ is the mean of the three stress components or the hydrostatic stress. The stress σ_{ij} can be regarded as a superposition of the two stress states $\tfrac{1}{3}\delta_{ij}\sigma_{kk}$ (hydrostatic stress) and the deviatoric stress or stress deviator represented by s_{ij} (total stress−hydrostatic stress) where

$$s_{ij} = \sigma_{ij} - \tfrac{1}{3}\delta_{ij}\theta, \; s_{kk=0}, \qquad (3.31a)$$

with

$$\theta \equiv \sigma_{kk}. \qquad (3.31b)$$

Similarly, we can separate the strain ε_{ij} into a mean strain $\varepsilon_{kk}/3$ (one-third of dilatation, Δ_v) and a deviatoric strain e_{ij} where

$$e_{ij} = \varepsilon_{ij} - \tfrac{1}{3}\delta_{ij}\Delta_v, \; e_{kk=0}, \tag{3.32a}$$

with

$$\Delta_v \equiv \varepsilon_{kk}. \tag{3.32b}$$

The six equations expressing Hooke's Law [Eqs. (3.7) or (3.15)] are equivalent to a simple expression

$$s_{ij} = 2Ge_{ij} \tag{3.33a}$$

with

$$\theta = (3\lambda + 2G)\Delta_v$$

$$= 3K\Delta_v, \tag{3.33b}$$

where K is the modulus of volume expansion or bulk modulus, to be discussed later in more detail.

Simple Loading Conditions

In order to gain some insight into the physical significance of the elastic constants entering Eqs. (3.28) and (3.29), we consider the behavior of isotropic elastic bodies subjected to pure shear, simple tension, and hydrostatic pressure.

Consider an elemental rectangular parallelepiped with the sides parallel to the coordinate axes and submitted to the action of pure shear characterized by σ_{12}

$$\sigma_{12} = \text{constant} \tag{3.34a}$$

and

$$\sigma_{11} = \sigma_{22} = \sigma_{33} = \sigma_{31} = \sigma_{23} = 0. \tag{3.34b}$$

From the substitution from Eq. (3.34) into Eq. (3.29) we get

$$2\varepsilon_{12} = \frac{\sigma_{12}}{G} = \sigma_{12}J \tag{3.35a}$$

and

$$\varepsilon_{11} = \varepsilon_{22} = \varepsilon_{33} = \varepsilon_{31} = \varepsilon_{23} = 0. \tag{3.35b}$$

Equation (3.35a) shows that a parallelepiped, whose faces are parallel to the coordinate planes, is sheared in the x_1–x_2 plane so that the right angle between the edges of the parallelepiped parallel to the x_1 and x_2 axes is diminished by the angle $\alpha_{12} = 2\varepsilon_{12} = \gamma_{12}$. Therefore

$$G = \frac{\sigma_{12}}{\alpha_{12}}. \tag{3.35c}$$

Thus, the term G represents the ratio of the shear stress to the change in angle α_{12} produced by the shear stress. For this reason, G is called the modulus of elasticity in shear and J is called shear compliance. We note that in pure shear, the change in shape is not accompanied by any change in volume $(e_{kk=0})$.

Next, consider the state of simple tension in which normal stress σ_{11} is uniformly distributed over two opposite sides such that

$$\sigma_{11} = \text{constant} \tag{3.36a}$$

and

$$\sigma_{22} = \sigma_{33} = \sigma_{12} = \sigma_{31} = \sigma_{23} = 0. \tag{3.36b}$$

From the substitution from Eq. (3.36) in Eq. (3.29) we get

$$\varepsilon_{11} = \frac{\sigma_{11}}{E} = \sigma_{11}D, \tag{3.37a}$$

$$\varepsilon_{22} = \varepsilon_{33} = -v\varepsilon_{11} \equiv -v\frac{\sigma_{11}}{E}, \tag{3.37b}$$

and

$$\varepsilon_{12} = \varepsilon_{31} = \varepsilon_{23} = 0, \tag{3.37c}$$

where

$$E = \frac{G(3\lambda + 2G)}{\lambda + G}$$

$$= 2G(1 + v)$$

$$= \frac{9GK}{G + 3K}, \tag{3.38}$$

K will be described later.

Thus, E represents the ratio of the tensile stress to the strain produced in the direction of stress applied, and is called Young's modulus of elasticity in tension. D is called tensile compliance and the term v denotes the ratio of the contraction of the linear elements perpendicular to the axis of the element to the longitudinal extension and is called the Poisson ratio. We note that if $\sigma_{11} > 0$, then a tensile stress will produce an extension in the direction of the axis of the element and a contraction in its cross section. We note that there is a volume change (increase) in elastic materials if $v < 0.5$ (true for all physical substances). For most polymeric materials $v \sim 0.5$ and there is negligible volume change similar to shear experiments.

Finally, consider a body of arbitrary shape to a hydrostatic pressure of uniform intensity p distributed over its surface (bulk compression or dilatation). The system of stresses are

$$\sigma_{11} = \sigma_{22} = \sigma_{33} = -p, \tag{3.39a}$$

or

$$\sigma_{11} + \sigma_{22} + \sigma_{33} = \theta = -3p,$$

and

$$\sigma_{12} = \sigma_{31} = \sigma_{23} = 0. \tag{3.39b}$$

From Eqs. (3.29) and (3.39) we get

$$\varepsilon_{11} = \varepsilon_{22} = \varepsilon_{33} = \frac{-p}{3\lambda + 2G} = \frac{-p(1 - 2v)}{E} \qquad (3.40a)$$

and

$$\varepsilon_{12} = \varepsilon_{31} = \varepsilon_{23} = 0. \qquad (3.40b)$$

The volume expansion, the so-called dilatation Δ_v, is related to hydrostatic pressure as

$$\Delta_v = \varepsilon_{11} + \varepsilon_{22} + \varepsilon_{33} = -p/K$$

$$= -pB, \qquad (3.41)$$

where

$$K = \lambda + \tfrac{2}{3}G = \frac{E}{3(1 - 2v)}. \qquad (3.42)$$

Thus, K represents the ratio of compressive stress to the cubical compression and is called the modulus of volume expansion or bulk modulus. B is called bulk compliance.

Since K is positive for all physical substances, it follows that

$$v < 0.5. \qquad (3.43)$$

For most metals $v \sim \tfrac{1}{3}$ and for amorphous polymers $v \sim \tfrac{1}{2}$.

3.1.2. Viscous Liquids

The classical theory of hydrodynamics deals with properties of linear viscous liquids, for which, in accordance with Newton's Law, the stress is always directly proportional to rate of strain but independent of the strain itself (Flugge, 1975). We consider a state of simple tension where tensile stress is applied in the x_1 direction

$$\sigma_{11} = F\dot{\varepsilon}_{11}. \qquad (3.44)$$

Here and elsewhere we shall employ the dot to designate the ordinary and partial derivative with respect to time. The quantity $\dot{\varepsilon}_{11}$ $(=d\varepsilon_{11}/dt)$ is called the strain rate. For a state of pure shear with nonzero stress and strain components σ_{12} and $\dot{\varepsilon}_{12}$

$$\sigma_{12} = 2\eta\dot{\varepsilon}_{12}, \qquad (3.45)$$

where η is the absolute viscosity (Pa.s) of the liquid. The stress–strain behavior of a linear viscous liquid is shown in Fig. 3.2(b).

3.1.3. Viscoelasticity

For an ideally elastic solid in accordance with Hooke's Law, stress is always directly proportional to strain in small deformations but independent of the rate of strain. For an ideally viscous liquid in accordance with Newton's Law, the stress is always directly proportional to the rate of strain but independent of the strain itself. These categories are idealizations. The behavior of many solids approaches Hooke's Law for infinitesimal strains, and that of many liquids approaches Newton's Law for infinitesimal rates of strain, deviations in behavior are observed for other conditions. Even if both strain and rate of strain are infinitesimal, a material may exhibit behavior which combines liquidlike and solidlike characteristics. Materials whose behavior exhibits such characteristics are called viscoelastic. The viscoelastic material has a memory, i.e., the material response is not only determined by the current state of stress, but is also determined by the past states of stress. A similar situation exists if one considers the deformation being specified, and thus, the current stress depends upon the entire past history of deformation. If the ratio of stress to strain is a function of time (or frequency) alone, and not of stress magnitude, we have linear viscoelastic behavior (Bland, 1960; Tobolsky, 1960; Gurtin and Sternberg, 1962; McCrum et al., 1967; Christensen, 1971; Flugge, 1975; Norwick and Berry, 1972; Sternstein, 1977; Ferry, 1980; Plazek, 1980; Ward, 1983; Tschoegl, 1989).

Anelasticity and *viscoplasticity* denote two special types of viscoelasticity. Anelastic material will eventually stop deforming and assume a final equilibrium strain proportional to the applied stress. Total eventual recoverability on unloading is an implied reciprocal aspect of anelasticity. A viscoplastic material, on the other hand, will continue to deform without limit under a constant stress, and on unloading exhibits no recovery of the creep strain. Viscoplasticity is a characteristic of an internal flow process; anelasticity is a characteristic of an internal relaxation process. In the remaining section, we present constitutive stress–strain relationships for anelasticity, however, we will refer to it by a more general term of viscoelasticity.

3.1.3.1. Constitutive Equation

The constitutive equation for an isothermal, linear viscoelastic material is based on the Boltzmann linear superposition principle (Boltzmann, 1874), that the effects of mechanical history (sequential changes in strain or stress) are linearly additive. We will assume constant temperature and humidity conditions, so that the only independent variable will be time. For a strain history shown in Fig. 3.7, the total stress at any time is given by

$$\sigma(t) = \sum_{i=0}^{n} C(t - t_i)\Delta\varepsilon_i\delta(t - t_i), \tag{3.46}$$

where $\Delta\varepsilon_i$ is the strain increment imposed at time t_i, $C(t)$ is the stress relaxation modulus, and $\delta(t)$ is a step function such that $\delta(t) = 0$ if $t < 0$ and $\delta(t) = 1$ if $t > 0$. If the strain history is a continuous function of time, the stress is described as a function of the rate of strain history or, alternatively, the strain is described as a function of the history of the rate of change of stress. The generalized constitutive equation for an anisotropic linear viscoelastic

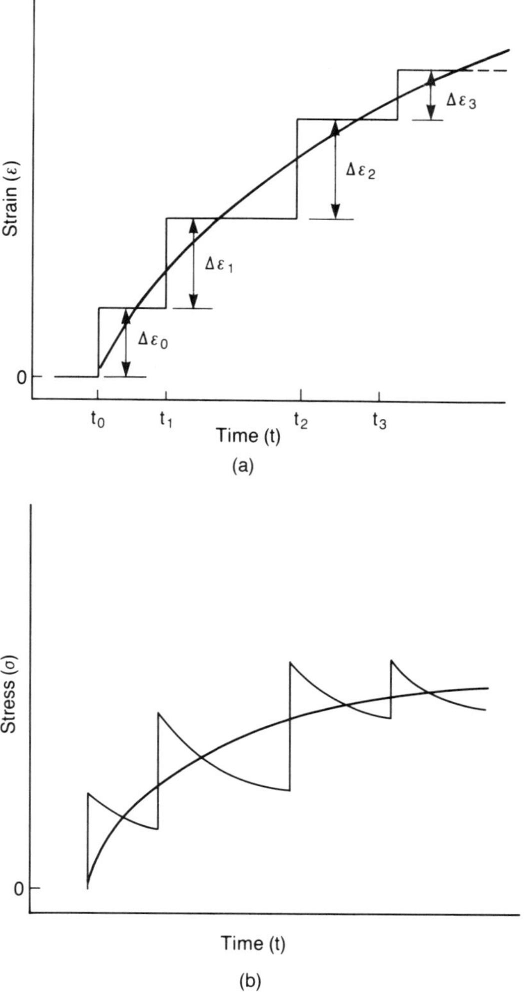

Fig. 3.7. (a) Incremental strain history as a superposition of steps, and (b) corresponding stress history.

material is given as (Christensen, 1971)

$$\boldsymbol{\sigma}(t) = \int_0^t \mathbf{C}(t - \tau)\dot{\boldsymbol{\varepsilon}}(\tau)\,d\tau \tag{3.47}$$

or

$$\sigma_{ij}(t) = \int_0^t C_{ijkl}(t - \tau)\dot{\varepsilon}_{kl}(\tau)\,d\tau,$$

where functions $C_{ijkl}(t)$ are termed relaxation functions. Although $\varepsilon_{ij}(t) = 0$

for $t < 0$, the lower limit in Eq. (3.47) may be changed from 0 to $-\infty$ through a shift of the time scale, as long as $\varepsilon_{ij}(t) \to 0$ as $t \to -\infty$.

An alternative form of the stress–strain relation may be obtained by reversing the roles of stress and strain in Eq. (3.47), in such a way that the current strain is determined by the current value and past history of stress

$$\varepsilon(t) = \int_{-\infty}^{t} \mathbf{S}(t - \tau)\dot{\sigma}(\tau)\, d\tau, \tag{3.48}$$

where functions $S_{ijkl}(t)$ are termed creep functions and $S_{ijkl}(t) = 0$ for $-\infty < t < 0$. Obviously, Eqs. (3.47) and (3.48) are not independent relations. It follows that the creep and relaxation functions must be related.

For the most general loading condition, there are six stress and six strain components. The simplifications of $C_{ijkl}(t)$ and $S_{ijkl}(t)$ in Eqs. (3.46) and (3.47) are made for various loading situations, similar to that of C_{ijkl} and S_{ijkl} in the constitutive equations for elasticity, Eqs. (3.7) and (3.8). For the most general case of an anisotropic viscoelastic body, the number of independent viscoelastic constants is 21. For an orthotropic material the number of constants is reduced to nine, and for an isotropic material the number of constants is reduced to two. The constitutive relations of viscoelastic materials for special cases can be developed parallel to the elasticity treatment presented earlier.

Following a procedure parallel to that in elasticity, whereby the deviatoric components of stress s_{ij} and strain e_{ij} are introduced, Eq. (3.47) reduces to the constitutive equation for an *isotropic material* (Christensen, 1971)

$$s_{ij} = \int_{-\infty}^{t} 2G(t - \tau)\dot{e}_{ij}(\tau)\, d\tau \tag{3.49a}$$

and

$$\sigma_{kk}/3 = \int_{-\infty}^{t} K(t - \tau)\dot{\varepsilon}_{kk}(\tau)\, d\tau, \tag{3.49b}$$

where $G(t)$ and $K(t)$ are independent isotropic relaxation functions. In a similar way, from Eq. (3.48) we get

$$2e_{ij} = \int_{-\infty}^{t} J(t - \tau)\dot{s}_{ij}(\tau)\, d\tau \tag{3.50a}$$

and

$$\varepsilon_{kk} = \int_{-\infty}^{t} \tfrac{1}{3} B(t - \tau)\dot{\sigma}_{kk}(\tau)\, d\tau, \tag{3.50b}$$

where $J(t)$ and $B(t)$ are the two isotropic creep functions. We note that $G(t)$ and $J(t)$ are the relaxation and creep functions appropriate to states of shear, while $K(t)$ and $B(t)$ are defined relative to states of dilatation.

Obviously, Eqs. (3.49a) and (3.49b) are not relations which are independent of Eqs. (3.50a) and (3.50b). If one function is given analytically, the other can be calculated. To display the relationship between these functions, it is expedient to introduce the Laplace transformation. Let $f(t)$ be a continuous function on $0 \le t \le \infty$ and let it be of exponential order as $t \to \infty$. Then the Laplace transform $\bar{f}(s)$ of $f(t)$ is defined as

$$\bar{f}(s) = \int_0^\infty f(t)e^{-st}\, dt. \tag{3.51}$$

The Laplace transform of Eqs. (3.49) and (3.50) gives

$$\bar{s}_{ij}(s) = 2s\bar{G}(s)\bar{e}_{ij}(s), \tag{3.52a}$$

$$\bar{\sigma}_{kk}(s) = 3s\bar{K}(s)\bar{\varepsilon}_{kk}(s), \tag{3.52b}$$

$$\bar{e}_{ij}(s) = s\bar{J}(s)\bar{s}_{ij}(s)/2, \tag{3.53a}$$

$$\bar{\varepsilon}_{kk}(s) = s\bar{B}(s)\bar{\sigma}_{kk}(s)/3. \tag{3.53b}$$

It follows from Eqs. (3.52) and (3.53) that

$$\bar{G}\bar{J} = 1/s^2 \tag{3.54}$$

and

$$\bar{K}\bar{B} = 1/s^2, \tag{3.55}$$

that is,

$$\int_0^t G(t-\tau)J(\tau)\, d\tau = \int_0^t G(\tau)J(t-\tau)\, d\tau = t \tag{3.56}$$

and

$$\int_0^t K(t-\tau)B(\tau)\, d\tau = \int_0^t K(\tau)B(t-\tau)\, d\tau = t. \tag{3.57}$$

From which it follows that $G(t)J(t) \le 1$ and $K(t)B(t) \le 1$ (Tschoegl, 1989). Only for an elastic solid, where both G and J or K and B are not time-dependent, does one obtain $GJ = KB = 1$.

3.1.3.2. Description of Time-Dependent Deformation Experiments

The constitutive equations of the preceding section can be used to describe the response of an *isotropic linear viscoelastic material* to various kinds of time-dependent patterns of stress and strain. Here, we describe experiments in pure shear, simple tension, and hydrostatic pressure. Tensile experiments are often easy to perform compared to the shear experiments, but have the disadvantage that simultaneous changes in both shape and volume make the behavior more difficult to interpret on a molecular basis; moreover, perceptible volume changes significantly modify the relaxation and creep functions. However, in polymeric systems, v is very close to $\frac{1}{2}[K(t) \gg G(t)]$ and the volume change is negligible in comparison with the change in shape. (These types of polymeric materials are called "incompressible" or "soft" elastic solids.) The simple extension gives the same information as pure shear, and the results of the two experiments are interconvertible.

Dynamic Modulus and Dynamic Compliance

In a periodic or dynamic experiment used to provide information corresponding to very short times, if the stress is varied periodically at a frequency ω ($= 2\pi f$ where f is in cps or Hz)[2], it is found that the strain will also alternate sinusoidally but will be out of phase with stress (Fig. 3.8).

In a pure shear experiment with nonzero stress and strain components σ_{12} and ε_{12}, let sinusoidally varying shear strain as

$$\varepsilon_{12}(\omega) = \varepsilon_{12}^0 \sin \omega t, \tag{3.58}$$

where ε_{12}^0 is the maximum amplitude of the strain. Substituting in Eq. (3.49), we have

$$\sigma_{12}(\omega) = 2\varepsilon_{12}^0 [G'(\omega) \sin \omega t + G''(\omega) \cos \omega t], \tag{3.59}$$

thereby defining two frequency-dependent functions—the shear storage modulus $G'(\omega)$ and the shear loss modulus $G''(\omega)$.

It is instructive to write the stress in an alternative form displaying the amplitude $\sigma_{12}^0(\omega)$ of the stress and the phase angle $\delta(\omega)$ between stress and strain. From the trigonometric relations

$$\sigma_{12}(\omega) = \sigma_{12}^0 \sin(\omega t + \delta)$$
$$= \sigma_{12}^0 \cos \delta \sin \omega t + \sigma_{12}^0 \sin \delta \cos \omega t. \tag{3.60}$$

Comparison of Eqs. (3.59) and (3.60) shows that

$$G'(\omega) = (\sigma_{12}^0/2\varepsilon_{12}^0) \cos \delta, \tag{3.61a}$$

$$G''(\omega) = (\sigma_{12}^0/2\varepsilon_{12}^0) \sin \delta, \tag{3.61b}$$

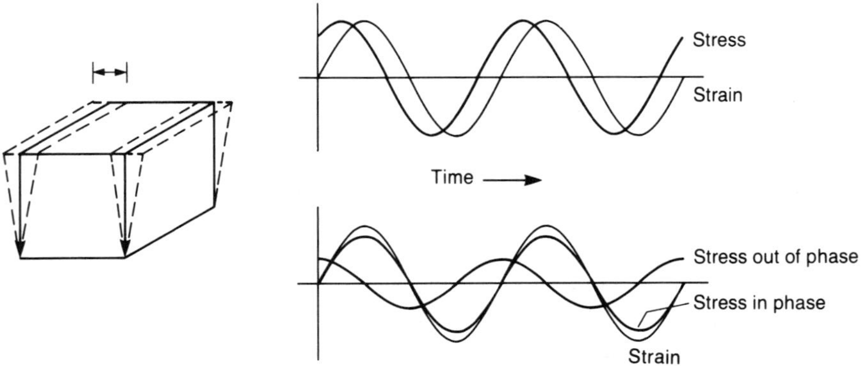

Fig. 3.8. Geometry and time profile of a shear experiment with sinusoidally varying shear.

---

[2] A periodic experiment at frequency ω is qualitatively equivalent to a transient experiment at time $t = 1/\omega$.

and

$$G''(\omega)/G'(\omega) = \tan \delta. \tag{3.61c}$$

The $\tan \delta$ is called the loss tangent or tangent delta or damping factor. It is evident that each periodic or dynamic measurement at a given frequency provides simultaneously two independent quantities, either G' and G'' or else $\sigma_{12}^0/\varepsilon_{12}^0$, the ratio of peak stress to peak strain.

It is usually convenient to express the sinusoidally varying stress as a complex quantity, then the modulus is also complex, given by

$$\sigma_{12}^*(\omega)/2\varepsilon_{12}^*(\omega) = G^*(\omega) = G'(\omega) + iG''(\omega), \tag{3.62a}$$

and

$$|G^*| = \sigma_{12}^0/2\varepsilon_{12}^0 = [|G'|^2 + |G''|^2]^{1/2}, \tag{3.62b}$$

where G^* is a shear complex modulus comprising real and imaginary components—shear storage modulus, G', and shear loss modulus, G'', respectively. However, G'' is a real quantity and is the coefficient of the imaginary term in Eq. (3.61). G^*, G', and G'' are functions of frequency (or time or strain rate) and temperature. The G' is the ratio of the stress in phase with the strain to the strain, whereas G'' is the ratio of the stress 90° out of phase with the strain to the strain. The ratio G''/G' $(=\tan \delta)$ determines the phase relationship between σ_{12} and ε_{12}. The loss angle δ ranges from 0 to $\pi/2$ and therefore $\tan \delta$ takes on only positive values between 0 and ∞. The stress always leads the strain.

The data from sinusoidal experiments can be expressed in terms of complex compliance

$$2\varepsilon_{12}^*(\omega)/\sigma_{12}^*(\omega) = J^*(\omega) = J'(\omega) - iJ''(\omega) = 1/G^*(\omega) \tag{3.63a}$$

and

$$|J^*|^2 = J'^2 + J''^2, \tag{3.63b}$$

and

$$\tan \delta = J''/J' = G''/G'. \tag{3.63c}$$

The storage compliance J' is the ratio of the strain in the phase with the stress to the stress, whereas the loss compliance J'' is the ratio of the strain 90° out of phase with the stress to the stress.

The shear storage modulus G' and compliance J' are so named because they are directly proportional to the average energy storage in a cycle of deformation. The shear loss modulus, G'', and loss compliance, J'', are directly proportional to the average viscous dissipation or loss of energy as heat in a cycle of deformation. Figure 3.9 shows the interrelationships of the quantities defined by Eqs. (3.61)–(3.63). It is clear from this figure that the complex modulus describes how the stress leads the strain with a phase angle δ, whereas the complex compliance illustrates how the strain lags the stress by the same angle.

Although $J^*(\omega) = 1/G^*(\omega)$, their individual components are not recipro-

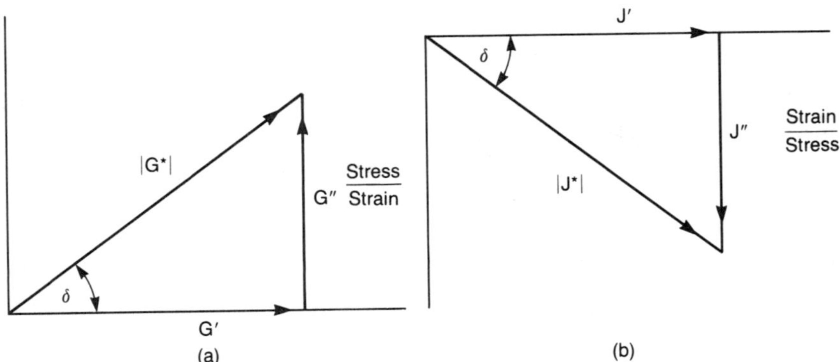

Fig. 3.9. Vectorial resolution of the component of (a) complex modulus and (b) compliance in sinusoidal shear deformation.

cally related but are connected by the following equations:

$$J'(\omega) = \frac{G'(\omega)}{G'(\omega)^2 + G''(\omega)^2} = \frac{1/G'(\omega)}{1 + \tan^2 \delta}, \tag{3.64a}$$

$$J''(\omega) = \frac{G''(\omega)}{G'(\omega)^2 + G''(\omega)^2} = \frac{1/G''(\omega)}{1 + \tan^{-2} \delta}, \tag{3.64b}$$

$$G'(\omega) = \frac{J'(\omega)}{J'(\omega)^2 + J''(\omega)^2} = \frac{1/J'(\omega)}{1 + \tan^2 \delta}, \tag{3.65a}$$

$$G''(\omega) = \frac{J''(\omega)}{J'(\omega)^2 + J''(\omega)^2} = \frac{1/J''(\omega)}{1 + \tan^{-2} \delta}. \tag{3.65b}$$

As an alternative to $G^*(\omega)$, the phase relationships can equally well be expressed by a complex viscosity

$$\eta^* = \eta' - i\eta'', \tag{3.66a}$$

which is frequently used to describe viscoelastic liquids. The ratio of stress in phase with the rate of strain to the rate of strain is η', and η'' is the stress 90° out of phase with the rate of strain divided by the rate of strain. Thus the phase relations are the opposite of those for G^*, and the individual components are given by

$$\eta' = G''/\omega, \tag{3.66b}$$

$$\eta'' = G'/\omega. \tag{3.66c}$$

The in-phase or real component η' for a viscoelastic liquid approaches the steady-flow viscosity η_0 as the frequency ω approaches zero. Equations (3.61)–(3.66) are written for simple shear. Analogous relations hold for a simple extension with the G's replaced by E's and the J's by D's, and for bulk compression or dilatation with the G's replaced by K's and the J's by B's.

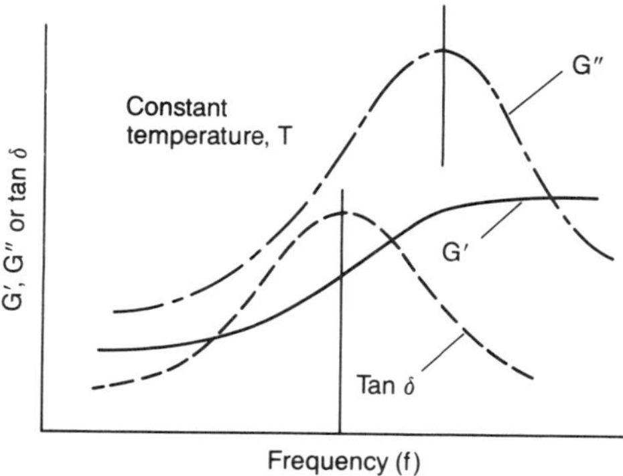

Fig. 3.10. Storage (G'), loss moduli (G''), and tangent modulus (tan δ) as functions of frequency.

Periodic measurements can be made at frequencies from 10^{-15} Hz to 10^8 Hz from cryogenic temperatures ($\sim -100°$ C) to few hundred degrees centigrade.

It is known that the loss modulus G'' and the loss tangent tan δ for real materials have pronounced maxima at certain frequencies, and that the storage modulus G' increases with increasing frequency in the manner shown in Fig. 3.10. These characteristic curves apply at a particular temperature T, and it has been shown experimentally that an increase in temperature effectively moves these curves to the right (using the time–temperature superposition, to be discussed later) so that (in the case of G', G'', and tan δ) the peaks occur at higher frequencies.

Stress Relaxation

In a stress relaxation test, the material is instantaneously loaded by a stress to some fixed deformation that is then held constant. Immediately, the material starts to relax, and the stress originally imposed continues to diminish over a period of time. The relaxation modulus at a given time is lower at higher temperatures (time–temperature superposition, to be discussed later).

In a pure shear loading, suppose a shear strain ε_{12} ($= \varepsilon_{12}^0$) is imposed within a brief period of time t_0 by a constant rate of strain $\dot{\varepsilon}_{12} = \varepsilon_{12}^0/t_0$ for a total period of $t = t_1$ (Fig. 3.11). The constitutive equation (3.47) can then be expressed as

$$\sigma_{12}(t) = \int_0^{t_0} 2G(t - t')\frac{\varepsilon_{12}^0}{t_0}\, dt', \qquad (3.67)$$

since the rate of strain is zero both before and after the interval represented by this integral (Gottenberg and Christensen, 1964; Bhushan and Dauer, 1978). By the Theorem of the Mean, Eq. (3.67) can be written as

$$\sigma_{12}(t) = 2\varepsilon_{12}^0 G(t + nt_0), \qquad 0 \le n \le 1, \qquad (3.68a)$$

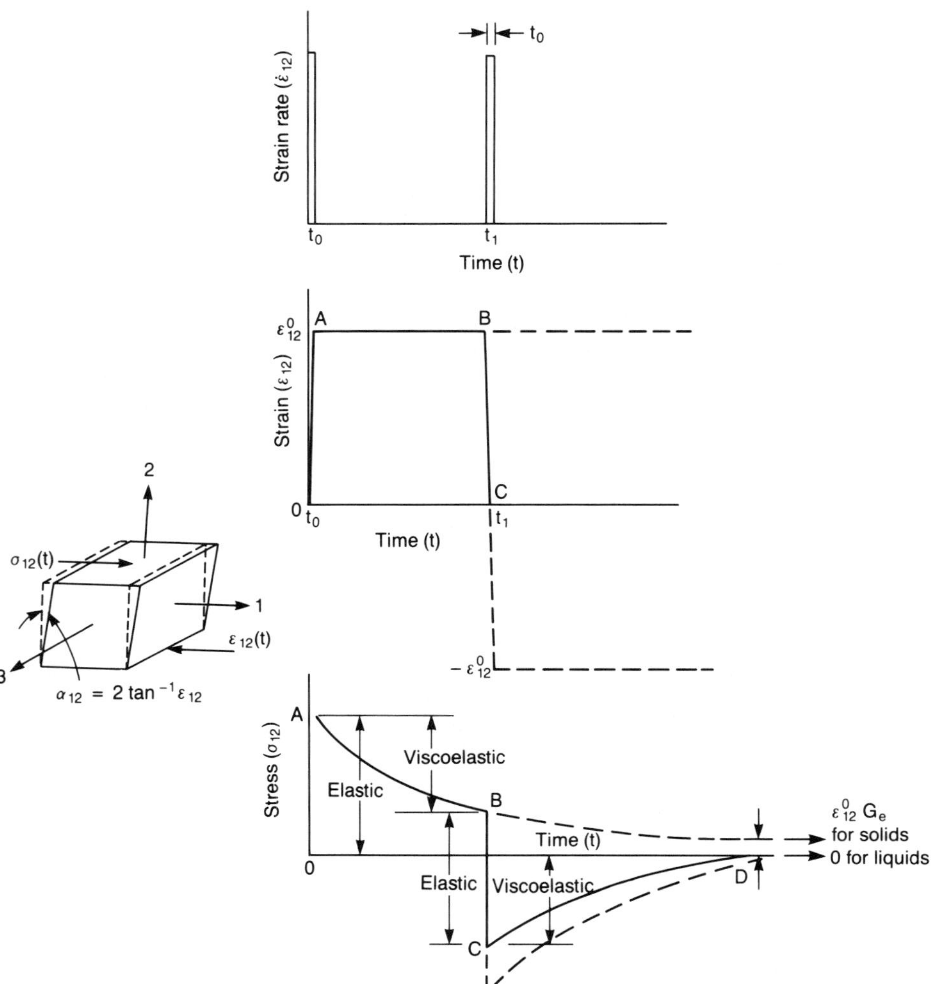

Fig. 3.11. Geometry and time profile of a shear stress relaxation experiment. Dotted lines after time $= t_1$ show the strain profiles, ε_{12}^0 applied at $t = 0$, and $-\varepsilon_{12}^0$ applied at $t = t_1$ and corresponding stress relaxation profiles whereas firm lines show the net profiles. Also shown are values of σ_{12} $(t \to \infty)$ during loading; σ_{12} $(t \to \infty) \to 0$ during unloading.

and for times long compared with t_0, the loading interval, this is indistinguishable from

$$\sigma_{12}(t) = 2\varepsilon_{12}^0 G(t), \qquad (3.68b)$$

where $G(t)$ is the relaxation modulus, and for a perfectly elastic solid the equilibrium shear modulus G is defined as $\sigma_{12}/2\varepsilon_{12}$, and $G(t)$ is its time-dependent analog as measured in an experiment with this particular time pattern. The rate of decrease in $G(t)$ is a monotonically decreasing function of time. The initial time derivative is the greatest, and the ultimate rate is zero for both viscoelastic solids and liquids. For liquids, $G(t)$ approaches zero as $t \to \infty$; for solids, $G(t)$ approaches a constant finite value (G_e).

A given continuously varying strain history, as shown in Fig. 3.7, can be considered as a succession of increments, and the resulting stresses can be added according to the Boltzmann linear superposition principle. As a general case, if there exists a sequence of finite strain changes $(\varepsilon_{12})_i$ (with $i = 1, \ldots, n$) each at a time point t_i ($-\infty \leq t_i \leq t$), then

$$\sigma_{12}(t) = \sum_{i=0}^{n} 2(\varepsilon_{12})_i G(t - t_i)\delta(t - t_i). \tag{3.69}$$

If now the load is suddenly removed after time t_1 (at the point B in Fig. 3.11), there is an immediate elastic recovery to C (which may be below zero, depending on the initial deformation). The residual stress then causes further relaxation and the body returns more or less towards its initial state. The Boltzmann linear superposition principle can be used to predict the stress history during the recovery period. The removal of stress in the unloading of a relaxation experiment is equivalent to applying an additional strain $-\varepsilon_{12}^0$ at $t = t_1$. We take the strain as being specified in terms of a unit step function $\delta(t)$ (the effect of t_0 is neglected here)

$$\varepsilon_{12}(t) = \varepsilon_{12}^0\delta(t), \qquad t \leq t_1,$$
$$= \varepsilon_{12}^0\delta(t) - \varepsilon_{12}^0\delta(t - t_1), \qquad t \geq t_1. \tag{3.70}$$

The relaxation stress after unloading is given by

$$\sigma_{12}(t) = 2\varepsilon_{12}^0[G(t) - G(t - t_1)], \qquad t \geq t_1. \tag{3.71}$$

The second term is the mirror image of the first term across the horizontal axis displaced horizontally by the time t_1 (the dotted lines in Fig. 3.11). We note that σ_{12} $(t \to \infty) \to 0$ both for solids and liquids.

Equations (3.67)–(3.71) are written for pure shear. Analogous relations hold for a simple extension with all $2G(t)$'s replaced by $E(t)$'s. In a simple extension experiment, if an isotropic cubical element is elongated in one direction (say $\varepsilon_{11} = \varepsilon_{11}^0$), the two mutually orthogonal directions will contract with an equal amount in the two directions

$$\varepsilon_{22}(t) = \varepsilon_{33}(t) = -v(t)\varepsilon_{11} \equiv -v(t)\varepsilon_{11}^0. \tag{3.72}$$

Although ε_{11} is constant by definition during the stress relaxation experiment of a viscoelastic material, ε_{22} and ε_{33} will, in general, change. The stresses are

$$\sigma_{11}(t) = \sigma_T(t) - p_a,$$
$$\sigma_{22} = -p_a, \tag{3.73}$$
$$\sigma_{33} = -p_a,$$

where σ_T is the tensile stress resulting from the applied force and p_a is the ambient pressure. An operation analogous to Eq. (3.68) then gives

$$\sigma_T(t) = \varepsilon_{11}^0 E(t), \tag{3.74}$$

where $E(t)$ is the tensile relaxation modulus or time-dependent Young's modulus. From Eq. (3.38) we get a relationship between $E(t)$ and $G(t)$ for a polymer with $v(t) \sim 0.5$

$$E(t) = 3G(t). \tag{3.75}$$

Finally, if the dimensions of a cubical element are increased or decreased uniformly by the application of normal forces on all faces (bulk compression or dilatation)

$$\sigma_{11} = \sigma_{22} = \sigma_{33} = -p(t) - p_a, \tag{3.76}$$

where $-p$ is the hydrostatic pressure and p_a is the ambient pressure. The three nonzero strain components ($\varepsilon_{11}, \varepsilon_{22}, \varepsilon_{33}$) are equal. Analogous to Eqs. (3.68) and (3.74) we have

$$-p(t) = \Delta_v K(t), \tag{3.77}$$

where $K(t)$ is termed the bulk relaxation modulus and Δ_v is the volume compression or dilatation. For a solid or liquid in which there is no time-dependent response, or for any material at equilibrium under pressure, $K(t)$ and $K^*(\omega)$ (complex dynamic modulus) become the equilibrium bulk modulus K_e.

Creep

In a creep test (which is the opposite of a stress relaxation test), the material is instantaneously loaded to some stress that is maintained at a constant level. This stress results in a strain (point A), which is an elastic response. As the stress σ_{12} ($= \sigma_{12}^0$) is maintained, the sample deforms or creeps viscoelastically (Fig. 3.12). The total creep at a given time is higher at higher temperatures (time–temperature superposition, to be discussed later). In a pure shear experiment the dependence of the creep strain ε_{12} on time for a viscoelastic solid can be derived similarly to that for a relaxation experiment, with the result

$$2\varepsilon_{12}(t) = \sigma_{12}^0 J(t). \tag{3.78}$$

The effect of t_0 is neglected here. $J(t)$ is the creep compliance which has the dimensions of a reciprocal modulus and is a monotonically nondecreasing function of time. For a perfectly elastic solid $J = 1/G$. However, for a viscoelastic material $J(t) \neq 1/G(t)$, because of the difference between the two experimental time patterns.

For linear and branched amorphous polymers that are viscoelastic liquids, $J(t)$ is the sum of three contributions

$$2\varepsilon_{12}(t)/\sigma_{12}^0 \equiv J(t) = J_g + J_d\psi(t) + t/\eta_0, \tag{3.79a}$$

where a J_g glassy (or glasslike) compliance arises from an apparently instanta-

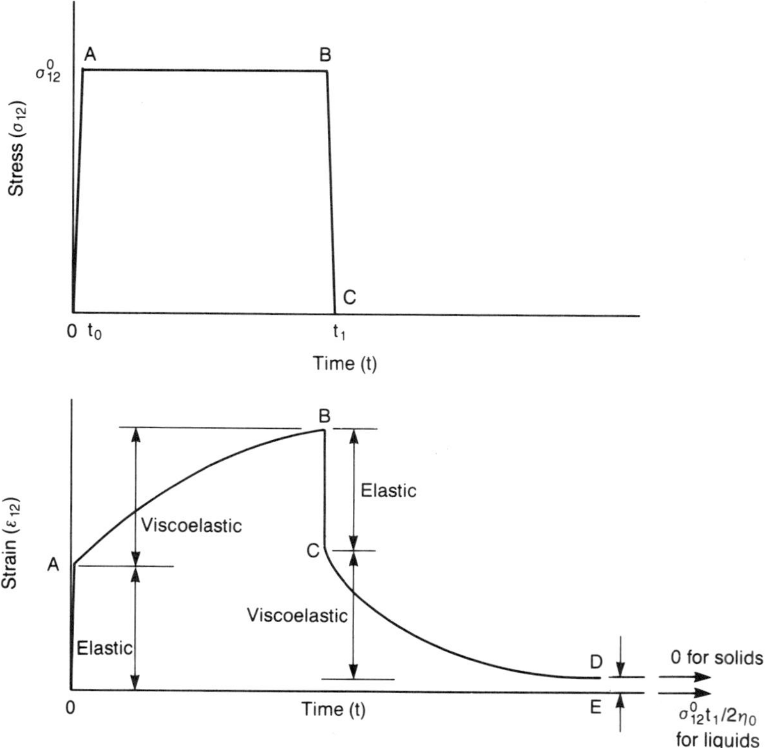

Fig. 3.12. Time profile of a shear creep experiment. Shown are values for ε_{12} ($t \to \infty$) during unloading, the values for ε_{12} ($t \to \infty$) during loading are $\sigma_{12}^0 J_e$ for solids and $\sigma_{12}^0 J_e + \sigma_{12}^0 t_1 / \eta_0$.

neous contribution to the deformation; $\psi(t)$ is the normalized recoverable creep compliance function, which is zero when t is zero (i.e., the time at which the stress is applied to the sample) and is equal to one when $t = \infty$; J_d is the normalization constant, the compliance; and t/η_0 represents the viscous flow and η_0 is the limiting low-rate-of-shear viscosity. The first two terms on the right-hand side of Eq. (3.79a) are proportional to the recoverable portion of the deformation accumulated during creep, extending to steady-state conditions which are obtained when $\psi(t)$ becomes equal to one. The sum of the two recoverable compliance terms $J_g + J_d$ is called J_e, the *equilibrium recoverable shear compliance*, a measure of elastic deformation during steady flow. It is obvious from the preceding statements that the remaining third term in Eq. (3.79a) is proportional to the *nonrecoverable* (permanent) deformation accumulated during creep. Under constant stress, the strain rate approaches a limiting value, and a situation of steady-state flow is eventually attained, governed by the Newtonian viscosity, η_0. The steady-state compliance

$[\psi(t) = 1]$ is given as

$$J(t) \equiv J_e^0 = J_g + J_d + t/\eta_0$$

$$\equiv J_e + t/\eta_0, \tag{3.79b}$$

and

$$\frac{dJ(t)}{dt} = 1/\eta_0. \tag{3.79c}$$

That is to say, the creep velocity measured in steady state is simply the reciprocal of viscosity.

For viscoelastic solids (such as a cross-linked polymer), all of the deformation is found to be ultimately recoverable, i.e., no viscous flow is observed or the third term in Eq. (3.79a) is equal to zero (η is operationally infinite). Instead of reaching a limiting creep velocity, the creep of a viscoelastic solid decelerates from the initial highest velocity to zero velocity and an equilibrium compliance J_e is reached under steady state.

If the load (σ_{12}^0) is instantaneously removed from the sample after time t_1, there is an immediate elastic recovery to C followed by a final gradual viscoelastic recovery. The strain after unloading (Fig. 3.12) based on Boltzmann's superposition principle is given by

$$2\varepsilon_{12}(t) = \sigma_{12}^0[J(t) - J(t - t_1)], \qquad t \geq t_1. \tag{3.80}$$

For a viscoelastic solid, the creep compliance during recovery is given by

$$J(t) - J(t - t_1) = J_d[\psi(t) - \psi(t - t_1)], \qquad t > t_1. \tag{3.81a}$$

It approaches zero as $t \to \infty$. For a viscoelastic liquid, the creep compliance during recovery is given by

$$J(t) - J(t - t_1) = J_d[\psi(t) - \psi(t - t_1)] + t_1/\eta_0, \qquad t > t_1, \tag{3.81b}$$

and it approaches a final value (DE) of t_1/η_0. This is a measure of the viscous flow and is a completely nonrecoverable response.

Equations (3.78)–(3.81) are written for pure shear. Analogous relations hold for a simple extension with the J's replaced by D's. A sudden tensile stress $\sigma_{11} = \sigma_{11}^0$ produces a time-dependent strain

$$\varepsilon_{11}(t) = \sigma_{11}^0 D(t), \tag{3.82}$$

where $D(t)$ is the tensile creep function which is related to shear and bulk creep as

$$D(t) = J(t)/3 + B(t)/9$$

$$\sim J(t)/3 \qquad \text{for} \quad v \sim 0.5. \tag{3.83}$$

If the uniform pressure $-p$ is applied suddenly on all faces and the volume change is followed as a function of time, the bulk creep experiment is described by the analog of Eq. (3.78)

$$\Delta_v(t) = -pB(t), \tag{3.84}$$

Table 3.3. Summary of moduli and compliances

Deformation	Pure shear	Simple extension[a]	Bulk compression
Complex modulus	$G^*(\omega)$	$E^*(\omega)$	$K^*(\omega)$
Storage modulus	$G'(\omega)$	$E'(\omega)$	$K'(\omega)$
Loss modulus	$G''(\omega)$	$E''(\omega)$	$K''(\omega)$
Complex compliance	$J^*(\omega)$	$D^*(\omega)$	$B^*(\omega)$
Storage compliance	$J'(\omega)$	$D'(\omega)$	$B'(\omega)$
Loss compliance	$J''(\omega)$	$D''(\omega)$	$B''(\omega)$
Relaxation modulus	$G(t)$	$E(t)$	$K(t)$
Creep compliance	$J(t)$	$D(t)$	$B(t)$
Equilibrium modulus	G_e	E_e	K_e
Glasslike modulus	G_g	E_g	K_g
Equilibrium compliance	J_e	D_e	$B_e \ (=\beta)$
Glasslike compliance	J_g	D_g	B_g
Steady-state compliance	J_e^0	D_e^0	—
Steady-flow viscosity[b]	η_0	$\bar{\eta}_0$	—
Dynamic viscosity	$\eta'(\omega)$	$\bar{\eta}'(\omega)$	$\eta_v'(\omega)$

[a] For this type of deformation each modulus may be called a Young's modulus (relaxation Young's modulus, storage Young's modulus, etc.).
[b] At vanishing shear rate.

where $B(t)$ is the bulk creep compliance. For a solid or liquid in which there is no time-dependent response, or for any material at equilibrium under pressure, $B(t)$ and $B^*(\omega)$ (complex dynamic compliance) become the thermodynamic compressibility $\beta = -(1/V)(\partial V/\partial p)_T$ where V, p, and T are volume, pressure, and temperature, respectively.

Summary of Moduli and Compliances

The various moduli and compliances which have been introduced for infinitesimal deformations are summarized in Table 3.3. All moduli have the dimensions of stress (Pa) and all compliances have the dimensions of reciprocal stress (Pa^{-1}).

3.1.3.3. Mechanical Model Analogies of Linear Viscoelastic Behavior

The time-dependence of viscoelastic functions such as $G(t)$ or $J(t)$, or of the frequency dependence of viscoelastic functions such as $G'(\omega)$, $G''(\omega)$, $J'(\omega)$, or $J''(\omega)$, can be imitated by the behavior of a mechanical model with a sufficient number of perfectly elastic elements (springs) and viscous elements (dashpots imagined as pistons moving in a cylinder containing a viscous lubricant) (Bland, 1960; Flugge, 1975). Each spring element is assigned a stiffness (force/displacement) analogous to a modulus contribution, G_i, and each dashpot is assigned a frictional resistance (force/rate of displacement) analogous to a viscous contribution, η_i. Force applied to the terminals of the model is analogous to stress, the relative displacement of the terminals is analogous to strain, and the rate of displacement is analogous to strain rate. The dimensions do not correspond (force/displacement is N/m and modulus is N/m^2) and the geometry looks like extension rather than shear in the case of shear loading, but the mathematical analogy is satisfactory. Various viscoelastic behaviors can be represented by Maxwell, Voigt–Kelvin, and their combinations thereof.

Maxwell Models

The simplest mechanical model analogous to a viscoelastic system is one spring (stress is proportional to strain) combined with one dashpot (stress is proportional to strain rate and independent of the strain), either in series (Maxwell) or in parallel (Voigt–Kelvin). Figure 3.13(a) represents a Maxwell model which consists of a spring and a dashpot in series. We choose a simple shear (with nonzero stress and strain components σ_{12} and ε_{12}) as the type of deformation, though any other deformation would do as well. At all times, the stress on each element is the same and the total strain is additive of the strains of each element, i.e., $\varepsilon_{12} = \varepsilon_{12}^{\text{elastic}} + \varepsilon_{12}^{\text{viscous}}$. Expressing this as the differ-

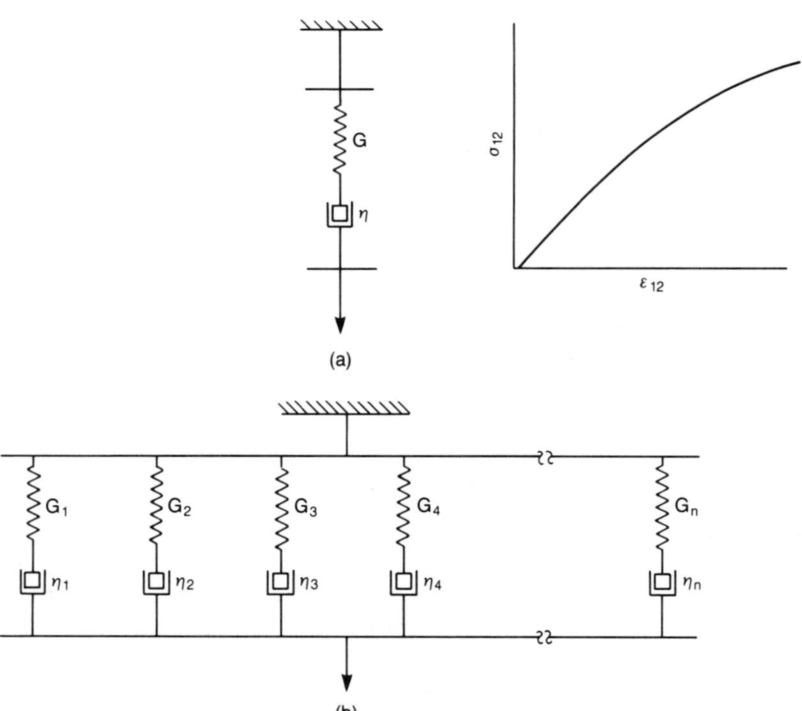

Fig. 3.13. (a) Simple Maxwell model and its stress–strain behavior, and (b) the generalized Maxwell model.

ential equation leads to the equation of motion of a Maxwell unit

$$\frac{\sigma_{12}(t)}{\tau} + \dot{\sigma}_{12}(t) = G\dot{\varepsilon}_{12}(t), \tag{3.85}$$

and

$$\tau = \eta/G, \tag{3.86}$$

where the spring corresponds to a shear rigidity $G = 1/J$ and the dashpot to a viscosity η.

To understand what this equation implies, the Maxwell element is subjected to a *stress relaxation* experiment in which a strain ε_{12}^0 is applied as a step function and is held constant (Fig. 3.11). With $\varepsilon_{12} = \varepsilon_{12}^0$ or $\dot{\varepsilon}_{12} = 0$, Eq. (3.85) is a homogeneous differential equation for the σ_{12} and has the solution

$$\sigma_{12}(t) = \sigma_{12}^0 \exp(-t/\tau), \tag{3.87a}$$

or the relaxation modulus

$$\frac{\sigma_{12}(t)}{\varepsilon_{12}^0} = G(t) = G \exp(-t/\tau), \tag{3.87b}$$

where G is the relaxation modulus at time equal to zero and when the constant strain is applied and τ is called the relaxation time that determines the rate of decay. We note that at $t = \tau$, the stress is reduced to $1/e$ times the original value. We note that stress decreases exponentially under constant strain (i.e., the material relaxes) and approaches zero at $t \to \infty$. Within the realm of a linear constitutive equation [Eq. (3.85)], the material shows a typical property of a fluid: its capability of unlimited deformation under finite stress. The material described by Eq. (3.85) is known as *Maxwell fluid*. Expressions for various viscoelastic functions exhibited by the Maxwell element are presented in Table 3.4. A typical stress–strain curve predicted by the Maxwell model is also shown in Fig. 3.13(a). Under conditions of constant stress, a Maxwell body shows instantaneous elastic deformation first, followed by viscous flow.

Voigt–Kelvin Model

Figure 3.14(a) represents a Voigt–Kelvin (or simply Voigt or Kelvin) model which consists of a spring and a dashpot in parallel. At all times, the strain of the two elements is the same, and the total stress will be split into spring and dashpot in whichever may be necessary to make strain the same. We get

$$\varepsilon_{12}(t) + \tau\dot{\varepsilon}_{12}(t) = J\sigma_{12}(t), \tag{3.88}$$

where

$$\tau = \eta J = \eta/G. \tag{3.89}$$

The Voigt–Kelvin element is subjected to a *creep* experiment in which a stress σ_{12}^0 is applied as a step function and is held constant (Fig. 3.12). For this

Table 3.4. Viscoelastic functions for various mechanical models in pure shear[a]

Maxwell model		Voigt–Kelvin model	
Single element	Infinite elements[b]	Single element	Infinite elements[c]
$G(t) = G_i \exp(-t/\tau_i)$ $J(t) = J_i + t/\eta_i$ $G'(\omega) = G_i \omega^2 \tau_i^2 (1 + \omega^2 \tau_i^2)$ $G''(\omega) = G_i \omega \tau_i/(1 + \omega^2 \tau_i^2)$ $\eta'(\omega) = \eta_i/(1 + \omega^2 \tau_i^2)$ $J'(\omega) = J_i$ $J''(\omega) = J_i/\omega \tau_i = 1/\omega \eta_i$ $\tan \delta = 1/\omega \tau_i$	$G(t) = G_e + \int_{-\infty}^{\infty} H(\tau) \exp(-t/\tau) d(\ln \tau)$ $G'(\omega) = G_e + \int_{-\infty}^{\infty} [H(\tau)\omega^2 \tau^2/(1 + \omega^2 \tau^2)]d(\ln \tau)$ $(G_e = 0$ for a viscoelastic liquid) $G''(\omega) = \int_{-\infty}^{\infty} [H(\tau)\omega \tau/(1 + \omega^2 \tau^2)]d(\ln \tau)$ $\eta' = \int_{-\infty}^{\infty} [H(\tau)\tau/(1 + \omega^2 \tau^2)]d(\ln \tau)$ $\eta_0 = \int_{-\infty}^{\infty} H(\tau)\tau d(\ln \tau)$	$G(t) = G_i$ $J(t) = J_i[1 - \exp(-t/\tau_i)]$ $G'(\omega) = G_i$ $G''(\omega) = G_i \omega \tau_i = \omega \eta_i$ $\eta'(\omega) = \eta_i$ $J'(\omega) = J_i/(1 + \omega^2 \tau_i^2)$ $J''(\omega) = J_i \omega \tau_i/(1 + \omega^2 \tau_i^2)$ $\tan \delta = \omega \tau_i$	$J(t) = J_g + \int_{-\infty}^{\infty} L(\tau)[1 - \exp(-t/\tau)]d(\ln \tau) + t/\eta_0$ $J'(\omega) = J_g + \int_{-\infty}^{\infty} [L(\tau)/(1 + \omega^2 \tau^2)]d(\ln \tau)$ $J''(\omega) = \int_{-\infty}^{\infty} [L(\tau)\omega \tau/(1 + \omega^2 \tau^2)]d(\ln \tau) + 1/\omega \eta_0$

[a] Similar expressions can be written for the cases of simple extension and bulk compression.

[b] G_e is zero for a viscoelastic liquid.

[c] η_0 is infinity for a viscoelastic solid.

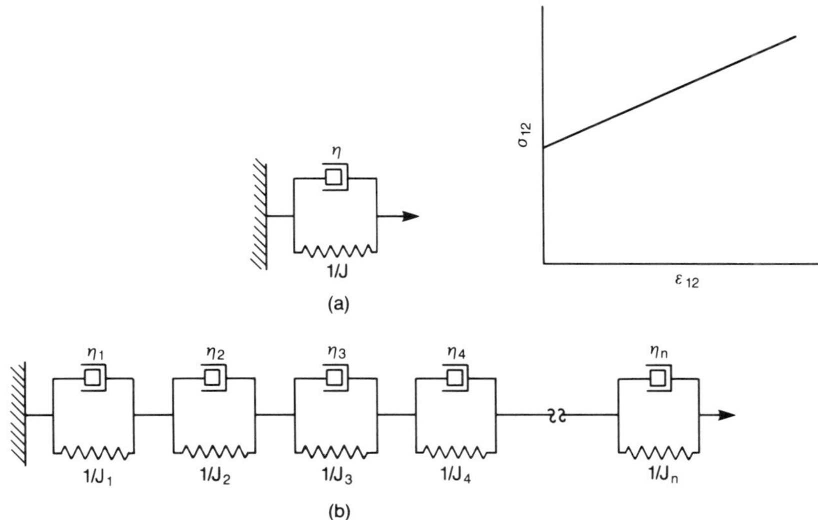

Fig. 3.14. (a) Simple Voigt–Kelvin model and its stress–strain behavior, and (b) the generalized Voigt–Kelvin model.

case, Eq. (3.88) reduced to

$$\frac{\varepsilon_{12}(t)}{\sigma_{12}^{0}} = J(t) = J[1 - \exp(-t/\tau)], \tag{3.90}$$

where J is the creep compliance at time equal to zero when constant stress is applied and τ is called the retardation time which determines the rate of decay of the first derivative of $J(t)$. τ represents the length of time the model takes to attain $(1 - 1/e)$ or 0.632 of the equilibrium strain. We note that strain increases and approaches a limit. This is almost the behavior of an elastic solid. The material defined by Eq. (3.90) is known as the *Voigt–Kelvin solid*. Expressions for various viscoelastic functions for the Voigt–Kelvin model are presented in Table 3.4.

Generalized Models

Any number of Maxwell elements in series have the properties of a single Maxwell element with $G = \sum G_i$ and $1/\eta = \sum(1/\eta_i)$; any number of Voigt–Kelvin elements in parallel have the properties of a single Voigt–Kelvin element with $J = \sum J_i$ and $\eta = \sum \eta_i$. However, Maxwell elements in parallel or Voigt–Kelvin elements in series, as in Figs. 3.13(b) and 3.14(b), exhibit multiple discrete viscoelastic spectra.

A generalized Maxwell model [Fig. 3.13(b)] consists of a group of Maxwell elements in parallel. It represents a discrete spectrum of relaxation times, each time τ_i being associated with a spectral strength G_i. Since in a parallel arrangement, the stresses are additive, it can readily be shown that the viscoelastic functions are obtained simply by summing the expressions in Eq. (3.87) over all the parallel elements for the relaxation experiment in pure shear; thus, if there are n elements,

$$G(t) = \sum_{i=1}^{n} G_i \exp(-t/\tau_i) \tag{3.91}$$

and

$$\tau_i = \eta_i/G_i. \tag{3.92}$$

Expressions for other viscoelastic functions can be obtained similarly.

A generalized Voigt–Kelvin model consists of a group of Voigt–Kelvin elements in series. It represents a discrete spectrum of retardation times, each time τ_i being associated with a spectral compliance magnitude J_i. Since in a series arrangement the strains are additive, it turns out that for the generalized Voigt–Kelvin model, the viscoelastic functions are obtained by summing the expressions in Eq. (3.90) over all series elements for a creep experiment in pure shear. Thus

$$J(t) = \sum_{i=1}^{n} J_i[1 - \exp(-t/\tau_i)] \tag{3.93}$$

and

$$\tau_i = \eta_i J_i = \eta_i/G_i. \tag{3.94}$$

Expressions for other viscoelastic functions can be obtained similarly (Ferry, 1980).

Any experimentally observed stress relaxation curve which decreases monotonically can be fitted with any desired degree of accuracy to a series of terms as in Eq. (3.91) by taking n sufficiently large, and this would amount to determining the discrete spectrum of "lines" each with a location τ_i and intensity G_i. Similarly, fitting creep data to Eq. (3.93) would amount to experimental determination of the discrete retardation spectrum (Tschoegl, 1989).

The models of Figs. 3.13 and 3.14 are equivalent to an appropriate assignment of parameters, subject to certain requirements which depend on whether the system is a viscoelastic liquid or a viscoelastic solid. All viscoelastic materials are intermediate between solids and liquids. In a relaxation experiment: for liquids, $G(t)$ approaches zero as $t \to \infty$; for solids, $G(t)$ approaches a constant finite value (G_e). Conversely, in a creep experiment: for liquids, $J(t)$ continues to grow and does not have an equilibrium compliance but has a steady-state compliance $(J_e^0 = J_g + J_d + t/\eta_0)$; for solids, $J(t)$ approaches a constant finite value $(J_e = J_g + J_d)$. For a liquid, all viscosities in the generalized Maxwell model [Fig. 3.13(b)] must be finite and one spring in the generalized Voigt–Kelvin model [Fig. 3.14(b)] must be zero (to add a term t/η_0). For a solid, one of the relaxation times (or viscosity) in the generalized Maxwell model must be infinite (resulting in an added constant term G_e) and all springs in the generalized Voigt–Kelvin model must be nonzero (see Fig. 3.26 to be presented later).

We finally note that complex models involving both Maxwell and Voigt–Kelvin elements can be produced for certain viscoelastic materials (Flugge, 1975).

Continuous Viscoelastic Spectra

If the number of elements in the Maxwell model is increased without limit, the result is a continuous spectrum in which each infinitesimal contribution to rigidity, $F(\tau)\tau$, is associated with relaxation times whose logarithms lie in the range between τ and $\tau + d\tau$, a measure of the population of relaxation mechanisms with relaxation times in this interval. Experience has shown that a logarithmic time scale is much more convenient than a linear time scale; accordingly, the continuous relaxation spectrum is defined as $H(\tau)d(\ln \tau)$, the contribution to rigidity associated with relaxation times whose logarithms lie in the range between $\ln(\tau)$ and $\ln(\tau) + d(\ln \tau)$, a measure of the population of relaxation mechanisms with relaxation times in this interval. [Evidently, $H(\tau) = F(\tau)\tau$.] For the continuous spectrum, Eq. (3.91) becomes

$$G(t) = G_e + \int_{-\infty}^{\infty} H(\tau) \exp(-t/\tau)d(\ln \tau), \qquad (3.95)$$

which may alternatively be taken as a mathematical definition of $H(\tau)$ in terms of the relaxation function $G(t)$. The constant equlibrium modulus G_e is added to allow for a discrete contribution to the spectrum with $\tau = \infty$, for viscoelastic solids; for viscoelastic liquids, of course, $G_e = 0$.

In an entirely analogous manner, if the Voigt–Kelvin model is made infinite in extent, it represents a continuous spectrum of retardation times, $L(\tau)$, alternatively defined by the continuous analog of Eq. (3.93)

$$J(t) = J_g + \int_{-\infty}^{\infty} L(\tau)[1 - \exp(-t/\tau)]d(\ln \tau) + t/\eta_0. \qquad (3.96)$$

The glassy compliance J_g which arises from an apparently instantaneous contribution of the distribution must be added to allow for the possibility of a discrete contribution with $\tau = 0$.

Expressions for other viscoelastic functions are presented in Table 3.4.

Andrade Creep

In creep experiments at the long-time end of the plateau zone, Henderson (1951), Kennedy (1953), Van Holde (1957), and Plazek (1960, 1966) have shown that creep compliance for some crystalline and amorphous polymers can be described by the relation

$$J(t) = J_g + \beta t^{1/3} + t/\eta_0. \qquad (3.97)$$

The constants β and J_g are determined by plotting $J(t) - t/\eta_0$ after η_0 has been determined from the steady-state flow, against $t^{1/3}$. An equation similar to this (but without the t/η_0 term) was originally introduced by Andrade

(1910, 1962) to describe the creep response of metals. The coefficient β has the temperature dependence; when referred to a standard reference temperature T_0, the ratio $\beta T_0 \rho_0 / \beta T \rho$ is found to be identical to $a_T^{1/3}$ (shift factor, to be discussed later). Equation (3.97) would imply that the creep compliance in excess of the steady-flow term increases without limit instead of approaching the customary limit J_e, and cannot be valid at extremely long times.

The retardation spectrum corresponding to Eq. (3.97) is (Plazek, 1960)

$$L(\tau) = 0.246\beta\tau^{1/3}. \tag{3.98}$$

3.1.3.4. Time (or Frequency) and Temperature Effects

The mechanical properties of viscoelastic materials are highly temperature- and strain-rate-dependent. The effects of temperature and strain rate are *interdependent* and for "thermorheologically simple" materials (which exhibit, for a constant change of temperature, a pure shift in the viscoelastic function when plotted against the logarithm of time, Schwarzl and Staverman, 1952), a time (or frequency) temperature transformation is feasible for properties such as ultimate tensile properties, dynamic modulus, damping factor, relaxation modulus, dynamic compliance, creep compliance, viscosity, and relaxation and retardation time spectra. The interdependence is often expressed as a "shift factor," a_T, which implies the horizontal shift needed for the data points (representing the mechanical properties plotted on the ordinate against frequency or time on the abscissa on a log-log paper) at a temperature to correspond to the data points at some other temperature.

A relationship originally derived empirically but later verified theoretically, expresses the extent of the horizontal shift with the change in temperature for an amorphous rubberlike ($T > T_g$) polymer. The temperature dependence of $\log_{10} a_T$ is given in the temperature range $T_g < T < (T_g + 100 \text{ K})$ by the WLF equation (Williams et al., 1955)

$$\log_{10} a_T = -\frac{8.86(T - T_0)}{101.6\,\text{K} + T - T_0}, \tag{3.99}$$

where T_0 is an arbitrary reference temperature and has to be chosen suitably. The appropriate value of the reference temperature is usually governed by the equation

$$T_0 \sim T_g + (50 \pm 5 \text{ K}). \tag{3.100}$$

We note that $\log a_T$ decreases with increasing temperature. It is positive if $T < T_0$, is negative if $T > T_0$, and is, of course, unity at $T = T_0$. The principle of superposition (transformation) requires that all relaxation times of various molecular mechanisms have the same temperature dependence. Thus, time–temperature equivalence can be applied to any given polymer, provided a change in structure, relaxation mechanisms, or modes of motion do not occur over the temperature range of interest. This is satisfied strictly by the amorphous phase of polymers only. Therefore, the WLF equation is applicable to

amorphous and low-crystallinity, rubberlike polymers over a wide temperature range but may not be too successful if appreciable crystallinity exists
in the sample (Smith, 1958; Morland and Lee, 1960; Nakayasu et al., 1961;
Bahadur and Ludema, 1972; Engel and Lasky, 1977; Sternstein, 1977; Ferry,
1980).

By applying Eq. (3.99), the value of any viscoelastic property obtained at
temperature T and frequency f (or time, t, or τ) can be related to the reference
temperature T_0 by the following equation:

$$\text{modulus} \qquad C(\omega, T) = \frac{T\rho}{T_0\rho_0} C(\omega a_T, T_0), \qquad (3.101a)$$

$$\text{compliance} \quad S(\omega, T) = \frac{T_0\rho_0}{T\rho} S(\omega a_T, T_0), \qquad (3.101b)$$

where ρ and ρ_0 are the material densities at T and T_0, respectively. For time
t or τ instead of frequency ω, replace ωa_T by t/a_T or τ/a_T. Temperatures and
density ratios are due merely to thermal expansion. (It is generally sufficient
to approximate the ratio $T\rho/T_0\rho_0$ by the temperature ratio T/T_0.) Thus the
effects of a change from T to T_0 is to multiply the ultimate tensile properties,
modulus (C), viscosity (η), and the relaxation time spectra by $T_0\rho_0/T\rho$, and
multiply the frequency scale by a_T or divide the time scale by a_T; for compliance (S) and retardation time spectra (L) multiply by $T\rho/T_0\rho_0$ and with the
same frequency (or time) adjustment; and for the damping factor (tan δ)
to only make the same frequency (or time) adjustment. Data from tests
conducted on a given material at various temperatures T and frequencies ω,
can thus be used to construct a single master curve at temperature T_0 with a
so-called reduced viscoelastic function as a function of the reduced frequency
ωa_T or the reduced time t/a_T or τ/a_T. The master curve is generally plotted on
a log-log scale.

Once such master curves have been constructed, it is likewise possible to
predict the values of viscoelastic functions over wide frequency or temperature ranges, although the predictions may become less reliable as the upper
and lower limits of the frequencies and temperatures considered become
further removed from those at which the measurements were originally made.

When time–temperature equivalence is valid, then the shift factors can be
related to an apparent activation energy of the relaxation process, by assuming that the viscoelastic deformation process has an Arrhenius temperature
dependence [which is very similar to the earlier form of the WLF equation
(3.99)]

$$\log_{10} a_T = \log_{10}(t/t_0) = \frac{-\Delta H_a}{2.303R}\left(\frac{1}{T_0} - \frac{1}{T}\right), \qquad (3.102)$$

where ΔH_a is the activation energy in kJ/mole (kcal/mole), R is the universal
gas constant ($= 8.313$ kJ/mole K or 1.986 kcal/mole K), and t and t_0 are the
times at absolute temperatures T and T_0, respectively, when strains are equal.

The activation energy can be obtained from the slope of a straight line on a plot of $\log_{10} a_T$ versus $1/T$.

Bhushan and Connolly (1986) have presented evidence of time–humidity transformation very similar to time–temperature transformation. This data will be presented later in the chapter. Using the time–humidity superposition, viscoelastic functions at any frequency (time) and humidity can be predicted from the master curve.

3.2. Dynamic Modulus Data

3.2.1. Measurement Techniques

Various instruments, called dynamic viscoelastometers, are employed to make dynamic mechanical measurements (Ferry, 1980; Plazek, 1980). Most commonly used instruments are Rheovibron (also called Autovibron), Dyna Stat (manufactured by Imass Inc., Hingham, Massachusetts), and Dynamic mechanical thermal analyzer (DMTA, manufactured by Polymer Laboratories, Loughborough, U.K.). The main deformation modes are tension/compression, bending beam (flexure), and shear (torsion) (Sternstein, 1983; Weber et al., 1983; Wetton et al., 1983). They can be used at a fixed frequency ranging from about 0.03 Hz to over 100 Hz and at temperatures ranging from cryogenic temperature of about $-150°$ C to $300°–500°$ C. They can be used for both thin (~ 10 μm) and thick films; the deformation mode for thin films is primarily tension. The typical sample size is 5–10-mm wide and 15–20-mm long. Another instrument dynamic mechanical analysis (DMA) system (manufactured by E. I. duPont Company, Wilmington, Delaware) is used for thick films (typically greater than 100 μm in thickness) to measure dynamic properties in bending at temperatures ranging from $-150°$ to $500°$ C, however, measurements can be made only at the resonant frequency of the sample at the test temperature. The instruments use microprocessor-controlled instrumentation so that the measurements can be made automatically. Here we describe the principles of operation of DMTA and DMA instruments.

3.2.1.1. DMTA

We describe the principles of operation of a typical dynamic instrument (DMTA) subjected to deformation in the tensile mode and operable at various fixed oscillating frequencies. A preset force (typically 2–20 N) is applied to a sample held at each end in the measurement head in the tensile mode and this force is maintained at a steady level. An oscillator unit (voice coil or stepping motor) then applies a sinusoidally varying force to the sample to achieve a given peak strain value. The oscillating stress experienced by the specimen is proportional to the drive current from the oscillator. The strain produced is converted to a proportional voltage by a nonloading transducer.

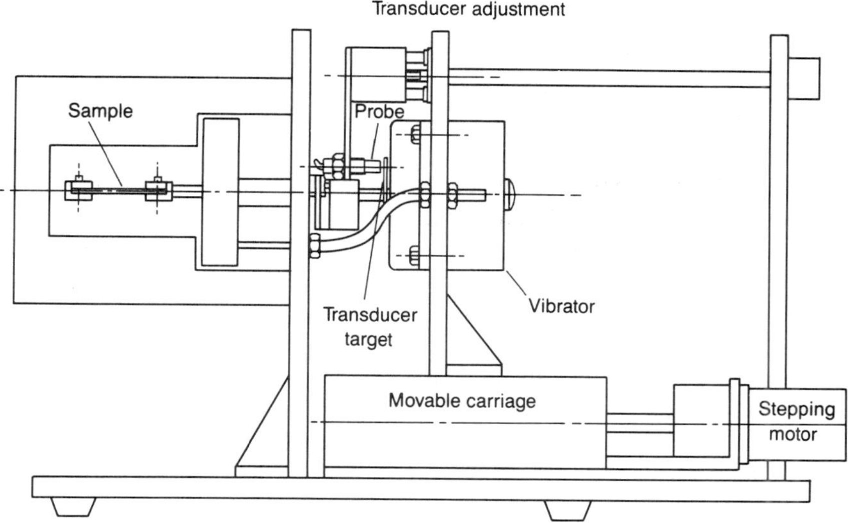

Fig. 3.15. Schematic of the DMTA tensile system for the measurement of the dynamic modulus and damping factor of thin viscoelastic films (Polymer Laboratories, Loughborough, U.K.).

The dynamic analyzer unit compares the phase and amplitude of stress and strain and computes absolute values of log (storage modulus) and tan δ (Wetton et al., 1983).

A schematic of a DMTA system in the tensile mode is shown in Fig. 3.15. The sample is mounted on two clamps with one of the clamps mounted on the drive shaft. The drive shaft is supported on a series of springs. The tension is applied in the form of DC fed to the vibrator coil. The entire vibrator assembly is mounted on a movable carriage and this carriage is moved to allow the springs within the system to return to their neutral position (i.e., the force from the coil and the sample tension are opposite and equal to one another; the springs do not contribute as they are in the neutral position).

The displacement transducer has a target rigidly coupled to the drive shaft and a probe mounted on the movable carriage. The position of the springs is registered by following the average signal from the displacement transducer. The displacement transducer signal is therefore monitored over a whole number of cycles and, if the average displacement differs from the set value by more than a given amount, the tensile control commands the carriage to move to bring the average signal back to its set value. This movement may be in either direction.

The dynamic force is applied by the movement of the vibrator assembly carriage performed by a stepping motor. The electronics send the appropriate number of steps to the motor and, by counting these steps, displays the displacement of the carriage from its initial position.

3.2.1.2. DMA

A dynamic mechanical analysis (DMA) system is used to measure dynamic properties (E^*, tan δ) as a function of the temperature ranging from $-150°$ to $500°$ C but only at the natural frequency of the sample (Bhushan, 1990). In the DMA, a viscoelastic material is physically stressed and then released. When the stress is removed, deformation energy is converted to mechanical oscillation at a resonant frequency characteristic of the test material. Because deformation energy is dissipated by internal molecular motion (heat), the amplitude of oscillation decreases, or is damped with time. The rate of damping is proportional to the rate at which a material can dissipate energy internally.

The sample is sinusoidally oscillated at its resonant frequency with an amplitude selected by the operator. An amount of energy, equal to that dissipated by the sample, is added on each cycle to maintain a constant amplitude of oscillation. This make-up energy is a direct measure of sample damping (energy dissipation) and is expressed as a logarithmic function (in decibels) or as an absolute linear function that can be subsequently converted to tan δ. Digital values of frequency (f) and damping are continuously displayed.

A schematic diagram of the DMA system (model 981 manufactured by E. I. duPont Co.) is shown in Fig. 3.16. To make a measurement, a material of known dimensions is clamped between the two sample arms. The sample-arm pivot system is oscillated at its resonance frequency by an electromechanical transducer. The frequency and amplitude of this oscillation are detected by a linear variable differential transducer (LVDT) positioned at the opposite ends of the active arm. The LVDT provides a signal to an electromechanical

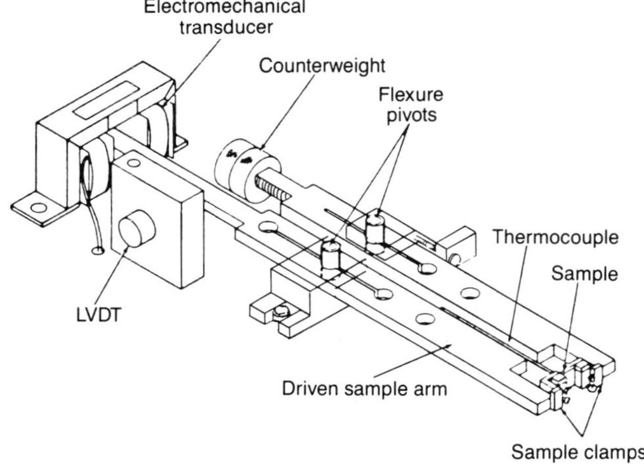

Fig. 3.16. Schematic of the DMA system for the measurement of the dynamic modulus and damping factor of viscoelastic films (E. I. duPont Co.).

transducer, which in turn keeps the sample oscillating at constant amplitude. The complex modulus (E^*) and tan δ of the sample can be obtained from the resonant frequency (f) and the damping signal (V), respectively, by the following relations:

$$E^* = \frac{4\pi^2 f^2 J - K}{2w[(L/2) + D]^2}\left(\frac{L}{T}\right)^3, \tag{3.103}$$

where E^* is the complex modulus (Pa), f is the DMA frequency (Hz), J is the moment of inertia of the arm (N·m^2), K is the spring constant of pivot (Nm/rad), D is the clamping distance, w is the sample width, T is the sample thickness, and L is the sample length. Also

$$\tan \delta = CV/f^2, \tag{3.104}$$

where V is the DMA damping signal (mV) and C is the system constant (~ 0.25 Hz^2/mV).

3.2.2. Experimental Results

The effect of temperature on storage and loss moduli of biaxially-oriented PET film (Mylar A, 92 DB) was measured using Rheovibron at a fixed frequency of 35 Hz and at temperatures from $-90°$ to $160°$ C, Fig. 3.17. Specimens were cut at $0°$, $-45°$, and $90°$ to the MD; for these specimens, extinction angles were at about $-45°$ and $+45°$ to the MD. Except at elevated temperatures, specimens cut at $-45°$ to the MD had the largest

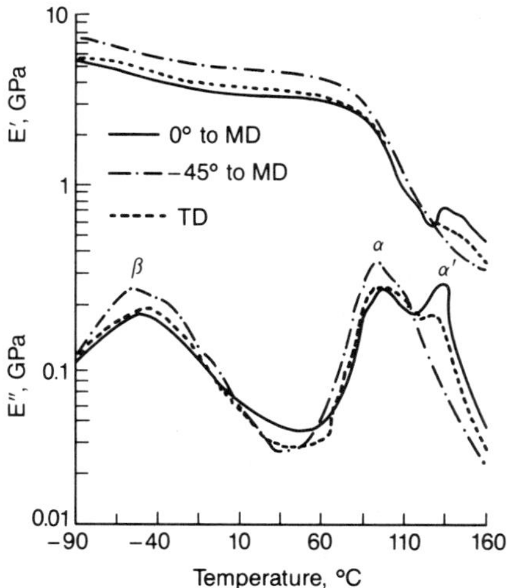

Fig. 3.17. Storage (E') and loss (E'') moduli as a function of temperature measured at 35 Hz on 36.1-μm-thick PET (Mylar A) films for which the major axis (slow optical axis) is about $-45°$ to the MD. Direction of stress application was $0°$, $-45°$, and $90°$ to the MD (Chu and Smith, 1973).

moduli. The temperature at which a significant drop in storage modulus occurs corresponds to the glass transition temperature of the PET film. The storage modulus continues to drop at a steady rate up to the melting point of PET (Chu and Smith, 1973).

A molecular view of a loss modulus curve is that for a given maximum (β- or α-relaxation) more and more large segments are free to use up energy as the temperature is increased. The loss will continue to increase with temperature as long as an increase in temperature will free more segments of a particular type to engage in the mechanical loss process. The maximum is past when the next temperature increment releases fewer segments than the last lower temperature increment. A minimum between two loss peaks is reached when all of the segments in the sample have been released to participate in the process and a temperature increment will release no more of that type. A new loss maximum (α') is started when the enthalpy of the solid has increased to the point that a new type of segment rotation has become possible. Progressive temperature increments will release more segments.

The effect of frequency on the dynamic modulus of the PET films at various test temperatures was studied using Dyna Stat (Smith, 1983–84), Fig. 3.18. [The measurements were also made on a coated tape A (23.4-μm-thick PET coated with a 4-μm-thick calendered frontcoat and a 4-μm-thick uncalendered backcoat) and the absolute values of the storage moduli were within 5% of the PET film.] The storage modulus versus frequency at different temperatures were superimposed, giving the master curve shown in Fig. 3.19, for which the reference temperature is 25° C. We expected the frequency shift factor (a_T) to have an Arrhenius temperature dependence. The activation energy (ΔH_a) is then calculated from a straight line on a plot of log a_T versus $1/T$ (Fig. 3.20) whose slope gives an activation energy of

$$\Delta H_a \sim 240 \text{ kJ/mole (58 kcal/mole)}. \tag{3.105}$$

We will compare this value later with that of creep data (activation energy ~ 200 kJ/mole).

The effect of temperature on the complex modulus and tan δ of free films of two magnetic coatings, about 0.2 mm in thickness and designated A and D, was measured using the DMA system at about 44 Hz from about 0° to 90° C, Fig. 3.21. (For the coating compositions, see Section 3.4.) The temperature at which a transition in the complex modulus occurs corresponds to the melting point of the polyester (soft segment) of the copolymer polyester–polyurethane (Bhushan et al., 1984a; Bhushan, 1990). The magnetic coating D had a slightly higher modulus and a substantially higher inflection temperature in the modulus (60° C) because coating D had a higher proportion of the hard segment of polyurethane compared to that of coating A. We note that the modulus of composite polymers containing rigid fillers such as magnetic coating depends on many factors, notably the volume fraction of the filler, its geometrical arrangement, the degree of adhesion to the binder, and the extent of the porosity that may also be present as a third component of the structure (Nielsen, 1974).

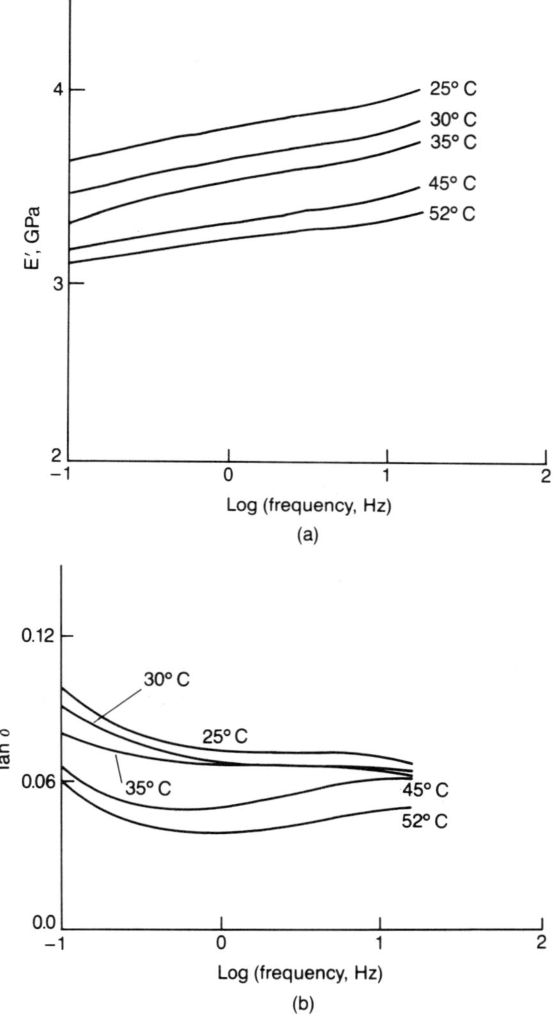

Fig. 3.18. Storage modulus and tan δ as a function of log frequency at various temperatures and a relative humidity of 40% RH on 23.4-μm-thick PET (Mylar A) film along its major axis (slow optical axis).

3.3. Tensile Relaxation and Creep Data

A considerable amount of experimental data has been published on the viscoelastic properties of unoriented, amorphous, or partially crystalline PET films (Thompson and Woods, 1956; Ward, 1964; Illers and Breuer, 1963; Pinnock and Ward, 1966; McCrum et al., 1967; Tajiri et al., 1970), uniaxially oriented PET fibers (Murayama et al., 1968; Gupta et al., 1984), and biaxially-oriented semicrystalline films used for flexible magnetic-media substrate

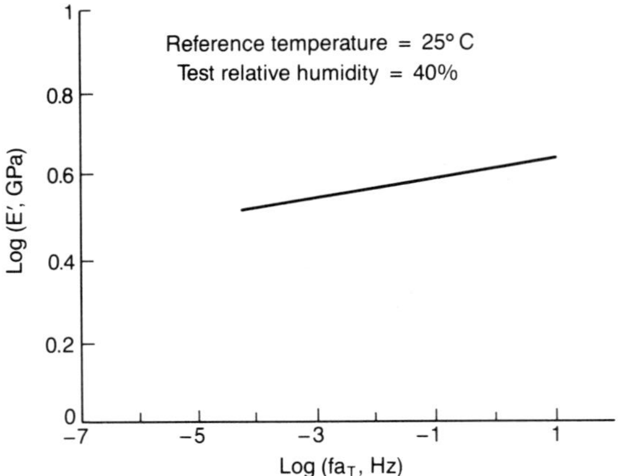

Fig. 3.19. Master curve of log (storage modulus) as a function of log (frequency) of 23.4-μm-thick PET film (Mylar A) along its major axis reduced to 25° C and 40% RH.

(Chen and Raj, 1977; Bogy et al., 1979; Hawthorne, 1981; Bhushan et al., 1984c; Eshel et al., 1984; Vallat et al., 1986; Bhushan and Connolly, 1986; Vallat et al., 1988a, b).

The molecules in the oriented-PET film are in the nonequilibrium state. During manufacturing of PET, as the polymer is cooled at a finite rate through the glass transition, the molecules are not able to reach their equilibrium conformation with respect to temperature, due to a rapid increase in viscosity and an associated decrease in molecular mobility in the T_g region.

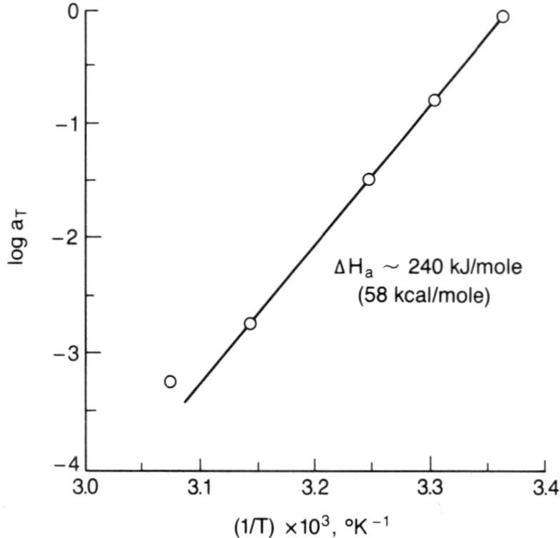

Fig. 3.20. Temperature dependence of shift factors, log a_T, obtained from the master curves in Fig. 3.19.

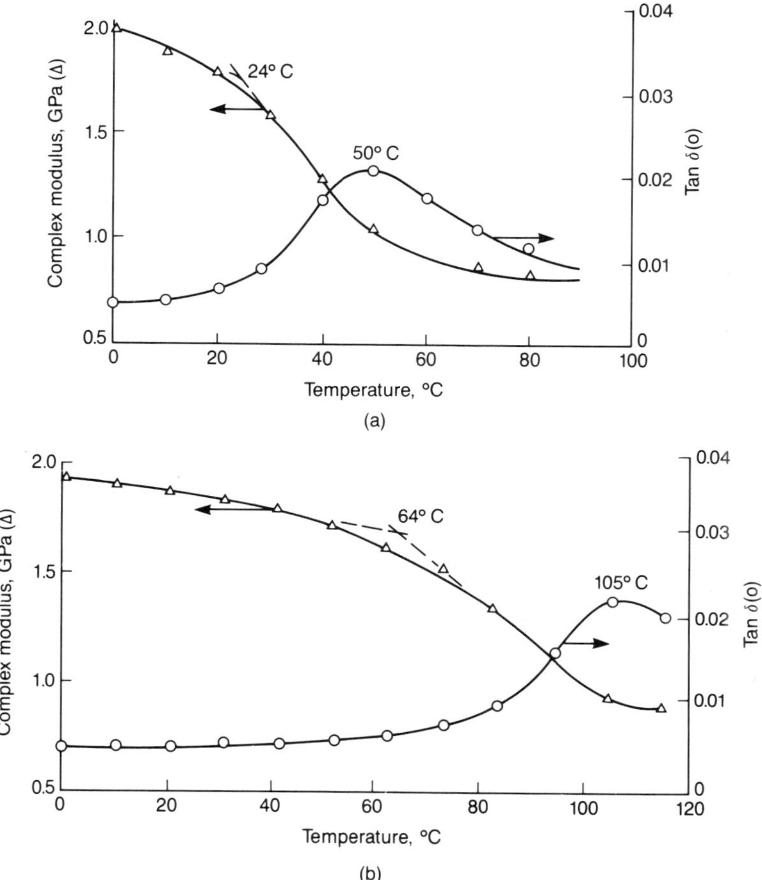

Fig. 3.21. Complex modulus and loss tangent (tan δ) measured at 44 Hz on free film of (a) magnetic coating A and (b) magnetic coating D (Bhushan, 1990).

As the temperature is further decreased, the polymer molecules are essentially "frozen" into a nonequilibrium state by undergoing further packing. PET films exhibit dramatic structural and property changes upon thermal treatments below the glass-transition temperature–physical aging (Tant and Wilkes, 1981; Vallat et al., 1986, 1988a, b) and above the glass-transition temperature-annealing (Groeninckx et al., 1976, 1980a, b; Vallat et al., 1986, 1988a, b).

Investigations of PET specimens with different degrees of crystallinity show that completely amorphous linear polymers are viscoelastic liquids which flow, while even small amounts of crystallinity preclude flow. The crystallites act as crosslinks because some polymer molecules participate in the formation of more than one crystallite. The rigid crystallites also act as active fillers and the strength of the softening dispersion (i.e., the decrease in modulus from the glassy level down to the rubbery level) diminishes as the rubbery level increases and the viscoelastic response broadens on the time or frequency scales (Illers and Breuer, 1963; Tajiri et al., 1970; Groeninckx et al., 1976; Tant and Wilkes, 1981). This is typical behavior for crystallizable polymers.

Studies of the stress relaxation of uniaxially-oriented PET fibers (Murayama et al., 1968) and biaxially-oriented PET films (Hawthorne, 1981; Bhushan et al., 1984c; Bhushan and Connolly, 1986) show clearly that the effect of orientation is to increase the magnitude of the relaxation modulus and to decrease the temperature-dependence of the viscoelastic response. Logarithmic plots of the shift factor a_T for both the uni- and biaxially-oriented PET's as a function of temperature have a signficantly lower slope than that obtained on unoriented material. Since the morphology of oriented and unoriented samples with the same percent crystallinity is unavoidably different it is not possible to isolate the cause. It is also difficult to ascertain the influence of the orientation on the breadth of the viscoelastic response, since the oriented samples are also crystalline. Both the unoriented crystalline and the oriented crystalline PET have broad responses (Murayama et al., 1968; Hawthorne, 1981; Bhushan et al., 1984c).

Most of the published viscoelastic data have been taken for relatively short durations—a few minutes to a few hours. Bhushan et al. (1984c, 1986) have conducted creep and relaxation tests of 23.4-μm-thick biaxially-oriented PET films (Mylar A, 92 DB) along three film orientations at various temperatures and humidities for a duration of several hundred hours. They have obtained the constitutive relationships for the orthotropic viscoelastic behavior of the PET films under plane stress conditions for various temperature and humidity conditions. By using the time–temperature and time–humidity superposition principles, they determined shift factors for temperature and humidity, respectively. They have studied the creep and relaxation behavior of PET film under time-dependent temperature fields (nonisothermal viscoelasticity) as magnetic media may be stored at more than one temperature before use. They have also studied the creep and relaxation behavior of magnetic tapes and the free films of magnetic coatings. The effect of the thermal treatment of PET film on the creep behavior also has been studied to obtain a conditioning thermal history which would produce a stabilized film (low shrinkage and low creep/relaxation) (Vallat et al., 1986, 1988a, b). Details of these studies follow.

3.3.1. Descriptions of Creep and Relaxation Test Apparatuses

3.3.1.1. Creep

In a creep experiment, a stress (or tension) is applied and held constant and the resulting strain (or elongation) is measured as a function of time. Various designs to measure the creep of PET films have been used. Here we present the details of three apparatuses designed to conduct simple-extension creep tests. The data will be presented later.

Creep Apparatus at IBM Corp., Tucson

A schematic of a creep apparatus is shown in Fig. 3.22 (Bhushan et al., 1984c).
It was placed inside a commercially available environmental chamber in
order to conduct tests at various temperatures and humidities. Test samples
were 12.7-mm wide and 0.2-m long, and clamped on both sides. Alignment of
the samples was important to eliminate out-of-plane distortions and to apply
a uniform stress across the width of the sample. Each clamp was tightened
with a screw that pressed a small plate against the PET film. On the top
clamps, the plates were saddle-shaped to reduce twist.

Linear variable-displacement transducers (LVDTs) were not used to measure creep displacement because of their sensitivity to temperature and
humidity changes. Since the test apparatus was placed directly in the oven,
proximity probes were used instead. However, to decrease the minimum
measurable strain increment for these probes, amplification was required. By
using a balance beam-loading mechanism, a 2 : 1 amplification was obtained.
Since the fixture and samples had to stabilize prior to loading, application
and removal of the loads had to be controlled remotely. Figure 3.22 shows
one beam loaded in the fixture. Each beam was 0.26-m long, with a balance
weight (counterweight) on the outer end. Tension from 3.5 MPa to 14 MPa
could be applied to samples using weights of different sizes. Two air cylinders
raised and lowered the weights suspended from the crossbar. During the test,

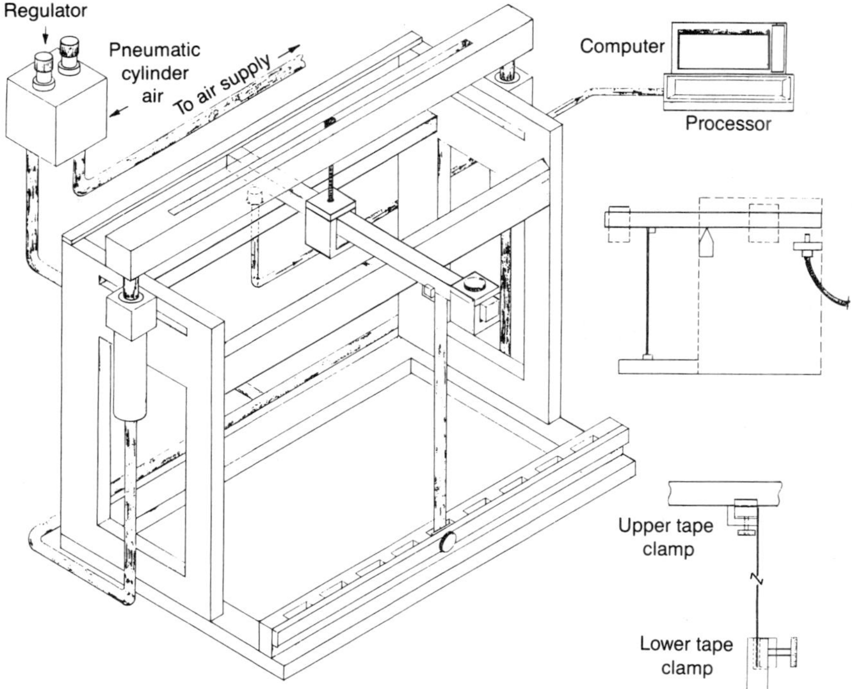

Fig. 3.22. Schematic of a tensile creep apparatus (Bhushan et al., 1984c).

a shield was placed around the samples to minimize the effect of air flow in the chamber. The output of the proximity probes was fed to a printer.

The experiments proceeded in the following way: after the samples were clamped, the apparatus was placed in the chamber. The samples were under a light load (about 140 mN) until equilibrium was reached (about 4 h). This light load was applied by the balance weights. Creep deformation was initialized. Then the test load of typically 2.08 N (excluding the initial 140-mN load) or 7.0 MPa was applied remotely without opening the environmental chamber. Typically, a creep test was conducted for 1 week and a recovery test was conducted for 4 days. At the end of the week, the loads were removed and only the light load (about 140 mN) remained.

Physical aging of PET film is known to result in some shrinkage. A sample in each test was loaded only to a small load of 140 mN (with a relatively small contribution in creep) to measure shrinkage, so that the total creep in an experiment could be calculated by adding measured creep to shrinkage. We found that shrinkage was small ($<2\%$ of the total strain) within the range of measurement accuracy.

Creep Apparatus at the University of Pittsburgh

A schematic of a differential elongational creep apparatus is shown in Fig. 3.23 (Vallat et al., 1986). (A similar instrument has also been used by Raj et al., 1974.) High sensitivity is achieved by measuring the differential displacement between a loaded sample and a reference sample, thus compensating for fluctuations caused by thermal expansion. Twin samples (4)[3] and (5) of matched length (about 6-mm wide and 150-mm long) were hung in parallel holes in an aluminum rod whose temperature was controlled by temperature-controlled silicone oil circulating through a surrounding jacket. The length of each was monitored with a LVDT (9) (Schaevitz coil 050HR with a LPM-21 signal conditioning unit). The LVDT cores (8) were supported on a quartz rod. Only one sample (5) was loaded during creep runs. The outputs of the two LVDT's were connected in opposition so that changes in length caused by thermal fluctuations or material shrinkage cancel, and only the creep deformation of sample (5) is observed. Weights (10) were loaded onto sample (5) by lowering a supporting pan (11) with a solenoid. The solenoid core is (13) and its coil is (14). The bellows (7) which was part of the vacuum chamber provided thermal isolation for the LVDT's from the sample thermostat. The standoffs which held the bellows at constant length were Invar tubes. The micrometers were used to center the LVDT cores. The output from the LVDT monitoring the slightly loaded reference sample was used to follow its length while it was in the apparatus. In this manner, the instrument served as a *differential elongational creep apparatus and a linear dilatometer*. Because of the sizable expansion of the aluminum sample cell walls, it was necessary to

---

[3] Numbers refer to similarly numbered components in Fig. 3.23.

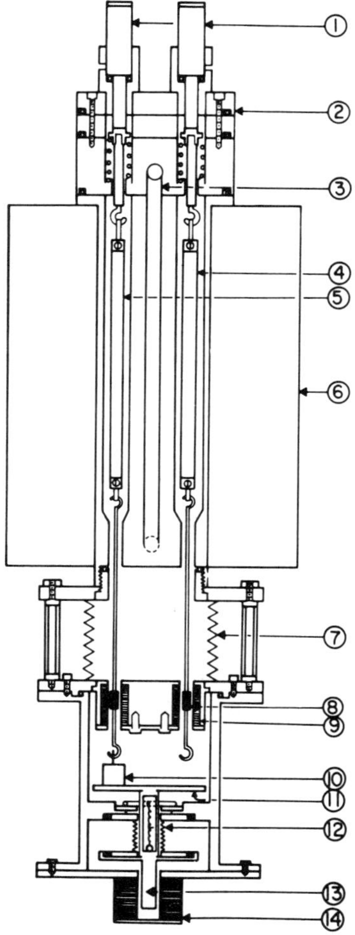

Fig. 3.23. Schematic of the cross section of the differential elongation creep apparatus showing: (1) micrometers for LVDT core adjustment; (2) the O-ring seal; (3) the circulation hole in the aluminum sample chamber for pumping thermostated silicone oil (Dow Corning DC 200, 100 cSt); (4) the reference sample; (5) the creep sample; (6) the insulation around the aluminum sample chamber; (7) bellows for thermal isolation; (8) the LVDT core; (9) the LVDT coil; (10) the weight; (11) the support platform; (12) the bellows vacuum seal; (13) the solenoid core; and (14) the solenoid coil (Vallat et al., 1986).

calibrate the relative displacement readings to extract absolute changes in sample length. Thin rods of quartz, alumina, and aluminum were used in the calibration. It was necessary to account for the temperature dependences of the expansivities of the calibration materials even in the limited temperature range 25°–160° C to obtain a single calibration curve.

The resolution of displacements with LVDT's is often claimed to approach

10 nm. For constrained motion of the core along a fixed path, the claim is probably justified. In the present case where friction is avoided, even after careful alignment, constraints were not possible and slight lateral motions alter the output so that the resolution was of the order of a micron. Hooks were chosen to allow for self-alignment of the film samples. To eliminate the shifting, epoxy adhesive was applied to the hooks after the samples are clamped in place. Clamps and quartz rods hung from the samples. The rods supported the LVDT cores. These attached parts totalling about 70 mN constituted a preload on the sample that creates a stress of about 0.5 MPa in the samples. A preload was necessary to straighten out the films. Deformations caused by this preload were not significant during the creep measurements.

Creep Apparatus at IBM Corporation, San Jose

This apparatus at the IBM Corporation in San Jose was used to measure longitudinal and lateral strains as a function of time at desired temperatures and humidities, in order to obtain creep properties as well as the Poisson ratio (Bogy et al., 1979). Since the change in width is much less than the change in length, the measurement of width change must be performed with much higher accuracy than the measurement of the length change. A schematic of the creep apparatus is shown in Fig. 3.24(a). In determining the longitudinal strain, the elongation of a tape sample of about 0.9-m length was measured from the position at which it was clamped to the position at which the load was applied. The LVDT measured the travel of the loaded end of the specimen, which was used to obtain the longitudinal strain.

The laser-scanning technique was used for the measurement of the lateral contraction of specimens. Figure 3.24(b) shows a schematic of the width measurement apparatus. A 3-mW He–Ne laser was directed onto a rotating mirror by means of the two fixed mirrors M_1 and M_2 oriented at 45° to the beam direction. The angular speed of the rotating mirror was 1800 rpm, and for each revolution the fixed mirrors M_3 and M_4 were scanned by the laser beam. When these mirrors were scanned, their reflected beams swept across the active area of the photodiodes PD1 and PD2. A test piece of PET and calibration slits of known width (0.356 mm on the left side and 0.350 mm on the right side) were arranged in front of each of the photodetectors so as to create two vertical gaps which can be swept horizontally by the laser beam. As the beam swept each gap, it produced a gate pulse whose width was proportional to the gap width. The gate pulse and the 20-MHz oscillator signal were combined, and the resulting output was sent to an electronic counter. The output from the counter was displayed by numerical readouts for each of the gaps. At the beginning of a test, a strip of PET was placed on the apparatus and slightly loaded (<2.5 N) to take out slack. Then the counter readings were recorded for each of the four gaps. If C_1 and C_4 denote counter readings corresponding to the gap between the edge of the fixed frame and the edge of the strip on the left- and right-hand sides, and if C_2 and

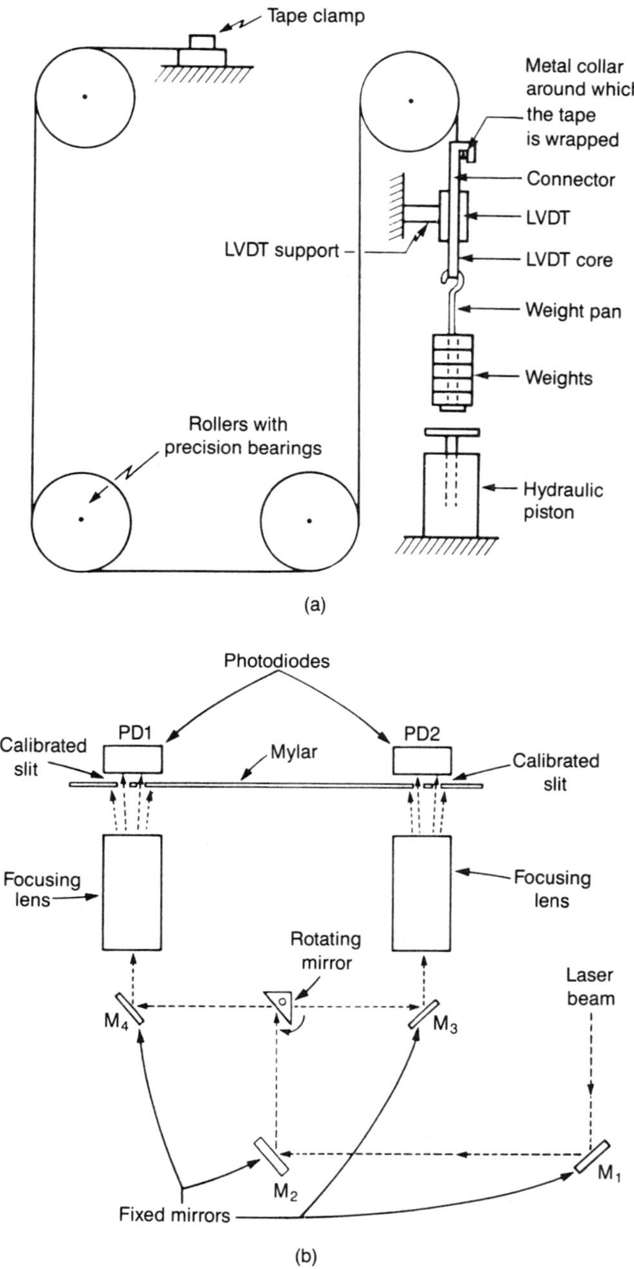

Fig. 3.24. Schematic of a tensile creep and lateral strain apparatus (a) experimental setup for longitudinal strain measurement and (b) scanning laser system for lateral strain measurement (Bogy et al., 1979). © 1979 International Business Machines Corporation; reprinted with permission.

C_3 correspond to the calibration slit readings on the left- and right-hand sides, respectively, then the sum S of widths of the gaps between the fixed edges of the frame and the left and right edges of the strip is given by

$$S = \frac{0.356C_1}{C_2} + \frac{0.350C_4}{C_3}. \tag{3.106}$$

If $S(0)$ and $S(t_0)$ are the sums of the two gaps between the fixed frame and the contracting PET edges at $t = 0$ and $t = t_0$, respectively. The lateral strain, experienced by the specimen in the time interval from $t = 0$ to $t = t_0$, is the $-[S(t_0) - S(0)]$ divided by the original width of the strip.

To obtain the instantaneous Poisson ratio, the uniaxial loading of the PET strips was increased in increments of about 5 N, and the longitudinal and lateral strain changes from an initially slightly loaded (2.5 N) reference configuration were measured for each loading. These two strain measurements were each plotted against applied stress. The ratio of the slope of the lateral strain versus stress to the slope of the longitudinal strain versus stress gave the Poisson ratio.

3.3.1.2. Relaxation

In a relaxation experiment, strain (or elongation) is applied and held constant and the resulting stress (or tension) is measured as a function of time. A schematic of the relaxation test apparatus at the IBM Corp., Tucson, is shown in Fig. 3.25 (Bhushan et al., 1984c). A sample, 12.7-mm wide and 0.24-m long, was clamped on both sides. Proper clamping of the sample is extremely important, therefore, the clamp design was selected so that the sample is straight and uniformly loaded. One end of the sample was clamped to a cantilevered beam with strain gauges. The tension of the PET film as a function of time was measured by the strain gauges. The cantilevered beam was mounted on a linear slide. The initial strain was applied by the slide, which was driven by a dc linear motor. The movement of the slide was measured by a proximity probe against a fixed reference. During the test, the apparatus was placed inside an environmental chamber and the motor controller was placed outside the chamber. The displacement of the sample could be controlled and measured from the outside without opening the chamber.

The linear slide used at the upper end of the sample made it possible for the load not to be supported by the springs (in the slide) during relaxation. The cantilevered beam was very stiff at 400 N/mm and did not change (less than 1% of the sample elongation) during the test. Sample elongation was continuously monitored by the proximity probe.

For preparation of the test, the sample was properly mounted but not loaded. The test apparatus was placed inside an environmental chamber. The apparatus and the test sample were allowed to stabilize, which took 2–4 h. The desired initial strain was applied by a motor controller. A probe reading was taken at 140 mN tension and used as a reference. The desired elongation

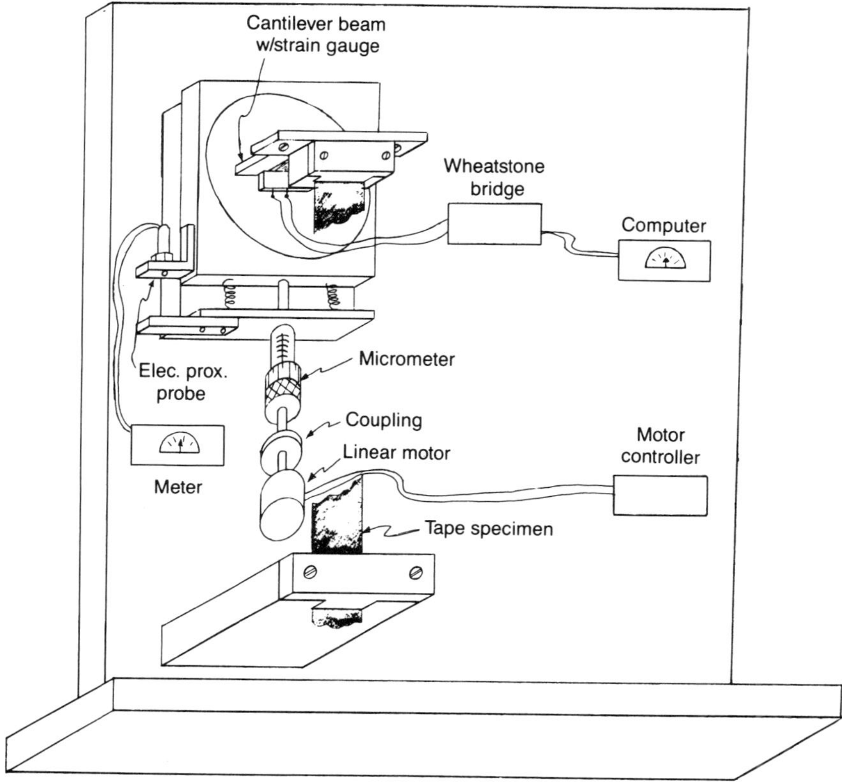

Fig. 3.25. Schematic of a stress relaxation apparatus (Bhushan et al., 1984c).

of 450 μm (strain $\sim 2 \times 10^{-3}$) was applied and the additional tension was measured. Once the test had started, the tension was measured by strain gauges and periodically the information was stored on a computer.

3.3.2. Constitutive Relations for Analysis of Isothermal Experimental Data

Biaxially-oriented PET film can be modeled as an orthotropic, linear-visco-elastic material. The film subjected to the simultaneous action of loads confined to its plane (relevant for the experimental data to be presented here), represents a plane-stress condition. The characteristics of the material within the plane are presented in Table 3.2. For this case, there are three stresses (σ_1, σ_2, σ_6) and three strains (ε_1, ε_2, ε_6) nonzero components and four elastic constants (C_{11}, C_{12}, C_{22}, and C_{66}) are required to specify the in-plane material response. The nonzero components of the stress and strain vectors with respect to a plane orthonormal coordinate system can be written in simplified vector form

$$\boldsymbol{\sigma}(t) = \begin{bmatrix} \sigma_{11} \\ \sigma_{22} \\ \sigma_{12} \end{bmatrix}, \qquad \boldsymbol{\varepsilon}(t) = \begin{bmatrix} \varepsilon_{11} \\ \varepsilon_{22} \\ 2\varepsilon_{12} \end{bmatrix}. \tag{3.107}$$

The elastic tensor $\mathbf{C}$ in matrix form can be written as

$$\mathbf{C}(t) = \begin{bmatrix} C_{11} & C_{12} & 0 \\ C_{12} & C_{22} & 0 \\ 0 & 0 & C_{66} \end{bmatrix}, \tag{3.108}$$

where C_{11} and C_{22} are the tensile elastic constants in the principal directions, C_{66} is the shear elastic constant, and C_{12} is related to the tensile elastic constants through the Poisson ratio (i.e., $C_{12} = v_{12}C_{22} = v_{21}C_{11}$). These elastic constants are time-dependent for a viscoelastic material. Similarly, we can write the compliance tensor $\mathbf{S}(t)$ in the matrix form.

Further, we can write the stress–strain relations using the engineering constants in a plane rectangular coordinate system, and which are coincident with the material's principal axes:

$$\begin{bmatrix} \sigma_{11} \\ \sigma_{22} \\ \sigma_{12} \end{bmatrix} = \begin{bmatrix} E_{11} & 0 & 0 \\ 0 & E_2 & 0 \\ 0 & 0 & G_{12} \end{bmatrix} \begin{bmatrix} \varepsilon_{11} \\ \varepsilon_{22} \\ 2\varepsilon_{12} \end{bmatrix}, \tag{3.109}$$

so that

$$\begin{bmatrix} \varepsilon_{11} \\ \varepsilon_{22} \\ 2\varepsilon_{12} \end{bmatrix} = \begin{bmatrix} D_1 & 0 & 0 \\ 0 & D_2 & 0 \\ 0 & 0 & J_{12} \end{bmatrix} \begin{bmatrix} \sigma_{11} \\ \sigma_{22} \\ \sigma_{12} \end{bmatrix}. \tag{3.110}$$

$E_1(t)$ and $E_2(t)$ are Young's moduli and $G_{12}(t)$ is the shear modulus. Similarly, $D_1(t)$ and $D_2(t)$ are the tensile creep compliances and $J_{12}(t)$ is the shear creep compliance. The components of the $\mathbf{C}(t)$ and $\mathbf{S}(t)$ tensors are related to measured creep and relaxation functions, $D_1(t)$, $D_2(t)$, $J_{12}(t)$, $E_1(t)$, $E_2(t)$, and $G_{12}(t)$, by the following expressions:

$$\begin{aligned} D_1(t) &= S_{11}(t) \quad &\text{and} \quad E_1(t) &= C_{11}(t)[1 - v_{12}(t)v_{21}(t)], \\ D_2(t) &= S_{22}(t) \quad &\text{and} \quad E_2(t) &= C_{22}(t)[1 - v_{12}(t)v_{21}(t)], \\ J_{12}(t) &= S_{66}(t) \quad &\text{and} \quad G_{12}(t) &= C_{66}(t). \end{aligned} \tag{3.111}$$

We will take the principal axes of the material as the spatial axes. If this is not the case, a rotation must be performed that will, in general, make all the components of the matrix $\mathbf{C}(t)$ nonzero.

Four independent relationships are needed to determine the components of matrix $\mathbf{C}$, three of which were obtained from the sample-extension creep experiments, and the fourth was provided by the assumed value of the Poisson ratio (v_{12}). The simple-extension creep experiments were performed under conditions of plane stress by loading material strips having a constant stress σ_0 and measuring the resulting elongation. The samples were cut from a 0.6-m-wide PET roll at three different (optical axes) orientations α, namely, $0°$, $45°$, and $90°$, with respect to the material's principal axis. The results can be written as relationships of the form

$$D_\alpha(t) = \frac{\varepsilon_\alpha(t)}{\sigma_0},$$ (3.112)

with $\alpha = 0°$, $45°$, and $90°$. The creep data were modeled using a generalized Kelvin–Voigt model, and were reduced to the form of the tensile-creep compliance function, $D_\alpha(t)$, $\alpha = 0°$, $45°$, and $90°$,

$$D_\alpha(t) = D_\alpha^0 + \sum_{k=1}^{I} D_\alpha^k [1 - \exp(-t/\tau_\alpha^k)],$$ (3.113)

where τ_α are the discrete retardation times. The experimental data were fitted using the nonlinear least-squares fit Marquardt algorithm.

The retardation spectrum gives the *continuous* spectrum of retardation times (Ferry, 1980). For numerical simplicity, the creep compliance data were fitted to discrete retardation times. For a viscoelastic solid that has a crystalline phase, one of the retardation times in a creep test is approximately zero, and the corresponding compliance contribution is D_α^0, the equilibrium compliance. This condition can be represented by the mechanical behavior of a spring (only) in series (or a zero viscosity of one of the Voigt–Kelvin elements which represents the contribution of the crystalline phase) in a generalized Voigt–Kelvin model [see Fig. 3.26(a)]. If the viscoelastic solid has only an amorphous phase (which is liquidlike), the compliance always continues to increase with time and the equilibrium compliance does not exist. A similar analysis can be developed for the stress–relaxation mode. For a viscoelastic solid that has a crystalline phase present, one of the relaxation times must be infinite, and the corresponding modulus contribution is the equilibrium (relaxed) modulus. This condition can be represented by a spring (only) in parallel (or an infinite viscosity of one of the Maxwell elements which represents the contribution of the crystalline phase) in a generalized Maxwell model [Fig. 3.26(b)].

In order to transform the measured creep compliance functions into stress relaxation functions, the orientation transformation of $D_\alpha(t)$ to the components $D_1(t)$, $D_2(t)$, and $J_{12}(t)$ of the creep compliance tensor is required. The resulting relationships are

$$D_1(t) = D_0(t),$$

$$D_2(t) = D_{90}(t),$$ (3.114)

$$J_{12}(t) = 4D_{45}(t) - [D_0(t) + D_{90}(t)] + 2v_{12}(t)D_0(t),$$

where v_{12} is the Poisson ratio $(= -\varepsilon_{22}/\varepsilon_{12}$ when a pure tensile or compressive load is applied in the 1-direction).

The creep-compliance functions in Eq. (3.114) were then transformed into

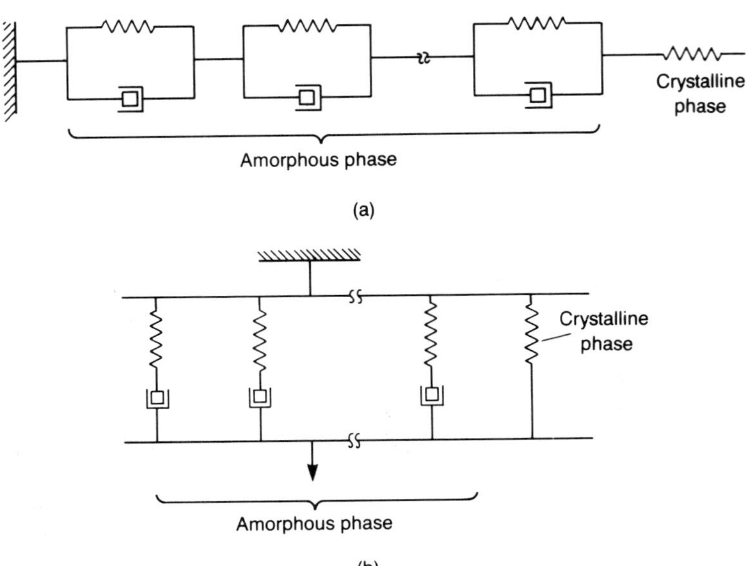

Fig. 3.26. Mechanical model of a semicrystalline material subjected to (a) creep experiment—spring in series in a generalized Voigt–Kelvin model and (b) relaxation experiment—a spring in parallel in a generalized Maxwell model.

relaxation functions via the convolution equation[4] [see Eq. (3.56)]

$$E_i(t)D_i(0) - \int_0^t E_i(t)\frac{dD_i(t-\tau)}{d\tau}\,d\tau = 1, \qquad i = 1, 2 \text{ (no summation of } i),$$

and   (3.116)

$$G_{12}(t)J_{12}(0) - \int_0^t G_{12}(t)\frac{dJ_{12}(t-\tau)}{d\tau}\,d\tau = 1.$$

---

[4] We should mention that simpler approximate expressions are available for the inversion of creep-compliance functions into relaxation functions. In particular, a widely used approximation is given by (Ferry, 1980)

$$D_i(t)E_i(t) \simeq \frac{\sin m\pi}{m\pi}, \qquad i = 1, 2 \text{ (no summation on } i),$$

and   (3.115)

$$J_{12}(t)G_{12}(t) \simeq \frac{\sin m\pi}{m\pi},$$

where m is the slope of the doubly logarithmic plot of $D_i(t)$ or $J_{12}(t)$ versus log t. This relationship is valid only if $D_i(t)$ or $J_{12}(t)$ is a linear function of log t over a moderate interval of the time scale, and the slope m is much less than 1. However, none of these conditions are fulfilled by the experimental data of PET films. In particular, the log-log representation of the measured creep functions has the behavior of a quadratic function that deviates significantly from a straight line (see data, to be discussed later) and the more complex inversion is necessary.

An appropriate procedure for the numerical integration is proposed by Taylor (1964). This procedure is extremely accurate and allows for large time increments to be used before a loss of accuracy becomes evident. The algorithm has the following form (Taylor, 1964) with $E = E_1$, E_2, and G_{12}, $D = D_1$, D_2, and J_{12}:

$$E^n = \frac{2}{D^0 + D^1}\left[1 + \tfrac{1}{2}E^0(D^{n-1} - D^n) + \frac{1}{2}\sum_{i=1}^{n-1} E^i(D^{n-i-1} - D^{n-i+1})\right] \quad (3.117)$$

and

$$E^0 = \frac{1}{D^0}, \quad (3.118)$$

where $E^k = E(t_k)$, $D^k = D(t_k)$, $k = 0, \ldots, n$, $t_k = k\,\Delta t$, and, for simplicity, we have assumed that the time increment Δt is a fixed constant. For shorter tests (~ 50 h) we used $\Delta t = \tfrac{1}{2}$ h, and for longer tests (~ 500 h) we used $\Delta t = 2\tfrac{1}{2}$ h.

The components of the tensile-relaxation matrix $\mathbf{C}$ now follow (Table 3.2)

$$C_{11}(t) = E_1^2(t)/B(t), \quad (3.119a)$$

$$C_{12}(t) = v_{12}(t)E_1(t)E_2(t)/B(t)$$

$$= v_{12}(t)C_{22}(t), \quad (3.119b)$$

$$C_{22}(t) = E_1(t)E_2(t)/B(t)$$

$$= C_{12}(t)/v_{12}(t), \quad (3.119c)$$

$$C_{66}(t) = G_{12}(t), \quad (3.119d)$$

where $B(t) = E_1(t) - v_{12}^2(t)E_2(t)$. We also note that

$$v_{21}/v_{12} = C_{22}/C_{11} = E_2/E_1. \quad (3.119e)$$

Calculations were made only for C_{11}, C_{12}, and C_{66}. C_{22} was calculated from C_{12} and v_{12} [Eq. (3.119c)].

The functions $C_{ij}(t)$ in Eq. (3.119) are obtained in numerical form. For use in finite-element stress analyses to be presented in the next chapter, they were fitted by a Prony series of the form [Eq. (3.91)]

$$C_{ij}(t) = C_{ij}^0 + \sum_{k=1}^{I} C_{ij}^k \exp(-t/\tau_{ij}^k). \quad (3.120)$$

Using the Marquardt algorithm and two relaxation times τ^k ($k = 1, 2$). The resulting relaxation times τ_{11}^k, τ_{12}^k, τ_{22}^k, and τ_{66}^k were averaged, and the average values were then used to calculate new coefficients of C_{ij}^k in Eq. (3.120) where τ^k are the relaxation times. The number of relaxation times I to be used in the representation will depend on the duration of the creep experiments and the frequency of measurements made during the early part of the experiments.

3.3.3. Experimental Results

Bhushan et al. (1984c) and Bhushan and Connolly (1986) conducted creep and relaxation tests on 12.7-mm-wide, 23.4-μm-thick, and 0.2-m-long PET films (Mylar A, 92 DB) cut from a 6-m-wide PET roll at three different orientations, 0°, 45°, and 90°, relative to the fast optical axis, i.e., material axis with minimum Young's modulus of elasticity (which was found to be as the extinction orientation nearest the machine direction, for more details see Chapter 2). Their data were found to be very reproducible (within a few percent).

Bhushan and Connolly (1986) conducted an initial set of experiments to determine if the PET material behaved as a linearly viscoelastic material, which requires that creep compliance as a function of time be independent of stress. They also ran tests to study the effect of PET-film thickness on creep compliance.

The next set of creep experiments was conducted at constant humidity and different temperatures, and at constant temperature and different humidities. From the measured creep compliances [$D_0(t)$, $D_{45}(t)$, $D_{90}(t)$] and an assumed constant value of the Poisson ratio (v_{12}), Bhushan et al. (1984c) predicted the relaxation functions. They also conducted the stress relaxation tests and compared the data with the inverted relaxation functions obtained from creep tests. Verification would imply that the derived constitutive relationship for PET film and the associated assumption of linear viscoelasticity are valid.

The data at various temperatures and humidities were used to determine the shift factors by time–temperature and time–humidity superpositions. The superpositions are valid for thermorheologically simple materials that exhibit a pure shift along the logarithmic time axis in the creep function, relaxation modulus, and other characteristic functions of the material, when plotted against the logarithm of time for a constant change of temperature or humidity. An inherent assumption in the superposition principle is that the unrelaxed (at $t = 0+$) and relaxed (at $t = \infty$) viscoelastic functions are the same at different temperatures (McCrum and Morris, 1964). To accomplish this, a slight vertical shift is needed before the viscoelastic functions can be shifted along the log time axis.

Bhushan and Connolly (1986) studied the creep and relaxation behavior of PET film under time-dependent temperature fields (nonisothermal viscoelasticity). Tests were also conducted on coated PET films (magnetic tapes) and free films of magnetic coatings to compare their results with those of tests on uncoated PET films. Vallat et al. (1986, 1988a, b) studied the effect of thermal treatments of PET film on creep behavior.

3.3.3.1. Linearity of PET Material Response

Bhushan and Connolly (1986) tested the 23.4-μm-thick PET film along the 0° orientation at 51.7° C and 30% RH and at four loads of 0.97, 2.08, 3.19, and 4.30 N (excluding the initial 140 mN), corresponding to normal stresses of 3.3, 7.0, 10.8, and 14.5 MPa, respectively. Creep-compliance data for all four tests

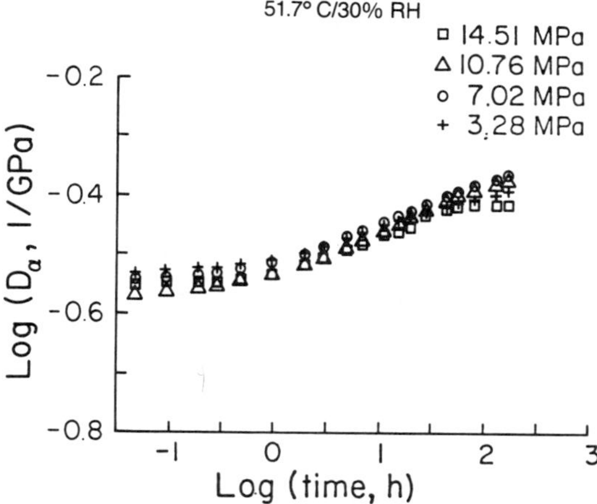

Fig. 3.27. Log (tensile-creep compliance) as a function of log (time) for a 23.4-μm-thick film (Mylar A) along 0° orientation at 51.7° C and 30% RH. Tests were conducted at different stresses (Bhushan and Connolly, 1986).

are shown in Fig. 3.27. Note that the data of the different tests superimpose quite well indicating that total deformation is a linear function of load in the range tested. Similar results are also reported by Vallat et al. (1986).

If the material is linearly viscoelastic, it should, in addition, follow the Boltzmann superposition principle. This comparison was made by using the creep-recovery data. If $D(t)$ is the tensile-creep compliance, the tensile compliance during recovery is given as follows:

$$D_\alpha^r(t) = D_\alpha(t) - D_\alpha(t - t_1), \qquad (3.121)$$

where the load is removed at time t_1. The measured $D_\alpha^r(t)$ were compared with those predicted in Eq. (3.121). $D_\alpha(t)$ beyond the test duration was predicted from extrapolation of the straight line obtained by curve fitting of the log (creep compliance) versus log (time) data. A sample comparison is shown in Fig. 3.28. We found some discrepancy after about 100 h of recovery. The Boltzmann superposition principle strictly holds for a short time (a couple of hours), therefore, the PET material has restricted linearity.

3.3.3.2. Effect of PET Film Thickness

Three films of thicknesses, 14.5, 23.4, and 36.1 μm, were selected to study the effect of film thickness on creep compliance, and the results are presented in Fig. 3.29. Note that there are differences between films of different thicknesses, which we believe are real and are probably caused by differences in crystallinity from the manufacturing process. Thinner films exhibit superior creep resistance. We note that thinner films had also exhibited a higher Young's modulus, see Chapter 2.

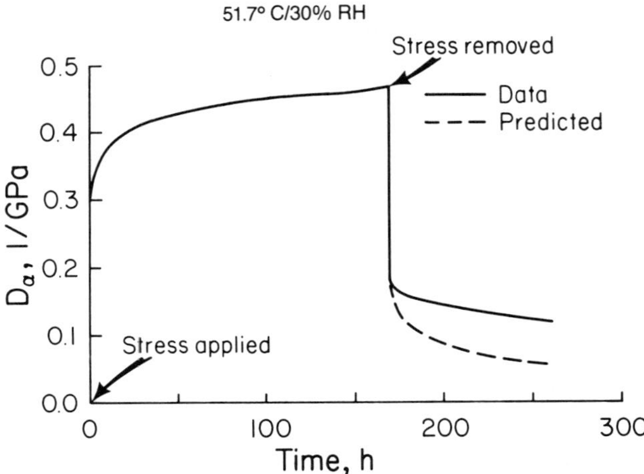

Fig. 3.28. Tensile-creep compliance as a function of time for a 23.4-μm-thick PET film (Mylar A) along 0° orientation at 51.7° C and 30% RH. Data show comparisons between actual recovery data and those predicted from the Boltzmann superposition principle (Bhushan and Connolly, 1986).

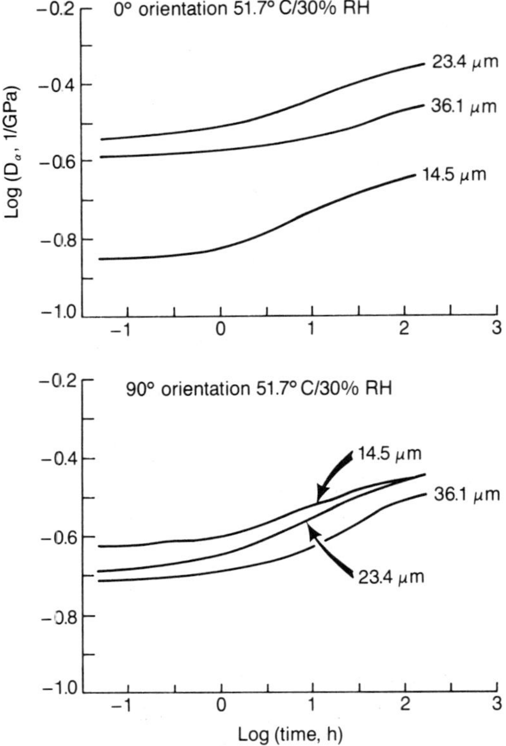

Fig. 3.29. Log (tensile-creep compliance) as a function of log (time) for a PET film (Mylar A) in two orientations at 51.7° C and 30% RH for different PET film thicknesses (Bhushan and Connolly, 1986).

3.3.3.3. Effects of Temperature and Humidity in PET Films

Creep tests were conducted on 23.4-μm-thick PET films for eleven different combinations of temperature and humidity, with temperatures below the glass-transition temperature. The experiments were divided into two groups: one for constant humidity and varying temperatures, the other for constant temperature and varying humidities. One experiment was common to both groups. All tests were conducted at a normal stress of 7.0 MPa.

The measured creep compliances as a function of time for samples in three orientations ($\alpha = 0°$, 45°, and 90°) are shown in Figs. 3.30–3.34. Note the effects of temperature and humidity in the samples at 45° to the material's axis (Figs. 3.30–3.32). For low temperatures and humidities, the compliance function remains bounded by those at 0° and 90°, however, for temperatures at

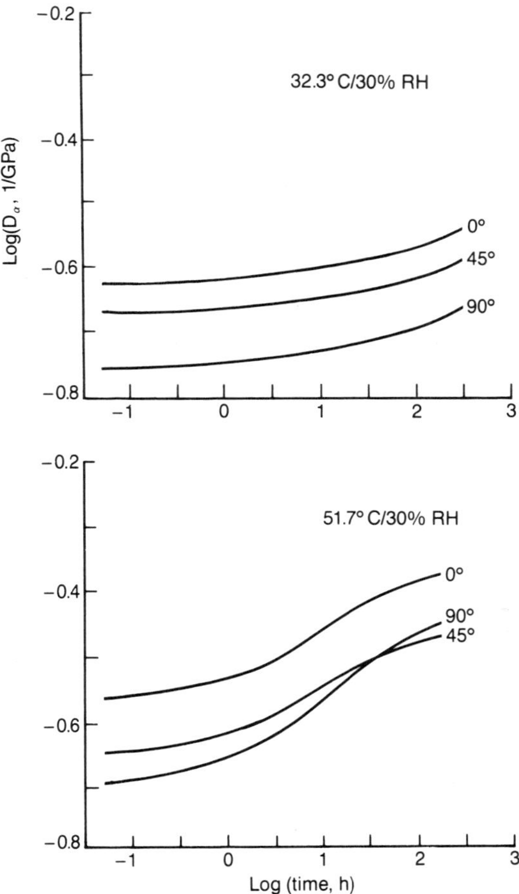

Fig. 3.30. Log (tensile-creep compliance) as a function of log (time) for a 23.4-μm-thick PET film in three orientations at two different temperatures (Bhushan et al., 1984c).

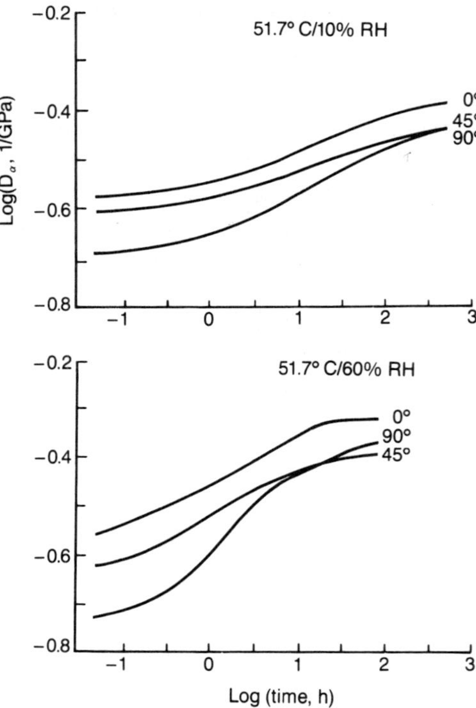

Fig. 3.31. Log (tensile-creep compliance) as a function of log (time) for a 23.4-μm-thick PET film (Mylar A) in three orientations at two different humidities (Bhushan et al., 1984c).

or above 51.7° C and all humidities at 51.7° C, the curves for 45° and 90° cross over after some time, and the compliance for the 45° orientation becomes lower than that for 90°. The time taken for this crossover to take place is also found to be dependent on temperature and humidity (Bhushan et al., 1984c). The results in Figs. 3.33 and 3.34 show a strong dependence on both temperature and humidity. As expected, the creep compliance decreases with decreasing temperature and humidity. Saturation is also observed to occur faster at higher temperatures and humidities. Creep compliance and its recovery data are presented in Fig. 3.35 for tests at 32.3° C/30% RH and at 51.7° C/30% RH in three orientations.

Constitutive Stress–Strain Relationships

The measured creep-compliance functions $[D_\alpha(t)]$ are now transformed into relaxation functions $[C_{ij}(t)]$ using the procedure described earlier. To do this, the creep data is first fitted with a Prony series [Eq. (3.113)]. The components of the relaxation tensor **C** obtained from Eq. (3.119) were fitted using a Prony series [Eq. (3.120)]. The time scales should be selected so as to resolve the

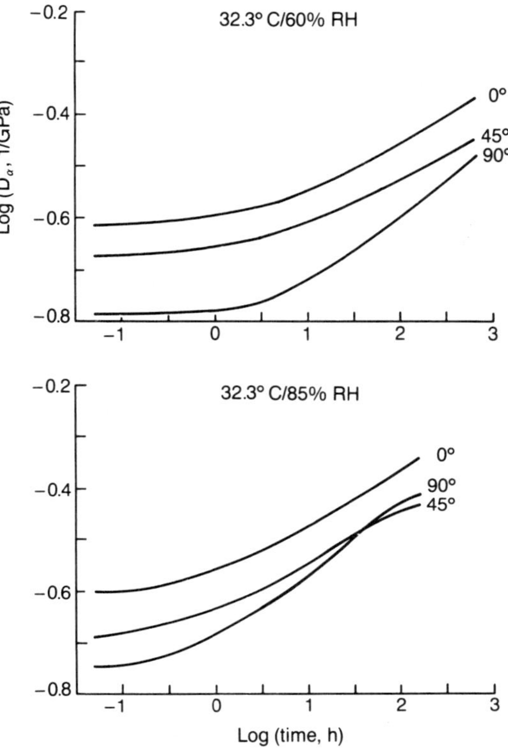

Fig. 3.32. Log (tensile-creep compliance) as a function of log (time) for a 23.4-μm-thick PET film (Mylar A) in three orientations at two different humidities (Bhushan and Connolly, 1986).

initial sequence of measurements, be separated by about a decade, and be much lower than the duration of the experiments. For the experimental duration (~ 150 h), it was found that a three-time scales fit was not possible, so only two-time scales were used. In Fig. 3.36, a comparison of the experimental, creep-compliance data at 0° orientation at 51.7° C and 30% RH is made between curves fitted by using only one and two retardation times. Note the inadequacy of a fit using only one time. We recognize that if tests are conducted for longer durations, additional relaxation (and retardation) times will be picked, because most materials have continuous relaxation (or retardation) time spectra having numerous dominant times. Relaxation (or retardation) due to these times will be an important factor at longer times.

To calculate the components of the creep compliance tensor and the relaxation tensor in addition to three tensile creep tests we need the Poisson ratio, v_{12}. Bogy et al. (1979) simultaneously measured the longitudinal extension Δl and the lateral contraction Δw of the PET films and coated tapes loaded in the longitudinal direction. (See Section 3.3.1 on the description of creep and

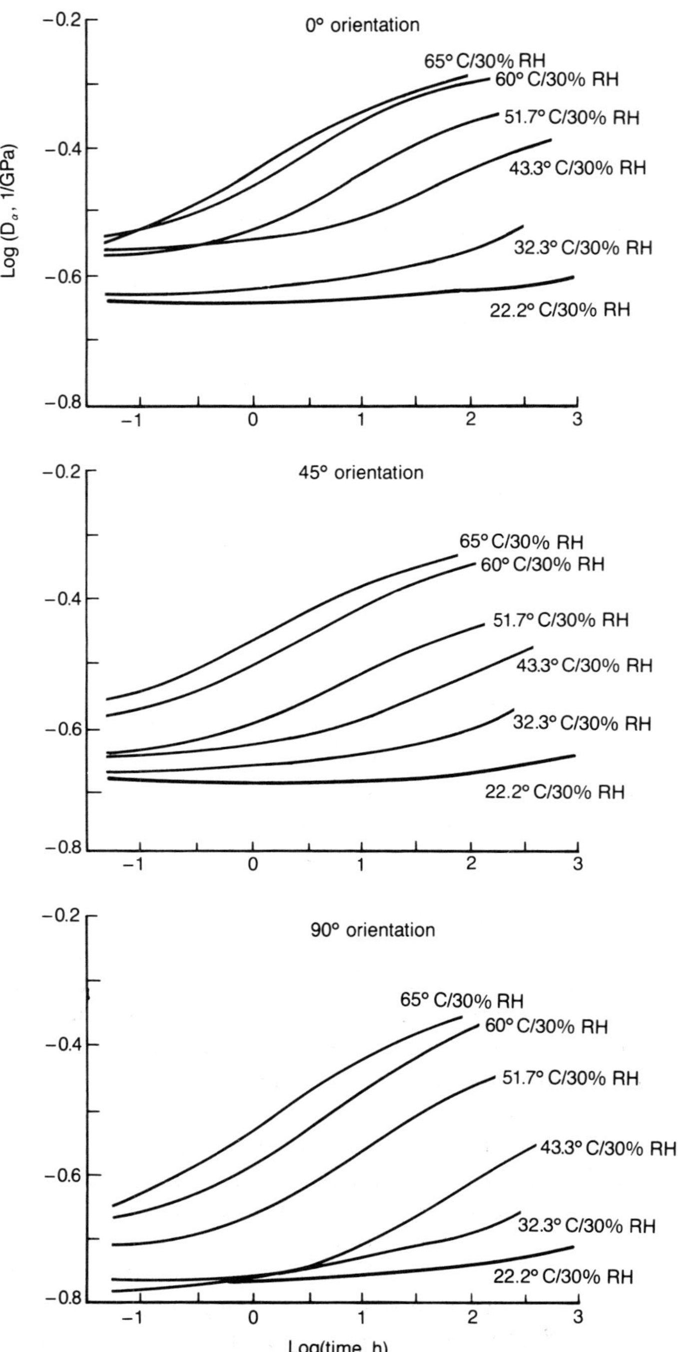

Fig. 3.33. Log (tensile-creep compliance) as a function of log (time) for a 23.4-μm-thick PET film (Mylar A) in three orientations at six different temperatures and 30% RH (Bhushan et al., 1984c).

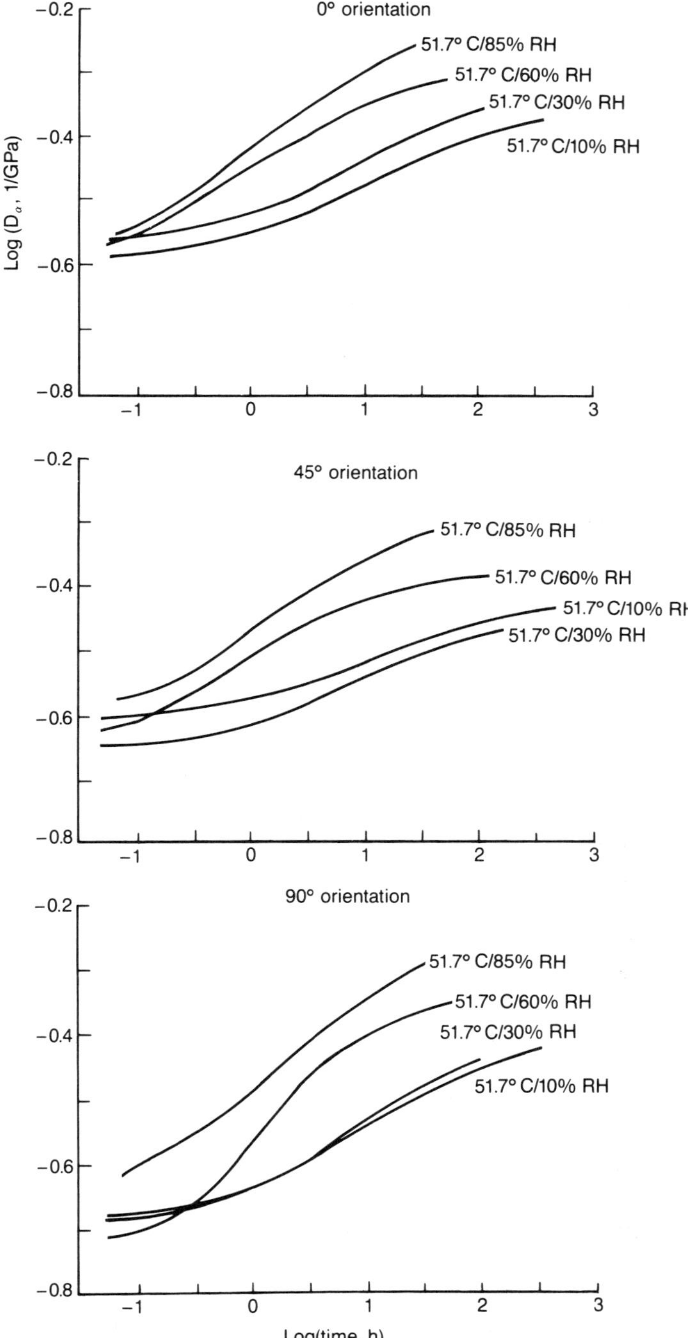

Fig. 3.34. Log (tensile-creep compliance) as a function of log (time) for a 23.4-μm-thick PET film (Mylar A) in three orientations at four different humidities (Bhushan et al., 1984c).

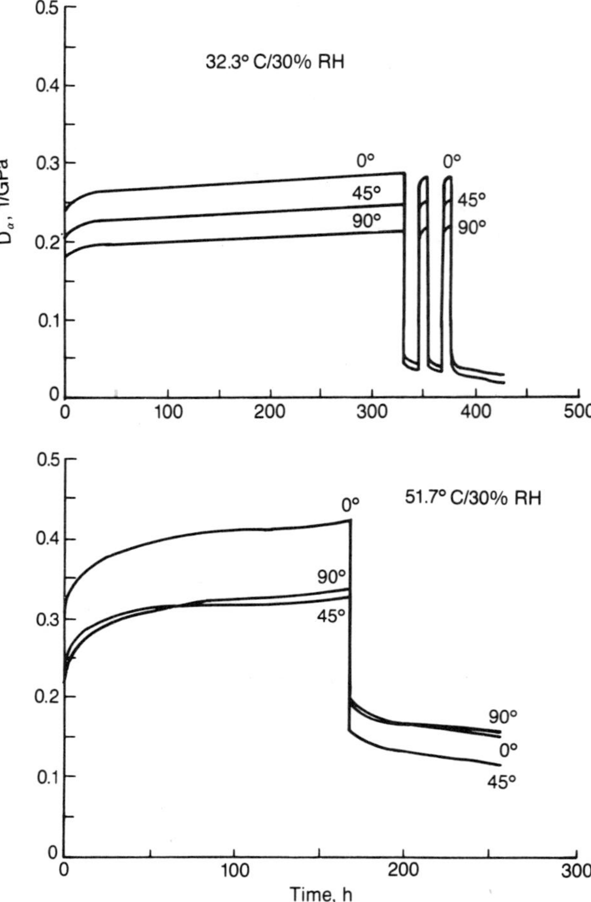

Fig. 3.35. Log (tensile-creep compliance) as a function of log (time) for a 23.4-μm-thick PET film (Mylar A) in three orientations including recovery data (Bhushan et al., 1984c).

relaxation apparatuses for details.) They found that the Poisson ratio for a 23.4-μm-thick PET film at 45° and 55° C was about 0.28 and to a first approximation, time-independent, Fig. 3.37. The Poisson ratio for a 38.1-μm-thick PET film at temperatures ranging from 22° to 55° C was about 0.23 and again time-independent. The Poisson ratios for the coated and uncoated tapes were comparable. This result that the Poisson ratio is time-independent is of significance since it greatly simplifies analytical calculations and also reduces the experimental effort. We have assigned the value of $v_{12} = 0.3$ throughout our study. Our experimentation with several values of v_{12}, namely, 0.2, 0.3, and 0.4, did not show a significant effect on the calculated value of **C**.

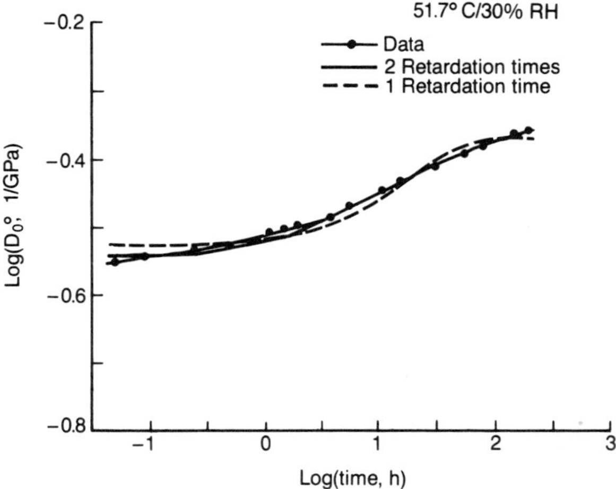

Fig. 3.36. Log (tensile-creep compliance) as a function of log (time) for a 23.4-μm-thick PET film (Mylar A) along 0° orientation. Experimental data and curve fits by a Prony series are shown (Bhushan et al., 1984c).

The resulting coefficients and relaxation and retardation times are given in Tables 3.5 and 3.6. Note that the relaxation and retardation times are different. Predicted values of log $E_\alpha(t)$ as a function of log t at 0° and 90° orientations for different temperatures and humidities are plotted in Figs. 3.38–3.40. Values for $E_{45}(t)$ can be calculated using Eqs. (3.25).

Direct stress-relaxation tests were conducted to verify the predictions obtained from creep data. Tests were conducted at 0° orientation of PET films at 43.3°, 51.7°, and 60° C, and at 30% RH. Results are presented in Fig. 3.41

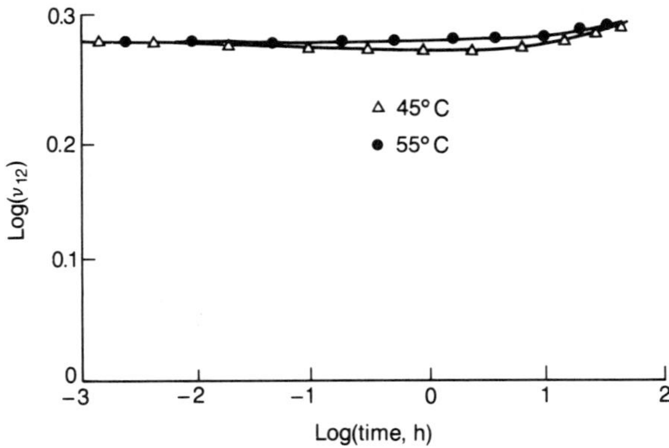

Fig. 3.37. The Poisson ratio as a function of log (time) for a 23.4-μm-thick PET film (Mylar A) at two temperatures, normal stress = 13 MPa.

Table 3.5. Coefficients for tensile creep compliance (D_α) and retardation times (τ_α) of PET films (Mylar A) and coated magnetic tapes

α, deg.	°C/RH%	D_α^0, 1/GPa	D_α^1, 1/GPa	D_α^2, 1/GPa	τ_α^1, h	τ_α^2, h	D_α^r, 1/GPa x^a	y^a
				(a) *PET film*				
0	22.3/30	0.243	0.0117	0.0368	32.73	1461.7	—	—
45		0.220	0.0082	0.0314	13.69	1196.7	—	—
90		0.189	0.0074	0.0295	14.52	1259.8	—	—
0	32.3/30	0.238	0.0209	0.1063[a]	11.47	978.0[b]	0.0602	0.0393
45		0.213	0.0176	0.1217[a]	11.13	1352.7[b]	0.0542	0.0319 (93)
90		0.174	0.0182	0.1056[a]	9.67	1104.1[b]	0.0593	0.0409
0	43.3/30	0.283	0.0468	0.0689	17.97	253.5	0.1360	0.0951
45		0.225	0.0307	0.0619	7.76	141.4	0.1139	0.0775 (142)
90		0.170	0.0323	0.0758	9.96	157.1	0.1212	0.0900
0	51.7/30	0.286	0.0627	0.0916	3.75	44.52	0.1920	0.1244
45		0.228	0.0555	0.0625	3.73	41.83	0.1436	0.0817 (90)
90		0.206	0.0588	0.0903	3.83	41.93	0.1858	0.1281
0	60.0/30	0.290	0.1055	0.0950	2.03	32.75	0.2257	0.1434
45		0.259	0.0839	0.0837	1.93	31.97	0.2127	0.1043 (138)
90		0.218	0.0929	0.1051	2.61	38.54	0.1664	0.1521
0	65.0/30	0.279	0.1074	0.1025	0.78	19.21	0.2522	0.1269
45		0.267	0.0899	0.0827	0.76	18.00	0.2268	0.1012 (138)
90		0.223	0.0999	0.1054	0.90	19.28	0.2578	0.1433
0	51.7/10	0.271	0.0673	0.0770	5.67	91.62	0.1809	0.1318
45		0.253	0.0770	0.0815	5.21	88.42	0.1624	0.1157 (118)
90		0.211	0.0770	0.0815	7.02	122.90	0.2043	0.1624
0	51.7/60	0.269	0.0882	0.1253	0.37	8.16	0.2627	0.1978
45		0.233	0.1030	0.0767	0.71	13.56	0.2159	0.1560 (50)
90		0.188	0.1508	0.0969	1.72	23.29	0.2539	0.1951
0	51.7/85	0.279	0.1068	0.1707	0.49	10.97	0.3084	0.1614
45		0.259	0.1067	0.1284	0.86	13.49	0.2735	0.1273 (100)
90		0.227	0.1223	0.1582	0.83	12.86	0.3370	0.1737
0	32.3/60	0.246	0.0620	0.1153	12.50	204.5	0.1995	0.1488
45		0.209	0.0388	0.0917	5.36	137.7	0.1641	0.1206 (157)
90		0.161	0.0620	0.1120	17.03	326.0	0.1758	0.1466
0	32.3/85	0.251	0.0733	0.1293	5.58	52.99	0.2237	0.1690
45		0.208	0.0621	0.1167	3.76	63.12	0.1835	0.1343 (74)
90		0.176	0.0554	0.1441	1.55	35.74	0.1687	0.1267
				(b) *Tapes A and D*				
Tape A	22.2/30	0.199	0.0524	0.0466	2.16	491.4	—	—
Tape D (without a back coat)	22.2/30	0.161	0.0257	0.0620	11.02	938.7	—	—

[a] x is the value after unloading; y is the value after the number of hours indicated in parentheses.
[b] Since the retardation time is larger than the test duration, the absolute values of the data are questionable.

Table 3.6. Coefficients for the relaxation modulus (C_{ij}) and the average relaxation times (τ_i) of PET films (Mylar A)

°C/RH%	C_{ij}	C_{ij}^0, GPa	C_{ij}^1, GPa	C_{ij}^2, GPa	τ_1, h	τ_2, h
22.2/30	C_{11}	3.90	0.20	0.56		
	C_{12}	1.50	0.07	0.23		
	C_{22}	4.99	0.22	0.75	18.02	1114.00
	C_{66}	1.43	0.06	0.19		
32.3/30	C_{11}	2.90	0.41	1.49[a]		
	C_{12}	1.07	0.19	0.71[a]	9.87	818.00[a]
	C_{22}	3.55	0.63	2.36[a]		
	C_{66}	1.12	0.12	0.48[a]		
43.3/30	C_{11}	2.94	0.56	0.71		
	C_{12}	1.22	0.41	0.45	9.02	131.83
	C_{22}	4.08	1.36	1.51		
	C_{66}	1.19	0.19	0.22		
51.7/30	C_{11}	2.56	0.76	0.67		
	C_{12}	0.95	0.42	0.30	2.98	33.75
	C_{22}	3.15	1.41	1.01		
	C_{66}	1.15	0.33	0.19		
60.0/30	C_{11}	2.28	1.09	0.54		
	C_{12}	0.81	0.53	0.28	1.56	26.90
	C_{22}	2.69	1.65	0.94		
	C_{66}	0.91	0.33	0.17		
65.0/30	C_{11}	2.28	1.16	0.59		
	C_{12}	0.78	0.49	0.25	0.56	14.85
	C_{22}	2.60	1.64	0.83		
	C_{66}	0.88	0.32	0.16		
51.7/10	C_{11}	2.68	0.91	0.60		
	C_{12}	0.91	0.43	0.27	4.43	79.62
	C_{22}	3.03	1.45	0.91		
	C_{66}	1.06	0.23	0.15		
51.7/60	C_{11}	2.29	1.47	0.52		
	C_{12}	0.77	0.79	0.29	0.47	12.33
	C_{22}	2.58	2.65	0.98		
	C_{66}	0.97	0.45	0.14		
51.7/85	C_{11}	1.98	1.26	0.76		
	C_{12}	0.65	0.57	0.27	0.54	9.56
	C_{22}	2.18	1.91	0.91		
	C_{66}	0.81	0.39	0.25		
32.3/60	C_{11}	2.67	1.07	1.03		
	C_{12}	1.05	0.61	0.55	8.26	155.80
	C_{22}	3.50	2.02	1.83		
	C_{66}	1.12	0.30	0.27		
32.3/85	C_{11}	2.44	1.23	0.99		
	C_{12}	0.86	0.65	0.39	2.68	40.81
	C_{22}	2.88	2.16	1.30		
	C_{66}	1.04	0.35	0.43		

[a] Since relaxation times are larger than test durations, the absolute values of the data are questionable.

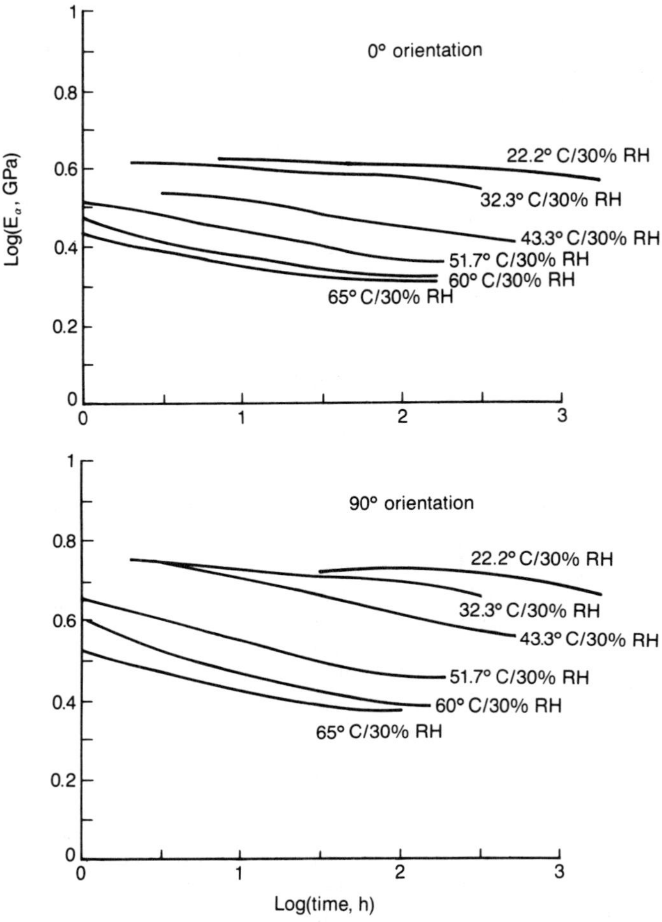

Fig. 3.38. Log (tensile-relaxation modulus) as a function of log (time) for a 23.4-μm-thick PET film (Mylar A) in two orientations at five different temperatures (Bhushan et al., 1984c).

for comparisons with the directly measured data. We have normalized the initial values of the relaxation modulus with those at 51.7° C and 30% RH of creep data. A much better agreement can be observed in the normalized case, which is due to the fact that measurements of the initial displacements in the creep data are subject to some error, because of the difficulty in loading the samples instantaneously. The normalized data eliminates this discrepancy. Very good agreement can be observed in Fig. 3.41 between the experimentally measured stress-relaxation data and that obtained using the inverted function C_{ij}, validating our approach and assumption of linear viscoelasticity.

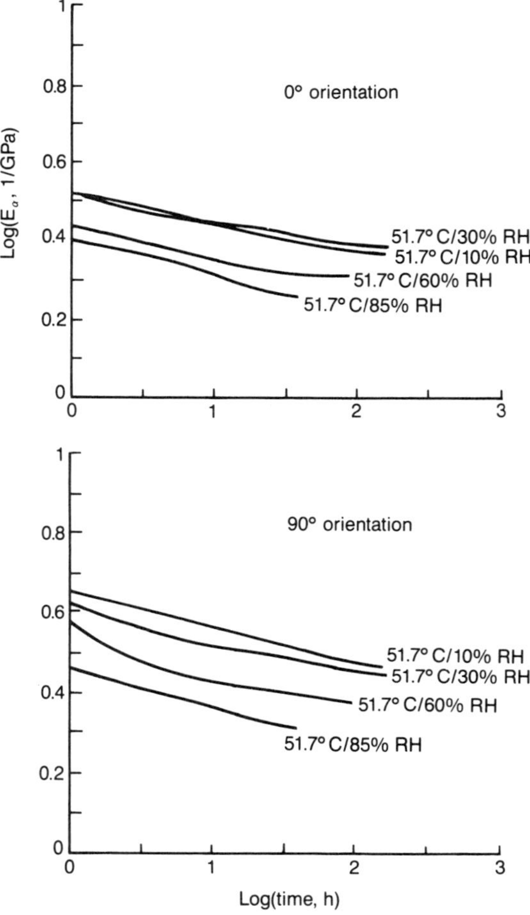

Fig. 3.39. Log (tensile-relaxation modulus) as a function of log (time) for a 23.4-μm-thick PET film (Mylar A) in two orientations at four different humidities (Bhushan et al., 1984c).

Time–Temperature Superposition

Log (creep compliance) as a function of log (time) at different temperatures were superimposed, giving the master curves shown in Fig. 3.42, for which the reference temperature is 51.7° C. To get the best superposition, it was necessary to shift the curves by small amounts (commonly designated log c_T) along the log $D_\alpha(t)$ axis. An inherent assumption in the superposition principle is that the unrelaxed (at $t = 0+$) and relaxed compliance or moduli (at $t = \infty$) are the same at different temperatures (McCrum and Morris, 1964). Therefore, the curves had to be shifted vertically to offset the difference in the

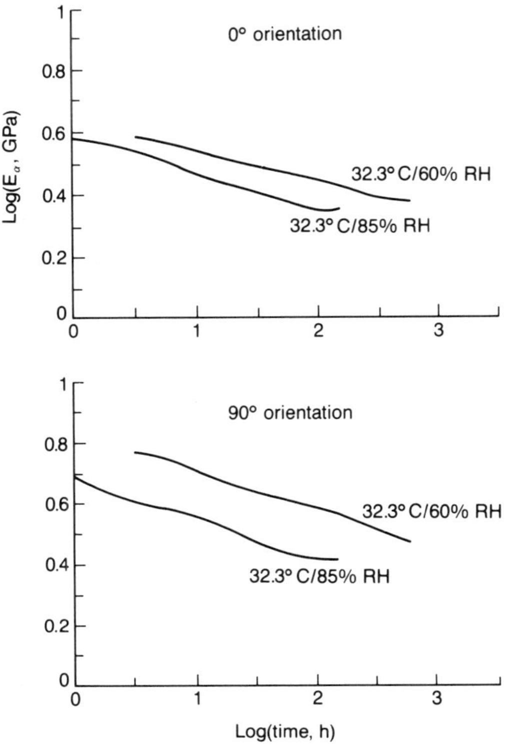

Fig. 3.40. Log (tensile-relaxation modulus) as a function of log (time) for a 23.4-μm-thick PET film (Mylar A) in two orientations at different humidities (Bhushan and Connolly, 1986).

unrelaxed and relaxed moduli at different temperatures. Also, $D_\alpha(t)$ data at different temperatures reflected run-to-run variability, especially the initial elongation during loading (near time zero). The vertical shift offset these errors. Log c_T had no fixed pattern with temperature; maximum log c_T was about 0.05 (log GPa^{-1}), which provided some uncertainty in the shift distances along the log t axis giving values of log a_T, the shift factor. An uncertainty in a_T of 10% was introduced because of log c_T.

We note that the short time compliance level for the less oriented direction (0°) with a lower modulus of elasticity is considerably greater than in the 90° direction. However, at long reduced times the compliance curves begin to merge. The compliance curves for the 45° directions are intermediate.

We expected a_T to have an Arrhenius temperature dependence, i.e.,

$$a_{T_2}/a_{T_1} = t_2/t_1 = \exp[(-\Delta H_a/R)(1/T_1 - 1/T_2)], \qquad (3.122)$$

where ΔH_a is the activation energy in kJ/mole (kcal/mole), R is the universal gas constant (equal to 8.313 kJ/mole K or 1.986 kcal/mole K), T is the

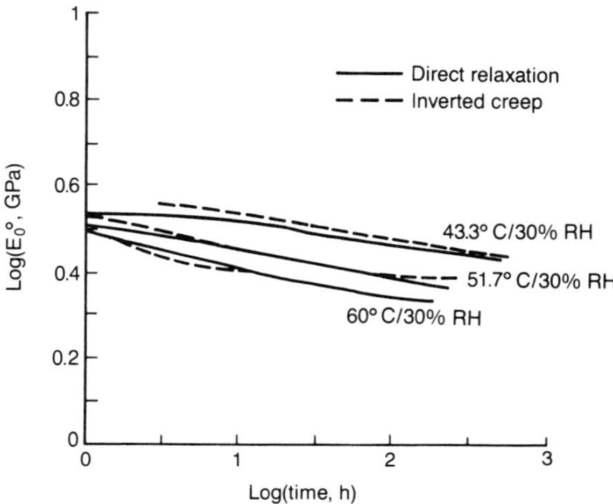

Fig. 3.41. Comparisons of log (tensile-relaxation modulus) as a function of log (time) for a 23.4-μm-thick PET film (Mylar A) along the 0° orientation of predictions obtained from creep data and as directly measured (Bhushan et al., 1984c).

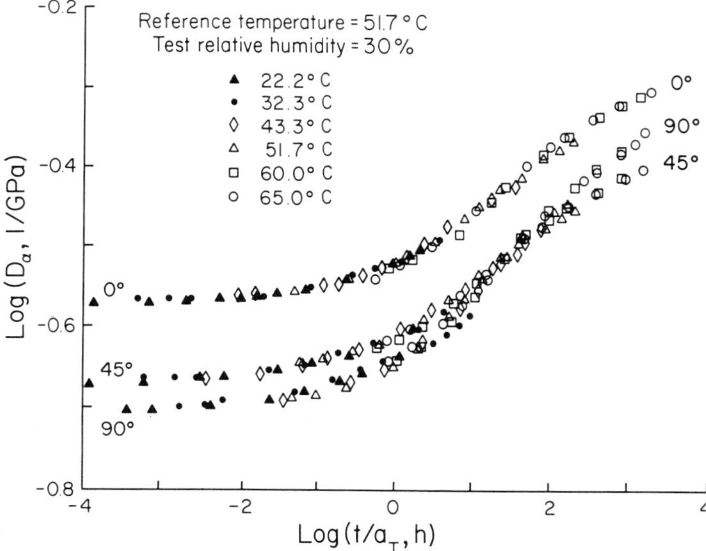

Fig. 3.42. Master curves of log (tensile-creep compliance) as a function of log (time) for a 23.4-μm-thick PET film (Mylar A) in three orientations reduced to 51.7° C and 30% RH. All the data were taken at 30% RH (Bhushan and Connolly, 1986).

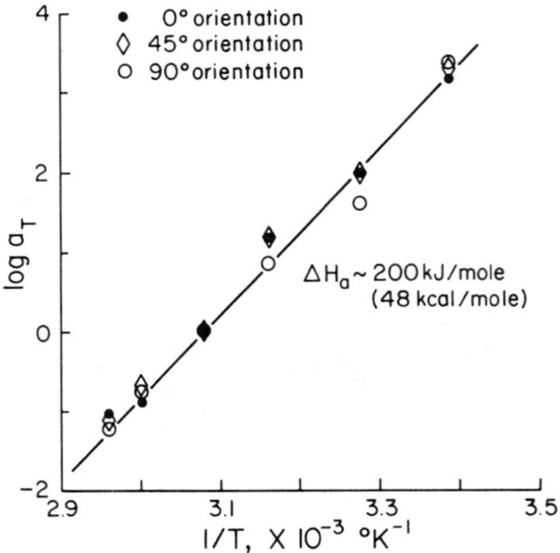

Fig. 3.43. Temperature dependence of shift factors, log a_T, obtained from constructing the master curves in Fig. 3.42 (Bhushan and Connolly, 1986).

absolute temperature in K, and t is time. The activation energy is calculated from a straight line on a plot of log a_T versus $1/T$ (Fig. 3.43) whose slope gives an activation energy of

$$\Delta H_a \sim 200 \text{ kJ/mole (48 kcal/mole).}$$

This activation energy is comparable to the value of 240 kJ/mole (58 kcal/mole) for the storage modulus of PET film, as reported earlier [Eq. (3.105)]. This activation energy also compares well with 192 kJ/mole (46 kcal/mole) for PET fiber reported by Murayama et al. (1968). Log a_T data were also plotted as a function of T in Fig. 3.44. The data conformed rather well with the equation

$$\log a_T \sim -0.110(T - 324.7)$$
$$-0.110(\Delta T). \tag{3.123}$$

Equation (3.123) compares very well with the data of Murayama et al. (1968) and Smith (1983–84).

Note that the apparent activation energy of PET film is constant in the range of the test conditions (below glass transition temperature), which implies that the shift factor can be determined from an Arrhenius equation. We note that a single activation energy does not imply a single relaxation time (Ferry, 1980).

Time–Humidity Superposition

Tests were conducted at four relative humidities of 10, 30, 60, and 85% at 51.7° C and at three humidities of 30, 60, and 85% at 32.3° C. The log (creep compliance) as a function of log (time) data at different humidities were

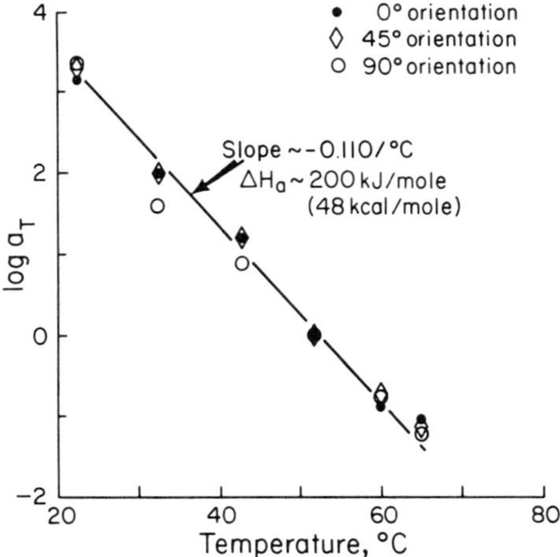

Fig. 3.44. Temperature dependence of shift factors, log a_T, obtained from constructing the master curves in Fig. 3.42 (Bhushan and Connolly, 1986).

superimposed, giving the master curves shown in Figs. 3.45 and 3.46. The reference humidity for both temperatures (51.7° and 32.3° C) was 60%. To get the best superposition, it was necessary to vertically shift the curves by small amounts (log c_{RH}) along the log D_α axis. The vertical shift was partly caused by the variability in the initial elongation during loading, and its magnitude was in the range 0–0.02 (log GPa), which was less than that needed in a time–temperature superposition. Log c_{RH} had no fixed relationship to humidity. The time–humidity superposition held quite well.

We derived a relationship between the shift factor (a_{RH}) as a function of relative humidity (RH) or the amount of water (W) present. The log (shift factor) as a function of relative humidity was plotted on linear curves and the results are shown in Figs. 3.47 and 3.48. The data fit reasonably well with a straight line. We found the following relationships:

$$\log a_{RH} \sim 0.025/RH \quad \text{at } 51.7° \text{ C,}$$
$$\log a_{RH} \sim 0.047/RH \quad \text{at } 32.3° \text{ C.} \tag{3.124}$$

To compare the relationships between the shift factor and the effect of humidity at different temperatures, it is better to relate the shift factor to the amount of water (W) present. Therefore, an analysis was carried out to convert RH to W. We know that

$$\text{RH, \%} = PP \times 100/VP, \tag{3.125}$$

where PP is the partial pressure of water and VP is vapor pressure; and at RH = 100%, PP = VP. VP follows an Arrhenius equation and values at any temperature are given in many handbooks (Weast, 1972). From VP and RH at a temperature, PP can thus be calculated using Eq. (3.125). PP is related

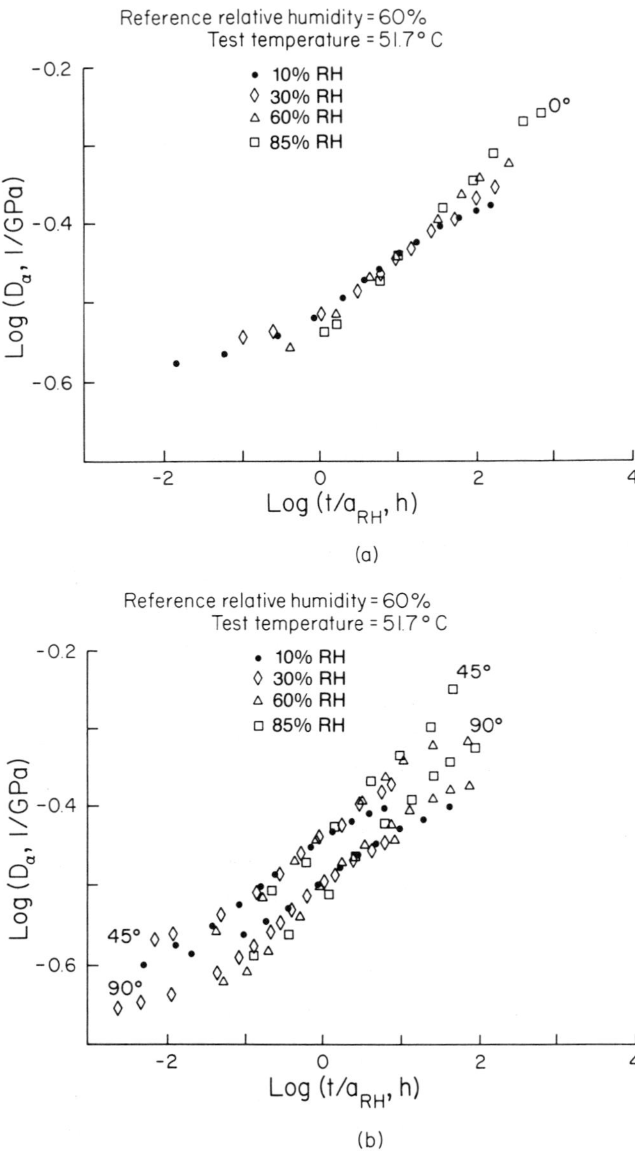

Fig. 3.45. Master curves of log (tensile-creep compliance) as a function of log (time) for a 23.4-μm-thick PET film (Mylar A) in three orientations reduced to 51.7° C and 60% RH. All the data were taken at 51.7° C: (a) 0° orientation and (b) 45° and 90° orientations (Bhushan and Connolly, 1986).

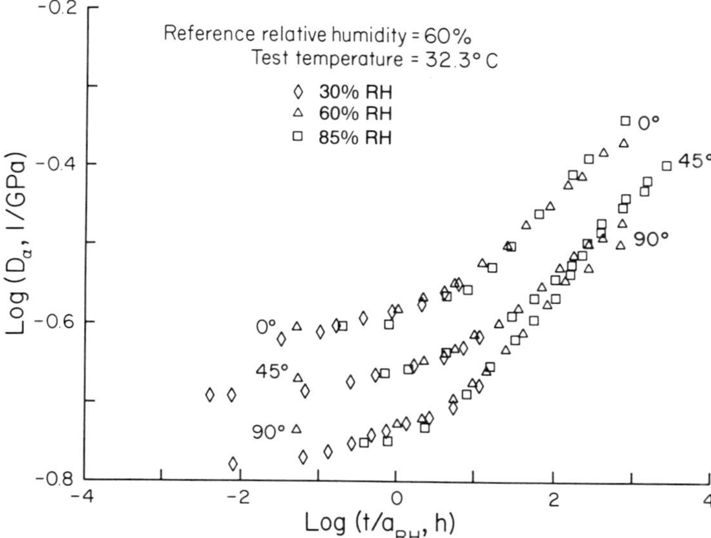

Fig. 3.46. Master curves of log (tensile-creep compliance) as a function of log (time) for a 23.4-μm-thick PET film (Mylar A) in three orientations reduced to 32.3° C and 60% RH. All the data were taken at 32.3° C (Bhushan and Connolly, 1986).

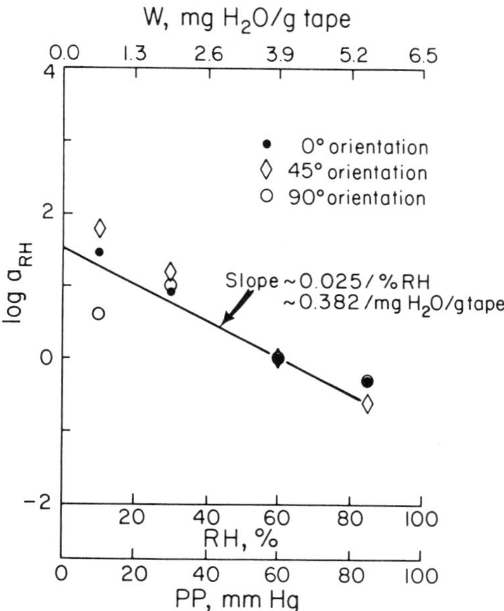

Fig. 3.47. Humidity dependence of shift factors, log a_{RH}, obtained from constructing the master curves at 51.7° C and 60% RH in Fig. 3.45 (Bhushan and Connolly, 1986).

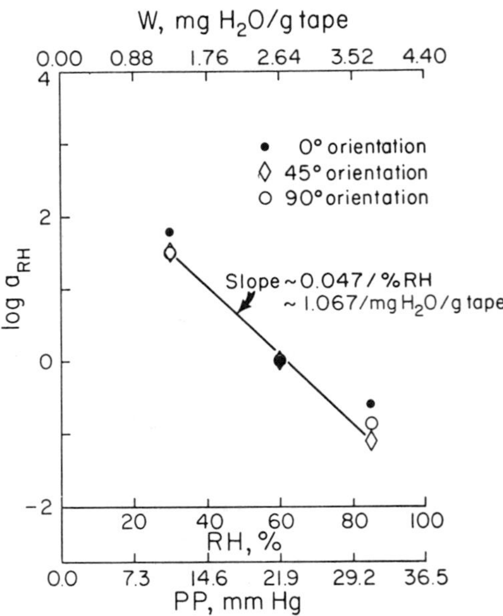

Fig. 3.48. Humidity dependence of shift factors, log a_{RH}, obtained from constructing the master curves at 32.3° C and 60% RH in Fig. 3.46 (Bhushan and Connolly, 1986).

to W by the following relationship:

$$W = K(PP), \qquad (3.126)$$

where K follows an Arrhenius equation with a negative activation energy (ΔH_a)

$$K = K_0 \exp(+\Delta H_a/RT). \qquad (3.127)$$

Cuddihy (1976, 1984) has measured the amount of water absorbed by a commercial, computer magnetic tape[5] as a function of temperature at $-5°$, $20°$, and $55°$ C. He found the following values of the constants in Eq. (3.127):

$$K_0 = 4.360 \times 10^{-6} \text{ mg H}_2\text{O/(g of tape)(mm Hg)}, \qquad (3.128a)$$

$$\Delta H_a = 26 \text{ kJ/mole (6.2 kcal/mole)}. \qquad (3.128b)$$

Using Eqs. (3.127) and (3.128), we get

$$\begin{aligned} K_{32.3°C} &= 0.1206 \text{ mg H}_2\text{O/(g of tape) (mm Hg)}, \\ K_{51.7°C} &= 0.0654 \text{ mg H}_2\text{O/(g of tape) (mm Hg)}. \end{aligned} \qquad (3.129)$$

---

[5] We have assumed that the amount of water absorbed by a magnetic tape is comparable to that absorbed by PET film.

From Weast (1972), vapor pressures are

$$VP_{32.3°C} = 36.48 \text{ mm Hg},$$
$$VP_{51.7°C} = 100.1 \text{ mm Hg}. \tag{3.130}$$

Using the data reported in Eqs. (3.128) and (3.129), we can now convert RH to W at the two temperatures using Eqs. (3.125) and (3.126). Converted W values are also plotted in Figs. 3.47 and 3.48. The relationships between $\log a_{RH}$ versus PP and W are given as follows:

$$\log a_{RH} \sim 0.025/\text{mm Hg} \qquad \text{at } 51.7° \text{ C}$$

$$= 0.382/\text{mg H}_2\text{O/g tape}, \tag{3.131a}$$

$$\log a_{RH} \sim 0.129/\text{mm Hg} \qquad \text{at } 32.3° \text{ C}$$

$$= 1.067/\text{mg H}_2\text{O/g tape}. \tag{3.131b}$$

We believe that at higher humidities, water vapor plasticizes the PET film, which lowers the glass-transition temperature (T_g) and increases creep. From Eq. (3.131), note that $\log a_{RH}$ is not independent of temperature. The effect of water vapor on creep behavior is more pronounced at lower temperatures. The humidity–acceleration factor is related to T_g. At a given water-vapor level, T_g is fixed; but at a lower temperature there is increased crowding of the molecules in the material, resulting in a higher acceleration factor. Temperature sensitivity drops off at higher temperatures, therefore, we expect that at temperatures (T) below T_g, the larger the $T - T_g$, the larger the a_{RH} and at T greater than T_g, the larger the $T - T_g$, the smaller the a_{RH}.

The shift factor $a_{T,RH}$ at any temperature and humidity can now be calculated by combining the temperature and humidity shift factors given by Eqs. (3.123) and (3.124), as follows:

$$\log a_{T,RH} = \log a_T + \log a_{RH}. \tag{3.132}$$

Summary

We have obtained constitutive stress–strain relationships for the orthotropic, linear viscoelastic behavior of the PET film under plane-stress conditions. The range of application of the derived constitutive relationships in the time domain is also important. As a general rule, calculations should never be performed for times longer than four times the largest relaxation time scale τ_1. The magnitude of such a scale is dependent on the duration of the experiments.

Long-term creep tests data were presented on PET film which was used to determine the shift factors for temperature and humidity. The activation energy of PET film was found to be about 200 kJ/mole (48 kcal/mole) and the log shift factor ($\log a_T$) was related to temperature (T) as follows:

$$\log a_T = -0.110(\Delta T).$$

We found that the humidity has a significant effect on the creep behavior of the film; the effect of water vapor on creep behavior was more pronounced at lower temperatures. Shift factors (a_{RH}) were related to the amount of water vapor as follows:

$$\log a_{RH} \sim 0.382/\text{mg } H_2O/\text{g tape at } 51.7° \text{ C} \quad \text{and}$$

$$1.067/\text{mg } H_2O/\text{g tape at } 32.3° \text{ C}.$$

Using the temperature and humidity shift factors, the shift factor at any temperature and humidity can be predicted.

3.3.3.4. Thermoviscoelastic Behavior of PET Films

Models to predict the viscoelastic response of thermorheologically-simple materials subjected to time-dependent temperature fields have been proposed by several researchers (Morland and Lee, 1960; Stouffer and Wineman, 1971; Stouffer, 1972). Morland and Lee (1960) have extended the time–temperature superposition, which is commonly used for isothermal viscoelastic behavior to nonisothermal conditions. Their extension to transient environments rests on the assumption that the amount of creep (or relaxation) that occurs in an infinitesimal interval of time $(t, t + \Delta t)$ depends only on the value of the temperature in that increment of time, independent of temperature history. The temperature history in the time interval $(0, t)$ has no influence on the response other than to determine the value of the creep (or relaxation) function at time t using the shift factors. Furthermore, the temperature history cannot reverse the creep (or relaxation) process; it can only vary the rate of response.

Illustrative examples are shown in Figs. 3.49(a)–(c). In the sample shown in Fig. 3.49(a), the temperature is changed after time t. The creep response from 0 to t is given by the creep curve from the isothermal creep test (A–B). At time t_1, as the temperature changes, new locations on the creep curves (C or D) are determined from isothermal tests at the current temperatures (T_3 or T_2) using the shift factors. The material then follows the corresponding curves (C–E or D–E). The slight change in creep compliance at time t_1 is caused by changes in the elastic deflections at different temperatures. The creep compliance at $t \geq t_1$ is given by the following equation:

$$D(t)|_{T_2, T_3} = D(t_1 a_{T_2, T_3} + t - t_1)|_{T_2, T_3} + D(0)|_{T_2, T_3} - D(0)|_{T_1}. \quad (3.133)$$

Figure 3.49(b) shows a condition where the temperature is changed and the material is unloaded. Note that the rate of creep recovery depends on the temperature. Figure 3.49(c) shows a condition where the temperature and stress are simultaneously changed. We have used Boltzmann's superposition in the conditions shown in Figs. 3.49(b) and (c). Stress superposition is done at a *constant* temperature.

Stouffer (1972) developed a so-called "thermal hereditary" law, which is a generalization of the concept of the development of the Boltzmann superposi-

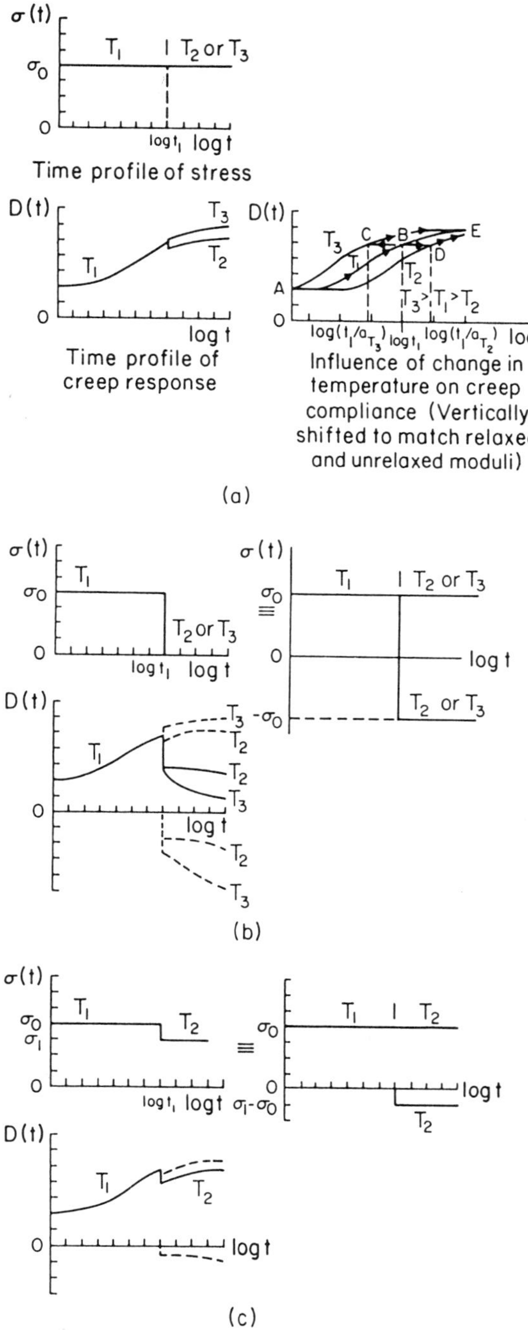

Fig. 3.49. (a) Creep response when the temperature is changed during the test; (b) creep response when the temperature is changed at the stress unloading; and (c) creep response when both the stress and temperature are changed simultaneously during the test (based on Morland and Lee's theory).

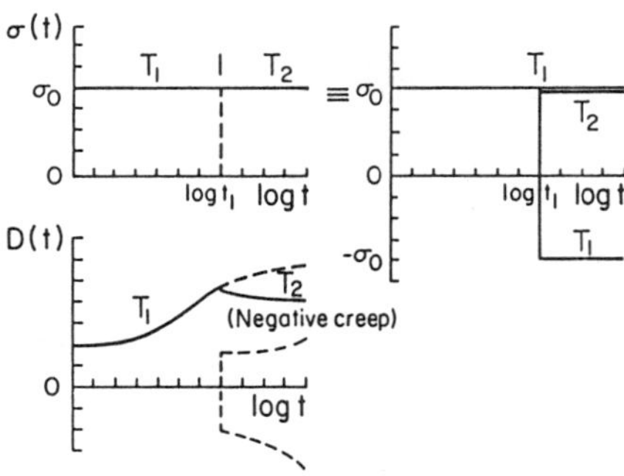

Fig. 3.50. Creep response when the temperature is changed during the test (based on Stouffer's theory).

tion integral of linear viscoelasticity. The main difference is that the roles of stress and temperature are interchanged, which is incorrect because the creep response is not linear in the temperature history. An illustrative example is shown in Fig. 3.50. Creep compliance at $t > t_1$ is given as

$$D(t)|_{T_2} = D(t)|_{T_1} - D(t - t_1)|_{T_1} + D(t - t_1)|_{T_2}. \tag{3.134}$$

Stouffer did not realize that *responses of stresses at different temperatures are not additive.* Depending on the $D(t)$ profile, T_1, T_2, and t_1, the creep process can be reversed (Fig. 3.50). This was found *not* to be true for PET films, see the following data.

Bhushan and Connolly (1986) conducted creep and relaxation tests to determine which of the two theories was applicable to PET film. After testing at a temperature for a certain period, the temperature was decreased at a constant relative humidity and the test continued. During the test, as the temperature decreased, some drift might have been introduced in the measurement gauges. No drift was measured in the proximity probes used in the creep tester because of the temperature variation from 51.7° to 22° C. In the strain gauges used in the relaxation tester, the drift was negligible ($<0.5\%$). As the temperature decreased during the test, the thermal expansion mismatch between the PET film and the steel frame of the tester affected the data, which needed to be corrected. It took about 20 min for the temperature of the temperature/humidity chamber (the same temperature as that of the PET film) to cool from 51.7° to 32.2° C, and the tester frame took about 1 h. The coefficients of the thermal expansion of PET film and steel are $17 \times 10^{-6}/°C$ and $11.7 \times 10^{-6}/°C$, respectively. Therefore, for a 0.2-m-long sample in the creep test, an error of about 31 μm in elongation, or 1.6×10^{-4} in the strain, would be introduced when it is cooled from 51.7° to 22.2° C, which is negligible. (The load, of course, is not affected.) In the relaxation tests, the initial strain was kept constant as the temperature was lowered and the correction in load was made. An error of 31 μm in elongation corresponded to about a

40-mN load, which was about 1–2% of the load measured. Although these errors were small, the data were adjusted accordingly.

During their experiments, the test was initiated at 51.7° C and 30% RH for 40 h, then the temperature was changed to either 43.3° C and 30% RH or 22.2° C and 30% RH for an additional 800 h. Creep-compliance and relaxation-moduli responses after the temperature reduction were predicted using Morland and Lee's and Stouffer's analyses, and the predicted data were compared with the experimental data, Figs. 3.51 and 3.52. Both analyses predicted an

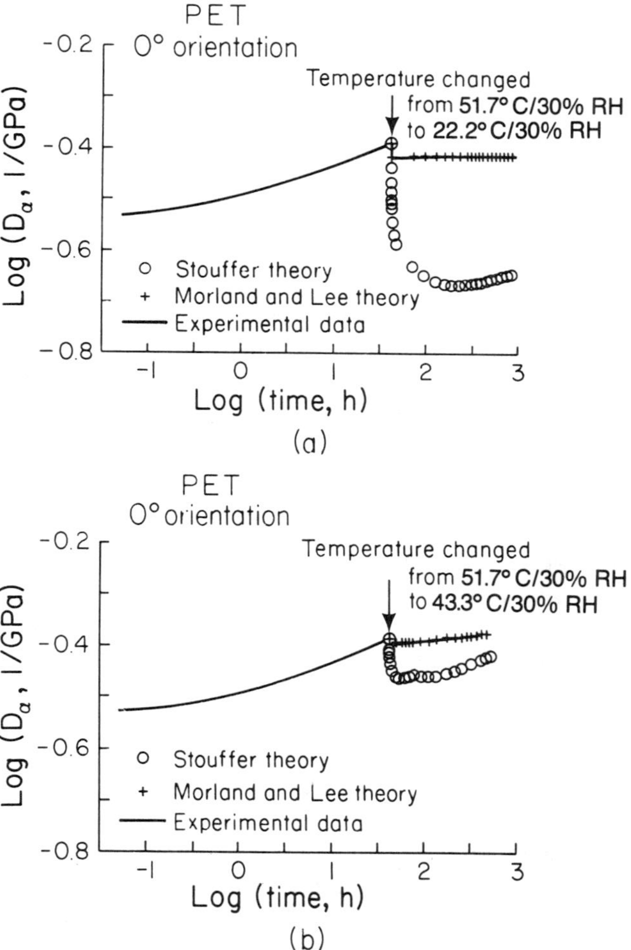

Fig. 3.51. Log (tensile-creep compliance) as a function of log (time) for a 23.4-μm-thick PET film (Mylar A) along 0° orientation at two temperatures changes: (a) from 51.7° to 22.2° C, and (b) from 51.7° to 43.3° C at a constant relative humidity of 30% RH (Bhushan and Connolly, 1986).

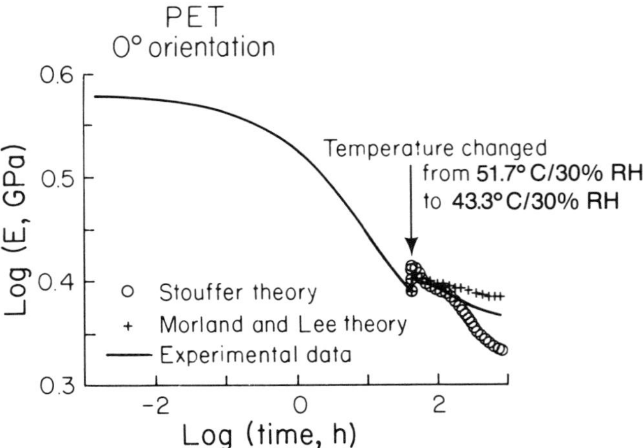

Fig. 3.52. Log (tensile-relaxation modulus) as a function of log (time) for a 23.4-μm-thick PET film (Mylar A) along 0° orientation at the temperature change from 51.7° to 43.3° C at a constant relative humidity of 30% RH (Bhushan and Connolly, 1986).

abrupt change in creep or relaxation behavior as the temperature changed because of changes in elastic moduli. Stouffer's analysis predicted a reversal in the creep or relaxation behavior, which we found to be untrue. *Morland and Lee's analysis predicted the creep or relaxation behavior of PET film reasonably well.*

3.3.3.5. Viscoelastic Behavior of Coated Tapes and Magnetic Coatings

Tensile creep tests under a constant load were performed on the coated tapes. It must be recognized at once that the creep of a multilayer tape represents a more complicated situation than does the creep of a single-layer tape, where (for small strains) a constant load produces a constant stress. Although a constant load imposes a constant nominal stress on the composite, the *actual* stress on each component may change substantially as the creep takes place, decreasing on the layers which relax most rapidly, and increasing on those which relax more slowly. As a consequence of the continual repartitioning of the load between the different layers, a constant load experiment on a multilayered tape represents neither creep at constant stress nor stress relaxation at constant strain. Despite these complications, considerable insight into the viscoelastic behavior of the tape can be obtained by a comparison of creep measurements performed on the multilayered tape and on comparison samples of tape substrate and calendered and uncalendered coatings.

Tape A (23.4-μm-thick PET film with a 4-μm-thick calendered frontcoat and a 4-μm-thick uncalendered backcoat), tape D (with a coating formulation designated D and with no backcoat), and free films of magnetic coatings A and D (for coating compositions, see Section 3.4) were used for the study

(Bhushan et al., 1984c; Bhushan and Connolly, 1986; Bhushan, 1990). The magnetic coating D had a slightly higher modulus and a substantially higher inflection temperature in the modulus (60° C) compared to that of coating A, Fig. 3.21 (Bhushan, 1990). The thickness of the free films of the magnetic coatings was about 0.2 mm, and they were as cast (and not calendered), unless otherwise specified.

Creep experiments were conducted on tape A, tape D with no backcoat (NBC), and PET films at two conditions: 22.2° C and 30% RH and 51.7° C and 30% RH, and their results are presented in Fig. 3.53 and Table 3.5. We find that the creep rates of the composite tapes are higher than the bare PET substrate at 22° C, however, the creep behavior of tape A and PET films in the three orientations at 51.7° C is comparable.

Relaxation experiments on coating A (uncalendered and calendered), tape A, stripped tape A (PET film with an unknown orientation), and PET film at

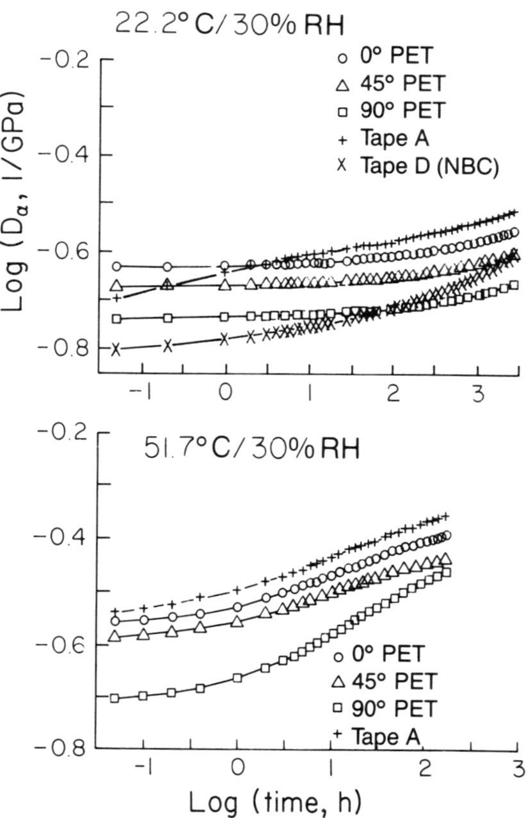

Fig. 3.53. Comparisons of log (tensile-creep compliance) as a function of log (time) for a 23.4-μm-thick PET film (Mylar A) along three orientations and tapes A and D at (a) 22.2° C and 30% RH and (b) 51.7° C and 30% RH (Bhushan and Connolly, 1986).

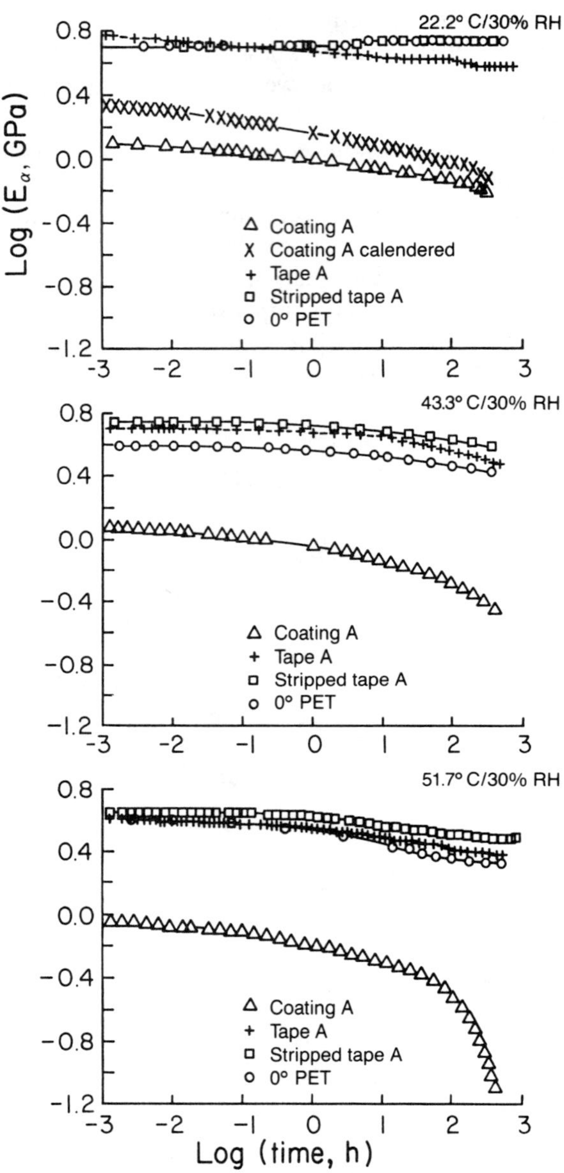

Fig. 3.54. Comparisons of log (tensile-relaxation moduli) as a function of log (time) of tape A, stripped tape A, 23.4-μm-thick PET film (Mylar A) along 0° orientation and magnetic coating A (uncalendered and calendered) at various temperatures (Bhushan and Connolly, 1986).

Table 3.7. Coefficients for the relaxation modulus (C) and the relaxation times (τ_i) of coatings and tapes (the tests were conducted at 30% RH)

Material	Temp., °C	C^0, GPa	C^1, GPa	C^2, GPa	τ_1, h	τ_2, h
Coating A	22.2	0.69	0.24	0.22	0.11	26.18
	32.3	0.59	0.37	0.24	0.24	56.60
	43.3	0.38	0.36	0.41	0.25	72.98
	51.7	0.08	0.33	0.46	0.27	124.01
Coating A, calendered	2.22	0.81	0.82	0.15	0.56	46.51
Coating D	22.2	0.97	0.30	0.44	0.40	87.49
	51.7	0.33	0.79	0.60	0.08	95.45
Tape A	22.2	3.81	1.35	0.58	0.16	154.49
Stripped tape A	22.2	4.04	0.28	0.61	14.25	957.37
Tape A	43.3	3.04	0.46	1.38	1.22	77.81
	51.7	2.37	0.59	0.87	0.79	38.73
Stripped tape A	51.7	2.95	0.73	0.67	2.23	73.27

0° orientation were conducted at 22.2°, 43.3°, and 51.7° C, and their results are presented in Fig. 3.54. The data were fitted to a Prony series. The elastic constants obtained from the fit are presented in Table 3.7. There was no difference in the relaxation behavior of the PET film and the stripped tape, as expected. The initial modulus of the coated and stripped tapes was comparable within experimental error. We again found that at room temperature the relaxation behavior of PET film is substantially slower than that of the coated tape (Tables 3.5 and 3.7) because the rate of relaxation of PET film is substantially slower than that of the magnetic coatings. At higher temperatures, there was little difference in the relaxation behavior of PET film and the coated tape; at high temperatures, the coating relaxes to a very low modulus after a short period (< 1 h at 51.7° C), thus shedding most of its load, then the coating rides on the PET film as a grease, with no mechanical contribution and *there is hardly any difference between the relaxation behavior of the tape and PET film.* Smith (1983–84) has shown that the shift factors in the relaxation studies at temperatures ranging from 25° to 52° C for magnetic tape A are comparable to those of the PET substrate.

Relaxation data of coatings A and D (the modulus of D being higher than that of A, Fig. 3.21) at different temperatures are presented in Fig. 3.55 and Table 3.7. The rate of relaxation of coating D was slower than that of coating A. The relaxation times of coating A increased with temperature because the modulus dropped rapidly at longer times for higher temperatures. The opposite effect had been seen in the PET film when the modulus dropped rapidly at shorter times for higher temperatures. We observe that the calendered coating A had a lower rate of relaxation than that of the uncalendered coating A. Calendered coatings also had a higher initial elastic modulus (Bhushan, 1985; Chapter 2). Coating D and tape D had lower rates of relaxation than coating A and tape A, respectively, as expected. The decreased viscoelastic response (higher modulus and a lower rate of relaxation) of the

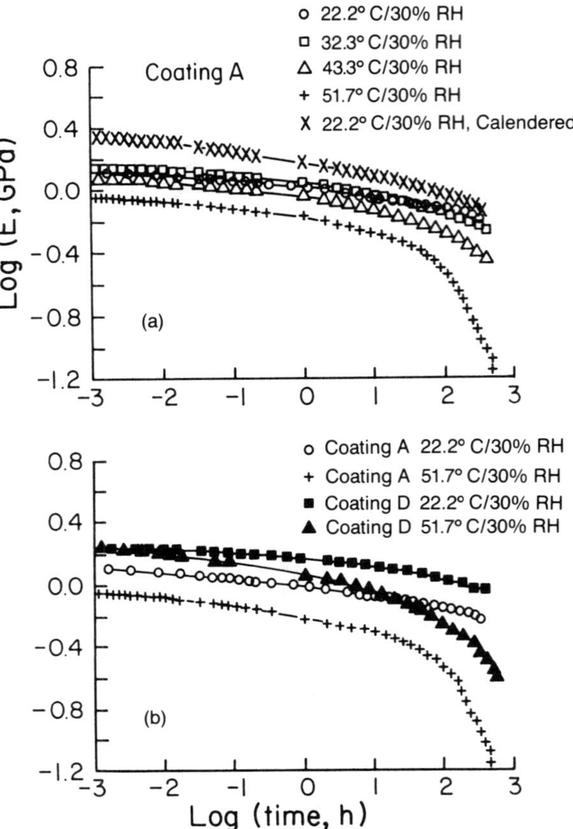

Fig. 3.55. Log (tensile-relaxation modulus) as a function of log (time): (a) of the magnetic coating A at various temperatures and a calendered coating A at 22.2° C and 30% RH; and (b) of magnetic coatings A and D at 22.2° C and 30% RH and 51.7° C and 30% RH (Bhushan and Connolly, 1986).

magnetic coating may also be obtained with a "burn-in procedure" in which the finished magnetic tape reel is heated for a period of time (e.g., at 52° C for 40 h).

The stresses are asymmetric in a (composite) tape wound onto a reel with high stresses at the outer layers—magnetic coating and back coating (if present) (Timoshenko, 1958). Bending produces the greatest strains at the surfaces of the tape which accentuates the viscoelastic effect of the magnetic coating. Thus, despite the large relative thickness of the PET substrate, the viscoelastic behavior of the tapes in a wound reel is strongly influenced by the magnetic coating (Appendix 3.A). In order to better simulate the viscoelastic deformation of a wound tape reel, Berry and Pritchet (1988) proposed a so-called "mandrel test" for flexural stress-relaxation and recovery studies where a strip of tape is bent around a cylindrical mandrel, Fig. 3.56(a). When

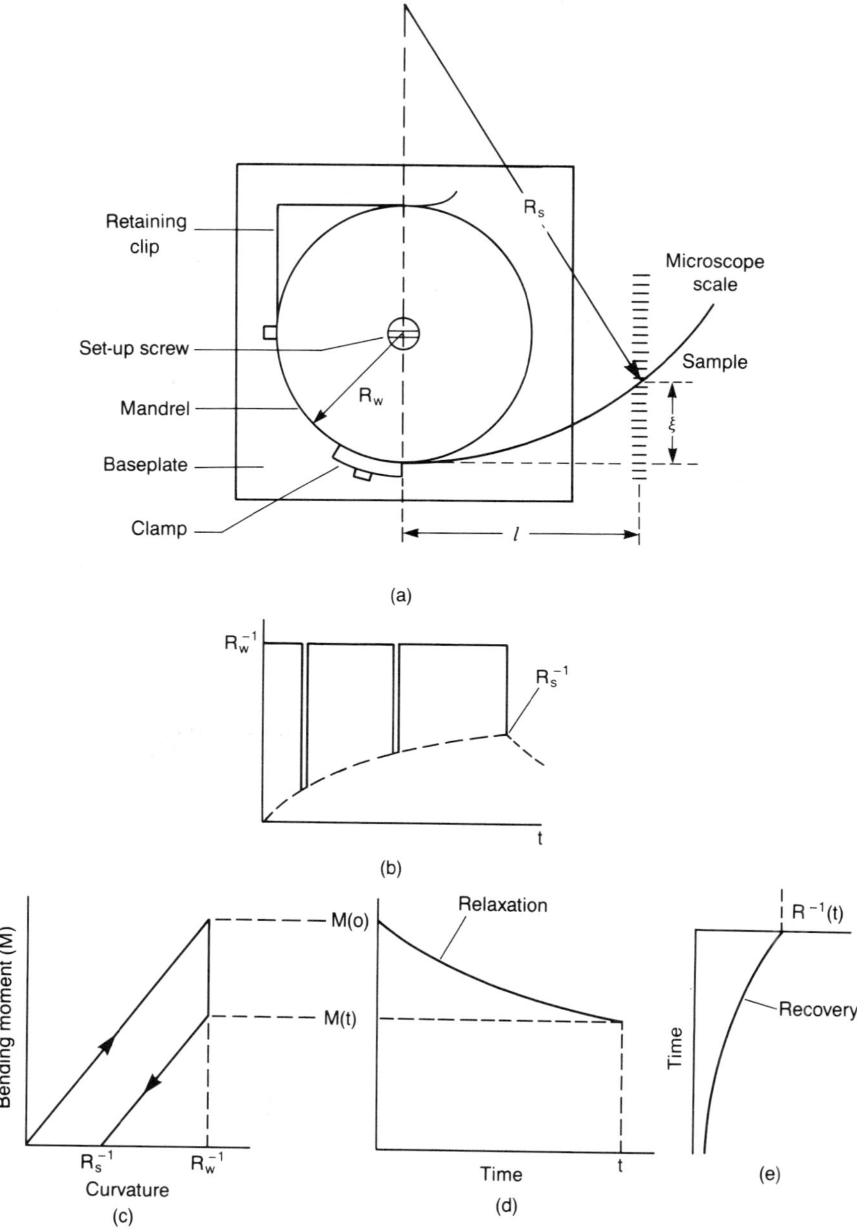

Fig. 3.56. (a) Schematic of the mandrel test for flexural stress relaxation and recovery studies; (b) growth of the initial springback curvature R_s^{-1}, tracked by periodic momentary releases from the wound-up position; (c) bending moment-curvature diagram for windup, and springback under unrelaxed conditions; (d) relaxation at constant curvature; and (e) recovery at zero net moment (Berry and Pritchet, 1988). © 1988 International Business Machines Corporation; reprinted with permission.

desired, the elastic springback curvature R_s of the strip can be obtained by measuring the deflection ξ at a known distance l. One end of a strip of a tape, approximately 25-mm long and 3-mm wide, is clamped to a squat copper mandrel fitted with a retaining clip to hold the tape in the wound-up position. The mandrel is attached by a centered set-up screw to a square base copper baseplate. With the baseplate clipped in a reference position against guides on the stage of a measuring microscope, the set-up screw is tightened during initial assembly after the mandrel has been oriented on the baseplate so that the sample is initially perpendicular to the graduated eyepiece scale. To facilitate sample set-up, including measurement of the viewing length l, the microscope is carried out on graduated slides giving micrometer control of the x, y, and z coordinates.

In the first, or relaxation, phase of the mandrel test, the sample is wound around the mandrel and secured, to maintain a fixed curvature $1/R_w$, Fig. 3.56(a). To obtain surface strains comparable to those employed in the tension tests (reported earlier), mandrels of 12.7-mm radius were employed. Periodically, the sample is momentarily released for observation of the initial springback by the optical microscope, and then rewound, Fig. 3.56(b). If the tape were ideally elastic, the tape would always spring back to its initial shape, no matter how prolonged the windup time. A viscoelastic tape, on the other hand, undergoes during windup a progressive relaxation with time t of the bending moment $M(t)$ or the bending stiffness $K(t) \equiv M(t)/(1/R_w)$, required to maintain the curvature $1/R_w$ and correspondingly does not spring back to its initial position upon release, Figs. 3.56(c) and (d). The kinetics of relaxation of the flexural bending stiffness can thus be followed by periodically releasing the strip, noting the elastic springback curvature $1/R_s$, and rewinding the tape. At any desired point this sequence may be discontinued and the springback curvature observed instead as a function of time, so as to ascertain the extent and kinetics of any subsequent creep recovery (flexural aftereffect), Fig. 3.56(e).

The springback curvature $1/R_s$ is related to the measured deflection ξ of an initially straight sample [Fig. 3.56(a)] by simple geometry

$$1/R_s = 2\xi/(\xi^2 + l^2). \tag{3.135}$$

Since the initial windup is presumed to take place rapidly under unrelaxed conditions, the slope of the windup line in Fig. 3.56(c) corresponds to the unrelaxed bending stiffness $K(0)$. Since the springback also occurs under unrelaxed conditions, it is evident that both the windup and springback lines of Fig. 3.56(c) are parallel. From similar triangles, we thus find

$$K(t)/K(0) = M(t)/M(0) = 1 - (R_w/R_s). \tag{3.136}$$

Thus, the normalized bending stiffness (K) parameter is equivalent to the bending moment (M) relaxation function for a multilayered sample.

For the special case of an unilayer (substrate identified by a subscript s), we

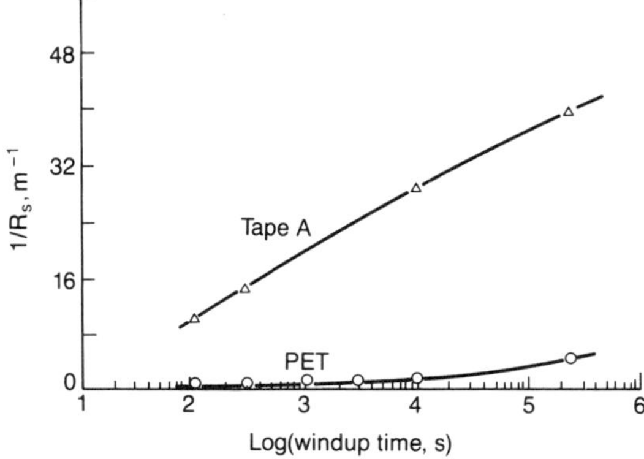

Fig. 3.57. Springback curvature as a function of log (windup time) for 23.4-μm-thick PET substrate (Mylar A) and Tape A bent around a mandrel of 12.7-mm radius at 22° C. The springback curvature was measured 10 s after release (Berry and Pritchet, 1988). © 1988 International Business Machines Corporation; reprinted with permission.

have the considerable simplification that the neutral surface of bending remains in a constant position at the center of the strip. As a consequence, the normalized flexural stress–relaxation modulus is equal to the normalized bending stiffness (Appendix 3.A),

$$E_s(t)/E_s(0) = K(t)/K(0). \tag{3.137}$$

The interpretation of the normalized bending stiffness for the composite sample (magnetic tape) is more complicated, since it involves the combined stress–relaxation behavior of all layers of the tape.

The growth of springback curvature $R_s(t)$ with windup time for PET and a trilayer tape A at room temperature (22° C) is shown in Fig. 3.57. We replot the data of Fig. 3.57 at 22° C and additional data at 52° C in the form of normalized bending stiffness for PET and coated tape A. At room temperature, the relaxation of PET occurs slowly. By comparison, the relaxation of PET was strongly accelerated at 52° C. We note the striking difference between the room-temperature relaxation behavior of bare PET and the trilayer tape A. The relaxation at 22° C is at least eight times larger in tape A than in the bare PET substrate. It is abundantly clear, merely from an inspection of Figs. 3.57 and 3.58, that the surface coatings exerted an overriding influence on the room temperature relaxation behavior of the tape, and the presence of the coatings cannot be ignored in predicting the behavior of the tape under service conditions (Berry and Pritchet, 1988).

To gain additional insight into the viscoelastic behavior of the tape, the recovery experiments were conducted on the tape A and PET after various windup times, Fig. 3.59. For the windup periods employed (3 days or less), the curvature developed in Fig. 3.58 by both bare PET and the tape proved to be

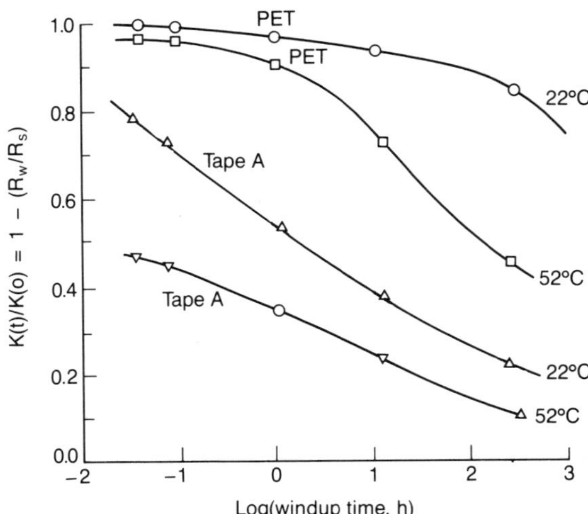

Fig. 3.58. Normalized bending-stiffness as a function log (windup time) of tape A and bare 23.4-μm-thick PET substrate (Mylar A) bent around a mandrel of 12.7-mm radius. The springback curvature was measured 10 s after release (Berry and Pritchet, 1988). © 1988 International Business Machines Corporation; reprinted with permission.

almost totally recoverable. The time required for recovery increased with the prior windup time, in the manner expected for an anelastic system governed by a wide distribution of relaxation times. It must be borne in mind, however, that the observation of recoverable behavior did not prove that the surface coatings were themselves behaving in an intrinsically anelastic (recoverable) manner. This follows from the fact that the recoverable behavior of the PET core would enforce pseudo-anelastic behavior on the composite tape, even if the coatings themselves were totally viscoplastic. Finally, it should be noted that the recoverable behavior seen in these experiments did not continue indefinitely, but increasingly gave way to viscoplastic behavior as the time and temperature of the deformation were increased.

In common with bimetallic strips utilized in thermostats and other devices, a multilayer composed of materials having different coefficients of thermal expansion has a general tendency to bend or curl when subjected to a change in temperature (Appendix 3.B). Thermal curling has been exploited by Berry and Pritchet (1988) as a useful tool for the study of recording tape. The tape samples were clamped at one end in the manner of a cantilever, and mounted on a temperature-controlled platform inside an environmental chamber containing dry helium. Changes in the curvature of a sample were calculated from deflection measurements made with a microscope of the type shown in Fig. 3.56. Using this technique, Berry and Pritchet predicted the coefficient of linear thermal expansion of CrO_2-particle magnetic coating to be about $1.5 \times 10^{-5}/°C$.

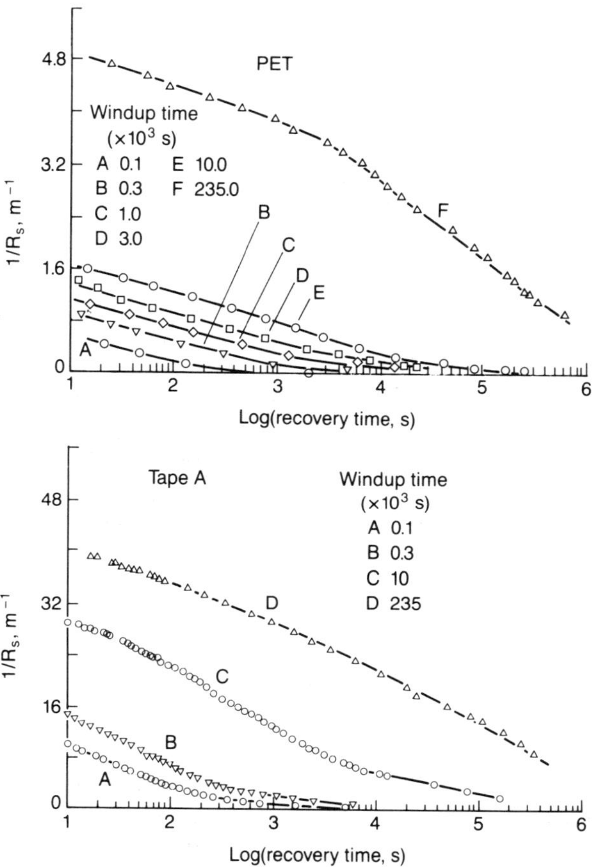

Fig. 3.59. Creep recovery (flexural aftereffect) in 23.4-μm-thick PET substrate (Mylar A) and tape A after being bent around a mandrel of 12.7-mm radius at 22° C for various windup times (Berry and Pritchet, 1988). © 1988 International Business Machines Corporation; reprinted with permission.

Summary

The relaxation and creep rates of the magnetic coatings were much faster than that of the PET film. Thus, the relaxation and creep rate of coated tape was much faster than that of the PET film at room temperature; however, at a high temperature after a short time (or a low temperature after a long time), the relaxation and creep rates of coated tape were comparable to those of the PET film. At room temperature, tape A is an asymmetrical trilayer in which surface coatings exert a major effect on both the initial stiffness and subsequent relaxation behavior of the tape. However, at high temperatures, the coating relaxes to a very low modulus after a short period, thus shedding most of its load, then the coating rides on the PET film as a grease, with no mechanical contribution and there is hardly any difference between the viscoelastic behavior of the tape and PET film. Shift factors for coated tapes were comparable to those of the PET films.

3.3.3.6. Effects of Thermal Treatment of PET Films

The biaxially-oriented PET (Mylar A) samples in three orientations (0°, 45°, and 90°) were tested. Their measured density at 25° C was 1397.6 kg/m^3. All samples were dried *in situ* at room temperature in a vacuum overnight in the University of Pittsburgh's differential elongational creep apparatus before tensile creep measurements were initiated. After drying, the sample chamber was filled with nitrogen to aid in maintaining uniform temperatures in the specimens. During various thermal conditioning treatments the samples were unconstrained or virtually so. When such treatments took place in the elongational creep instrument, the strips of film were under a tensile stress of 0.5 MPa which kept the samples straight but added negligible change in length after an initial deformation of about 7×10^{-5} % during the first 3 days after installation. Long-term thermal treatments were carried out on loosely rolled film which was held at constant temperatures in glass cylinders immersed in a temperature-controlled silicone oil bath.

Two different kinds of thermal effects involving different mechanisms will be distinguished: one is the result of the samples residence time at temperatures above T_g. Storage time at temperatures greater than T_g will be called *annealing*; if the film is heated at a high temperature ($>150°$ C) for a short time, it will be designated as *high temperature heat set*. On the other hand, storage of the polymer below T_g which will be designated as *physical aging* (Struik, 1978), results in similar, but distinguishable effects. For the PET samples held at temperatures below 150° C, the degree of crystallinity, crystalline size, and composition do not change significantly, however, the degree of crystallinity increases appreciably if the PET samples are held at temperatures above about 150° C but below its melting temperature.

Thermal Effects

While samples of the material "as-received" in three orientations were heated, the length was monitored in the tensile creep apparatus with only the preload attached (Vallat and Plazek, 1988a). They were sequentially: (1) heated from room temperature up to 140° C at about 1° C per minute; (2) held at 140° C for 3 days; and (3) cooled down from 140° C at about 0.5° C per minute. During this treatment the films were kept taut with a small tensile stress of 0.5 MPa which caused negligible elongation during the 3 days (less than 10 microns).

The behavior during the initial heating is shown in Fig. 3.60. The specimen along the 0° orientation displayed the highest initial expansion coefficient. At temperatures designated T_i (inversion temperature) initial shrinkage is observed during this first heating. The 0° sample starts shrinking at the lowest temperature 64° C with the 45° oriented sample close behind at 65° C. The

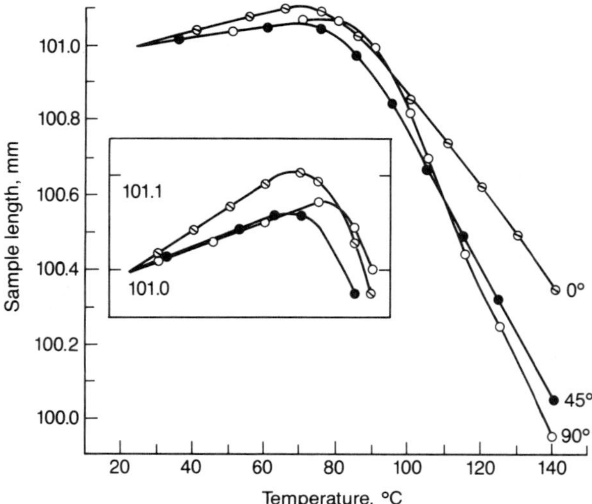

Fig. 3.60. The sample length as a function of temperature for the 23.4-μm-thick PET films (Mylar A), as-received, during initial heating from room temperature at 1° C per minute. Inset shows the initiation of irreversible shrinkage with an expanded length scale (Vallat and Plazek, 1988a).

90° sample continues expanding for an additional 10° C before starting its reversal. During this first heating, the rate of shrinkage is highest for the most oriented direction 90° and slowest for the least, 0°, so that even though the 90° sample starts shrinking last it shrinks the most (about 1.1%) by the time 140° C is reached. During the shrinkage, the orientation of the noncrystalline chains decreases (Samuels, 1974), therefore the most oriented sample should shrink more. For this material, the 90° oriented film is more oriented than the 0° in agreement with this deduction.

The isothermal shrinkage at 140° C is presented in Fig. 3.61 where all of the specimens show slightly more than an additional 0.5% shrinkage. The only significant difference is that the 0° sample and to a lesser extent the 45° sample indicate a leveling out as a function of the logarithmic time of annealing after 1 day (roughly 10^5 s).

The thermal contraction behavior after 3 days of annealing at 140° C is seen in Fig. 3.62. Thermal contraction coefficients are greatest above T_g and lowest below T_g for the 90° direction. The contraction coefficient of the 45° sample is closer to that of the 0° sample above T_g and close to that of 90° below T_g. Note that the T_g (the temperature at which, during cooling, a rapid decrease in the rate of change of volume with temperature occurs) of the 90° oriented material (in the direction of highest orientation) identified by the intersection of the high- and low-temperature straight lines is 8° C higher than that of the 0° material. The T_g for 45° is in between the two but closer to that for 0°. The initial cooling displayed in Fig. 3.62 was followed by several

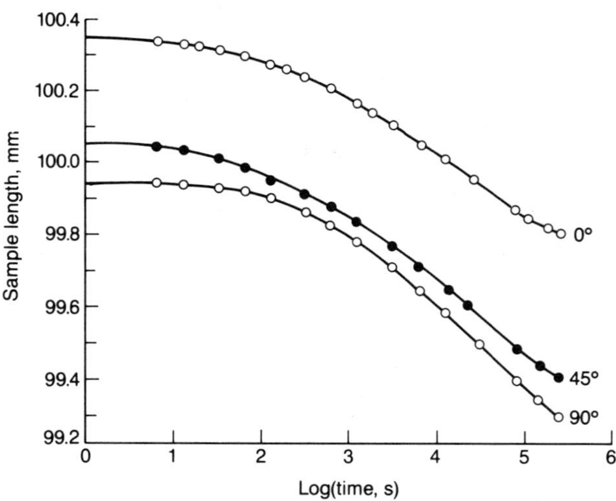

Fig. 3.61. The sample length as a function of time for the 23.4-μm-thick PET films (Mylar A) at 140° C for 3 days (annealing) following the initial heating shown in Fig. 3.60 (Vallat and Plazek, 1988a).

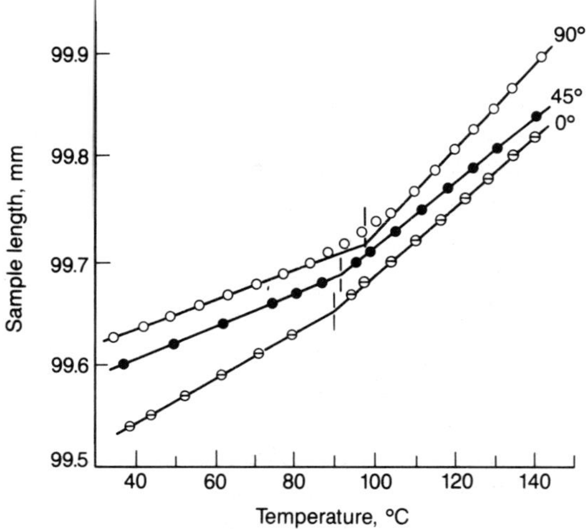

Fig. 3.62. The sample length as a function of temperature for the 23.4-μm-thick PET films (Mylar A) during cooling at about 1° C per minute after annealing at 140° C for 3 days. Glass temperatures, T_g, are indicated at the intersections of the drawn lines. Thermal contraction coefficients are given in Table 3.8 (Vallat and Plazek, 1988a).

Table 3.8. Anisotropic thermal parameters of
the biaxially-oriented PET film (Mylar A)

Film orientation	0°	45°	90°
T_i, °C	64	65	75
T_g, °C	89	91	97
Shrinkage, % after 10^5 s	1.42	1.68	1.77
$\alpha_{c,g}$, $\times 10^{-5}/°C$, 1st heating	2.4	1.4	1.4
$\alpha_{c,g}$, $\times 10^{-5}/°C$	2.2	1.7	1.5
$\alpha_{c,l}$, $\times 10^{-5}/°C$	3.1	3.2	4.1

heating and cooling cycles between 30° and 140° C. The values of the thermal contraction coefficients $\alpha_{c,g}$ and $\alpha_{c,l}$ shown in Table 3.8 are average values obtained during the cycling. The subscripts c, g, and l identify the presence of crystalline, glassy, and liquidlike material. Table 3.8 also presents the T_g's and other quantitative parameters obtained from Figs. 3.60–3.62. (We note that the thermal properties of these thermally-treated samples presented in Table 3.8 are different than the typical properties of the "as-received" PET samples presented in Chapter 2.) Based on linear dilatometric measurements, Vallat et al. (1986) reported an increase in T_g with annealing at 140° C.

To investigate the effects of increased anisotropy, a "tensilized" Mylar was prepared from a 23.4-μm-thick Mylar A sheet 120-mm long in the 90° direction and 150-mm long in the 0° direction by stretching it in the 90° direction by 8% at 150° C. The stretching rate was 4×10^{-2} mm/min. The film was then heat-set for 5 min at 190° C while maintaining the tension.

The enhancement of the orientation in the 90° direction by tensilizing has dramatic consequences. Specimens cut parallel to the 90° and 0° directions were measured with the following thermal histories. The 90° tensilized specimen was heated from room temperature to 140° C at about 1° C/min. The initial temperature of shrinkage occurred at 53° C but the subsequent shrinkage was so great and fast that it could not be followed with the sensitive but short range (~ 100 μm) LVDT that was used; see Fig. 3.63. The sample which was held at 140° C for 2 h before cooling to room temperature had to be removed from the tensile creep apparatus to determine the extent of the initial loss of length (on the order of 5%). After reinstallation in the apparatus the second heating up to and down from 140° C after the 2 h of annealing at that highest temperature yielded the curves shown in Fig. 3.63. The T_g found from the second cooling was 97° C which is, within experimental uncertainty, the same as that determined on the stock film after 3 days of annealing at 140° C. Substantial shrinkage ($\sim 0.6\%$) occurred at 140° C during the second 2-h annealing period. The tensilization clearly reduced the thermal expansivities both above and below T_g in the 90° direction.

However when treated similarly with 1° C/min heating and 0.5° C/min cooling, the tensilized film increased in the 0° direction in length irreversibly by 1% during the initial heating (see Fig. 3.64) and thereafter indicated a

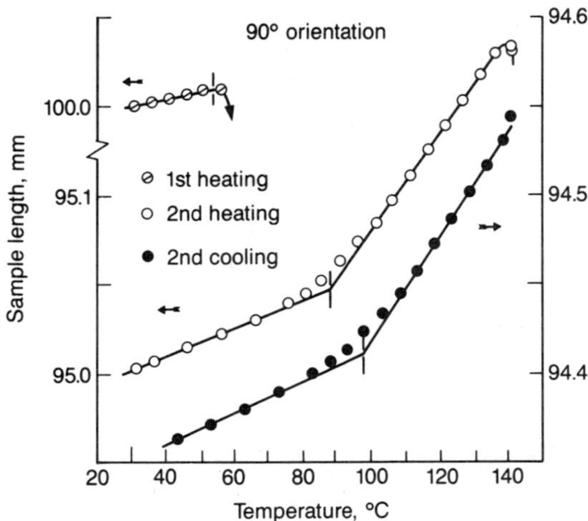

Fig. 3.63. The sample length as a function of temperature for the 23.4-μm-thick tensilized PET film (produced by additional stretching in the 90° orientation of the biaxially-stretched PET film by 8% at 150° C and heat setting at 190° C) along 90° orientation during initial heating (1° C/min), second heating (1° C/min) and second cooling (0.5° C/min) (Vallat and Plazek, 1988a).

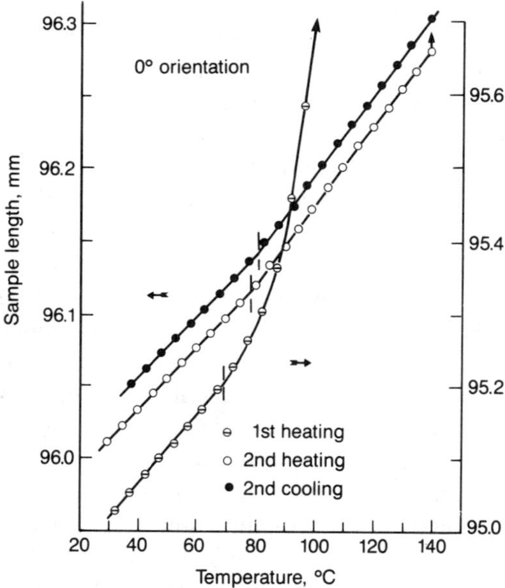

Fig. 3.64. The sample length as a function of temperature for the 23.4-μm-thick tensilized PET film along 0° orientation during initial heating (1° C/min), second heating (1° C/min) and second cooling (0.5° C/min) (Vallat and Plazek, 1988a).

Table 3.9. Anisotropic thermal parameters
of "tensilized" PET film (Mylar A)

Orientation	0°	90°
T_i, °C	69	53
T_f, °C, 2nd heating	78.5	88
T_g, °C	81	97.5
$\alpha_{c,g}$, $\times 10^{-5}/$°C, 1st heating	2.5	0.42
$\alpha_{c,g}$, $\times 10^{-5}/$°C, 2nd heating	2.2	0.82
$\alpha_{c,l}$, $\times 10^{-5}/$°C, 2nd heating	2.8	3.0
$\alpha_{c,g}$, $\times 10^{-5}/$°C, 2nd cooling	2.2	0.95
$\alpha_{c,l}$, $\times 10^{-5}/$°C, 2nd cooling	2.8	3.2

depressed T_g of 81° C. The signal was lost again during the first heating and was out of range during the subsequent cooling. In the following heating and cooling cycle, $\alpha_{c,g}$ reverted to a value of 2.2 $\times$ $10^{-5}/$°C which is the same as that of the untensilized film during repeated cycling. The results obtained from Figs. 3.63 and 3.64 are presented in Table 3.9. More data on shrinkage is presented in Chapters 2 and 6.

Differential scanning calorimetric (DSC) thermograms were obtained on the PET (Mylar A) as received and on the physically altered specimens described as being "tensilized." The samples in an unsealed aluminum pan and under a dry nitrogen atmosphere were heated at 10° C/min from 35° to 300° C. It can be seen in Fig. 3.65 (top curve) that the as-received material exhibits a small but definite exothermic shift near 82° C during an initial heating which reflects the shrinkage that starts at this temperature. Vallat et al. (1986) have reported that subsequent heatings show the usual endothermic shift near 82° C which yields the fictive temperature T_f. Glass temperatures T_g can only be obtained from cooling measurements. The initial exothermic shift is believed to reflect the shrinkage seen in Fig. 3.60. The melting peak temperature is 256° C. The second curve in Fig. 3.65 was determined on the tensilized film. Upon first heating there is only a hint of an exothermic shift near 84° C. In the melting range, two endothermic peaks are seen that appear to be remnants of the usual melting peak which has been split by a super-imposed exothermic process, which we assume is the release upon melting of high-energy molecular chain conformations which were formed during the tensilization (Gupta et al., 1984). The melting peak of the as-received PET has been scaled down in proportion to the relative masses of polymer being measured, and is shown as a short dashed line superimposed on the thermo-gram of the same film after tensilization. The exothermic process decreases to a minimum at 257° C which is well below the approaching baseline. The tensilized film could not have a lower degree of crystallinity than the as-received film. Therefore, the smaller apparent latent heat of fusion here is evidence against arguments which invoke melting and recrystallization to

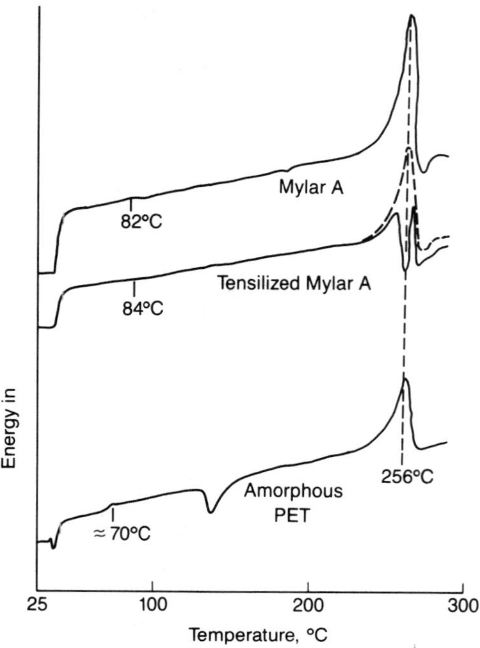

Fig. 3.65. First heating thermograms from DSC measurements for the as-received 23.4-μm-thick PET film (Mylar A); tensilized PET film; and quenched unoriented (amorphous) PET film (Vallat and Plazek, 1988a).

explain double melting peaks. Likewise, additional crystallization during the tensilization process is not a possible explanation. At still higher temperatures (265° C) in the two top thermograms there appears to be an additional exothermic peak which could also be reflecting recovery processes in the polymer melt. (The as-received PET curve was obtained with 62 μN of sample while 38 μN of the tensilized sample was used.) For comparison, an isotropic quenched film of PET, which was thought to be largely amorphous, was measured. The resulting curve is also shown in Fig. 3.65 where a fictive temperature slightly under 70° C is seen. The relative areas of the crystallization (∼ 130° C) and melting peaks (∼ 258° C) indicate that substantial crystallization had occurred prior to the DSC scan.

Creep Measurements

Creep measurements were made on film samples with the three orientations between room temperature and 140° C and a normal stress of 28.3 MPa (Vallat et al., 1986, 1988b). Before the measurements, samples were allowed to stabilize.

(a) Conditioning by Annealing

The influence of annealing for a PET sample along 45° orientation is shown in
Fig. 3.66. Each curve is identified by the temperature at which the creep is
performed and the time in minutes at which the sample was at this given
temperature before starting the creep run. The number in brackets gives the
order in which they were made. The sequence of measurements for the $D(t)$
curves in Fig. 3.66 is 141° C, 130° C, 120° C, 110° C, 141° C, 141° C, 78° C,
and 21° C. The creep run number 5 was started 120 min after the temperature
was reached the second time; after the recovery run, a second creep run
(number 6) was made without changing the temperature. In this case, the
annealing time includes that of creep number 5. The measurement at 141° C
was a check run, which clearly indicates a lower compliance level and/or a
lower rate of creep than was measured earlier at nearly the same temperature
141° C. There is a strong influence of the temperature on the initial creep
compliance.

By plotting these same results on a linear scale of $D(t)$ versus $t^{1/3}$ (Fig. 3.67),
two different regions of linearity called Andrade creep appear: one below T_g
and the other above 120–130° C. The rate of creep is much higher above T_g.
The same type of behavior is observed for samples along 0° and 90° orienta-

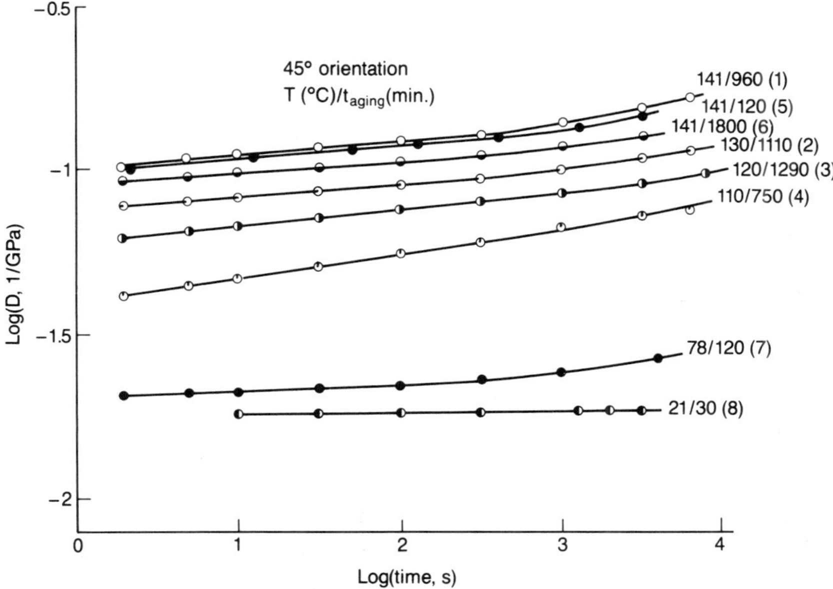

Fig. 3.66. Log (tensile-creep compliance) as a function of log (time) for a 23.4-μm-thick
PET sample (Mylar A) along 45° orientation at various temperatures. The numbers
in brackets indicate the order in which the creep runs were made (Vallat et al.,
1986).

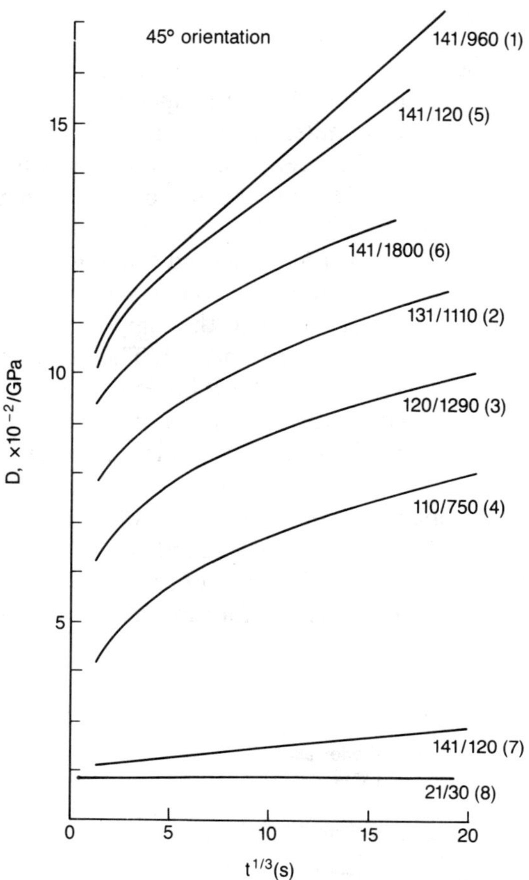

Fig. 3.67. The tensile-creep compliances of Fig. 3.66 plotted as a function of $t^{1/3}$ (Andrade plots). Between 22° and 141° C there are two different regions of Andrade creep: one below T_g and the other one above T_g (Vallat et al., 1986).

tions (Vallat et al., 1988b). When an amorphous polymer sample is heated through its T_g, the density is less than its equilibrium value, but tends toward its higher equilibrium value with time. For a semicrystalline polymer, the behavior is more complex because the noncrystalline chains are constrained by the crystalline part of the polymer and are not in equilibrium even above T_g. Below T_g, the amorphous part of the polymer should also exhibit densification causing physical aging which should be reflected in the mechanical properties of the material. This effect will be shown in the following sections.

(b) Conditioning by Annealing and Aging

In an attempt to stabilize the biaxially-oriented PET morphology, a PET film along 90° orientation was annealed at 140° C for 12 days and physically aged

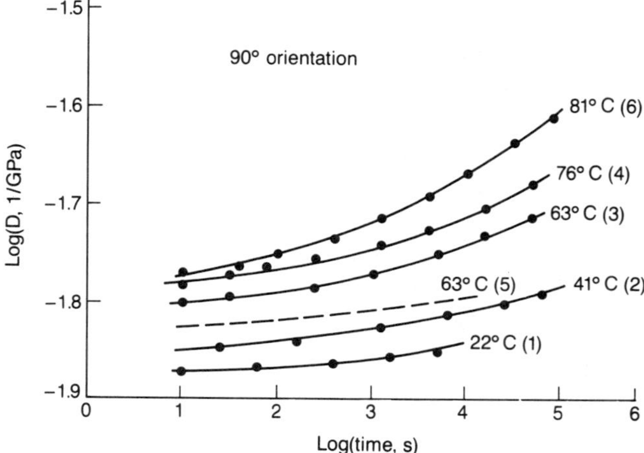

Fig. 3.68. Log (tensile-creep compliance) as a function of log (time) for a 23.4-μm-thick PET sample (Mylar A) along 90° orientation for six temperatures from 22° to 81° C after 12 days of annealing at 140° C followed by 24.5 days of aging at 85° C. Sequence of measurements (shown by the numbers in brackets) with increasing temperature except for the fifth run, which was a check at 63° C (Vallat et al., 1988b).

at 85° C for 24.5 days. The annealing temperature of 140° C (above T_g) was chosen so that the degree of crystallinity and crystallite size would not be altered appreciably but stresses in the noncrystalline chain could be relieved. To stabilize the amorphous regions of the PET for measurements below T_g it is desirable generally to hold the film about 15° C below T_g. Creep compliance curves obtained following the thermal treatment are shown in Fig. 3.68. The measurements were made successively at increasing temperatures with the exception of the check run at 63° C. A 5-day interval passed between the two runs at 63° C with 3 days between the start of the first run at 63° C and that at 76° C. More than 1 day of recovery following each creep run was needed to avoid serious errors stemming from the sample's previous stress history. Clearly additional aging occurred during the series of runs. The rates of creep at 76° C are virtually the same as those seen in the first run at 63° C, whereas those seen during the check run are substantially lower. The log $D(t)$ curve is clearly shifted to longer times. Measurements were restricted to temperatures below the aging temperatures, but obviously the aging was insufficient to prevent further aging during the series of runs.

(c) Conditioning by Aging

The effect of physical aging is shown in Fig. 3.69 for a PET sample along a 90° orientation. The creep compliance $D(t)$ for a sample quenched from 120° to 85° C is presented as a function of $t^{1/3}$. (The sample was heated to 120° C to erase the effects of previous thermal history.) For each creep run, three

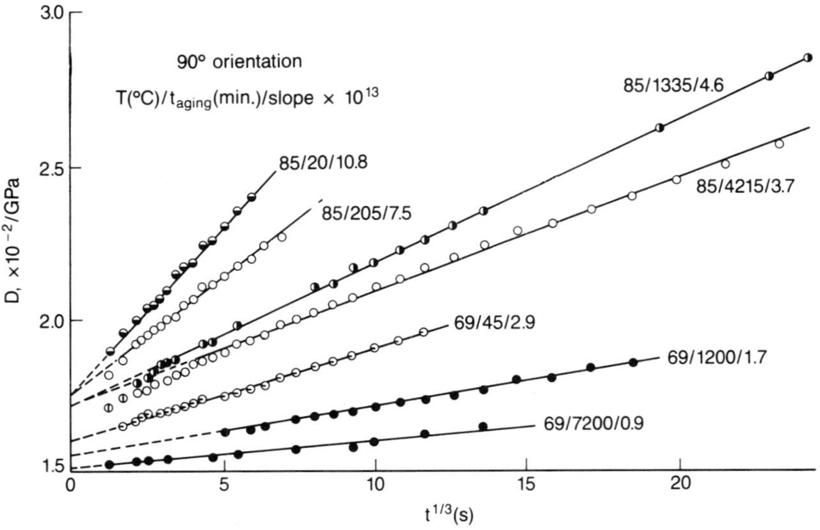

Fig. 3.69. Andrade plots of the tensile-creep compliance $D(t)$ as a function of $t^{1/3}$ for 23.4-μm-thick PET samples (Mylar A) along 90° orientation showing the effect of aging at two different temperatures: 69° and 85° C. In both cases, samples were quenched from 120° C (Vallat et al., 1986).

numbers are given, the first one is the temperature, the second one is the aging time in minutes at this temperature when the creep is started, and the third one is the slope of the Andrade plot. The rate of creep decreased by a factor of 25 as the aging time increased from 20 min to 70 h. A second set of curves for the same sample was obtained after a heating to 120° C which erases the aging that occurred at 85° C. The second series followed a quenching to 69° C. Between 45 min and 5 days at 69° C, the creep rate decreased by 33-fold. These decreases are generally attributable to the spontaneous increase in density of the amorphous regions of the sample. (The amorphous phase is liquidlike with a nonequilibrium density.)

The *reversibility of the effect of aging below* T_g is illustrated in Fig. 3.70 and in Table 3.10. Two pairs of runs on a sample along the 90° orientation taken after 1 h and 17 h of aging at 90° C yielded nearly identical results following cooling from above 110° C. The decrease in the rate of creep (aging effects) is erased by heating above the T_g, where an equilibrium density of the amorphous material is achieved. Table 3.10 gives other evidence of the reversibility of aging for a sample along 45° orientation quenched from 131° to 78° C. The slopes of the Andrade plot are given as a function of aging time at 78° C. Again, the slope decreases as the aging time increases. The sample was then heated to 130° C and quenched again to 78° C. After an aging time of 120 min, the slope of the Andrade plot is nearly identical to that obtained after the first quenching.

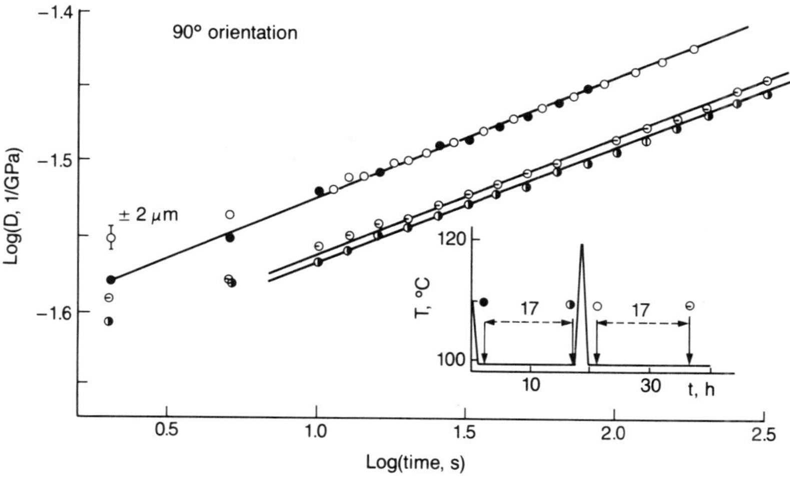

Fig. 3.70. Log (tensile-creep compliance) as a function of log (time) at 90° C for a 23.4-μm-thick PET sample (Mylar A) along 90° orientation. The onset temperature as a function of time plot gives the thermal history of the sample and the corresponding symbols used in the $D(t)$ curves (Vallat et al., 1986).

Figure 3.71 compares the creep behavior of the "as-received" PET sample in the two main orientations 0° and 90° at 42° C. It can be seen that the sample along the 90° orientation, which corresponds to the most oriented direction, has a lower compliance than that of the 0° orientation, as well as a lower creep rate as indicated by the slope of the Andrade plot, 0.95×10^{13} compared to 1.01×10^{13} for the sample along the 0° orientation. The temperature was then taken to 90° C, which is slightly above the initial T_g of the PET and this temperature was maintained for about 1 h to erase any previous aging. After the 1 h of annealing above T_g, the temperature was decreased to 42° C at which creep measurements were made as a function of aging time. After this thermal treatment, the film along the 90° orientation showed a higher compliance indicating some loss of orientation. With the same thermal treatment, the film along the 0° orientation presented a slightly lower compliance, which means that the behavior of the biaxially-oriented PET film tends

Table 3.10. Reversibility of aging for a PET film (Mylar A) along 45° orientation at 78° C

	After first cooling				After second cooling
Aging time, min	120	1200	1650	2800	120
Slope Andrade Plot ($\times 10^{13}$)	4.4	2.9	2.5	2.3	4.3

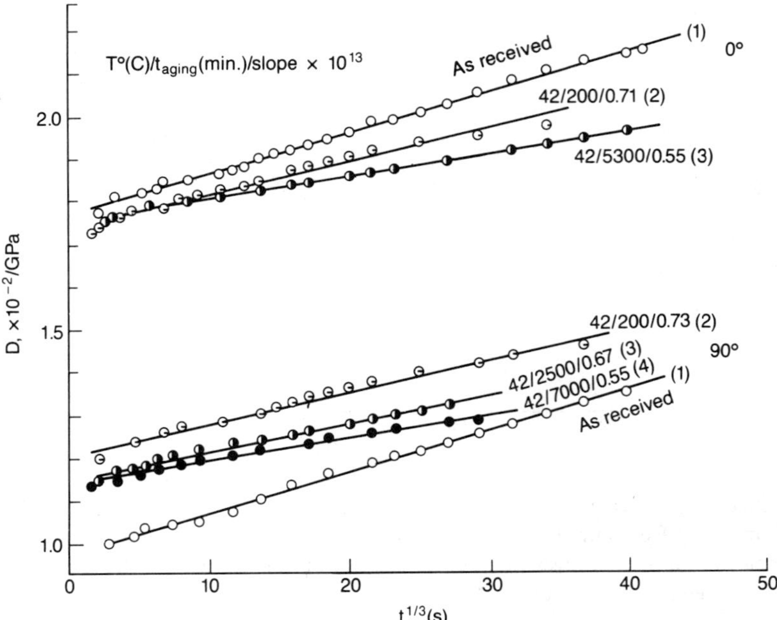

Fig. 3.71. Andrade plots of the tensile-creep compliance as a function of $t^{1/3}$ at 42° C for a 23.4-μm-thick PET sample (Mylar A) along 0° and 90° orientations. The curves with open circles correspond to the films as we received them, the slopes of the Andrade plots are 1.01×10^{-13} and 0.95×10^{-13} for films with 0° and 90° orientations, respectively. The other curves show the effect of aging at 42° C after a heat treatment of about 1 h at 90° C (which is slightly above the initial T_g of PET) (Vallat et al., 1986).

to become more isotropic. As the material is maintained at 42° C it ages, and the slope of the Andrade plot decreases for both orientations.

(d) High Temperature Heat-Set

A PET film was heated to 190° C for 15 min and subsequently cooled to 41° C where a creep run identified in Fig. 3.72 as "no aging" at 60° C was carried out. The 15 min at 190° C "heat set" to the specimen reduces the strength of the viscoelastic dispersion (Gupta et al., 1984). Subsequent heating to 60° C with aging times of 2, 19, and 765 h yielded the $D(t)$ curves, also shown in Fig. 3.72. Following 1 day of recovery at 60° C and cooling to 41° C a final determination of $D(t)$ was determined after a total of 790 h of aging at 60° C. The $D(t)$ results are presented as functions of $t^{1/3}$ and Andrade creep is observed within experimental uncertainties. Compared to the creep compliance curve initially obtained at 60° C, labeled as "as-received," the decrease in the rate of creep after 765 h of aging is 140-fold. At 41° C the initial $D(t)$ curve measured with no aging at 60° C displays rates of creep that are twenty-two times faster than those measured after the 790 h of aging at 60° C.

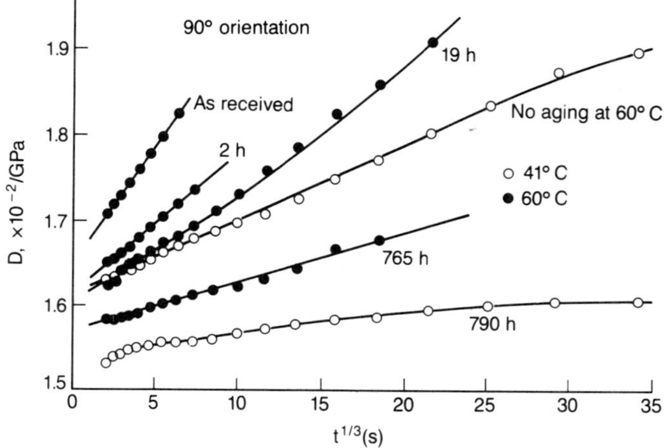

Fig. 3.72. Andrade plots of the tensile-creep compliance as a function of $t^{1/3}$ for a 23.4-μm-thick PET sample (Mylar A) along 90° orientation measured at 60° and 41° C after an additional "heat set" of 15 min at 190° C and aging at 60° C for different times. Results obtained at 60° C on an untreated "as-received" sample are shown for comparison. Aging times at 60° C are indicated (Vallat et al., 1988b).

(e) Role of Crystallinity

Tant and Wilkes (1981) treated the amorphous films of PET by heating under vacuum at various temperatures to induce different levels of crystallinity. Table 3.11 shows the level of crystallinity obtained (based on density measurements), along with the time and temperature of the crystallization of each. These samples were annealed at a temperature above their T_g for 10 min to erase any previous aging. After annealing, the samples were then immediately quenched below T_g in an ice-water bath, removed, and stored at ambient conditions. The stress relaxation tests were performed on the amorphous and semicrystalline samples as a function of sub-T_g aging time. The percent relaxation of stress during the first 10 min of the experiment was calculated for each sample. The stress relaxation data for the amorphous PET and the various

Table 3.11. Level of crystallinity obtained with various treatments of amorphous PET

Crystallization temperature, °C	Crystallization time, h	Percent crystallinity by density
102	12	12
102	48	22
113	3	29
163	24	39
205	24	51

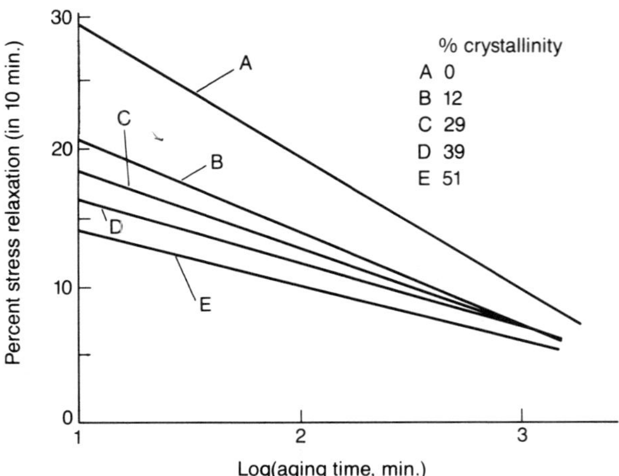

Fig. 3.73. Percent stress relaxation in 10 min as function of log (sub-T_g aging time) for a PET of various degrees of crystallinity (Tant and Wilkes, 1981).

degrees of semicrystalline PET are given in Fig. 3.73. In general, these curves move to lower values of percent stress relaxation with an increasing degree of crystallinity. Both the overall extent and rate of physical aging in PET decrease with increasing percent crystallinity.

(f) Reduced Creep Compliance Curves

Vallat et al. (1988b) generated master curves using time–temperature superposition for the PET films with different thermal histories. They found the shift factors to be nonunique. It results from the density changes occurring below T_g in the noncrystalline regions of the polymer and from the broadening of the viscoelastic dispersion that occurs at temperatures above T_g, presumably due to increasing restraints on the noncrystalline chain segments. Hence, the entire thermal history of the material plays a role in determining its response.

Discussion and Conclusions

Two different effects of heating to temperatures above ambient on biaxially-oriented PET film have been demonstrated. An irreversible shrinkage of a few percent, which occurs in the temperature range between the glass temperature T_g and the melting temperature T_m (annealing), is known to represent a partial loss of orientation in the noncrystalline molecules while the crystallites formed in the oriented state preserve most of the orientation (Judge and Stein, 1961; Krigbaum and Roe, 1964; Oono et al., 1973; Samuels, 1974; Gupta et al., 1984). Concomitant with the observed shrinkage during this annealing above T_g is some deceleration of the rate of creep. Limited data from the pressurized bellows volume dilatometer (Zoller et al., 1976) indicate that an increase in density does occur under conditions where the film length is decreasing (Vallat et al., 1988a). The decrease in the rate of creep can be attributed to the increase in density of the amorphous phase since it is believed there is no change in the degree of crystallinity. Since the film was anisotropic in its plane, different amounts and rates of shrinkage and values of T_g were observed along with differing thermal expansion coefficients in various directions relative to the primary optic axis. The anisotropy in the properties is increased further by the increase in anisotropy due to the tensilization process. Linear dilatometric measurements on the PET film showed an increase in T_g with annealing at 140° C (Vallat et al., 1986).

When a biaxially oriented film is cooled from a temperature above T_g to below T_g, a deceleration of mechanical response, commonly called physical aging, is also observed. Physical aging in most, if not all, cases can be traced to time-dependent densification (increases in the density) of the unstable amorphous phase (with nonequilibrium density) of the total amorphous as well as partially crystalline polymers (Kovacs et al., 1963; Plazek and Magill, 1966; Struik, 1978; Tant and Wilkes, 1981; Richardson and Savill, 1977). Creep rates were found to decrease significantly with the residence time at temperatures below T_g. (The effect is much greater than that of residence times at temperatures above T_g.) As with purely amorphous polymers, this deceleration was reversible in that every time a specimen was taken to a temperature above T_g and then cooled, the deceleration previously accumulated at temperatures below T_g was erased. Heating to temperatures above T_g results in an equilibrium value for the amorphous density. Subsequent cooling below T_g reinitiates the reversible time-dependent densification of the amorphous material and the deceleration of all rate processes which also decreases the rate of the creep.

Since all the creep rates measured are sensitive to the sample density, the viscoelastic mechanisms contributing to the creep deformations must arise from the amorphous portions of the PET film. We found two regions of viscoelastic response that are dominated by Andrade creep (Fig. 3.63). The deformation contributing to $D(t)$ at temperatures below T_g (which is between 90° and 100° C for most of the specimens used here for thermal studies) can be described by the Andrade cube root of time relation. At temperatures above T_g (110°–130° C) the $D(t)$ versus $t^{1/3}$ curves have negative second derivatives. In this region of the viscoelastic response it is certain that the mechanisms involved with the usual softening process are contributing. In a completely amorphous polymer, these would be the mechanisms that populate the primary softening peak in the viscoelastic spectrum. At sufficiently high temperatures, 140° C in this case, the Andrade creep linearity reappears. There is no known reason to indicate that the two Andrade regions are related.

Measurements made after various thermal treatments which were intended to stabilize the PET's physical state were shown to be not totally effective. For example, samples were annealed at 140° C for 12 days and aged at 95° C for over 24 days before measurement, but aging was insufficient to prevent further aging. Nevertheless, the creep compliance of the PET material can be significantly reduced (by an order of magnitude or more) by the physical aging of the amorphous structure on the order of 15° C below T_g. This reduction in creep compliance can be accomplished in the commercial films by the physical aging of a jumbo roll wound at a relatively low tension at a temperature of the order of 70° C for a period of the order of 1 day. Since some shrinkage occurs during aging, it needs to be removed in the free form (under extremely low tension) before aging, and this is accomplished by exposing the commercial PET film to a temperature above T_g (e.g., 140° C for a few seconds) during the winding/unwinding cycle.

3.4. Compressive Creep Data

It is relatively easy to measure creep in tension; however, few efforts have been made to measure creep in compression. Gillen (1978) has used a commercially available thermomechanical analysis system (TMS) to measure the creep compliance of relatively thick polymers (7.6–12.7 mm). Bhushan and Smith (1985) have used this technique to measure the creep compliance of thin PET films and cast films of magnetic coatings.

3.4.1. Description of Creep Test Apparatus

A schematic diagram of the TMS apparatus is shown in Fig. 3.74. The test sample was placed on the platform of a fixed quartz sample tube. An appropriate quartz probe was fitted to the probe assembly, which consisted of a shaft upon which the core of a linear voltage differential transducer (LVDT) was mounted. Any change in the position of the core with respect to the LVDT changed the cylindrical-transformer output. Therefore, any motion of the probe caused by penetration into the sample, or by a change in sample size, was transmitted with very high sensitivity as an electrical signal to a recorder.

The probe assembly included the quartz probe, core rod (including the LVDT core), and the weight tray, which permitted a choice of loads on the sample surface. The probe assembly was supported by a plastic float (rigidly attached) fixed to the shaft and totally immersed in a high-density fluorocarbon fluid. This method had the important advantage that the true load on the sample was essentially independent of the probe position over the range where the plastic float remains totally immersed. Therefore, the sample load was independent of sample thickness or any change in the sample thickness during the experiment. The lower portion of the fixed quartz sample tube (with walls partially cut away to permit insertion of the sample) was placed inside a cylindrical furnace. The furnace was embedded in a block of aluminum that serves as a heat sink. For cryogenic experiments, the unit was cooled by adding liquid nitrogen to the Dewar flask.

Test samples for testing were taken from PET films and cast films of

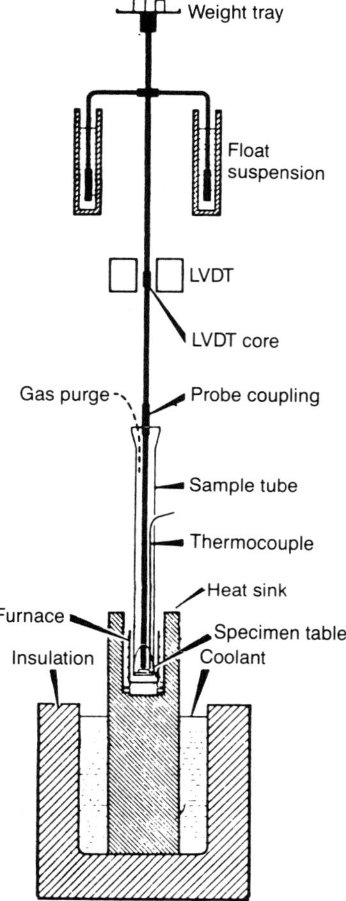

Fig. 3.74. Schematic of thermomechanical analysis system (TMS).

magnetic coating by punching out disks using a single-hole paper punch. This yielded a flat, round sample 7 mm in diameter. The test procedure consisted of placing a test sample on the platform on a fixed sample tube, then lowering the probe into the sample under a minimum load of about 1 mN (contact load). The sample was allowed to stabilize for 2 min, after which the desired test load (in most tests it was 10 mN) was applied and the deflection of the probe recorded versus time. The zero-load condition was established by incrementally loading the probe while monitoring the probe position. The zero-load condition was reached when the probe just touched the sample. They estimated that the sample had a load of 1 mN at the initial zero-load condition (contact load). The load was applied electromechanically. Note that the initial contact stress on the surface with a 10-mN load is of the order of 35–50 MPa. However, due to viscoelastic flow, the contact stress in the creep experiments after a short duration would be much lower ($\sim$1 MPa).

For a spherically-tipped quartz probe of radius R placed on a polymeric sample of thickness h under load, the creep compliance in compression $D(t)$ for the sample is given as (Bhushan and Smith, 1985)

$$D(t) = \frac{1.78 R^{1/2} \rho(t)^{3/2}}{W} \{1 + 0.252[R\rho(t)]^{1/2}/h\}, \qquad (3.138)$$

where $\rho(t)$ is the penetration in the sample, t is time, and W is the weight of the probe. The smallest radius of the probe should be selected such that the maximum penetration in the sample is as low as possible to eliminate the influence of the backup plate. Therefore, the smallest radius of commercially available quartz probes of 0.457-mm radius was selected.

3.4.2. Experimental Results

3.4.2.1. PET Films and Cast Films of Magnetic Coatings

Tests were conducted on a biaxially-oriented PET (Mylar A) substrate and five cast films of the magnetic coatings were designated A to E. Coatings A to D were constructed with a polyester–polyurethane with a similar chemistry, except that the difference in the polyurethane (hard segments) contents in coatings A and D were to be 32% and 40%, respectively. Tapes B and C were polyester–polyurethanes that were hardened by compounding with poly(vinyl chloride) and poly(vinylidene chloride), respectively. Coating E was made with a different grade of polyester–polyurethane. All coatings were prepared with CrO_2 magnetic particles at a 50% particle-volume concentration (PVC). The films of magnetic ink were made by casting the ink slurry on PTFE-coated glass plates. All free films used were uncalendered (as coated), except those referred to as calendered. The thickness of the magnetic coating was typically 0.2–0.4 mm, and of the PET film was 0.325-mm thick. Although the PET film of 23.4 μm or less and 76.2 μm in thickness as a substrate for tape and flexible disk, respectively, and coating thicknesses of about 3–5 μm are used in most magnetic media, thicker films were selected because meaningful data on thinner films cannot be obtained.

Most tests were conducted at 22° C (ambient temperature) and 52° C (maximum temperature during media shipment). Limited tests were conducted at $-20°$ and 60° C. The duration of the tests was approximately 16 h. Creep compliance $D(t)$ for the PET substrate and for coatings A and E is plotted in Figs. 3.75 and 3.76. Selected data of normal penetration $\rho(t)$ and creep compliance $D(t)$ for the PET substrate and all the coatings are presented in Table 3.12. The general nature of creep curves is similar to that in tension, however, the creep compliance is higher in compression than that in tension. As seen in tensile creep tests, coating D in compressive creep tests also exhibited lower creep than the coating A.

3.4.2.2. Calendered Versus Uncalendered Magnetic Coatings

Since the recording surface in a magnetic tape is calendered (in tapes having a backcoat, the backcoat is not generally calendered), it is of interest to study

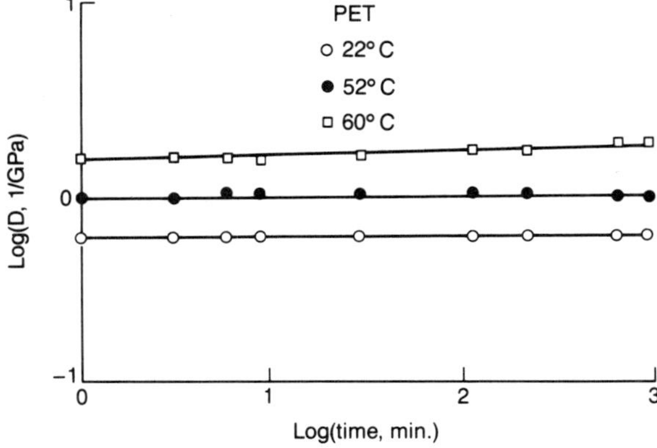

Fig. 3.75. Log (compressive-creep compliance) as a function of log (time) for a 0.325-mm-thick PET sample (Mylar A) at various temperatures (Bhushan and Smith, 1985).

the influence of calendering on creep compliance. Coating A was calendered using a small, laboratory, two-roll calender with a face width of 190 mm. The steel roll was 114 mm in diameter and the highly compressed paper roll was 153 mm in diameter. The coating was calendered twice at 74° C and at a load of 385 N/mm. Assuming a modulus and a Poisson ratio of the paper roll of 2.07 GPa and 0.30, respectively, the normal pressure was calculated to be roughly 72 MPa. Creep tests on calendered and uncalendered films were conducted at 22° and 52° C. Results are presented in Table 3.13. Note that calendered films have a lower penetration (or creep compliance) than un-calendered films. The difference is greater at 52° C. Calendering makes the film denser and improves the mechanical properties (higher modulus). This is due to decreased porosity of the films after calendering. However, creep rates in both films are about the same. The surface roughnesses of uncalendered and calendered films as measured by a noncontact optical profiler were 100 nm and 18 nm rms, respectively. Because penetration is significantly greater than surface roughness, we believe that surface roughness does not influence creep. In calendered films, creep compliance decreased at 52° C after some time. This may be due to the removal of residual stresses introduced from calendering.

3.4.2.3. Recovery Experiments

Recovery experiments were conducted on the PET film and coating A at 22° and 52° C. Samples were loaded for 5 min and then unloaded. Penetration during recovery was measured for 16 h. Results are presented in Figs. 3.77 and 3.78. At 22° C, the instantaneous recovery in creep compliance (BC) during unloading is comparable to the elastic component of creep compliance during loading at time zero (0A). Further recovery is very small. At 52° C, instantaneous recovery (FG) appears greater than instantaneous elastic compliance (OE). This anomaly may be due to a rather large creep at

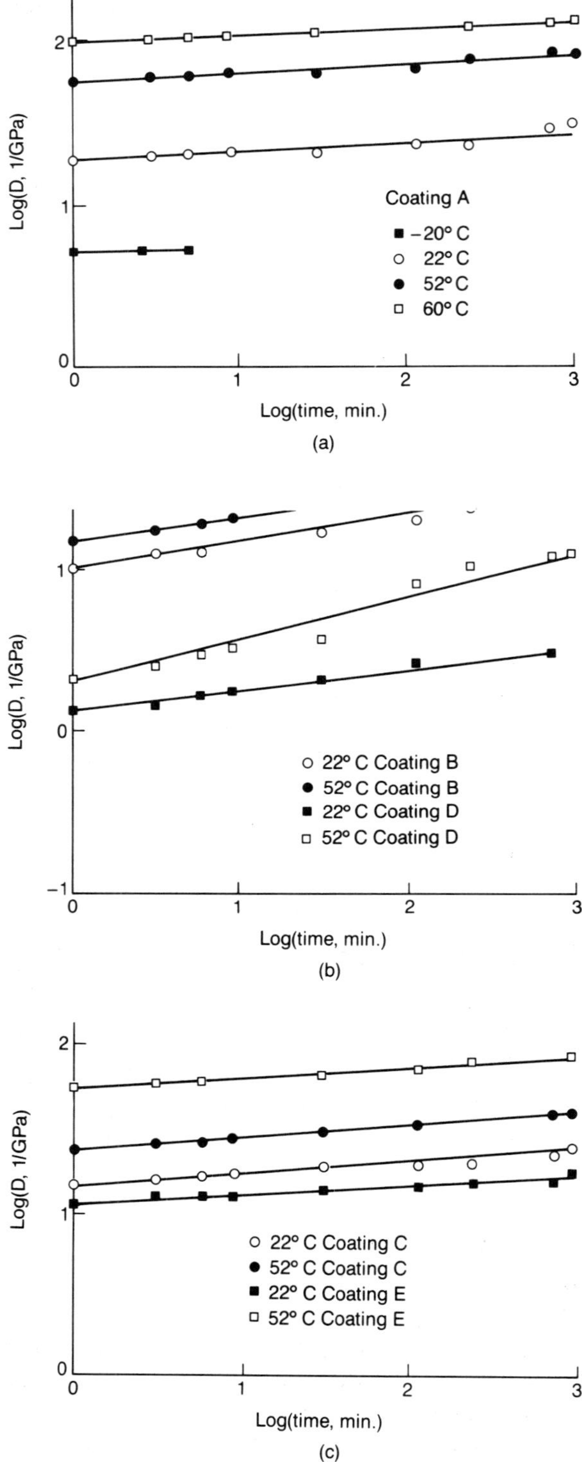

Fig. 3.76. Log (compressive-creep compliance) as a function of log (time) for magnetic coatings A to E at various temperatures (Bhushan and Smith, 1985).

Table 3.12. Compressive creep data for 0.325-mm-thick PET film (Mylar A) and different magnetic coatings

Specimen	Test temperature, °C	$\rho(t)-\rho(30\ s)$, μm		$D(t)$, 1/GPa		
		5 min	16 h	30 s	5 min	16 h
PET substrate	22	0.01	0.03	0.6	0.6	0.7
(thickness	52	0.01	0.05	0.9	1.0	1.3
direction)	60	0.008	0.10	1.6	1.7	2.1
Coating A	−20	0.20		5.7	7	
	22	0.16	1.03	20	22	32
	52	0.30	1.60	59	64	89
	60	0.28	1.18	106	112	133
Coating B	22	0.12	1.30	13	15	30
	52	0.30	1.95	15	18	39
Coating C	22	0.10	0.70	16	17	27
	52	0.31	0.98	23	27	37
Coating D	22	0.05	0.35	1.1	1.6	3.3
	52	0.15	1.34	2.2	3.0	13
Coating E	22	0.07	0.37	13	14	16
	52	0.32	1.70	58	63	89

Table 3.13. Compressive creep data for calendered and uncalendered magnetic coating A

Calendered (C)/ uncalendered (UC)	Test temperature, °C	$\rho(t)-\rho(30\ s)$, μm		$D(t)$, 1/GPa		
		5 min	16 h	30 s	5 min	16 h
UC	22	0.16	1.03	20	22	32
C	22	0.18	1.19	11	12	22
UC	52	0.30	1.60	59	64	89
C	52	0.34	1.42	26	30	43

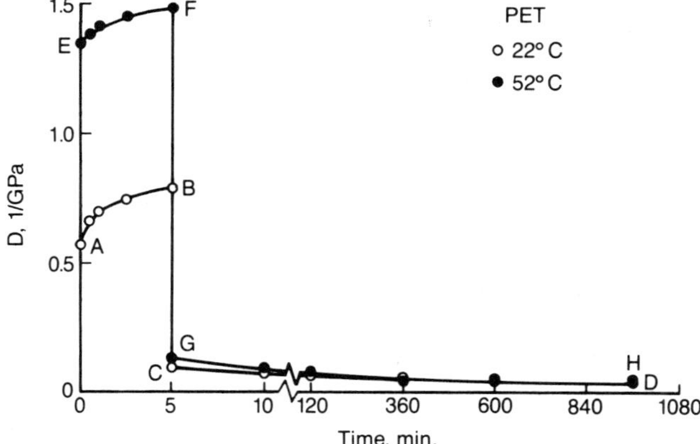

Fig. 3.77. Compressive-creep compliance as a function of time during loading and recovery for a 0.325-mm-thick PET sample (Mylar A) (Bhushan and Smith, 1985).

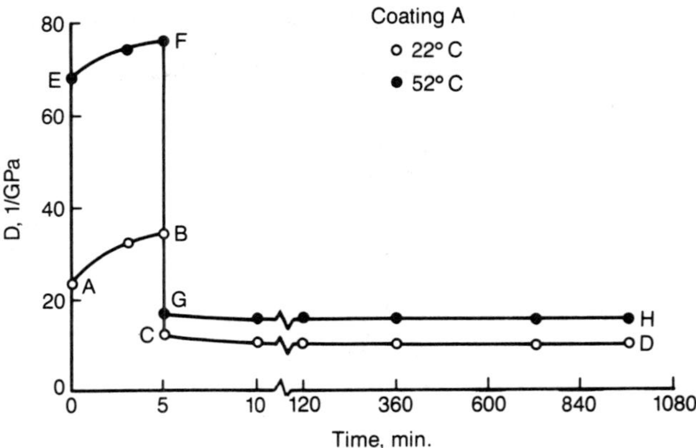

Fig. 3.78. Compressive-creep compliance as a function of time during loading and recovery of magnetic coating A (Bhushan and Smith, 1985).

time 0+, which makes the measurement of the elastic component of creep compliance difficult. Note that a fair portion of penetration does not recover instantaneously.

3.4.2.4. Summary

Creep compliance measurements in compression were made on a 0.325-mm-thick PET substrate and five typical magnetic coatings. By comparing the creep data in compression with that of in tension presented earlier, we note that creep compliance of the PET and magnetic coatings is higher in compression than in tension. Calendered films showed lower creep compliance than uncalendered films. The difference was greater at 52° C than at 22° C. Creep rates of both calendered and uncalendered films were comparable. Recovery experiments were conducted on the PET substrate and the coating A. The elastic component was recovered instantaneously during unloading, but a significant portion of the remaining strain did not recover after 16 h.

References

Andrade, E. N. da C. (1910). *Proc. Roy Soc.* **A84**, 1.
Andrade, E. N. da C. (1962). The validity of the $t^{1/3}$ law of flow of metals. *Phil. Mag.* **7**, 2003–2014.
Bahadur, S. and Ludema, K. C. (1972). Time–temperature superposition of large-strain shear properties of the ethylene–propylene copolymer system. *J. Appl. Poly. Sci.* **16**, 361–380.
Berry, B. S. and Pritchet, W. C. (1984). Bending-cantilever method for the study of moisture swelling in polymers. *IBM J. Res. Develop.* **28**, 662–667.
Berry, B. S. and Pritchet, W. C. (1988). Elastic and viscoelastic behavior of a magnetic recording tape. *IBM J. Res. Develop.* **32**, 682–694.
Bhushan, B. (1985). Anisotropic mechanical properties of biaxially-oriented poly

(ethylene terephthalate) films. In "Tribology and Mechanics of Magnetic Storage Systems" (B. Bhushan and N. S. Eiss, eds.), Vol. 2, pp. 119–126. SP-19 ASLE, Park Ridge, Illinois.

Bhushan, B. (1990). "Tribology and Mechanics of Magnetic Storage Devices." Springer-Verlag, New York.

Bhushan, B., Bradshaw, R. L., and Sharma, B. S. (1984a). Friction in magnetic tapes II: Role of physical properties. *ASLE Trans.* **27**, 89–100.

Bhushan, B., Hahn, F. W., Sharma, B. S., and Connolly, D. (1984b). Long-term reliability of magnetic tapes for digital recording. In "Tribology and Mechanics of Magnetic Storage Systems" (B. Bhushan et al., eds.) Vol. 1, pp. 132–147. SP-16, ASLE, Park Ridge, Illinois.

Bhushan, B., Heinrich, J. C., and Connolly, D. (1984c). Orthotropic viscoelastic behavior of polyethylene terephthalate film under plane-stress conditions. In "Tribology and Mechanics of Magnetic Storage Systems" (B. Bhushan et al., eds.), Vol. 1, pp. 158–171. SP-16, ASLE, Park Ridge, Illinois.

Bhushan, B. and Connolly, D. (1986). Viscoelastic properties of poly(ethylene terephthalate). *ASLE Trans.* **29**, 489–499.

Bhushan, B. and Dauer, F. W. (1978). Experimental determination of the relaxation modulus of a linear–isotropic–viscoelastic material. *Exp. Mech.* **18**, 421–425.

Bhushan, B. and Smith, D. R. (1985). Measurement of creep properties in compression and their influence on the friction of magnetic tapes. *ASLE Trans.* **28**, 325–335.

Bland, D. R. (1960). "The Theory of Linear Viscoelasticity." Pergamon Press, New York.

Bogy, D. B., Bugdayci, N., and Talke, F. E. (1979). Experimental determinations of creep functions for thin orthotropic polymer films. *IBM J. Res. Develop.* **23**, 450–458.

Boltzmann, L. (1874). Zur theorie der elastischen nachwirkung. *Sitzungsber. Math. Naturwiss. Kaiserl. Akad. Wiss.* **70** (2), 275.

Chen, Y. T. and Raj, R. (1977). "Creep Properties of Mylar PB 92 and PB 300." Technical Report No. TR 44.0316, March, IBM Corp., Boulder, Colorado.

Christensen, R. M. (1971). "Theory of Viscoelasticity," 2nd ed. Academic Press, New York.

Chu, W. H. and Smith, T. L. (1973). Rodlike superstructures in and mechanical properties of biaxially oriented polyethylene terephthalate films. In "Structure and Properties of Polymer Films" (R. W. Lenz and R. S. Stein, eds.), Plenum Press, New York.

Cuddihy, E. F. (1976). Hygroscopic properties of magnetic recording tape. *IEEE Trans. Mag.* **MAG-12**, 126–135.

Cuddihy, E. F. (1984). Personal communications.

Engel, P. A. and Lasky, R. C. (1977). Mechanical response and heat build up in repetitively impacted elastomers. *Exp. Mech.* **17**, 97–105.

Eshel, A., Baker, S., Hartman, A., and Orcutt, F. K. (1984). Mechanical aspects of archival storage of magnetic tape. In "Tribology and Mechanics of Magnetic Storage Systems" (B. Bhushan et al., eds.), Vol. 1, pp. 148–157. SP-16, ASLE, Park Ridge, Illinois.

Ferry, J. D. (1980). "Viscoelastic Properties of Polymers," 3rd ed. Wiley, New York.

Flugge, W. (1975). "Viscoelasticity." Springer-Verlag, New York.

Gillen, K. T. (1978). Use of a thermomechanical analyzer to estimate the tensile compliance of polymeric films. *J. Appl. Poly. Sci.* **22**, 1291–1302.

Goldstein, H. (1953). "Classical Mechanics." Addison-Wesley, Reading, Massachusetts.

Gottenberg, W. G. and Christensen, R. M. (1964). An experiment for determination of the mechanical properties in shear for a linear, isotropic viscoelastic solid. *Int. J. Eng. Sci.* **2**, 45.

Greenberg, H. J., Stephens, R. L., and Talke, F. E. (1977). Dimensional stability of floppy disks. *IEEE Trans. Magn.* **MAG-13**, 1397–1399.

Greenberg, H. J., Stephens, R. L., and Talke, F. E. (1978). Measurement of dimensional changes in thin polymer films. *Experimental Mechanics* **18**, 115–120.

Groeninckx, G., Berghmans, H., and Smets, G. (1976). Morphology and modulus-temperature behavior of semicrystalline poly(ethylene terephthalate) (PET). *J. Poly. Sci.: Poly. Phys. Ed.* **14**, 591–602.

Groeninckx, G., Reynaers, H., Berghmans, H., and Smets, G. (1980a). Morphology and melting behavior of semicrystalline poly(ethylene terephthalate) (PET). I. Isothermally crystallized PET. *J. Poly. Sci.: Poly. Phys. Ed.* **18**, 1311–1324.

Groeninckx, G. and Reynaers, H. (1980b). Morphology and melting behavior of semicrystalline poly(ethylene terephthalate). II. Annealed PET. *J. Poly. Sci., Poly. Phys. Ed.* **18**, 1325–1341.

Gupta, V. B., Ramesh, C., and Gupta, A. K. (1984). Structure–property relationship in heat-set poly(ethylene terephthalate) fibers. III. Stress–relaxation behavior. *J. Appl. Poly. Sci.* **29**, 4203–4218.

Gurtin, M. E. and Sternberg, E. (1962). On the linear theory of viscoelasticity. *Arch. Rational Mech. Anal.* **11**, 291.

Hawthorne, J. M. (1981). Stress relaxation behavior of biaxially oriented poly(ethylene) terephthalate. *J. Appl. Poly. Sci.* **26**, 3317–3324.

Hearmon, R. F. S. (1961). "An Introduction to Applied Anisotropic Elasticity." Oxford University Press, London.

Henderson, C. (1951). The application of Boltzmann's superposition theory to materials exhibiting reversible β flow. *Proc. Roy. Soc.* (*London*) **A206**, 72–86.

Illers, K. H. and Breuer, H. (1963). Molecular motions in polyethylene terephthalate. *J. Colloid Sci.* **18**, 1–31.

Judge, J. T. and Stein, R. S. (1961). Growth of crystals from molten crosslinked oriented polyethylene. *J. Appl. Phys.* **32**, 2357–2363.

Kennedy, A. J. (1953). On the generality of the cubic creep function. *J. Mech. Phys. Solids* **1**, 172–181.

Kovacs, A. J., Stratton, R. A., and Ferry, J. D. (1963). Dynamic mechanical properties of polyvinyl acetate in shear in the glass transition temperature range. *J. Phys. Chem.* **67**, 152–161.

Krigbaum, W. R. and Roe, R. J. (1964). Diffraction study of crystallite orientation in stretched polychloroprene vulcanizates. *J. Poly. Sci.* **A-2**, 2, 4391–4414.

Lekhnitskii, S. G. (1968). "Anisotropic Plates." Gordon and Breach, London, U.K.

Lekhnitskii, S. G. (1981). "Theory of Elasticity of an Anisotropic Body." Mir, Moscow, USSR.

Love, A. E. H. (1944). "A Treatise on the Mathematical Theory of Elasticity." Dover, New York.

McCrum, N. G. and Morris, E. L. (1964). On the measurement of the activation energies for creep and stress relaxation. *Proc. Roy. Soc. London Ser. A* **281**, 258–273.

McCrum, N. G., Read, B. E., and Williams, G. (1967). "Anelastic and Dielectric Effects in Polymeric Solids." Wiley, New York.

Morland, L. W. and Lee, E. H. (1960). Stress analysis for linear viscoelastic materials with temperature variation. *Trans. Soc. Rheology* **4**, 233–263.

Murayama, T., Dumbleton, J. H., and Williams, M. L. (1968). Viscoelasticity of oriented poly(ethylene terephthalate). *J. Poly. Sci.* **6**, 787–793.

Nakayasu, H., Markovitz, and Plazek, D. J. (1961). The frequency and temperature dependence of the dynamic mechanical properties of a high density polyethylene. *Trans. Soc. Rheo.* **5**, 261–283.

Nielsen, L. E. (1974). "Mechanical Properties of Polymers and Composites," Vol. 2. Marcel Dekker, New York.

Norwick, A. S. and Berry, B. S. (1972). "Anelastic Relaxation in Crystalline Solids." Academic Press, New York.

Oono, R., Miyasaka, K., and Ishikawa, K. (1973). Crystallization kinetics of biaxially stretched natural rubber. *J. Poly. Sci.: Poly. Phys. Ed.* **11**, 1477–1488.

Pinnock, P. R. and Ward, I. M. (1966). The temperature dependence of viscoelastic behavior in polyethylene terephthalate. *Polymer* **7**, 255–266.

Plazek, D. J. (1960). Dynamic mechanical and creep properties of a 23% cellulose nitrate solution: Andrade creep in polymeric systems. *J. Colloid Sci.* **15**, 50–75.

Plazek, D. J. (1966). Effect of crosslink density on the creep behavior of natural rubber vulcanizates. *J. Poly. Sci.* **A-2**, 4, 745–763.

Plazek, D. J. (1980). Viscoelastic and steady-state rheological response. In "Methods of Experimental Physics," Vol. 16C, Chapter 11. Academic Press, New York.

Plazek, D. J. and Magill, J. H. (1966). Physical properties of aromatic hydrocarbons. I. Viscous and viscoelastic behavior of $1:3:5$-tri-α-naphthyl benzene. *J. Chem. Phys.* **45**, 3038–3050.

Raj, R., Chen, Y. T., and Van Baren, F. (1974). High sensitivity creep tester. *Rev. Sci. Instrum.* **45**, 1502–1503.

Richardson, M. J. and Savill, N. G. (1977). Enthalpy relaxation in amorphous polymers. *Polymer* **18**, 413–414.

Samuels, R. J. (1974). "Structured Polymer Properties." Wiley, New York.

Schwarzl, F. and Staverman, A. J. (1952). Time–temperature dependence of linear viscoelastic behavior. *J. Appl. Phys.* **23**, 838–843.

Smith, T. L. (1958). Dependence of the ultimate properties of a GR–S rubber on strain rate and temperature. *J. Poly. Sci.* **32**, 99–113.

Smith, T. L. (1983–84). Personal communications.

Sokolnikoff, I. S. (1956). "Mathematical Theory of Elasticity," 2nd ed., McGraw-Hill, New York.

Sternstein, S. S. (1977). Mechanical properties of glassy polymers. In "Properties of Solid Polymeric Materials," Part B (J. M. Schultz, ed.), pp. 541–598. Academic Press, New York.

Sternstein, S. S. (1983). Transient and dynamic characterization of viscoelastic solids. In "Polymer Characterization: Spectroscopic, Chromatographic, and Physical Instrumental Methods" (C. D. Craver, ed.), Advances in Chemistry Series No. 203, pp. 123–147. American Chemical Society, Washington, D.C.

Stouffer, D. C. (1972). A thermal–hereditary theory for linear viscoelasticity. *J. Appl. Math. Phys.* **23**, 845–851.

Stouffer, D. C. and Wineman, A. S. (1971). Linear viscoelastic materials with environmental dependent properties. *Int. J. Eng. Sci.* **9**, 193–212.

Struik, L. C. E. (1978). "Physical Aging in Amorphous Polymers and Other Materials." Elsevier, New York.

Tajiri, K., Fujii, Y. Makoto, A., and Kawai, H. (1970). Linear viscoelastic properties of polyethylene terephthalate and its related polymers. *J. Macromol. Sci.-Phys.* **B4**, 1–38.

Tant, M. R. and Wilkes, G. L. (1981). Physical aging studies of semicrystalline poly(ethylene terephthalate). *J. Appl. Poly. Sci.* **26**, 2813–2825.

Taylor, R. L. (1964). Creep and relaxation. *AIAA. T. N.* **2**, 1659–1660.

Taylor, R. L., Pister, K. S., and Goudreau, G. L. (1970). Thermomechanical analysis of viscoelastic solids. *Inst. J. Num. Meth. Eng.* **2**, 45–59.

Thompson, A. B. and Woods, D. W. (1956). The transitions of polyethylene terephthalate. *Trans. Faraday Soc.* **52**, 1383–1397.

Timoshenko, S. P. (1958). "Strength of Materials, Part I: Elementary Theory and Problems." Van Nostrand, New York.

Timoshenko, S. P. and Goodier, J. N. (1970). "Theory of Elasticity." McGraw-Hill, New York.

Tobolsky, A. V. (1960). "Properties and Structure of Polymers." Wiley, New York.

Tschoegl, N. W. (1989). "The Phenomenological Theory of Linear Viscoelastic Behavior." Springer-Verlag, New York.

Vallat, M. F., Plazek, D. J. and Bhushan, B. (1986). Effects of thermal treatment of biaxially oriented poly(ethylene terephthalate). *J. Poly. Sci. Part B: Poly. Phys.* **24**, 2123–2134.

Vallat, M. F. and Plazek, D. J. (1988a). Effects of thermal treatment of biaxially oriented poly(ethylene terephthalate). II. The anisotropic glass temperature. *J. Poly. Sci., Part B: Poly. Phys.* **26**, 545–554.

Vallat, M. F., Plazek, D. J., and Bhushan, B. (1988b). Effects of thermal treatment of biaxially oriented poly(ethylene terephthalate). III. Creep behavior following various thermal histories. *J. Poly. Sci., Part B: Poly Phys.* **26**, 555–567.

Van Holde, K. (1957). A study of the creep of nitrocellulose. *J. Poly. Sci.* **24**, 417–427.

Ward, I. M. (1964). The temperature dependence of extensional creep in polyethylene terephthalate. *Polymer* (*London*) **5**, 59.

Ward, I. M. (1983). "Mechanical Properties of Solid Polymers," 2nd ed. Wiley, New York.

Weast, R. C. (1972). "Handbook of Chemistry and Physics." CRC Press, Cleveland, Ohio.

Weber, S. M., Manson, J. A., and Lang, R. W. (1983). Dynamic mechanical spectroscopy using the autovibron DDV-III-C. In "Polymer Characterization: Spectroscopic, Chromatographic, and Physical Instrumental Methods" (C. D. Craver, ed.), Advances in Chemistry Series 203, 109–122. American Chemical Society, Washington, D.C.

Wetton, R. E., Croucher, T. G., and Fursdon, J. W. M. (1983). The PL-dynamic mechanical thermal analyzer and its application to the study of polymer transitions. In "Polymer Characterization: Spectroscopic, Chromatographic, and Physical Instrumental Methods" (C. D. Craver, ed.), ACS Advances in Chemistry Series 203, pp. 95–108. American Chemical Society, Washington, D.C.

Williams, M. L., Landel, R. F., and Ferry, J. D. (1955). The temperature dependence of relaxation mechanisms in amorphous polymers and other glass-forming liquids. *J. Amer. Chem. Soc.* **77**, 3701–3707.

Zoller, P., Bolli, P., Pahud, V., and Ackerman, H. (1976). Apparatus for measuring pressure–volume–temperature relationships of polymers to 350° C and 2200 kg/cm^2. *Rev. Sci. Instrum.* **47**, 948–952.

Appendix 3.A. Analysis of Flexural and Tensile Stress Relaxation of a Multilayered Tape

Here we consider a beam of magnetic tape composed of three layers of equal width w and thicknesses t_b, t_s, t_f, respectively, t_b being the thickness of the backcoat, t_s being the thickness of the tape substrate, and t_f being the thickness of the frontcoat. The corresponding stress relaxation moduli are denoted by $E_b(t)$, $E_s(t)$, and $E_f(t)$.

3.A.1. Flexural Relaxation at Constant Curvature

The composite beam is bent to a radius of R_w by a pure moment couple $M(t)$, Fig. 3.A.1 (Berry and Pritchet, 1988). For the bending analysis of a composite beam, we make the usual assumptions that plane cross sections remain plane on bending, and the radius R_w is much larger than the thickness of the tape. The strain $\varepsilon(z)$ at a distance z from the inner reference surface $z = 0$ is then

$$\varepsilon(z) = (z - \bar{z})/R_w, \tag{3.A.1}$$

where $\bar{z}$ defines the position of the neutral surface (centroid). Since the sample is bent by a pure couple M, the location of the neutral surface can be obtained from the condition that the stress distribution (with positive stresses above

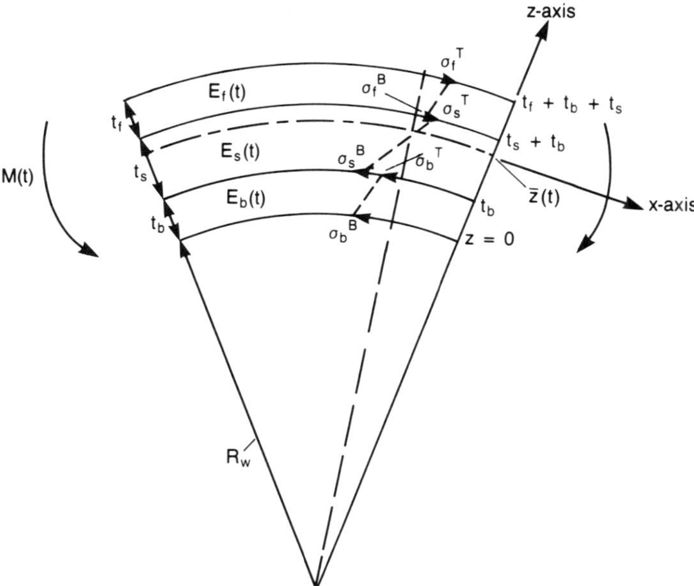

Fig. 3.A.1. Flexural stress profile of a trilayer magnetic tape bent on a cylinder of radius R_w due to a bending moment $M(t)$ (Berry and Pritchet, 1988). © 1988 International Business Machines Corporation; reprinted with permission.

the neutral axis and negative stresses below the neutral axis)

$$\sigma(z, t) = E(z, t)\varepsilon(z)$$

$$= E(z, t)(z - \bar{z})/R_w, \tag{3.A.2}$$

produces no net axial force. Hence, we obtain

$$(w/R_w) \int_0^{t_b+t_s+t_f} E(z, t)(z - \bar{z})\, dz = 0, \tag{3.A.3}$$

where w is the cross-sectional width and $E(z, t)$ is the stress–relaxation modulus at position z and time t after the initial windup. Equation (3.A.3) for the trilayer leads to

$$\bar{z}(t) = \frac{E_b(t)t_b^2 + E_s(t)[(t_b + t_s)^2 - t_b^2] + E_f(t)[(t_b + t_s + t_f)^2 - (t_s + t_b)^2]}{2[E_b(t)t_b + E_s(t)t_s + E_f(t)t_f]}, \tag{3.A.4}$$

where $E_b(t)$, $E_s(t)$, and $E_f(t)$ represent the stress–relaxation moduli of the component layers.

The magnitude of the bending moment $M(t)$ in equilibrium with the stress distribution is given by

$$M(t) = (w/R_w) \int_0^{t_b+t_s+t_f} E(z, t)(z - \bar{z})^2\, dz. \tag{3.A.5}$$

The bending stiffness $K(t)$ of the trilayer thus follows as

$$K(t) \equiv M(t)/(1/R_w) = E_b(t)w[\tfrac{1}{3}t_b^3 - \bar{z}(t)t_b^2 + \bar{z}^2(t)t_b]$$

$$+ E_s(t)w\{\tfrac{1}{3}[(t_b + t_s)^3 - t_b^3] - \bar{z}(t)[(t_b + t_s)^2 - t_b^2]$$

$$+ \bar{z}^2(t)t_s\}$$

$$+ E_f(t)w\{\tfrac{1}{3}[(t_b + t_s + t_f)^3 - (t_b + t_s)^3]$$

$$- \bar{z}(t)[(t_b + t_s + t_f)^2 - (t_b + t_s)^2] + \bar{z}(t)^2 t_f\}. \tag{3.A.6}$$

This analysis is general and can be applied to any number of layers.

For a unilayer ($t_b = t_f = 0$) Eq. (3.A.6) reduces to

$$K(t) = (wt_s^3/12)E_s(t). \tag{3.A.7a}$$

In the dimensionless form obtained by normalizing to the initial unrelaxed values $K(0)$ and $E_s(0)$, Eq. (3.A.7a) becomes

$$K(t)/K(0) = E_s(t)/E_s(0). \tag{3.A.7b}$$

The interpretation of the normalized bending stiffness parameter $K(t)/K(0)$ for the composite sample is more complicated, since it involves the combined stress–relaxation behavior of all layers of the tape.

From Eq. (3.A.2), we can write expressions for the flexural stresses at various locations on the beam. For a point on the top surface of the front coat, the flexural stress is

$$\sigma_f^{T}(t) = E_f(t)(t_b + t_s + t_f - \bar{z})/R_w, \tag{3.A.8a}$$

while for a point in the same section at the frontcoat–substrate interface, the flexure stress is

$$\sigma_f^{B}(t) = E_f(t)(t_b + t_s - \bar{z})/R_w. \tag{3.A.8b}$$

For a point in the substrate at the substrate–frontcoat interface

$$\sigma_s^{T}(t) = E_s(t)(t_b + t_s - \bar{z})/R_w, \tag{3.A.9a}$$

at the substrate–backcoat interface in the substrate

$$\sigma_s^{B}(t) = E_s(t)(t_b - \bar{z})/R_w, \tag{3.A.9b}$$

at the substrate–backcoat interface in the backcoat

$$\sigma_b^{T}(t) = E_b(t)(t_b - \bar{z})/R_w, \tag{3.A.10a}$$

and at the bottom of the backcoat

$$\sigma_b^{B}(t) = -E_b(t)\bar{z}/R_w. \tag{3.A.10b}$$

Table 3.A.1 shows the various flexural stress values at time $=0$, for the tape substrate, and for the backcoated and unbackcoated tapes. We note that for cases shown in Table 3.A.1, the greatest flexural stress occurs in the substrate, rather than in the lower modulus coating—which is further from the centroid. Note that for the cases where the coating and PET substrate are of equal modulus, the greatest flexural stress would occur in the coating.

With time, the bending moment or flexural stress in the bent tape relaxes. The stresses in the coating relax at a much faster rate than those in the PET. At the same time the initial elastic flexural strain becomes partially viscoelastic. Since this occurs at a rate proportional to the stress relaxation, the viscoelastic strain in coating will be greater than that in the PET. Thus when the tape is unloaded it will not return immediately to its undeformed configuration.

3.A.2. Tensile Relaxation at Constant Elongation

The total axial force exerted on the multilayer to maintain a constant elongation strain ε_0 applied at $t = 0$ is the sum of forces on the component layers (or the tensile stiffness = modulus of elasticity $\times$ cross-sectional area of a parallel arrangement of springs is the sum of their individual spring constants), Fig. 3.A.2. The composite relaxation modulus $\bar{E}(t)$ is given as

$$\bar{E}(t) = \frac{t_b E_b(t) + t_s E_s(t) + t_f E_f(t)}{t_b + t_s + t_f}. \tag{3.A.11}$$

In the case of magnetic tapes, the substrate thickness is substantially larger than that of front- and backcoats, therefore, viscoelastic deformation of the

Table 3.A.1. Flexural stresses of a composite magnetic tape (substrate thickness, $t_s = 23.4\ \mu$m) bent to a 12.7-mm radius as wrapped (time = 0)

	t_b, μm	t_f, μm	E_b, GPa	E_s, GPa	E_f, GPa	$\bar{z}$, μm	σ_f^T, MPa	σ_f^B, MPa	σ_s^T, MPa	σ_s^B, MPa	σ_b^T, MPa	σ_b^B, MPa
Backcoated tape A	3.6	4	1.1	3.7	2.2	16.0	2.6	1.9	3.2	−3.6	−1.07	−1.38
Unbackcoated tape A	0	4	—	3.7	2.2	13.0	2.4	1.8	3.0	−3.8		
PET substrate	0	0		3.7		11.7			3.4	−3.4		

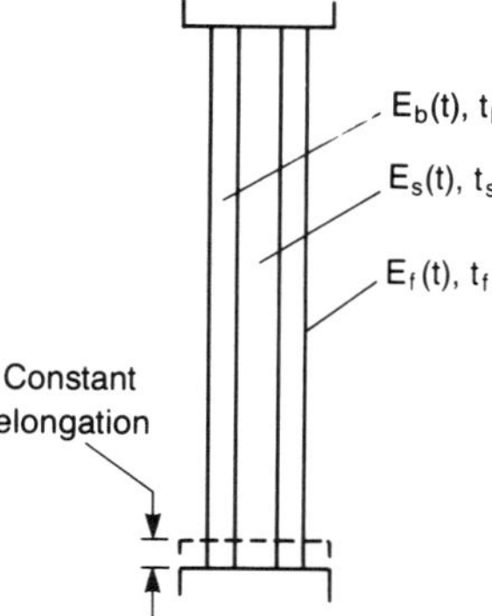

Fig. 3.A.2. A trilayer magnetic tape under constant elongation, as imposed by stretching between rigid crossheads.

tape in a tensile test is primarily dominated by the properties of PET as long as the relaxation modulus of the coating is equal to or less than of the PET.

Appendix 3.B. Analysis of Thermal Curling of a Multilayered Magnetic Tape in the Elastic Regime

Any multilayered strip composed of material of different coefficients of thermal expansions has a general tendency to bend or curl when subjected to a change in temperature (Timoshenko and Goodier, 1970; Berry and Pritchet, 1984, 1988). Magnetic tapes (both bilayered and trilayered) curl when exposed to heat. The change in curvature of the curled tape can be measured using the mandrel test described in this chapter. Curvature changes as much as 20 m^{-1} for a temperature change of $10°$ C have been measured. These measurements can be used to predict some mechanical or thermal properties of the materials.

We present an analysis of thermal curling for a trilayered strip, as shown in Fig. 3.B.1 (Berry and Pritchet, 1988). We denote the layered thicknesses by t_b, t_s, and t_f for backcoat, substrate, and frontcoat, respectively. The respective moduli and coefficients of linear thermal expansion are E_i and α_i ($i = $ b, s, f). The trilayer is assumed to be flat and stress-free at a reference temperature T_0 and to curl to a curvature $1/R$ in response to a temperature change $\Delta T = T - T_0$. The conditions for equilibrium involve: (i) a balance of internal axial forces; (ii) a balance of internal bending moments; and (iii) dimensional matching at the interfaces. If we suppose, that the forces P_b and P_s are the compressive and P_f is tensile, the force balance may be written

$$P_b + P_s = P_f, \tag{3.B.1}$$

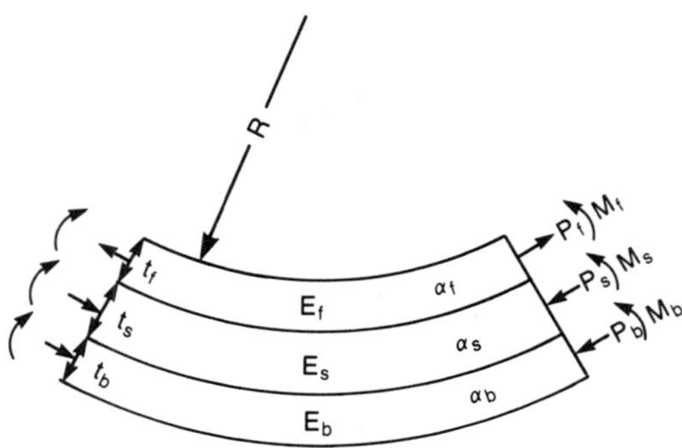

Fig. 3.B.1. Force and moment diagram for the thermal curling of a trilayer (Berry and Pritchet, 1984).

and the moment balance follows as

$$\sum M_i = M_b + M_s + M_f = P_b(t_b + t_s)/2 + (P_b + P_s)(t_s + t_f)/2. \quad (3.B.2)$$

With the usual assumption that plane cross sections remain plane on bending, the bending moment on the ith layer is $M_i = E_i I_i/R$. We thus obtain

$$(1/R)(E_b I_b + E_s I_s + E_f I_f) = P_b(t_b + t_s)/2 + (P_b + P_s)(t_s + t_f)/2, \quad (3.B.3)$$

where $I_i = wt_i^3/12$ is the second moment of area of the layer i about its center, and w is the width of the cross section. Further, recognizing that the radius R is much greater than the thicknesses of the component layers, the requirement of dimensional matching at the interfaces leads to the equations

$$\alpha_b \Delta T - P_b/E_b A_b - t_b/2R = \alpha_s \Delta T - P_s/E_s A_s + t_s/2R, \quad (3.B.4)$$

and

$$\alpha_s \Delta T - P_s/E_s A_s - t_s/2R = \alpha_f \Delta T - P_f/E_f A_f + t_f/2R, \quad (3.B.5)$$

where A_i denotes the cross-sectional area wt_i of the ith layer. After the elimination of P_f from Eq. (3.B.5) by the use of Eq. (3.B.1), rearrangement of Eqs. (3.B.4) and (3.B.5) yields

$$X_{bs} = (P_b/S_b) - (P_s/S_s), \quad (3.B.6)$$

$$X_{sf} = (P_b/S_f) + P_s[(1/S_s) + (1/S_f)], \quad (3.B.7)$$

where the tensile stiffnesses S_i are defined by $S_i \equiv E_i A_i$ and

$$X_{bs} \equiv (\alpha_b - \alpha_s)\Delta T - (t_b + t_s)/2R, \quad (3.B.8)$$

and

$$X_{sf} \equiv (\alpha_s - \alpha_f)\Delta T - (t_s + t_f)/2R. \quad (3.B.9)$$

Solving Eqs. (3.B.6) and (3.B.7) for P_b and P_s, we obtain

$$P_b = S_b\{(S_s + S_f)X_{bs} + S_f X_{sf}\}/(S_b + S_s + S_f) \quad (3.B.10)$$

and

$$P_s = S_s\{S_f X_{sf} - S_b X_{bs}\}/(S_b + S_s + S_f).\tag{3.B.11}$$

When these results are inserted into Eq. (3.B.3), we finally obtain

$$(1/R)\{[2(S_b + S_s + S_f)(E_b I_b + E_s I_s + E_f I_f)] + [S_b(S_s + S_f)(t_b + t_s)^2/2]$$
$$+ [S_b S_f(t_b + t_s)(t_s + t_f)] + [S_f(S_b + S_s)(t_s + t_f)^2/2]\}$$
$$= (\alpha_b - \alpha_s)\Delta T S_b[(t_b + t_s)(S_s + S_f) + (t_s + t_f)S_f]$$
$$+ (\alpha_s - \alpha_f)\Delta T S_f[(t_b + t_s)S_b + (t_s + t_f)(S_b + S_s)].\tag{3.B.12}$$

It is of interest to examine Eq. (3.B.12) for some simpler cases. For the case of the totally symmetrical trilayer we have $\alpha_b = \alpha_f$, $E_b = E_f$, $S_b = S_f$, and $t_b = t_f$. For this case, we obtain $1/R = 0$, meaning that a symmetrical trilayer does not exhibit thermal curling, as is already evident from the postulated symmetry. For the geometrically symmetrical but elastically asymmetrical trilayer which approximately represents the backcoated magnetic tape, Eq. (3.B.12) shows only a modest simplification. The case of the bilayer is obtained by assigning any one of the layers a zero thickness. For example, if $t_b = 0$, Eq. (3.B.12) becomes

$$(1/R)\{[2/(t_s + t_f)](1/S_s + 1/S_f)(E_s I_s - E_f I_f) + (t_s + t_f)/2\} = (\alpha_s - \alpha_f)\Delta T,$$
$$\tag{3.B.13}$$

in accord with previously available results (Timoshenko and Goodier, 1970).

It is helpful to rearrange Eq. (3.B.13) for a bilayer so as to disentangle, to some extent, the effects of geometry and elasticity on the curling of the bilayer.

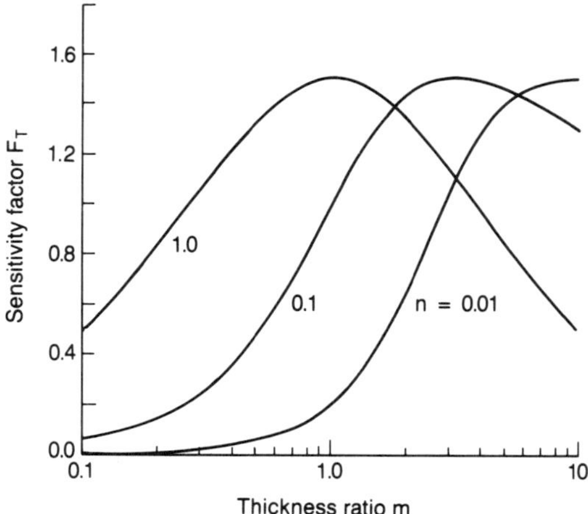

Fig. 3.B.2. The dimensional sensitivity factor F_T as a function of the bilayer thickness ratio $m = t_s/t_f$, for different values of the modulus ratio $n = E_s/E_f$.

To this end, we note that Eq. (3.B.13) can be written in the form

$$1/R = F_T(\alpha_s - \alpha_f)\Delta T/(t_s + t_f), \qquad (3.B.14a)$$

where F_T is a dimensionless sensitivity factor given by

$$F_T = 2/\{[(1 + mn)(1 + nm^3)/3mn(1 + m)^2] + 1\}, \qquad (3.B.14b)$$

where m denotes the thickness ratio t_s/t_f and n denotes the ratio of elastic moduli E_s/E_f. Since F_T depends only on the ratios m and n, we may note first that if these quantities are held constant, the slope of the curvature–temperature plot will vary inversely with the total thickness $t_s + t_f$. We thus begin to understand why thermal curling can be a prominent attribute of thin multilayers. Further insight is provided by Fig. 3.B.2, which shows how F_T varies with the thickness ratio m for several widely different values of the modulus ratio n. For equal moduli ($n = 1$), F_T follows a symmetrical bell-shaped curve with a maximum value of 1.5 for $t_s/t_f = 1$. As E_s/E_f is changed, the position of the maximum shifts, the curve becomes asymmetrical, but the maximum sensitivity factor remains equal to 1.5. In brief, the slope of the temperature–curvature plot for a bilayer with substantial geometrical asymmetry (such as those of interest for magnetic tapes) is quite strongly dependent on the modulus ratio n. The slope can be large if the minor component has a relatively high stiffness, but becomes much reduced if relative thinness is also combined with a low relative modulus. With these considerations in mind it becomes somewhat easier to understand how the slopes of the experimental thermal curling curves ($1/R$ as a function of ΔT) depend on the different expansion coefficients, thicknesses, and elasticities of the two layers. If the E_s/E_f is known for a bilayer then ($\alpha_s - \alpha_f$) can be obtained from the slopes of thermal curling curves.

CHAPTER 4

Stress Analysis of Flexible Media

4.1. Wound Magnetic Tape Reels

The analyses of the stresses generated during the winding of thin flexible webs, such as magnetic tape, film, and paper have been presented by several authors (Anonymous, 1959; Tramposch, 1963; Altmann, 1968; Owen, 1971; Monk et al., 1975; Umanskii et al., 1978; Bertram and Eshel, 1980; Yagoda, 1980a, b; Connolly and Winarski, 1984; Eshel et al., 1984). A knowledge of these initial stresses, how they can be altered, and how they produce defects in the wound material provide design and winding guidelines that help minimize the adverse effects of such stresses.

During the tape's winding process, when tape is wound onto a hub of some compliance and with a certain winding tension, a stress field develops. During winding, in addition, the tape can be subjected to applied deformations when it is wound on extraneous particles, irregularities in the hub surface, entrapped air, etc., which induce additional stresses in the tape. The most important components of the stress field are the hoop (also called circumferential) stress, which is caused by the winding tension applied over the cross-sectional area of the tape, and the radial interlayer pressure, which is caused by the radial component of the winding stress. As more and more wraps are added, the radial stress in the inner wraps accumulates continuously and the hoop stress (wound in tension) decreases. The hub compliance relieves the hoop stress, since the radial displacement of the tape is smaller and so is the hoop strain. At the same time, it reduces the buildup of radial pressure. However, if the hub is too rigid, the tensile hoop stress generated during winding remains in the wound material near the hub. For materials that have across-the-width variations in mechanical properties, such as magnetic tape, film, or paper, these high hoop stresses produce tension bands or lanes in the tape (Pfeiffer, 1966; Frye, 1967; Monk et al., 1975; Bhushan et al., 1984a; Eshel et al., 1984). (The effects of the Poisson ratio tend to reduce the tensile hoop stress with increasing radial pressure.) If the hub is too compliant, its function is transferred to the wound tape, which produces compressive circumferential stresses in the tape. Sufficient, negative, hoop stresses can cause the wound tape to buckle. Furthermore, the radial pressure during winding must be sufficient to prevent interlayer tape slippage on acceleration. To ensure that the wound tape remains undistorted during storage, the hub must be of uniform compliance, which requires as close a uniform hub thickness as possible. A mismatch in the coefficient of thermal or hygroscopic expansion between the hub and the tape will lead to further stresses induced in the wound tape when there is a temperature or humidity change. These stresses must be considered when selecting hub material and geometry. Stress variations across the reel radius can be reduced by tapered winding tension (or constant torque winding) rather than constant tension winding.

The viscoelastic nature of the tape material causes these stresses to relax with time and a viscoelastic deformation then forms that can, eventually, cause loss of data (Tramposch, 1965, 1967; Pfeiffer, 1966; Bhushan et al., 1984a; Eshel et al., 1984; Bhushan and Connolly, 1986; Heinrich et al., 1986, Chapter 5). This process is accelerated at higher temperatures and humidities. The relaxation of a stored wound reel necessitates rewinding to re-establish reel tension and protect reel integrity. The qualitative effects of various winding, mechanical properties, and environmental parameters on stress distributions in the reel are shown in Fig. 4.1.

In the stress analysis of wound magnetic tape reels, the hub is assumed to be either an elastic or viscoelastic material and the tape reel itself is usually assumed to be either elastic or viscoelastic and anisotropic. We first present elastic analyses used to predict the initial stress field generated during the winding of tape onto a hub, based on the work by Connolly and Winarski (1984). Then an axisymmetric finite-element model for the analysis of stress relaxation in a wound magnetic tape reel is presented. It is used to predict viscoelastic deformation in the tape caused by the initial winding stresses and geometric deformations imposed on it (Heinrich et al., 1986).

4.1.1. Initial Stress Field

Analyses used to predict the initial stress field generated during the tape's winding process are presented. A comparison of the predicted and measured radial stresses in a reel of tape is presented. The analysis is then applied to provide guidelines in the selection of hub material and geometry, as well as to the winding tension profile. This analysis is further used to calculate the maximum amount of tape that can be stored on a given hub without adversely affecting the wound tape. We will then include an analysis of the thermal stresses caused by the mismatch in the coefficients of thermal expansion between the hub and the tape, and between the radial and circumferential directions in the tape. These analyses are based on the work by Connolly and Winarski (1984).

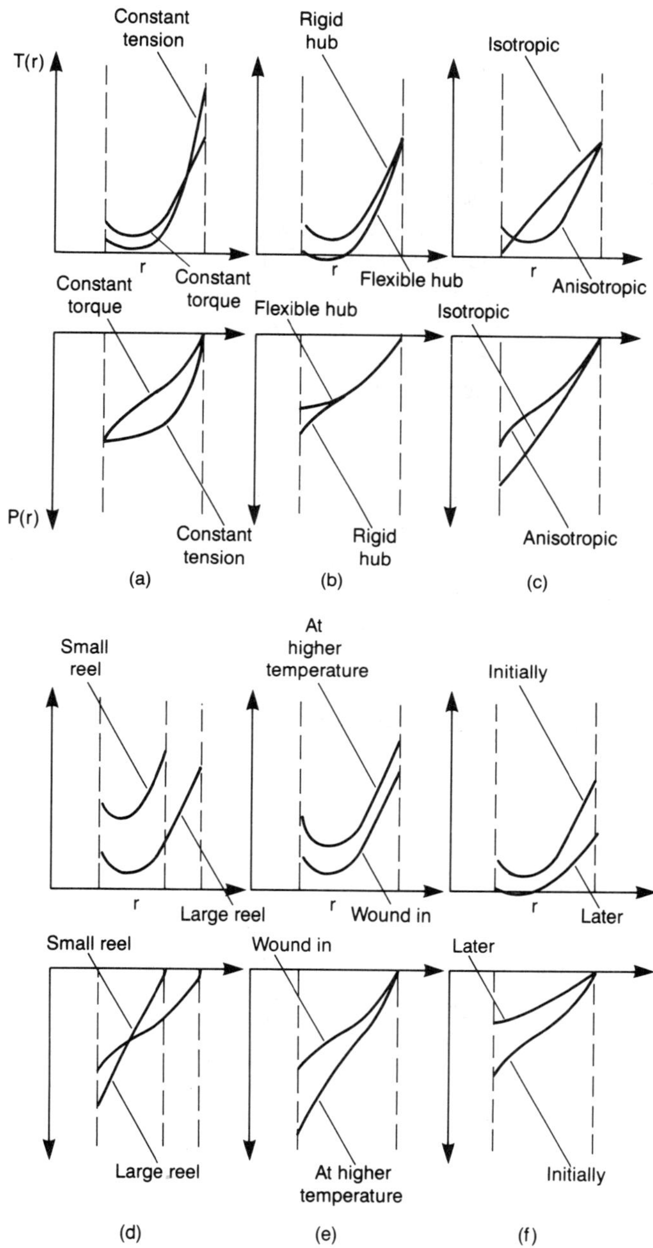

Fig. 4.1. Schematic description of the effect of various parameters on the stress distribution in a reel: (a) effect of winding pattern; (b) effect of hub flexibility; (c) effect of reel anisotropy; (d) effect of reel radius; (e) effect of temperature rise or humidity reduction; and (f) effect of stress relaxation (Eshel et al., 1984).

4.1.1.1. Analytical Techniques

Although the actual winding process is a continually-progressing operation, it can be represented realistically by the incremental addition of single layers of tapes, each of which is assumed a closed, thin-walled ring with known internal tension. Since it can be shown that the variation in tape tension along the tape length is small, the shear stresses between layers and in radial planes can be neglected. Consequently, the winding pattern can be presented by the model shown in Fig. 4.2. A portion of the applied winding stress of the tape is resisted by the centrifugal acceleration of the web, and only a portion of the applied winding stress acts to compress the underlying reel structure (Bertram and Eshel, 1980; Yagoda, 1980b). For the sake of simplicity, the centrifugally-induced stresses are neglected here.

The following assumptions were made for a perfectly wound reel of magnetic tape:

(1) The hub is a right-circular cylinder and remains so during and after winding. The relationship between the hub deflection and the radial pressure exerted on it by the tape is linear.
(2) The tape is of uniform thickness, which is small in comparison with its width, and offers no resistance to bending.
(3) During and following winding, the tape reel is considered to be a homogeneous and orthotropic elastic cylinder. Young's modulus in the radial direction, E_r, is smaller than that in the circumferential direction, E_θ. This is due to the softening effects of the tape's coating and the entrapped air during winding.

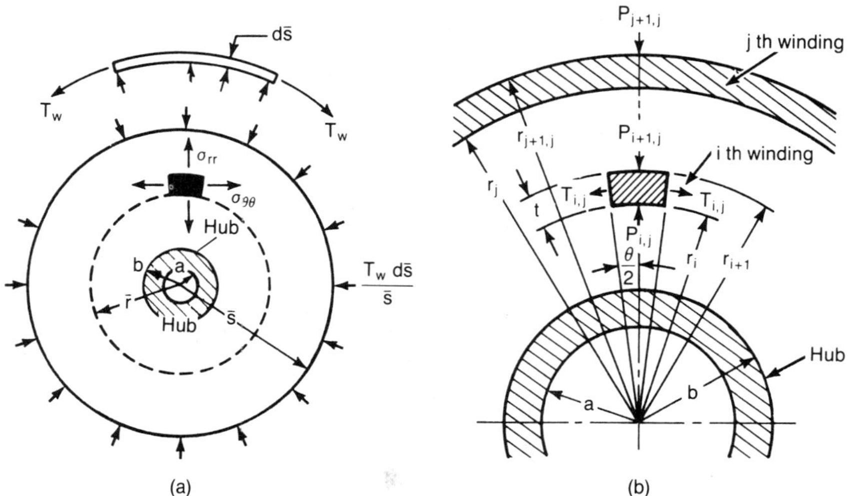

Fig. 4.2. Schematic of stresses acting on a wound tape reel for (a) Altmann and (b) Tramposch models.

(4) Although the winding process is continuous, it can be modeled as the successive addition of closed rings with a known internal tension.
(5) Shear stresses and relative motion between the layers can be neglected.
(6) The stress components at a point in the tape are independent of its width and those in the width direction are negligible, therefore, plane-stress conditions exist.

The assumptions have thus reduced the formulation to a *one-dimensional, plane-stress axisymmetric problem*, with variations occurring only in the radial direction. We now need to determine the radial- and hoop-stress components at any radial location in the wound tape. In the literature, two solution techniques exist, one proposed initially by Anonymous (1959), which was subsequently used by many others (Altmann, 1966; Owen, 1971; Umanskii et al., 1978; Bertram and Eshel, 1980; Yagoda, 1980a), and another proposed by Tramposch (1963). We will take the one presented by Altmann (1968) as typical of the first approach and contrast it with that of Tramposch (1963) and will discuss their respective advantages, and then the agreement of both approaches is verified.

In the remainder of this section, we will use the following notation to refer to the equations shown in Tables 4.1–4.3: 4.1A-3 refers to Table 4.1, Altmann column, Eq. [3]; 4.1T-2 refers to Table 4.1, Tramposch column, Eq. [2]; and

Table 4.1. Problem formulation for the analysis used to predict the initial stress field generated during the tape's winding process (Connolly and Winarski, 1984)

Altmann	Tramposch

Stress equation of equilibrium

$$r\frac{d\sigma_{rr}}{dr} + \sigma_{rr} - \sigma_{\theta\theta} = 0. \qquad (1)$$

$$T_{i,j} + P_{i,j} + (r_i + t)\frac{P_{i+1,j} - P_{i,j}}{t} = 0. \qquad (1)$$

Stress–displacement relationship

$$\sigma_{rr} = \frac{E_r}{(1 - v_{r\theta}v_{\theta r})}\left(\frac{du}{dr} + v_{\theta r}\frac{u}{r}\right),$$
$$\sigma_{\theta\theta} = \frac{E_\theta}{(1 - v_{r\theta}v_{\theta r})}\left(\frac{u}{r} + v_{r\theta}\frac{du}{dr}\right). \qquad (2)$$

$$P_{i,j} = \frac{-E_r}{(1 - v_{r\theta}v_{\theta r})}\left(\frac{u_{i+1,j} - u_{i,j}}{t} + v_{\theta r}\frac{u_{i,j}}{r_i}\right),$$
$$T_{i,j} = \frac{E_\theta}{(1 - v_{r\theta}v_{\theta r})}\left(\frac{u_{i,j}}{r_i} + v_{r\theta}\frac{u_{i+1,j} - u_{i,j}}{t}\right). \qquad (2)$$

Displacement equation of equilibrium

$$\frac{d^2u}{dr^2} + \left(1 + v_{\theta r} - v_{r\theta}\frac{E_\theta}{E_r}\right)\frac{1}{r}\frac{du}{dr} - \frac{E_\theta}{E_r}\frac{u}{r^2} = 0. \qquad (3)$$

$$\frac{u_{i+2,j} - 2u_{i+1,j} + u_{i,j}}{t^2}$$
$$+ \left(1 + v_{\theta r} - v_{r\theta}\frac{E_\theta}{E_r}\right)\frac{1}{r_i}\frac{u_{i+1,j} - u_{i,j}}{t}$$
$$- \frac{E_\theta}{E_r}\frac{u_{i,j}}{r_i^2} = 0, \qquad i = 1,\dots,j-1. \qquad (3)$$

Table 4.2. Boundary conditions for the problem formulation in Table 4.1 (Connolly and Winarski, 1984)

	Altmann		Tramposch	

At the hub

$$u(1) = \frac{\sigma_{\mathrm{rr}}^{(1)}}{E_{\mathrm{c}}}, \qquad (1)$$

where

$$E_{\mathrm{c}} = \frac{\left(1 - \dfrac{a^2}{b^2}\right)E_{\mathrm{h}}}{(1 + \nu_{\mathrm{h}})\dfrac{a^2}{b^2} + (1 - \nu_{\mathrm{h}})},$$

$E_{\mathrm{c}} \ldots$ Hub elasticity.

$$u_{1,j} = fP_{1,j}, \qquad (1)$$

where

$$f = \frac{b}{E_{\mathrm{c}}},$$

$f \ldots$ Hub compliance.

At the "current" outer wrap

$$\sigma_{\mathrm{rr}}(s) = \frac{T_{\mathrm{w}}}{s}\,ds. \qquad (2)$$

Note:
$$\sigma_{\mathrm{rr}}(s + ds) = 0.$$
See Fig. 4.1(a)

$$P_{j,j} = \frac{T_{j,j}t}{r_j}. \qquad (2)$$

Note:
$$P_{j+1,j} = 0.$$
See Fig. 4.1(b)

so on. The tables show the equations in tabular form for easy comparison and quick reference (Connolly and Winarski, 1984).

Altmann (1968)

Consider a hub, whose inner radius is a and outer radius is b (see Fig. 4.2), onto which tape is wound. With the addition of each wrap, wound under tension, there is a change induced in the stress state existing in the already-wound wraps. Altmann evaluates the incremental change in the stress state induced at an inner radial location r (normalized with respect to b, $r = \bar{r}/b$, where $\bar{r}$ is the actual radial location), when the current outer wrap is added [see Fig. 4.2(a)]. At equilibrium, the stress component increments σ_{rr} and $\sigma_{\theta\theta}$, namely, the radial and hoop stress, are related through the stress equation of equilibrium 4.1A-1. Then, using the stress–displacement relationships for an orthotropic, linearly elastic solid under plane-stress conditions 4.1A-2, he obtains the displacement equation of equilibrium 4.1A-3. This gives the change in radial displacement u (again normalized with respect to b) at a radial location r caused by the addition of the current outer wrap at a (normalized) radial location s; u is fully determined when two boundary conditions are specified, one at the hub and the other at the current outer wrap. The boundary condition at the hub, 4.2A-1, relates the radial deflection of the outer hub radius ($r = 1$) to the pressure exerted on it by the wound tape. We have defined E_{c}, in 4.2A-1, as the hub elasticity (Yagoda, 1980a), for a hub whose geometry is thick, hollow, right-circular cylindrical and whose compo-

Table 4.3. Solution for the problem described in Tables 4.1 and 4.2 (Connolly and Winarski, 1984)

Altmann	Tramposch

Displacement equation of equilibrium gives

Altmann	Tramposch
$$u(r) = Ar^{\gamma-\delta} + Br^{-(\gamma+\delta)}, \qquad (1)$$	$$-a_i u_{i-1,j} + D_i u_{i,j} - u_{i+1,j} = 0; \quad i = 2,\dots,j, \qquad (1)$$

where (Altmann) / where (Tramposch)

$$\beta^2 = \frac{E_\theta}{E_r},$$

$$\delta = \frac{1}{2}\left(v_{\theta r} - v_{r\theta}\frac{E_\theta}{E_r}\right), \qquad \gamma = \sqrt{\delta^2 + \beta^2}$$

$$a_i = 1 - \left(1 + \frac{E_\theta}{E_r}\frac{\bar{t}}{1 + (i-2)\bar{t}}\right)\frac{\bar{t}}{1 + (i-2)\bar{t}},$$

$$\bar{t} = \frac{t}{b}, \qquad D_i = 2 - \frac{\bar{t}}{1 + (i-2)\bar{t}}.$$

Hub boundary condition gives

$$A + B = \frac{1}{E_c}[a_r A + b_r B], \qquad (2)$$

$$D_1 u_{1,j} - u_{2,j} = 0, \qquad (2)$$

where (Altmann):

$$a_r = (v_{\theta r} + \eta)\frac{E_r}{(1 - v_{r\theta}v_{\theta r})}, \qquad b_r = (v_{\theta r} - \rho)\frac{E_r}{(1 - v_{r\theta}v_{\theta r})}$$

$$\eta = \gamma - \delta, \qquad \rho = \gamma + \delta$$

where (Tramposch):

$$D_1 = 1 - \left(v_h - \frac{1}{F}\right)\bar{t},$$

$$F = \frac{fE_r}{1 - v_{r\theta}v_{\theta r}}\frac{1}{b}$$

"Current" outer wrap boundary conditions gives

$$[a_r As^{\eta} + b_r Bs^{-\rho}]\frac{1}{s} = \frac{-T_w}{s}ds \qquad (3)$$

$$-a_{j+1}u_{j,j} + D_{j+1}u_{j+1,j} = b_{j+1} \qquad (3)$$

where (Tramposch)

$$a_{j+1} = 1 - v_h\frac{\bar{t}}{1 + (j-1)\bar{t}}, \qquad D_{j+1} = 1$$

$$b_{j+1} = -\left(\frac{1 - v_{r\theta}v_{\theta r}}{E_r}\right)\frac{t\bar{t}}{1 + (j-1)\bar{t}}T_{j,j}$$

Solving for A and B gives

$$\sigma_{rr}(r) = \frac{1 + mr^{-2\gamma}}{r^{1-\eta}}\frac{s^{1-\eta}}{(1 + ms^{-2\gamma})}\frac{T_w}{s}ds,$$

$$\sigma_{\theta\theta}(r) = \frac{-(\eta - \rho mr^{-2\gamma})}{r^{1-\eta}}\frac{s^{1-\eta}}{(1 + ms^{-2\gamma})}\frac{T_w}{s}ds, \qquad (4)$$

where (Altmann)

$$m = \frac{\gamma + \mu - E_\theta/E_c}{\gamma + \mu + E_\theta/E_c}, \qquad \mu = \frac{1}{2}\left(v_{\theta r} + v_{r\theta}\frac{E_\theta}{E_r}\right).$$

This gives a matrix equation for $u_{i,j}$, $1 \le i \le j + 1$,

$$\begin{bmatrix} D_1 & -1 & 0 & 0 & \cdots & 0 & 0 \\ -a_2 & D_2 & -1 & 0 & & & \\ 0 & -a_3 & D_3 & -1 & & & \\ \vdots & & & & & \vdots & \vdots \\ & & & & & D_j & -1 \\ 0 & 0 & 0 & 0 & \cdots & -a_{j+1} & D_{j+1} \end{bmatrix}, \qquad (4)$$

$$\begin{Bmatrix} u_{1,j} \\ \vdots \\ u_{j,j} \\ u_{j+1,j} \end{Bmatrix} = \begin{Bmatrix} 0 \\ \vdots \\ 0 \\ b_{j+1} \end{Bmatrix}.$$

Table 4.3 (*continued*)

Altmann	Tramposch

Solve simply

$$u_{j+1,j} = \frac{b_{j+1}}{D_{j+1} - a_{j+1}s_j} \equiv z_{j+1}$$

$$\text{for } i = j, j-1, \ldots, 1,$$

$$u_{i,j} = s_i u_{i+1}; \qquad s_k = \frac{1}{D_k - a_k s_{k-1}},$$

$$k = 2, \ldots, j. \tag{5}$$

The total radial and hoop stress distribution at a radial location r in the fully wound reel is

$$P(r) = \frac{1 + mr^{-2\gamma}}{r^{1-\eta}} \int_r^R \frac{s^{1-\eta}}{1 - ms^{-2\gamma}} \frac{T_w}{s} ds,$$

$$T(r) = T_w - \frac{(\eta - \rho mr^{-2\gamma})}{r^{1-\eta}} \int_r^R \frac{s^{1-\eta}}{1 + ms^{-2\gamma}} \frac{T_w}{s} ds. \tag{5}$$

The total radial and hoop stress distribution in the ith wrap for a total of N wraps wound is

$$P_{i,N} = P_{i,i} - \sum_{j=i+1}^{N} \left[\frac{u_{i+1,j} - u_{i,j}}{t} + v_{\theta r} \frac{u_{i,j}}{r_i} \right],$$

$$T_{i,N} = T_{i,i} - \sum_{j=i+1}^{N} \left[v_{\theta r} \frac{u_{i+1,j} - u_{i,j}}{t} + \frac{E_\theta}{E_r} \frac{u_{i,j}}{r_i} \right]. \tag{6}$$

sition is homogeneous, isotropic, and linearly elastic (Bertram and Eshel, 1980; Timoshenko and Goodier, 1970). E_c is a function of the hub stiffness and geometry, where E_h is Young's modulus of the hub, v_h is the Poisson ratio of the hub, and t_h is the hub thickness. For orthotropic hubs, used by some manufacturers, Lekhnitskii (1981) has a corresponding formula. The current outer-wrap boundary condition, 4.2A-2, is a statement of equilibrium for the applied radial pressure, which is caused by the addition of that wrap, of incremental thickness, ds, with internal hoop stress, T_w, caused by the winding tension (T_w is equal to the winding tension divided by the tape's cross-sectional area, wt, with w as web or tape width and t as the wrap thickness) at a (normalized) radial location s. Note that a variable, winding-tension profile can be incorporated into the analysis by considering T_w as a function of s in this boundary condition.

An exact solution of the displacement equation of equilibrium, 4.1A-3, which is a second-order, ordinary differential equation, is given in 4.3A-1, with two unknown constants of integration. From the hub compliance's boundary condition, 4.2A-2, we obtain, using 4.3A-1 and 4.1A-2, one equation, 4.3A-2, that relates the unknown constants A and B; from the current outer-wrap boundary condition 4.2A-2, we get a second equation, 4.3A-3, again using 4.3A-1 and 4.1A-2, that relates A and B. From these two equations, 4.3A-1 and 4.3A-2, we can fully determine u and, using 4.1A-2, also σ_{rr} and $\sigma_{\theta\theta}$. The final stress distribution in the fully wound tape, at a radial location r, is calculated by an integration over all the increments added on that radial location to give the total radial and circumferential stress $P(r)$ and

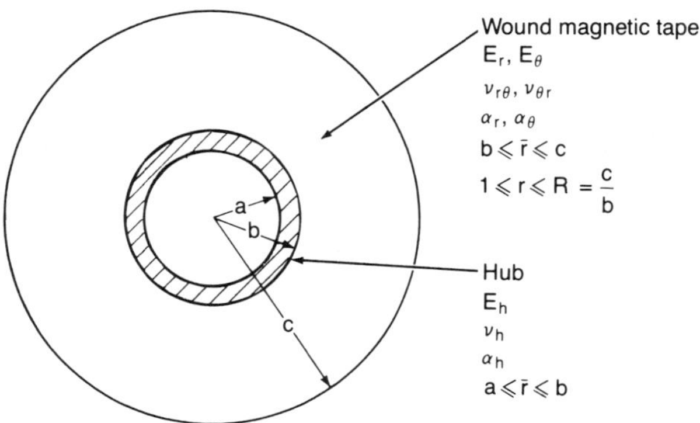

Fig. 4.3. Schematic of fully wound tape geometry.

$T(r)$, respectively, in 4.3A-4. There, R is the (normalized) outer-wrap radius of the fully wound tape reel equal to c divided by b where c is the outer wrap radius (see Fig. 4.3).

Tramposch (1963)

Unlike Altmann, Tramposch in his analysis assumes that

$$\nu_{r\theta} = \nu_{\theta r}(E_r/E_\theta). \tag{4.1}$$

In the plane-stress, axisymmetric, wound tape, we have two Young's moduli, E_θ and E_r. E_θ is Young's modulus of the magnetic tape in the circumferential direction, in tension (assumed to be identical to Young's modulus of the tape in tension), E_r is Young's modulus in the radial direction, in compression, and $\nu_{r\theta}$ and $\nu_{\theta r}$ are the two Poisson ratios. $\nu_{r\theta}$ is the radial Poisson ratio, i.e., the ratio of the circumferential strain to the radial strain for an element under pure radial stress and $\nu_{\theta r}$ is the circumferential Poisson ratio, i.e., the ratio of the radial strain to the circumferential strain for an element under pure circumferential stress. We note that these four material constants cannot all be independent and they require that Eq. (4.1) hold. A measured value of $\nu_{\theta r} = 0.30$ is reported by Bogy et al. (1979). A measured value of $\nu_{r\theta} = 0.05$ is presented by Umanskii et al. (1978), who then use Eq. (4.1) to get $\nu_{\theta r}$. Those who do not accept Eq. (4.1) use

$$\nu_{r\theta} = \nu_{\theta r}, \tag{4.2}$$

for which there appears to be no theoretical or experimental justification.

While Altmann's analysis was presented in the more general case, Tramposch's analysis is presented, as he did, with Eq. (4.1) holding. Moreover, the normalized radial distance and radial displacement is not used, as in the previous analysis. Tramposch, with a view to incorporating linear variations in the width direction of tape thickness, hub diameter, and hub flexibility, obtains a discretized (on the wrap thickness) stress equation of equilibrium, 4.1T-1, relating the induced increment in radial pressure, $T_{i,j}$, and hoop stress, $P_{i,j}$, in the ith wrap caused by the addition of the jth or the current outer wrap. Using the discretized stress–displacement relationship, 4.1T-2, he obtains a three-point, finite-difference, displacement equation of equilibrium, 4.1T-3, relating the displacements produced in three adjacent inner wraps, caused by the addition of the jth wrap, which is the current outer wrap. The boundary conditions are again the effect of hub compliance, 4.2T-1, where f, the hub compliance, is defined for the same cylindrical hub described in Altmann's analysis, and the current outer-wrap pressure, 4.2T-2. Again note that a variable, winding-tension profile can be incorporated into this analysis by considering $T_{j,j}$, the winding stress in the current outer wrap, as a function of j in this boundary condition.

Rewriting the displacement equation of equilibrium, 4.1T-3, in the form given in 4.3T-1, the hub's boundary condition, 4.2T-1, as given in 4.3T-2 using 4.1T-2, and the current outer-wrap boundary condition, 4.2T-2, as given in 4.3T-3, again using 4.1T-2, produces a matrix equation, 4.3T-4, for the radial displacements $u_{i,j}$, $1 \leq i \leq j + 1$, in the wound wraps, which can be solved algebraically. Recall at this stage that $u_{j+1,j}$ is defined as the displacement of the outer surface of the current outer wrap, as illustrated in Fig. 4.2(b). This approach differs slightly from that of Tramposch's. The solution for $u_{i,j}$ is given in 4.3T-5. A summation over all the wraps wound on top of a specific wrap gives the total displacement for that wrap. Using the stress–displacement relationship, 4.1T-2, Tramposch obtains the change in stress in the ith wrap caused by the addition of the jth wrap. A summation over all the wraps wound on a specific wrap gives the final stress distribution in that wrap, for each wrap in the fully wound reel of tape, 4.3T-6.

Agreement

Thus, under the stipulated conditions, we can evaluate the initial stress conditions in tape wound onto a reel. The above analyses were programmed by Connolly and Winarski (1984) and both were found to agree. An immediate implication is that, since Altmann's model does not involve tape thickness but only the amount of tape stored, which is tape thickness multiplied by the number of wraps, tape thickness does not affect the initial stress field. This fact is borne out using Tramposch's model. From a programming point of view, Altmann's analysis is represented by a few lines of code involving numerical integration, while that of Tramposch, even though it is only algebraic, is more complicated. Tramposch does, however, present an analysis to account for the across-the-width variations mentioned earlier, which offers reasonable extendibility to his approach.

On completion of the winding of a wrap, the internal stresses begin to relax, because of the viscoelastic nature of the magnetic tape (Tramposch, 1965,

1967; Mukherjee, 1974; Bhushan et al., 1984a). This process is accelerated at higher temperatures and humidities.

4.1.1.2. Measurement of Radial Stresses in a Wound Tape Reel

The radial-stress distribution in wound tape reel can be measured by inserting force sensing resistors or sandwiched/uncovered pull tabs or by using strain gages (sandwiched pull tabs are steel pull tabs sandwiched between brass covers). The friction tab technique has been used by Pfeiffer (1966), Umanskii et al. (1978), Bertram and Eshel (1980), Bhushan et al. (1984a), Connolly and Winarski (1984), and Bhushan (1990). Tabs of magnetic tape or stainless steel (typically 25-μm thick, 12.7-mm wide, and 50-mm long) are inserted axially into the reel periodically during winding. Their strength, smoothness, dimensional stability, and thinness (to avoid disturbing the reel structure as much as possible) are considerations in this selection. Once wound, these tabs are pulled out of the reel and the tab pull force is measured with a spring gauge. Measurements of the pull force, along with separate measurements of the combined coefficients of friction between the tab and both the magnetic coating and backcoating (~ 0.6 for tape tabs) are then used to determine the interlayer pressure. The combined coefficient of friction is obtained independently by loading a tab between several layers of tape on a flat table, applying a load and measuring the force required to pull the tab. There may be variability in the friction between tabs and different tape layers. Sandwiched pull tabs avoid this variability. An experiment to measure the hoop stress inside rolls of paper by splicing an especially made strain gauge directly into the web of paper is described by Hussain et al. (1962). Such an experiment on magnetic tape has not been attempted.

To compare the theoretical and experimental results, Connolly and Winarski (1984) ran Altmann's model under the conditions of both Eqs. (4.1) and (4.2), with $v_{\theta r} = 0.3$, $E_\theta = 3.45$ GPa, and E_r in the range of 0.17–0.28 GPa (Chapter 2), together with the geometry of the experimental tape reel. The value of E_r is very uncertain. It is dependent to a large extent on the amount of air entrapped during winding. A lower value of E_r is obtained immediately after winding, while the higher value is obtained from a tape reel that had been wound for a day or so, because entrapped air between the layers seeps out (Pfeiffer, 1967; Bhushan et al., 1984a). A comparison of the predicted radial stress and that measured experimentally is given in Fig. 4.4. Much better agreement is obtained with Eq. (4.1) and we have used this.

4.1.1.3. Role of Reel Geometry and Winding Parameters

Equipped with our analytical technique and the necessary materials' properties, we can look at the effect of the various variables, such as pack geometry, hub geometry and stiffness, the winding tension profile, and the winding speed, on the stresses generated during winding, and how they can be altered

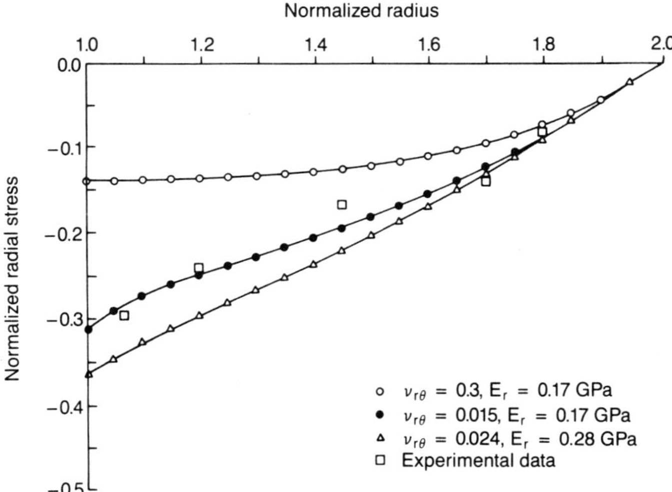

Fig. 4.4. Comparison of the theoretical and experimental radial stress in a tape reel. Comparison of the effect of the value of the Poisson ratio $v_{r\theta}$ on the predicted radial stress. The analysis used $v_{\theta r} = 0.3$ and $E_\theta = 3.45\,\text{GPa}$ (Connolly and Winarski, 1984).

to minimize the probability of creating a stress profile that leads to magnetic tape defects. In the numerical examples we present (Fig. 4.3)

$$a = 50.8\,\text{mm}, \qquad b = 69.85\,\text{mm}, \qquad c = 125.73\,\text{mm},$$

$$E_\theta = 3.45\,\text{GPa}, \qquad v_{\theta r} = 0.3, \qquad E_r = 0.17\,\text{GPa},$$

$$E_h = 6.89\,\text{GPa}, \qquad v_h = 0.33, \qquad t_h = 19.05\,\text{mm},$$

together with Eq. (4.1) holding, since these properties resemble those of common, magnetic tape reels found today. Then we show the influence of tape-reel geometry, as well as hub geometry and material property variations on the tape's stresses. Reel stresses are normalized with the winding stress, T_w.

The outer hub radius, b, must be increased to permit an increase in the amount of tape stored. This is so because the inner wraps of tape, acting as a hub, go into compression circumferentially as more wraps are added. This effect can be seen in Fig. 4.5, where we look at the effect of storing more tape on our selected hub. If we increase R, we increase the radial stress as expected, but we also put the inner wraps into compression circumferentially. Thus, long lengths of tape are stored on larger diameter hubs.

The hub should be of uniform stiffness across its width (Waites, 1982). Otherwise, it will cause a nonuniform collapse along the width, producing a tension gradient in the tape and a nonuniform permanent distortion across the width of the tape with time. Our analysis assumes a hub of uniform stiffness.

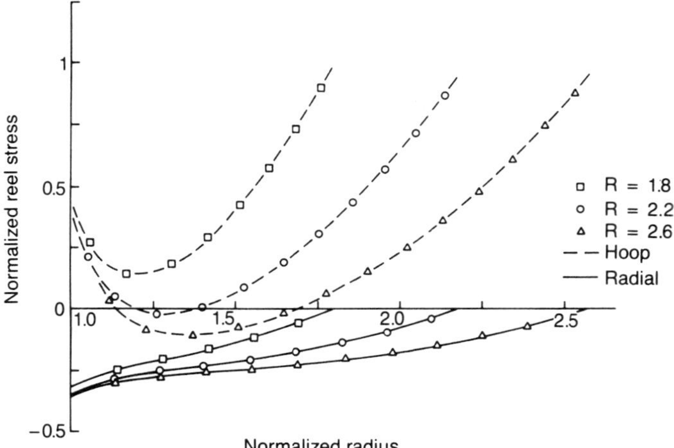

Fig. 4.5. Predicted effect on the tape's circumferential and radial stresses of the fully wound tape's outer wrap radius (Connolly and Winarski, 1984).

Hub thickness and material contribute to the hub's compliance. As mentioned in the introduction, if the hub is too compliant, it shifts its function to the inner wraps and puts them into compression circumferentially. If this compression becomes so high as to overcome the friction between the layers, the tape will fold back on itself and buckle. To avoid this, we must increase the hub's stiffness by increasing its thickness or changing to a material with a higher Young's modulus. When compression occurs farther out in the pack, then we must increase the outer hub's radius or decrease the amount of tape in the pack, as mentioned earlier.

If the hub is too stiff, the high, tensile hoop stresses caused by winding remain in the tape near the hub. This goes unnoticed in narrow tape, but in wide tape, which has significant variations in mechanical properties across its width, these high hoop stresses produce tension bands or lanes in the tape. With time, as the stresses relax, these distortions become permanent and adversely affect the tape's performance. This effect has also been found in paper (Tramposch, 1965; Frye, 1967) and cellophane (Pfeiffer, 1966). Figure 4.6 shows the effect on the tape's hoop and the radial stresses of a hub, i.e.: too compliant, $E_r = 0.689$ GPa; too stiff, $E_r = 68.9$ GPa; and intermediate, $E_r = 6.89$ GPa. Figure 4.7 shows the same effect for different hub thicknesses, i.e., too compliant, $t_h = 6.35$ mm; reasonable, $t_h = 19.05$ mm; and more rigid, $t_h = 31.75$ mm. To avoid tension ridges, we advise not to use a too-rigid hub, thereby keeping the amplitude of the tension ridges small and protecting the tape from more severe distortions.

For a constant winding tension, we see from 4.3A-5 that the resultant stress is linearly proportional to the winding stress. For wide rolls, it is common to use decreasing winding tension with increasing tape winding radius. The

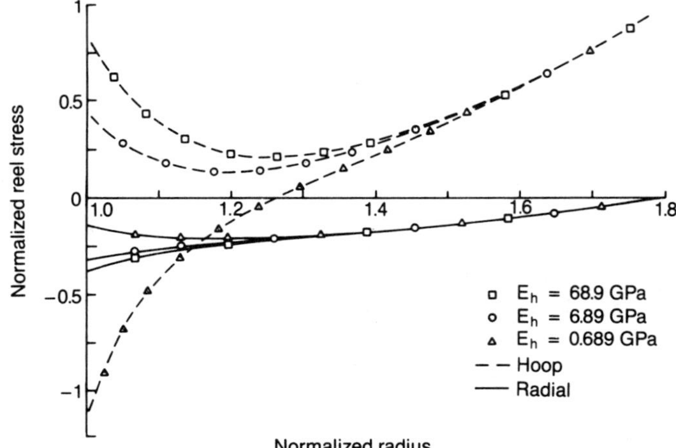

Fig. 4.6. Predicted effect on the tape's circumferential and radial stresses of Young's modulus of the hub (Connolly and Winarski, 1984).

effect on the hoop-stress profile of a linearly decreasing (or tapered) winding tension, which reduces to 0.6 times its initial value during tape winding, is shown in Fig. 4.8. Thus, the stress variations (and consequently, the variations of time base errors) across the reel radius can be reduced by tapered winding tension (or constant torque winding) rather than constant tension winding (Mukherjee, 1974).

Changing the winding tension and speed has an effect on the amount of air entrapped and thereby changes the tape's radial Young's modulus. Higher

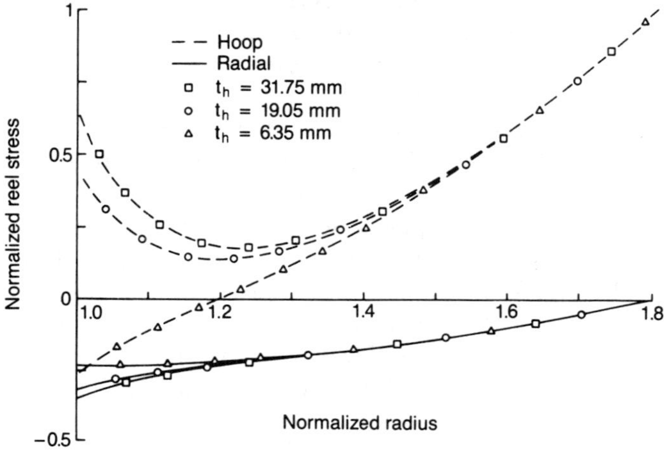

Fig. 4.7. Predicted effect on the tape's circumferential and radial stresses of the thickness of the hub (Connolly and Winarski, 1984).

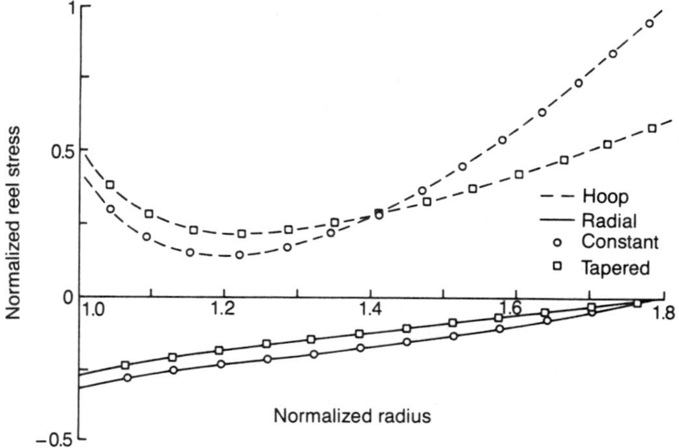

Fig. 4.8. Predicted effect on the tape's circumferential and radial stresses of a tapered winding tension (Connolly and Winarski, 1984).

winding tension and low winding speeds result in higher E_r (Chapter 2). The effect of varying the tape's radial Young's modulus is shown in Fig. 4.9. There, with Eq. (4.1) holding, we look at the predicted radial and hoop stresses in our basic tape reel as a function of E_r. Note the significant difference between the hoop-stress distribution in the orthotropic tape reel ($E_r = 0.17$ GPa or 0.345 GPa) as opposed to the isotropic one ($E_r = E_\theta = 3.45$ GPa). Thus, the assumption of orthotropy leads to significantly different predicted behaviors than that of isotropy.

The radial pressure of the wound tape must be sufficient to avoid slippage

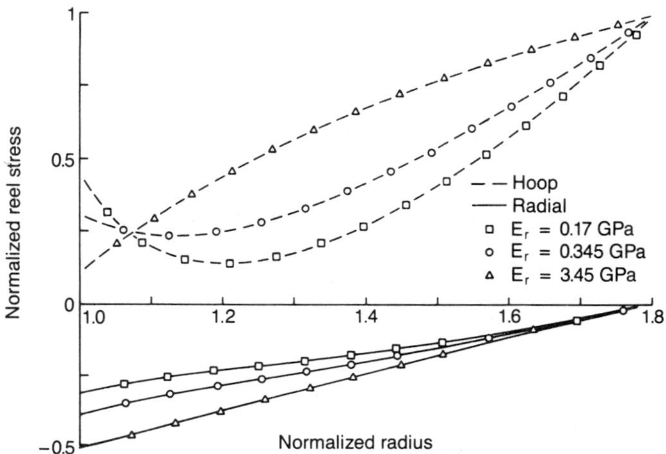

Fig. 4.9. Predicted effect on the tape's circumferential and radial stresses of the tape's radial modulus (Connolly and Winarski, 1984).

in the layers of tape during acceleration. Note from Figs. 4.6 and 4.7 that there is little increase in the radial pressure when the stiffness of the hub is increased. Thus, other considerations, such as the thermal-stress effects, have to be taken into account to limit the reduction in the radial stress of the wound tape. This is considered in the next section.

4.1.1.4. Environmental Stresses

This section presents a model to predict the stress change induced in a reel of tape when it is subject to a temperature or humidity change from that at which it was wound. The stresses result from the mismatch in the thermal or hygroscopic coefficient of expansion in (a) the hub and the tape and (b) the tape in the radial and the circumferential directions (Tramposch, 1965; Umanskii et al., 1981; Connolly and Winarski, 1984). The analysis for a change in humidity is analogous to that for a change in temperature if we use the appropriate coefficients of hygroscopic expansion rather than those of thermal expansion. Therefore, we present only the thermal-stress analysis.

The following undesirable effects of thermal stresses have been noted in the literature (Yagoda, 1980a):

1. An increased circumferential compression in the inner layers, which may cause wrinkling of the tape near the hub.
2. An increased circumferential tension in the outer wraps of tape, which causes increased distortion.
3. Accelerated stress relaxation; any associated viscoelastic deformation occurs at a faster rate.

A temperature drop causes a decrease in the radial stress, which increases the chance of interlayer slip on reel acceleration. Also, there is a critical stress at which the radial stress at a radial location can become zero, which has the danger of "layering" (which is a radial separation of the layers) the tape.

Assumptions and Problem Formulation

In addition to the assumptions made in the winding analyses, the following assumptions are made here:

1. The reel of tape behaves as a continuum, which is assumed to be an orthotropic, linearly elastic solid.
2. The materials' properties are independent of temperature and humidity in the range considered here.
3. The temperature (and humidity) change does not vary through the wound tape.

The problem is modeled as two concentric, thin, right-circular cylinders, as shown in Fig. 4.3. Since neither the tape nor the hub is constrained in the axial direction, there are no thermal stresses in that direction. Because the model is axisymmetric, the in-plane components of the displacement (u_r, u_θ) reduce to

$$u_r = u(r), \qquad u_\theta = 0. \tag{4.3}$$

The plane-stress, stress-displacement relationship for an orthotropic, linearly thermoelastic solid (Mukherjee, 1974) is

$$du/dr - \alpha_r \Delta T = (\sigma_{rr} - v_{r\theta}\sigma_{\theta\theta})/E_r, \tag{4.4}$$

$$u/r - \alpha_\theta \Delta T = (\sigma_{\theta\theta} - v_{\theta r}\sigma_{rr})/E_\theta, \tag{4.5}$$

where α_r and α_θ are the radial and circumferential coefficients of thermal expansion, respectively, and ΔT is the temperature change.

The stress equation of equilibrium is

$$d\sigma_{rr}/dr + (\sigma_{rr} - \sigma_{\theta\theta})/r = 0. \tag{4.6}$$

Boundary Conditions

The boundary conditions are that there are no applied tractions at the inner hub radius, $r = a$, or at the outer radius of the wound tape, $r = c$, while at the interface, $r = b$, the displacement u and the normal stress σ_{rr} are continuous. We write these conditions as

$$\sigma_{rr}^h(a) = 0, \tag{4.7}$$

$$\sigma_{rr}^h(b) = \sigma_{rr}^t(b), \qquad u^h(b) = u^t(b), \tag{4.8}$$

$$\sigma_{rr}^t(c) = 0, \tag{4.9}$$

where superscripts h and t refer to the hub and tape, respectively.

Problem Solution

Substituting for the radial displacement u, using the stress-displacement in Eqs. (4.4) and (4.5) in the stress equation of equilibrium of Eq. (4.6), we get the displacement equation of equilibrium

$$(d^2u/dr^2) + (1/r)(du/dr) - (E_\theta/E_r)(u/r^2) = [(\alpha_r + v_{\theta r}\alpha_\theta) - (E_\theta/E_r)(\alpha_\theta + v_{r\theta}\alpha_r)], \tag{4.10}$$

which governs the radial displacement $u(r)$.

For an isotropic hub

$$E_r = E_\theta \equiv E_h, \qquad v_{r\theta} = v_{\theta r} \equiv v_h, \qquad \alpha_r = \alpha_\theta \equiv \alpha_h. \tag{4.11}$$

So, a solution to the displacement equation of equilibrium in Eq. (4.10) for the hub is

$$u^h(r) = C_1^h r + C_2^h/r \qquad (a \le r \le b), \tag{4.12}$$

where C_1^h and C_2^h are constants to be determined from the boundary conditions, Eqs. (4.7)–(4.9).

For the orthotropic tape, a solution to the displacement equation of equilibrium, Eq. (4.10), is

$$u^t(r) = C_1^t r^\beta + C_2^t/r^\beta \qquad (b \le r \le c), \qquad (4.13)$$

where C_1^t and C_2^t are constants to be determined from the boundary conditions in Eqs. (4.7)–(4.9) and $\beta^2 = E_\theta/E_r$.

The case $\beta = 1$ and $\alpha_r^t \ne \alpha_\theta^t$, i.e., where magnetic tape is anisotropic only with respect to its thermal expansitivity, is not considered, since it does not seem physically realizable.

Considering the boundary conditions in Eqs. (4.7)–(4.9) and the stress–displacement relationship in Eqs. (4.4) and (4.5), we find that the radial and circumferential components of the thermal stress in the tape are

$$\sigma_{rr}^t = [E_r/(1 - v_{r\theta}v_{\theta r})]\{(\beta + v_{\theta r})C_1^t r^{\beta-1} - (\beta - v_{\theta r})(C_2^t/r^{\beta+1})$$
$$+ [(1 + v_{\theta r})\Omega - \kappa]\Delta T\}, \qquad (4.14)$$

$$\sigma_{\theta\theta}^t = [E_\theta/(1 - v_{r\theta}v_{\theta r})]\{(1 + \beta v_{r\theta})C_1^t r^{\beta-1} + (1 - \beta v_{r\theta})(C_2^t/r^{\beta+1})$$
$$+ [(1 + v_{r\theta})\Omega - \lambda\kappa]\Delta T\}, \qquad (4.15)$$

where

$$\kappa = \alpha_r v_{\theta r} + \alpha_\theta; \qquad \lambda = (\alpha_\theta + v_{r\theta}\alpha_r)/\kappa,$$
$$\Omega = [\kappa(1 - \lambda\beta^2)/(1 - \beta^2)] \qquad (\beta \ne 1),$$

and

$$C_1^t = [1/(\beta + v_{\theta r})][\beta - v_\theta](C_2^t/C^{2\beta}) + \gamma^t],$$
$$C_2^t = -[(1 - v_{r\theta}v_{\theta r})/(\beta - v_{\theta r})]\{[\xi_1\xi_5 + (E_h/E_r)\xi_3]/$$
$$[\xi_1\xi_4 - (E_h/E_r)\xi_2]\}b^{2\beta+1},$$

where

$$\xi_1 = (1 - v_h)(b/a)^2 + (1 + v_h),$$
$$\xi_2 = (1 - v_{r\theta}v_{\theta r})\{[1/(\beta + v_{\theta r})(b/c)^{2\beta}] + [1/(\beta - v_{\theta r})]\},$$
$$\xi_3 = (\alpha_h - \Omega)\Delta T - [1/(\beta + v_{\theta r})](b/c)^{\beta-1}\gamma^t,$$
$$\xi_4 = [(b/c)^{2\beta} - 1]/[(b/a)^2 - 1],$$
$$\xi_5 = [1/(1 - v_{r\theta}v_{\theta r})\gamma^t]\{[(b/a)^{\beta-1} - 1]/[(b/a)^2 - 1]\},$$
$$\gamma^t = [\kappa - (1 + v_{\theta r})\Omega]\Delta T.$$

Numerical Results

Little information, in general, is known about α_r. In Yagoda (1980b), we find that $\alpha_r = 50 \times 10^{-6}/°C$. Looking at the effect of variations in this value, and with

$$a = 50.8 \text{ mm}, \qquad b = 69.85 \text{ mm}, \qquad c = 125.73 \text{ mm},$$
$$E_\theta = 3.45 \text{ GPa}, \qquad v_{\theta r} = 0.3, \qquad \alpha_\theta = 18 \times 10^{-6}/°C,$$
$$E_h = 6.89 \text{ GPa}, \qquad v_h = 0.33, \qquad \alpha_h = 27 \times 10^{-6}/°C,$$

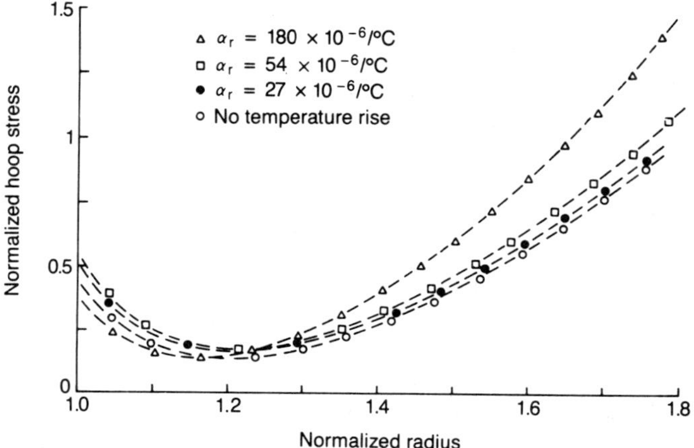

Fig. 4.10. Predicted effect on the tape's circumferential stress of a temperature rise of 25° C for various tape's radial coefficients of thermal expansion (Connolly and Winarski, 1984).

and Eq. (4.1) holding, we calculate the thermal-stress field induced in wound tape.

We plot in Fig. 4.10 the change in the hoop stress, due to winding, of a temperature increase of $\Delta T = 25°$ C for $\alpha_r = 27$, 54, and $180 \times 10^{-6}/°$C. We plot in Fig. 4.11 the change in the radial stress, due to winding, of a temperature decrease of 25° C for these values of α_r. Also, the change in the radial

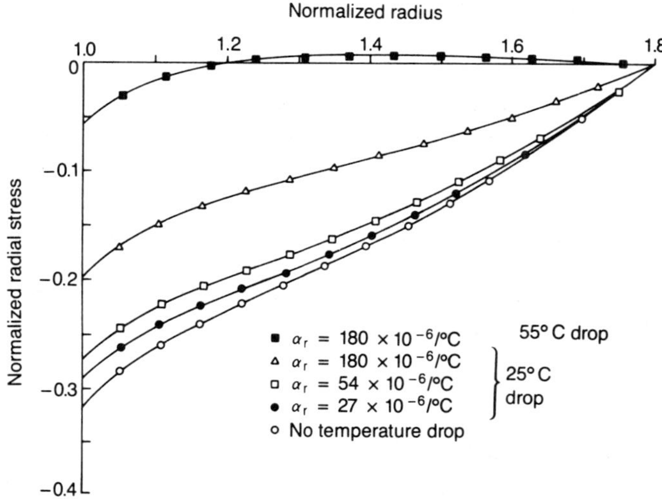

Fig. 4.11. Predicted effect on the tape's radial stress of a temperature drop of 25° C for the various tape's radial coefficients of thermal expansion, and 55° C for tape's radial coefficient of thermal expansion $\alpha_r = 180 \times 10^{-6}/°$C (Connolly and Winarski, 1984).

stress, due to winding, of a temperature decrease of 55° C for $\alpha_r = 180 \times 10^{-6}/°C$ is shown as an extreme case to illustrate the fact that the radial stress can indeed be reduced to zero on cooling. This phenomenon is, of course, dependent on hub material and geometry and, therefore, presents serious consideration on their selection. Storage of tape reels at higher temperatures and humidities, that result in significant stress relaxation of the tape, would produce the same effect with a lesser hub-tape thermal mismatch or temperature change.

Interlayer slip is the tangential motion of the tape relative to the hub. It is caused by rotational acceleration after exposure of the tape to environmental stress for a period of time. As we have seen, hub stiffness above a certain level has little effect on the radial stress, but any mismatch in the thermal coefficient of expansion between the hub and the tape can have a far greater effect on its value. In extensive testing, we found that a hub with adequate compliance and the least thermal mismatch allowed us to attain the maximum rotational acceleration before tape-to-tape slippage occurred (Graham et al., 1981; Richard and Winarski, 1982).

4.1.1.5. Summary

In this section, we have contrasted the existing analytical techniques to calculate the initial stress field generated during the winding of tape onto a hub. We show that while the solution methods differ in ease of use and extendibility, their predictions nonetheless agree. Using a pull-tab test, we measured the radial stresses found inside the wound tape. These measured data agree very well with the analytical predictions when we restrict the ratio of the tape reel's radial Poisson ratio to the circumferential Poisson ratio to be that of the tape reel's radial Young's modulus to the circumferential Young's modulus.

On the basis of this analysis, guidelines for hub design and winding-tension profiles that can prevent the occurrence of commonly known winding defects are given. An analysis of the thermal stress field created in the wound tape by a change in temperature has been described. The lack of a reasonable value for the radial coefficient of thermal expansion for the wound tape allows only a study of the induced thermal-stress field. However, this is sufficient to indicate the application of this analysis to hub geometry and material selection to minimize the adverse effects of thermal stresses on the wound magnetic tape.

4.1.2. Stress Relaxation

To analyze the response of the PET film to applied stresses and deformations, a constitutive equation is needed that relates the state of stress to a corresponding state of strain at all points in time for the given environmental conditions. In Chapter 3, we have shown that biaxially-oriented PET behaves as an orthotropic, linearly viscoelastic material at temperatures below those of the glass transition, which is 65–70° C. We derived constitutive relationships from experimental data.

We now extend the method to obtain constitutive stress–strain relationships for axisymmetric viscoelastic solids from experimental measurements. We introduce a finite-element model for the quasi-static analysis of deformations and stresses in an axisymmetric, orthotropic, viscoelastic solid composed of PET. We then apply it to the analysis of reels of PET subject to geometric deformations and prestressing caused by the winding dynamics.

Finally, we present examples of numerical simulations that show the behavior of the wound PET under different conditions of temperature and humidity. The analysis reported here can be utilized for reliability predictions of magnetic tape in storage (Tramposch, 1965, 1967; Pfeiffer, 1966; Eshel et al., 1984; Heinrich et al., 1986).

4.1.2.1. Constitutive Relationships

We derive constitutive stress–strain relationships for an axisymmetric, orthotropic, viscoelastic solid composed of PET. Since the model is axisymmetric (every cross section at any θ remains the same), the displacement components are as follows: $u_r(r, z)$, $u_\theta = 0$, and $u_z(r, z)$, which leads to $\sigma_{r\theta} = \sigma_{\theta z} = 0$. However, $\sigma_{\theta\theta} \neq 0$ as strain in the r direction creates strain in the θ direction.

The second-order stress and strain tensors are written, with respect to the principal material's axis, in simplified vector form as

$$\boldsymbol{\sigma} = \begin{bmatrix} \sigma_{rr} \\ \sigma_{\theta\theta} \\ \sigma_{zz} \\ \sigma_{rz} \end{bmatrix}, \qquad \boldsymbol{\varepsilon} = \begin{bmatrix} \varepsilon_{rr} \\ \varepsilon_{\theta\theta} \\ \varepsilon_{zz} \\ 2\varepsilon_{rz} \end{bmatrix}, \tag{4.16}$$

and the fourth-order symmetric tensor $\mathbf{C}$ is written, with respect to the principal material's matrix, in simplified matrix form as

$$\mathbf{C} = \begin{bmatrix} C_{rr} & C_{r\theta} & C_{rz} & 0 \\ C_{r\theta} & C_{\theta\theta} & C_{\theta z} & 0 \\ C_{rz} & C_{\theta z} & C_{zz} & 0 \\ 0 & 0 & 0 & C_{ss} \end{bmatrix}, \tag{4.17}$$

with seven, independent, nonzero functions $C_{\alpha\beta}(t)$, since the material is orthotropic and $\sigma_{r\theta}$ and $\sigma_{\theta z}$ are assumed to be zero. C_{ss} describes the material's mechanical response in shear, while the other coefficients $C_{\alpha\beta}$ characterize the response in tension.

Seven independent measurements are needed to determine the functions $C_{\alpha\beta}(t)$. Four of these are provided by creep measurements and three are derived from the Poisson ratios $v_{r\theta}(t)$, $v_{rz}(t)$, and $v_{\theta z}(t)$ (see Chapter 3).

The creep-compliance functions measured experimentally are denoted by:

$D_1(t)$ = radial creep compliance,

$D_2(t)$ = creep compliance at $0°$ orientation in the θ–z plane,

$D_3(t)$ = creep compliance at $90°$ orientation in the θ–z plane,

$D_4(t)$ = creep compliance at $45°$ orientation in the r–z plane,

where the orientation is measured with respect to the material's principal axis (see Chapter 3).

A calculation requiring reorientation of the compliance tensor is then used to relate the measured creep functions D_i ($i = 1, 2, 3, 4$) to the components D_α (α = r, θ, z, s) of the tensile-creep-compliance tensor $\mathbf{D}$ (see Chapter 3). This results in

$$D_r(t) = D_1(t), \tag{4.18a}$$

$$D_\theta(t) = D_2(t), \tag{4.18b}$$

$$D_z(t) = D_3(t), \tag{4.18c}$$

$$D_s(t) = 4D_4(t) - [D_1(t) + D_3(t)] + 2v_{rz}(t)D_1(t). \tag{4.18d}$$

The Poisson ratios are defined as

$$v_{r\theta}(t) = -\varepsilon_{\theta\theta}(t)/\varepsilon_{rr}(t) \quad \text{for applied stress in the r direction,} \tag{4.19a}$$

$$v_{rz}(t) = -\varepsilon_{zz}(t)/\varepsilon_{rr}(t) \quad \text{for applied stress in the r direction,} \tag{4.19b}$$

$$v_{\theta z}(t) = -\varepsilon_{zz}(t)/\varepsilon_{\theta\theta}(t) \quad \text{for applied stress in the } \theta \text{ direction.} \tag{4.19c}$$

The compliance functions $D_\alpha(t)$ can be transformed into the corresponding relaxation functions $E_\alpha(t)$ through the convolution integral (see Chapter 3)

$$E_\alpha(t)D_\alpha(0) - \int_0^t E_\alpha(\tau)[dD_\alpha(t - \tau)/d\tau]\, d\tau = 1, \tag{4.20}$$

and, finally, the components of the stress tensor $\mathbf{C}$ are obtained from

$$C_{rr}(t) = E_r(t)\{1 - [E_z(t)/E_\theta(t)]v_{\theta z}^2(t)\}A(t), \tag{4.21a}$$

$$C_{r\theta}(t) = [E_\theta(t)v_{r\theta}(t) + E_z(t)v_{rz}(t)v_{\theta z}(t)]A(t), \tag{4.21b}$$

$$C_{rz}(t) = E_z(t)[v_{r\theta}(t)v_{\theta z}(t) + v_{rz}(t)]A(t), \tag{4.21c}$$

$$C_{\theta\theta}(t) = E_\theta(t)\{1 - [E_z(t)/E_r(t)]v_{rz}^2(t)\}A(t), \tag{4.21d}$$

$$C_{\theta z}(t) = [E_z(t)/E_r(t)][E_r(t)v_{\theta z}(t) + E_\theta(t)v_{r\theta}(t)v_{rz}(t)]A(t), \tag{4.21e}$$

$$C_{zz}(t) = E_z(t)\{1 - [E_\theta(t)/E_r(t)]v_{r\theta}^2(t)\}A(t), \tag{4.21f}$$

$$C_{ss}(t) = E_s(t), \tag{4.21g}$$

where

$$A^{-1}(t) = 1 - 2[E_z(t)/E_r(t)]v_{r\theta}(t)v_{\theta z}(t)v_{rz}(t)$$
$$- v_{rz}^2(t)[E_z(t)/E_r(t)] - v_{rz}^2(t)[E_z(t)/E_\theta(t)]$$
$$- v_{r\theta}^2(t)[E_\theta(t)/E_r(t)].$$

Also note that

$$v_{\alpha\beta}(t) = v_{\beta\alpha}(t)[E_\alpha(t)/E_\beta(t)]. \tag{4.22}$$

4.1.2.2. Relaxation Matrix from Experimental Measurements

Throughout this work, the basic assumption is that a stack of wound tape can be modeled as a linearly viscoelastic solid. A more detailed analysis, introducing the layered nature of the stack in the radial direction, cannot be justified, since there are too many uncertainties in the data needed for such an analysis. In fact, the lack of data needed to fully characterize the material's properties will cause some difficulties even when the solid model is assumed. However, we shall show later that we can reduce the degree of uncertainty to dependence on only three parameters, whose effect can be effectively studied by means of the finite-element model.

To determine the functions $D_i(t)$, three creep experiments were performed. Two simple extension measurements were obtained from tape strips cut at $0°$ and $90°$ orientation in the $\theta-z$ plane with respect to the material's principal axis. These experiments are described in detail in Chapter 3. A third measurement, for $D_1(t = 0)$, was obtained from compression tests on a reel of tape for the r direction (see Chapter 3). The time-dependence of $D_1(t)$ is assumed to be the same as that of E_θ. A fourth experiment is needed to determine $D_4(t)$ from which to derive $D_s(t)$, ultimately needed to get $C_{ss}(t)$, which relates the shear stress $\sigma_{rz}(t)$ to the shear strain $\varepsilon_{rz}(t)$. However, this would require measurements at an angle ϕ (typically $\phi = 45°$) in the r–z plane. Experimentally, this is not attainable because of the layered nature of the stack. Instead, we will derive an approximation to the shear modulus $C_{ss}(t)$ knowing the shear modulus in the $\theta-z$ plane, in a manner to be described later.

The values of the Poisson ratios $v_{r\theta}$, v_{rz}, and $v_{\theta z}$ are also unknown. We will assume that the value of the Poisson ratios are independent of time (see Chapter 3), and also that $v_{r\theta} = v_{rz}$. Variations of these parameters over a range of plausible values will be used to assess their influence on stress distribution.

The functions $D_i(t)$ ($i = 1, 2, 3$) obtained by experimental procedures are fitted by a Prony series as discussed in Chapter 3. Since it is not possible to obtain the function $D_4(t)$ from experimental measurements, we will construct the function $D_s(t)$ as

$$D_s(t) = 4\bar{D}_4(t) - [D_2(t) + D_3(t)] + 2v_{\theta z}D_2(t), \tag{4.23}$$

which is related to the shear modulus in the $\theta-z$ direction by Eq. (4.20), where

$\bar{D}_4$ is the compliance at 45° orientation in the θ–z plane and $D_s(t)$ is fit by a Prony series.

The functions $D_\alpha(t)$ are then inverted following the steps given in Chapter 3. The components $C_{ij}(t)$ of the relaxation tensor $\mathbf{C}$ are fitted using a Prony series. It has been shown in Chapter 3 that two time scales give excellent agreement with experimental measurements for times comparable to the duration of the experiments. The main limitation imposed by the choice of two time scales is that calculations carried for long periods of time will not be valid, due to saturation, after the simulation has reached roughly $t = 4\tau_2$, τ_2 being the largest time scale.

To account for the softer behavior expected in the shear coefficient $C_{44}(t)$, we do not define it using Eq. (4.21g) but through the following expression:

$$C_{44}(t) = \{E_r(0)/\tfrac{1}{2}[E_\theta(0) + E_z(0)]\}\, E_s(t) \equiv G E_s(t). \tag{4.24}$$

This expression relates the shear modulus $E_s(t)$ in the θ–z plane to the shear modulus $C_{44}(t)$ in the r–z plane through the ratio of the moduli. The value of the coefficient of $E_s(t)$ gives a factor of about $\tfrac{1}{13}$, which reflects the difference between radial and tangential properties. The coefficient of $E_s(t)$ has been varied over a range of values to assess its influence on the stress distribution.

The coefficients C_{ij}^k and relaxation times τ_k for the cases under consideration are given in Table 4.4 (Bhushan et al., 1984b; Bhushan and Connolly, 1986).

4.1.2.3. Axisymmetric Finite-Element Model

The basic assumptions that the material is linearly viscoelastic and undergoes quasi-static deformations in an axisymmetric geometry imply that at each time t, in the absence of body force, we must satisfy the equilibrium equations

$$\left.\begin{aligned}(1/r)(\partial/\partial r)(r\sigma_{rr}) + (\partial\sigma_{rz}/\partial z) - (1/r)\sigma_{\theta\theta} = 0,\\(1/r)(\partial/\partial r)(r\sigma_{rz}) + (\partial\sigma_{rr}/\partial z) = 0,\end{aligned}\right\} \tag{4.25}$$

over the region Ω in the r–z plane corresponding to an axisymmetric section of the body, whose boundary we denote by Γ.

The boundary conditions associated with Eq. (4.25) considered in this work can be of the form

$$\mathbf{u} = \bar{\mathbf{u}} \quad \text{on } \Gamma_u, \tag{4.26a}$$

$$\mathbf{t} = \bar{\mathbf{t}} \quad \text{on } \Gamma_t, \tag{4.26b}$$

where

$$\mathbf{u} = \begin{bmatrix} u_r \\ u_z \end{bmatrix} \quad \text{and} \quad \mathbf{t} = \begin{bmatrix} \sigma_{rr}n_r + \sigma_{rz}n_z \\ \sigma_{rz}n_r + \sigma_{zz}n_z \end{bmatrix} = \begin{bmatrix} t_r \\ t_z \end{bmatrix} \tag{4.27}$$

Table 4.4. Coefficients for the relaxation moduli ($C_{\alpha\beta}$) and average relaxation times (τ_k) of 23.4-μm-thick biaxially-oriented PET (Mylar A) films

°C/RH%	$C_{\alpha\beta}$	$C_{\alpha\beta}^0$, GPa	$C_{\alpha\beta}^1$, GPa	$C_{\alpha\beta}^2$, GPa	τ_1, h	τ_2, h
43.3/30	C_{rr}	0.220	0.053	0.066		
	$C_{r\theta}$	0.042	0.010	0.012		
	C_{rz}	0.013	0.004	0.005		
	$C_{\theta\theta}$	2.948	0.557	0.714	8.93	126.20
	$C_{\theta z}$	1.237	0.401	0.468		
	C_{zz}	4.098	1.316	1.542		
	C_{ss}	0.088	0.014	0.017		
51.7/30	C_{rr}	0.187	0.068	0.066		
	$C_{r\theta}$	0.036	0.013	0.011		
	C_{rz}	0.010	0.004	0.003		
	$C_{\theta\theta}$	2.622	0.854	0.727	2.72	35.17
	$C_{\theta z}$	0.987	0.412	0.333		
	C_{zz}	3.279	1.361	1.101		
	C_{ss}	0.085	0.025	0.015		
65.0/30	C_{rr}	0.161	0.059	0.067		
	$C_{r\theta}$	0.031	0.011	0.012		
	C_{rz}	0.008	0.003	0.004		
	$C_{\theta\theta}$	2.335	0.800	0.867	0.612	15.00
	$C_{\theta z}$	0.807	0.330	0.381		
	C_{zz}	2.674	1.099	1.261		
	C_{ss}	0.066	0.016	0.018		
51.7/85	C_{rr}	0.134	0.098	0.056		
	$C_{r\theta}$	0.026	0.019	0.011		
	C_{rz}	0.007	0.006	0.003		
	$C_{\theta\theta}$	1.979	1.272	0.793	0.420	9.28
	$C_{\theta z}$	0.657	0.585	0.293		
	C_{zz}	2.182	1.904	0.966		
	C_{ss}	0.060	0.028	0.019		

are the displacements and traction vectors, respectively, and $\mathbf{n} = [n_r, n_z]^T$ denotes the outwards unit normal to the surface. The portion of the boundary Γ_u and Γ_t are such that $\Gamma_u \cup \Gamma_t = \Gamma$ and $\Gamma_u \cap \Gamma_t = \phi$, the empty set.

To seek a finite-element approximation to Eqs. (4.26) and (4.27) using the Ritz method, we reformulate the problem in the following variational form: at each time t, the mechanical state that satisfies Eq. (4.25), subject to the boundary conditions in Eq. (4.26), is the one that makes the variation

$$\delta U = 0,$$

where

$$U(t) = \frac{1}{2} \int_{-\infty}^{\infty} \left[\int_{\Omega} \varepsilon^T \mathbf{C}(t - \tau)(d\varepsilon/d\tau) \, d\Omega \right] d\tau - \int_{\Gamma} \mathbf{u}^T \mathbf{t} \, d\Gamma \qquad (4.28)$$

over all the continuous displacement fields $\mathbf{u}$ that satisfy Eq. (4.26a) in Γ_u.

Using the strain-displacement relationships

$$\begin{bmatrix} \varepsilon_{rr} \\ \varepsilon_{\theta\theta} \\ \varepsilon_{zz} \\ 2\varepsilon_{rz} \end{bmatrix} = \begin{bmatrix} (\partial/\partial r) & 0 \\ (1/r) & 0 \\ 0 & (\partial/\partial z) \\ (\partial/\partial z) & (\partial/\partial r) \end{bmatrix} \begin{bmatrix} u_r \\ u_z \end{bmatrix}. \tag{4.29}$$

The total potential energy U can be expressed in terms of the displacement field $\mathbf{u}$ only, which will be approximated using a rectangular mesh of four noded bilinear elements (see Zienkiewicz, 1977). The displacements are hence approximated by an expression of the form

$$\mathbf{u} \simeq \sum_{k=1}^{NN} N^i(r, z)\mathbf{u}^k, \tag{4.30}$$

where NN is the total number of nodal points in the region, $N^i(r, z)$ are piecewise, bilinear, polynomial functions, and $\mathbf{u}^k$ is the approximation to u at nodal point k.

Replacing Eqs. (4.29) and (4.30) into Eq. (4.28) and differentiating with respect to the nodal parameters u_r^k and u_z^k, we obtain the Euler equations, which give the necessary condition for a minimum potential energy to occur. These equations will consist of a semidiscrete system of linear, first-order, integro-differential equation in the unknown parameters, of the form

$$\int_0^t \mathbf{K}(t - \tau)\dot{\mathbf{u}}\, d\tau = \mathbf{F} - \mathbf{K}(t)\mathbf{u}^0, \tag{4.31}$$

where $\mathbf{u} = [u_r^1 u_z^1 u_r^2 u_z^2 \ldots u_r^{nn} u_z^{nn}] = [u_1 u_2 \ldots u_{2n}]$ is the vector containing the unknown nodal parameters, the dot indicates time differentiation, and $\mathbf{u}^0$ is the vector of initial elastic displacements. The vector $\mathbf{F}$ is given by

$$f_i = \begin{bmatrix} \displaystyle\int_{\Gamma_t} N^k t_r\, d\Gamma \\[1em] \displaystyle\int_{\Gamma_t} N^k t_z\, d\Gamma \end{bmatrix}, \tag{4.32}$$

and the matrix $\mathbf{K}(\xi)$ is written as

$$\mathbf{K}(\xi) = C_{11}(\xi)\mathbf{A} + C_{12}(\xi)\mathbf{B} + C_{13}(\xi)\mathbf{C} + C_{22}(\xi)\mathbf{D}$$
$$+ C_{23}(\xi)\mathbf{E} + C_{33}(\xi)\mathbf{F} + C_{44}(\xi)\mathbf{G}. \tag{4.33}$$

The constant matrices $\mathbf{A} - \mathbf{G}$ in Eq. (4.33) are geometry-dependent only and their components are given by

$$a_{ij} = \begin{cases} \displaystyle\int\int_\Omega r(\partial N^i/\partial r)(\partial N^j/\partial r)\, dr\, dz & \text{for } i \text{ odd and } j \text{ odd,} \\[1em] 0 & \text{otherwise,} \end{cases} \tag{4.34a}$$

$$b_{ij} = \begin{cases} \displaystyle\iint_\Omega [(\partial N^i/\partial r)N^j + N^i(\partial N^j/\partial r] \, dr \, dz & \text{for } i \text{ odd and } j \text{ odd,} \\ 0 & \text{otherwise,} \end{cases} \tag{4.34b}$$

$$c_{ij} = \begin{cases} \displaystyle\iint_\Omega r(\partial N^i/\partial r)(\partial N^j/\partial z) \, dr \, dz & \text{for } i \text{ odd and } j \text{ even,} \\ \displaystyle\iint_\Omega r(\partial N^i/\partial z)(\partial N^j/\partial r) \, dr \, dz & \text{for } i \text{ even and } j \text{ odd,} \\ 0 & \text{otherwise} \end{cases} \tag{4.34c}$$

$$d_{ij} = \begin{cases} \displaystyle\iint_\Omega (1/r)N^i N^j \, dr \, dz & \text{for } i \text{ odd and } j \text{ odd,} \\ 0 & \text{otherwise,} \end{cases} \tag{4.34d}$$

$$e_{ij} = \begin{cases} \displaystyle\iint_\Omega N^i(\partial N^j/\partial z) \, dr \, dz & \text{for } i \text{ odd and } j \text{ even,} \\ \displaystyle\iint_\Omega (\partial N^i/\partial z) N^j \, dr \, dz & \text{for } i \text{ even and } j \text{ odd,} \\ 0 & \text{otherwise,} \end{cases} \tag{4.34e}$$

$$f_{ij} = \begin{cases} \displaystyle\iint_\Omega r(\partial N^i/\partial z)(\partial N^j/\partial z) \, dr \, dz & \text{for } i \text{ even and } j \text{ even,} \\ 0, & \text{otherwise,} \end{cases} \tag{4.34f}$$

$$g_{ij} = \begin{cases} \dfrac{1}{2}\displaystyle\iint_\Omega r(\partial N^i/\partial z)(\partial N^j/\partial z) \, dr \, dz & \text{for } i \text{ odd and } j \text{ odd,} \\ \dfrac{1}{2}\displaystyle\iint_\Omega r(\partial N^i/\partial r)(\partial N^j/\partial z) \, dr \, dz & \text{for } i \text{ even and } j \text{ odd,} \\ \dfrac{1}{2}\displaystyle\iint_\Omega r(\partial N^i/\partial z)(\partial N^j/\partial r) \, dr \, dz & \text{for } i \text{ odd and } j \text{ even,} \\ \dfrac{1}{2}\displaystyle\iint_\Omega r(\partial N^i/\partial r)(\partial N^j/\partial r) \, dr \, dz & \text{for } i \text{ even and } j \text{ even.} \end{cases} \tag{4.34g}$$

The stiffness matrix $\mathbf{K}(\xi)$ has been written in the form shown in Eq. (4.33) to make the time integration more efficient. This is now explained in the context of the first component of $\mathbf{K}(\xi)$, i.e., we consider the integral

$$\mathbf{v}(t) = \int_0^t C_{11}(t - \tau)\mathbf{Au}(\tau) \, d\tau. \tag{4.35}$$

Equation (4.35) represents a set of Volterra integral equations of the second kind. A standard numerical technique to solve this type of equation is obtained by the forward approximation of the time-derivative term, which is

$$\dot{\mathbf{u}}(\tau) \simeq (\mathbf{u}^i - \mathbf{u}^{i-1})/\Delta t_i, \tag{4.36}$$

where $\Delta t_i = t_i - t_{i-1}$. This is equivalent to using a linear Lagrangian interpolation scheme in the time dimension [see also Taylor et al. (1970)]. Using a Prony series for C_{11}, the exponential terms can be integrated exactly over each time interval, and in Eq. (4.25) each component of $\mathbf{v}$ can be written as

$$v_i(t) = a_{ij} \left\{ \sum_{n=1}^{N} C_{11}^0 (u_j^n - u_j^{n-1}) + \sum_{k=1}^{I} C_{11}^k e^{-t/\tau_k} \right.$$
$$\left. \times \left[\sum_{n=1}^{N} \tau_k (e^{t_n/\tau_k} - e^{t_{n-1}/\tau_k})(u_j^n - u_j^{n-1})/\Delta t_n \right] \right\}, \qquad (4.37)$$

where $t_n = t$ is the current time. For computational purposes, we rewrite $v_i(t)$ as

$$v_i(t) = a_{ij} \left\{ C_{11}^0 + (1/\Delta t_N) \left[\sum_{k=1}^{I} C_{11}^k \tau_k (1 - e^{-\Delta t_N/\tau_k}) \right] \right\} (u_j^N - u_j^{N-1})$$
$$+ a_{ij} \left\{ C_{11}^0 (u_j^{N-1} - u_j^0) + \sum_{k=1}^{I} C_{11}^k \tau_k e^{-t/\tau_k} \right.$$
$$\left. \times \left[\sum_{n=1}^{N-1} (1/\Delta t_n)(e^{t_n/\tau_k} - e^{t_{n-1}/\tau_k})(u_j^n - u_j^{n-1}) \right] \right\}. \qquad (4.38)$$

The following computational advantages are obtained from Eq. (4.38):

(1) The second term in the right-hand side is the "memory" term and can now be constructed recursively, each time step requiring only the addition of newly calculated terms, by defining

$$\alpha_j^k(t_N) = \sum_{n=1}^{N-1} (1/\Delta t_n)(e^{t_n/\tau_k} - e^{t_{n-1}/\tau_k})(u_j^n - u_j^{n-1}),$$
$$\alpha_j^k(t_1) = 0, \qquad (4.39)$$

so we have

$$\alpha_j^k(t_{N+1}) = \alpha_j^k(t_N) + (1/\Delta t_N)(e^{t_N/\tau_k} - e^{t_{N-1}/\tau_k})(u_j^N - u_j^{N-1}). \qquad (4.40)$$

(2) The storage requirements are proportional to the number of nodes in the region and the number of Maxwell elements I used in the viscoelastic model, but independent of the number of time steps.

(3) The coefficients of u_i^n in the first term of Eq. (4.38) are constant for a constant time step Δt. As long as the time step remains unchanged, only one matrix inversion is required in the transient solution.

This integration procedure is applied to each component of the stiffness matrix $\mathbf{K}$ defined in Eq. (4.33). The resulting system of linear equations is then solved using a banded, symmetric-equation solver based on a Cholesky decomposition. If the time step is not variable, the matrix needs to be decomposed only once, at the first time step, and the solution is advanced in time by updating the right-hand side and performing one forward and one backward substitution at each time step.

4.1.2.4. Numerical Results

A numerical study of the parameters has been carried out to determine the range of the unknown parameters in the constitutive relationships. The experimental data for the condition at 51.7° C and 30% RH was used to obtain constitutive relationships using the values for $v_{12} = v_{13}$, v_{23}, and G given in Table 4.5.

The results of a geometry presented later were then analyzed for qualitative and quantitative behaviors and the following criteria set forth to evaluate the results:

(1) Negative hoop stress caused by the presence of bump geometry cannot develop in the tape.
(2) The shear stress σ_{rz} should be small when compared to the radial and hoop stresses, σ_{rr} and $\sigma_{\theta\theta}$, caused by the geometric deformation.

Based on these criteria, we found that only the values, $v_{12} = v_{13} = 0.01$ and $G = 0.025$, were adequate, since all but cases 1 and 5 in Table 4.5 produced stress fields that were acceptable.

For subsequent calculations, the values $v_{12} = v_{13} = 0.01$ and $v_{23} = 0.3$ were chosen. The coefficient G, as calculated by Eq. (4.24), gives $G = 0.074$, and further experimentation using these values have produced acceptable stress fields, so G has been calculated using Eq. (4.24).

In Fig. 4.12(a), a sketch of a radial cross section of a reel of tape is given, where the dimensions and the discretized region are shown. We have assumed that the tape has been wrapped under tension (2.2 N or 5.3 MPa), which causes a prestressing of the tape, and that the stress field is calculated using the method outlined in the previous section and included in the simulations, as shown in Figs. 4.12(b) and (c). Since the tape is modeled as a solid body, we assume that a deformation is applied at the base of the hub, simulating a bump in the hub, which is of a sinusoidal shape given by

$$u_r(1, z) = \begin{cases} \delta \sin(\pi/l)(z - l_1) & \text{if } l_1 \leq z \leq l + l_1, \\ 0 & \text{otherwise,} \end{cases} \qquad (4.41)$$

Table 4.5. Range of variables used in the parametric study at 51.7° C and 30% RH

Case	$v_{12} = v_{13}$	v_{23}	G
1	0.01	0.2	0.025
2	0.01	0.2	0.125
3	0.1	0.2	0.025
4	0.1	0.2	0.125
5	0.01	0.4	0.025
6	0.01	0.4	0.125
7	0.1	0.4	0.025
8	0.1	0.4	0.125

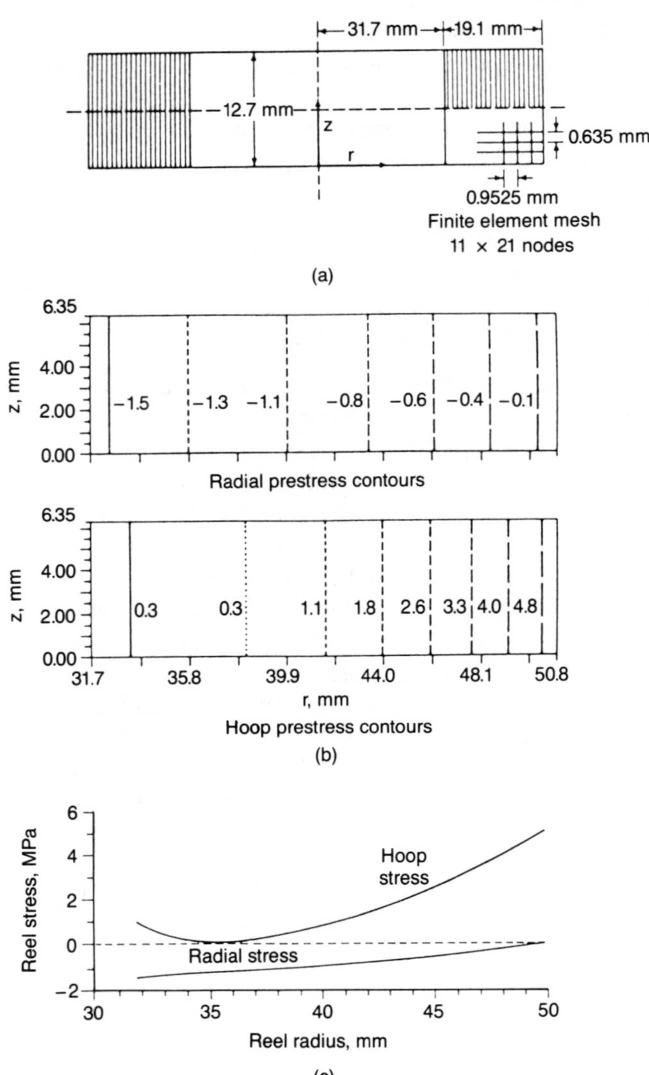

Fig. 4.12. (a) Schematic of a tape reel showing the finite-element mesh. (b) Radial and hoop prestressing contours caused by winding at 22° C and 30% RH. Stresses are in MPa. (c) Radial and hoop prestressing caused by winding at 22° C and 30% RH as a function of radius (Heinrich et al., 1986).

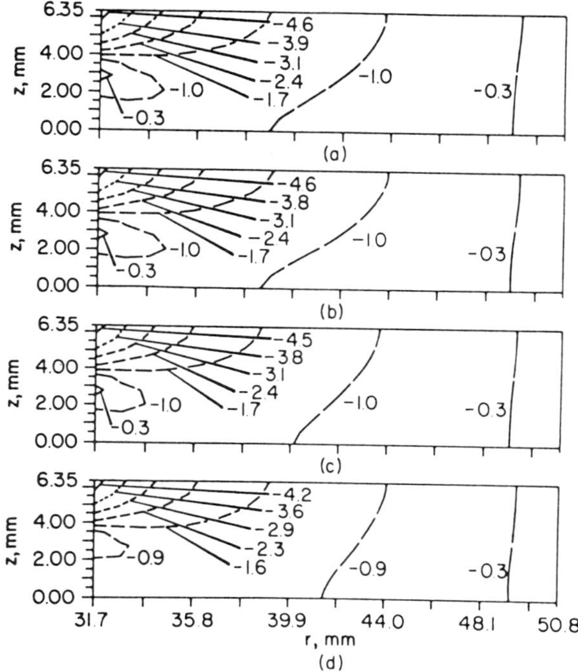

Fig. 4.13. Radial stress contours at time = 0 h with prestressing: (a) 43.3° C and 30% RH; (b) 65° C and 30% RH; (c) 51.7° C and 30% RH; and (d) 51.7° C and 85% RH. Stresses are in MPa (Heinrich et al., 1986).

where δ is the amplitude, l is the distortion length of the bump, and l_1 is the phase factor. A symmetric case with $\delta = 0.508$ mm, $l = 6.4$ mm, and $l_1 = 3.2$ mm was chosen to represent a typical bump.

Calculations were performed under these conditions for all four cases shown in Table 4.4. In Figs. 4.13 and 4.14, the normal radial and hoop stresses, σ_{rr} and $\sigma_{\theta\theta}$, respectively, are given for time $t = 0$, before any relaxation has taken place. The stress fields are qualitatively and quantitatively similar, as expected; however, the maximum differences are not negligible, and are of the order of 10%. Figure 4.15 shows the normal axial stress σ_{zz} and the shear stress distribution σ_{rz} at time $t = 0$ and $t = 170$ h for case 2 in Table 4.4. In Fig. 4.16, the normal stresses σ_{rr} and $\sigma_{\theta\theta}$ are given for case 2 after 170 h and for case 4 after 40 h. The effect of the environmental conditions in the relaxation times can be observed. Finally, in Fig. 4.17, the maximum normal stresses, σ_{rr}, $\sigma_{\theta\theta}$, and σ_{zz} and the shear stress σ_{rz}, have been plotted as a function of log t for all cases in Table 4.4.

Bhushan et al. (1984a) reported an example of the change in interlayer pressure (σ_{rr}) caused by stress relaxation at room temperature, Fig. 4.18. Measurements were made on a 12.7-mm-wide and 40-μm-thick data-processing

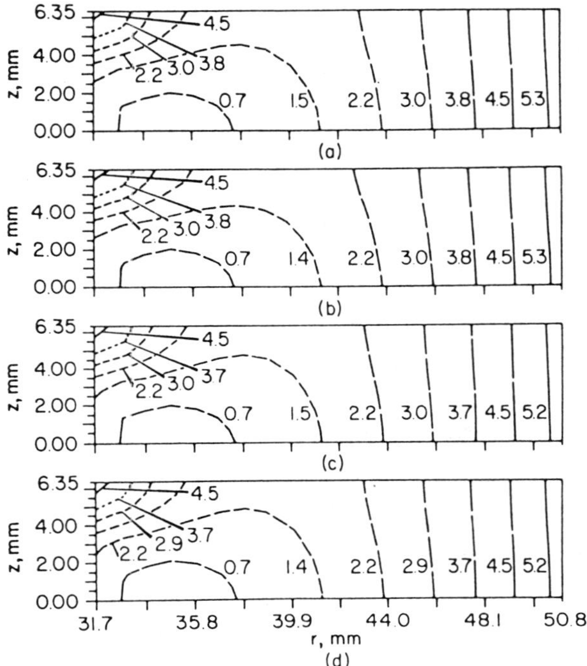

Fig. 4.14. Hoop stress contours at time = 0 h with prestressing: (a) 43.3° C and 30% RH; (b) 65° C and 30% RH; (c) 51.7° C and 30% RH; and (d) 51.7° C and 85% RH. Stresses are in MPa (Heinrich et al., 1986).

tape wound at a tension of 2.2 N. Eshel et al. (1984) also reported that stress relaxation occurred with storage. The effect was accelerated at higher temperatures and humidities.

The ultimate objective of these analyses is to relate the stress relaxation inside the tape to a viscoelastic geometric deformation (Bhushan et al., 1984a; Eshel et al., 1984). Unfortunately, there is no direct way to do this, so we must introduce two additional assumptions, which are:

(1) The percentage of viscoelastic deformation is equal to the percentage of stress relaxation that has taken place at a given time for given environmental conditions. Thus, the process is governed by the same time constants as the stress relaxation.

(2) In the deformed state, the percentage of deformation which is viscoelastic is the same everywhere.

We can then write, for the viscoelastic deformation $u_r^a(r, z, t)$ in the r direction,

$$u_r^a(r, z, t) = u_r(r, z, 0)[1 - (A_0 + A_1 e^{-t/\tau_1} + A_2 e^{-t/\tau_2})], \qquad (4.42)$$

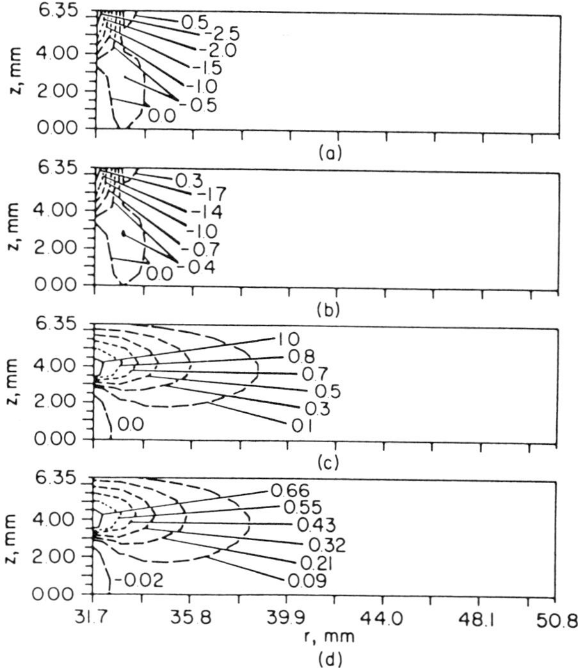

Fig. 4.15. Axial stress contours with prestressing at 51.7° C and 30% RH: (a) at time = 0; (b) at time = 170 h. Shear stress σ_{rz} with prestressing at 51.7° C and 30% RH; (c) at time = 0 h; and (d) at time = 170 h. Stresses are in MPa (Heinrich et al., 1986).

where the coefficients A_i are obtained from the hoop-stress equation $\sigma_{\theta\theta}$ by rewriting it in the form

$$\sigma_{\theta\theta}(t) = \sigma_{\theta\theta}(A_0 + A_1 e^{-t/\tau_1} + A_2 e^{-t/\tau_2}). \tag{4.43}$$

In Fig. 4.19(a), we show the maximum viscoelastic deformation at the innermost wrap plotted against $\log t$ for the calculations just discussed. Figure 4.19(b) shows how the distortion in the hub propagates throughout the tape layers for case 1 in Table 4.4.

Finally, we examine the effect of varying the distortion length l of the applied deformation. For this purpose, we reduced the value of l of the applied deformation. For this purpose, we reduced the value of l by one-half to 3.2 mm, set $l_1 = 4.8$ mm, and kept the amplitude the same, so as to produce a much sharper bump. In Fig. 4.20, we show the displacement u_r as a function of radius for the two bump geometries under consideration at $z = 6.35$ mm. Figure 4.21 shows the stress fields, σ_{rr} and $\sigma_{\theta\theta}$, for this second geometry, under the conditions of case 2 in Table 4.4. These should be compared with Figs. 4.13(a) and 4.14(a).

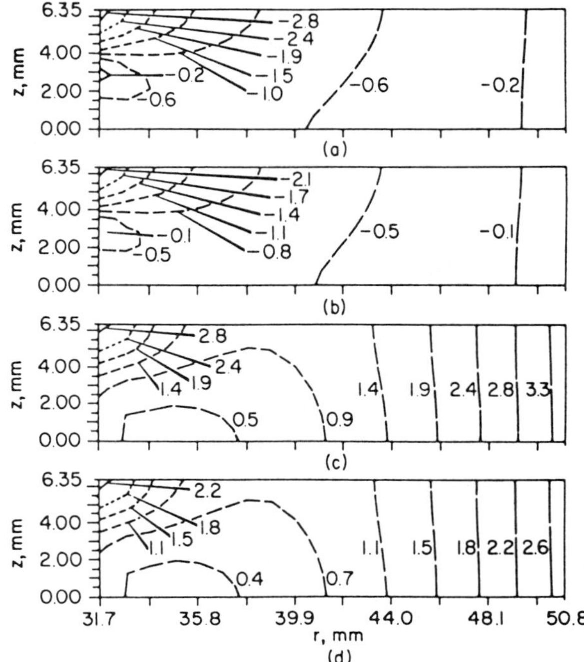

Fig. 4.16. Radial stress contours with prestressing at time = 170 h: (a) 51.7° C and 30% RH; (b) 51.7° C and 85% RH. Hoop stress contours with prestressing at time = 170 h; (c) 51.7° C and 30% RH; and (d) 51.7° C and 85% RH. Stresses are in MPa (Heinrich et al., 1986).

4.1.2.5. Summary

We have presented an axisymmetric, finite-element model for the prediction of stress relaxation and viscoelastic deformation induced by the geometric distortion introduced in reels of tape. The analysis performed shows no significant differences in the initial stress fields for the same geometric distortion caused by changes in temperature or relative humidity affecting the constitutive relationships (change in the elastic modulus at time zero). The greatest difference in the maximum stresses for the cases discussed here is less than 7%. We did not consider stresses induced by thermal and hygroscopic expansivity mismatches (Section 4.1.1.4).

Significant differences are observed, however, in the relaxation times caused by varying conditions of temperature and relative humidity. For example, in case 1 in Table 4.4 (43.3° C and 30% RH) the maximum hoop stress relaxes only 29% in 500 h. On the other hand, in case 4 of Table 4.4 (51.7° C and 85% RH) we get a 51% relaxation in the maximum hoop stress in only 42 h. In terms of viscoelastic deformation for the sinusoidal bump of

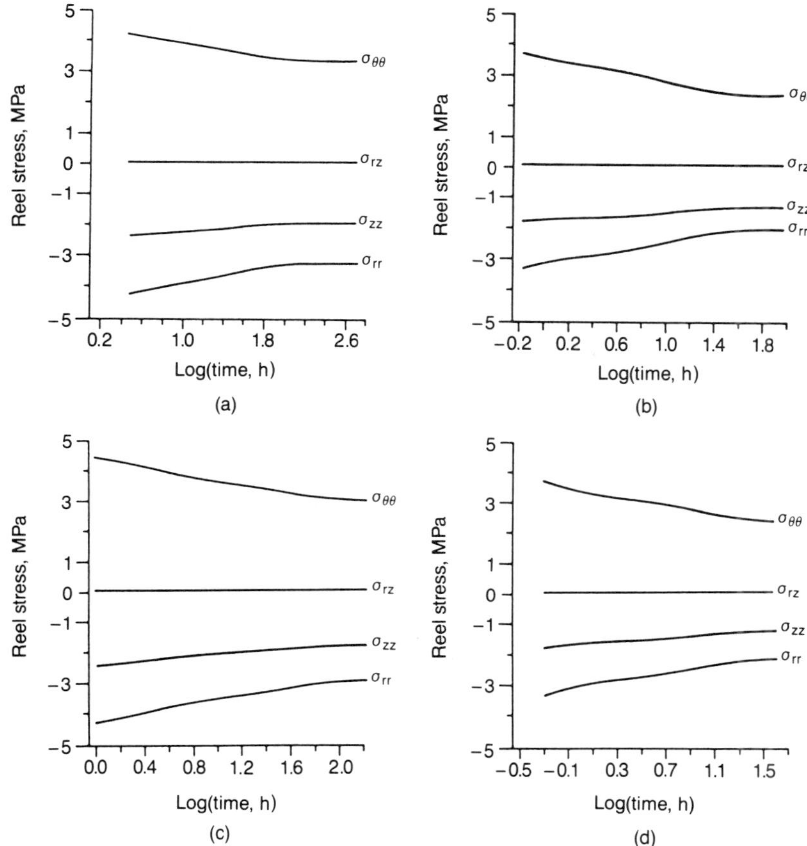

Fig. 4.17. Relaxation of maximum normal stresses and shear stress with prestressing as a function of log time: (a) 43.3° C and 30% RH; (b) 65° C and 30% RH; (c) 51.7° C and 30% RH; and (d) 51.7° C and 85% RH (Heinrich et al., 1986).

amplitude 0.0508 mm, this gives a maximum deformation of 0.015 mm in case 1 and 0.026 mm in case 4. That is, approximately twice the deformation in one-twelfth of the time. In general, the rate at which stresses relax increases with an increase in temperature and relative humidity.

The stresses developed in the tape, because of an applied geometric deformation (bump) of the type given in Eq. (4.41), are directly proportional to the amplitude of the bump. When the distortion length is reduced, keeping the amplitude constant so that a sharper deformation is induced, the radial stresses increase as expected; however, the displacement decreases faster in the radial direction, as shown in Fig. 4.20. This results in a smaller hoop stress, in this case by as much as 18%. Note also that the normal stress in the axial direction increases significantly by 25% in this example.

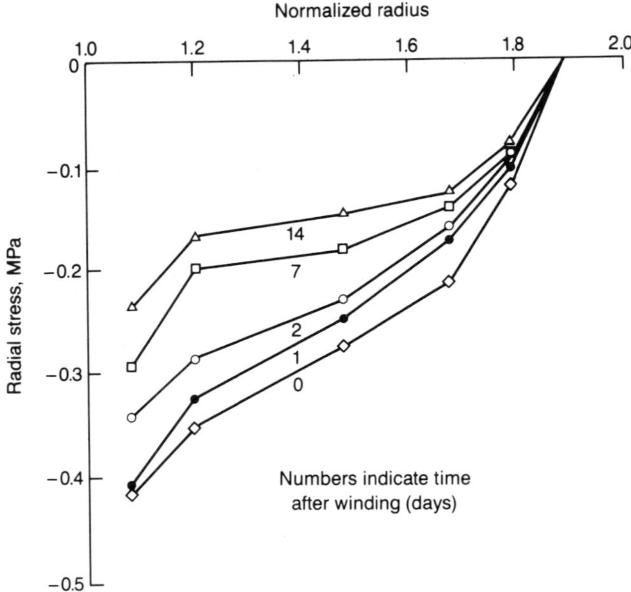

Fig. 4.18. Relaxation of interlayer pressure with time. A 12.7-mm-wide and 40-μm-thick tape was wound at 2.2 N and 2 m/s (winding stress = 4.3 MPa) (Bhushan, 1990).

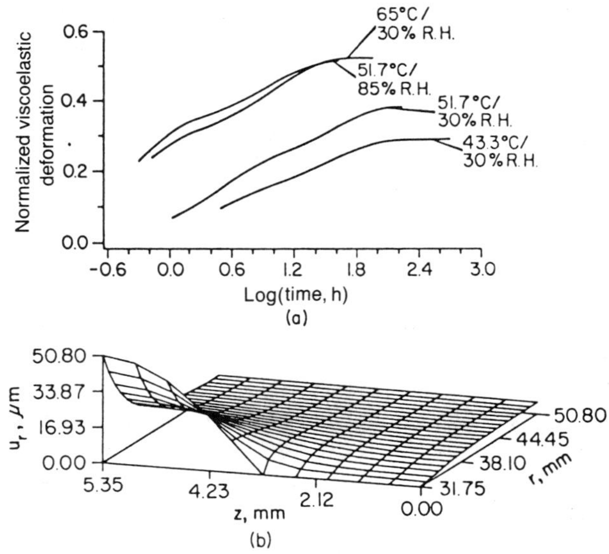

Fig. 4.19. (a) Normalized, radial, viscoelastic deformation in the innermost wrap at the tape reel's centerline as a function of log (time) for various temperatures and relative humidities. (b) Total radial deformation u_r as a function of reel radius r and reel width z (Heinrich et al., 1986).

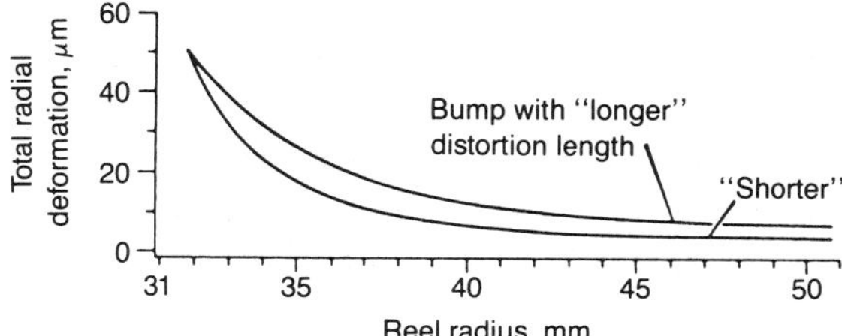

Fig. 4.20. Total radial deformation as a function of tape-reel radius for two different bump geometries (Heinrich et al., 1986).

It should be pointed out that the data used in this work pertain to uncoated PET only. For coated magnetic tape, we expect quantitative differences that should be small at higher temperatures and humidities, but could be significant at room-ambient conditions (see Chapter 3). The overall trends should, however, remain the same. Finally, we note that the validity of the model is restricted to time scales related to the duration of the experiments and the number of relaxation times used in the Prony series are representations of the various functions. The use of the model for longer periods of time will yield erroneous results.

4.2. Flexible Disks

The dimensional changes in the biaxially-oriented PET substrate are partially due to temperature and humidity changes, as well as to creep recovery (shrinkage) from stresses of manufacture. Additional deformations are apt to result from the stresses due to spinning over a long period of time. This is

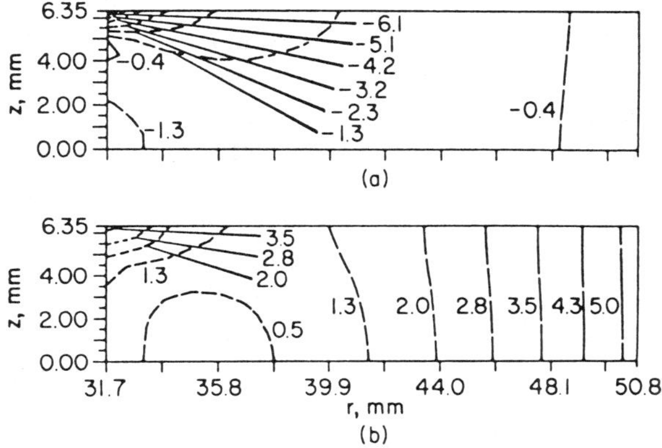

Fig. 4.21. Stress contours at time = 0 h with prestressing at 43.3° C and 30% RH: (a) radial and (b) hoop. Stresses are in MPa (Heinrich et al., 1986).

because PET tends to "flow" under stress even at room temperature if the time of observation is long enough. If the flexible disk is expected to rotate constantly at few hundred rpm, once installed, the creep due to the centrifugal stresses of rotation of a few Pa (generally very small) could possibly become an important factor over the life of the disk (5–8 years) especially at high temperatures and/or high humidity. Actually, the concern is not so much that the dimensions may change, but rather that the change will not be axisymmetric. Then an initially circular data track would creep to a noncircular one and could require high resolution servo techniques for tracking. This is because the PET sheet is anisotropic, i.e., it has different mechanical, thermal, and hygroscopic properties in different directions, caused by the manufacturing process. The situation is further complicated by the fact that the axes of material symmetry do not have the same directions at different transverse locations on a roll of PET. The orientation angle of the material symmetry axes may vary up to 45° across a 6.1-m-wide roll of PET (Chapter 2).

The problem of finding the stress distributions in rotating circular disks made of isotropic and anisotropic materials has been studied exhaustively (Green and Zerna, 1954; Halpin and Pagano, 1968; Tang, 1969; Timoshenko and Goodier, 1970; Lekhnitskii, 1968, 1981; Darlington and Saunders, 1974; Bogy and Talke, 1984). In the stress analysis of spinning a flexible disk, the disk is assumed as an orthotropic elastic material to obtain initial stresses, and is assumed as an orthotropic viscoelastic material to predict the creep that would result from the stresses of rotation for a few years. We present here both analyses, as advanced by Bogy and Talke (1984).

4.2.1. Elasticity Solution

We wish to determine the noncircular elastic deformation due to spinning of an initially circular disk made of an anisotropic material. The stress field for a complete disk can be found in the literature (Lekhnitskii, 1968), and the displacement field follows from a simple integration of the associated strains. The solution of an annular disk that is fixed at its inner boundary to a rigid shaft, for the case of cylindrical anisotropy for which the deformation remains axisymmetric, is presented by Tang (1969). In the more general anisotropic cases, and also for the orthotropic case, the deformation is not axisymmetric and the solution technique makes use of two complex variables that depend on the particular material. These variables affect a transformation that carries circles into ellipses. By the use of conformal mapping, one ellipse can be mapped back to a circle, but two cannot. For this reason, it is relatively easy to solve the general problem for a complete disk or for a circular hole in the infinite plane, since only one boundary is involved (Lekhnitskii, 1968; Green and Zerna, 1954).

In the analysis presented here, the circular disk with a central hole is treated by a superposition technique. An infinite system of equations is generated for deriving coefficients that appear in a series representation of the solution. This procedure is exact but does not appear to lead to a practical solution for the purpose of numerical calculations. An alternate iterative scheme is then adopted in which the first and second approximations are derived. This scheme should give satisfactory results when the ratio of inner radius to outer radius is small in comparison to unity. Numerical results based on this scheme are presented in order to evaluate the effect of the central clamping on the solution. Such calculations require all four orthotropic plate material constants.

4.2.1.1. Orthotropic Solid Disk

Bogy and Talke (1984) considered a continuous disk of radius b spinning in its plane about its center at constant angular speed ω. The disk is assumed to be homogeneous with material symmetry about a set of perpendicular axes, as indicated by the dashed lines in Fig. 4.22. A set of Cartesian coordinates x_1, x_2 is fixed in the disk, aligned with the material symmetry axes. With reference to this frame, the equations of equilibrium for steady spin appear as

$$\frac{\partial \sigma_{11}}{\partial x_1} + \frac{\partial \sigma_{12}}{\partial x_2} + \rho\omega^2 x_1 = 0,$$

$$\frac{\partial \sigma_{12}}{\partial x_1} + \frac{\partial \sigma_{22}}{\partial x_2} + \rho\omega^2 x_2 = 0,$$

(4.44)

where ρ is the density of the disk material. The constitutive relations and the strain displacement relations take the form (see Chapter 3)

$$\varepsilon_{11} = S_{11}\sigma_{11} + S_{12}\sigma_{22}, \qquad \varepsilon_{22} = S_{12}\sigma_{11} + S_{22}\sigma_{22}, \qquad \varepsilon_{12} = \tfrac{1}{2}S_{66}\sigma_{12},$$

(4.45)

$$\varepsilon_{11} = \frac{\partial u_1}{\partial x_1}, \qquad \varepsilon_{22} = \frac{\partial u_2}{\partial x_2}, \qquad \varepsilon_{12} = \frac{1}{2}\left(\frac{\partial u_1}{\partial x_2} + \frac{\partial u_2}{\partial x_1}\right).$$

(4.46)

The four elastic compliances, S_{11}, S_{12}, S_{22}, S_{66}, which characterize the two-dimensional (plate extension) deformation of an orthotropic material are

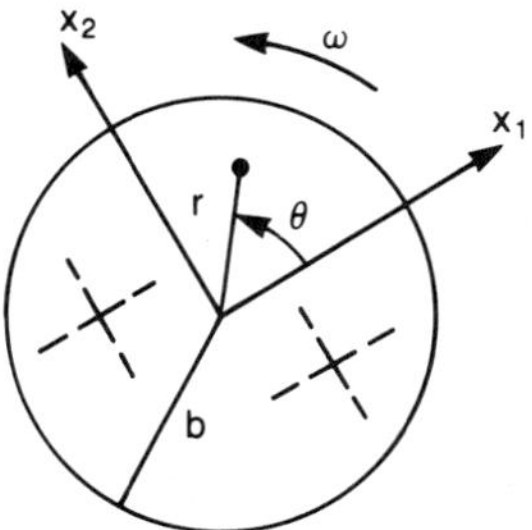

Fig. 4.22. Schematic of a solid disk, a set of perpendicular axes as indicated by the dashed lines show material symmetry (Bogy and Talke, 1984).

related to the more familiar material constants as shown in Chapter 3. The boundary conditions are no traction at $r = b$, i.e.,

$$\sigma_{rr} = \sigma_{r\theta} = 0 \qquad \text{at} \quad r = b. \tag{4.47}$$

The stress field of the solution for Eqs. (4.44)–(4.47) is (Lekhnitskii, 1968)

$$\sigma_{11} = \frac{\rho\omega^2}{2}[(1 - A)(b^2 - x_1^2 - x_2^2) + 2Ax_2^2],$$

$$\sigma_{22} = \frac{\rho\omega^2}{2}[(1 - A)(b^2 - x_1^2 - x_2^2) + 2Ax_1^2], \tag{4.48}$$

$$\sigma_{12} = -\rho\omega^2 Ax_1 x_2,$$

in which

$$A = \frac{S_{11} + 2S_{12} + S_{22}}{3S_{11} + 2S_{12} + S_{66} + 3S_{22}}. \tag{4.49}$$

The stresses for a spinning flexible disk are negligibly small (~ 1 Pa or less). However, stress exposure at high temperature and high humidity may result in some creep deformation.

The strain field results from Eqs. (4.45) and (4.48) and then the displacement field follows from the integration of Eq. (4.46). We get

$$
\begin{aligned}
u_1 &= \frac{\rho\omega^2 b^3}{2}\left\{(S_{11} + S_{12})(1 - A)\left(\frac{x_1}{b} - \frac{x_1^3}{3b^3} - \frac{x_1 x_2^2}{b^3}\right)\right. \\
&\quad \left. + 2A\left(S_{12}\frac{x_1^3}{3b^3} + S_{11}\frac{x_1 x_2^2}{b^3}\right)\right\}, \\
u_2 &= \frac{\rho\omega^2 b^3}{2}\left\{(S_{12} + S_{22})(1 - A)\left(\frac{x_2}{b} - \frac{x_2^3}{3b^3} - \frac{x_1^2 x_2}{b^3}\right)\right. \\
&\quad \left. + 2A\left(S_{12}\frac{x_2^3}{3b^3} + S_{22}\frac{x_1^2 x_2}{b^3}\right)\right\}.
\end{aligned}
\tag{4.50}
$$

In order to indicate the nonaxisymmetric nature of the deformation, we need the polar components u_r, u_θ which we obtain from Eq. (4.50) as follows:

$$u_r + iu_\theta = e^{-i\theta}(u_1 + iu_2), \tag{4.51}$$

so that

$$
\begin{aligned}
u_r(b, \theta) = \frac{\rho\omega^2 b^3}{2}\left\{\tfrac{1}{4}(S_{11} + 2S_{12} + S_{22}) + \frac{(1 - A)}{3}(S_{11} - S_{22})\cos 2\theta \right. \\
\left. + \left[\tfrac{1}{12}(S_{11} + 2S_{12} + S_{22}) - \frac{A}{3}(S_{11} + S_{22})\right]\cos 4\theta\right\}.
\end{aligned}
\tag{4.52}
$$

This gives the θ-dependence of the radial displacement at the disk boundary $r = b$ due to spinning. In the isotropic limit, $S_{11}, S_{22} \to 1/E$, $S_{12} \to -v/E$, $S_{66} \to 2(1 + v)/E$ and we find, with the use of Eq. (4.49), that the coefficients of cos 2θ, cos 4θ in Eq. (4.52) vanish while $\frac{1}{4}(S_{11} + 2S_{12} + S_{22}) \rightarrow (1 - v)/2E$. Some general features of θ-dependence can be pointed out. The term with no θ-dependence contributes circular expansion, the cos 2θ term contributes an elliptical expansion, and the cos 4θ term contributes a four-lobed "cloverleaf" expansion. Therefore, we conclude that the deformed shape cannot be elliptical unless the cos 4θ term vanishes, which would require that the relation $2S_{12} + S_{66} = S_{11} + S_{22}$ holds.

4.2.1.2. Annular Disk: Fixed Inner Boundary

Bogy and Talke (1984) next considered the problem of a spinning annular disk of outside radius b and inside radius a, where the disk is attached to a rigid shaft. This configuration is depicted in Fig. 4.23. We again restrict our attention to the case of orthotropic symmetry and align the coordinate axes with the symmetry axes so that Eqs. (4.44)–(4.46) remain valid but the boundary conditions are replaced by

$$\sigma_{rr} = \sigma_{r\theta} = 0 \qquad \text{at} \quad r = b,$$
$$u_1 = u_2 = 0 \qquad \text{at} \quad r = a. \tag{4.53}$$

The solution for the complete disk satisfies the first of these conditions but not the second. Clearly, the solution sought presently approaches that given by Eqs. (4.48) and (4.49) in the limit as $a/b \rightarrow 0$, since Eq. (4.49) satisfies $u_1 = u_2 = 0$ at $r = 0$. Therefore, the complete disk solution must be a good approximation in the outer regions of the disk to the solution satisfying Eq. (4.53) whenever $a/b \ll 1$.

If we denote the complete disk solution by S^0 and the annular disk solution by S, then we may write

$$S = S^0 + S'. \tag{4.54}$$

It follows that the difference solution, $S' = S - S^0$, satisfies Eqs. (4.44)–(4.46) and the following boundary conditions

$$\sigma'_{rr} = \sigma'_{r\theta} = 0 \qquad \text{at} \quad r = b, \qquad u'_1 = -u^0_1, \qquad u'_2 = -u^0_2 \qquad \text{at} \quad r = a. \tag{4.55}$$

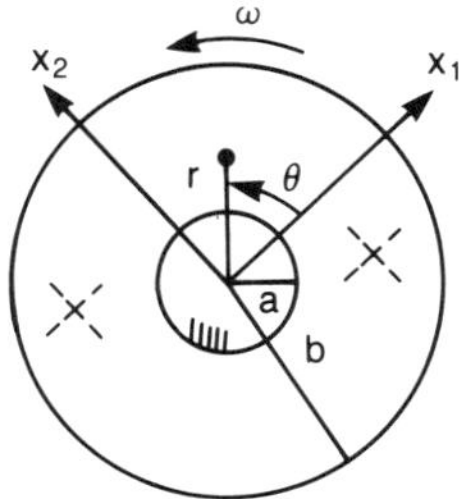

Fig. 4.23. Schematic of an annular disk, a set of perpendicular axes as indicated by the dashed lines show material symmetry (Bogy and Talke, 1984).

Thus the S' solution leaves the outer boundary traction free and annuls the displacement of the S^0 solution at the inner boundary. In view of Eq. (4.50) the displacement conditions in Eq. (4.55) have the form

$$u_1'(a, \theta) = P_1(e^{i\theta} + e^{-i\theta}) + P_3(e^{i3\theta} + e^{-i3\theta}),$$
$$u_2'(a, \theta) = Q_1(e^{i\theta} - e^{-i\theta}) + Q_3(e^{i3\theta} - e^{-i3\theta}), \tag{4.56}$$

where

$$P_1 = -\frac{\rho\omega^2 b^3}{2}\left[-(S_{11} + S_{12})(1 - A)\left(\frac{a^3}{4b^3} - \frac{a}{2b}\right) + \frac{A}{4}(S_{11} + S_{12})\frac{a^3}{b^3}\right],$$

$$P_3 = -\frac{\rho\omega^2 b^3}{2}\left[(S_{11} + S_{12})(1 - A)\frac{a^3}{12b^3} + \frac{A}{4}\left(\frac{S_{12}}{3} - S_{11}\right)\frac{a^3}{b^3}\right]$$

$$Q_1 = -\frac{i\rho\omega^2 b^3}{2}\left[(S_{12} + S_{22})(1 - A)\left(\frac{a^3}{4b^3} - \frac{a}{2b}\right) - \frac{A}{4}(S_{12} + S_{22})\frac{a^3}{b^3}\right],$$

$$Q_3 = -\frac{i\rho\omega^2 b^3}{2}\left[(S_{12} + S_{22})(1 - A)\frac{a^3}{12b^3} + \frac{A}{4}\left(\frac{S_{12}}{3} - S_{22}\right)\frac{a^3}{b^3}\right]. \tag{4.57}$$

The complex representation of plane solutions for anisotropic materials has the form

$$u_1 = \sum_{\alpha=1}^{2} [p_\alpha \Phi_\alpha(z_\alpha) + \bar{p}_\alpha \overline{\Phi}_\alpha(\bar{z}_\alpha)],$$

$$u_2 = \sum_{\alpha=1}^{2} [q_\alpha \Phi_\alpha(z_\alpha) + \bar{q}_\alpha \overline{\Phi}_\alpha(\bar{z}_\alpha)], \tag{4.58}$$

$$\sigma_{11} = \sum_{\alpha=1}^{2} [\mu_\alpha^2 \Phi_\alpha'(z_\alpha) + \bar{\mu}_\alpha^2 \overline{\Phi}_\alpha'(\bar{z}_\alpha)],$$

$$\sigma_{22} = \sum_{\alpha=1}^{2} [\Phi_\alpha'(z_\alpha) + \overline{\Phi}_\alpha'(\bar{z}_\alpha)], \tag{4.59}$$

$$\sigma_{12} = -\sum_{\alpha=1}^{2} [\mu_\alpha \Phi_\alpha'(z_\alpha) + \bar{\mu}_\alpha \overline{\Phi}_\alpha'(\bar{z}_\alpha)],$$

where $\Phi_1(z_1)$, $\Phi_2(z_2)$ are analytic functions of the complex variables z_1, z_2 defined by

$$z_1 = x_1 + \mu_1 x_2, \qquad z_2 = x_1 + \mu_2 x_2, \tag{4.60}$$

and primes denote differentiation while the superposed bar denotes complex conjugate. The complex material constants p_1, p_2, q_1, q_2, μ_1, and μ_2 appearing in Eqs. (4.58)–(4.60) are defined in the case of orthotropic materials in terms of the four real constants S_{11}, S_{22}, S_{12}, S_{66} by

$$\left. \begin{aligned} p_1 &= S_{11}\mu_1^2 + S_{12}, & p_2 &= S_{11}\mu_2^2 + S_{12}, \\ q_1 &= S_{12}\mu_1 + \frac{S_{22}}{\mu_1}, & q_2 &= S_{12}\mu_2 + \frac{S_{22}}{\mu_2}, \end{aligned} \right\} \tag{4.61}$$

μ_1, μ_2 are roots of

$$S_{11}\mu^4 + (2S_{12} + S_{66})\mu^2 + S_{22} = 0.$$

The roots μ_1, μ_2 take one of the following forms:

$$\mu_1 = \beta_1 i, \qquad \mu_2 = \beta_2 i,$$
$$\mu_1 = \alpha + \beta i, \qquad \mu_2 = -\alpha + \beta_i. \tag{4.62}$$

In order to obtain the analytic functions $\Phi_1(z_1)$, $\Phi_2(z_2)$ needed in Eqs. (4.58) and (4.59) for the difference solutions S' satisfying Eqs. (4.55) and (4.56), Bogy and Talke (1984) made use of two known solutions; one is the solution defined in the unbounded exterior region $r > a$ and capable of satisfying arbitrary prescribed boundary displacements at $r = a$, the other is the solution defined on the interior region $r < b$ and capable of satisfying arbitrary prescribed boundary tractions at $r = b$. These two solutions have the following forms for the problem at hand (Lekhnitskii, 1968):

$$\Phi_\alpha^{(1)}(z_\alpha) = \sum_{m=0}^{\infty} A_{\alpha m}[\zeta_\alpha(z_\alpha)]^{-m}, \qquad \alpha = 1, 2,$$

$$\zeta_\alpha(z_\alpha) = \frac{z_\alpha + \sqrt{z_\alpha^2 - a^2(1 + \mu_\alpha^2)}}{a(1 - i\mu_\alpha)}, \tag{4.63}$$

$$\Phi_\alpha^{(2)}(z_\alpha) = \sum_{m=0}^{\infty} B_{\alpha m} P_{\alpha m}(z_\alpha), \qquad \alpha = 1, 2,$$

$$P_{\alpha 0}(z_\alpha) = 1,$$

$$P_{\alpha m} = -\left\{ \left(\frac{z_\alpha + \sqrt{z_\alpha^2 - b^2(1 + \mu_\alpha^2)}}{b(1 - i\mu_\alpha)} \right)^m + \left(\frac{z_\alpha - \sqrt{z_\alpha^2 - b^2(1 + \mu_\alpha^2)}}{b(1 - i\mu_\alpha)} \right)^m \right\},$$

$$m = 1, 2, 3, \ldots. \tag{4.64}$$

The functions $\zeta_\alpha(z_\alpha)$ and $P_{\alpha m}(z_\alpha)$ are defined so that

$$\zeta_\alpha^{-m} = e^{-im\theta} \qquad \text{at} \quad r = a,$$
$$P_{\alpha m} = e^{im\theta} - t_\alpha e^{-im\theta} \qquad \text{at} \quad r = b, \tag{4.65}$$

where

$$t_\alpha = (1 + i\mu_\alpha)/(1 - i\mu_\alpha). \tag{4.66}$$

We can represent ζ_α^{-m} at $r = b$ and $P_{\alpha m}$ at $r = a$ by the Fourier expansions

$$\zeta_\alpha^{-m} = \sum_{n=0}^{\infty} (a_n^{\alpha m} e^{in\theta} + b_n^{\alpha m} e^{-in\theta}) \qquad \text{at} \quad r = b,$$

$$P_{\alpha m} = \sum_{n=0}^{\infty} (c_n^{\alpha m} e^{in\theta} + d_n^{\alpha m} e^{-in\theta}) \qquad \text{at} \quad r = a. \tag{4.67}$$

Next we define the difference solution S' by

$$\Phi_\alpha^{(\prime)}(z_\alpha) = \Phi_\alpha^{(1)}(z_\alpha) + \Phi_\alpha^{(2)}(z_\alpha) \tag{4.68}$$

and determine $A_{\alpha m}$, $B_{\alpha m}$ in Eqs. (4.63) and (4.64) so that the boundary conditions in Eqs. (4.55) and (4.56) are satisfied. The integrated form of traction boundary conditions at $r = b$ in terms of $\Phi_\alpha(z_\alpha)$ is (Lekhnitskii, 1978)

$$\sum_{\alpha=1}^{2} [\Phi_\alpha(z_\alpha) + \overline{\Phi}_\alpha(\bar{z}_\alpha)] = -\int_0^s X_2 \, ds,$$

$$\sum_{\alpha=1}^{2} [\mu_\alpha\Phi_\alpha(z_\alpha) + \bar{\mu}_\alpha\overline{\Phi}_\alpha(\bar{z}_\alpha)] = \int_0^s X_1 \, ds,$$

$$(4.69)$$

where X_1, X_2 represent the traction components along the boundary, which by Eq. (4.55) are zero at $r = b$ for the solution S'. The general form for displacement boundary conditions in terms of $\Phi_\alpha(z_\alpha)$ is, from Eq. (4.58),

$$\sum_{\alpha=1}^{2} [p_\alpha\Phi_\alpha(z_\alpha) + \bar{p}_\alpha\overline{\Phi}_\alpha(\bar{z}_\alpha)] = u_1^* = \sum_{m=0}^{\infty} (\alpha_m e^{im\theta} + \bar{\alpha}_m e^{-im\theta}),$$

$$\sum_{\alpha=1}^{2} [q_\alpha\Phi_\alpha(z_\alpha) + \bar{q}_\alpha\overline{\Phi}_\alpha(\bar{z}_\alpha)] = u_2^* = \sum_{m=0}^{\infty} (\beta_m e^{im\theta} + \bar{\beta}_m e^{-im\theta}),$$

$$(4.70)$$

where u_1^*, u_2^* represent the components of displacement at the boundary, which are given by Eqs. (4.56) and (4.57) for the solution S'. By use of Eqs. (4.62)–(4.67) and equating the powers of $e^{ik\theta}$, we obtain the following system of equations for determining $A_{\alpha k}$, $B_{\alpha k}$, $k = 0, 1, 2, \ldots$,

$$\sum_{\alpha=1}^{2} \left\{ \sum_{m=0}^{\infty} A_{\alpha m} a_k^{\alpha m} + B_{\alpha k} + \sum_{m=0}^{\infty} \overline{A}_{\alpha m} \bar{b}_k^{\alpha m} + \overline{B}_{\alpha k} \bar{t}_\alpha \right\} = 0,$$

$$\sum_{\alpha=1}^{2} \left\{ \mu_\alpha \left[\sum_{m=0}^{\infty} A_{\alpha m} a_k^{\alpha m} + B_{\alpha k} \right] + \bar{\mu}_\alpha \left[\sum_{m=0}^{\infty} \overline{A}_{\alpha m} \bar{b}_k^{\alpha m} + \overline{B}_{\alpha k} \right] \right\} = 0,$$

$$\sum_{\alpha=1}^{2} \left\{ p_\alpha \sum_{m=0}^{\infty} B_{\alpha m} c_k^{\alpha m} + \bar{p}_\alpha \left[\overline{A}_{\alpha k} + \sum_{m=0}^{\infty} \overline{B}_{\alpha m} \bar{d}_k^{\alpha m} \right] \right\} = \alpha_k,$$

$$\sum_{\alpha=1}^{2} \left\{ q_\alpha \sum_{m=0}^{\infty} B_{\alpha m} c_k^{\alpha m} + \bar{q}_\alpha \left[\overline{A}_{\alpha k} + \sum_{m=0}^{\infty} \overline{B}_{\alpha m} \bar{d}_k^{\alpha m} \right] \right\} = \beta_k.$$

$$(4.71)$$

These equations together with the four obtained by conjugating them, provide an infinite system of eight equations for determining the eightfold infinity of unknowns contained in the real and imaginary parts of $A_{\alpha k}$, $B_{\alpha k}$; $\alpha = 1, 2$; $k = 0, 1, 2 \ldots$.

We observe that all the Eqs. (4.71) are coupled through the Fourier coefficients of ζ_α^{-m} at $r = b$ and $P_{\alpha m}$ at $r = a$ given in Eq. (4.67). It is this coupling which causes the major complication from the point of view of solving the system for $A_{\alpha m}$, $B_{\alpha m}$. Various approaches are possible for obtaining approximate results. We could merely truncate the system at some finite value of $k = K$ and obtain an approximate solution for $A_{\alpha k}$, $B_{\alpha k}$, $k = 0, 1, \ldots, K$. The coefficients $a_k^{\alpha m}$, $b_k^{\alpha m}$, $c_k^{\alpha m}$, $d_k^{\alpha m}$ would first be computed explicitly then the system could be solved numerically. A rational criterion for deciding where to truncate the series can be based on the order of terms in the parameter a/b. In fact, if $a/b \ll 1$ it makes sense to expand the equations in powers of a/b and neglect all terms of $O[(a/b)^N]$ for some chosen N. By direct expansion it can be shown from Eq. (4.65) that

$$p_{\alpha m}(z_\alpha)|_{r=a} = R_{\alpha m} + S_{\alpha m}(e^{i\theta} + t_\alpha e^{-i\theta})\frac{a}{b} + O\left[\left(\frac{a}{b}\right)^2\right], \qquad (4.72)$$

where

$$R_{\alpha m} = -\left[\left(\frac{i}{2}\right)^m + \left(\frac{-i}{2}\right)^m\right][\sqrt{4(1 + i\mu_\alpha)}]^m,$$

$$S_{\alpha m} = \frac{m}{2}\left[\left(\frac{i}{2}\right)^{m-1} + \left(\frac{-i}{2}\right)^{m-1}\right][\sqrt{4(1 + i\mu_\alpha)}]^{m-1}, \qquad (4.73)$$

$$c_0^{\alpha m} = R_{\alpha m}, \quad c_1^{\alpha m} = S_{\alpha m}\frac{a}{b}, \quad d_1^{\alpha m} = S_{\alpha m}t_\alpha\frac{a}{b}c_k^{\alpha m}, \quad d_k^{\alpha m} = O\left(\frac{a^2}{b^2}\right), \quad k > 1. \quad (4.74)$$

Likewise, from Eq. (4.64),

$$[\zeta_\alpha(z_\alpha)]^{-1}|_{r=b} = \frac{a}{b}\frac{1}{e^{i\theta} + t_\alpha e^{-i\theta}} + O\left(\frac{a^2}{b^2}\right). \qquad (4.75)$$

Expressions (4.72) and (4.75) could be used in Eqs. (4.69) and (4.70) to derive the first-order equations in a/b corresponding to Eq. (4.71). Some simplification is obtained in this manner but the resulting system would still not be uncoupled. An alternate approximate scheme to be described next appears to be more suitable for practical calculations where $a/b \ll 1$.

4.2.1.3. Approximate Scheme for Annular Disk: $a/b \ll 1$

In this approach, Bogy and Talke (1984) viewed the solution S as the sum of an infinite number of solutions

$$S = S^0 + S^{(1)} + S^{(2)} + \cdots, \qquad (4.76)$$

where S^0 is the solution for the complete disk $r < b$, $S^{(1)}$ is the solution for the exterior region $r > a$ that satisfies

$$\underline{u}^{(1)}(a) = -\underline{u}^0(a), \qquad (4.77)$$

and $S^{(2)}$ is the solution for the complete disk $r < b$ that satisfies

$$\underline{T}^{(2)}(b) = -\underline{T}^{(1)}(b), \qquad (4.78)$$

etc. This is not truly a perturbation expansion in the parameter a/b since $S^{(1)}$ contains $O(a/b)$ and $O(a^3/b^3)$ terms because Eq. (4.57) contains these terms. The form of $S^{(1)}$ is quite simple and easy to obtain from Eqs. (4.63) and (4.70) with Eqs. (4.56) and (4.57). We find that

$$\alpha_1 = \bar{\alpha}_1 = P_1, \qquad \alpha_3 = \bar{\alpha}_3 = P_3, \qquad \text{other } \alpha_m = 0,$$
$$\beta_1 = -\bar{\beta}_1 = Q_1, \qquad \beta_3 = -\bar{\beta}_3 = Q_3, \qquad \text{other } \beta_m = 0, \tag{4.79}$$

which gives

$$\Phi_1^{(1)}(z_1) = \frac{1}{D}[(P_1 q_2 + Q_1 p_2)\zeta_1^{-1} + (P_3 q_2 + Q_3 p_2)\zeta_1^{-3}],$$

$$\Phi_2^{(1)}(z_2) = -\frac{1}{D}[(P_1 q_1 + Q_1 p_1)\zeta_2^{-1} + (P_3 q_1 + Q_3 p_1)\zeta_2^{-3}], \tag{4.80}$$

$$D = p_1 q_2 - p_2 q_1.$$

The displacement field $u_1^{(1)}$, $u_2^{(1)}$ follows from the use of Eq. (4.80) in Eq. (4.58)

$$u_1^{(1)} + iu_2^{(1)} = \frac{p_1 + iq_1}{D}\left[\frac{P_1 q_2 + Q_1 p_2}{\zeta_1} + \frac{P_3 q_2 + Q_3 p_2}{\zeta_1^3}\right]$$
$$- \frac{p_2 + iq_2}{D}\left[\frac{P_1 q_1 + Q_1 p_1}{\zeta_2} + \frac{P_3 q_1 + Q_3 p_1}{\zeta_2^3}\right]$$
$$+ \frac{\bar{p}_1 + i\bar{q}_1}{\bar{D}}\left[\frac{P_1 \bar{q}_2 - Q_1 \bar{p}_2}{\bar{\zeta}_1(\bar{z}_1)} + \frac{P_3 \bar{q}_2 - Q_3 \bar{p}_2}{\bar{\zeta}_1(\bar{z}_1)^3}\right]$$
$$- \frac{\bar{p}_2 + i\bar{q}_2}{\bar{D}}\left[\frac{P_1 \bar{q}_1 - Q_1 \bar{p}_1}{\bar{\zeta}_2(\bar{z}_2)} + \frac{P_3 \bar{q}_1 - Q_3 \bar{p}_1}{\bar{\zeta}_2(\bar{z}_2)^3}\right]. \tag{4.81}$$

We observe from Eq. (4.52) that $u_r^0(b, \theta)$ is $O(1)$ in a/b. The $O(a^2/b^2)$ correction to $u_r(b, \theta)$ can be obtained from Eq. (4.81). We see from Eq. (4.57) that

$$P_1 = \frac{\rho\omega^2 b^3}{2}\left[(S_{11} + S_{12})(1 - A)\frac{a}{2b} + O\left(\frac{a^3}{b^3}\right)\right],$$

$$P_3 = O(a^3/b^3),$$

$$Q_1 = \frac{i\rho\omega^2 b^3}{2}\left[(S_{12} + S_{22})(1 - A)\frac{a}{2b} + O\left(\frac{a^3}{b^3}\right)\right], \tag{4.82}$$

$$Q_3 = O(a^3/b^3),$$

and also from Eq. (4.75) that

$$\frac{1}{\zeta_\alpha(b)} = \frac{a}{b}\frac{1}{e^{i\theta} + t_\alpha e^{-i\theta}} + O\left(\frac{a^2}{b^2}\right),$$

$$\frac{1}{\zeta_\alpha^3(b)} = O\left(\frac{a^3}{b^3}\right), \tag{4.83}$$

giving, with Eq. (4.81)

$$[u_r^{(1)} + iu_\theta^{(1)}]|_{r=b} = e^{-i\theta}[u_1^{(1)} + iu_2^{(1)}]|_{r=b}$$

$$= \frac{\rho\omega^2 b^3}{2}\frac{a^2}{b^2}\left[\frac{p_1 + iq_1}{D}(a_1 q_2 + b_1 p_2)\frac{1}{e^{i2\theta} + t_1}\right.$$

$$- \frac{p_2 + iq_2}{D}(a_1 q_1 + b_1 p_1)\frac{1}{e^{i2\theta} + t_2} \qquad (4.84)$$

$$- \frac{\bar{p}_1 + i\bar{q}_1}{\bar{D}}(a_1 \bar{q}_2 - b_1 \bar{p}_2)\frac{1}{1 + \bar{t}_1 e^{i2\theta}}$$

$$\left. - \frac{\bar{p}_2 + i\bar{q}_2}{\bar{D}}(a_1 \bar{q}_1 - b_1 \bar{p}_1)\frac{1}{1 + \bar{t}_2 e^{i2\theta}}\right],$$

where

$$a_1 = -\tfrac{1}{2}(S_{11} + S_{12})(1 - A),$$

$$b_1 = \frac{i}{2}(S_{12} + S_{22})(1 - A). \qquad (4.85)$$

Thus, we see that the first order correction is $O(a^2/b^2)$ instead of $O(a/b)$.

In order to evaluate the effect of the inner clamping on the solution, Bogy and Talke (1984) calculated $u_r(b, \theta)$ for a centrally clamped disk [Eq. (4.84)] as well as for a complete disk [Eq. (4.52)] for the assumed values of elastic constants. Based on the results obtained, they made the following observations. *The effect of the central clamping is unimportant in the outer region of the disk for $a/b \leq \frac{1}{7}$* (it would become more important for larger a/b). By neglecting the effect of the central clamping, they obtained an upper bound on the angular variation of the radial displacement. Thus, we know the actual variation for the centrally clamped disk will not exceed the values obtained for the unclamped disk. Therefore, we base the remainder of the analysis on the simple results in Eqs. (4.50) and (4.52), including the entire viscoelastic analysis.

4.2.1.4. Estimates Based on Incomplete Knowledge of Elastic Compliances

As seen in Eq. (4.45), we need to determine experimentally four elastic constants in order to completely characterize the in-plane deformation of an orthotropic sheet (see Chapter 3). They are the tensile moduli E_1, E_2 in the two principal directions, the in-plane shear modulus G_{12}, and one of the Poisson ratios, v_{12}, v_{21}, giving in-plane lateral contraction with tension in the two principal directions. There is considerable difficulty in determining experimentally G_{12}, v_{12}, or v_{21}. But, since these quantities are related to the components of the elasticity tensor, the tensor transformation law for the change of coordinates may be used to derive (see Chapter 3)

$$\frac{1}{E(\phi)} = \frac{\cos^4 \phi}{E_1} + \left(\frac{1}{G_{12}} - \frac{2v_{12}}{E_1}\right)\sin^2 \phi \cos^2 \phi + \frac{\sin^4 \phi}{E_2}, \qquad (4.86)$$

where ϕ is the angle between the axes x_1 and x_1'. By performing a simple tension experiment on a specimen cut with its tension direction at 45° from the E_1 direction, we get

$$\frac{1}{G_{12}} - \frac{2v_{12}}{E_1} = \frac{4}{E_{(45)}} - \frac{1}{E_1} - \frac{1}{E_2}. \qquad (4.87)$$

This relation, together with E_1, E_2, $E_{(45)}$ obtained from tension tests on specimens cut parallel to $0°$, $90°$ and $45°$, leaves only one undetermined constant, say v_{12}. If we are unable to measure v_{12}, then the following question arises. How good of an estimate can be made based on this incomplete information on material constants? There are various ways to proceed. Two ways which we can also extend to the viscoelasticity analysis are described below.

Assumption on the Poisson ratio

Bogy and Talke (1984) assumed they knew E_1, E_2, $E_{(45)}$. They further assumed the Poisson ratios v_{12}, v_{21}, lie in the range $0.2 < v_{12}, v_{21} < 0.5$ (these values bracket published values for polymers). We know that $v_{12}/E_1 = v_{21}/E_2$, so if $E_1 > E_2$, then $v_{12} > v_{21}$. We can choose $v_{21} = 0.2$, then calculate v_{12} for a lower choice on v_{12}; then we can choose $v_{12} = 0.5$ for an upper choice on v_{12}. We then make our computation for these two choices of v_{12} and assume this brackets the actual result. This procedure was used for the following values of PET:

$$E_1 = 3.47 \text{ GPa}, \qquad E_2 = 5.3 \text{ GPa}, \qquad E_{(45)} = 4.07 \text{ GPa}.$$

$$\text{Case I.} \quad v_{21} = 0.31, \quad v_{12} = 0.20. \qquad (4.88)$$

$$\text{Case II.} \quad v_{21} = 0.50, \quad v_{12} = 0.33.$$

These values with Eq. (4.52) yield the results shown in Fig. 4.24. As shown there, this rather crude method of bracketing the Poisson ratio yields a quite satisfactory bracket for u_r.

Assumption of Elliptical $E(\phi)$ Curve

Results presented in Chapter 3 indicate that the experimental instantaneous elastic $E(\phi)$ curve for PET is essentially elliptical. This means that

$$\frac{2}{E_{(45)}^2} = \frac{1}{E_1^2} + \frac{1}{E_2^2}. \qquad (4.89)$$

It was previously shown, in the discussion after Eq. (4.52), that $u_r(b, \theta)$ is

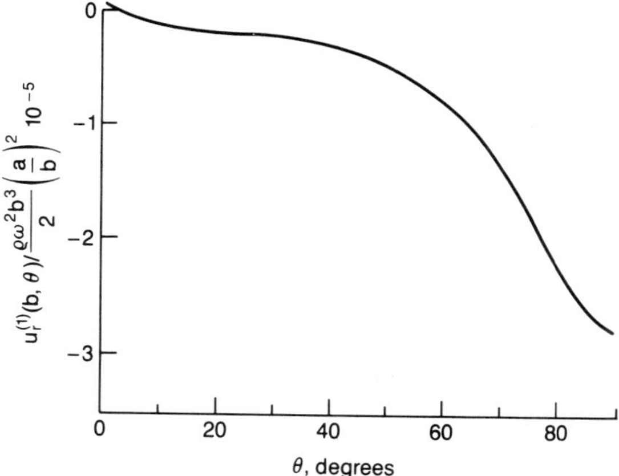

Fig. 4.24. First-order correction $u_r^{(1)}(b, \theta)$ due to inner clamping of spinning orthotropic disk (Bogy and Talke, 1984).

elliptical if $2S_{12} + S_{66} = S_{11} + S_{22}$ or, from Lekhnitskii (1968), if

$$\frac{1}{G_{12}} - \frac{2\nu_{12}}{E_1} = \frac{1}{E_1} + \frac{1}{E_2}, \tag{4.90}$$

which, in view of Eq. (4.87), is the condition

$$\frac{2}{E_{(45)}} = \frac{1}{E_1} + \frac{1}{E_2}. \tag{4.91}$$

Now Eqs. (4.89) and (4.91) are not the same, so it follows that an exactly elliptical $E(\phi)$ curve does not produce an exactly elliptical deformation $u_r(b, \theta)$. However, $E_{(45)}$ computed from Eq. (4.89) is quite close to that computed from Eq. (4.91). For example, if $E_1 = 1.5$, $E_2 = 1$, then Eq. (4.89) gives $E_{(45)} = 1.18$ and Eq. (4.91) gives $E_{(45)} = 1.20$. It follows that materials for which the $E(\phi)$ curve is elliptical will have an almost elliptical $u_r(b, \theta)$ curve as computed from Eq. (4.52). If we assume that the $u_r(b, \theta)$ curve is elliptical, then the maximum deviation in u_r is obtained from the values at $\theta = 0$, $\pi/2$, and Eq. (4.52) gives

$$\max_\theta \Delta u_r(b, \theta) = \frac{\rho \omega^2 b^3}{3} \left(\frac{1}{E_1} - \frac{1}{E_2} \right)(1 - A), \tag{4.92}$$

where A is given in Eq. (4.49), but can also be written as

$$A = I_1/(3I_1 + I_2),$$

$$I_1 = S_{11} + S_{22} + 2S_{12} = \frac{1}{E_1} + \frac{1}{E_2} - \frac{2\nu_{12}}{E_1}, \tag{4.93}$$

$$I_2 = S_{66} - 4S_{12} = \frac{1}{G_{12}} + \frac{4\nu_{12}}{E_1},$$

where I_1, I_2 are invariants of the elasticity tensor under a change of coordinates (Lekhnitskii, 1981). I_1 is associated with total volume change and it can be shown that $I_1 > 0$. Clearly, $I_2 > 0$ also, hence $A < \frac{1}{3}$.

From Eq. (4.49), we also have

$$A = \left(\frac{1}{E_1} + \frac{1}{E_2} - \frac{2v_{12}}{E_1}\right) \bigg/ \left[3\left(\frac{1}{E_1} + \frac{1}{E_2}\right) - \frac{2v_{12}}{E_2} + \frac{1}{G}\right], \tag{4.94}$$

$$A = \frac{1}{4} - \frac{v_{12}}{2(1 + E_1/E_2)}. \tag{4.95}$$

Based on this expression, we find

$$\tfrac{1}{8} < A < \tfrac{1}{4}. \tag{4.96}$$

The values of E_1 and E_2 in Eq. (4.88) were used with Eqs. (4.92) and (4.96) to compute $\max_{\theta} \Delta u_r(b, \theta)$ with the results

$$3.40 < (2 \times 10^7) \max_{\theta} \Delta u_r(b, \theta)/\rho\omega^2 b^3 < 3.96. \tag{4.97}$$

The corresponding values from Fig. 4.25 based on the assumptions on the Poisson ratios are 3.66 for Case I and 3.54 for Case II, both within the range obtained here.

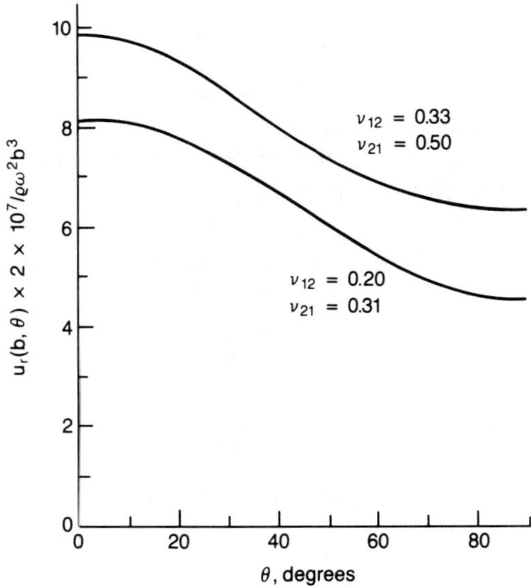

Fig. 4.25. Bounds on $u_r(b, \theta)$ using E_1 (=3.47 GPa) and E_2 (=5.31 GPa) values for PET with bounding assumptions on the Poisson ratio with bounding v_{12}, v_{21} (Bogy and Talke, 1984).

4.2.2. Viscoelasticity Solution

The problem analyzed by Bogy and Talke (1984) is the quasi-static creep of the spinning viscoelastic disk of radius b, which has orthotropic symmetry. We again use a coordinate frame that is fixed in the disk and aligned with the symmetry axes. With reference to this frame, the stress equations of motion are

$$\frac{\partial \sigma_{11}(t)}{\partial x_1} + \frac{\partial \sigma_{12}(t)}{\partial x_2} + \rho \omega^2 x_1 \delta(t) = \rho \frac{\partial^2 u_1(t)}{\partial t^2},$$

$$\frac{\partial \sigma_{12}(t)}{\partial x_1} + \frac{\partial \sigma_{22}(t)}{\partial x_2} + \rho \omega^2 x_2 \delta(t) = \rho \frac{\partial^2 u_2(t)}{\partial t^2}, \tag{4.98}$$

in which $\delta(t)$ is the step function. The constitutive relations for a linearly viscoelastic material and time-dependent strain–displacement relations are presented in Chapter 3. The Laplace transform of the equations of motion, constitutive relations, and strain–displacement relations are

$$\frac{\partial \bar{\sigma}_{11}(s)}{\partial x_1} + \frac{\partial \bar{\sigma}_{12}(s)}{\partial x_2} = -\frac{\rho \omega^2 x_1}{s},$$

$$\frac{\partial \bar{\sigma}_{12}(s)}{\partial x_1} + \frac{\partial \bar{\sigma}_{22}(s)}{\partial x_2} = -\frac{\rho \omega^2 x_2}{s}, \tag{4.99}$$

$$\bar{\varepsilon}_{11}(s) = s\bar{S}_{11}(s)\bar{\sigma}_{11}(s) + s\bar{S}_{12}(s)\bar{\sigma}_{22}(s),$$

$$\bar{\varepsilon}_{22}(s) = s\bar{S}_{12}(s)\bar{\sigma}_{11}(s) + s\bar{S}_{22}(s)\bar{\sigma}_{22}(s), \tag{4.100}$$

$$\bar{\varepsilon}_{12}(s) = \tfrac{1}{2}s\bar{S}_{66}(s)\bar{\sigma}_{12}(s),$$

$$\bar{\varepsilon}_{11}(s) = \frac{\partial \bar{u}_1(s)}{\partial x_1}, \qquad \bar{\varepsilon}_{22}(s) = \frac{\partial \bar{u}_2(s)}{\partial x_2},$$

$$2\bar{\varepsilon}_{12}(s) = \frac{\partial \bar{u}_1(s)}{\partial x_2} + \frac{\partial \bar{u}_2(s)}{\partial x_1}. \tag{4.101}$$

Because of the orthotropic symmetry and our special choice of coordinates, only four independent $\bar{S}_{11}(s)$, $\bar{S}_{12}(s)$, $\bar{S}_{22}(s)$, and $(1/2)\bar{S}_{66}(s)$ enter into the two-dimensional equations (4.100).

4.2.2.1. Solution

For quasi-static viscoelasticity boundary value problems, the method of solution makes use of the elastic–viscoelastic correspondence principle. The correspondence is between the elasticity solution of the problem and the Laplace transform of the viscoelasticity solution. Some care must be exercised in solving Eqs. (4.99)–(4.101) for $\bar{\sigma}_{ij}$, $\bar{\varepsilon}_{ij}$, $\bar{u}_i$. In particular, we cannot obtain $\bar{\sigma}_{ij}$ from Eqs. (4.48) and (4.49) by merely replacing S_{ij} by $s\bar{S}_{ij}(s)$. We must retrace the derivation of Eqs. (4.48)–(4.50).

Equation (4.99) is satisfied identically if we introduce a function ϕ by

$$\bar{\sigma}_{11}(s) = \frac{\partial^2 \phi(s)}{\partial x_1^2} - \frac{\rho \omega^2}{2s}(x_1^2 + x_2^2 - b^2),$$

$$\bar{\sigma}_{22}(s) = \frac{\partial^2 \phi(s)}{\partial x_2^2} - \frac{\rho \omega^2}{2s}(x_1^2 + x_2^2 - b^2), \qquad (4.102)$$

$$\bar{\sigma}_{12}(s) = -\frac{\partial^2 \phi(s)}{\partial x_1 \partial x_2}.$$

The $\bar{\varepsilon}_{ij}(s)$ expressed in terms of $\phi(s)$ follow from the use of Eq. (4.102) in Eq. (4.100). In order to determine ϕ, the resulting strain components are substituted into the strain compatibility equation

$$\frac{\partial^2 \bar{\varepsilon}_{11}(s)}{\partial x_2^2} + \frac{\partial^2 \bar{\varepsilon}_{22}(s)}{\partial x_1^2} = 2\frac{\partial^2 \bar{\varepsilon}_{12}(s)}{\partial x_1 \partial x_2}, \qquad (4.103)$$

to yield

$$s\bar{S}_{22}\frac{\partial^4 \phi}{\partial x_1^4} + s(2\bar{S}_{12} + \bar{S}_{66})\frac{\partial^4 \phi}{\partial x_1^2 x_2^2} + s\bar{S}_{11}\frac{\partial^4 \phi}{\partial x_2^4} = \rho\omega^2(\bar{S}_{11} + 2\bar{S}_{12} + \bar{S}_{22}). \qquad (4.104)$$

The particular solution of this equation is

$$\phi = \frac{\rho\omega^2}{8s} B(s)(x_1^2 + x_2^2 - b^2)^2, \qquad (4.105)$$

where

$$B(s) = \frac{\bar{S}_{11}(s) + 2\bar{S}_{12}(s) + \bar{S}_{22}(s)}{3\bar{S}_{11}(s) + 2\bar{S}_{12}(s) + \bar{S}_{66}(s) + 3\bar{S}_{22}(s)}. \qquad (4.106)$$

The use of Eq. (4.105) in Eq. (4.102) gives

$$\bar{\sigma}_{11}(s) = \frac{\rho\omega^2}{2s}\{[1 - B(s)](b^2 - x_1^2 - x_2^2) + 2B(s)x_2^2\},$$

$$\bar{\sigma}_{22}(s) = \frac{\rho\omega^2}{2s}\{[1 - B(s)](b^2 - x_1^2 - x_2^2) + 2B(s)x_1^2\}, \qquad (4.107)$$

$$\sigma_{12}(s) = -\frac{\rho\omega^2}{s} B(s)x_1 x_2.$$

Upon comparing Eqs. (4.100), (4.101), (4.106), (4.107) with Eqs. (4.45), (4.44), (4.49), (4.48), we can record $\bar{u}_i(s)$ immediately as

$$\bar{u}_1(s) = \frac{\rho\omega^2 b^3}{2}\left\{[\bar{S}_{11}(s) + \bar{S}_{12}(s)][1 - B(s)]\left(\frac{x_1}{b} - \frac{x_1^3}{3b^3} - \frac{x_1 x_2^2}{b^3}\right)\right.$$

$$\left. + 2B(s)\left[\bar{S}_{12}(s)\frac{x_1^3}{3b^3} + \bar{S}_{11}(s)\frac{x_1 x_2^2}{b^3}\right]\right\},$$

$$\bar{u}_2(s) = \frac{\rho\omega^2 b^3}{2}\left\{[\bar{S}_{12}(s) + \bar{S}_{22}(s)][1 - B(s)]\left(\frac{x_2}{b} - \frac{x_2^3}{3b^3} - \frac{x_1^2 x_2}{b^3}\right)\right.$$

$$\left. + 2B(s)\left[\bar{S}_{12}(s)\frac{x_2^3}{3b^3} + \bar{S}_{22}(s)\frac{x_1^2 x_2}{b^3}\right]\right\}. \tag{4.108}$$

The physical displacement components $u_\alpha(\underline{x}, t)$ are obtained from the Laplace inversion formula and Eq. (4.108)

$$u_\alpha(\underline{x}, t) = \frac{1}{2\pi i}\int_{c-i\infty}^{c+i\infty} \bar{u}_\alpha(\underline{x}, s)e^{st}\, ds, \qquad \alpha = 1, 2, \tag{4.109}$$

where c is chosen so that the path of integration is to the right of all singularities of $\bar{u}_\alpha(\underline{x}, s)$ in the complex s-plane.

The creep compliance expression corresponding to Eq. (4.86) is

$$S(t; \phi) = S_{11}(t)\cos^4\phi + [S_{66}(t) - 2S_{12}(t)]\sin^2\phi\cos^2\phi + S_{22}(t)\sin^4\phi. \tag{4.110}$$

If we assume $S(t, \theta)$ is elliptical, we obtain, corresponding to Eq. (4.92),

$$\max_\theta \Delta u_r(b, \theta, t) = u_1(b, 0, t) - u_2(0, b, t)$$

$$= \frac{\rho\omega^2 b^3}{6\pi i}\int_{c-i\infty}^{c+i\infty} [\bar{S}_{11}(s) - \bar{S}_{22}(s)][1 - B(s)]e^{st}\, ds. \tag{4.111}$$

It is apparent from Eqs. (4.106) and (4.111) that the computation of $\max \Delta u_r(t)$ requires the knowledge of the four creep compliances $S_{11}(t)$, $S_{22}(t)$, $S_{12}(t)$, $S_{66}(t)$.

4.2.2.2. Estimates for Large Times

A suitable mathematical form of the creep functions $S_{11}(t)$, $S_{22}(t)$ for PET is

$$S_{ij}(t) = S_{ij}^0 + k_{ij}t^{p_{ij}}, \qquad ij = 11, 22. \tag{4.112}$$

We assume the form Eq. (4.112) also describes $S_{12}(t)$ and $S_{66}(t)$. Then, the Laplace transforms are (Doetsch, 1961)

$$\bar{S}_{ij}(s) = \frac{S_{ij}^0}{s} + K_{ij}\Gamma(p_{ij} + 1)\frac{1}{s^{p_{ij}+1}}, \tag{4.113}$$

where $\Gamma(\ )$ is the gamma function. Let

$$m_{ij} = k_{ij}\Gamma(p_{ij} + 1), \tag{4.114}$$

then $B(s)$ in Eq. (4.106) has the form

$B(s)$

$$= \frac{S_{11}^0 + 2S_{12}^0 + S_{22}^0 + m_{11}s^{-p_{11}} + 2m_{12}s^{-p_{12}} + m_{22}s^{-p_{22}}}{3S_{11}^0 + 2S_{12}^0 + S_{66}^0 + 3S_{22}^0 + 3m_{11}s^{-p_{11}} + 2m_{12}s^{-p_{12}} + m_{66}s^{-p_{12}} + m_{66}s^{-p_{66}} + 3m_{22}s^{-p_{22}}}, \tag{4.115}$$

and

$$\bar{S}_{11}(s) - \bar{S}_{22}(s) = (S_{11}^0 - S_{22}^0)s^{-1} + m_{11}s^{-p_{11}-1} - m_{22}s^{-p_{22}-1}. \quad (4.116)$$

Now let $I(s)$ be defined as

$$I(s) = [\bar{S}_{11}(s) - \bar{S}_{22}(s)][1 - B(s)]. \quad (4.117)$$

Then from Eqs. (4.115)–(4.117), we have the following possibilities for the limit of $I(s)$ as $s \to 0$ ($p_{11} > p_{22}$ by choice of axes)

$$I(s) \to \begin{cases} \frac{2}{3}m_{11}s^{-p_{11}-1} & \text{if } p_{11} > p_{12}, p_{66}, p_{22}, \\[2mm] \dfrac{m_{11}^2}{m_{12}}s^{p_{12}-2p_{11}-1} & \text{if } p_{12} > p_{11} > p_{66}, p_{22}, \\[2mm] m_{11}s^{-p_{11}-1} - \dfrac{m_{11}^2}{m_{66}}s^{p_{66}-2p_{11}-1} & \text{if } p_{66} > p_{11} > p_{12}, p_{22}. \end{cases} \quad (4.118)$$

From this limit and Doetsch (1961), we obtain from Eq. (4.111) as $t \to \infty$

$$\frac{\max \Delta u_r}{\rho\omega^2 b^3/3}$$

$$\to \begin{cases} \frac{2}{3}k_{11}t^{p_{11}} & \text{if } p_{11} > p_{12}, p_{66}, p_{22}, \\[3mm] \dfrac{[k_{11}\Gamma(p_{11}+1)]^2}{k_{12}\Gamma(p_{12}+1)\Gamma(-p_{12}+2p_{11}+1)}t^{2p_{11}-p_{12}} & \text{if } p_{12} > p_{11} > p_{66}, p_{22}, \\[3mm] k_{11}t^{p_{11}} - \dfrac{[k_{11}\Gamma(p_{11}+1)]^2}{k_{66}\Gamma(p_{66}+1)}t^{2p_{11}-p_{66}} & \text{if } p_{66} > p_{11} > p_{12}, p_{22}. \end{cases}$$

$$(4.119)$$

From the limited experimental evidence on creep of oriented polymers, it is expected that $0 < p_{ij} < 1$ and also $p_{11} > p_{12}, p_{66}$, in which case,

$$\frac{\max \Delta u_r}{\rho\omega^2 b^3/3} \to \frac{2}{3}k_{11}t^{p_{11}} \qquad \text{as} \quad t \to \infty. \quad (4.120)$$

Typical values for the principal creep compliances in tension for PET obtained at the constant temperature of 36° C are (in units of 1/MPa for t in minutes)

$$\begin{aligned} S_{11}^0 &= 2.42 \times 10^{-4}, & k_{11} &= 3.68 \times 10^{+10}, & p_{11} &= 0.310, \\ S_{22}^0 &= 1.85 \times 10^{-4}, & k_{22} &= 7.98 \times 10^{-10}, & p_{22} &= 0.243. \end{aligned} \quad (4.121)$$

The quantities S_{ij}^0, p_{ij} are independent of temperature and the dependence of k_{ij} on temperature can be represented by

$$k_{ij}(T) = k_{ij}(36) \times 10^{0.1 p_{ij}(T-36)}. \quad (4.122)$$

Finally, we use Eqs. (4.120) and (4.121) to make the $t = 5$ years estimate of $\max_\theta \Delta u_r(b, \theta, t)$ for the typical configuration for which

$$\rho = 1400 \text{ kg/m}^3, \qquad \omega = 1500 \text{ rpm}, \qquad b = 178 \text{ mm}. \quad (4.123)$$

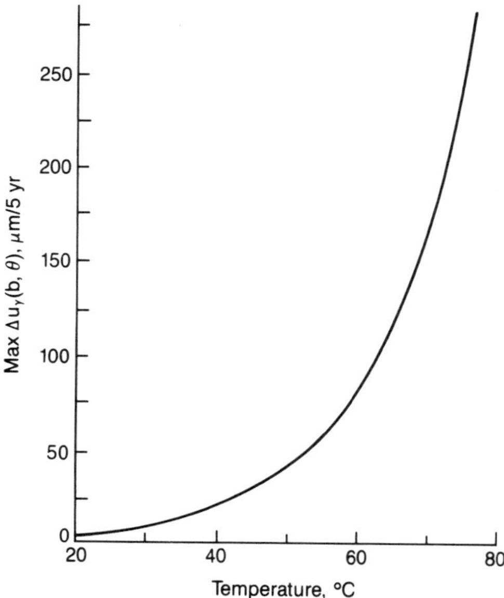

Fig. 4.26. Maximum creep runout as a function of temperature for a flexible disk with $b = 175$ mm and spinning at 1500 rpm (Bogy and Talke, 1984).

We obtain these values

$$\max_{\theta} \Delta u_r(b, \theta, t) \text{ (in } \mu\text{m)} \sim 6.254 \times 10^6 \, k_{11} t^{p_{11}}, \qquad (4.124)$$

which, on setting $t = 5$ years $- 2.62 \times 10^6$ min gives the results shown in Fig. 4.26.

4.2.3. Summary

Based on the work by Bogy and Talke (1984), an analysis is presented to predict the nonsymmetric creep over 5 years of an initially circular data track on a biaxially oriented PET annular disk spinning at a constant angular speed. We first obtained the elastic deformation of an orthotropic *continuous* disk and computed the nonsymmetric radial displacement for a particular orthotropic material of known elastic constants. Then, an approximate solution was obtained for a similar *annular* disk clamped at its inside radius. It was found that the deformation at the outer region of the disk was diminished by central clamping as expected, and the effect of the central clamping could be neglected for inner to outer radius ratios of less than $\frac{1}{7}$. Next, the elastic solution for the continuous disk was studied in order to obtain reliable estimates when only partial information is available for the elastic constants.

The steady isothermal creep solution was then obtained for an orthotropic viscoelastic continuous disk. A large time one-term estimate of the maximum radial runout at the outer radius of the disk was derived, which required for its evaluation only the two principal creep functions. This estimate was evaluated at $t = 5$ years for various fixed temperatures and the results are shown in Fig. 4.26. *It shows the creep runout to be less than* 25 μm *for temperatures less than* 40° C.

It should be realized that the results presented in Fig. 4.26 do not include distortions that may occur in the disk as a result of humidity changes, temperature changes, or the creep shrinkage due to the relaxation of the previously unrelieved stresses caused by the biaxial stretching of the PET (Chapter 6). These distortions also need to be carefully evaluated for the disk configuration. Various annealing processes appear to be available for controlling the latter, if necessary (see Chapters 2 and 3).

References

Altmann, H. C. (1968). Formulas for computing the stresses in center wound rolls. *Tech. Assoc. Pulp Paper Ind.* **51** (4), 176–179.

Anonymous (1959). "Magnetic Tape Study." General Kinetics, Inc., Contract No. D18-119-SC42, Arlington, Virginia.

Bertram, N. and Eshel, A. (1980). "Recording Media Archival Attributes (Magnetic)." Final Technical Report No. RADC-TR-80-123. Rome Air Development Center, Rome, New York.

Bhushan, B. (1990). "Tribology and Mechanics of Magnetic Storage Devices." Springer-Verlag, New York.

Bhushan, B. and Connolly, D. (1986). Viscoelastic properties of poly(ethylene terephthalate). *ASLE Trans.* **29**, 489–499.

Bhushan, B., Hahn, F. W., Sharma, B. S., and Connolly, D. (1984a). Long-term reliability of magnetic tapes for digital recording. In "Tribology and Mechanics of Magnetic Storage Systems" (B. Bhushan et al., eds.), pp. 132–147. SP-16, ASLE, Park Ridge, Illinois.

Bhushan, B., Heinrich, J. C., and Connolly, D. (1984b). Orthotropic viscoelastic behavior of polyethylene terephthalate film under plane-stress conditions. In "Tribology and Mechanics of Magnetic Storage Systems" (B. Bhushan et al., eds.), pp. 158–171. SP-16, ASLE, Park Ridge, Illinois.

Bogy, D. B., Bugdayci, N., and Talke, F. E. (1979). Experimental determination of creep functions for thin orthotropic films. *IBM J. Res. Develop.* **23**, 450–458.

Bogy, D. B. and Talke, F. E. (1984). Creep of a rotating orthotropic polymer circular disk. In "Tribology and Mechanics of Magnetic Storage Systems" (B. Bhushan et al., eds.), pp. 115–125. SP-16, ASLE, Park Ridge, Illinois.

Connolly, D. and Winarski, D. J. (1984). Stress analysis of wound magnetic tapes. In "Tribology and Mechanics of Magnetic Storage Systems" (B. Bhushan et al., eds.), pp. 172–182. SP-16, ASLE, Park Ridge, Illinois.

Darlington, M. W. and Saunders, D. W. (1974). The anisotropy of creep behavior in oriented thermoplastics. *Poly. Eng. Sci.* **14**, 616–620.

Doetsch, G. (1961). "Guide to the Applications of Laplace Transforms." D. Van Nostrand, New York.

Eshel, A., Baker, S., Hartman, A., and Orcutt, F. K. (1984). Mechanical aspects of archival storage of magnetic tape. In "Tribology and Mechanics of Magnetic Storage Systems" (B. Bhushan et al., eds.), pp. 148–157. SP-16, ASLE, Park Ridge, Illinois.

Frye, K. G. (1967). Winding variables and their effect on roll hardness and roll quality. *TAPPI* **50** (7).

Graham, S., Richard, M., and Winarski, D. (1981). Tape reel hub. *IBM Tech. Disclosure Bull.* **24** (5), October.

Green, A. E. and Zerna, W. (1954). "Theory of Elasticity." Oxford University Press, Oxford, U.K.

Halpin, J. C. and Pagano, N. J. (1968). Observation on linear anisotropic viscoelasticity. *J. Composite Mat.* **2**, 68–80.

Heinrich, J. C., Connolly, D., and Bhushan, B. (1986). Axisymmetric, finite-element analysis of stress relaxation in wound magnetic tapes. *ASLE Trans.* **29**, 75–84.

Hussain, S. M., Farrell, W. R., and Gunning, J. R. (1962). Most paper in the roll is in unstable condition. *Can. Pulp Paper Ind.* **21**, 52.

Lekhnitskii, S. G. (1968). "Anisotropic Plates." Gordon and Breach, New York.

Lekhnitskii, S. G. (1981). "Theory of Elasticity of an Anisotropic Body." Mir, Moscow, USSR.

Monk, D. W., Lautner, W. K., and McMullen, J. F. (1975). Internal stresses within rolls of cellophane. *TAPPI* **58** (8), 152–155.

Mukherjee, S. (1974). Time base errors in video tape packs. *ASME J. Appl. Mech.* **96**, 625–630.

Owen, R. J. (1971). "Magnetic Head/Tape Interface Study for Satellite Tape Recorders." Technical Report, Vol. 2, IITRI Proj. No. E6134, Contract No. NAS5-11622, NASA Goddard Space Flight Center, Greenbelt, Maryland, pp. 227–242.

Pfeiffer, J. D. (1966). Internal pressures in a wound roll of paper. *TAPPI* **49** (8), 342–347.

Pfeiffer, J. D. (1967). The mechanics of a rolling nip on paper webs. Technical Report. Beloit-Eastern Technical Center, Downingtown, Pennsylvania.

Richard, M. and Winarski, D. (1982). "Tape Reel Hub." U.S. Patent 4,350,309, September 28.

Tang, S. (1969). Elastic stresses in rotating anisotropic disks. *Int. J. Mech. Sci.* **11**, 509–517.

Taylor, R. L., Pister, K. S., and Goudreau, G. L. (1970). Thermomechanical analysis of viscoelastic solids. *Inst. J. Num. Meth. Eng.* **2**, 45–59.

Timoshenko, S. P. and Goodier, J. N. (1970). "Theory of Elasticity," 3rd ed. McGraw-Hill, New York.

Tramposch, H. (1963). "Internal Forces in a Wound Reel of Magnetic Tape." Technical Rept. TR00.973, IBM Data Systems Division, Poughkeepsie, New York.

Tramposch, H. (1965). Relaxation of internal forces in a wound reel of magnetic tape. *ASME J. Appl. Mech.* **32**, 865–873.

Tramposch, H. (1967). Anisotropic relaxation of internal forces in a wound reel of magnetic tape. *ASME J. Appl. Mech.* **34**, 888–894.

Umanskii, E. S., Kryuchov, V. V., and Rakovskii, V. A. (1978). Determination of the stressed state of a coil of magnetic tape. *Problemy Prochnosti.* **3**, March, 83–85.

Umanskii, E. S., Kryuchov, V. V., and Shidlovskii, N. S. (1981). Estimating the effect of the temperature factor on the bearing capacity of a reel of magnetic tape. *Problemy Prochnosti.* **8**, August, 62–65.

Waites, J. B. (1982). Care, handling, and management of magnetic tape. In "Magnetic Tape Recording for the Eighties," pp. 45–59. NASA, Washington, D.C.

Yagoda, H. P. (1980a). Resolution of a core problem in wound rolls. *ASME J. Appl. Mech.* **47**, 847–856.

Yagoda, H. P. (1980b). Centrifugally induced stresses within center-wound rolls— Part I. *Mech. Res. Comm.* **7** (3), 181–193.

Zienkiewicz, O. C. (1977). "The Finite Element Method," 3rd ed. McGraw-Hill, London.

CHAPTER 5

Long-Term Reliability of Magnetic Tapes

In high-density, high data-rate tape drives, a magnetic tape is transported over a stationary or rotating (audio, video, and data-processing) read/write head at head–tape separations ranging from 0.1 μm to 0.2 μm. For optimum magnetic-recording performance, the entire tape surface must be maintained at a fixed head-tape separation. Any instantaneous variations in this separation will lead to undesirable modulation and possible extinction of the magnetic signal. Localized increases in head-tape separation are the principal cause of permanent or temporary read and write errors. Furthermore, it is desirable that a constant and uniform tension (along the length and across the width of the tape) and speed be maintained while the tape is running in a tape drive. Any local variations in tension and/or speed can adversely affect the read/write performance (reduction in magnetic amplitude[1] because of an increase in the head–tape separation) of the tape drive. The machine designer can sense these variations and design appropriate servo mechanisms to maintain constant tape tension and speed with some success. The overall reliability of the machine can be improved if the mechanical integrity of the recording tape while it is manufactured, stored, transported, and operated is maintained.

Magnetic tapes are wound onto a reel under a constant or tapered tension, which causes circumferential (hoop) and radial (interlayer pressure) stresses in the tape (Chapter 4). The tape may be at elevated temperatures, up to 52° C, during shipping, and typically at 22° C and 50% relative humidity (RH) during storage in a computer room. Some archival tapes may be stored for 10 years or longer. Due to the viscoelastic nature of magnetic tapes, the

---

[1] The magnetic signal amplitude on a given track has to drop to roughly 15–25% of its nominal amplitude to cause an error in an IBM 3480/3490 data-processing drive. Furthermore, the signal drop has to occur for a certain number (typically 8) of bits before it is counted as an error. Most drives are equipped with automatic gain control (AGC) to boost signal, if necessary, therefore most errors are temporary errors. No hard (nonrecoverable) errors are expected to occur in the life of a drive.

stresses in wound tape relax with time. Storage at high temperatures and humidities and thinner tapes accelerate this effect (Bhushan et al., 1984a; Bhushan and Connolly, 1986; Heinrich et al., 1986, Chapter 3), which results in low hoop stress and interlayer pressure. If the interlayer pressure (radial stress) has relaxed significantly and there is low friction between the wraps, it is likely that tape slippage (tangential motion of the tape relative to the hub) will occur during rewinding or handling. This is referred to as interlayer slip (ILS) and involves a dynamic balance between the rewinding tension and the relaxed pressure through the frictional forces between the wraps. Another manifestation of stress relaxation inside the wound tape is the creation of loose wraps. During normal machine operation, under constant tape tension, the tape may undergo an instantaneous speed variation (ISV) caused by sudden tape acceleration in the region of these loose wraps. Another source of ISVs is tape vibrations excited by a damaged edge of the tape or a tape-path component.

If, as the tape is wound, nonuniform stresses occur in localized regions, localized viscoelastic tape distortions will form with time, which we will call out-of-plane distortions (OPD). (A failure rate model to assess viscoelastic-deformation-related defects is presented in Appendix 5.C.) These distortions result in a nonuniform tension across the width of the tape. If these localized viscoelastic deformations are excessive, they can lead to a permanent data loss due to the increased head–tape separation. For given winding conditions, a nonuniform stress field may be generated due to (a) an uneven tape-stack profile and (b) the mechanical print-through (from an irregular hub surface or entrapped contaminants). Under uniform applied tension, the tape does exhibit some viscoelastic elongation, whose magnitude is usually quite small and does not cause undesirable effects.

Finally, tape-stacking irregularities, such as stack shift or staggered wrap, may lead to damage of the tape-edge tracks caused by the bending of the unsupported regions of tape. To understand how staggered wraps are formed, we must know the winding dynamics, particularly the role of the hydro-dynamic air film at the inlet zone, which facilitates the wrap shift. Then, once they are formed, a consideration of the stresses they have undergone, and of the rate at which these stresses relax leading to viscoelastic deformation in the tape, is necessary to quantify the resulting damage.

The type of problem that might occur depends on the tape and reel, manufacturing process, and the subsequent handling of the tape. The rate and severity of the problems also depend on the viscoelastic behavior (deformation kinetics) of the tape. Proper selection of magnetic-tape materials and storage conditions (temperature and humidity) can minimize viscoelastic deformation hence magnetic errors. Tape-reliability problems are discussed in detail under the following five categories: (a) interlayer slip, (b) instantaneous speed variations, (c) uneven tape-stack profile, (d) mechanical print-through, and (e) staggered wraps.

The function of a tape reel is to protect the tape from damage and contamination during handling. Some design guidelines for the tape reel will be presented in a separate section.

5.1. Interlayer Slip (ILS)

When a carefully wound reel of tape is stored, the temperature and humidity may vary with time. These variations may relax the tape's original winding-stress profile with time (because of the viscoelastic nature of the magnetic tape material) and generate areas of low interlayer pressure in the tape stack. In extreme cases, physical voids are created, as illustrated in Fig. 5.1. If the interlayer pressure has relaxed significantly and there is low friction between the wraps, it is likely that interlayer slip, commonly referred to as ILS (tangential motion of the tape relative to the hub), will occur. During subsequent processing of these reels on a tape drive, tensioning of the tape and acceleration of the reel can result in ILS. Although tangential ILS is more common, axial slip may also occur. The magnitude of the tangential tape slip can range from a few micrometers to several centimeters. The control of tape velocity, as well as of tape tension, is complicated by this unexpected motion. Drives that use vacuum columns to regulate tape tension and decouple tape spooling may escape the effects of ILS (Tramposch, 1965, 1967; Owen, 1971; Mukherjee, 1974, 1975; Bertram and Eshel, 1980; Anonymous, 1983; Bhushan et al., 1984b).

Low interlayer pressures, which predispose the tape stack to ILS, are created as follows: the winding of successive layers of tape onto a reel increases the radial compressive stresses in the wraps of tape at and near the hub. The continued inward radial deflection of the hub converts the circumferential tensile strain in the tape, originally caused by winding, to a compressive strain. At the end of winding, the distribution of hoop stress and radial interlayer pressure can approach that shown in Fig. 5.2. The tape, hub, and winding parameters sufficient to obtain this condition can be determined using the analysis presented in Chapter 4. A highly compressible hub, high outer-wrap winding tension, low inner-wrap winding tension, the length of tape stored on a reel, and the entrapment of air during high-speed winding are strong contributors to low interlayer pressure.

The interlayer pressure from winding is further reduced by temperature and humidity cycling and/or storage. An example of the change in interlayer pressure caused by stress relaxation at room conditions was presented in Chapter 4. The interlayer pressure can be reduced when the wound tape is subjected to a temperature or humidity change, and is dependent on the relative value of the coefficients of thermal and hygroscopic expansion of the hub and the tape. This effect is aggravated not only by the magnitude of this differential mismatch, but also by the mismatch in the tape's radial and circumferential coefficients of thermal and hygroscopic expansion (Chapter 4).

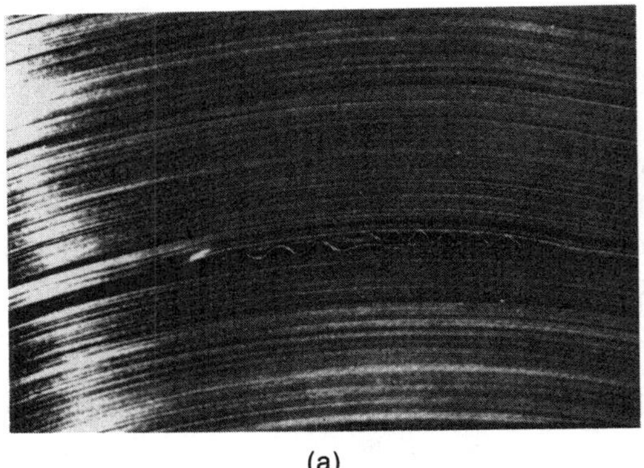

(a)

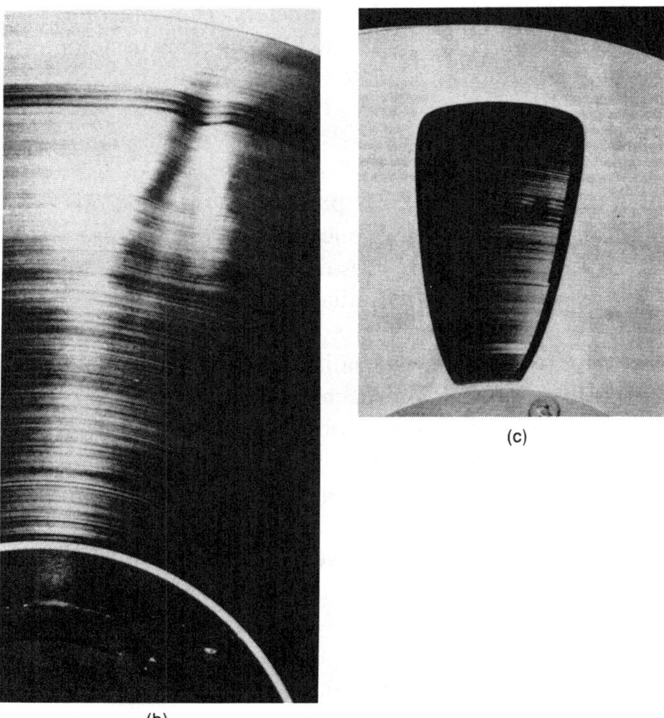

(b)

(c)

Fig. 5.1. Voids in the tape stack leading to (a) cinching, note the complete foldover of one tape strand, (b) spoking, and (c) windowing.

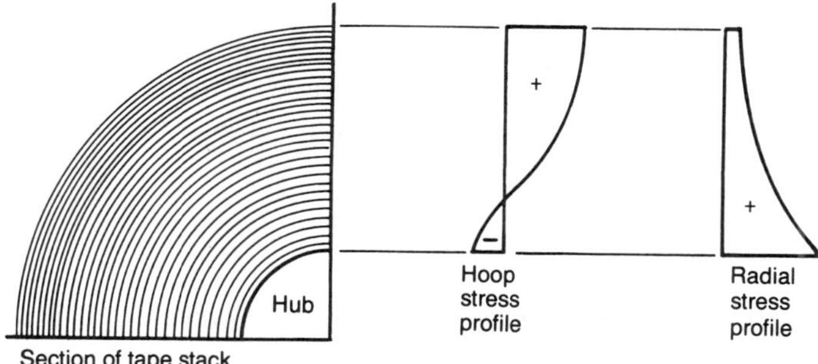

Fig. 5.2. Stress distribution in wound tape.

5.1.1. Various Forms of Tape Distortions

Low interlayer pressures inside the wound tape can also lead to various other forms of tape damage and distortions. Terms that describe these effects are cinching, spoking, windowing, and telescoping.

5.1.1.1. Cinching

When physical voids exist between layers of tape, the upper layer can buckle to fill the void, as shown in Fig. 5.1(a). During ILS, each wave in the tape folds over itself to produce a cyclical series of creases.

5.1.1.2. Spoking

A sudden mechanical shock can cause buckling of the tape stack along several radial lines corresponding to spokes, Fig. 5.1(b). This buckling is facilitated by the combined conditions of low interlayer pressure and compressive hoop stress. Compressive hoop stresses occur when the hub is too compliant and shifts its function to the inner wraps, which are put in compression (Chapter 4). Spoking generally occurs in areas where the hub is of variable thickness and thus compresses nonuniformly. A similar phenomenon on wound paper rolls is called a "star" by that industry (Pfeiffer, 1966; Frye, 1967).

5.1.1.3. Windowing

Loose winding can lead to voids or "windows" in the tape pack, Fig. 5.1(c), especially those packs that are later subjected to temperature and humidity extremes.

5.1.1.4. Telescoping

Although tangential ILS is more common, axial slip (also called pack slip) may take place during tape handling. This relative axial shifting of the following layers in an increasing manner is referred to as telescoping. Possible tape distortions caused by this telescoping motion of the tape stack are edge damage caused by the tape stack striking the reel flange and nonuniform hoop stresses caused by the axial offset. Tape distortions in the form of creases, edge damage, and waviness can produce large head–tape separations that may make recording impossible.

5.1.2. Model for Reel Slippage

We derive the condition for reel slippage. As noted previously, the final stress pattern from time–temperature–humidity effects can include an area of low interlayer pressure. The force preventing ILS between two layers is the product of the interlayer pressure, the contact area, and the coefficient of friction. Consider a reel rotating at an angular speed ω_f with a known tension and radial pressure distributions—$T(r, \omega_f)$ and $P(r, \omega_f)$. If the reel is subjected to an angular acceleration $\dot{\omega} = \alpha$, the reel is expected to slip if there exists some value of the radius r in the range $r_i \leq r \leq r_f$ such that (Bertram and Eshel, 1980)

$$\mu P(r, \omega_f)(2\pi r^2) \pm T(r, \omega_f)r/w \leq \rho\pi(r_f^4 - r^4)\alpha/2, \tag{5.1}$$

where μ is the coefficient of friction between layers; $P(r, \omega_f)$ is the radial interlayer pressure at a particular radius r; $T(r, \omega_f)$ is the circumferential tension at a particular radius r; r is the radius of the tape stack at which slip occurs (the minimum radius that satisfies the equation), r_f and r_i are the outer and inner radii of the tape stack, w is the tape width, and ρ is the tape density. Here the plus applies when the deceleration tends to tighten the reel and the minus applies when the deceleration tends to loosen the reel.

This expression defines the importance of designing systems that have a low ratio of outer to inner hub radii, low reel accelerations, and high interlayer friction. With respect to the latter, tapes are sometimes backcoated with a relatively rough coating that has a higher friction than that of an uncoated polyester substrate.

5.1.3. Methods to Contain ILS

Tapes that show less stress-relaxation behavior and have high interlayer friction and appropriate winding stresses should have a lower tendency toward ILS. Fast winding and/or low tension winding generally cause reel looseness due to air entrapment. One possible means to eliminate interlayer locked air is the use of a packing roller to load the tape against the reel during fast winding.

Several methods which have been proposed to prevent or contain the basic ILS problem in a tape reel are as follows (Geller, 1983):

- Rewind all new tapes prior to use.
- Rewind tapes prior to long-term storage.

- Rewind tapes prior to use if they have been exposed to another environment or stored for a long period of time.
- Rewind periodically (annually or semiannually). Rewind every 3.5 years if the temperature and humidity are controlled.
- Rewind tapes several times before use if they appear to be cinched or spoked.
- Monitor interlayer pressures through tab-pull measurements and rewind accordingly.

The rewindings recommended above are intended to be full length and at low speeds conducive to even windings.

5.1.4. Estimate of Periods Between Rewinds

Based on some experiments, Bertram and Eshel (1980) found that pressure reduction to about 50% of the original level brought the reels to the threshold of slippage for a deceleration of about 400 rad/s^2. Therefore, they suggested that the time to rewind occurs when reel interlayer pressure relaxes to 50% of the original interlayer pressure. They assumed that interlayer pressure relaxes proportionally to the tension level, and the time for the tension to relax 50% of its original value is roughly the same as the time for creep strain in a constant-tension creep test to reach 50% of the elastic strain. Finally, based on creep data, they assumed that total strain in a creep test behaves with time as

$$\varepsilon \sim t^{0.07}. \tag{5.2a}$$

Using the above and assuming that if the creep strain $\Delta\varepsilon_1$ is known at a given time t_1, then the time $t_{50\%}$ for the creep strain to reach 50% of the elastic strain, at the same temperature, may be found by the extrapolation

$$t_{50\%} = t_1 \left[\frac{(1/E) + (0.5/E)}{(1/E) + \Delta\varepsilon_1/\sigma} \right]^{1/0.07}, \tag{5.2b}$$

where $E\ (=\varepsilon(t)/\sigma)$ is the relaxation modulus and σ is the applied stress in a creep test. Various experimental results have been extrapolated according to Eq. (5.2b). The resulting (conservative) estimates are shown in Fig. 5.3. Based on this figure, it is safe to retain a reel wound at 18° C (65° F) for at least 3.5 years without rewinding, provided that the environment is maintained at 18° C and constant relative humidity. At 24° C (75° F), the safe period is reduced to about 10 months.

5.1.4.1. Environment for Archival Storage

Bertram and Eshel and others (Bhushan, 1990) have shown that a polyester–polyurethane binder system commonly used for media goes through a hydrolytic degradation which is accelerated at high temperature/humidity,

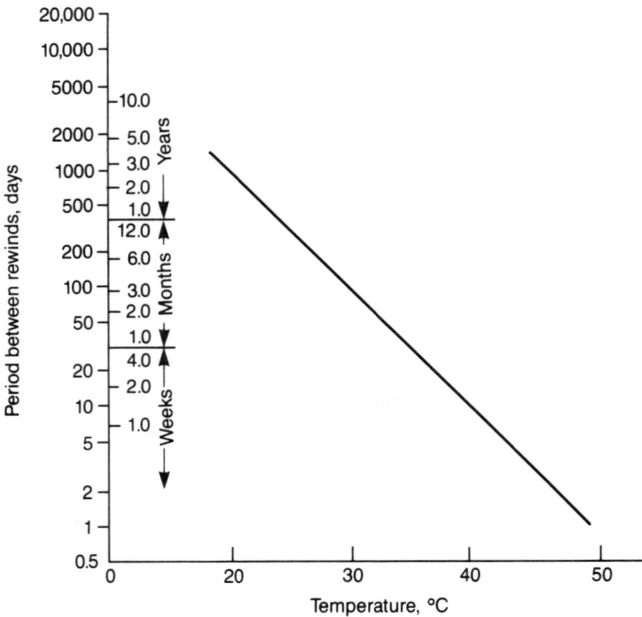

Fig. 5.3. Estimated period between rewinds as a function of storage temperature based on a nonlinear extrapolation of time for the total strain to reach 150% of the elastic strain ($\varepsilon \sim t^{0.07}$) (Bertram and Eshel, 1980).

Fig. 5.4. In order to ensure minimal hydrolysis at 18° C, a humidity lower than 40% must be maintained. Therefore, Bertram and Eshel (1980) recommended the environment for tape archival storage as $18 \pm 2°$ C ($65 \pm 3°$ F) and $40 \pm 5\%$ RH. If this environment is carefully maintained, a rewind frequency of once in 3.5 years is recommended. In the case of environmental fluctuations, it is recommended that unrecorded test reels of the type and size used in the particular archival storage facility be stored semimonthly with embedded friction tabs. A semimonthly check of the stored test reels would indicate the date range of reels requiring tension restoration.

5.2. Instantaneous Speed Variations (ISV)

Large variations in tape speed having millisecond wavelengths are called "instantaneous speed variations" (ISV), also called "high frequency flutter" or "frequency modulation (FM) noise." The magnitude of these speed variations ranges from 5% to 40% of nominal. ISVs occur most often when there is a loose wrap of tape in the reel or when the tape is excited into a vibration mode by a damaged edge of the tape or a tape path component.

Layers in a wound tape reel which are stress relaxed during storage can get

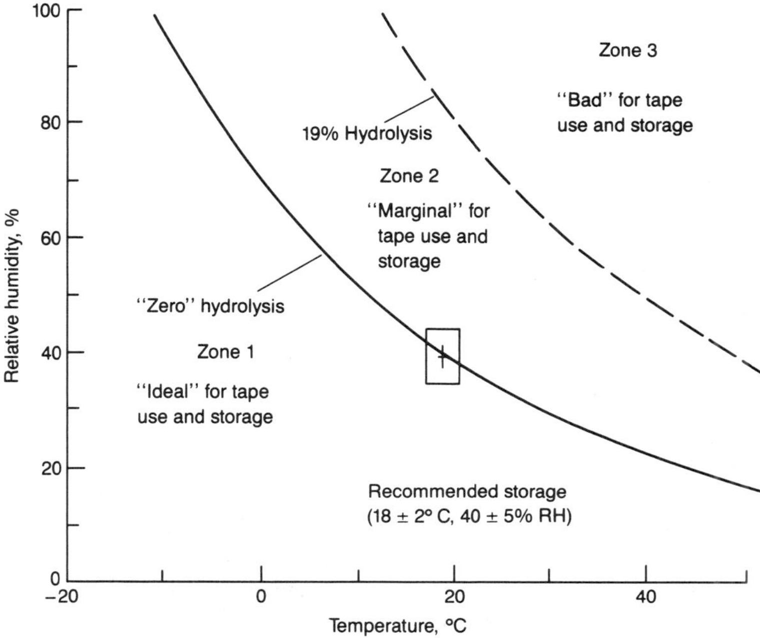

Fig. 5.4. Tape storage environment recommendation on the basis of the hydrolysis of the polyester–polyurethane binder system commonly used in the tape coatings (Bertram and Eshel, 1980).

pinned by high interlayer friction areas. During normal machine operation, under constant tape tension, the pinned areas break loose; allowing relaxed, lower tension tape (or loose wraps) to enter the tape path, they are accelerated over the head (ISV) while it is being written (or read). Then, at a later time when the tape is being read, what was written on the part of the tape where the ISV occurred cannot be read because the read detect card cannot track the data. The same problem can occur if the damaged tape edge or a tape path component excites the tape. If either of these conditions cause the tape to vibrate with a large enough amplitude as the tape passes over the head, an ISV could be flagged. An ISV would manifest itself as temporary and, finally, permanent read or write errors. ISVs are measured against the maximum permissible longitudinal acceleration (as the quotient of a 0.05% change in velocity divided by the time for one flux change in the case of the IBM 3480/ 3490 drive; the drive has a 15 dB safety margin with respect to tape vibrations). The higher the acceleration is in the tape, the lower is the signal-to-noise ratio. Any deviations perpendicular to the track result in tracking errors.

The most severe form of ISV is referred to as "popping" or "blow out," because of the popping sound heard each time the unwinding tape slips on the supply tape pack. Popping occurs in the forward direction near the beginning of the tape, in the reverse direction near the end of the tape, or when the tape speed slows down.

5.2.1. Loose Wraps

ISVs created by loose wraps are a form of low-amplitude ILS at or near the point of unwinding. They generally occur at the outer winding radii. The ISV type of interlayer slip might be unique in that it appears to take place only at or near the point of unwinding when the interlayer pressure is zero. The common belt equation can be used to describe the transition zone between drive and residual tape tension as

$$T_D = T_R \exp(\mu\theta), \tag{5.3}$$

where T_D is the applied drive tension, T_R is the residual relaxed tape tension in the reel at the point of unwinding, μ is the coefficient of friction between layers, and θ is the angular length of the transition zone between T_D and T_R.

If a random discontinuity is encountered such that either μ or T_R is suddenly lowered, then ILS occurs. Some or all of the tape in the transition zone is then elongated by exposure to the higher tension applied by the drive. An example of the velocity increase at the recording head caused by the propagation of this motion is shown in Fig. 5.5.

The basic hypothesis for ISV has been verified by intentionally creating a low-tension discontinuity. This was done by inserting tabs of the type previously described for measuring the interlayer pressure. When these tabs are removed, the tension in the upper layers of tape is reduced. In subsequent drive processing, ISVs were produced at most locations where the tab had been pulled.

The ISV mechanism might also be different from the common ILS in that

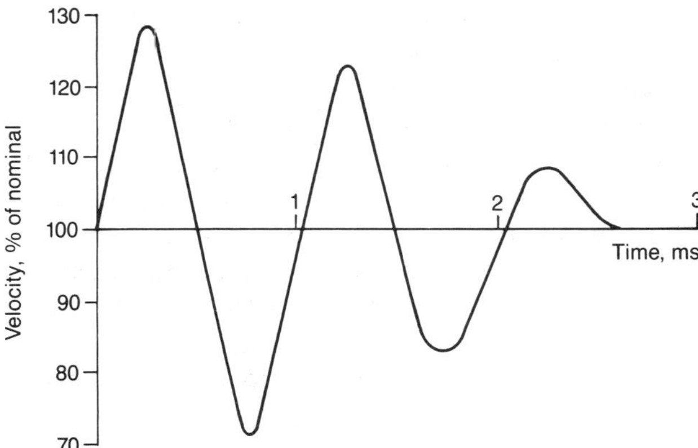

Fig. 5.5. Example of instantaneous speed variation. Measurements were made by discriminating the read signal of a tape prerecorded at constant frequency (Bhushan et al., 1984b).

stress relaxation is not obligatory. Experiments in which tape was wound at higher than normal tension and then unwound immediately at normal tension show the same basic phenomenon. The velocity profile at the head is opposite in this case, however, as tape in the transition zone must contract when it is exposed to the lower tension applied by the drive. The results of this experiment duplicate those of a 6.35-mm tape drive, where the design tension during winding is higher than that during unwinding (Anonymous, 1982).

The methods used to prevent ISV caused by loose wraps are similar to those described for ILS.

5.2.2. Tape Vibrations

Tape can be excited into several modes of vibration, e.g., longitudinal-compression and shear (Crandall and Dahl, 1959; Skurdrzyk, 1968; Snowdon, 1968). Fluctuations in the frictional forces and stick-slip effects, as the tape moves through the tape drives, tension variations across the tape width and tape nonuniformities (such as cupping and curvature) can excite longitudinal tape vibrations. These vibrations are generally found to be of a random character and have a spectrum dominated by the fundamental longitudinal resonant frequency (Prager, 1959; von Behren and Youngquist, 1955; Walraven, 1969; Baumann, 1972; Mukherjee, 1974; Eshel and Kennedy, 1979; Fell, 1980; Smith and Sievers, 1985; Smith, 1986). Longitudinal tape vibrations can also arise from the acoustic resonance of the tape path components.

A wear scar on a tape guide can excite the tape into a shear mode of vibration. When the wear scar reaches a significant depth (a few microns), the tape is excited in the shear mode by the surface traction between the tape edge and the groove sides (Bhushan, 1987). These two modes of vibrations, longitudinal-compression and shear, coexist in the tape and have characteristic frequencies and independent mode shapes.

For the analysis of vibration modes of the tape, we only consider in-plane motion and the tape is assumed to be a linear, elastic, and isotropic thin film. Since the tape is a lightly damped system, damping is neglected. For a tape in sliding contact with the guides, the longitudinal waves are modeled by integrating the following second-order differential equation:

$$(E/\rho)\left(\frac{d^2u}{dx^2}\right) = \frac{d^2u}{dt^2}, \tag{5.4a}$$

where E is Young's modulus of elasticity of the tape, ρ is the mass density of the tape, and u is the longitudinal displacement the tape (along the x direction). The boundary conditions are the fixed ends of the tape at each of the drive's reels (Fig. 5.6), i.e, $u(0, t)$ and $u(L, t)$ are zero. The eigenvalues for the longitudinal waves are given as

$$f_i^c = \left(\frac{2i - 1}{2L}\right)(E/\rho)^{1/2}, \tag{5.4b}$$

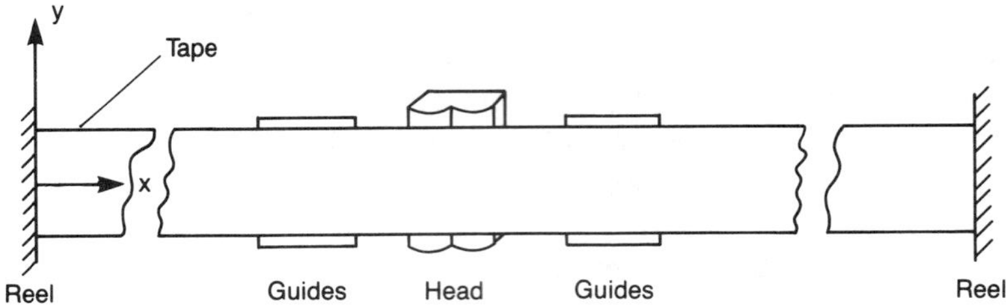

Fig. 5.6. Model of tape path in a tape drive.

where L is the tape length between the two drive's reels and is the vibrating tape span. The fixed boundary conditions at each end of the tape result in odd harmonics of the fundamental frequency. The propagation velocity for the longitudinal wave is

$$V_c = (E/\rho)^{1/2}. \tag{5.4c}$$

For a $(L=)$ 0.6-m-long tape

$$f_1^c \sim 1.5 \text{ kHz}. \tag{5.4d}$$

Subsequent eigenvalues are 4.5, 7.5, 10.5, 13.5, and 16.5 kHz, ... The set of 1.5–16.5 kHz longitudinal waves were all seen in the IBM 3480 tape path, Fig. 5.7(a) (Winarski, 1991). This empirical measurement was obtained by differentiating the signal read from a tape written with an all-ones data pattern. Damping within the tape prevented longitudinal frequencies higher than 16.5 kHz.

The shear waves are modeled by integrating the following second-order differential equation:

$$(G/\rho)\left(\frac{d^2u}{dy^2}\right) = \frac{d^2u}{dt^2}, \tag{5.5a}$$

where G is the shear modulus of elasticity of the tape and y is the coordinate along the width of the tape (Fig. 5.6). The boundary conditions were that the reels of tape are so far away that the ends of the tape are essentially free and that the edges of the tape are free to oscillate. The eigenvalues for the shear waves are given as

$$f_i^s = \left(\frac{i}{2w}\right)(G/\rho)^{1/2}, \tag{5.5b}$$

where w is the width of tape. And the propagation velocity of the shear wave is

$$V_s = (G/\rho)^{1/2}. \tag{5.5c}$$

For a 12.7-mm-wide tape,

$$f_1^s \sim 50 \text{ kHz}. \tag{5.5d}$$

Subsequent eigenvalues are 100, 150 kHz, ... For a more complete analysis,

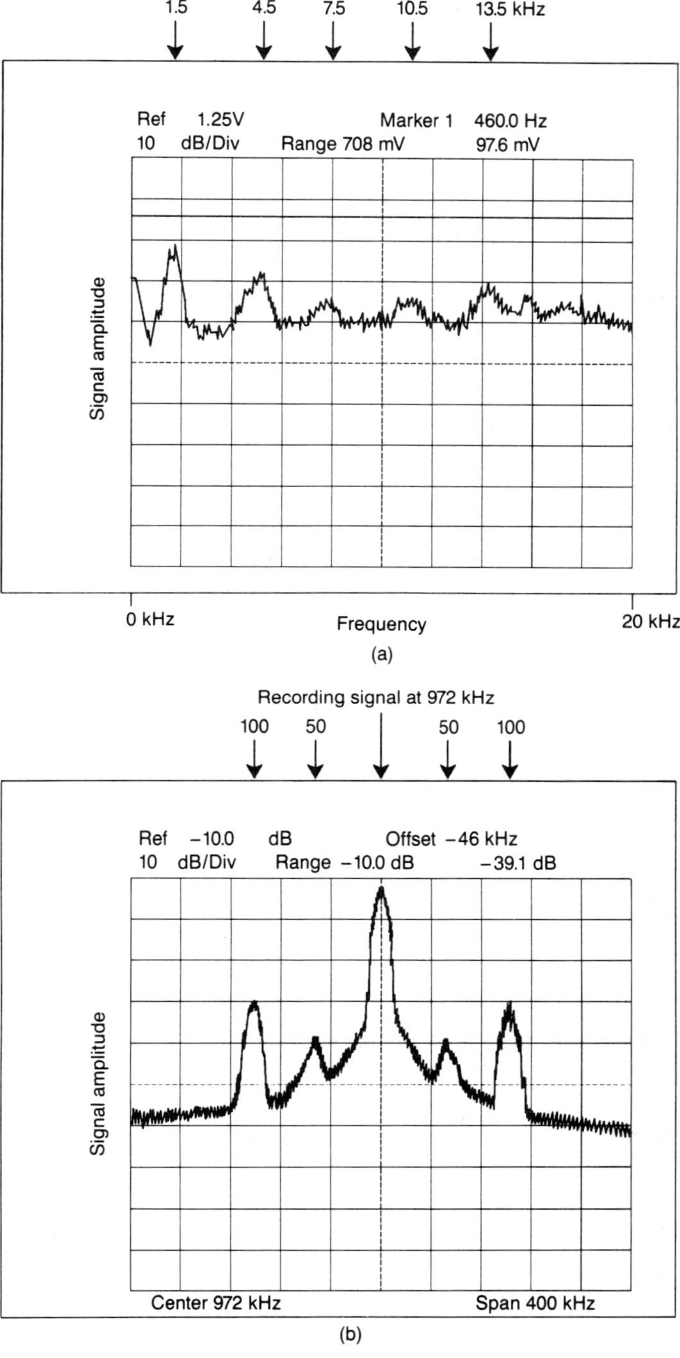

Fig. 5.7. (a) Longitudinal waves (1.5, 4.5, 7.5, 10.5, 13.5 kHz) in an IBM 3480 tape drive at a tape velocity of 1 m/s, and (b) shear waves (50 and 100 kHz) in an IBM 3480 drive at a tape velocity of 1 m/s (Winarski, 1991).

dynamic modulus and damping should be included. The tape stiffness (EI, where I is the moment of inertia) has an effect on the ability to excite vibrations and the vibration amplitude. A tape with a higher stiffness (e.g., thicker tape) may be less susceptible to vibrations.

An example of shear waves for the IBM 3480 tape drive, excited by a worn compliant guide, is shown in Fig. 5.7(b) (Winarski, 1991). There were two mode shapes associated with these shear waves. The spectrum shows 50-kHz and 100-kHz "side lobes" around the 972-kHz data signal. At 50 kHz, the amplitude of the wave varies linearly across the width of the tape with a node at the centerline of the tape. This mode shape was consistent with one edge of the tape being excited. At 100 kHz, the amplitude of the wave had a triangular profile, with nodes between the centerline of the tape and the outer edges. This second mode shape is consistent with both edges of the tape being excited simultaneously. These shear waves can affect data reliability because their direction of oscillation is parallel to the direction in which data is recorded.

To eliminate ISVs caused by tape vibrations, sources of tape excitation must be eliminated. This may require the redesign and selection of different materials and coating for tape path components and a flat tape with smooth edges.

5.3. Uneven Tape-Stack Profile (Hardband)

In the magnetic-tape manufacturing process, polyethylene terephthalate (PET) substrate webs, herein referred to as jumbos (typically a 3350-m-long and 600-mm-wide web for data-processing applications), are coated on one or both sides. Coating speeds are typically 2.5 m/s at a web tension of about 175 N/m. Coated webs are rewound at a tapered web tension of about 90–60 N/m (lower web tension at larger roll diameter). The wound rolls are subsequently slit into tape widths (e.g., 12.7 mm) at fairly high speeds (typically 2.5 m/s). The total length of tape is then wound onto intermediate reels, which we will call pancakes, prior to being wound onto drive reels (or cartridges) (see Chapter 1). If the profile of a tape stack's outer wrap is measured on the coated jumbos, the intermediate pancakes and machine reels, deviations from flatness across the tape are sometimes found. These are present as circumferential bumps or ridges that occur in a cylindrical wound reel (along the tape length). Due to the cumulative effect of the distortions on the wraps, these bumps become worse (in the sense of increasing bump height, but not necessarily in the sense of bump height to bump distortion length ratio) as we move away from the hub. If the tape is allowed to be stored with the bumps, formed by any source, the distorted region undergoes nonuniform stresses. These relax with time, which causes the initial elastic strain to become partially viscoelastic. When the tape is unwound, the viscoelastic component of strain does not recover instantaneously. The rate of viscoelastic deformation

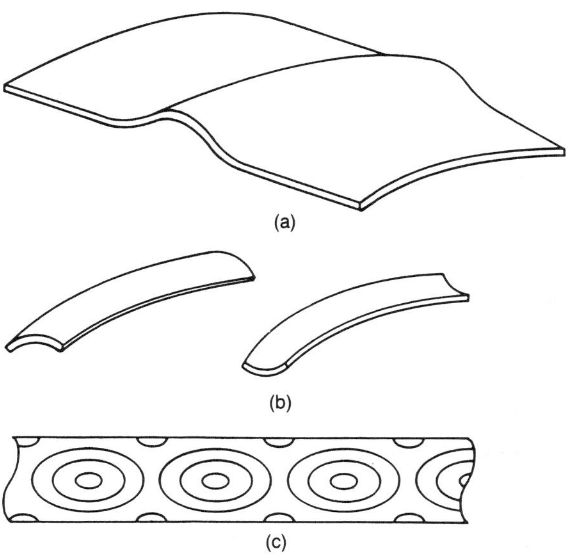

Fig. 5.8. Schematics of a typical distortion on a web (a) distorted 12.7-mm-wide tape cut from distorted area and (b) contours of buckled tape after the tape is straightened.

is higher at higher temperatures and humidities. When the distorted (cupped) tapes are straightened with little tension, the internal stresses can produce buckling (Fig. 5.8). Such buckling can be observed by simply laying the tape on a flat surface and observing the distorted reflections from the smooth recording surface. An application of a large enough tension, obviously, flattens the tape. (This tension is referred to as "tension to flatten.") These distortions affect data reliability by increasing the head–tape separation in the distortion zone caused by the lower tape tension in this zone.

In the magnetic tape industry, the term hardband is used where these ridges are generally formed by thicker sections that exist in a continuous manner along the length of the web either from substrate or coating thickness variations. The result of similar problems in the paper industry has been termed bagging, stretch-outs, tires, balloons, or bubbles (Pfeiffer, 1966).

5.3.1. Sources and Methods of Preventing Distortions

Potential sources of these distortions include (a) substrate or coating-thickness variations (if the PET substrate or coating-thickness variations are consistent along the same line, these will add up in a reel and can result in a more severe bump) and (b) the wrapping-in of entrapped air or scratches in the substrate or the coating. This process is aided by the presence of tension ridges in the moving web at the jumbo level, and tape cupping at the pancake and machine reel levels. At the typical speeds used in tape manufacturing, hydrodynamic air-film thicknesses of 10–25 μm can be generated at the inlet zone. Interlayer friction prevents the collapse of these air pockets under the radial compressive stresses. The formation of air pockets is likely at a coating station, because of the large width of the web. Pancake or machine reel hub deformation can also result in tape distortions. Differential hub deformation due to the radial winding stresses across the hub width (convex-shaped contour) can be significant for some plastic hub materials.

5.3.1.1. Substrate and Coating-Thickness Variations

Polyester films should be uniform in thickness across the width and along the length, and variations should be less than 0.025 μm. Any thickness variation across the width should be staggered along the length to minimize accumulation. This can be accomplished by controls during manufacturing, e.g., by oscillating the extrusion gate or more commonly the web rewind station by roughly 25–50 mm at a frequency of about one-fourth cycle/min.

Similar requirements exist for coated tapes as well. A coating uniformity with less than 0.1 μm variation is achievable. Some coating thickness variations come from the backcoat because it is not calendered. A burnishing operation to remove streaks can be added in the coating line. This involves using a burnishing tool (typically made of tungsten carbide) rotating at 30,000 rpm and contacting the web under light pressure. A coating head due to its inherent design may also produce a nonuniform coating thickness. If the ink resides in the head reservoir too long, particle separation may occur leading to coating nonuniformity. This can be improved by one of the following techniques: multidrip system in a coating head, stirrer in the ink pool, or a flow-through head. Any coating thickness variations across the width should also be staggered. This can be accomplished by either oscillating the coater head or the jumbo rewind station. Oscillation of the coater head may create instability in the coating process; therefore, oscillation of the jumbo rewind station is preferred.

Figure 5.9 shows two examples of uneven stack profiles on a machine reel: one where the variation is caused by coating-thickness variations, and the other where it is not. The two tape reels had about 3300 wraps. The thicknesses of the PET substrate and the coated tapes were measured using a Mahr 1234 IC/Z differential transformer gauge. (The upper, movable probe had a radius of 1.47 mm and exerted a 0.75 N load on the sample. The resolution of the instrument was 0.025 μm. Repeatability of any given measurement was usually ± 0.050 μm.) Each tape sample was measured at three places along the length, using a series of ten measurements across the width (12.7 mm). The plotted results are an average of three readings referenced to the measurements at one edge. A stylus profiler trace of the stack profile was also made and accompanies each thickness plot.

5.3.1.2. Entrapped Air Pockets, Tension Ridges, and Scratches

Variations in the tape-stack profile can also be caused by the winding of the web or tape over air pockets, tension ridges, scratches, or hub anomalies.

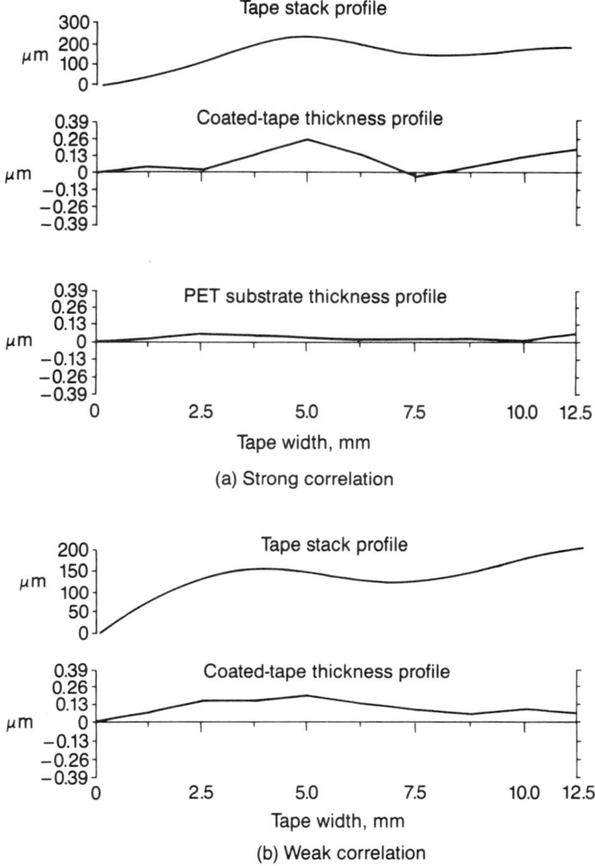

Fig. 5.9. Comparisons of coating and substrate thickness profiles with a tape-stack profile of a machine reel (a) strong correlation and (b) weak correlation (Bhushan et al., 1984b).

Interlayer friction prevents the collapse of air pockets or the flattening of tension ridges. Air can be entrapped at the jumbo, pancake, or machine reel level. Due to the large width of the web at jumbo level, there is less likelihood of an edge leakage of the air film.

During the winding process both the incoming tape and that already wound onto the rotating roll pump air into the nip, as shown in Fig. 5.10. The hydrodynamic air-film thickness, at the nip, is given as (Blok and Van Rossum, 1953)

$$h(0)/R = 0.426(12\eta Vw/T)^{2/3}, \tag{5.6}$$

where η is the absolute viscosity of air, V is the winding speed, w is the width of the web or tape, $T =$ the winding tension, and R is the current winding radius of the roll. Equation (5.6) illustrates the dependence of the air-film

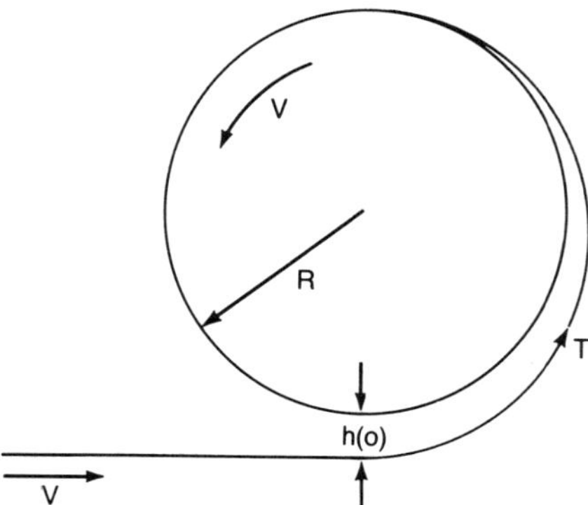

Fig. 5.10. Schematic of tape (or web) being wound showing hydrodynamic air film generated.

thickness on the winding parameters. Part of this air is squeezed out as the next wrap is wound but part of it is locked in. For jumbo rewinding conditions: $R = 150–225$ mm, $w = 600$ mm, $T/w = 90$ mN/mm, and $V = 2.5$ m/s, the air-film thickness $h \sim 25–35$ μm. We note that the air film is very thick and a very small entrapped portion can cause wound-in ridges. As mentioned earlier, a nonuniformity of about 50 nm over the entire length would give an unacceptable ridge height on the outside diameter.

Tension ridges along the web length are present in the process line between support rollers. These ridges are very visible and may have peak-to-peak amplitudes as high as 5 mm with a wavelength on the order of 100 mm across the web. The cause of tension ridges is a lack of adequate web support complicated by substrate flexibility, high web tension, and anisotropic mechanical properties (see Chapter 1). In addition, a coated web with tension ridges in the drying oven would result in viscoelastic deformation. This deformation would be nonuniform across the cross section due to the different rates of stress relaxation of the magnetic coating and PET, Fig. 5.11. It would also be nonuniform across the width of web due to the nonuniform stress distribution. The nonuniform viscoelastic deformation would result in tension ridges and cupping, even at lower tensions at the rewind station. Backcoating introduces additional stresses. Curing of the coating also introduces additional stresses which results in cupping (Hiratsuka et al., 1980). The tension ridges and other distortions in the web can be pinned at the rewind station. Although calendering tends to reduce tape distortions following coating, at least some viscoelastic deformations persist through subsequent winding, and result in air pockets. Finally, scratches in the substrate or the coating would cause an uneven stack profile.

Tension ridges are believed to be formed by lateral compression buckling. Euler buckling for a wide column (the web length) in compression (across the web width) of a 23.4-μm-thick and 600-mm-wide web gives a critical stress of

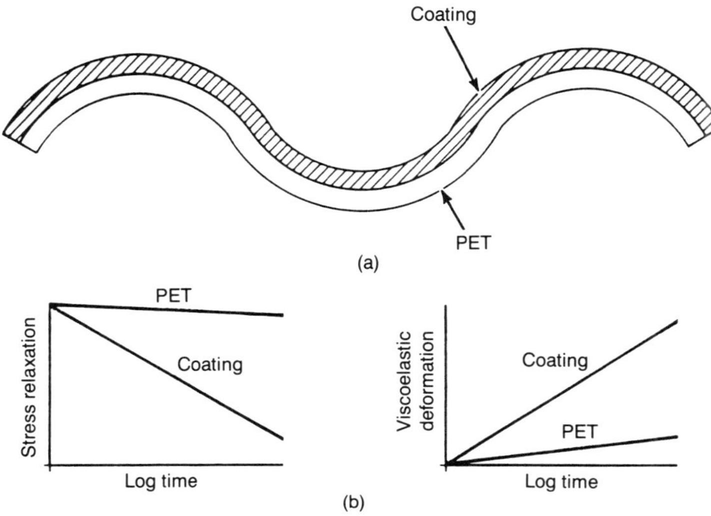

Fig. 5.11. (a) Schematic of coating curling on a distorted web, and (b) schematic of the viscoelastic deformation of a distorted web.

about 5 Pa; for the observed tension ridge wavelength of 100 mm, the stress is about 1 kPa; and for the 12.7-mm tape, the stress is about 12 kPa. The web stress analysis shows that a lateral compressive stress does exist in the center region of the web, which will produce buckling. Further analysis of this case showed that buckling would first occur at a tension of about 15 mN/mm (3.5 kPa maximum compression). Using the finite-element analysis, the first few buckling modes of the web with one clamped boundary and uniaxial tension are shown in Fig. 5.12. We note that uniaxial tension on each boundary of a web made of the orthotropic material (PET) also gives significant lateral compression in the center of an unsupported length of the web.

Conditions for Wound-In Distortions

An analysis of circular-distortion geometry was carried out to find out what is the minimum interlayer friction needed for air pockets to form and not collapse, and tension ridges to be wrapped. The schematic of a distortion in wound tape is shown in Fig. 5.13. We assume that the distortion has a radius R in the web before rewinding. If the friction is extremely high, the ends of the distortion (A and B) will be held in place (or pinned) and they will not slide. The objective here is to analyze the conditions that cause the distortions to be pinned. W is the load applied on the web as a result of the tension in the web and H is the horizontal reaction at its ends. The value of $2H/W$, which determines whether the distortion will be pinned or not, can be obtained by solving the problem for the deflection of a distortion with pinned ends.

From Walowit et al. (1973), the distortions will not be pinned in place if the

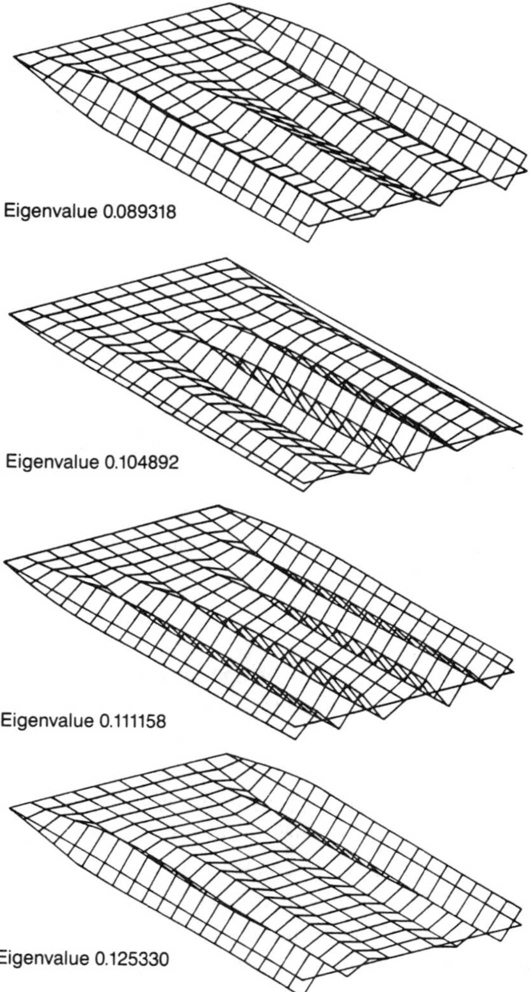

Fig. 5.12. Buckling mode shapes of a 600-mm-wide PET web with left clamped boundary and uniaxial tension.

coefficient of friction μ is

$$\mu < (2H/W)$$

$$= \frac{\sin^2 \theta_0 - 2 \cos \theta_0(\theta_0 \sin \theta_0 + \cos \theta_0 - 1) - 2\alpha \cos \theta_0(1 - \cos \theta_0)}{3 \sin \theta_0 \cos \theta_0 + \theta_0(2 \cos^2 \theta_0 + 1) + 2\alpha \sin \theta_0 \cos \theta_0}, \quad (5.7)$$

where $\alpha = \frac{1}{12}(t/R)$, $\alpha \to 0$, for most cases, and t is the thickness of the tape. A plot of $2H/W$ as a function of θ_0 is given in Fig. 5.14. We note that if $\theta_0 \to 0$, pinning would not occur in most of these cases unless friction is close to

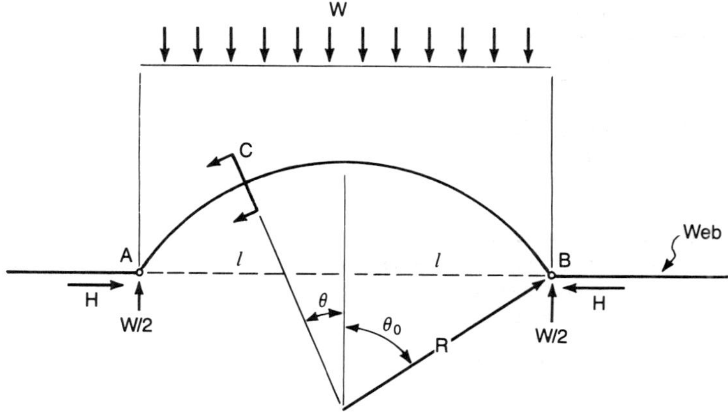

Fig. 5.13. Schematic of a distortion in a wound tape.

infinity and if $\theta_0 \to \pi/2$, μ should be less than $2/\pi$ or 0.63 for pinning not to occur.

Friction measurements were conducted on PET substrates, frontcoated tapes and front- and backcoated tapes. The test consisted of pulling the recording side of the tape over the back side of the tape, which was wrapped onto a Ni–Zn ferrite rod (Bhushan, 1990). Results are reported in Table 5.1. Note that PET and frontcoated tapes have lower interlayer friction than backcoated tapes. This suggests that the latter have a higher propensity for

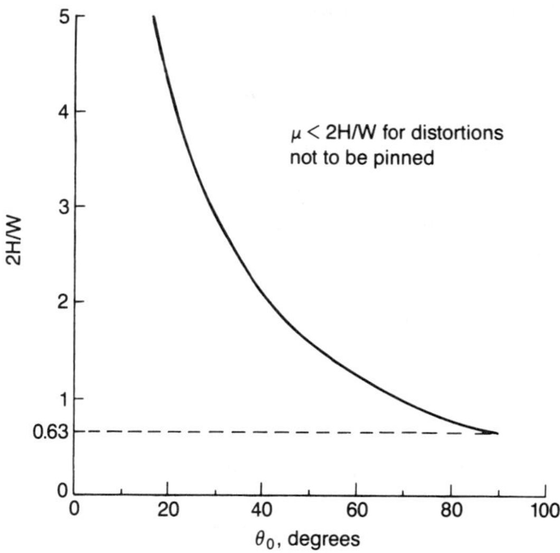

Fig. 5.14. The ratio $2H/W$ as a function of θ_0 for a distortion in a wound tape, shown in Fig. 5.13.

Table 5.1. Interlayer friction results on tapes and PET substrate

Specimens	Special treatment[a]	Coefficient of friction[b]
PET vs. itself	—	0.37
PET vs. recording side	—	0.38
Backcoat vs. recording side	—	0.47
Backcoat vs. recording side	With THF solvent after 15 s	0.54
Backcoat vs. recording side	After 5 min	0.49
Backcoat vs. recording side	After 15 min	0.49
Backcoat vs. recording side	After 45 min	0.48
Backcoat vs. recording side	After 2 h	0.47

[a] Solvents were applied using felt on the front- and backcoats of the tape.
[b] Measured at a wrap angle of 90° and an initial tension of 650 mN.

air-pocket formation and tension ridges to be wrapped. Further tests conducted on backcoated tapes, after applying a solvent to simulate the rewinding of a freshly coated web having retained some solvents, showed higher interlayer friction. Thus, the friction, particularly for backcoated tapes, is comparable to that needed for air pockets and tension ridges to be "pinned," and hence, the potential for air entrapment and tension ridges to be wrapped exists. This can be avoided by using a squeeze roller (see a later section).

If the distortions are pinned, the compressive bending stresses acting on the web at a point C in Fig. 5.13 for a uniformly distributed load are

$$\sigma_c = -\{(H/t)\cos\theta_0 + (W/t)[\tfrac{1}{2} - (\theta_0 - \theta)/2\theta_0]\sin\theta\}. \tag{5.8}$$

The $H/2W$ relationship can be obtained from Eq. (5.7). The maximum stress occurs at the ends ($\theta = \theta_0$)

$$(\sigma_c)_{max} = -[(H/t)\cos\theta_0 + (W/2t)\sin\theta_0]$$

$$= -(2W/t)[(H/2W)\cos\theta_0 + \tfrac{1}{4}\sin\theta_0]. \tag{5.9}$$

Now we calculate the compressive stresses for a typical case of the tension per unit tape width during a web or tape rewinding of 88 mN/mm and a radius of the web roll of 20 mm. For this case, a plot of $(\sigma_c)_{max}$ as a function of θ_0 is given in Fig. 5.15, which demonstrates that high compressive stresses (>0.5 MPa), added to the stress field caused by normal winding, can be incorporated in wound tape.

Stress Variations in the Outer Wraps of Tape Caused by a Circumferential Bump in the Stack Profile

We carry out the stress analysis of different bump geometries. We consider the outer wrap of a tape as a thin, cylindrical shell subjected to a uniform, internal stress (T/tw) caused by winding. Decomposing the radial displace-

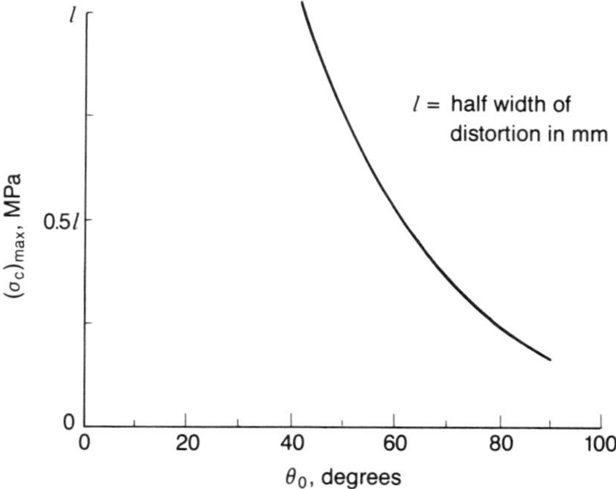

Fig. 5.15. Maximum compressive stress $(\sigma_c)_{max}$ as a function of θ_0 for a distortion in a wound tape, shown in Fig. 5.13.

ment $u(z)$, radius $R(z)$, and the hoop stress $T(z)$ into

$$u(z) = u_0 + u_1(z),$$

$$R(z) = R_0 + R_1(z), \qquad (5.10)$$

$$T(z) = T_0 + T_1(z),$$

where u_0, R_0, and T_0/tw are the radial displacement, radius, and hoop stress in the outer wrap (winding tension) in the absence of a bump, and $u_1(z)$, $R_1(z)$, and $T_1(z)/tw$ are the changes produced in the radial displacement, radius, and hoop stress as a result of the bump. From Timoshenko and Woinowsky-Krieger (1959), the deformation u in the radial direction of a thin tape ring of constant thickness and composed of an isotropic, linearly elastic solid is

$$D[d^4u(z)/dz^4] + [Et/R^2(z)]u(z) = T(z)/wR(z), \qquad (5.11)$$

where the flexural rigidity of the tape is

$$D = Et^3/12(1 - v^2), \qquad (5.12)$$

E is Young's modulus, v is the Poisson ratio, and t is the tape thickness. From Eqs. (5.10) and (5.11), and assuming that

$$u_1(z) = R_1(z), \qquad (5.13)$$

then the deformation change in the wrap of tape caused by the bump is equal to the increase in the wrap's radius and

$$[1/R(z)] \sim (1/R_0), \qquad (5.14)$$

which is true for bump heights that are small in comparison with the outer

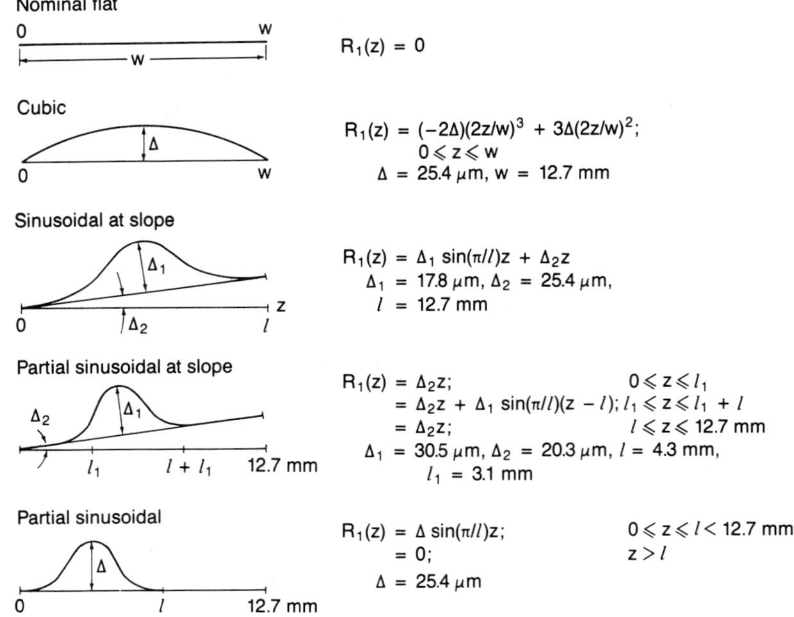

Fig. 5.16. Definition of bump geometries.

wrap radius, then the increase in hoop stress in the outer wrap caused by this known bump profile is

$$T_1(z)/tw = (E/R_0)R_1(z) + (DR_0/t)[d^4R_1(z)/dz^4], \qquad (5.15)$$

while the increase in radial pressure is

$$P_1(z) \sim -[T_1(z)/wR_0]. \qquad (5.16)$$

For typical bump profiles (Fig. 5.16), the calculations of $T_1(z)/tw$ and $P_1(z)$ are given in Table 5.2. Note that there is a substantial increase in radial and hoop stresses at the bump locations for certain bump geometries, which result in increased creep deformations.

Examples of an Uneven Tape-Stack Profile

Direct evidence of air entrapment was provided by sectioning pancakes [with an internal diameter (ID) of 100 mm and an outer diameter (OD) of 175 mm] with known problems in the tape-stack profile. The entire reel of tape (including the hub assembly) was cast in an epoxy, to maintain its integrity, and subsequently sectioned and polished. Figure 5.17 shows a low-magnification photograph of such a sectioned reel with an observable uneven tape-stack profile. Air entrapment is also visible. Note here that the cumulative effect of the defect lies in a line at a substantial angle to the radial direction from the hub. In this case, the defect disappeared at about two-thirds of the distance

Table 5.2. Variations in maximum hoop and radial stresses and tension as a function of bump geometry. Nominal tension = 2.2 N, T_0/tw = 5.3 MPa, $P_0 = -3.4$ kPa, $R_0 = 50.8$ mm, $t = 31$ μm, $w = 12.7$ mm, $E = 3.45$ GPa

Bump geometry	Max. $T_1(z)/tw$, MPa	Max. $P_1(z)$, kPa	Max. tension, N
Nominal flat	0	0	0
Cubic	1.7	-1.1	0.70
Sinusoidal at slope	1.4	-0.9	0.59
Partial sinusoidal at slope	2.5	-1.6	1.05
Partial sinusoidal	1.7	-1.1	0.70
	($l = 12.7$ mm)		
	2.8	-1.8	1.17
	($l = 2.54$ mm)		
	18.7	-12.2	7.84
	($l = 1.27$ mm)		

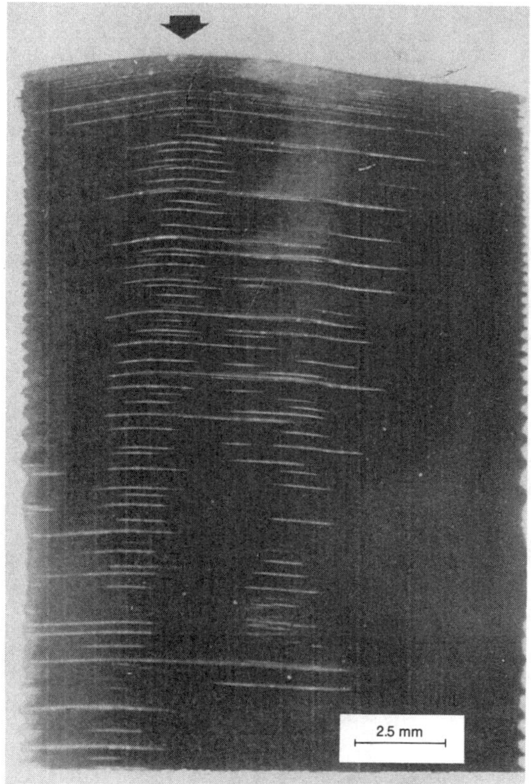

Fig. 5.17. Photograph of a cross section of a pancake. The top end is the outer wrap of the tape (Bhushan et al., 1984b).

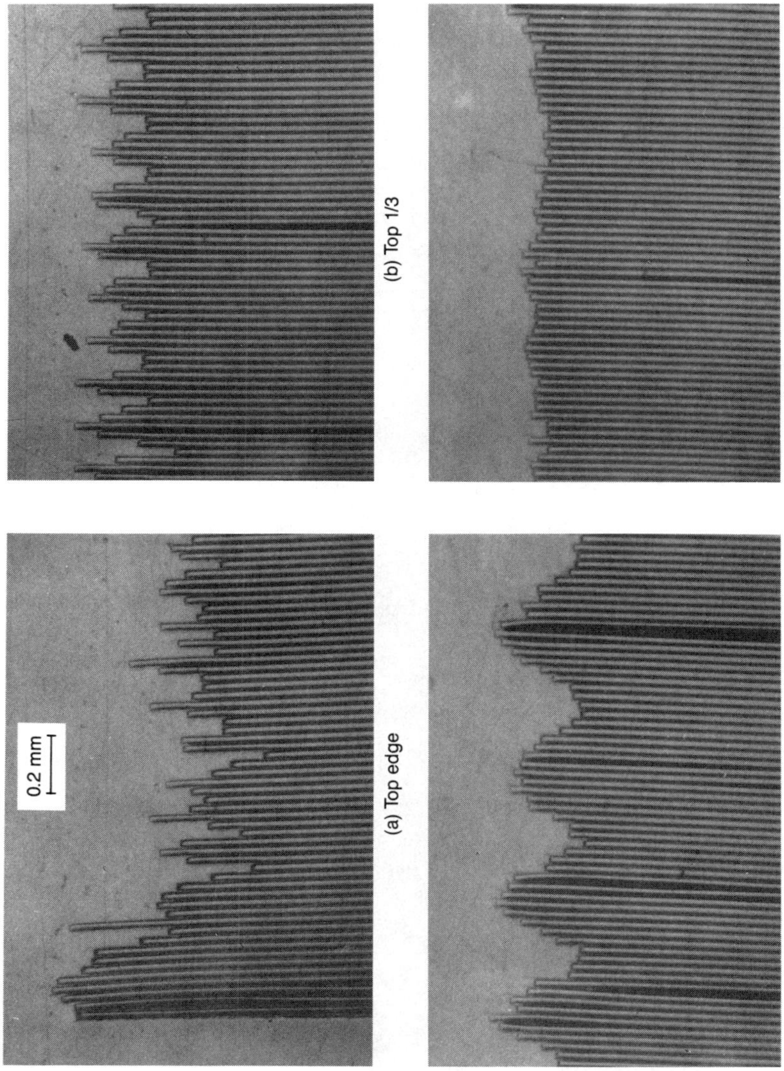

Fig. **5.18.** Edge views of a cross section of a pancake indicating staggered wraps and air gaps between the layers (Bhushan et al., 1984b).

from the hub to the outer diameter. The series of photographs in Fig. 5.18 exhibit areas of random tape droop, as well as of air gaps between the layers. Note also that the air pockets are associated with the crests of sinusoidal edge motion pictured in Fig. 5.17. The source of these air pockets could be air entrapment or tension ridges.

Figure 5.19 shows an example of an uneven stack profile on a 600-mm-wide jumbo roll. The variation in profile from hub to outer diameter is shown. Measurements were made with a replication technique in which an epoxy-coated metal plate is pressed against the jumbo surface. After curing, the replica is cut away from the jumbo. A profiler is then used to trace the replica.

Methods of Preventing Uneven Tape-Stack Profile

The uneven tape-stack profile problem can be eliminated in two ways:

(a) Minimize the entrapped air film and the pinning of tension ridges during rewinding of the web and slit tape by using a lay-on, packing, or squeeze roller at these stations (Anonymous; Bertram and Eshel, 1980). A squeeze roller also helps rewind the web at slightly lower tensions.
(b) Reduce tension ridges during unwinding, coating, drying, calendering, and rewinding of the web by keeping the overall tension low.

Industry practice is to use lay-on rollers made of rubber or soft plastic (e.g., NBR or EPDM with a hardness ~ 35–60 Shore A) at or close to the nip, to squeeze out the air film [Fig. 5.20(a)]. The important fact here is that the roller must be placed at the tangency point of the incoming web or tape, and

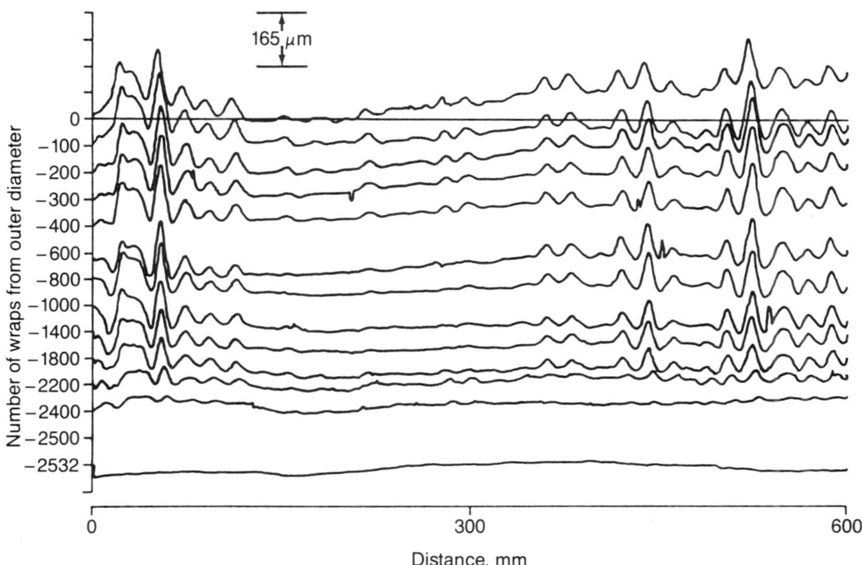

Fig. 5.19. An example of the variations in a tape-stack profile through a jumbo roll.

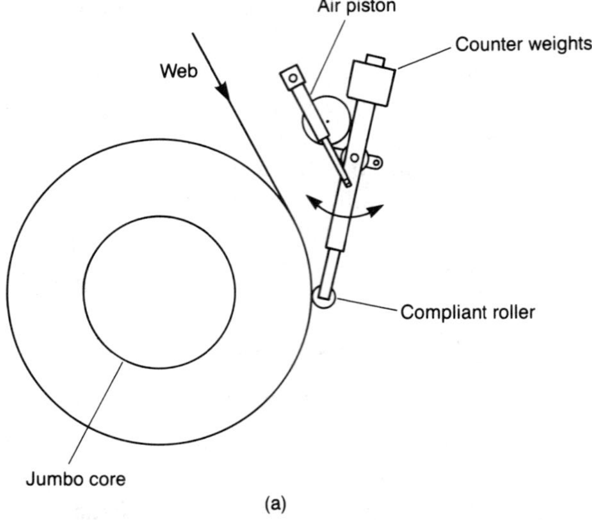

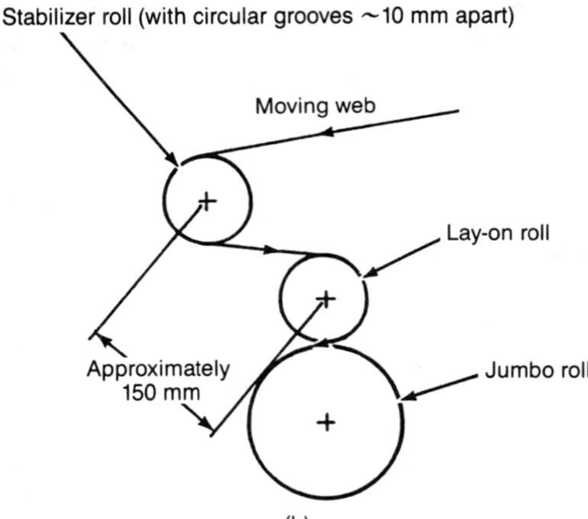

Fig. 5.20. Schematic of (a) a lay-on roll and (b) a stabilizer and lay-on rolls at a jumbo rewind station.

the reel at the rewind station. A stabilizer roll ahead of a lay-on roll are commonly used to stabilize the web, Fig. 5.20(b). In the case of jumbo rewind, the lay-on roller should be about 590-mm wide in the case of a 600-mm-wide web, so that it does not ride on the coating streaks on the ends. The two-edge segments of tape (12.7-mm wide for a 12.7-mm-wide tape) are thrown away. Therefore, edge segments need not be squeezed. A schematic of the pancake rewind station (after the tape web is slit) is shown in Fig. 5.21.

The load must be adequate to squeeze out the entrapped air. A good rule of thumb is that the external load applied on the roller and its elastic modulus

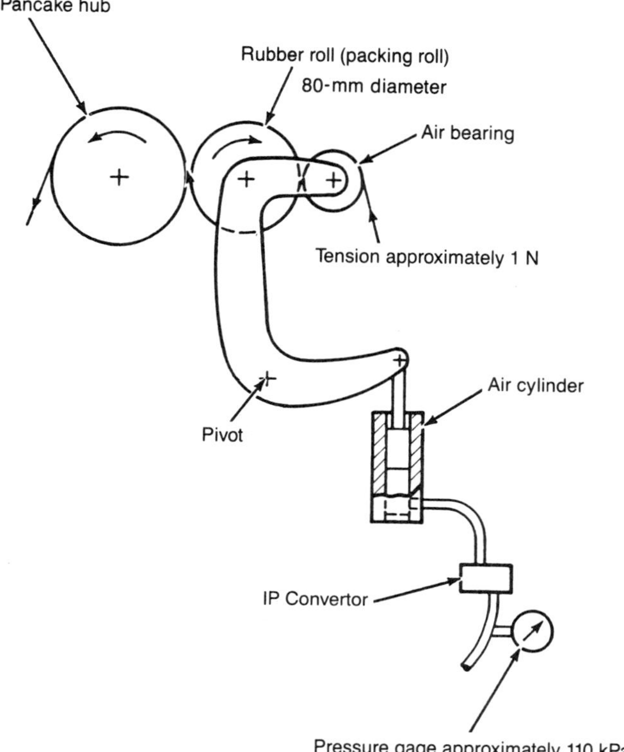

Fig. 5.21. Schematic of a lay-on roll at a pancake (slitter) rewind station.

should be selected so that the hydrodynamic air-film thickness is of the same order of magnitude as that of the composite surface roughness of the mating tape surfaces. A relationship between the normal load F and the hydrodynamic air-film thickness was derived from modification of the analysis presented by Blok and Van Rossum (1953). The roller was applied at the nip, the region where the web or tape begins to wrap around the reel it is being wound on, as shown in Fig. 5.22. Boundary conditions used for calculating the air-film thickness at the nip, as reported in Eq. (5.6), were modified to include the Hertz pressure exerted by the lay-on roller as follows (Winarski, 1984):

$$h(0)/R_1 = 0.426(12\eta Vw/TM)^{2/3}, \tag{5.17}$$

where $M = 1 + 0.564(R_1 w/T)[F(1/R_1 + 1/R_2)E'/w]^{1/2}$ and $1/E' = (1 - v_1^2)/E_1 + (1 - v_2^2)/E_2$; 1 and 2 refer to tape reel and lay-on roll, respectively.

Analysis was performed for the following parameters: $v_1 = 0.37$, $E_1 = 210$ MPa, $v_2 = 0.48$, $E_2 = 14$ MPa and variable, and $R_2 = 50$ mm. Additional jumbo winding parameters were: 150 mm $< R_1 <$ 225 mm, $T/w = 90$ mN/mm, $w = 600$ mm, and $V = 1.5$ m/s. Additional pancake winding pa-

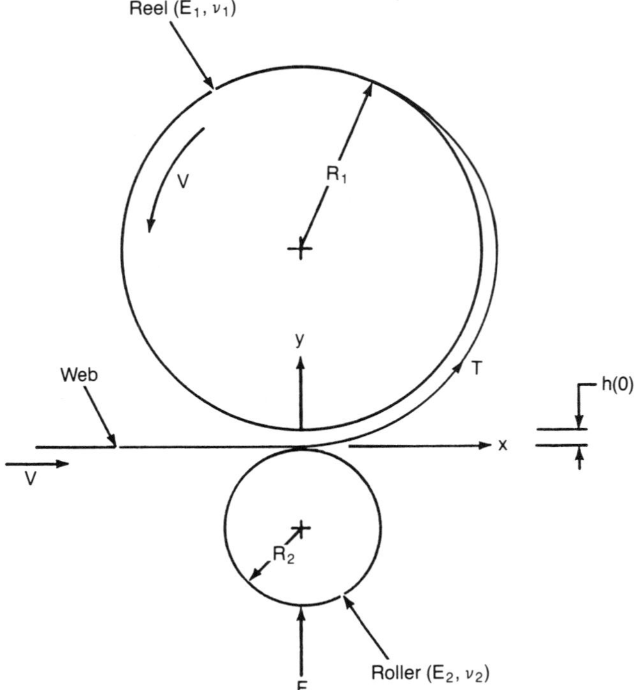

Fig. 5.22. Schematic of a tape (or web) being wound with a roller applied at the nip—the region where the tape begins to wrap around the reel.

rameters were: $100 \text{ mm} < R_1 < 175 \text{ mm}$, $T/w = 100 \text{ mN/mm}$, $w = 12.7 \text{ mm}$, and $V = 3.8 \text{ m/s}$. For the above parameters, for a given rubber, we present plots of the roller load F versus $h(0)$, the resulting air-film thickness for both the jumbo and pancake winding conditions, as shown in Figs. 5.23 and 5.24. Our analysis shows that a load of the order of 500 N should be used for a typically-used rubber material for jumbo rewind, and a load of the order of 50 N should be used for pancake rewind. We also note that the harder the roller material, the less the required load to achieve that prescribed air-film thickness.

Tension ridges can be minimized by: reducing the tension, increasing the web support, and using spreader or crowned rollers. Adequate web support in the ovens, and between various equipment in the coating process, should be applied. A reduction in the coating speed may also reduce the propensity of wound-in tension ridges.

5.3.2. How Does an Uneven Tape Stack Affect Data Reliability?

Winding tape over an out-of-plane distortion results, with time, in a viscoelastic tape deformation and, thus, in increased head-tape separation resulting in

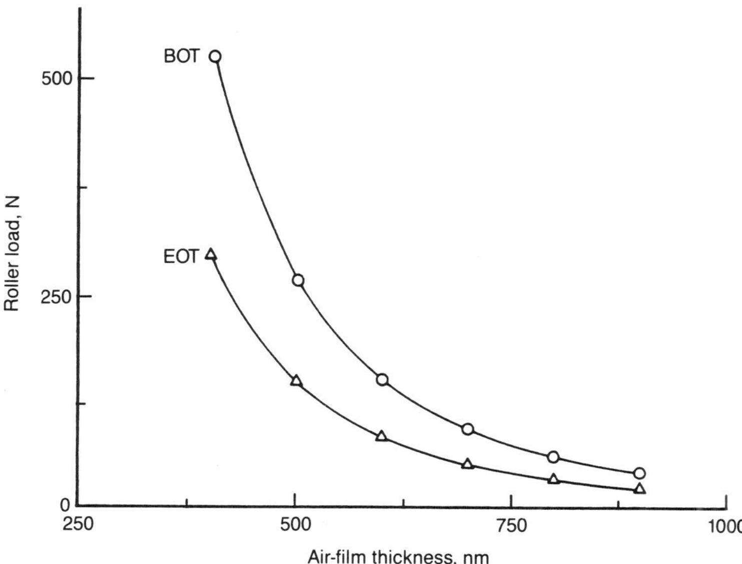

Fig. 5.23. Hydrodynamic air-film thickness as a function of roller load for jumbo rewind using a rubber roller of hardness ∼60 Shore A.

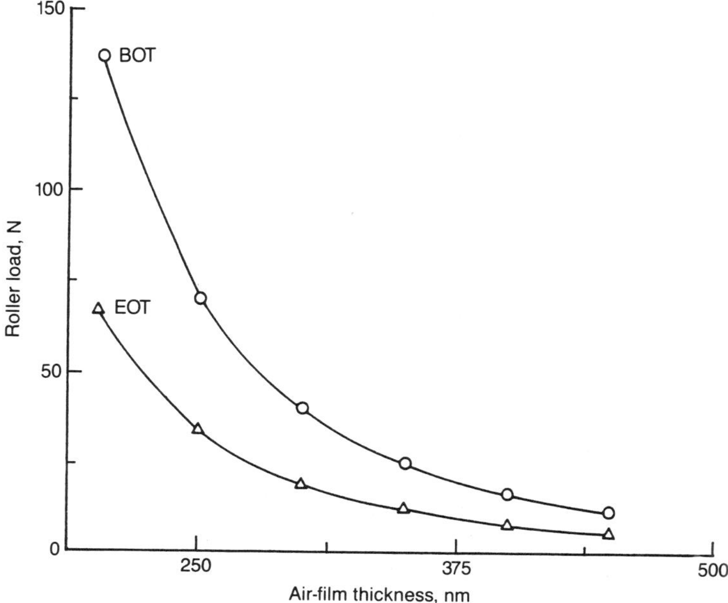

Fig. 5.24. Hydrodynamic air-film thickness as a function of roller load for pancake rewind using a rubber roller of hardness ∼60 Shore A.

magnetic errors. We describe here two experiments to demonstrate these effects.

The first experiment was conducted to show how a tape, wound over a distortion and stored for a period of time, deforms viscoelastically and forms an uneven stack. Three pancakes with a range of uneven profile severity were identified. The IBM 3480-type tape cartridges (with a 31-μm-thick, 12.7-mm-wide, and 165-m-long tape) were wound to represent each pancake. Typical stylus-profiler profiles of the cartridges are shown in Fig. 5.25. Each of the cartridges had been previously stored for 30 days at 52° C and 27% RH with several rewinds and was, therefore, in a dimensionally stable condition. Several meters of the flat tape were spliced onto the test cartridge. The cartridges were then tensioned on the 3480 tape drive (2.2 N nominal tension) and placed in storage at either 52° C and 27% RH, or at room conditions. Tape deformation, measured for various intervals of storage in terms of tension to flatten (TTF) in N, is plotted in Fig. 5.25. *A higher value of TTF signifies increased tape distortion* (see Appendix 5.A). We note that the elevated temperature has an accelerated effect and the bulk of the deformation occurs in about the first 100 h, as we would expect from viscoelastic data presented in Chapter 3.

We note that additional deformation would occur if the tape was retensioned and then stored again. To demonstrate this effect, the spliced tapes on three cartridges, already stored for 600 h at 52° C and 27% RH, were retensioned and then stored at 52° C and 27% RH for another 120 h. This final value of tension to flatten is also plotted in Fig. 5.25.

The cartridges in the previous experiment were also used to determine the deformation kinetics of the PET substrate by itself. Results using initially-flat PET samples, shown in Fig. 5.26, showed significantly less deformation than that of the coated tapes. This suggests a very important contribution from the coating, as expected from the viscoelastic data of Chapter 3. (The measured numbers of TTF reported in Fig. 5.26 can be predicted by the analysis presented in Appendix 5.A.)

The second experiment was conducted to show how localized viscoelastic deformation, resulting from being wound over a distortion, affects the data reliability of the tapes used on an IBM 3480 tape drive with an 18-track head. A sample of 50 cartridges was obtained to evaluate data reliability versus storage time at the extreme shipping condition of 52° C and 27% RH. The cartridges came from 50 different pancakes that included both hardband accepts and hardband rejects. The data reliability of the cartridges was measured in terms of mean-bits-to-failure (MBTF). A high MBTF number represents goodness, because on the average there are more good bits between errors. The significance of various MBTF levels is as follows: 0.4×10^7—failure; 0.4–1.0×10^7—marginal but acceptable, and 1.0×10^7 or greater—acceptable (see Appendix A). The acceptability of the 50 cartridges versus storage time is shown in Fig. 5.27. Cartridges that fail the MBTF criterion

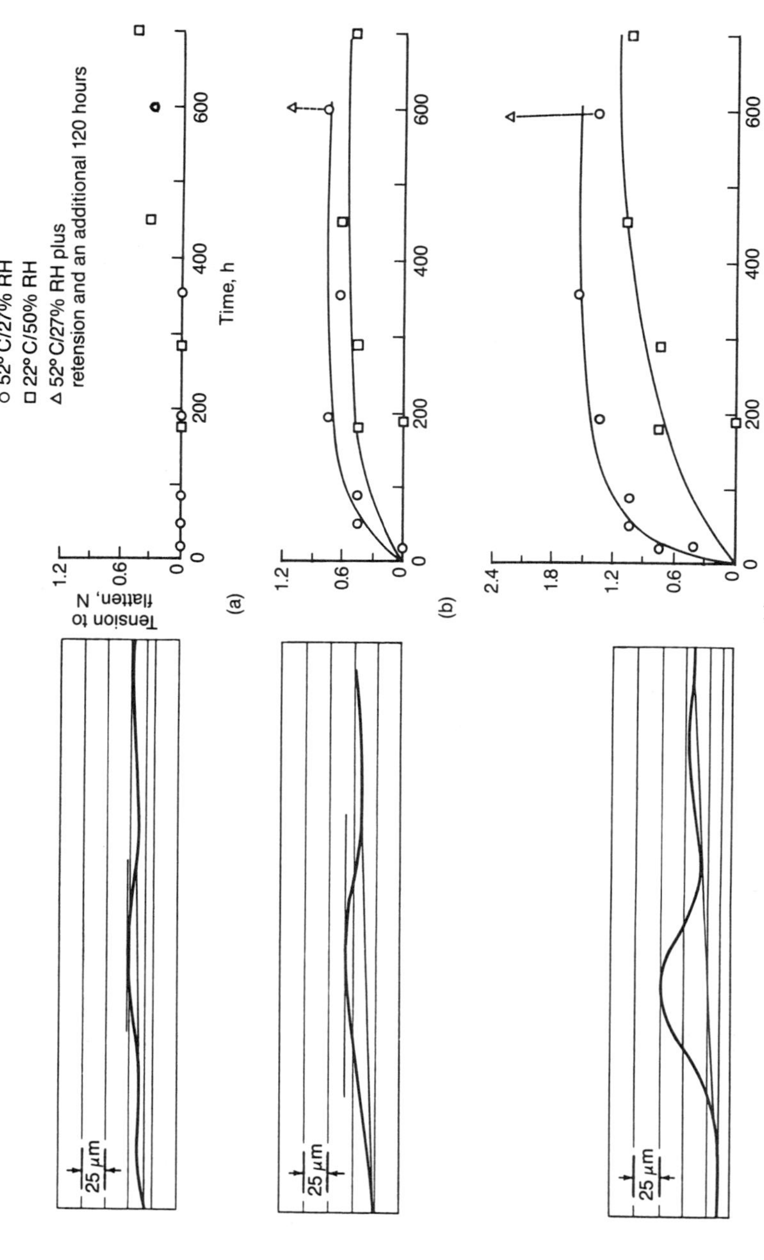

Fig. 5.25. Profile trace across a tape cartridge and tension-to-flatten as a function of storage time for three tape cartridges with different uneven profile severities: (a) 20-μm profile spread (small), (b) 30–35-μm profile spread (medium), and (c) 55–60-μm profile spread (severe).

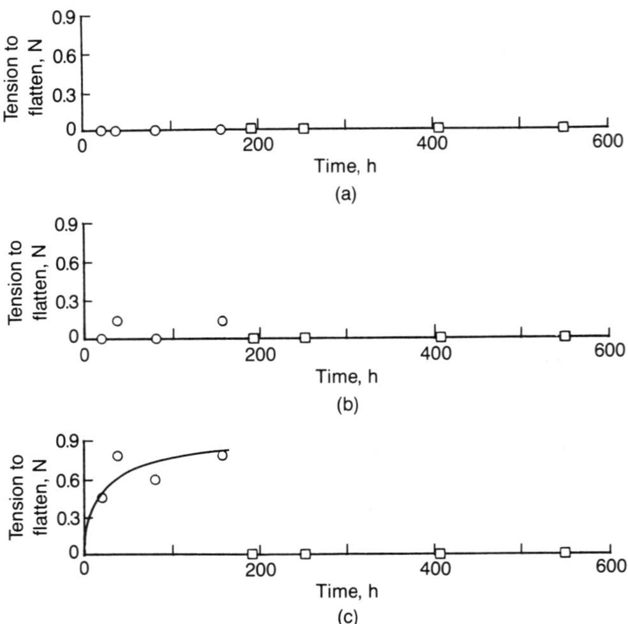

Fig. 5.26. Tension-to-flatten as a function of storage time for PET using cartridges with three different uneven profile severities: (a) 20-μm profile spread (small), (b) 30–35-μm profile spread (medium), and (c) 55–60-μm profile spread (severe).

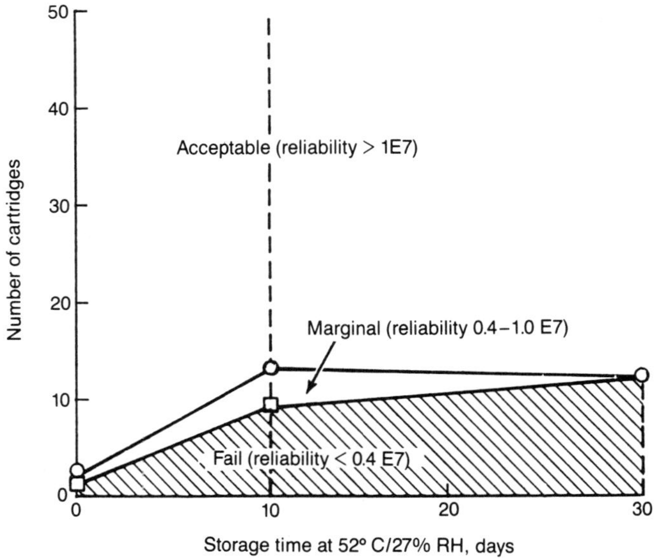

Fig. 5.27. Number of cartridge failures as a function of storage time at 52° C/27% RH.

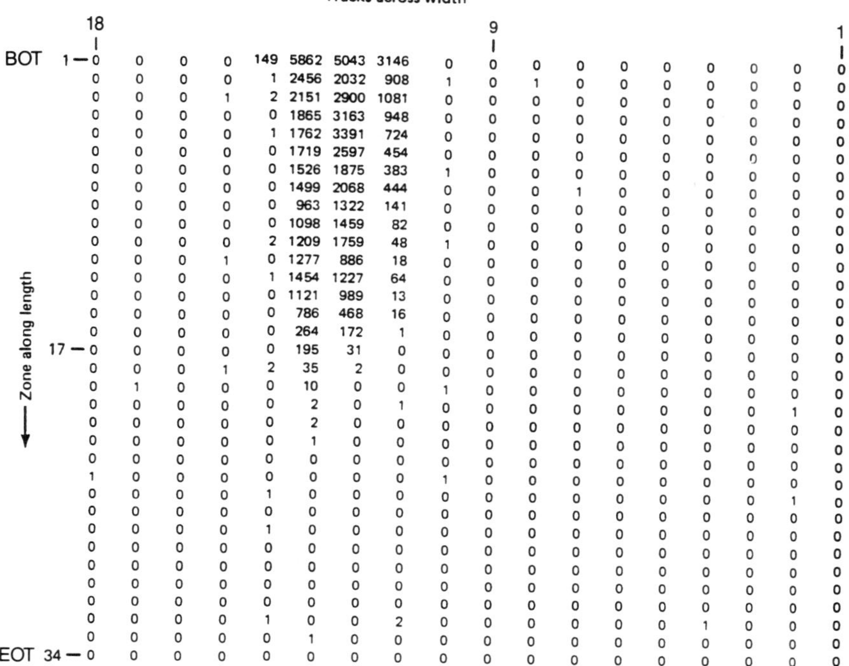

Zone	18	17	16	15	14	13	12	11	10	9	8	7	6	5	4	3	2	1
BOT 1	0	0	0	0	149	5862	5043	3146	0	0	0	0	0	0	0	0	0	0
2	0	0	0	0	1	2456	2032	908	1	0	1	0	0	0	0	0	0	0
3	0	0	0	1	2	2151	2900	1081	0	0	0	0	0	0	0	0	0	0
4	0	0	0	0	0	1865	3163	948	0	0	0	0	0	0	0	0	0	0
5	0	0	0	0	1	1762	3391	724	0	0	0	0	0	0	0	0	0	0
6	0	0	0	0	0	1719	2597	454	0	0	0	0	0	0	0	0	0	0
7	0	0	0	0	0	1526	1875	383	1	0	0	0	0	0	0	0	0	0
8	0	0	0	0	0	1499	2068	444	0	0	0	1	0	0	0	0	0	0
9	0	0	0	0	0	963	1322	141	0	0	0	0	0	0	0	0	0	0
10	0	0	0	0	0	1098	1459	82	0	0	0	0	0	0	0	0	0	0
11	0	0	0	0	2	1209	1759	48	1	0	0	0	0	0	0	0	0	0
12	0	0	0	1	0	1277	886	18	0	0	0	0	0	0	0	0	0	0
13	0	0	0	0	1	1454	1227	64	0	0	0	0	0	0	0	0	0	0
14	0	0	0	0	0	1121	989	13	0	0	0	0	0	0	0	0	0	0
15	0	0	0	0	0	786	468	16	0	0	0	0	0	0	0	0	0	0
16	0	0	0	0	0	264	172	1	0	0	0	0	0	0	0	0	0	0
17	0	0	0	0	0	195	31	0	0	0	0	0	0	0	0	0	0	0
18	0	0	0	1	2	35	2	0	0	0	0	0	0	0	0	0	0	0
19	0	1	0	0	0	10	0	0	1	0	0	0	0	0	0	0	0	0
20	0	0	0	0	0	2	0	1	0	0	0	0	0	0	0	0	1	0
21	0	0	0	0	0	2	0	0	0	0	0	0	0	0	0	0	0	0
22	0	0	0	0	0	1	0	0	0	0	0	0	0	0	0	0	0	0
23	0	0	0	0	0	0	0	0	0	0	0	0	0	0	0	0	0	0
24	1	0	0	0	0	0	0	0	1	0	0	0	0	0	0	0	0	0
25	0	0	0	0	1	0	0	0	0	0	0	0	0	0	0	0	1	0
26	0	0	0	0	0	0	0	0	0	0	0	0	0	0	0	0	0	0
27	0	0	0	0	1	0	0	0	0	0	0	0	0	0	0	0	0	0
28	0	0	0	0	0	0	0	0	0	0	0	0	0	0	0	0	0	0
29	0	0	0	0	0	0	0	0	0	0	0	0	0	0	0	0	0	0
30	0	0	0	0	0	0	0	0	0	0	0	0	0	0	0	0	0	0
31	0	0	0	0	0	0	0	0	0	0	0	0	0	0	0	0	0	0
32	0	0	0	0	1	0	0	2	0	0	0	0	0	0	1	0	0	0
33	0	0	0	0	0	1	0	0	0	0	0	0	0	0	0	0	0	0
EOT 34	0	0	0	0	0	0	0	0	0	0	0	0	0	0	0	0	0	0

Fig. 5.28. Error map of a failed cartridge, each zone is 4.8 m in length. The error tracks match the location of the uneven stack profile.

due to an uneven stack profile have the following characteristics:

- A large number of errors are present on one to five tracks matching the location of the uneven stack profile. The errors are greatest at the beginning of the tape where the distortion is most severe. This situation is illustrated by the error map in Fig. 5.28.
- There is reduced localized tape tension at the location of the errors. Tension profiles of the good tracks and the error tracks, as measured by five solid state pressure sensors mounted in a dummy glass head (see Appendix 5.B), are shown in Fig. 5.29.
- Low signal amplitude matches the location of errors, as shown in Fig. 5.30.

We also note that white-light interferometry [interference fringes observed through a simulated glass head against a moving tape, Bhushan (1990)] has been used to verify the suggested increase in head-to-tape separation at the location of error.

An example of the relationship between data reliability or probability of failure and tape distortion, in terms of tension to flatten, is shown in Fig. 5.31. The data indicates that there will be no failure if tension to flatten remains below 0.4 N and that 100% of the cartridges will fail when tension to flatten

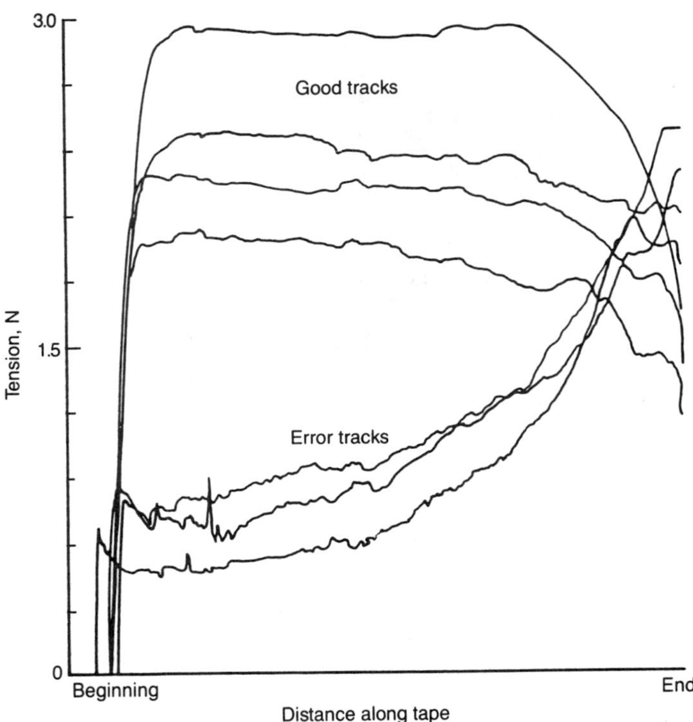

Fig. 5.29. Variation in tape tension through tape length along seven different tracks of a failed cartridge.

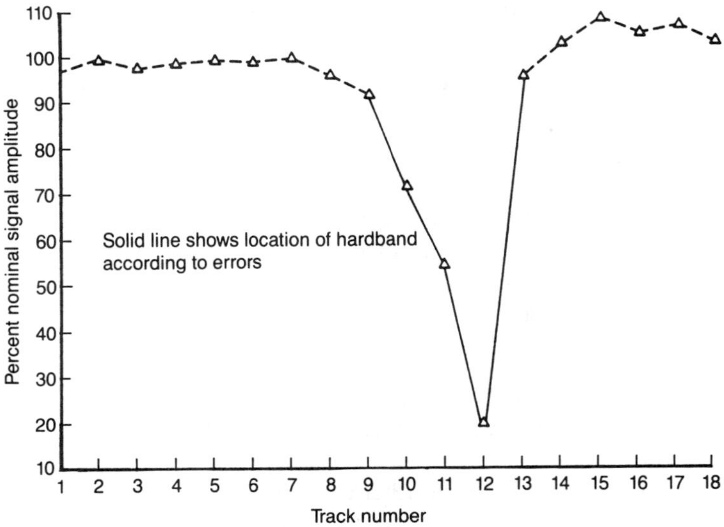

Fig. 5.30. Magnetic signal amplitude profile of a tape with magnetic errors at track numbers 10–13.

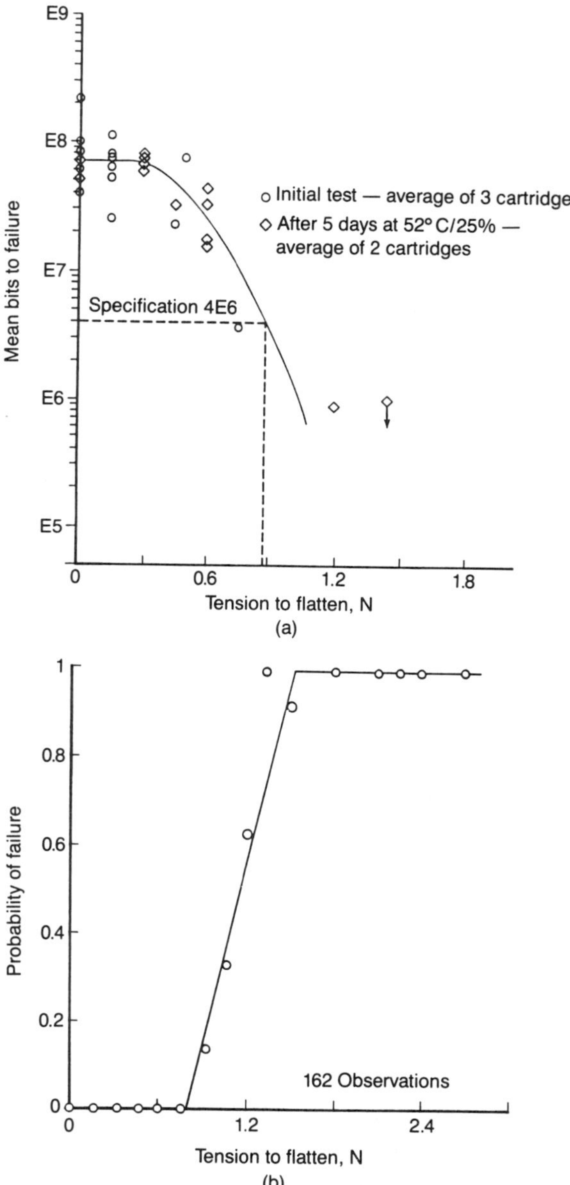

Fig. 5.31. (a) Data reliability, and (b) probability of failure as a function of the required tension-to-flatten.

reaches 0.75 N. Note that measurement errors, including differences in the 3480 drives used to determine data reliability, widen the sloped portion of the curve between 0.4 N and 0.75 N.

5.3.3. Summary

The out-of-plane deformation that results from winding successive layers of tape over a ridge is described and predicted. Sources of uneven stack profile, defined as localized circumferential ridges, include substrate or coating-thickness variations, air entrapment during winding, tension ridges, and scratches. Magnetic errors are shown to begin when the tape deformation achieves a tension-to-flatten value of about 0.7 N for an IBM 3480 drive. A limitation on the severity of uneven stack profile is found necessary in order to achieve the required data reliability for the tape drive. The resolution of the problem through process modifications that eliminate the cylindrical ridge is discussed. The recommended changes include uniformity in substrate and coating thickness, reduction of web tension at the coater head, the drying oven, and the jumbo rewind station; a reduction in coating speed; the use of spreader rollers and idler rollers; oven support; and the addition of a lay-on roll with an appropriate force at the rewind stations.

5.4. Mechanical Print-Through

Even when the operating machine conditions are kept constant, the tenting or spacing variations at the head–tape interface can arise because of the localized deformations in the magnetic tape. These deformations affect the magnetic performance in the same manner as the out-of-plane distortions described in the previous section.

When a tape is wound under tension onto a reel, the wraps conform to the profile of the wrap underneath. A flat profile is desirable but an uneven profile may be generated from the winding of tape over a contaminant particle, a rough hub surface, or an entrapped air bubble. The tape is then subjected to a localized, nonuniform, stress field and has a tension gradient across the defect. As this stress field relaxes during storage, the initial elastic strain becomes viscoelastic and a mechanical print-through is produced. High local stresses will produce a plastic deformation in the magnetic tape if they exceed its yield strength. The rate and the amount of the viscoelastic deformation are governed by the viscoelastic properties of the tape (PET substrate and magnetic coating), storage temperature, humidity, time, and stress gradient.

The data presented here are intended to place theoretical bounds on the size of the contaminant particles likely to cause magnetic errors in a tape drive. The experimental data presented on the kinetics of tape deformation is a result of an artificially induced defect. The rate of deformation is studied as a function of time, temperature, and the number of wraps away from the

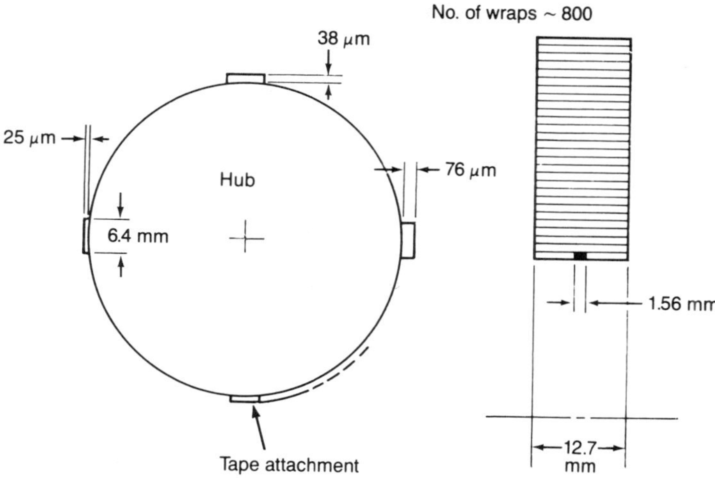

Fig. 5.32. Schematic showing the location of shims on a tape reel.

defect. Experiments were performed by winding tape on reels that have known defects at the hub. Shims made of PET were cut with nominal dimensions of 6.4 × 1.56 mm and thicknesses of 25, 38, and 76 μm. After attaching the 12.7-mm-wide and 31-μm-thick tape to the hub, shims of each thickness were placed on the first wrap, located 90° apart (see Fig. 5.32). The tape was wound on top of these artificial defects. All the tape reels were tested for magnetic performance, and those that had permanent magnetic errors at time zero were removed from the experiment. The tape reels were stored at different temperatures and for varying lengths of time. Then we recorded the number of wraps above the shim at which the first permanent error after storage occurred. The magnitude of the mechanical print-through at different wraps as a function of storage time was also measured. The maximum tape deformation of a single wrap in the print-through zone was recorded using a proximity probe while maintaining a nominal 0.5 N tension in the tape, since this deformation is strongly dependent on the applied nominal tension. The number of wraps away from the shim at which the first permanent error was observed is listed in Table 5.3. For the magnetic testing conditions used here we only observed magnetic errors in the location of a 76-μm shim. Even after 400 h of storage at 52° C, no magnetic errors were measured in the location of shims 25- and 38-μm thick.

To narrow down the critical thickness likely to cause a permanent error, more tapes were wound, but with only one shim in each reel. The shims were 25, 38, 51, 64, and 76-μm thick. Reels were stored at 52° C for varying lengths of time, up to a maximum of 80 h. The number of wraps exhibiting magnetic errors at 80 h as a function of shim thickness is plotted in Fig. 5.33. Again we found that shims 38-μm thick and less did not induce magnetic errors. Defor-

Table 5.3. Number of wraps where first permanent error occurred using a 76-μm-thick shim

		Storage time, h								
		3	5	10	20	30	40	80	120	400
Reel	Temp., °C	Number of wraps where first permanent error occurred								
1	52	66	65	55	87	54	67	93	84	91
2		51	57	70	78	70	72	—	83	125
1	43	ND	43	42	45	53	41	74	45	—
2		ND	42	41	ND	44	56	75	89	69
1	35	—	—	ND	ND	—	42	—	101	71
2		—	—	ND	ND	—	ND	—	74	69
1	22	—	—	ND	ND	—	ND	—	ND	ND
2		—	—	ND	ND	—	ND	—	ND	ND

ND—No error detected. Magnetically, the fortieth wrap is the first wrap that can be measured.

mations of the tape caused by shim print-through were measured for all reels, again using a proximity probe. Figure 5.34 shows maximum tape deformation versus a wrap number for 10 and 20 h of storage time at 52° C. Figure 5.35 shows a composite picture under an interference microscope of the 25-μm shim print-through across the tape width at about 50 wraps after storage at 52° C for 6 h. Tape deformation as a function of storage time for shims 76- and 38-μm thick is plotted in Fig. 5.36 for different storage temperatures. The relaxation times of tape deformation data are the same as that of stress relaxation data presented in Chapter 3 (see Appendix 5.C).

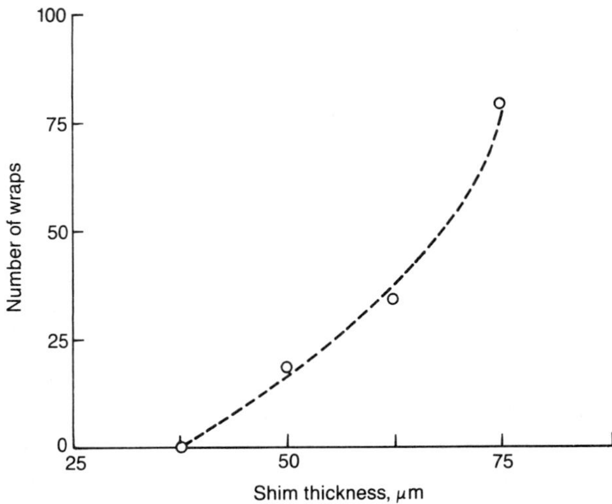

Fig. 5.33. Number of wraps at which the first magnetic error was observed (after storage at 52° C for 80 h) as a function of shim thickness (Bhushan et al., 1984b).

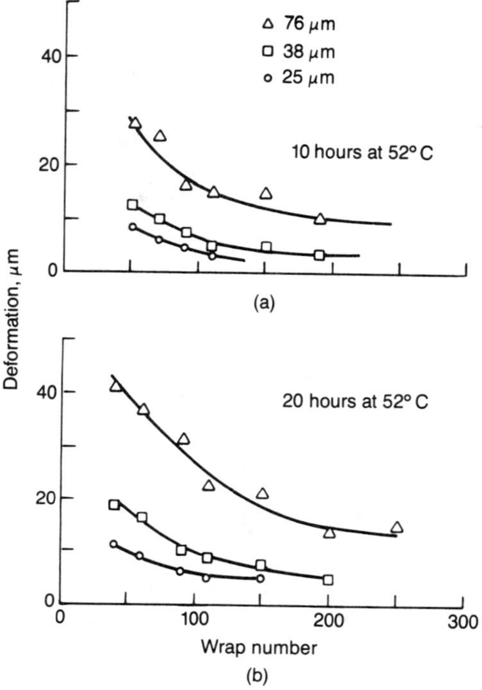

Fig. 5.34. Tape deformation as a function of the wrap number for shims 25, 38, and 76-μm thick at 52° C for (a) 10 h and (b) 20 h (Bhushan et al., 1984b).

These experiments show that viscoelastic/plastic deformation of the tape leading to magnetic errors can be caused by contaminant particles or tape debris getting wound in the tape or a glue dot sometimes used to attach the tape end to the reel hub during winding. (In most cases, water–Freon solution is used to attach the tape end to the hub in order to prevent any mechanical print-through as a result of tape attachment.) The number of affected wraps and the severity of the deformation are dependent on the size and shape of the particle, the tape thickness, and its deformation kinetics, storage temperature, and time.

5.5. Staggered Wraps

During the winding and the rewinding processes, the magnetic tape should stack perfectly, or damage occurs in the offset region of the tape. The severity of the damage depends on the tape's deformation kinetics and the winding and storage conditions. The result of the damage is magnetic errors at the edge tracks (Owen, 1971; Bhushan et al., 1984b). Staggered wraps generally droop radially inwards towards the hub, because of the radial pressure acting on the offset wraps. For a single staggered wrap, this radial pressure is the

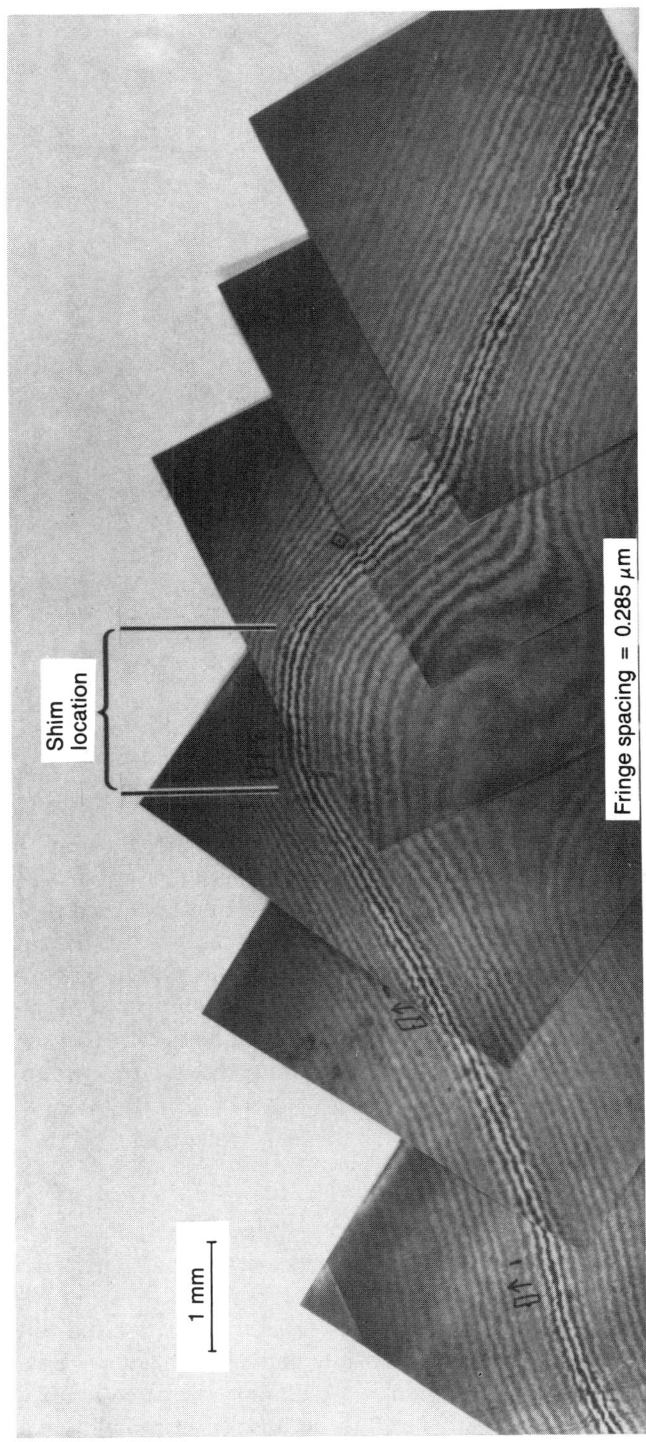

Fig. 5.35. Composite microscope picture of the 25-μm-thick shim print-through of a tape (Bhushan et al., 1984b).

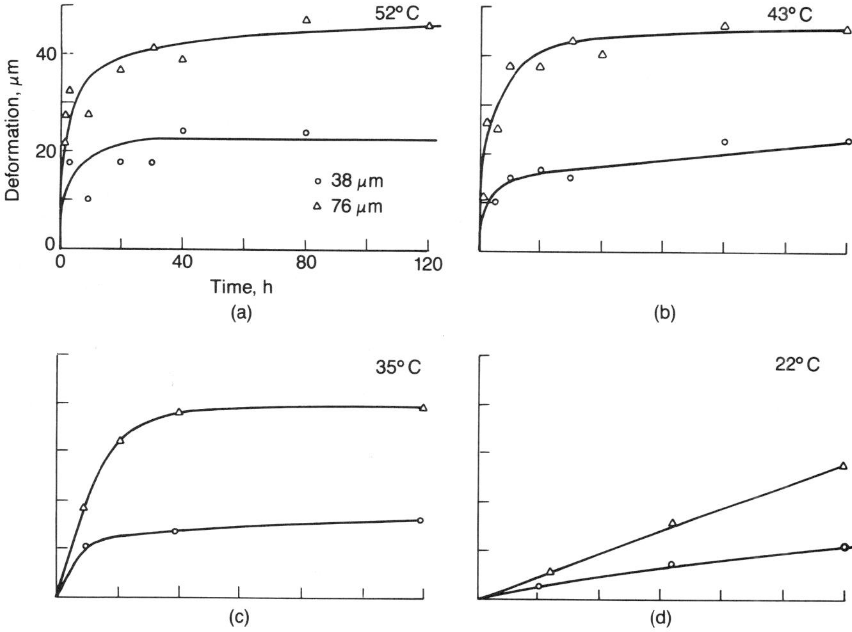

Fig. 5.36. Deformation at the fiftieth wrap as a function of storage time for shims 38 and 76-μm thick at (a) 52° C, (b) 43° C, (c) 35° C, and (d) 22° C (Bhushan et al., 1984b).

radial component of its hoop stress present caused by winding. When multiple wraps are offset, the cumulative effect of the radial pressure in all the outer offset wraps adds to the radial stress in a given offset wrap to produce droop. The droop is considerably greater in the latter case, since the total radial pressure can be two to three orders of magnitude greater than in the former. With time, the droop becomes partially viscoelastic/plastic and a crease is formed in the tape. This process is accelerated and aggravated by increasing temperatures and humidities. With the trend towards increased track density, a short overhang can, with time, produce a crease in the edge of the data track. With continued use of the tape, this crease produces an increased head–tape separation and a corresponding drop in signal amplitude. A severe crease will result in data loss. Staggered wraps can occur as a result of stack slip in wound tape or of the presence of some extraneous axial force during tape winding.

A destructive and complicated but quantitative method to study staggered wraps has been described earlier, in which a tape reel has to be potted and sectioned (Figs. 5.18 and 5.37). A less-involved method is to take shadow pictures of the sides of a reel. The technique is as follows: the tape reel is placed flat on a table, and a camera directly above. Blue light is shone on the reel at a 45° angle. This gives shadowing proportional to the height of the staggering. If more than one wrap is staggered at any location, they cannot

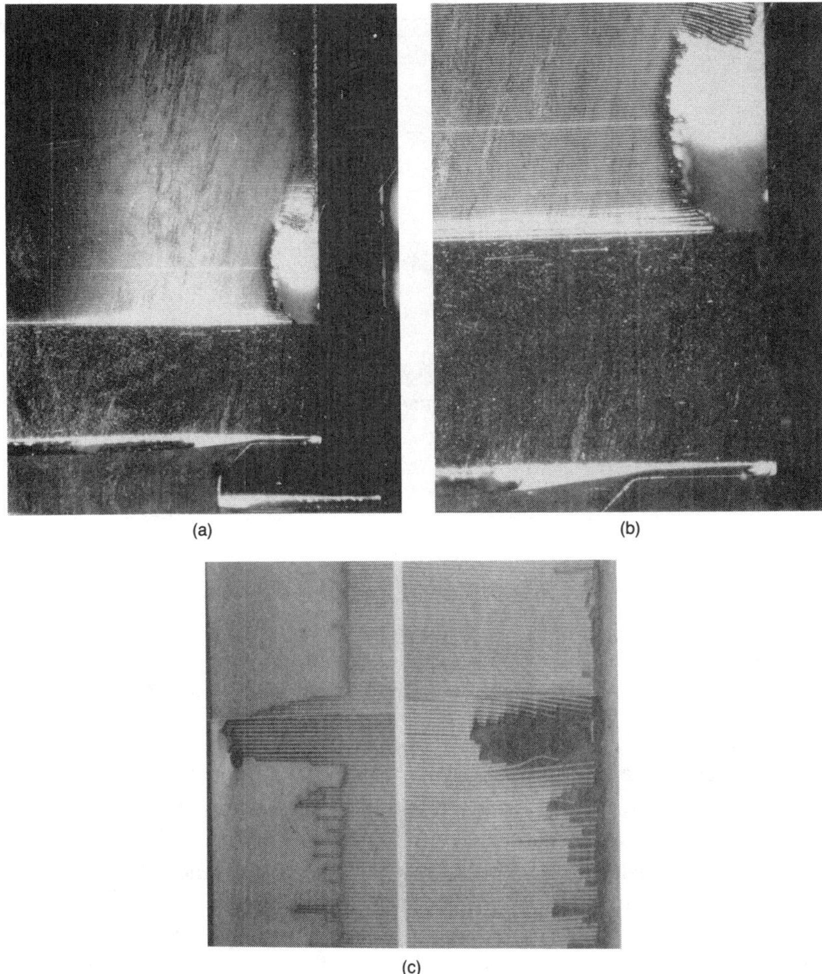

(a)

(b)

(c)

Fig. 5.37. Cross-sectional view of (a) a reel of tape with the tape stacked against one flange, (b) the same reel of tape as in (a) showing the area of maximum tape droop, and (c) another reel of tape with a stack slip in between.

be distinguished. However, this technique gives good semiquantitative information. An optical projection photograph of a staggered reel is shown in Fig. 5.38.

Various patterns appear in the staggered wraps. In Figs. 5.18(a) and (b), a totally irregular pattern is found, while in Fig. 5.18(c) the pattern appears to be periodic. All these figures show the same reel. In Fig. 5.37(a), we see that after a few wraps the tape slips to one flange and continues to stack up against this flange. In Fig. 5.37(b), we see a close-up view of the severe droop in the staggered wraps nearest the hub. Its severity decreases in the outer offset

Fig. 5.38. An optical projection (shadow) photograph of a side of the reel showing poor tape stacking (Bhushan et al., 1984b).

wraps; it is negligible when it is sufficiently far from the onset of staggering. Figure 5.37(c) shows a similar example in a stack. In Fig. 5.18, where a single wrap is offset, the amount of droop is negligible.

5.5.1. Sources of Tape Stagger

Tape stagger occurs either in the reel or at the entrance of the reel. Various sources can be identified in broad categories:

- Poor tape guiding in the drive causing tape wobble at the entrance of the reel.
- Poor tape quality (cupping, curvature, tension gradients across its width, poor slit edge).
- Elastohydrodynamic air-film generation during cartridge rewind facilitating axial interlayer slippage.
- Low winding tension or low reel tension after storage.

We believe that the mechanisms are interactive.

Using Eq. (5.6), the elastohydrodynamic air-film thickness, $h(0)$, for the IBM 3480 cartridge can be calculated for the following parameters: R varies from 24.6 mm to 48.3 mm, $T = 2.2$ N, and $V = 4$ m/s during rewind. The calculated air-film thickness is 2–4 μm which is substantially greater than the mating surface roughness and, thus, is sufficient to cause interlayer slippage. We note that air-film thickness decreases with an increase in tension and a decrease in the winding speed. Part of this hydrodynamically produced air film is squeezed out as the next wrap is wound but part of it is locked in. In the compression tests performed on wound tape, we found that it takes up to 20 h for all the air to seep out. Tape cupping and curvature, in addition to increased interlayer friction, help some of the air film to be locked in. Cross-sectional views of a reel of tape having entrapped air between the layers were shown in Figs. 5.17 and 5.18.

The tape and the drive must together, or separately, provide a forcing function to produce an axial motion in the tape. Hydrodynamic air film aids in the staggering of wraps but does not cause it on its own. Asymmetric tape nonuniformities, such as cupping, curvature, and tension gradients result in the preferential escape of the entrapped air film to one side of the tape, providing a reactive force to axial tape motion. Machine-reel wobble, poor tape guiding, and drive-motor misalignment with the machine reel's axis are sources of dynamic forcing functions that could produce staggered wraps. Rapid reel-motor acceleration and deceleration, as well as the presence of electrostatic fields, may also contribute to this problem.

Stack slip can occur in a tape reel, which is either wound with too low a tension or in which the radial stress has relaxed significantly with time when it is exposed to vibration or environmental stresses (Waites, 1982). Low winding tension produces a lower radial stress (Chapter 4); a temperature or humidity drop can reduce the radial pressure in the reel of tape to zero. Entrapped air, as shown in Fig. 5.18, helps stack slip to occur.

We conducted a matrix of tests to determine if the interlayer slippage is the only source of tape stagger. These tests investigated whether or not the amount of tape stagger is a function of the winding tensions and speeds that affect the air-film thickness. The tapes were wound at 2.2 N tension and winding speeds of 1, 2, 3, and 4 m/s on IBM 3480 tape reels. In addition, the tapes were wound at 1 m/s and at 1.1, 1.7, 2.2, 2.8, and 3.3 N of tension. The tape stagger in each reel was measured by the optical projection technique. We found that neither the speed nor the tension consistently affected the amount of stagger. We also found that for a given tape drive and winding conditions, different tapes stack differently. In addition, a tape staggers differently for a drive with different guiding methods. Therefore, we conclude that the elimination of interlayer slippage is not sufficient for the elimination of tape stagger. Attention should be paid to tape quality and tape drive guiding.

5.5.2. Effect of Winding Parameters and Storage Conditions

5.5.2.1. Analysis of Elastic Droop of a Staggered Wrap

A simple mathematical model for the elastic droop of a single staggered wrap and its experimental verification are now presented. The sensitivity of the tape

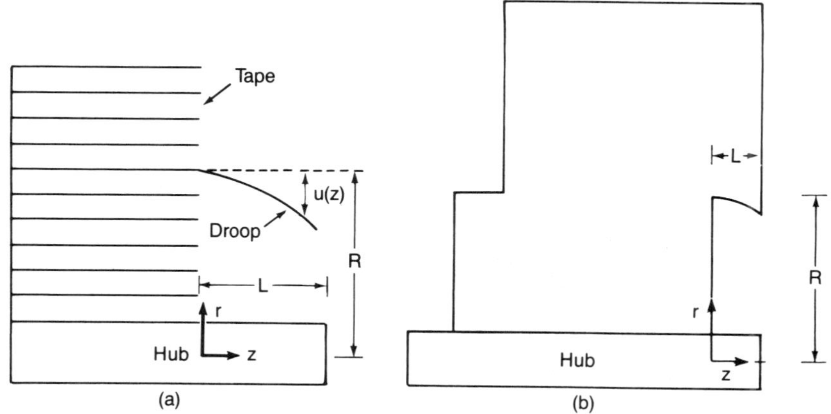

Fig. 5.39. Schematic of a cross-sectional view of a tape reel (a) with a single staggered wrap and (b) with a stack shift.

droop to overhang length, staggered-wrap radius, tape thickness, and radial stress or tape tension are shown. For a single staggered wrap with a length of overhang L, at a radius R [Fig. 5.39(a)], caused by some radial pressure $P(z; R)$, the droop $u(z)$ is given by (Timoshenko and Woinowsky-Krieger, 1959)

$$D(d^4u/dz^4) + (Et/R^2)u = P(z; R), \qquad 0 \leq z \leq L, \tag{5.18}$$

where D is the flexural rigidity of the tape, as defined earlier. One of the simplest, yet realistic assumptions that can be made for $P(z; R)$ is that

$$P(z; R) = [(T/tw) + (Eu/R)](t/R), \tag{5.19}$$

where T/tw is the (constant) circumferential (hoop) stress in the wrap caused by winding, while Eu/R is the relief in this stress caused by the droop of the tape. The radial stress (σ_{rr}) in the wrap caused by winding is equal to T/Rw. Substituting Eq. (5.19) into Eq. (5.18), u can be solved as

$$u(z) = e^{\beta z}(A \cos \beta z + B \sin \beta z) + e^{-\beta z}(C \cos \beta z + D \sin \beta z) - (TR/2twE), \tag{5.20}$$

where $\beta = [6(1 - v^2)/t^2R^2]^{1/4}$. Four boundary conditions are required to determine the four unknowns A, B, C, and D

$$u(z = 0) = 0, \qquad (du/dz)(z = 0) = 0, \tag{5.21}$$

$$(d^2u/dz^2)(z = L) = 0, \qquad (d^3u/dz^3)(z = L) = 0. \tag{5.22}$$

Equation (5.21) requires a zero droop and slope at $z = 0$, which is the fixed end, while Eq. (5.22) is a statement of zero shear and moment at $z = L$, which is the free end. Hence, in this case, $u(z)$ can be fully determined. For the case considered here, $E = 4.13$ GPa, $v = 0.30$, and $w = 12.7$ mm.

To verify the validity of the model, we designed an apparatus where we could easily alter the parameters and measure the resulting droop. The appa-

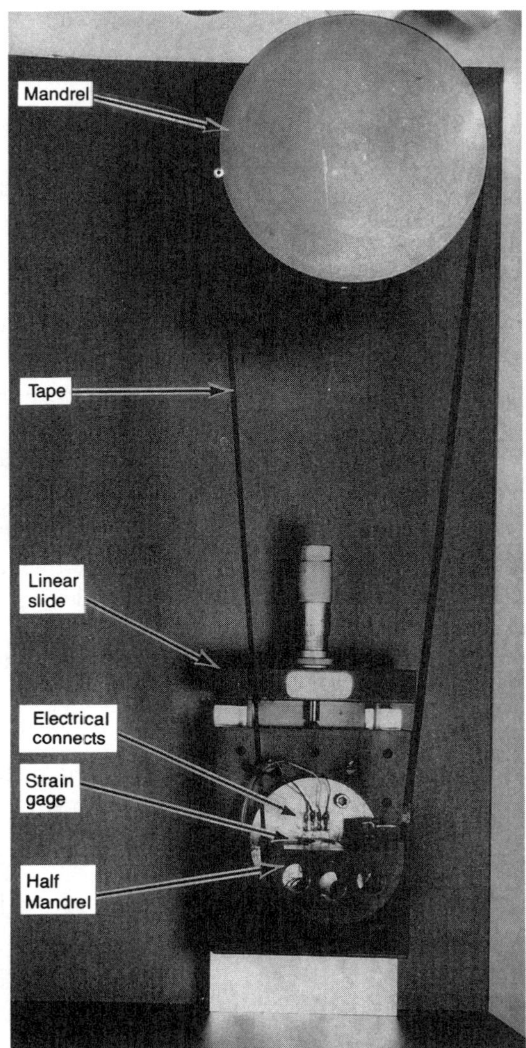

Fig. 5.40. Schematic of apparatus to measure the droop of an offset wrap (Bhushan et al., 1984b).

ratus, shown in Fig. 5.40, consists of a horizontal base plate on which a vertical back plate stands. To this back plate are attached an upper cylindrical mandrel of circular cross section and a lower semicircular cylindrical mandrel connected through a rectangular beam. On this rectangular beam, strain gauges are attached so that when a ring of tape is placed around both mandrels, a known tension can be placed in the tape using a sliding scale. On the upper mandrel, a cylindrical groove of smaller width than the tape was cut so that we could produce a staggered wrap. A set of mandrels of various radii allowed us to look at the effect of the wrap's radius. Using Moire

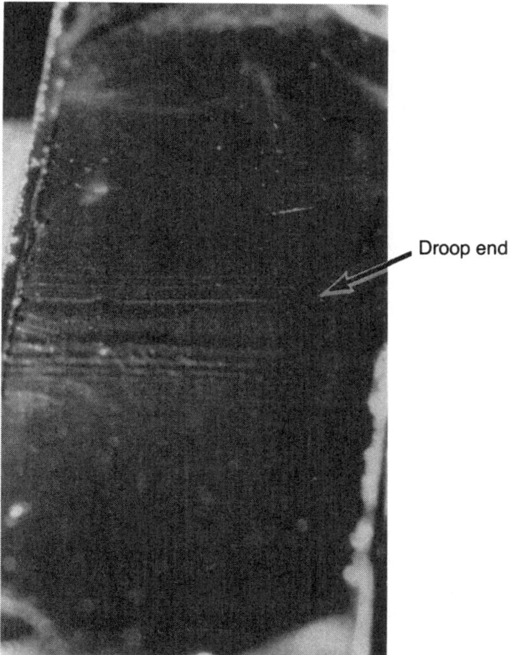

Fig. 5.41. Fringe pattern through diffraction grating caused by tape droop (Bhushan et al., 1984b).

interferometry, shining light on a diffraction grating at an angle, and taking a photograph of the diffracted light pattern, we measured the resulting droop (see Fig. 5.41). The detections obtained were in the range of 20–200 μm for tape tensions ranging from 3–5 N.

In Fig. 5.42 we show various comparisons of theory and experiment. The accuracy of the mathematical model was surprisingly good. The linearity between increasing tape tension and maximum tape droop is shown in Figs. 5.42(a) and (b) for various tape thicknesses at a given radius and for various radii at a given thickness. Note that for a constant tension and increasing radius there is a decreasing tape droop since there is a decreased radial pressure. In an actual tape reel, there is increasing hoop stress with increasing radius. Therefore, depending on the actual reel geometry, we can have a similar droop profile in an outer wrap as in an inner wrap. Of course, as $R \to \infty$ then $u(z) \to 0$, from Eq. (5.20). Figure 5.42(c) shows increasing droop with decreasing tape thickness for a given tension and radius. Increasing overhang lengths in the staggered wraps produce increasing droop, Fig. 5.42(d). The overhang length must be limited to about 0.5 mm for 12.7-mm-wide tape.

For multiple staggered wraps [see Fig. 5.39(b)], the radial pressure exerted on the innermost wraps can be two to three orders of magnitude greater than that given in Eq. (5.19). This is the cumulative effect of the radial stress in each

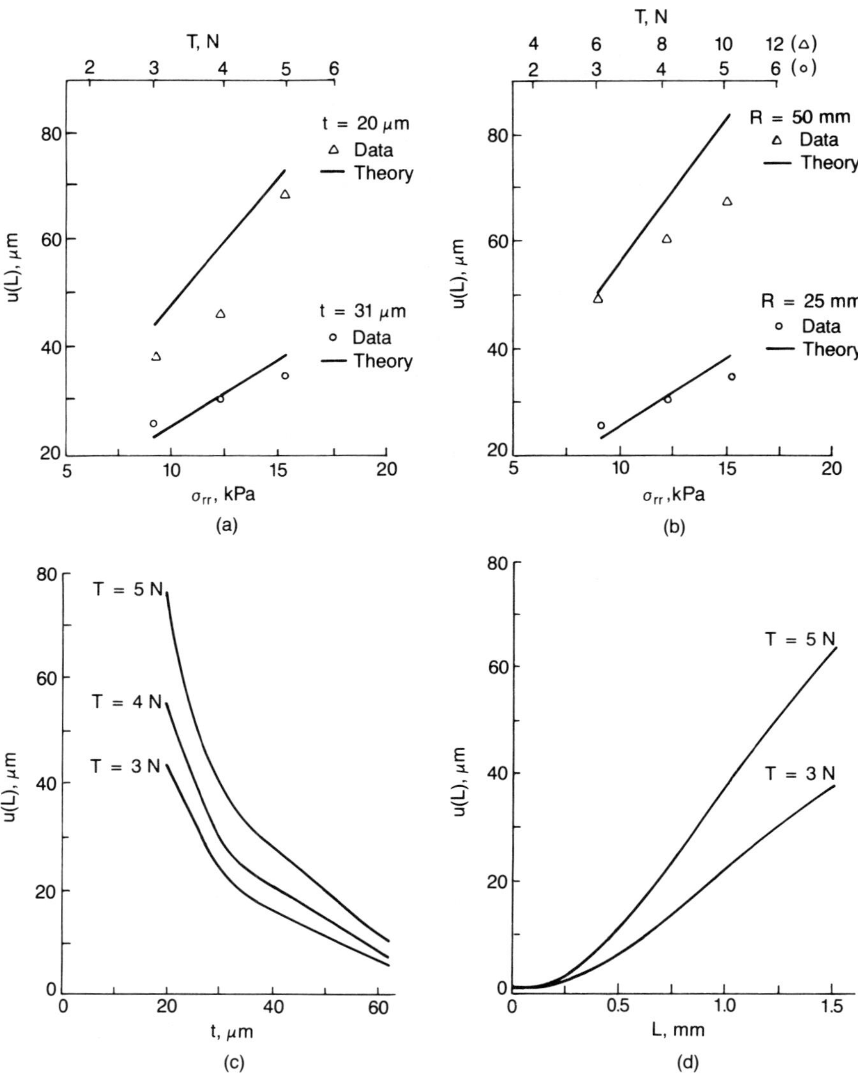

Fig. 5.42. Maximum tape droop (a) as a function of radial pressure for various tape thicknesses at radius 25 mm and overhang length 1 mm; (b) as a function of radial pressure for various wrap radii at a tape thickness of 31 μm and overhang length 1 mm; (c) as a function of tape thickness at radius 25 mm, overhang length 1 mm, and tensions of 3, 4, and 5 N; and (d) as a function of overhang length for a radial pressure of 12.26 kPa, radius 25 mm, and tape thickness of 31 μm (Bhushan et al., 1984b).

wrap on top of another wrap. An upper bound on the magnitude of the droop can be obtained by taking $P(z)$ as the radial pressure at that radius caused by winding (Frye, 1967). Once the droop has been determined, and knowing the stress relaxation and viscoelastic deformation kinetics, which depend strongly on temperature and humidity, we can predict the viscoelastic deformation in the tape right after unwinding from the reel.

5.5.2.2. Experimental Study

We present the results of an experimental study to investigate the effect of various winding parameters and storage conditions on the tape damage. Significant parameters are: overhang length, number of offset wraps, radial location of offset wraps, winding tension and speed, and storage conditions. A laboratory winder was used to wind reels with 12.7-mm-wide and 165-m-long tape having a select number of wraps at a select radial location of the reel offset (track 18—bottom of the reel, track 1—top of the reel) by a given length at different winding tensions and speeds. The reels were stored at select

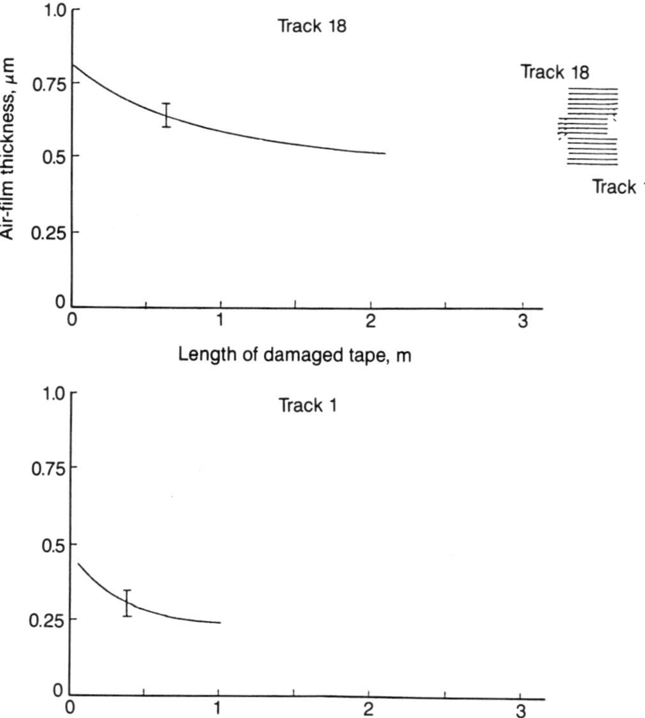

Fig. 5.43. Hydrodynamic air-film thickness in the offset zone as a function of length of damaged tape. Overhang length = 1 mm, number of offset wraps = 10 (track 18 side), winding tension = 2.2 N, winding speed = 2 m/s, radial location of first offset wrap = 38 mm, and storage condition = 52° C and 40 h.

conditions. The air-film thickness of the tape flying over a glass head was measured at 2.2 N and 2 m/s and was used as a measure of tape damage. Increased flying height suggested that the tape segment was damaged, a nominal flying height was about 0.1 μm.

Figure 5.43 shows the air-film thickness as a function of the length of damaged tape for the case when the ten wraps of a tape were offset on the track-18 side. There was, however, damage on the track-1 side as well, because offset on the track-18 side corresponds to offset on the wraps following to offset wraps on the track-1 side. Thus, offset on either side causes increased air-film thickness and the resulting magnetic errors on both sides. Damage on track 1 is different from damage on track 18 because of the differences in the number of offset wraps and the space available for offset wraps to bend over.

Figure 5.44(a) shows the mean air-film thickness of the first offset wrap, which is a maximum, and the length of damaged tape as a function of the number of offset wraps at both the track-18 and the track-1 sides. There was an increase in the film thickness and the length of damaged tape as a function of offset wraps up to about 50–100 wraps. The tape damage at tracks 1 and 18 are different and it can be explained by a model shown in Fig. 5.44(b). We found that damage for offset wraps less than ten on the track opposite to the offset side is greater than if the wraps are more than ten.

Figure 5.45 shows the effect that the overhang length (amount of offset) has on the air-film thickness. The air-film thickness increases as the amount of tape offset increases. The following observations were made:

Overhang length, mm	Tracks where damage occurred
Up to 0.5	None
Up to 1	1, 18
Up to 1.5	2, 17
Up to 2	3, 16

Figure 5.46 shows the effect of winding tension on the tape damage, which increases with the tension. Figure 5.47 shows that the tape damage increases as the radial location decreases because of increased radial pressure. This radial pressure is a result of an increased number of overlaying wraps, as discussed in Chapter 4.

The cartridges with ten offset wraps were exposed to various temperatures and times. The results of the exposures to the offset wraps are shown in Table 5.4. The tape damage occurs only when the cartridge with offset tape is exposed to a high temperature for a long period. At ambient, the damage is minimal. To compare the damage of one condition to another, the activation energy of the tape should be used (see Chapter 3). Since no damage occurred at room temperature, it implies that no plastic deformation occurred. Very high stresses can lead to crease or plastic deformation.

Finally, we ask if tape damage recovers if the damage tape is rewound perfectly and stored. A damaged cartridge with standard offset conditions

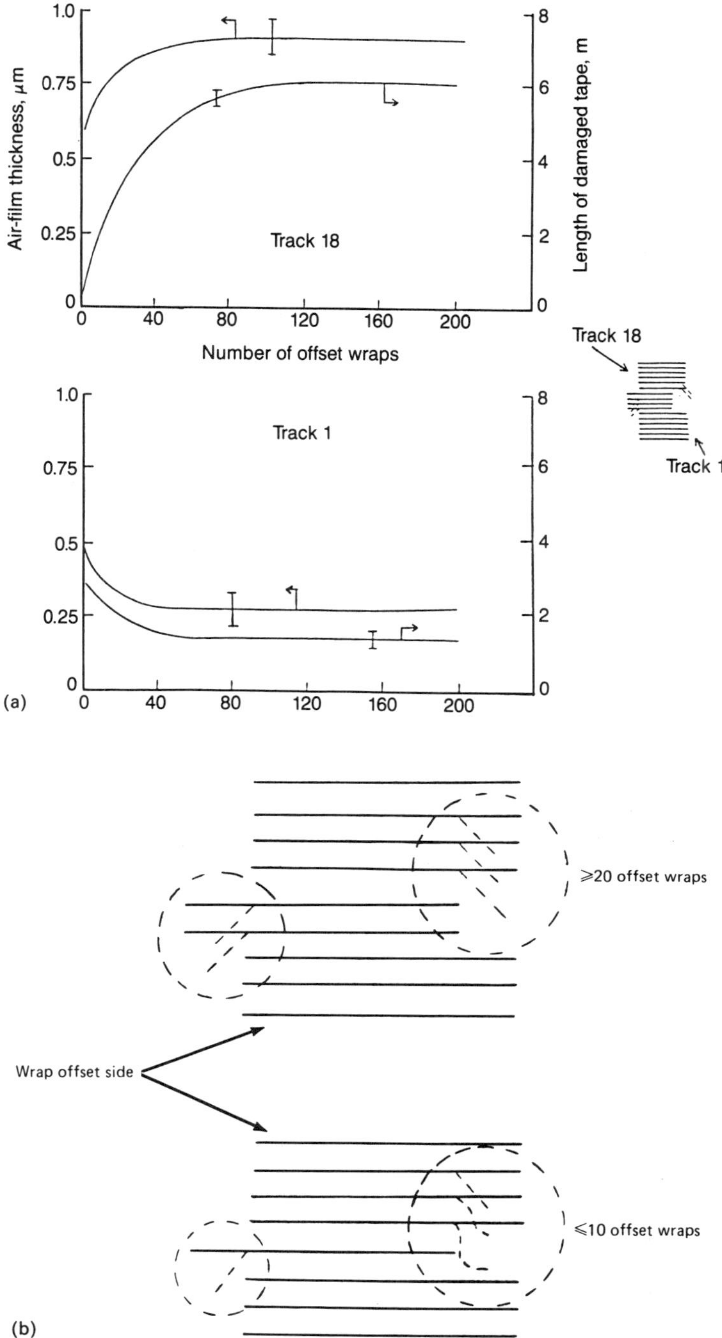

Fig. 5.44. (a) Hydrodynamic air-film thickness of the first offset wrap and length of damaged tape as a function of the number of offset wraps on the track 18 side. Overhang length = 1 mm, winding tension = 2.2 N, winding speed = 2 m/s, radial location of first offset wrap = 38 mm, and storage condition = 52° C and 40 h. (b) Schematic of the effect of the number of offset wraps on the tape deformation.

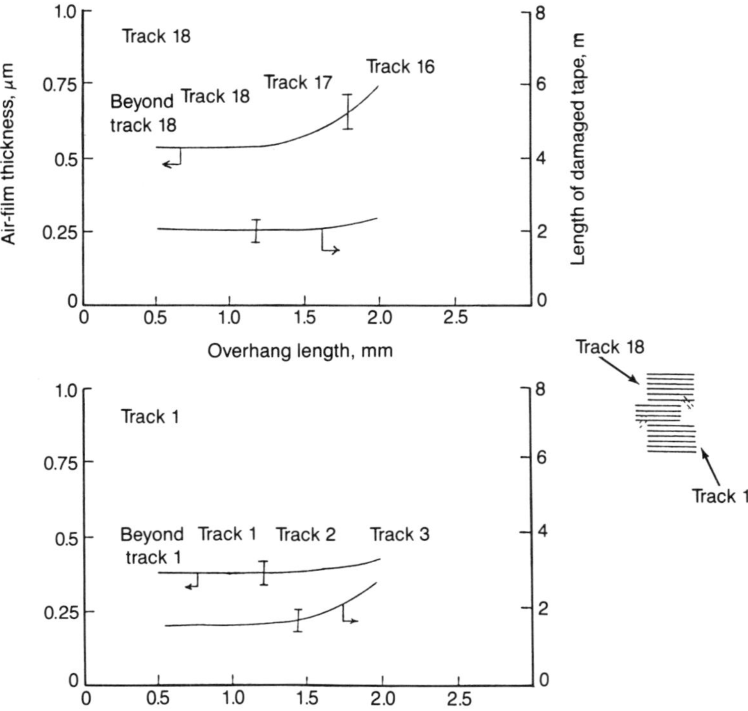

Fig. 5.45. Hydrodynamic air-film thickness of the first offset wrap and length of damaged tape as a function of overhang length on the track 18 side. Number of offset wraps = 10, winding tension = 2.2 N, winding speed = 2 m/s, radial location of the first offset wrap = 38 mm, and storage condition = 52° C and 40 h.

was exposed at 52° C for 40 h and did not recover when it was perfectly wound and stored at room temperature for 2 weeks. The cartridges recovered after storing at 52° C for 40 h. Thus, this procedure can be used to recover the damaged cartridges.

5.5.3. How Does Staggered Tape Cause Errors?

An IBM 3480 tape reel was wound with a select number of offset wraps at a select radial location of the reel by a given length. The reel was then stored at 52° C for 40 h. The damaged tape was placed over a glass head both in static and dynamic modes to measure its deformation pattern. To make static measurements, the creased tape was wrapped over a rod under a light tension of 0.55 N. A Moire grating with a grating spacing of 50 μm was placed over the tape so that the grating did not contact the tape. A fringe pattern was observed with a camera at 45° (β) to the Moire plane. The light source used

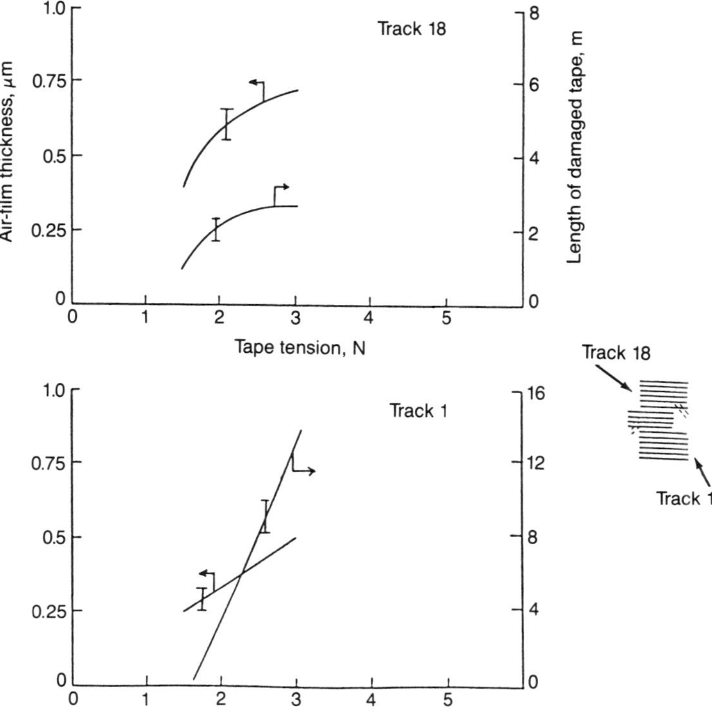

Fig. 5.46. Hydrodynamic air-film thickness of the first offset wrap and length of damaged tape as a function of winding tension. Overhang length = 1 mm, number of offset wraps = 10 (track 18 side), winding speed = 2 m/s, radial location of first offset wrap = 38 mm, and storage condition = 52° C and 40 h.

to illuminate the surface was at 35° (α) (Fig. 5.48). These measurements gave the Moire fringe spacing of:

$$(1/\text{number of gratings per unit distance})/(\tan \alpha + \tan \beta),$$

which corresponds to about 30.5 μm of spacing between the fringes. Figure 5.49 shows photographs of the Moire fringe patterns of creased tape. The tape lifts up at the crease about 50 μm at 0.55 N of tension.

Next, the air-film thickness of creased tapes was measured using the white light interferometry at 2.2 N and 2 m/s. We found an increase in the flying height in the creased zone. To further verify dynamic film thickness results, we made tension measurements in the damaged zone (near track 18) to see if there was a drop in tension, which could account for the increased flying. Tape-tension measurements were made on a glass head instrumented with five miniature pressure sensors that were equally spaced across the width of the glass head (see Appendix 5.B). Figure 5.50 shows variation in the tape

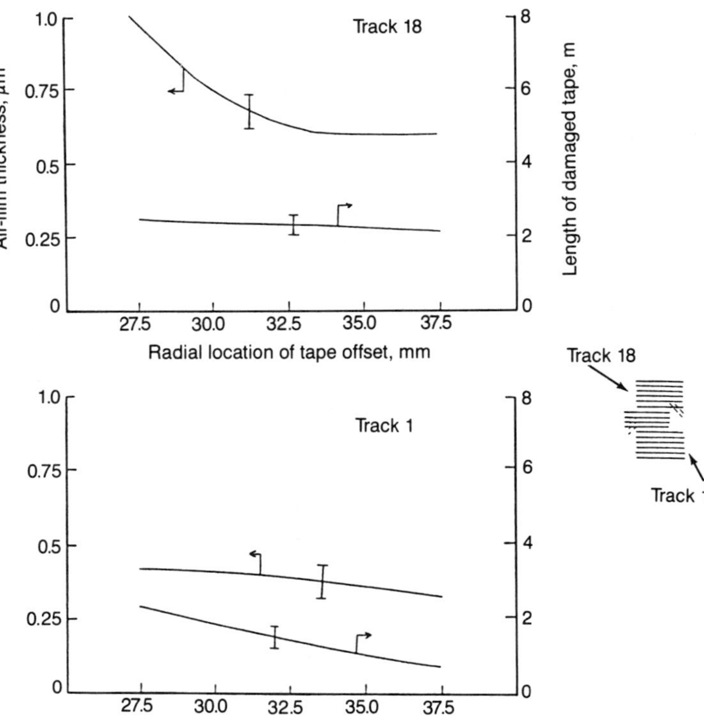

Fig. 5.47. Hydrodynamic air-film thickness of the first offset wrap and length of damaged tape as a function of the axial location of tape offset. Overhang length = 1 mm, number of offset wraps = 10 (track 18 side), winding tension = 2.2 N, winding speed = 2 m/s, and storage condition = 52° C and 40 h.

Table 5.4. Tape reels with offset wraps exposed to various temperatures for various times. Overhang length = 1 mm, number of offset wraps = 10, winding tension = 2.2 N, winding speed = 2 m/s and radial location of first offset wrap = 38 mm

Temperature, °C	Time	Tape tension, N	Tape damage occurred
22	2 days	2.2	None
22	2 days	9	None
52	1 h	2.2	None
52	5 min	2.8	Slight
52	40 h	2.2	Yes
60	2 h	2.2	Yes, comparable to 52° C and 40 h

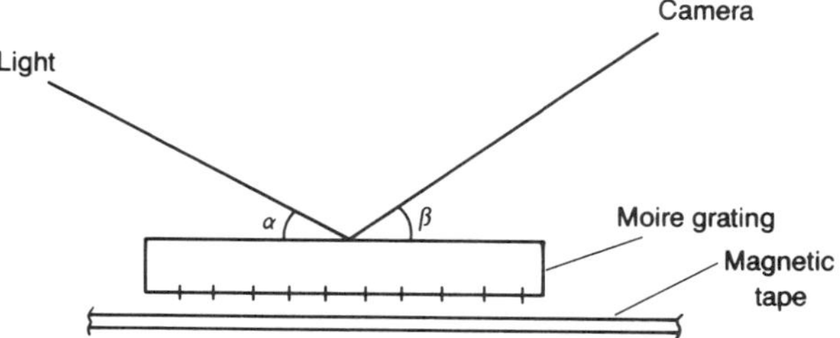

Fig. 5.48. Schematic of Moire grating method to measure out-of-plane distortion in a tape segment.

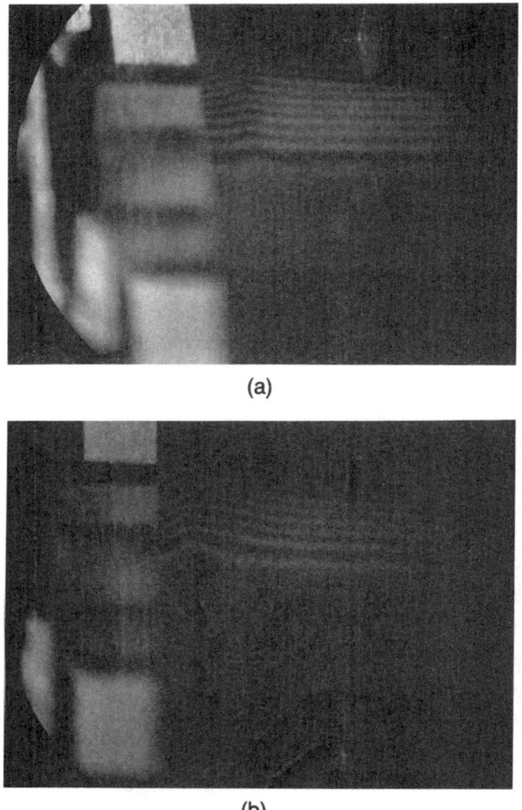

(a)

(b)

Fig. 5.49. Photographs of the Moire fringe patterns at a light tension of 0.55 N of creased tapes from an IBM 3480 cartridge wound at 2.2 N and 2 m/s with an overhang length of 1 mm at a radial location of (a) 38 mm and (b) 28 mm. Tape cartridges were exposed to 52° C for 40 h after winding. The inner and outer radial locations of the cartridge were 25.5 and 49 mm, respectively.

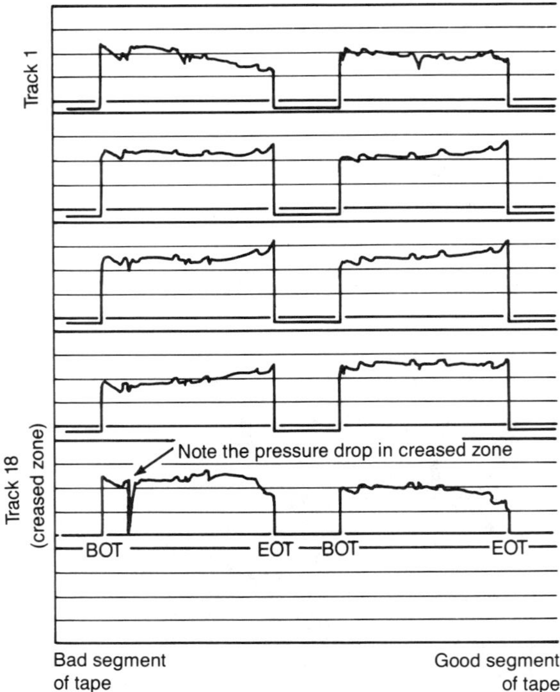

Fig. 5.50. Variation in the tape tension through the tape length along five tracks in bad and good segments of the tape. BOT—beginning of tape and EOT—end of tape.

tension for the five probe locations for the good and the bad segments of the tape. We found that the tape tension was very low in the creased zone, which must increase the flying height. Hence, from static and dynamic measurements, we conclude that there is an increased separation at the head-tape interface. This separation is a geometrical effect that results in increased flying height in the creased zone, which leads to magnetic errors.

Based on these observations, we propose a following model. The shift in the tape stack either in single or multiple layers [Fig. 5.51(a)] may lead to crease because of plastic deformation. Crease severity is affected by the quality of the edge of the tape underneath the offset stack, as shown in Fig. 5.51(b). Also, if the offset stack [Fig. 5.51(a)] is exposed to high temperature and humidity, viscoelastic deformation of the offset region occurs. This deformation is instantaneously nonrecoverable when the tape stack is unwound from the reel and traversed over the head right after winding. The plastic or viscoelastic deformation can either affect the flying height (geometrical effect), which results in a data-reliability problem, as shown in Fig. 5.51(c), or damages the portion of the tape at the offset location, which renders the tape incapable of writing.

Now we ask, how severe is the loss in magnetic signal in a creased tape and

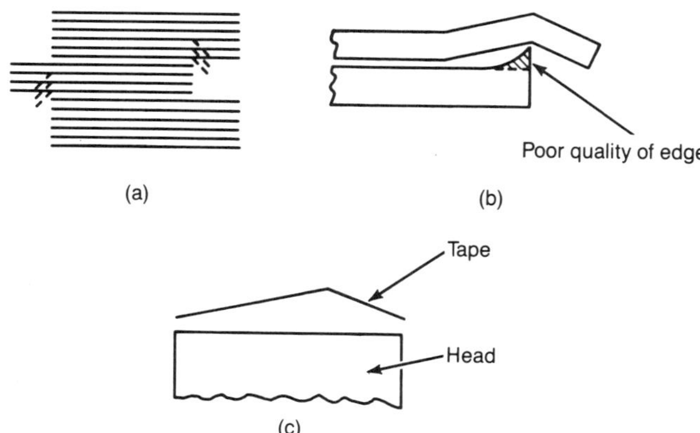

Fig. 5.51. Schematic of (a) shift in tape stack (single or multiple layers); (b) quality of edge can increase crease severity; and (c) viscoelastic deformation affects flying height (geometrical effect).

do magnetic errors in a creased tape occur in long lengths or short lengths? We developed a written tape with a visible crease in a colloidal suspension of magnetite (Fe_3O_4) and observed the drop-out areas. The unwritten areas become brownish. Magnetic signal attracts magnetite particles and makes them bluish. We found that the creased tape had a severe loss of amplitude which varies in the creased zone. Bit-by-bit information was obtained by measuring the magnetic amplitude of a 5-m-long creased tape at 976 kHz in a 3480 tape drive. Figure 5.52 shows the amplitude distribution of good and bad segments of the tape. Further dropout analysis showed that in less than 0.2% bits, the amplitude dropped below 15% of its nominal value and the longest dropout was 0.15 ms at a tape speed of 2 m/s (bit length $= \mu$m). Thus it appears that tape damage severity changes in the offset wraps and magnetic errors occur due to short length dropouts.

5.5.4. Methods of Preventing Tape Stagger

The solution to tape stagger lies in the tape drive and/or tape reel. Proper guiding in the drive probably should provide good stacking. A good stacking can be guaranteed only if the tape is guided inside the cartridge. However, if the tape is poorly guided in the drive, edge damage in guiding inside the reel can occur. Therefore, a perfect solution is properly guiding the tape in the drive and fine-tuning the guide inside the reel. Because a tape drive should perform on any competitive media, it should be insensitive to tape quality. Machine reel wobble, drive-motor misalignment with the machine reel's axis, and poor tape guiding must be avoided.

The most common causes of tape mistracking are: nonparallelism between the rollers' axes and a taper roller in the tape path and asymmetric tape

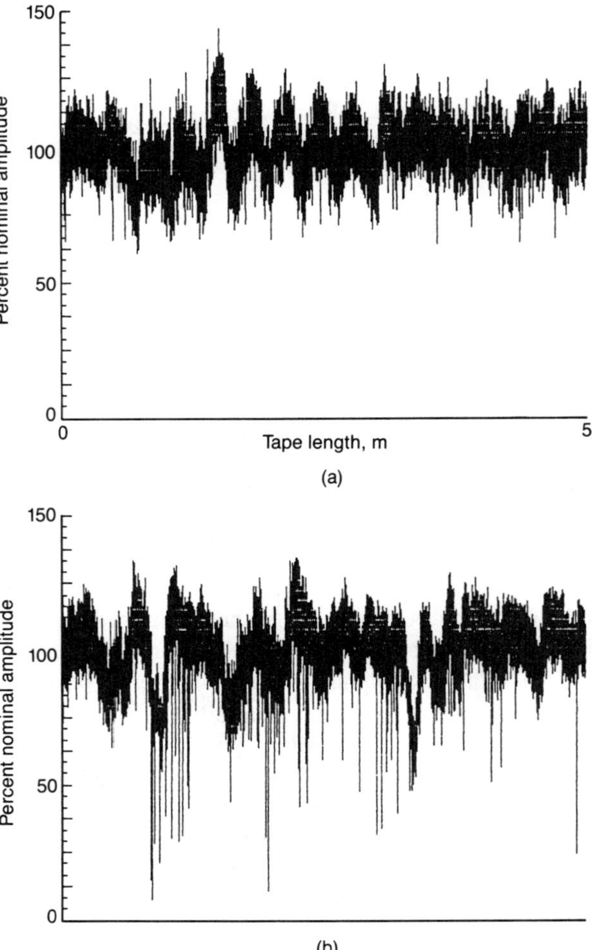

Fig. 5.52. Magnetic signal amplitude of (a) a good segment and (b) a bad (creased) segment of tape.

uniformities such as cupping, curvature, and tension gradients across its width. Figure 5.53(a) shows a tape approaching a roller on a tilted axis. The peripheral motion of the roller has a transverse component which will lead the tape off axis. Unintentional taper will induce greater speed and tension in the upper half of the tape width, with a consequent curvature of the tape, Fig. 5.53(b). The tape curvature will cause the tape to ride up the larger diameter of the rollers. Asymmetric tape such as a tape with curvature will cause the tape to ride off the intended center line as shown in Fig. 5.53(c).

Crowned rollers are commonly used to control the tracking of moving webs. Crowned rollers have the property of developing a restoring action on

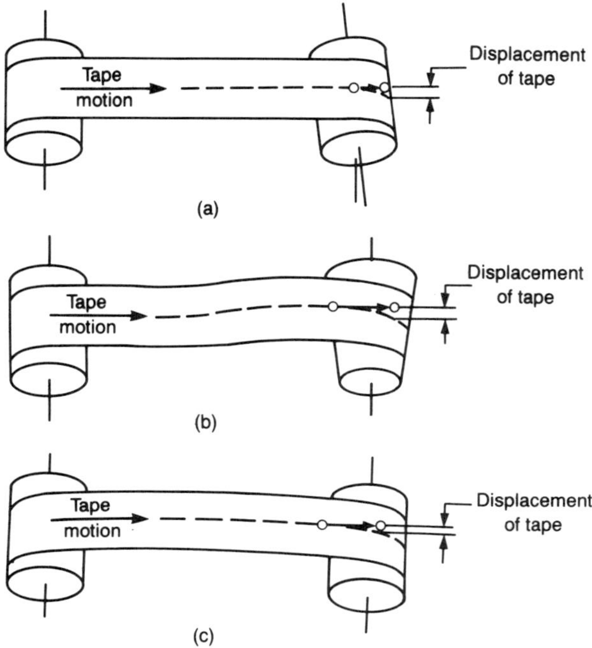

Fig. 5.53. Various causes of tape mistracking: (a) effect of tilted roller axis; (b) effect of tapered roller; and (c) effect of tape curvature.

a tape. Figure 5.54(a) shows a moving tape approaching a crowned roller. When the tape is on the roller surface with a tight frictional grip, it may be assumed to be moving with the velocity of the surface it touches. Since the crowned roller has its maximum diameter at the center plane, it also has its maximum velocity there. Where the roller diameter is decreased due to the crown contour, the velocity of the roller surface and the accompanying tape is also decreased. When the tape has a mistracking tendency, its center line will be displaced slightly from the center plane of the pulley, Fig. 5.54(b). However, the velocity and tension profiles are not displaced. The center of the tension, therefore, tends not to shift with the tape center line, and a bending moment is developed at the tape end approaching the roller and producing a curvature at that end. This curvature induces a counter-tendency for the tape to track to the opposite direction from that of the original mistracking, i.e., a restoring action is developed.

Edge guides are commonly used in the tape drive path and tape reels. Edge guidance may be considered to fall into two categories:

(1) two fixed guides which have a space between them which is slightly larger than the tape width;
(2) one fixed guide and one compliant (e.g., spring-loaded or soft) guide which exerts a continuous gentle pressure against one tape edge. This keeps the

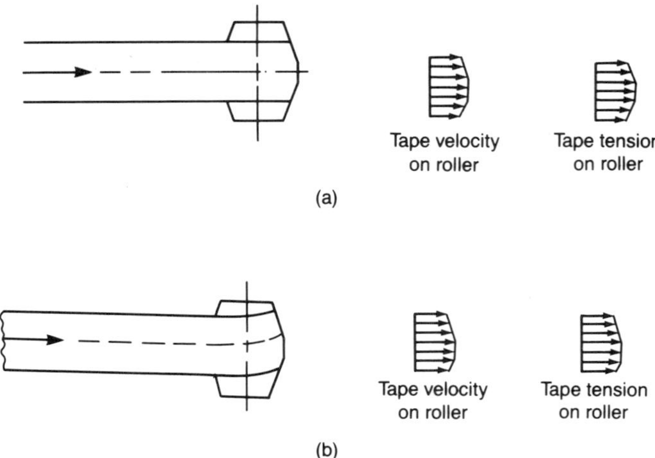

Fig. 5.54. The action of crowned roller (a) centered tape approaching roller, and (b) off-center tape approaching roller.

opposite tape edge in contact with the fixed guide, which is the reference position.

A third category of tape guidance control may be referred to as active guidance control. The active guidance technique involves the placement of one or more rollers (uncrowned) on a mounting which is pivotable. When the assembly is pivoted, the approaching tape span has its direction angle changed slightly while the departing tape span is merely twisted about its centered line. This action then is a deliberate misalignment which causes the tape to track to one side. By sensing some initial mistracking and introducing an approximate tilt to the rotor, the final mistracking can be compensated for and minimized. This action can be achieved either by a fully active servo system involving a powered transducer to implement the roller tilt and a tracking error detection circuit or by a nonpowered mechanical device which uses the offset tension of the displaced tape to produce the desired rotor tilt.

Going back to edge guides, to minimize tape-edge damage, the flange width spacing between two guides should be as large as possible compared to the tape width. However, for proper tape stacking, the flange width should be as close to the tape width as possible. The hydrodynamically generated air-film thickness should be small enough to allow guiding, but not too large to have unnecessary tape offset causing tape-edge damage. Since the reel flanges are tapered from the inner diameter (ID) to the outer diameter (OD) (see Chapter 1), more errors occur due to tape stagger at the beginning of the tape (BOT) zone and few in the middle of the tape (MOT) zone and almost none at the end of the tape (EOT) zone. To maintain no errors in the entire reel, the flange taper should be small (roughly 0.3 mm on each side for an IBM 3480 tape reel). To insure little stagger at EOT, the flange width near the hub should be only slightly larger than the tape width (about 0.3–0.4 mm larger than the tape width for a 12.7-mm tape, see Chapter 1). Furthermore, flanges should be of materials such as glass-filled polycarbonate to minimize its out-of-plane distortions especially near the OD. Use of a compliant disk inside the reel can be used to minimize stagger, e.g., a compliant disk such as a 75-μm-thick PET immediately adjacent to each flange of the reel and bonded near the ID of the flanges resulted in significant improvement in tape stagger.

Finally, a lay-on roller, as discussed for jumbo and pancake rewind to minimize uneven stack profile, may be used inside a tape reel to minimize interlayer slip. An example of a lay-on roller with flange guides is shown in Fig. 5.55. The flange guide can move freely over the pin as necessary when the

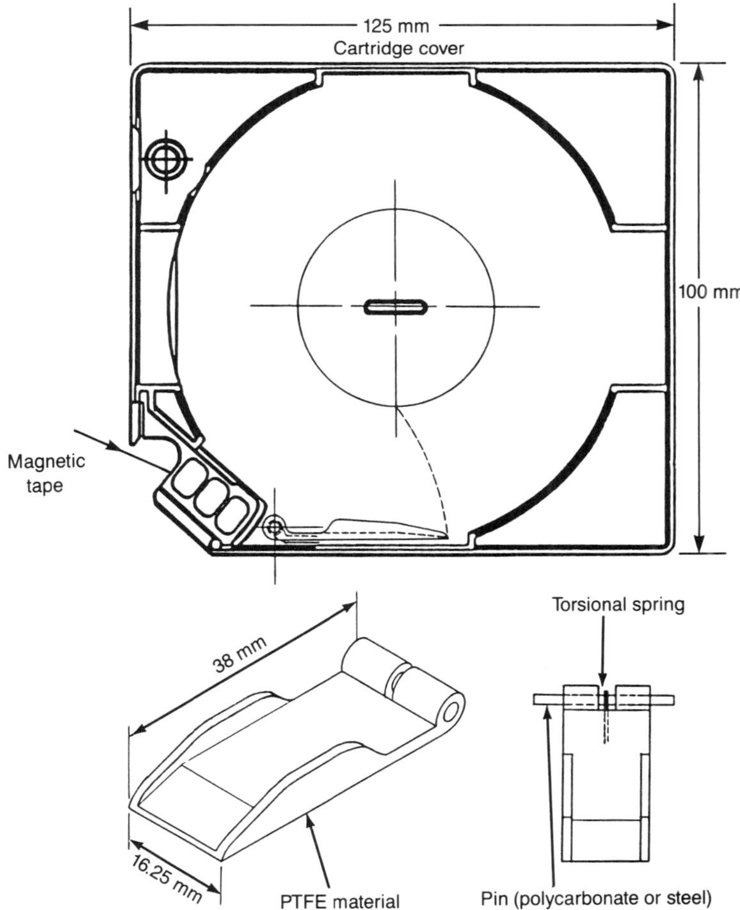

Fig. 5.55. Schematic of auto-lateral positioning: radial compliant flange for a 12.7-mm-wide tape reel cartridge.

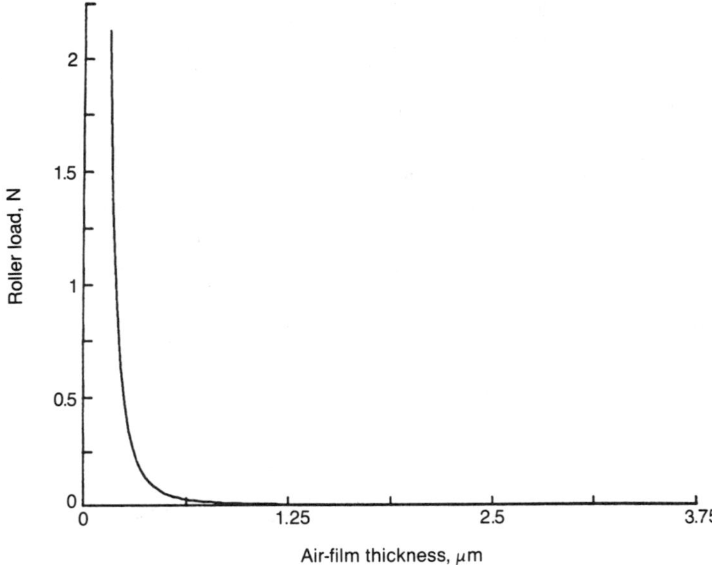

Fig. 5.56. Hydrodynamic air-film thickness as a function of roller load for a tape-reel rewind using a PTFE roller.

cartridge is unwound or rewound. A torsional spring keeps the flange pressed against the tape stack. The spring's stiffness should be selected to keep the desirable air-film thickness. The requirement is not to squeeze the air film entirely so that the tape stack can slip freely as is necessary for proper stacking and thus prevents edge damage. Figure 5.56 shows the hydrodynamic air-film thickness as a function of the total force applied on a lay-on roll made of PTFE ($E = 0.7$ GPa, $v = 0.48$). We note that a normal load of about 0.1 N reduces the film thickness to less than 0.5 μm, which is sufficient. An increase in load may result in a high roller and tape wear which would be a source of undesirable debris.

5.5.5. Summary

Poor tape guiding in the drive, poor tape quality, tension gradients across the tape width, and an elastohydrodynamic air-film generated at the nip during rewind result in a shift in the tape stack in either single or multiple layers. Continued exposure at high temperatures and humidity results in viscoelastic deformation, which does not recover at unloading. Very high stresses can lead to crease or plastic deformation. The quality of the tape-edge can increase crease severity. The viscoelastic deformation affects the flying height and increased head–tape separation results in magnetic errors. Crease severity changes in the offset wraps, and magnetic errors occur due to short length dropouts.

Winding parameters and storage conditions have a significant effect on tape damage. On the offset side, the flying height of the first offset wrap and length of creased tape increases with the number of offset wraps up to about 50–100 wraps. On the opposite side, the air-film thickness and length of creased tape decreases with an increase in the number of offset wraps up to about 50 wraps. The amount of tape offset does not have a significant effect on the flying height or the length of creased tape up to 1.5 mm of offset. Both the air-film thickness and the length of creased tape increased with an increase in tape tension. The radial film thickness and the length of creased tape increased as the radial location of the offset was moved toward the hub. An offset tape did not affect the air-film thickness when exposed to room temperature for 2 days or 52° C for 1 h. An offset tape exposed at 60° C for 2 h received the same damage as when exposed at 52° C for 40 h.

Elimination of staggered wraps requires that the tape be guided properly in the drive and on the reel. Since the amount of droop increases to the fourth power of the offset length, a reduction in the maximum attainable overhang significantly reduces the corresponding viscoelastic deformation. The overhang length in any event must be less than the distance between the tape edge and the outermost track (e.g., 0.7 mm for a 12.7-mm-wide tape for an IBM 3480 drive). Efforts must be made to reduce tape curvature, cupping, and tension gradients across the tape width. Crowned rollers, suitable edge guidance, or active guidance methods must be employed. Moreover, the use of lay-on or squeeze rollers where possible, to reduce the air-film thickness generated at the nip during winding, is beneficial.

5.6. Design of Tape Reels

The tape reel is an integral part of the tape drive transport system. A tape reel is made up of a hub and two flanges (see Chapter 1). The basic function of the flanges is to protect the tape from damage and contamination during handling and use. It is often the reel itself, damaged through mistreatment, that in turn damages the tape. Most modern tape drives require tape reels that are fabricated to close tolerances and have superior stability and flange rigidity. Some design guidelines follow.

5.6.1. Hub

Hub thickness and material contribute to the hub's compliance (Chapter 4). Increasing hub compliance reduces the hoop strain in the inner wraps, thereby reducing the hoop stress. If the hub is too compliant, its function is transferred to the wound tape, which produces compressive circumferential stresses in the tape. Sufficient negative, hoop stresses can cause the wound tape to buckle. Furthermore, the radial pressure during winding must be sufficient to prevent interlayer tape slippage on acceleration. To ensure that the wound tape remains undistorted during storage, the hub must be of uniform compliance, which requires as close a uniform hub thickness as possible. A mismatch in the coefficient of thermal or hygroscopic expansion between the hub and the tape will lead to further stresses induced in the wound tape when there is a temperature or humidity change. These stresses must be considered when selecting hub material and geometry (Connolly and Winarski, 1984, Chapter 4).

It is vital that the hub be as near a perfect cylinder as possible, both on its inner and outer surfaces. If the hub inner surface is imperfect, the reel will wobble on its spindle and feed the tape erratically onto the transport. This will result in tape edge damage and skewed tape travel across the transport head. If the hub outer surface is imperfect, the tape will nonuniformly stretch as it is stacked on the reel. The elastic deformation of the tape becomes viscoelastic as the tape is stored for a period of time, especially at high temperatures and humidity. When the deformed tape is replayed, signal distortion will occur.

Three specific geometrical parameters in the cylindrical section of the hub outer surface must be controlled: taper, straightness, and localized defects including foreign particles. If the hub outer surface is tapered, the nonuniform stretching will result in a tape curvature when unconstrained, which decreases from near the hub to the outer location on the reel, Fig. 5.57. There will be loss of tension on the long side of the tape. If the tape is curved (i.e., one edge longer than the other), straightening the tape, as is done in the drive tape path, will produce a bending moment (M) in the tape, and an uneven stress distribution. Uneven tension across the tape width causes tape vibration over the guide bearings, uneven spacing at the head–tape interface, and an uneven head wear pattern (Hu et al., 1984).

From simple geometry, the radius of curvature of the tape unwound from a tapered hub and fully deformed, is

$$R_1 = d_1 w/(d_2 - d_1), \tag{5.23}$$

where d_1 and d_2 are the diameters of the hub at the two ends and w is the width of the tape. And

$$h_{cur} = L^2/8R_1, \tag{5.24}$$

where h_{cur} the curvature for a tape length L (Fig. 5.57). Further, the stress (σ_{max}) or tension required to flatten (TTF) the tape is

$$\sigma_{max} = \pm \frac{E(d_2 - d_1)}{2d_1}, \tag{5.25}$$

or

$$TTF = \sigma_{max} wt, \tag{5.26}$$

where E is Young's modulus of elasticity of the tape and t is the tape thickness. For example, for $d_2 - d_1 = 80 \; \mu m$, $d_1 = 50$ mm, $E = 3.5$ GPa, $w = 12.7$ mm, and $t = 30 \; \mu m$, the tape tension TTF $= 1$ N.

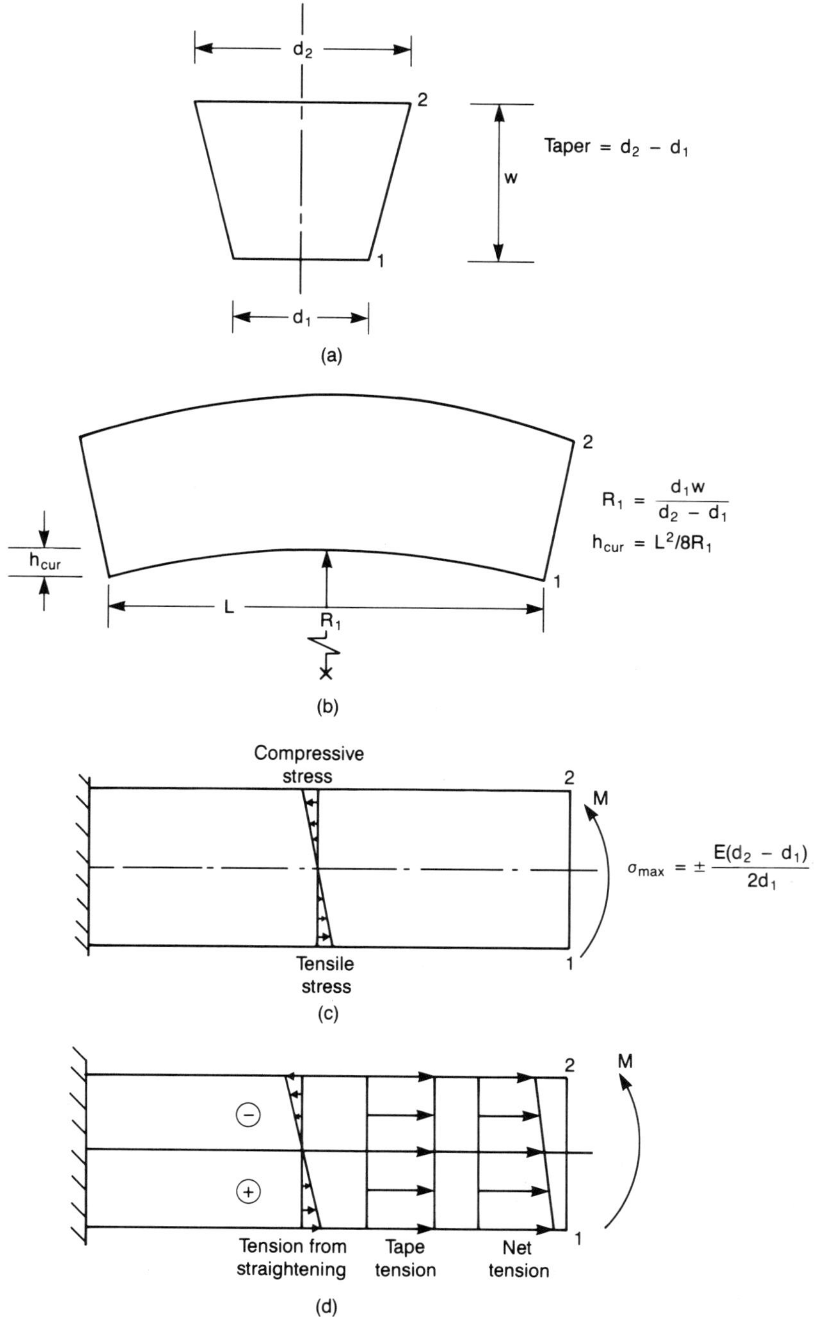

Fig. 5.57. Taper in the cylindrical section of a hub outer surface and its effect on the tape: (a) tapered hub; (b) unwound tape with curvature; (c) tape after straightening in drive; and (d) tape after tension added.

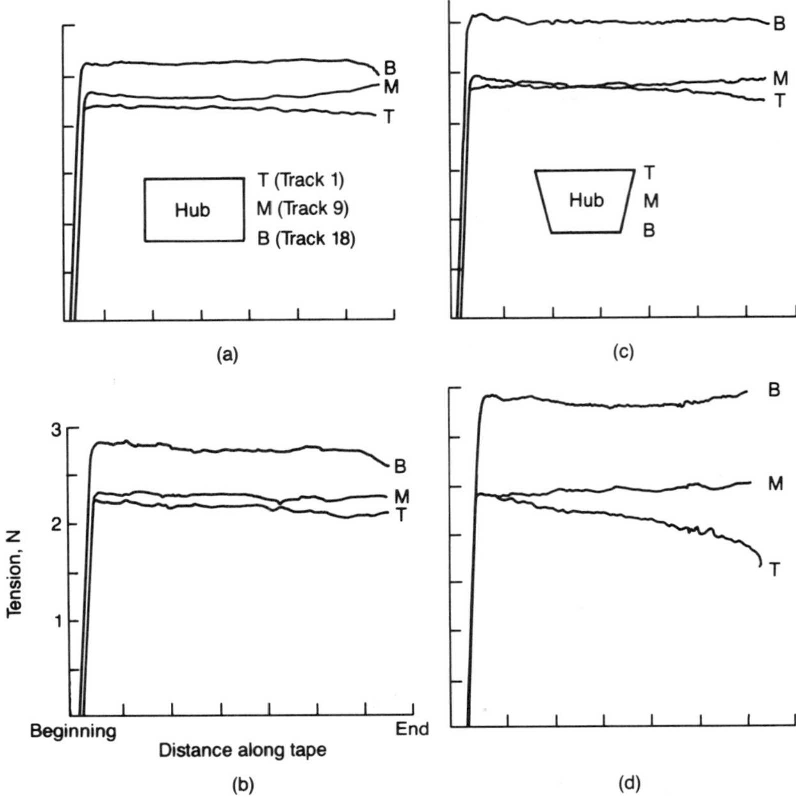

Fig. 5.58. Variation in tape tension through tape length along two edge tracks (top–T, bottom–B) and one middle (M) track for the 12.7-mm-wide tapes wound on (a) an untapered hub (average diameter = 51 mm) after as wound (no storage), (b) after storage at 52° C/30% RH for 24 h, (c) a tapered hub (diametral difference = 125 μm; average diameter = 51 mm) after as wound (no storage), and (d) after storage at 52° C/30% RH for 24 h. The initial tape curvature (h_{cur}) was 1.5 mm (bottom—B short).

Figure 5.58 shows the tension profiles at three tracks of the tapes wound onto untapered and tapered hubs after as-wound and after storage at 52° C/30% RH for 24 h. We note that there is a tension gradient of about 0.75 N because of the initial tape curvature, tape reel, and the tape drive used. After storage, there is no additional increase in the tension gradient for an untapered hub, however, for a tapered hub, tension of the tape edge with a larger hub radius decreased and of the tape end with smaller hub radius increased, as expected. Figure 5.59 shows the effect of hub taper on the tape curvature and the tension gradient across its width near the end of the tape, for a large number of reels with various hub taper after the wound reels were stored at 52° C/30% RH for 24 h. We find that hub taper has a significant effect on the

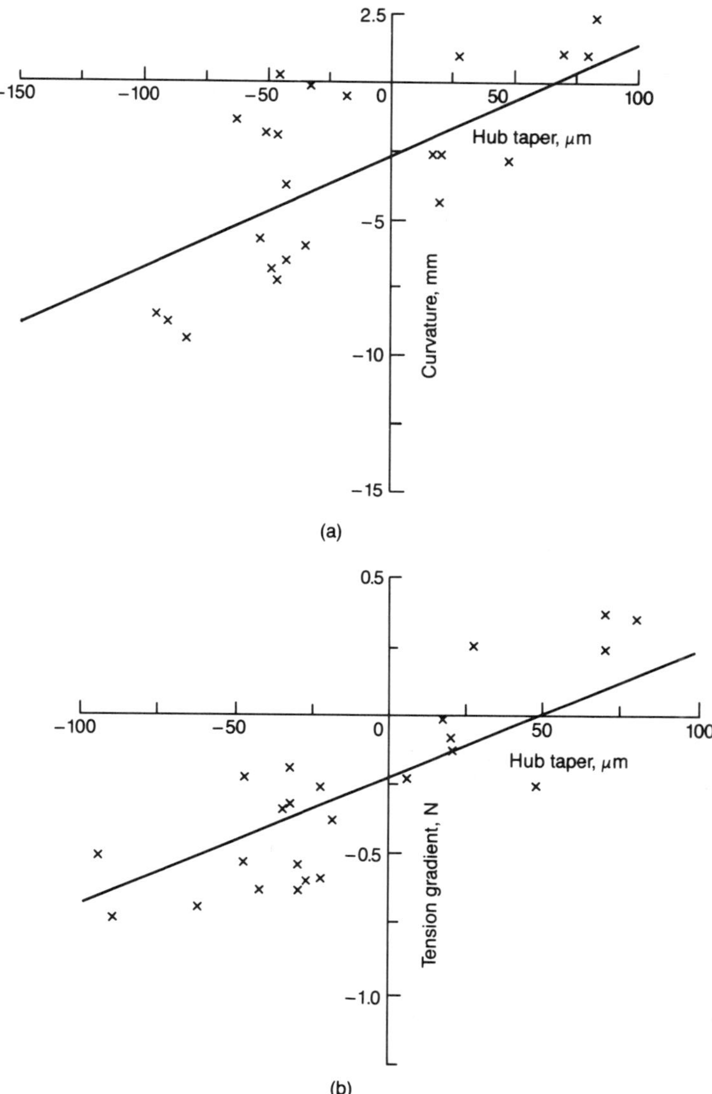

Fig. 5.59. Effect of hub taper on (a) the curvature and (b) the tension gradient across the width, the 12.7-mm-wide tape, near the end of tape, after the wound reels were stored at 52° C/30% RH for 24 h. The initial tape curvature was −2.5 mm and the tension gradient of the virgin tape on the drive was −0.25 N.

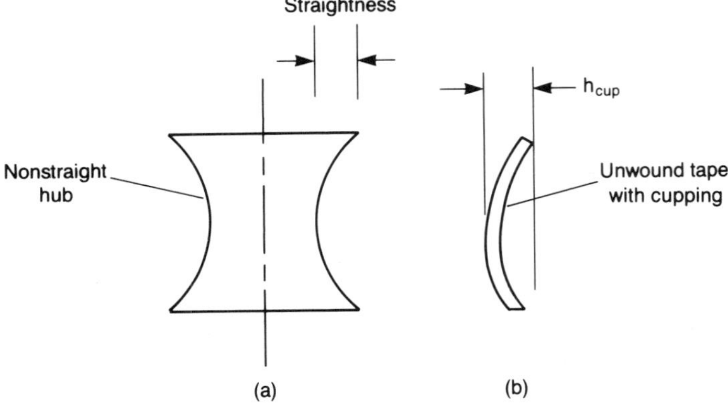

Fig. 5.60. (a) Nonstraightness in the cylindrical section of the hub outer surface, and (b) its effect on the tape straightness (known as cupping), which is defined as the difference between two parallel lines which contain the maximum excursions of the cylindrical surface; lines need not be parallel to the cylinder axis.

tape curvature and tension gradient, as expected. The bias in curvature and tension gradient of an untapered hub is because of the initial tape, the tape reel, and the tape drive used.

If the hub is nonstraight, again nonuniform stretching will create unevenly distributed tension within the reel pack and the tape will cup in the unconstrained condition, Fig. 5.60. (A cupped tape may also result in coating cracking (Hiratsuka et al., 1980).) The high tension regions of the tape will fly higher (resulting in low magnetic signal and magnetic errors) than the low tension regions when the tape is unwound and replayed. The tape damage from the hub straightness can be measured with the tension required to make the tape appear visually flat.

Normally, a hub taper and straightness of less than 50 μm is recommended for high-reliability data-processing applications. The curvature of a 12.7-mm-wide tape (h_{cur}, Fig. 5.57) is recommended to be less than 3.8 mm for a 1-m tape length, and cupping (h_{cup}, Fig. 5.60) to be less than 0.3 mm at any time during its use.

Foreign particles, scratches, and other defects on the reel-hub surface can cause the tape to deform locally (mechanical print-through, discussed earlier) and create unevenly distributed tension within the reel pack. This is why it is imperative that the reel be routinely and frequently inspected and cleaned.

5.6.2. Flanges

The reel must be designed to enable the tape to wind and unwind with minimal contact with the flanges. Any significant contact of the tape edge with the flange will usually result in tape-edge damage and loss of the edge recording track. The primary purpose of flanges is to protect the tape from damage and contamination during handling and use. However, it is desirable to get secondary tape guiding by the flanges. Therefore, the flange width should be larger but as close as possible to the tape width. For example, the flange width near the hub for a 12.7-mm (12.674 $\pm$ 0.025-mm)-wide tape for the IBM 3480 tape reel is 13.05 + 0.2/−0.1 mm (see Chapter 1). The flanges must be sufficiently rigid to withstand normal handling pressures and resist accidental nicking or gouging.

During normal tape transport operations, tape reels are subjected to rapid starts, stops, and direction reversals. These rapid changes of rotational movement can create large moments of inertia, especially at the outer circumference of a fast-spinning reel with thick flanges. Excessive reel inertia will also adversely affect rotating and braking components of the tape transport, shortening their service life. Consequently, reel manufacturers generally keep mass to a minimum after other design requirements are satisfied. The tapered flange design (thinner flanges near the outer diameter) in conventional reels gives it the desired low moment of inertia, while still providing adequate flange strength and stability. For example, the flange taper for the IBM 3480 tape reel is 0.35 mm on each flange, going from a radius of 25.5 mm to 49 mm (see Chapter 1).

5.6.3. Reel Materials

The tape reels are generally made of hard plastics such as glass-filled polycarbonate, phenolics, or polyvinyl chloride. Precision reels are classified by their flange material—metal or glass. Precision metal reels have precision fabricated metal flanges and accurately machined metal hubs. Precision glass reels have precision fabricated glass flanges and accurately machined metal hubs. The chemically strengthened glass is about ten times as strong as ordinary glass. Reel flanges made of this material are coated with a tough plastic material that will restrain glass fragments should failure occur.

References

Anonymous. "Techniques for Slitting and Winding." John Dusenbery Co. Inc., Randolph, New Jersey.

Anonymous (1982). "Instantaneous Speed Variation in the 6.30 mm Data Cartridge." ANSI Document X3B5/82-97, American National Standards Institute, December 9.

Anonymous (1983). "Tape/Head Interface Study." Contract No. NAS 5-26573 (RCA Govt. Systems Div., Camden, New Jersey). NASA, Greenbelt, Maryland.

Baumann, G. W. (1972). Viscoelastic behavior of computer tape subjected to periodic motion. *IBM J. Res. Develop.* **16**, 214–221.

Bertram, N. and Eshel, A. (1980). "Recording Media Archival Attributes (Magnetic)." Report No. RADC-TR-80-123. Rome Air Development Center, Rome, New York, April.

Bhushan, B., Heinrich, J. C., and Connolly, D. (1984a). Orthotropic viscoelastic behavior of polyethylene terephthalate film under plane-stress conditions. In "Tribology and Mechanics in Magnetic Storage Systems" (B. Bhushan et al., eds.), pp. 158–171. SP-16, ASLE, Park Ridge, Illinois.

Bhushan, B., Hahn, F. W., Sharma, B. S., and Connolly, D. (1984b). Long term reliability of magnetic tapes for digital recording. In "Tribology and Mechanics of Magnetic Storage Systems" (B. Bhushan et al., eds.), pp. 132–147. SP-16, ASLE, Park Ridge, Illinois.

Bhushan, B. (1987). Development of a wear test apparatus for screening bearing-flange materials in computer tape drives. *ASLE Trans.* **30**, 187–195.

Bhushan, B. (1990). "Tribology and Mechanics of Magnetic Storage Devices." Springer-Verlag, New York.

Bhushan, B. and Connolly, D. (1986). Viscoelastic properties of poly(ethylene terephthalate). *ASLE Trans.* **29**, 489–499.

Blok, H. and Van Rossum, J. J. (1953). The foil bearing—a new departure in hydrodynamic lubrication. *Lub. Eng.* **9**, 316–320.

Connolly, D. (1984). Personal communication.

Connolly, D. and Winarski, D. J. (1984). Stress analysis of wound magnetic tape. In "Tribology and Mechanics of Magnetic Storage Systems" (B. Bhushan et al., eds.), pp. 172–182. SP-16, ASLE, Park Ridge, Illinois.

Crandall, S. H. and Dahl, N. C. (1959). "An Introduction to the Mechanics of Solids." McGraw-Hill, New York.

Eshel, A. and Kennedy, T. G. (1979). Numerical simulation of longitudinal vibrations in foils including the effects of dry friction. *Proc. Summer Computer Simulation Conf.*, SCSC, Toronto, Canada, July 16–18.

Eshel, A., Baker, S., Hartman, S., and Orcutt, F. K. (1984). Mechanical aspects of archival storage of magnetic tape. In "Tribology and Mechanics of Magnetic Storage Systems" (B. Bhushan et al., eds.), pp. 148–157. SP-16, ASLE, Park Ridge, Illinois.

Fell, W. (1980). The influence of the elasticity of magnetic tape on some parameters of magnetic recording. *Radio and Electronics Eng.* **50**, 624–630.

Frye, K. F. (1967) Winding variables and their effect on roll hardness and roll quality. *TAPPI* **50** (7), 81A–86A.

Geller, S. B. (1983). "Care and Handling of Computer Magnetic Media." Report No. NBS SP500-101, National Bureau of Standards, Gaithersburg, Maryland, June.

Heinrich, J. C., Connolly, D., and Bhushan, B. (1986). Axisymmetric, finite-element analysis of stress relaxation in wound magnetic tapes. *ASLE Trans.* **29**, 75–84.

Hiratsuka, H., Hanafusa, H., Nakamura, K., and Hosokawa, S. (1980). Durability of magnetic recording tape for mass memory system. *Rev. Elec. Comm. Lab.* **28** (5–6), 449–458.

Hu, P. Y., Hollman, W., and Argumedo, A. J. (1984). Tension gradient measurement of magnetic tape. *IEEE Trans. Magn.* **MAG-20**, 921–923.

Meyer, P. L. (1973). "Introductory Probability and Statistical Analysis," 2nd ed. Addison-Wesley, Reading, Massachusetts.

Mukherjee, S. (1974). Time base errors in video tape packs. *J. Appl. Mech.* **96**, 625–630.

Mukherjee, S. (1975). Effect of multiple rewinds and temperature cycles on time base errors in video tape packs. *Int. J. Solids Structures* **11**, 887–894.

Nelson, W. (1982). "Applied Life Data Analysis." Wiley, New York.

Owen, R. J. (1971). "Magnetic Head/Tape Interface Study for Satellite Tape Recorders." Contract No. NAS5-11622 (IITRI, Chicago), NASA Greenbelt, Maryland.

Pfeiffer, J. D. (1966). Internal pressures in a wound roll of paper. *TAPPI* **49** (8), 342–347.

Prager, R. H. (1959). Time errors in magnetic tape recording. *JAES*, **7**, April.

Skurdrzyk, E. (1968). "Simple and Complex Vibratory Systems." Pennsylvania State University Press, University Park, Pennsylvania.

Smith, D. P. (1986). Magnetic tape speed variations. In "Tribology and Mechanics of Magnetic Storage Systems" (B. Bhushan and N. S. Eiss, eds.), Vol. 3, pp. 102–117. SP-21, ASLE, Park Ridge, Illinois.

Smith, D. P. and Sievers, J. A. (1985). Spatially coherent longitudinal vibrations in magnetic tape. In "Tribology and Mechanics of Magnetic Storage Systems" (B. Bhushan and N. S. Eiss, eds.), Vol. 2, pp. 80–86. SP-19, ASLE, Park Ridge, Illinois.

Snowdon, J. C. (1968). "Vibration and Shock in Damped Mechanical Systems." Wiley, New York.

Timoshenko, S., and Woinowsky-Krieger, S. (1959). "Theory of Plates and Shells," 2nd ed. McGraw-Hill, New York.

Tramposch, H. (1965). Relaxation of internal forces in a wound reel of magnetic tape. *J. Appl. Mech., Trans. ASME* **32**, 865–873.

Tramposch, H. (1967). Anisotropic relaxation of internal forces in a wound reel of magnetic tape. *J. App. Mech., Trans. ASME* **34**, 888–894.

von Behren, R. A. and Youngquist, R. J. (1955). Frequency modulation noise in magnetic recording. *JAES* **3**, January.

Waites, J. B. (1982). Care, handling, and management of magnetic tape. In "Magnetic Tape Recording for the Eighties" (F. Kalil, ed.), pp. 45–59. Reference Publication 1075, NASA, Washington, D.C.

Walowit, J. A., Murray, S. F., McCabe, J., Arwas, E. B., and Moyer, T. (1973). "Gas Lubricated Foil Bearing Technology Development for Propulsion and Power Systems." Report No. AFAPL-TR-73-92, Wright Patterson Air Force Base, Dayton, Ohio, December.

Walraven, A. (1969). The analysis of longitudinal tape vibrations. *IEEE Trans. Mag.* **MAG-5**, 452.

Winarski, D. J. (1984). personal communications.

Winarski, D. J. (1991). Use of ceramics for tape guiding in the IBM 3480 tape path. *Adv. Info. Storage Syst.* **1**, 37–48.

Appendix 5.A. Tension-to-Flatten Analysis for a Tape Reel with a Circumferential Bump

A *tension-to-flatten* (TTF) technique is commonly used to measure the tape distortion. It involves the measurement of the tension required to flatten the tape by a spring force gauge. The TTF measurement technique is illustrated schematically in Fig. 5.A.1. In the untensioned state shown in Fig. 5.A.1(a), the stretched area may be visualized as a "puckered" strand. As tension is applied, the pucker is reduced to the point where the strand finally becomes flat, Fig. 5.A.1(b). The tension required to achieve a visually flat condition is considered to be a measure of tape distortion.

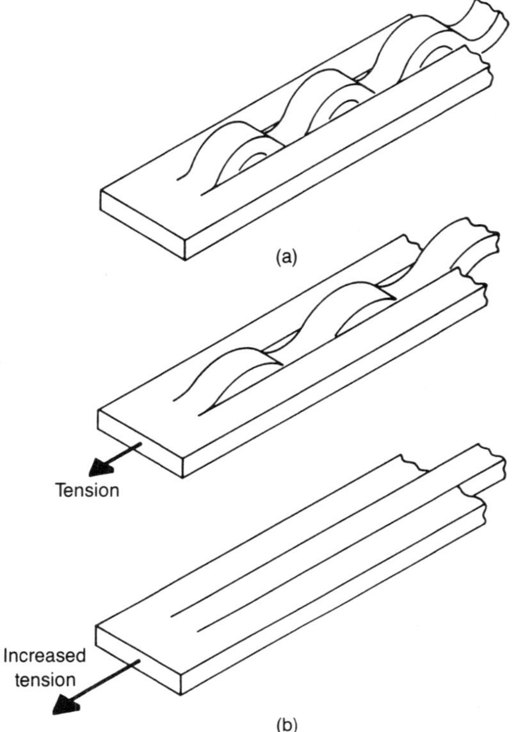

Fig. 5.A.1. Schematic of tape distortion (a) before and (b) after applied tension-to-flatten.

When tape is wound onto a spool, if a circumferential bump is generated in the stack, there is an increase in the circumferential (also called hoop) strain $\varepsilon_{\theta\theta}$ which is (see Chapter 4)

$$\varepsilon_{\theta\theta} = \frac{u(r, z)}{r}, \qquad 0 \le z \le w, \tag{5.A.1}$$

where $u(r, z)$ is the bump profile at a radius r as a function of the axial coordinate z, Fig. 5.A.2. Sources of such axisymmetric bumps are coating and substrate thickness variations, entrapped air pockets, tension ridges, scratches, and hub anomalies. There is also an increase in the radial and axial strains, but from the point of view of TTF these are not important. With time, this hoop strain becomes partially viscoelastic, a process which is accelerated with increasing temperature and humidity. When the tape is unwound, the radial viscoelastic strain prevents the tape from being flat. This strain should recover if it is unloaded for a sufficiently long time. Under machine operating conditions, this will not have time to occur.

Tension-to-flatten is the uniaxial tension one must apply to the tape to

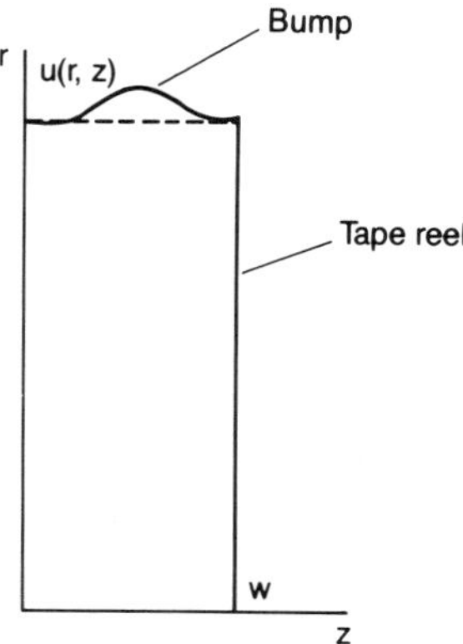

Fig. 5.A.2. Cross section of a reel with a bump.

make it flat. When the tape is flat, the total longitudinal strain is the same everywhere. Thus the applied tension has to supplement the viscoelastic hoop strain with an elastic longitudinal strain.

5.A.1. Tape Substrate

Consider, first, a piece of tape substrate deformed viscoelastically in six steps ($n = 6$), as shown in Fig. 5.A.3. The elastic longitudinal strain that must be applied to the tape to flatten it, when it is unwound off the reel where it was stored with axisymmetric "step" deformation, is

$$\delta_s^i(u/r) \quad \text{where } i = 1, 2, \ldots, 6, \tag{5.A.2}$$

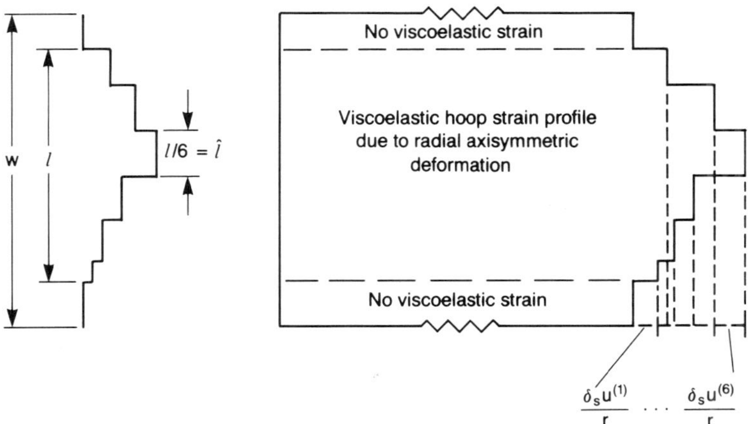

Fig. 5.A.3. Cross section and along-the-length strip of a tape with six equal-step axisymmetric deformations, wound on a reel.

where δ_s denotes the fraction of the initial elastic hoop strain that becomes viscoelastic and r is the radius of the bump in the wound reel. Then the TTF is simply

$$\text{TTF} = [(E_s t_s \delta_s)/r][(w - 6\hat{l})u^{(1)}(r, z) + (w - 5\hat{l})u^{(2)}(r, z) + \cdots + (w - \hat{l})u^{(6)}(r, z)]$$

$$= (E_s t_s \delta_s/r)\left[w \sum_{i=1}^{6} u^{(i)}(r, z) - \hat{l} \sum_{i=1}^{6} (7 - i)u^{(i)}(r, z) \right], \qquad (5.A.3)$$

where $l/6 = \hat{l}$. For a general n, we have

$$\text{TTF} = [(E_s t_s \delta_s)/r]\left\{ w \max_{0 \leq z \leq w} [u(r, z)] - \hat{l} \sum_{i=1}^{m} (n - i)u^{(i)} \right\}, \qquad (5.A.4)$$

where $\hat{l} = l/n$, l being defined in Fig. 5.A.3, δ_s is a function of time (t) the tape is loaded, temperature, and humidity, and E_s and t_s are Young's modulus of elasticity and thickness of the tape substrate, respectively. An $n \to \infty$, we have for a continuous bump profile $u(r, z)$

$$\text{TTF} = [(E_s t_s \delta_s)/r]\left\{ w \max_{0 \leq z \leq w} [u(r, z)] - \int_0^w u(r, z)\, dz \right\}. \qquad (5.A.5)$$

Thus, knowing the bump profile for the tape substrate and its relaxation kinetics, we can compute the TFF. For example, if

$$u(r, z) = u_0 \sin \pi[(z - l_1)/l], \qquad l_1 \leq z \leq l_1 + l \leq w, \qquad (5.A.6)$$

$E_s = 3.7$ GPa, $t_s = 23.4$ μm, $u_0 = 60$ μm, $r = 50$ mm, $w = 12.7$ mm, $l = 6.35$ mm, and $\delta_s = 0.38$ (e.g., PET at 52° C and 30% RH for $t = 170$ h, see Chapter 3), then

$$\text{TTF} = 3.34 \text{ N}.$$

These predicted numbers are very close to what was measured, Fig. 5.26, considering this is just a fictitious profile.

5.A.2. Composite Tape

Next we develop a model for a coated (composite) tape. As in the case of bending, the total hoop strain when wound is continuous across the substrate-coating interface. On unloading, due to the faster stress relaxation in the coating, there will be greater viscoelastic hoop strain in the coating than in the substrate. The elasticity of the substrate will tend to reduce the profile height. We need to calculate the proportion of the initial profile which is left on unloading.

We assume that the proportion of the initial profile left on unloading is the same as the proportion of the elastic bending moment lost $[M_R(0)]$ while the tape is bent. We call this factor $\delta_c(t)$

$$\delta_c(t) = \frac{M_R(0) - M_R(t)}{M_R(0)}. \qquad (5.A.7)$$

If the composite thickness is

$$t_c = t_b + t_s + t_f, \tag{5.A.8}$$

where t_b is the backcoat thickness and t_f is the frontcoat thickness and the composite Young's modulus of elasticity is

$$E_c = (E_b t_b + E_s t_s + E_f t_f)/(t_b + t_s + t_f), \tag{5.A.9}$$

where E_b = Young's modulus of the backcoat and E_f = Young's modulus of the frontcoat. Then, analogous to Eq. (5.A.5), we have for the composite

$$\text{TTF} = [(E_c t_c \delta_c)/r]\left\{ w \max_{0 \le z \le w} [u(r, z)] - \int_0^w u(r, z)\, dz\right\}. \tag{5.A.10}$$

For unbackcoated tape, E_b and t_b drop out from the analysis.

Appendix 5.B. Tension-Gradient Measurement Technique

The tension-gradient measurement device was developed to measure tension gradients both dynamically and statically across the width of the tape, as a function of tape length (Hu et al., 1984), Fig. 5.B.1. The basic system consists of a miniature pressure sensor having a diameter of 2 mm (Kulite CQ-080-10) epoxied in a ferrite or glass cylinder whose radius is equal to that of the magnetic head. (The heart of the sensor consists of a miniature silicon diaphragm to which a fully-active four-arm wheatstone bridge has been bonded.) A small hole (0.2 mm) was drilled to a depth of 0.5 mm at the apex of the cylinder. A 3-mm-diameter hole was drilled from the back side of the cylinder so that the hole just broke through to the 0.2-mm hole. The pressure sensor was placed into the 3-mm hole and the air bearing pressure developed at the head–tape interface is sensed through the 0.2-mm hole, which provides adequate sensing within the sensitivity range of the pressure transducer. This pressure is converted by the sensor into an electrical voltage, differentially amplified, and converted into tape tension (T) by the following relationship:

$$T = PRw, \tag{5.B.1}$$

where P is the air pressure at the head–tape interface, r is the cylinder radius, and w is the tape width.

The cylinder and sensor assembly is mounted on an x–y slide which is then placed on the drive in the tape path instead of on the magnetic head. The x–y slide is so mounted that the x direction of motion varies the tape-cylinder penetration, while the y direction of motion moves the cylinder perpendicularly to the direction of tape motion, Fig. 5.B.2. As stated earlier, the cylinder has the same radius of curvature as an actual head, and by penetrating the cylinder into the tape, similar to the head penetration used in an actual head assembly, the tape will fly over the cylinder similar to the manner in which it

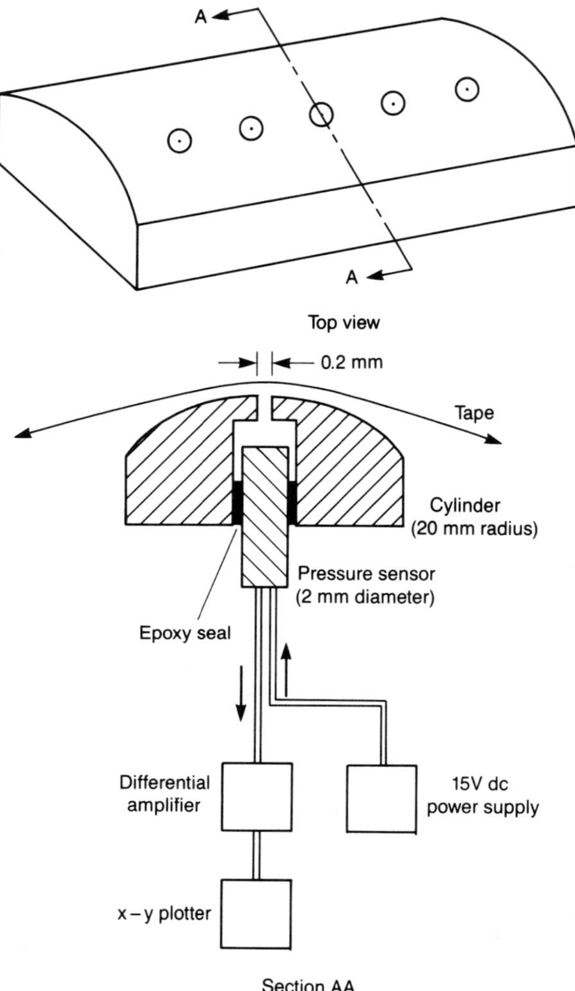

Fig. 5.B.1. Schematic of a tension-gradient measurement device (Hu et al., 1984).

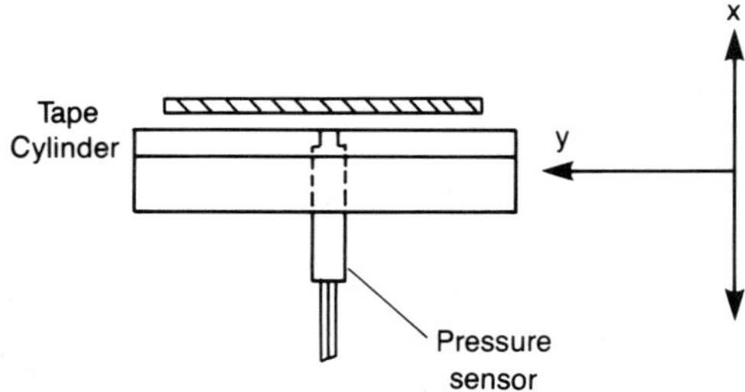

Fig. 5.B.2. Schematic of an x–y cylinder slide containing a tension transducer; only one pressure sensor is shown, though several sensors can be placed across the tape width.

flies over a magnetic head. By adjusting the device, the y motion of the sensor can be positioned at various track locations across the tape width, thereby giving track-by-track tension measurements as a function of tape length.

For tension measurements at several tracks simultaneously, several sensors (easily up to five) can be placed in the cylinder evenly across the width of the tape. Each of the sensors has its own amplifier.

Appendix 5.C. Instantaneous Failure Rate Model to Assess Failures Due to Viscoelastic Deformation-Related Defects

We propose an instantaneous failure rate (IFR) model for a tape-reel population in the field, due to viscoelastic deformation-related defects. The failure rate will depend on a *defect-size distribution* for the various types of defects which produce viscoelastic deformation in the magnetic tape, and *deformation kinetics* which prescribe the rate at which initially elastic deformation becomes viscoelastic as a function of time (Connolly, 1984).

The important features of the analysis included herein are:

1. For realistic defect size distributions one has a decreasing failure rate with time.
2. If one knows the exact defect-size distribution and the percent of the population failing during a period in the field, one can determine the critical defect size below which the defect will never cause a failure in that time period.
3. Then knowing this critical defect size and the defect size distribution from any other manufacturing lot, one can predict the percentage fall-out due to this defect type of the reel population in the field for any period of time.

In the analysis presented here standard statistical notation is used (see Appendix B for details). A random variable (r.v.) is denoted by a capital letter and a value of the random variable by a lowercase letter (Meyer, 1973; Nelson, 1982). Suppose that the defect size is a continuous r.v. D with a probability density function (p.d.f.) f satisfying

$$f(d) \geq 0 \qquad \text{for all } d \tag{5.C.1a}$$

and

$$\int_0^\infty f(d)\,dd = 1. \tag{5.C.1b}$$

For any defect sizes d_1 and d_2, such that $0 < d_1 < d_2 < \infty$, the cumulative probability distribution function is

$$P\{d_1 \leq D \leq d_2\} = \int_{d_1}^{d_2} f(d)\,dd, \tag{5.C.2}$$

which gives the probability of finding a defect in the size range (d_1, d_2).

By definition, the number of defects in a particular interval $(x, x + \Delta x)$ is

$$N(x) = \int_x^{x+\Delta x} f(d)\,dd \quad \text{(total number of defects)} \qquad (5.\text{C}.3a)$$

or

$$dN/dx = f(x) \quad \text{(total number of defects)}, \qquad (5.\text{C}.3b)$$

thus giving an interpretation to $f(d)$, as the rate of which the proportion of defects are changing with respect to their size.

Initially, at time $t = 0$, a distortion of severity D_E is produced in the media. This can be produced during winding by, for example, uneven tape stack profile, mechanical print through, or staggered wraps. At this initial instant, the amount of viscoelastic deformation, $d_v(t)$, due to this deformation is zero. As time progresses, the deformation severity remains D_E, but is no longer fully elastic: it has attained a degree of viscoelastic deformation, $d_v(t)$. This is acquired as the local stress field relaxes, a process which can be accelerated at elevated temperatures.

From the relaxation data presented in Chapter 3, stress relaxation can be expressed in the "Prony" series

$$\sigma(t) = \sigma_0[B_0 + B_1 \exp(-t/\tau_1) + B_2 \exp(-t/\tau_2)], \quad 0 \le t \le 3\tau_2, \qquad (5.\text{C}.4a)$$

where $B_0 + B_1 + B_2 = 1$. This restriction guarantees

$$\sigma(0) = \sigma_0, \qquad (5.\text{C}.4b)$$

the stress at time zero. Further, viscoelastic deformation is assumed to occur as follows:

$$d_v(t) = D_E[1 - (A_0 + A_1 \exp(-t/\tau_1) + A_2 \exp(-t/\tau_2)], \quad 0 \le t \le 3\tau_2, \qquad (5.\text{C}.5a)$$

where $A_0 + A_1 + A_2 = 1$ which gives

$$d_v(0) = 0. \qquad (5.\text{C}.5b)$$

Further

$$d_v(\infty) = D_E(1 - A_0), \qquad (5.\text{C}.5c)$$

giving the amount of viscoelastic deformation achieved in an "infinite" (long) time, which is some fraction of the elastic deformation. Thus, some portion of the deformation always remains elastic. We require that

$$B_2/B_1 = A_2/A_1 = \rho, \qquad (5.\text{C}.6)$$

which stipulates that the viscoelastic deformation occurs at a rate in proportion to the rate at which the stress relaxes. The tape deformation data (presented in this chapter which involved placing 38-μm and 76-μm-thick shims on a reel and exposing the tape reel to 52° C) have the same relaxation times as the stress relaxation data (presented in Chapter 3) and satisfy Eq. (5.C.6). Thus, our assumptions are verified.

We now assume that there is a critical defect size, D_F, below which a defect will not produce enough permanent or viscoelastic set to ever result in failure (by which we mean data loss due to a decrease in signal amplitude caused by increased head–tape separation as the viscoelastically deformed tape passes over the magnetic head). For each defect, whose size is greater than the critical size D_F, there is an associated time to fail t_f. From Eq. (5.C.5) and the definition of D_F, we get

$$D(t_f) = \frac{D_F(1 - A_0)}{1 - [A_0 + A_1 \exp(-t_f/\tau_1) + A_2 \exp(-t_f/\tau_2)]} \quad \text{for } D > D_F$$

$$= \frac{D_F}{1 - [\exp(-t_f/\tau_1) + \rho \exp(-t_f/\tau_2)]/(1 + \rho)}. \tag{5.C.7}$$

Notice that large defects have a short time to fail, and as the defect size approaches D_F the time to fail becomes infinite.

Equation (5.C.7) is a strictly monotone function and is differentiable. The associated probability density function, call it $g(t)$, is

$$g(t) = f(d)\frac{dd}{dt}, \tag{5.C.8}$$

where d is expressed in terms of t. Then the instantaneous failure rate (IFR) associated with the random variable T is given by

$$Z(t) = g(t). \tag{5.C.9}$$

The term dd/dt is small and exponentially decaying. Thus $Z(t)$ must decay with time unless $f(d)$, $d > D_F$, is exponentially growing, which is unlikely.

We now assume that the defect size r.v. D has a *Weibull p.d.f.* (see Appendix B), $f(d)$,

$$f(d) = \alpha\beta d^{\beta-1} \exp(-\alpha d^\beta), \qquad \alpha, \beta > 0. \tag{5.C.10}$$

We note that β has no unity and α has units of $(\text{length})^{-\beta}$. Then the instantaneous failure rate is given by Eq. (5.C.9) and is found to be

$$Z(t_f)$$
$$= \frac{\alpha\beta D_F^\beta[(1/\tau_1)\exp(-t_f/\tau_1)+(\rho/\tau_2)\exp(-t_f/\tau_2)]\exp\{-\alpha D_F/\{1-\{[\exp(-t_f/\tau_1)+\rho\exp(-t_f/\tau_2)]/(1+\rho)\}\}\}}{(1+\rho)\{1-[1/(1+\rho)][\exp(-t_f/\tau_1)+\rho\exp(-t_f/\tau_2)]\}^{\beta+1}}.$$
$$\tag{5.C.11}$$

We consider three different Weibull p.d.f.'s for three Weibull parameters, as shown in Fig. 5.C.1. In Fig. 5.C.2 are plots of the instantaneous failure rates for creep-related defects as a function of time $[Z(t_f)]$ at 52° C/30% RH ($\tau_1 = 2.8$ h, $\tau_2 = 36.8$ h, $A_0 = 0.36$, $A_1 = 0.34$, $A_2 = 0.32$, and $\rho = A_2/A_1 = 0.9$, data obtained by fitting Eq. (5.C.5) to the 76-μm-thick shim data in Fig. 5.36) assuming $D_F = 50$ (units of D_F are length). If we have a defect distribution with large defects (i.e., relative to D_F, the critical defect size to cause failure), such as when

$$\alpha = 0.001, \qquad \beta = 1.75,$$

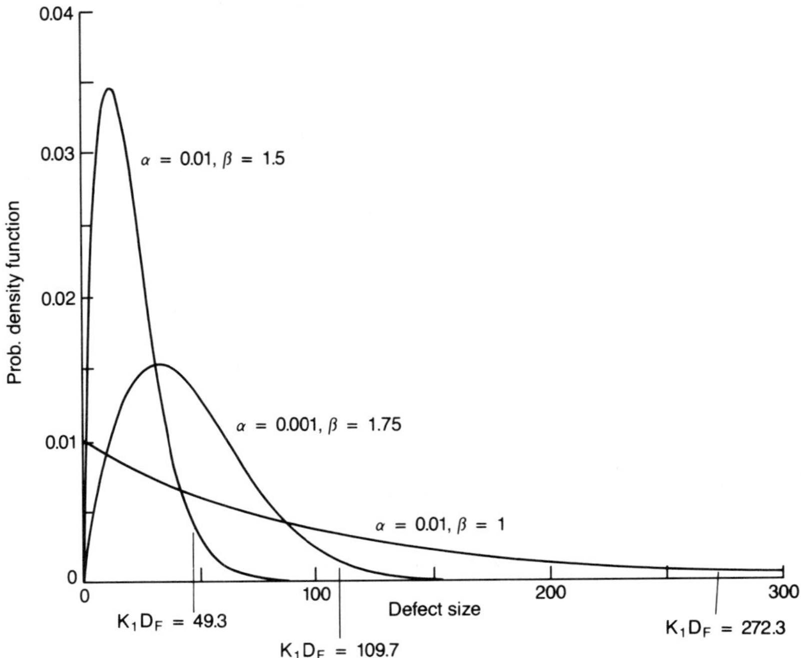

Fig 5.C.1. Weibull probability density functions for various Weibull parameters α, β. Arrows indicate the truncation (on the right-hand side) of the functions as a result of burn-in at 52° C/30% RH for 40 h.

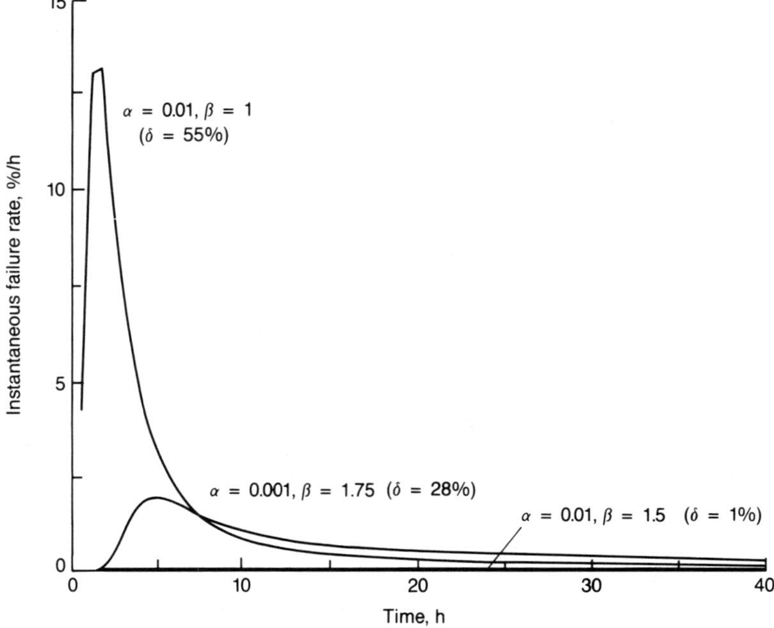

Fig. 5.C.2. Instantaneous failure rate %/h as a function of time for various Weibull probability density functions at 52° C/30% RH for $D_F = 50$.

and

$$\alpha = 0.01, \qquad \beta = 1,$$

then the instantaneous failure rate in the field is very large. On the other hand, if we have a defect distribution with few large defects (again relative to D_F), as when

$$\alpha = 0.01, \qquad \beta = 1.5,$$

then most defects will never lead to failure.

5.C.1. Effect of "Burn-In"

A "burn-in" approach (e.g., $52°$ C/30% RH for 40 h) can be used to accelerate the actual development of tape distortion at elevated temperatures. Bad tape reels are then eliminated by magnetic testing before shipment. After burn-in, the rate of change of the amount of viscoelastic deformation is small and exponentially decaying with time.

We first calculate a fraction of all potential creep-related media defects detected at the magnetic tester after burn-in. For burn-in at $52°$ C/30% RH, we have from Eq. (5.C.7)

$$D(t_f = 40\,\text{h}) = 1.19 D_F \equiv K_1 D_F,$$
$$D(t_f = 80\,\text{h}) = 1.057\, D_F \equiv K_2 D_F. \tag{5.C.12}$$

Equation (5.C.12) suggests that after a 40 h bake at $52°$ C/30% RH, defects whose size are greater than $1.19\, D_F$ will be detected and hence screened from the reel population before going to the field.

After a burn-in for 40 h, the defect distribution is truncated with no defects larger in size than $K_1 D_F$ (Fig. 5.C.1). This truncated r.v. has a Weibull p.d.f.

$$f(d) = \tilde{A}_0 \alpha \beta d^{\beta-1} \exp(-\alpha d^\beta), \qquad \tilde{A}_0, \alpha, \beta > 0, \tag{5.C.13}$$

where

$$\tilde{A}_0 \int_0^{K_1 D_F} \alpha \beta d^{\beta-1} \exp(-\alpha d^\beta)\,dd = 1,$$

or

$$\tilde{A}_0 = \frac{1}{1 - \exp[-\alpha(K_1 D_F)^\beta]}. \tag{5.C.14}$$

The proportion of the total population to fall-out in a further 40 h burn-in at $52°$ C or equivalently 5 years (based on the measured acceleration factors of 1000, reported in Chapter 3) in the field at $22°$ C is, call it δ,

$$\delta = \tilde{A}_0 \int_{K_2 D_F}^{K_1 D_F} \alpha \beta d^{\beta-1} \exp(-\alpha d^\beta)\,dd$$
$$= \frac{\exp[-\alpha(K_2 D_F)^\beta] - \exp[-\alpha(K_1 D_F)^\beta]}{1 - \exp[-\alpha(K_1 D_F)^\beta]}. \tag{5.C.15}$$

If, for one particular defect type, we know the exact defect size distribution (α, β) and the fall-out proportion δ, we can calculate from Eq. (5.C.15) the critical defect size D_F for that defect type. For example, if $\delta = 0.025$ and

$$\alpha = 0.01, \qquad \beta = 1.5 \to D_F = 41.2,$$
$$\alpha = 0.01, \qquad \beta = 1 \to D_F = 228.8,$$
$$\alpha = 0.001, \qquad \beta = 1.75 \to D_F = 92.2,$$

the units of D_F are length. Then if for subsequent manufacturing lots we know (α, β) and D_F for a particular defect type, we can calculate the percent fall-out in the 5-year period. For example, if $D_F = 50$, then Eq. (5.C.15) gives

$$\alpha = 0.01, \qquad \beta = 1.5 \to \delta = 1.1\%,$$
$$\alpha = 0.01, \qquad \beta = 1 \to \delta = 8.5\%,$$
$$\alpha = 0.001, \qquad \beta = 1.75 \to \delta = 10.5\%.$$

The instantaneous failure rate [Eq. (5.C.9)] then is

$$Z(t_f) = \frac{\alpha\beta\Lambda}{\tau_2} D_F^{\beta} \frac{\exp\{-\alpha\{D_F/[1 - \Lambda \exp(-t_f/\tau_2)]\}^{\beta} - t_f/\tau_2\}}{\{1 - \exp[-\alpha(K_1 D_F)^{\beta}]\}[1 - \Lambda \exp(-t_f/\tau_2)]^{\beta+1}}, \quad t \gg \tau_1,$$

$$(5.C.16)$$

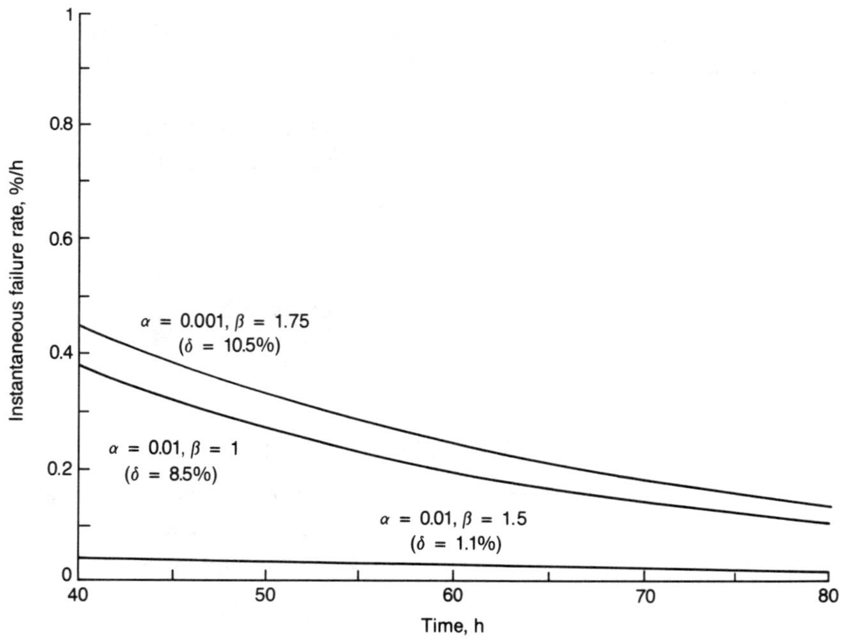

Fig. 5.C.3. Instantaneous failure rate %/h as a function of time for various truncated Weibull probability density functions at 52° C/30% RH for $D_F = 50$.

where $\Lambda = \rho/(1 + \rho) = 0.474$ for $52°$ C/30% RH. Figure 5.C.3 shows the plots of the instantaneous failure rates as a function of time $[Z(t_f)]$ for $D_F = 50$ for three Weibull distributions. It is interesting to compare scales in Figs. 5.C.2 and 5.C.3, burn-in reduces the failure rate considerably.

Since the tape continues to deform, though at a small rate, a certain amount of defects which potentially can cause an error may possibly escape the post-"burn-in" screen test. There are several possible alternatives by which the potential field defective escapes can be reduced to acceptable levels: adjust the signal amplitude failure level at the final magnetic tester to a higher clip level than that which causes failure in the drive system, and reduce the tape tension in the final magnetic tester by an appropriate amount such that allowance is made for the additional creep deformation that can occur in the field and that these reels do not pass the tester.

CHAPTER 6

Long-Term Reliability of Flexible Disks

For high areal densities, both linear and track densities need to be increased and the disk-surface defects need to be minimized (Hoagland et al., 1978). To accommodate high track densities without the need for expensive servo systems, the disk should be dimensionally stable; more specifically, that any dimensional changes do not affect the readback signal of a recorded track. In general, the substrate material must be as dimensionally stable as possible. It is acceptable, to an extent, for the disk to change its dimensions due to changes in the environment (temperature and humidity) and even to undergo time-dependent changes (shrinkage and creep due to the centrifugal forces of rotation), but it is crucial that these dimensional changes be the same (isotropic) or nearly so in all directions in the disk plane (Greenberg et al., 1977, 1978; Bogy and Talke, 1984). Mechanical misalignment, such as drive motor run-out and the repositioning of disks during use subsequent to initial write (Bhushan, 1990) also pose reliability problems. For a high recording-density system, it is necessary that a track which is recorded as a circle to be read as a circle with the same track width (typically 50–600 μm) within some run-out or track misregistration (TMR) limit (typically 15% of the track width). Commonly encountered long-term-reliability problems during usage and storage are the noncircular tracks, the nonuniform track width, and the changes in the spatial location of the tracks.

Servo techniques are used to accommodate the desired TMR for high recording density (for a track density of greater than 5.32 tracks/mm). Commonly used servo techniques include a sector servo with few (about three) outside servo tracks (see Chapter 1). As long as the radial deformations are linear with the radius, these servo techniques are satisfactory. However, if the properties of the disk substrate are not the same along the radius, or the temperature and humidity on the disk surface are not constant, the sector servo approach cannot be used satisfactorily. Even though, a suitable servo technique may be used, it is still desirable to have a dimensionally-stable disk so that servo requirements are minimal.

Polymers, by nature, deform to a finite amount as compared to many

438

other materials such as metals or ceramics. Anisotropic deformation (or track eccentricity) of the disks are the result of inherent anisotropic physical and chemical properties of polymer substrates [almost exclusively biaxially-oriented poly(ethylene terephthalate) or PET] used in the construction of the flexible disk. We have seen in Chapter 2 that it is feasible to obtain isotropic polymers with relatively low deformation characteristics, provided a compromise primarily in the cost and other desirable properties (such as Young's modulus of elasticity) can be made.

We analyze the effect of various environmental and operating parameters which affect the deformation of flexible disks, responsible for long-term reliability problems. Some data on measurements of in-plane deformation of an actual flexible disk as a function of the environmental conditions are also presented.

6.1. Analysis of Disk Deformation

We present analyses to study the effect of various environmental and operating parameters on the long-term reliability of flexible disks. Here we consider the effects of thermal and hygroscopic expansions, shrinkage (relaxation of built-in stresses during the manufacturing process), and elastic/viscoelastic (creep) deformations due to the centrifugal forces of rotation. Here, the thermal and hygroscopic deformations refer to recoverable deformations, i.e., deformations that are isolated from nonrecoverable shrinkage.

6.1.1. Thermal Expansion

We have seen in Chapter 2 that the coefficient of the linear thermal expansion of a biaxially-oriented PET is anisotropic. Typically, it is about $1 \times 10^{-5}/°C$ along the slow optical axis and $2 \times 10^{-5}/°C$ along the fast optical axis for Mylar A film. A maximum displacement $(u_r)_{max}$ of about 22 μm would occur at the outside diameter of a 90-mm disk (inner diameter = 25 mm) going from 10° to 52° C. Radial ellipsity or runout $(\Delta u_r)_{max}$ of about 11 μm would also be introduced for a 90-mm-diameter disk. The numbers would double for a 130-mm-diameter disk. The temperature rise results from the storage or operating environment as well as from the interface temperature rise, because of asperity contacts at the head–disk interface (Bhushan, 1990). The interface temperature rise is localized and its effect cannot be corrected by the sector servo approach.

6.1.2. Hygroscopic Expansion

We have seen in Chapter 2 that the coefficient of the linear hygroscopic expansion of a biaxially-oriented PET is much less anisotropic than the thermal expansion. Typically, it is about $0.5 \times 10^{-5}/\%$ RH along the slow optical axis and about $0.6 \times 10^{-5}/\%$ RH along the fast optical axis for Mylar A film. A maximum displacement $(u_r)_{max}$ of about 11 μm would occur at the outside diameter of a 90-mm-diameter disk going from 8% to 80% RH. A radial ellipsity or runout $(\Delta u_r)_{max}$ of 2 μm would also be introduced for a 90-mm-diameter disk. The numbers would double for a 130-mm-diameter disk.

6.1.3. Shrinkage

Biaxially-oriented PET is known to shrink with time at high temperatures and high humidities. Shrinkage is highly anisotropic, see Chapter 2. In some cases, there may even be shrinkage along one axis and expansion along the other. Typically, shrinkage is small below 60° C but, at and above 60° C, the shrinkage increases linearly with temperature and logarithmically with aging time. It is about 0.05% at 60° C for 100 h. This suggests that a maximum displacement of 13 μm would occur at 60° C for a 90-mm-diameter disk. The number would double for a 130-mm-diameter disk. We have seen in Chapter 2 that thin PET films appear to shrink more than thicker films. The shrinkage at and above 60° C is substantial and can be minimized by thermal treatments of the substrate (see Chapters 2 and 3). The operating and storage temperatures of the disk should be kept below 60° C to minimize the shrinkage.

The amount of shrinkage that a disk will experience depends upon the previous history of the material, i.e., the environmental conditions that the substrate has been exposed to. Therefore, the shrinkage may vary from one roll of film to another or from one disk to another.

6.1.4. Centrifugal Stresses and Displacements

The analytical expressions of centrifugal stresses and elastic displacements for a spinning orthotropic annular disk are presented in Chapter 4. For an orthotropic disk made of biaxially-oriented PET substrate, the maximum centrifugal stresses in the radial direction $(\sigma_{rr})_{max}$ are of the order of 1 Pa or less, at typical angular speeds (300–600 rpm) and at room temperature. Therefore, the elastic displacement in the radial direction (u_r) is of the order of a picometer or less. Thus the stresses and displacements are negligibly small. Stresses increase with the square of both angular velocity and the difference between outer and inner radii, and displacements increase with the square of angular velocity and the cube of radius (Chapter 4; Lekhnitskii, 1968; Tang, 1969).

Based on the analysis presented in Chapter 4, maximum creep runout (Δu_r) for 90-mm- and 130-mm-diameter disks made of biaxially-oriented PET (Mylar A), spinning at 300 rpm, is less than 50 nm and 125 nm, respectively, after use at 60° C for 5 years.

6.1.5. Summary

A summary of the typical deformations, for a disk made of biaxially-oriented PET and its parametric dependence on angular speed and radii, is presented in Table 6.1. We find that major sources of deformation leading to TMR are: thermal expansion, hygroscopic expansion, and shrinkage. Thermal expansion and shrinkage are highly anisotropic whereas hygroscopic expansion is isotropic to a first order. For a high recording density, servo techniques are needed to overcome the anisotropic deformation effects.

Table 6.1. Typical deformations of a 90-mm biaxially-oriented PET (Mylar A) disk (25-mm inner diameter) rotating at 300 rpm and their parametric dependence

		Dependence of	
Parameter	Typical value(s)	Angular speed (ω)	Inner and outer radii (a, b)
$(u_r)_{max}$, $(\Delta u_r)_{max}$ due to thermal expansion from 10° to 52° C	22 μm, 11 μm		($b - a$), linear with temperature
$(u_r)_{max}$, $(\Delta u_r)_{max}$ due to hygroscopic expansion from 8% to 80% RH	11 μm, 2 μm		($b - a$), linear with humidity
Maximum shrinkage at 60° C	13 μm		($b - a$), approximately linear with temperature
Maximum centrifugal stresses $(\sigma_{rr})_{max}$ at RT	<1 Pa	ω^2	$(b - a)^2$
Elastic $(u_r)_{max}$, $(\Delta u_r)_{max}$ due to centrifugal forces at RT	<1 pm	ω^2	b^3
$(\Delta u_r)_{max}$ due to creep at 60° C for 5 years	50 nm	ω^2	b^3

6.2. Measurements of Disk Deformation

We have seen in the preceding section that the significant disk deformation occurs because of thermal expansion, hygroscopic expansion and shrinkage, and little because of creep. We present here experimental data on the in-situ measurements of the time-dependent in-plane deformation of a rotating flexible disk as a function of temperature and humidity, based on the work by Greenberg et al. (1977, 1978). Izraelev (1983) has presented a recording method to make these deformation measurements.

6.2.1. Description of Disk Deformation Measurement Apparatuses

We describe here two measurement apparatuses. The measurement apparatus based upon a strobe lamp had an order of magnitude lower-resolution than the measurement apparatus based upon a scanning laser.

6.2.1.1. Stroboscopic-Disk Deformation Measurement Apparatus

The test apparatus to measure the in-plane deformation of a coated (non-transparent) or transparent rotating flexible disk using a strobe lamp is shown schematically in Fig. 6.1 (Greenberg et al., 1977). It consisted of a flat "Bernoulli-type" base plate on which the flexible disk rotated separated from the base plate by a thin self-acting air bearing (for the Bernoulli disk, see Chapter 1). In the case of a transparent disk, a narrow circumferential track was scratched in the disk at the start of a test using a pin located in the base plate. Thereafter, the disk was strobed at a fixed angular position during each rotation, while the position of the scratch was observed through a microscope relative to a fixed grid in the base plate. The trigger for the strobe lamp was controlled by a magnetic-sync transducer which could be time delayed through 360°; hence, measurements could be taken continuously around the entire circumference of the disk. Due to the high magnification and small depth of field, the cross hair in the microscope eyepiece had to be first aligned with the scratch on the disk surface and then refocused on the base-plate grid. For nontransparent disks, the procedure was the same except that now the

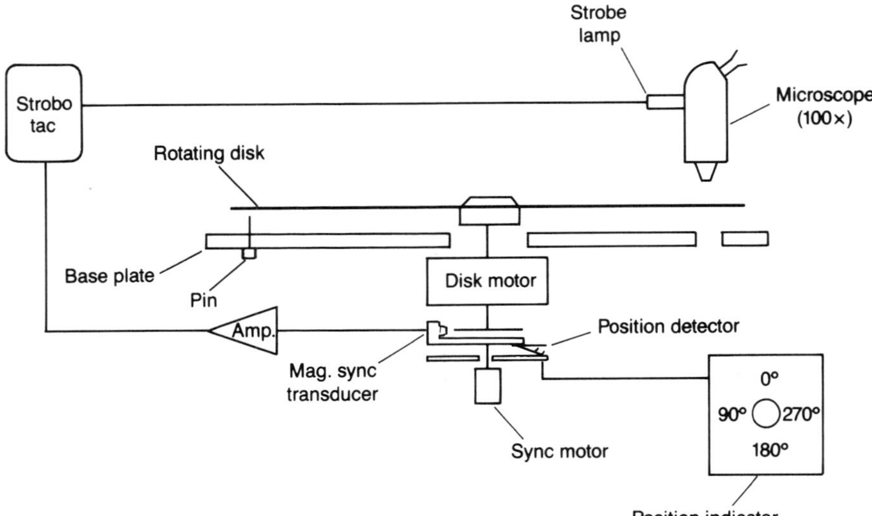

Fig. 6.1. Schematic of in-plane disk deformation measurement apparatus using a strobe lamp (Greenberg et al., 1977). © IEEE, 1977.

position of the disk edge was observed relative to the fixed grid rather than the position of the scratch.

Measurements made using the stroboscopic-disk tester required that the disk environment be disturbed. That is, the chamber had to be opened and a microscope positioned over the disk in order to take readings. Although the resolution of the stroboscopic-disk tester was only of the order of 10 μm, the stroboscopic-disk tester was found extremely useful as a "first-pass" coarse evaluation tool. It was used for long-term stability measurements of 1000 h or more (Greenberg et al., 1977).

6.2.1.2. Scanning-Laser-Disk Deformation Measurement Apparatus

The test apparatus to measure the in-plane deformation of a coated (non-transparent) rotating flexible disk using a scanning laser is shown schematically in Fig. 6.2 (Greenberg et al., 1977). From this figure we observe that a gap is formed between the rotating nontransparent disk and a reference edge fixed in the base plate on which the disk rotates. A mirror with 18 facets was mounted on the motor shaft concentrically with the flexible disk. Laser light was deflected from the rotating mirror so as to sweep radially across the gap between the disk and the reference edge. During this sweep, the light fell on a photodetector mounted over the gap, thereby producing a gate pulse whose width is proportional to the width of the gap. Using a high-frequency oscillator, Greenberg et al. (1977) determined the length of the gap in terms of how many wavelengths fit into the gate pulse. This gate-pulse-count N was then transmitted to an IBM System/7 computer. Eighteen such counts (every 20°) were received by the computer for each revolution of the disk corresponding to the 18 facets of the mirror; an optical position detector kept track of the sector being measured. The counts received by the computer were first converted into gap width and then into disk deformation by comparing them to those at the start of the test. In addition to measuring the total deformation of a disk from the start of the test to its end, we can measure the track distortion within any time interval Δt.

To obtain the gap width δ from the gate-pulse-count N, we note that

$$\delta = \frac{NV}{p}, \tag{6.1}$$

where V is the normal component of the scan velocity and p is the oscillator frequency. To ensure that the light beam will hit the same point on the edge of the disk, as the disk expands and contracts, the optics were designed so that the laser beam scans along a radius in the moving disk. That is, the tangential component of the scan velocity was matched to the speed of the edge of the disk, or

$$f\omega \cos \theta = R\Omega, \tag{6.2}$$

where f is the focal length of the lens, ω is the angular speed of the laser beam, θ is the angle between the scan direction and the tangent to the disk, R is the

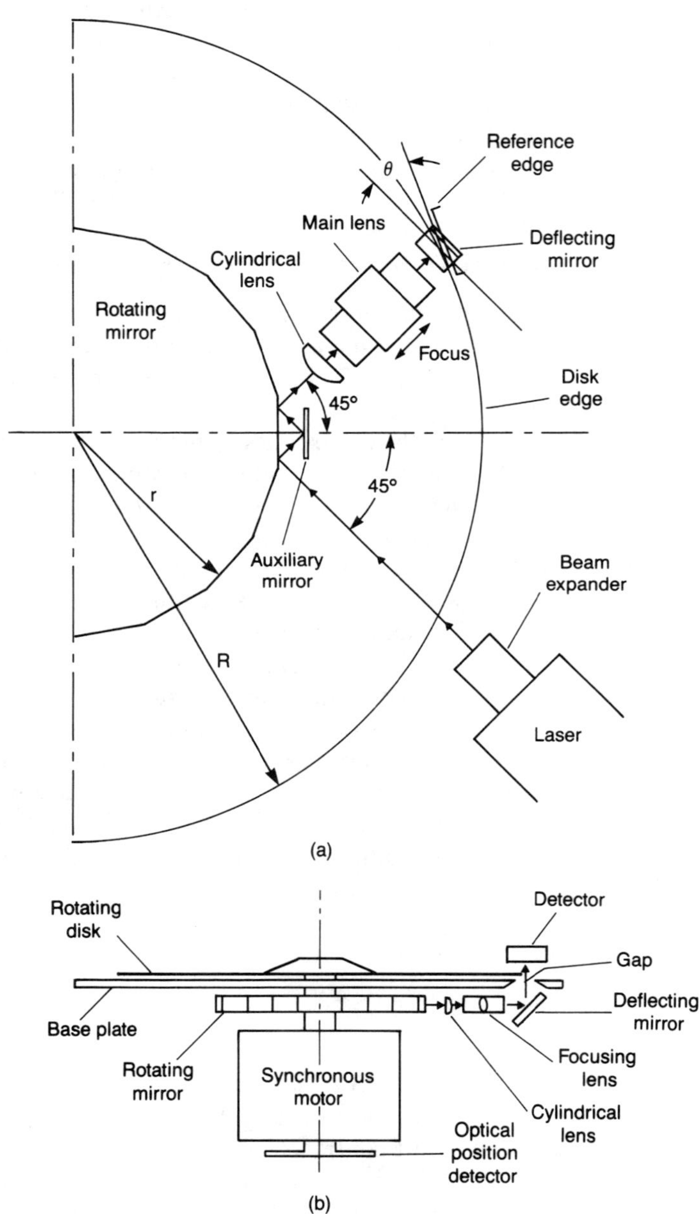

Fig. 6.2. Schematic of in-plane disk deformation measurement apparatus using a scanning laser (a) top view and (b) side view (Greenberg et al., 1977). © IEEE, 1977.

disk radius, and Ω is the rotational speed of the disk. Since the angular speed of the laser scan was selected as four times the rotational speed of the disk (i.e., $\omega = 4\Omega$) on account of the auxiliary mirror, we obtain from Eq. (6.2) that the focal length of the lens is given by

$$f = R/4 \cos \theta. \tag{6.3}$$

Noting that $\sin \theta = r/\sqrt{2}R$, where r is the radius of the 18-faceted mirror, we obtain

$$f = R/\{4[1 - (r/R)^2/2]^{1/2}\}. \tag{6.4}$$

Hence, using $V = 4f\Omega \sin \theta$, we have for the gap width

$$\delta = N\Omega r/\{p[2 - (r/R)^2]^{1/2}\}. \tag{6.5}$$

The oscillator-gate approach used had an inherent ± 1 count uncertainty, so it was necessary to keep the value of each count well under the desired resolution (about 2–5 μm for flexible-disk applications). For example, a typical setup to test 152.4-mm-diameter disks employed a mirror with a radius $r = 75$ mm rotating at 1800 rpm. The geometry of the system gave $\theta = 20°$, therefore, the appropriate lens for this configuration had a focal length of $f = 40$ mm, and the normal component of the scan speed was $V = 10.3$ m/s. A 60-MHz oscillator gave each count a value of 0.2 μm, i.e., less than 10% of the desired resolution. The cylindrical lens shown in Fig. 6.2 was provided to elongate the beam and, thereby, average out the effect of any roughness on the disk edge.

To test a disk, the equipment was brought to the desired test conditions (temperature and humidity) within the environmental chamber, and the disk was mounted. With the disk in place and spinning, the chamber was closed, and the first point was taken within the first two minutes. For the first few hours of a test, data was collected at frequent intervals since the early deformations (at elevated temperatures) were the most rapid (Greenberg et al., 1977).

6.2.2. Experimental Results

To measure thermal and hygroscopic deformations and shrinkage, the scanning-laser apparatus was used because of its high accuracy, and because of the fact that the environment does not have to be disturbed during measurement. To make the long-term shrinkage and creep measurements of a rotating disk, the stroboscopic apparatus was used. Measurements were made on 152.4-mm-diameter coated flexible disks [which were made of a 76.2-μm-thick biaxially-oriented PET (Mylar A) substrate with a 2.5 μm-thick particulate γ-Fe$_2$O$_3$ coating on both sides] as well as bare PET disks. We note that the deformation of a coated disk is primarily influenced by the properties of the PET-disk substrate.

6.2.2.1. Thermal and Hygroscopic Deformations

The reversible thermal and hygroscopic deformations of a disk were measured after equilibrating the disk for several hours at the test conditions so as to make the disk dimensionally stable. Then the temperature (humidity) was changed while holding the humidity (temperature) constant. Once this was achieved, the temperature (humidity) was changed to a new value without altering the humidity (temperature). Measurements were then taken until the disk had become dimensionally stabilized at the new conditions. The disk was equilibrated at the high temperature (humidity) and then reduced this temperature (humidity) to the lower value. This procedure eliminated the accelerated shrinkage associated with high temperatures (humidities).

Figure 6.3 shows the thermal deformation of a coated flexible disk in going from 58° to 46° C at 40% RH. In this plot, the shape of the disk at the reference time is displayed as a circle whose radius is of the order of magnitude of the expected deformation, and therefore the deformed shape becomes very exaggerated. Although the mapping to the highly exaggerated scale of Fig. 6.3 does not preserve shape, it depicts well the axes of preferential deformation. We note that thermal deformations are anisotropic with maximum and mini-

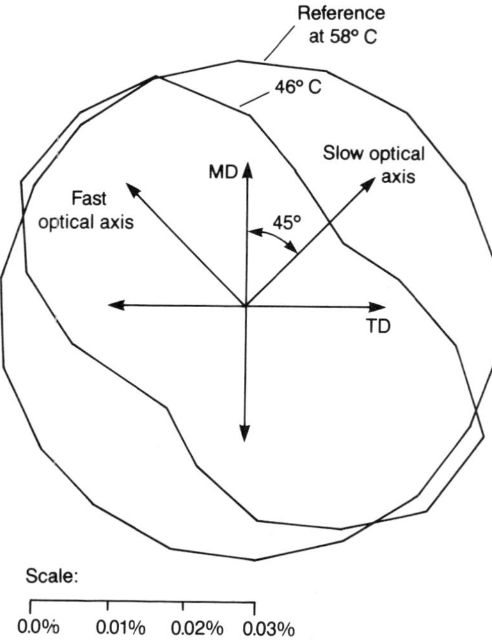

Fig. 6.3. Thermal deformation of a 152.4-mm-diameter flexible disk (76.2-μm-thick biaxially oriented PET, Mylar A substrate with a 2.5-μm-thick coating on both sides) in going from 58° to 46° C at 40% RH; MD—machine direction and TD—transverse direction (Greenberg et al., 1978).

mum values along the optical axes. For this disk, the optical axes were oriented at 45° from the machine direction. Anisotropy of the thermal expansion for a PET film has also been reported in Chapter 2.

Although the anisotropy and orientation of thermal deformation can be seen from Fig. 6.3, the magnitude of the deformation cannot be read directly because the aluminum base plate, which contains the reference edge, is also subjected to thermal deformation. If we assume that the coefficient of thermal expansion of aluminum is $\alpha_{Al} = 2.4 \times 10^{-5}/°C$, then the maximum and minimum thermal coefficients of the flexible disk can be calculated by subtracting the base-plate deformation, resulting in $\alpha_1 \sim 3.81 \times 10^{-5}/°C$ and $\alpha_2 \sim 2.2 \times 10^{-5}/°C$.

Figure 6.4 shows the hygroscopic deformation of a coated flexible disk (equilibrated at 60° C/40% RH for 2 weeks) when the humidity was lowered from 40% to 30% RH. Unlike thermal deformation (and shrinkage to be discussed later), hygroscopic deformation was found to be isotropic to first order as well as reversible. It appears that this phenomenon is related to the fact that the polymer absorbs moisture rather uniformly, regardless of structural orientation. This is consistent with the data presented in Chapter 2. The magnitude of the hygroscopic deformation can be seen directly from Fig. 6.4, since the base plate does not deform with changes in humidity. For the sample shown in Fig. 6.4, the linear coefficient of hygroscopic expansion is $\beta \sim 1.3 \times 10^{-5}/\%$ RH.

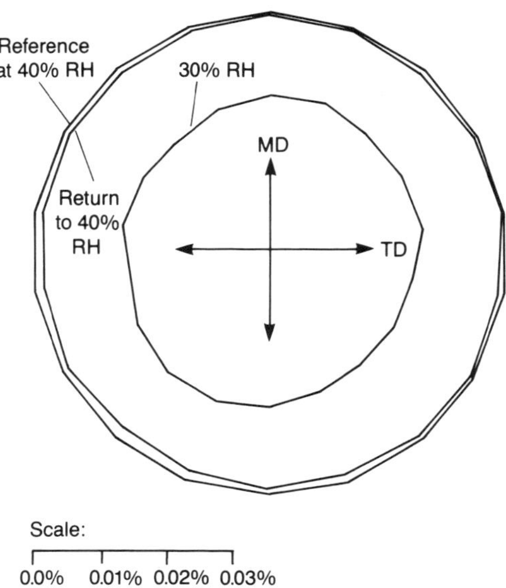

Fig. 6.4. Hygroscopic deformation of a 152.4-mm-diameter flexible disk (76.2-μm-thick biaxially-oriented PET, Mylar A substrate with a 2.5-μm-thick coating on both sides) in going from 40% to 30% RH and back to 40% RH at 60° C (Greenberg et al., 1978).

Greenberg et al. (1977) reported that annealing (heat treating) of the disk at 120° C for 1 h did not significantly alter the hygroscopic expansion.

6.2.2.2. Shrinkage

Shrinkage tests may be made on either the scanning-laser or the stroboscopic apparatus. To measure the shrinkage, the environmental chamber was brought to the desired temperature and humidity, and the disk was mounted. One hour was allowed for the disk to reach the equilibrium temperature and humidity, and then the first reading (reference) was taken. Computer-controlled readings were continued throughout the duration of the test at predetermined time intervals.

Figure 6.5 shows a typical result from the scanning-laser apparatus for the shrinkage of a coated flexible disk during 10 days at 60° C and 15% RH. We see that the maximum shrinkage occurred along the machine direction, while the minimum shrinkage occurred along the transverse direction. This is consistent with the data presented in Chapter 2.

Table 6.2 presents the long-term shrinkage data for PET disks using the stroboscopic apparatus. In Table 6.2, the first four entries show the accelerated shrinkage due to an increase in test temperatures at 50% RH, while entries 5 and 6 show the changes in shrinkage as a function of humidity. We note that temperatures higher than 60° C rapidly accelerate the shrinkage, though temperatures higher than 60° C are not realistic operating or storage

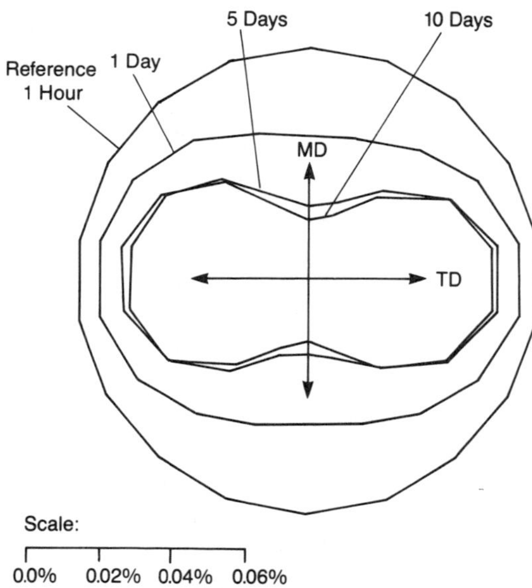

Fig. 6.5. Shrinkage of a 152.4-mm-diameter rotating flexible disk (76.2-μm-thick biaxially-oriented PET, Mylar A substrate with a 2.5-μm-thick coating on both sides) at 60° C/15% RH for 1 day, 5 days, and 10 days (Greenberg et al., 1978).

Table 6.2. Shrinkage in a 152.4-mm-diameter and 76.2-μm-thick biaxially-oriented PET (Mylar A) disk rotating at 1800 rpm (Greenberg et al., 1978)

					Shrinkage, %		
No.	Material	Temp., °C	RH, %	Time, h	max.	min.	anisotropic
1	Flexible disk	40	50	1000	0.02	0.01	0.01
2	Flexible disk	50	50	1000	0.03	0.01	0.02
3	Flexible disk	60	50	1000	0.07	0.03	0.04
4	Flexible disk	72	50	1000	0.13	0.05	0.08
5	Flexible disk	60	15	120	0.02	0	0.02
6	Flexible disk	60	40	120	0.03	0	0.03
7	Flexible disk	65	40	24	0.04	0	0.04
8	Flexible disk heated to 170° C for 10 min	65	40	24	~0	0	~0

temperatures. Entries 7 and 8 indicate that shrinkage can be reduced by annealing (heat treating) prior to testing.

Heat treating (annealing) may be done on the uncoated polymer film, the coated film before the disk is cut, or on the disk itself. It has been found that the best results are achieved if the film is annealed under light in-plane tension ($\sim$18 N/m) to prevent warping. The in-plane tension can conveniently be applied while the polymer film is in roll form or in disk form while it is rotating.

Greenberg et al. (1977) reported that a 23.4-μm-thick Mylar A shrank more than the 76.2-μm-thick Mylar A at 60° C/40% RH.

6.2.2.3. Creep

Creep, the viscoelastic deformation resulting from in-plane stresses caused by the rotation of the disk, is the most difficult deformation to observe since shrinkage is usually large and obscures the effect of creep. The stroboscopic apparatus described earlier allows the measurement of creep in a rotating disk in the following way. First, a stationary control disk was placed in the environmental chamber together with the rotating disk. Thereafter, deformation measurements as described earlier were made at fixed time intervals on both the continuously rotating disk as well as the stationary disk. The control disk, which was rotating only for a short period of time during initial scratching and each subsequent measurement, is essentially free of the forces of rotation for most of its lifetime and undergoes only shrinkage due to stress relaxation. The rotating disk, on the other hand, was experiencing creep in addition to shrinkage; thus, the difference in the deformation of the two disks corresponded to creep due to the forces of rotation.

Table 6.3 shows the creep results of PET disks for 1000 h at temperatures of 50° C, 60° C, and 72° C, respectively. We observe that the deformation of

Table 6.3. Creep in a 152.4-mm-diameter and 76.2-μm-thick biaxially-oriented PET (Mylar A) disk rotating at 1800 rpm (Greenberg et al., 1978)

		Maximum shrinkage, %			
Temp., °C	RH, %	spinning disk	control disk	Creep, %	Time, h
50	50	0.03	0.03	0	1000
60	50	0.07	0.07	0	1000
72	50	0.13	0.16	0.03	1000

the stationary and rotating disk are identical at the lower temperatures, and only insignificantly different as compared to shrinkage at the highest temperature of 72° C, suggesting that creep becomes increasingly important for temperatures at or above the glass transition temperature. However, in most recording applications, temperatures do not exceed 60° C, hence creep does not appear to be a concern (Greenberg et al., 1977, 1978; Bogy and Talke, 1984).

6.2.2.4. Summary

We find that the major sources of deformation on flexible biaxially-oriented PET disks are recoverable thermal and hygroscopic deformations and non-recoverable shrinkage. Creep due to rotational forces does not appear to be a significant source of deformation on flexible PET disks. Shrinkage is orthotropic with respect to the machine and transverse directions, while thermal expansion is orthotropic with respect to the optical axes; furthermore, hygroscopic expansion is, to first order, isotropic. Shrinkage is accelerated by both temperature and humidity, but can be reduced significantly by annealing.

References

Bhushan, B. (1990). "Tribology and Mechanics of Magnetic Storage Devices." Springer-Verlag, New York.

Bogy, D. B. and Talke, F. E. (1984). Creep of a rotating orthotropic polymer circular disk. In "Tribology and Mechanics of Magnetic Storage Systems" (B. Bhushan et al., eds.), Vol. 1, pp. 115–125. SP-16, ASLE, Park Ridge, Illinois.

Greenberg, H. J., Stephens, R. L., and Talke, F. E. (1977). Dimensional stability of floppy disks. *IEEE Trans. Magn.* **MAG-13**, 1397–1399.

Greenberg, H. J., Stephens, R. L., and Talke, F. E. (1978). Measurement of dimensional changes in thin polymer films. *Exp. Mech.* **18**, 115–120.

Hoagland, A. S., Oehme, W. F., and Talke, F. E. (1978). Narrow track defect studies on flexible media. *IEEE Trans. Magn.* **MAG-14**, 740–742.

Izraelev, V. (1983). On determination of thermal and hygroscopic expansion coefficients of PET floppy disk substrates by the recording method. *IEEE Trans. Magn.* **MAG-19**, 2253–2256.

Lekhnitskii, S. G. (1968). "Anistropic Plates." Gordon and Breach, New York.

Tang, S. (1969). Elastic stresses in rotating anisotropic disks. *Int. J. Mech. Sci.* **11**, 509–517.

APPENDIX A

Requirements and Supporting Test Methods for Magnetic Tapes and Tape Reels

This appendix presents the requirements and supporting test methods for magnetic tapes and tape-reels to meet performance and quality standards, and to enable the mechanical and magnetic interchangeability of tape reels between tape drives (Anonymous, 1975, 1977, 1979, 1983, 1984a, 1991). As an example, we present the requirements for 23.4-μm-thick tapes and tape cartridges to be used on the IBM 3480/3490 data-processing tape drives with an 18-track head (Anonymous, 1984a, b, 1991). Specifically, we present the environment and safety requirements, the requirements and supporting test methods for the physical and magnetic characteristics of the tape, tape magnetic-recording performance, tape quality, mechanical specifications of the tape cartridge, and cartridge quality. We also present methods for the care and handling of tape cartridges and tape drives. Finally, we present specifications for recording and track formats.

A.1. Environment and Safety

The conditions specified below refer to ambient conditions in the test or computer room and not to those inside the tape equipment.

A.1.1. Cartridge and Tape Testing Environment

Tests and measurements made on the tape, to check the requirements for an IBM 3480/3490 cartridge, will be carried out under following conditions:

Temperature: 23° C $\pm$ 2° C.
Relative humidity: 40–60%.
Conditioning before testing: 24 h.

A.1.2. Cartridge Operating Environment

Cartridges used for data interchange shall be operated under the following conditions:

452

Temperature: 15.6°–32.2° C. (The average temperature of the air surrounding the tape shall not exceed 40° C.)
Relative humidity: 20–80%.
Maximum wet bulb temperature: 25.6° C.

Conditioning before operating: If, during storage and/or transportation, a tape has been exposed to conditions outside the above values, it shall be conditioned for a period of at least 24 h.

A.1.3. Cartridge Storage Environment

Cartridges shall be stored under the following conditions:

Temperature: 4.4°–32.2° C.
Relative humidity: 5–80%.
Maximum wet bulb temperature: 26.7° C.

Continuous, extended storage of the tape cartridges at the maximum limits of the allowable temperature and humidity ranges is not recommended. However, if necessary, cartridges should be able to be stored in extreme conditions for up to *four weeks* without damaging the data or the cartridge. When tapes have been subjected to a storage environment outside the recommended ranges, reduced performance can result, if the tape cartridges are not reconditioned for a minimum of 24 h in the operating environment.

A.1.4. Transportation

A.1.4.1. Shipping Environment

(a) Unrecorded Cartridge

The packaged, unrecorded, tape cartridge must be capable of withstanding the following transportation environment:

Temperature: −23°–48.9° C.
Relative humidity: 5–100%.
Maximum wet bulb temperature: 26.7° C with no condensation in or on the cartridge.
Duration: 10 consecutive days.

(b) Recorded Cartridge

Temperature: 5.3°–32.2° C.
Relative humidity: 5–80%.
Maximum wet bulb temperature: 26.7° C with no condensation in or on the cartridge.

A.1.4.2. Hazards

Transportation of unrecorded tape cartridges involves three basic potential hazards.

(a) Impact Loads and Vibrations

The following recommendations should minimize damage to tape cartridges during transportation:

1. Avoid mechanical loads that would distort the cartridge shape.
2. Avoid dropping the cartridge more than 1 m.
3. Cartridges should be fitted into a rigid box containing adequate shock-absorbent material.
4. The final box must have a clean interior and a construction that provides sealing to prevent the ingress of dirt and water.
5. The orientation of the cartridges inside the final box should be such that their axes are horizontal.
6. The final box should be clearly marked to indicate its correct orientation.

(b) Extremes of Temperature and Humidity

Extreme changes in temperature and humidity should be avoided whenever possible. Whenever a cartridge is received it should be conditioned in the operating environment for a period of at least 24 h.

(c) Effects of Stray Magnetic Fields

A nominal spacing of not less than 80 mm should exist between the cartridge and the outer surface of the shipping container.

A.1.5. Safety

A.1.5.1. Safeness

The components of the tape and cartridge assembly must not constitute any safety or health hazard when used in the intended manner, or through any foreseeable misuse.

A.1.5.2. Flammability

The materials used in the external cartridge covers must have a flammability rating of at least 94V-2, as described in the Underwriters Laboratory UL94 *Standard for Safety—Test for Flammability of Plastic Materials.*

A.2. Physical and Magnetic Characteristics of Tape

The following is a description of the physical and magnetic characteristics of tape that are required for satisfactory operation. In addition to the properties described below, the following properties may also be specified for some tapes: Young's modulus of elasticity, yield strength, stress relaxation, edge smoothness, back surface reflectivity, and opacity of the tape.

A.2.1. Material

Definition. Materials used in the construction of magnetic tape (substrate, magnetic coating, and backcoat).

Requirement. The tape shall consist of a base material [oriented poly(ethylene terephthalate) film or its equivalent] coated on one side with a strong, yet flexible, layer of ferromagnetic material dispersed in a suitable binder. The back surface of the tape can also be coated.

A.2.2. Length

Definition. The minimum length of the tape wound around the tape reel hub.

Requirement. The minimum tape length shall be 165 m.

A.2.3. Width

Definition. The distance measured across the tape from tape-edge to tape-edge when the tape is under less than 0.28 N tension.

Requirement. The width of the tape shall be 12.650 mm $\pm$ 0.025 mm.

A.2.4. Thickness

Definition. The dimension measured perpendicular to the tape surface of the different layers of material that make up the tape as measured from the back to the front of the tape.

Recommendation. The thickness of the base material will be 92 gauge (23.4-μm thick) nominal. The total thickness of the tape will be between 25.9 μm and 33.7 μm.

A.2.5. Discontinuity

Definition. Any physical interruption in the tape, such as that produced by tape splicing or perforations.

Requirement. There will be no discontinuities.

A.2.6. Longitudinal Curvature

Definition. A departure from a straight line along the longitudinal dimension of the tape in the plane of the tape surface (see Chapter 5).

Requirement. Any deviation from a straight line will be gradual and will not exceed 3.8 mm for a 1-m span of tape. This corresponds to the minimum radius of curvature of 33 m when measured over an arc of circle.

Procedure. Allow a 1-m length of tape to unroll and measure its natural curvature on a flat smooth surface. Measure the maximum deviation of the concave edge of the tape from a straight line.

A.2.7. Out-of-Plane Distortions (OPDs)

Definition. OPDs are local deformations that cause portions of the tape to deviate from the plane of the surface of the tape. OPDs are most readily observed when the tape is lying on a flat surface under no tension (see Chapter 5).

Requirement. All visual evidence of out-of-plane distortion shall be removed when the tape is subjected to a uniform tension of 0.6 N.

A.2.8. Cupping

Definition. A departure across the width of tape (transverse to motion) from a flat surface (see Chapter 5).

Requirement. The departure across the width of the tape from a flat surface shall be less than 0.3 mm when placed concave side down on a smooth flat surface.

Procedure. Cut a 1 m $\pm$ 0.1 m sample of tape. Condition it for a minimum of 3 h in the test environment by hanging it so that the coated surface is freely exposed to the test environment. From the center portion of the acclimated tape, cut a sample 25 mm in length. Stand the cut sample on its end in a cylinder, which is 25-mm high and has an inside diameter of 13 mm $\pm$ 0.2 mm. With the cylinder standing on an optical comparator, measure the cupping by aligning the edges of the sample to the reticle and determining the distance from the aligned edges to the corresponding surface of the sample at its center.

A.2.9. Magnetic Coating Roughness

Definition. The average surface irregularity of the recording surface of the tape (Bhushan, 1990).

Requirement. The rms roughness of the coating surface shall be between 15 nm and 22 nm, in both the longitudinal and transverse directions, using a stylus profiler with a 90° conical diamond tip having a radius of 2.5 μm at a stylus force of 0.5 mN and a traverse length of about 1.3 mm. Gloss, if used as a process control, will be 95–110 when measured in the longitudinal direction with a 45° Gardner gloss meter (Bhushan, 1990).

A.2.10. Flexural Rigidity

Definition. The capability of the tape to resist bending in the longitudinal direction.

Recommendation. The flexural rigidity (Young's modulus of elasticity $\times$ the moment of inertia of the cross section with respect to the neutral axis) of the tape in the longitudinal direction will be between 0.06 N mm^2 and 0.16 N mm^2.

Procedure. Clamp a 180-mm-long tape sample in a universal testing machine, allowing a 100 mm separation between the machine jaws using a crosshead speed of 5 mm/min. Calculate Young's modulus of elasticity (E) using the slope of the stress–strain curve between 2.2 N and 6.7 N. The flexural rigidity is given by EI where $I = wt^3/12$, w is the tape width, and t is the tape thickness.

A.2.11. Coating (Peel) Adhesion

Definition. Pertains to the force required to peel any part of the coating from the tape base material (Bhushan, 1990; Bhushan and Gupta, 1991).

Requirement. The force required to peel any part of the coating from the tape base material will not be less than 1.5 N for a 12.7-mm-wide tape in a 180° peel test described below.

Procedure
1. Take a sample of the tape approximately 380-mm long and scribe a line through the coating across the width of the tape 125 mm from one end.
2. Using a double-sided pressure-sensitive tape, attach the full width of the sample to a smooth metal plate, with the recording surface facing the plate.
3. Fold the sample over 180° adjacent to, and parallel with, the scribed line. Attach the metal plate and the free end of the sample to the jaws of a universal testing machine (e.g., Instron), such that when the jaws are separated the tape is peeled. Set the jaw separation rate to about 250 mm/min.
4. Note the force at which any part of the coating first separates from the base material. If the sample peels away from the double-sided pressure-sensitive tape before the force exceeds the requirement, an alternative type of double-sided pressure tape will be used.

A.2.12. Frictional Drag Between the Recording Surface and the Back Surface

Definition. The resistance to motion between the tape coating and the back surface of the tape (Bhushan, 1990).

Requirement. The force required to move the recording surface, in relation to the back surface, shall not be less than 1.0 N.

Procedure. Wrap a tape sample around a 25-mm-diameter circular mandrel with the back surface of the tape sample facing outward. Place a second sample of tape, with the coated surface facing inward, around the first sample for a total wrap angle of 90°. Secure one end of the outer test tape to 0.65 N. Secure the other end of the test tape sample to a force gauge that is mounted on a motorized linear slide. Drive the slide at a speed of 1 mm/s. Connect the force gauge to a plotter to record the results.

A.2.13. Electrical Resistance of the Recording Surface

Definition. The electrical resistance is the resistance of the recording surface, measured in ohms per square.

Recommendation. The electrical resistance of the recording surface will not exceed 5×10^8 ohms per square for nonbackcoated tape, and 5×10^9 ohms per square for backcoated tape, but shall be greater than 10^5 ohms per square. The electrical resistance of any backcoat shall be less than 10^6 ohms per square.

Procedure. After a 24-h exposure to the test environment, two layers of the sample tape will be placed back-to-back between the strip electrodes, as shown in Fig. A.1, such that the recording surfaces are in contact with all of the electrodes. In mounting the sample for measurement, it is important that no conduction paths exist between the electrodes, except those through the sample. To ensure that the length of tape held between each strip electrode is the same, the sample shall be placed under 2.23 N $\pm$ 0.56 N tension as it is being clamped.

Measurements shall be made between each pair of adjacent electrodes, producing a total of five readings per sample. The resistance of the coating will be determined by means of a guarded circuit, as shown in Fig. A.1(b), using a 500 $\pm$ 10 V potential.

A.2.14. Magnetic Coercivity

Definition. The intrinsic coercive force of the magnetic coating (Chapter 1).

Requirement. The intrinsic coercive force of the coating shall be 41.4 $\pm$ 2.4 kA/m (520 Oe $\pm$ 30 Oe) when measured in the longitudinal direction.

Procedure. This test should be performed using a vibrating sample magnetometer or VSM (e.g., EG & G Princeton, Applied Research Model 155) to measure the magnetic characteristics of the tape. The magnetometer must be operated according to the manufacturer's operating procedures and set up according to the following specifications:

1. Calibrate for field strength H and for magnetization M against the appropriate standards.

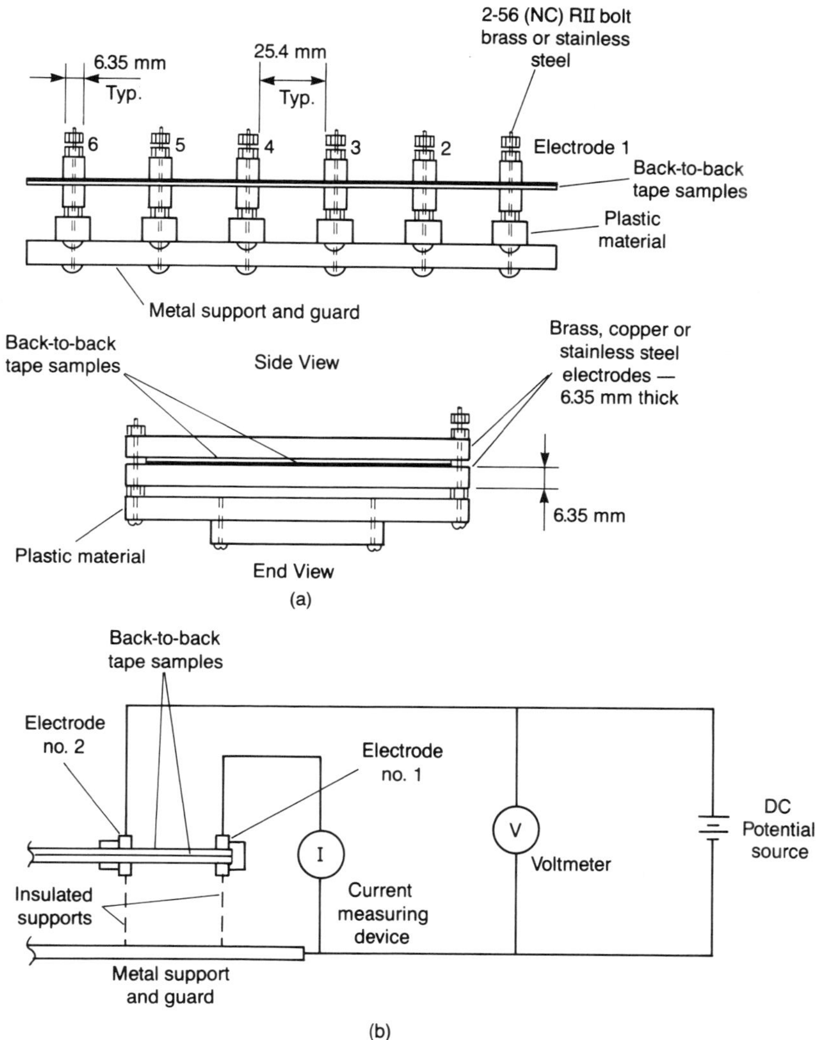

Fig. A.1. (a) Schematic arrangement of electrodes in the tape-resistance measurement setup, and (b) tape-resistance measurement circuit.

2. Make the lock-in amplifier time constant compatible with the sweep time.
3. Cycle the test sample at ± 350 kA/m (± 4400 Oe) (arbitrarily selected) at least four times before starting the actual test.
4. The sweep cycle time shall be 10 min; this is because H_c varies with the sweep time.
5. Magnetize to ± 350 kA/m ($+ 4400$ Oe), reduce H to zero, and increase H in a negative direction until the magnetization is reduced to zero.
6. Measure the negative value of field strength. This is magnetic coercivity value.

A.2.15. Magnetic-Particle Orientation Ratio

Definition. The squareness ratio is the longitudinal direction of the tape divided by the squareness ratio in the direction perpendicular to the tape edge. The squareness ratio is the ratio of the remnant flux density to the saturation density.

Requirement. The orientation ratio will be greater than or equal to 2 and the squareness ratio in the longitudinal direction will be greater than 0.80.

Procedure. The test should be performed with a vibrating sample magnetometer.

A.2.16. Remanent State

Definition. The remanent magnetic moment is the magnetic moment of the recording surface remaining in the absence of any magnetizing field. The maximum remanent of the recording surface is the remanent magnetic moment of the recording surface after subjecting the recording surface to a magnetizing field of 350 kA/m (4400 Oe).

Requirement. Prior to recording, the tape will have been anhysteretically erased (a process of erasure utilizing alternating fields of decaying level) to ensure that the remanent magnetic moment of the recording surface does not exceed 20% of the maximum remanent magnetic moment, and also that no low density transitions are present on the tape.

Procedure. The magnetic moments will be measured using a vibrating sample magnetometer (VSM), following these steps:

1. Use a 6–12-mm sample disk of tape.
2. Identify the longitudinal (coating) direction of the sample.
3. Mount the disk on the VSM sample holder with the longitudinal direction aligned from pole to pole.
4. Center the disk between the poles, according to the VSM manufacturer's instructions.
5. Measure the remanent magnetic moment of the disk.
6. Rotate the disk 180°, and repeat steps 4 and 5.
7. Cycle the disk four times around the hysteresis loop using a peak field by 350 kA/m (4400 Oe).
8. Measure the positive and negative values of the maximum remanent magnetic moment.
9. Compare the ratio of values obtained in steps 5 and 8.

A.2.17. Inhibitor Tape

Definition. Any tape that reduces the performance of the tape drive or subsequently used tapes.

Certain tape characteristics can contribute to poor tape-drive performance. These characteristics include high friction to tape-path components, high abrasivity, poor edge conditions, excessive tape wear residue products, electrostatic charge build up on the tape or tape-path components, and transfer of oxide coating to the back of the next tape layer. We only describe the tests for friction and abrasivity here.

Requirements. The tape shall not be an inhibitor tape.

A.2.17.1. Frictional Drag Between the Recording Surface of the Tape and Recording Head Surface

Definition. Pertains to the force required to move the recording surface of the tape over a ferrite surface as described in the following procedures.

Requirement. When the test is run according to the following procedures, the force required to move the tape at the beginning-of-tape (BOT) (at a point 1.3 m from the leader block of the cartridge) shall not exceed 1.5 N. The force required at the end-of-tape (EOT) (at a point 4.3 m from the junction of the tape with the cartridge hub) shall not exceed the force required at the BOT by more than a factor of 4.

Procedure
1. Wind a sample of tape on a 51-mm-diameter spool hub to an outside diameter of 97 mm. The wind tension shall be 2.2 N $\pm$ 0.2 N.
2. Repeat the following two steps five times:
   (a) Store at 50° C, 10–20% relative humidity for 48 h. (The total storage time at 50° C is 240 h.)
   (b) Acclimate at 23° C for 2 h and then unwind and rewind the tape on the spool hub at a tension to 2.2 N $\pm$ 0.2 N.
3. Acclimate the tape for 48 h at 30.5° C, 85% RH. At this environment, measure the force required to pull the recording surface of the tape over a 25.4-mm-diameter polished ferrite rod made of the material specified in the abrasivity test (90° wrap). The rod shall be polished to a 50 nm AA finish or better (N2 or ISO 1302).
4. Pull the sample over the rod at a speed of 1 mm/s while exerting a force of 0.64 N on the other end.
5. Take the measurements on the tape samples at the outer diameter of the reel and near the hub.

A.2.17.2. Tape Abrasivity (Head Wear)

Definition. Pertains to the tendency of the tape to wear the tape transport, including the head material.

Requirement. Head wear is simulated using a ferrite wear (Bhushan, 1990). The length of the wear pattern on the wear bar will not exceed 56 μm.

Procedure

1. Install a clean Ni–Zn ferrite square bar (4.5 mm × 4.5 mm × 18 mm) finished to a surface roughness of 50 nm AA (N2) or better, onto a holding fixture. The test edge facing upward should be unworn and free of chips. The radius of the test edge should not be greater than 13 μm.
2. Install the test fixture on a tape transport so that the wrap angle of the tape over the bar is 8° on each side, for 16° of total wrap.
3. Set the tape tension at the bar at 1.4 N.
4. With a tape speed of 1 m/s, make one pass of the tape over the wear bar. The length of tape passing over the wear bar should be 520 m $\pm$ 2.5 m. This length may be segmented into the appropriate number of cartridges.
5. Remove the holding fixture from the transport and measure the length of the flat worn on the wear bar. This measurement is most easily made using a microscope of known magnification, a camera, and a reference reticle. Magnification of 300 × or higher is recommended.
6. Measurements should be taken at the $\frac{1}{4}$, $\frac{1}{2}$, and $\frac{3}{4}$ points of the 12.65-mm width of the wear pattern. The three readings are averaged and the average length is used.

A.3. Tape Magnetic-Recording Performance

The magnetic-recording performance is defined by the testing requirements which follow. When performing the tests, the output or resultant signal will be measured on the same relative pass for both the reference tape and the tape under test (read-while-write, or on equipment without the read-while-write capability, on the first forward-read-pass) on the same equipment.

The following conditions will apply to all magnetic recording testing requirements, unless otherwise noted:

Tape condition: Pre-record condition.
Tape speed: ≤ 2.5 m/s.
Read-track: Within the written track.
Azimuth alignment: $\leq 6'$ between the mean write transitions and the read gap.
Write-gap length: 1.4 μm $\pm$ 0.2 μm.
Write-head saturation density: 0.34 T $\pm$ 0.03 T (3400 G $\pm$ 300 G).
Tape tension: 2.2 N $\pm$ 0.2 N.
Recording current: Standard measurement current.

A master standard reference tape is a tape selected as the standard for the reference field, signal amplitude, resolution, and overwrite. It is established at the NIST.

A.3.1. Typical Field

Requirement. The typical field of the tape shall be between 90% and 110% of the reference field of the master standard reference tape. Traceability to the reference field is provided by the calibration factors supplied with each secondary standard reference tape.

A.3.2. Signal Amplitude

Definition. Pertains to the average peak-to-peak value of the signal output measured over a minimum of 25 mm, exclusive of dropouts.

Requirement. The average signal amplitude at the physical recording density of 972 flux transitions/mm (ft/mm) will be between 70% and 140% of the standard reference amplitude.

A.3.3. Resolution

Definition. The ratio of the average signal amplitude at the physical recording density of 1458 ft/mm to that at the physical recording density of 972 ft/mm (Wang, 1966).

Requirement. The resolution of the tape will be between 80% and 120% of that for the master standard reference tape.

A.3.4. Overwrite

Definition. The ratio of the average amplitude of the residual of the fundamental frequency of the tone pattern, after being overwritten by the 972 ft/mm to the average signal amplitude of the 972 ft/mm signal. The average amplitude of the fundamental frequency of the tone pattern is the peak-to-peak amplitude of the sinusoidal signal with equal rms power.

Requirement. The overwrite ratio for the tape will be less than 120% of the same ratio for the master standard reference tape.

Procedure. A tone pattern is recorded at the standard measurement current (1.5 times the standard reference current that produces the reference field). A 972-ft/mm signal is then recorded over the tone pattern at the standard measurement current. Measure the average signal amplitude of the residual of the tone pattern fundamental and the average signal amplitude of the 972-ft/mm signal. Both amplitude measurements should be made using suitable filters.

A.3.5. Narrow-Band–Signal-to-Noise Ratio (NB–SNR)

Definition. Pertains to the ratio of the average read-signal power to the average integrated (side band) rms noise power in dB.

Requirement. The NB–SNR shall be equal to or greater than 30 dB when normalized to a track width of 410 μm. The normalization factor is $dB(410) = dB(w) + 10 \log(410/w)$, where w is the track width used when measuring $dB(w)$.

Procedure. The NB–SNR will be measured using a spectrum analyzer. The spectrum analyzer resolution bandwidth (RBW) will be 1 kHz and the video bandwidth (VBW) will be 10 Hz. The tape speed will be 762 mm/s.

1. Measure the read-signal amplitude of the 972-ft/mm signal using a spectrum analyzer, taking a minimum of 150 samples over a length of tape of 46-m minimum.
2. On the next pass (read only), measure the rms noise power over the same section of tape, and integrate the rms noise power (normalizing for the actual resolution bandwidth) over the range from 332 kHz to 366 kHz.

For other tape speeds the frequencies must be linearly scaled.

A.3.6. Velocity and Acceleration Profiles

Definition. Velocity and acceleration profiles of the tape during start/stop and streaming modes.

Requirement. The tape will reach the operating speed in a start distance of less than 17 mm and a start time of 18 ms (with an average acceleration of 110 m/s^2). The tape will stop in a stop distance of less than 18 mm and a stop time of 18 ms. During streaming, the tape velocity will vary a maximum of $\pm 7\%$ of the average operating speed. Longitudinal and lateral (shear) tape vibrations shall be within the specifications. Cartridges will be tested at the beginning, middle and end of the tape.

Procedure. Write an all ones pattern on the tape and then read it back. Measure start time, start distance, stop time, stop distance, and velocity. Using a spectrum analyzer, obtain the velocity and acceleration spectrums from the read-back signal. Examine the amplitude and frequency of the detected vibrations.

A.4. Tape Quality

The tape quality (including the effects of exposure to storage and shipping environments) is defined by the testing requirements which follow. The following conditions will apply to all quality testing requirements:

Environment: Operating environment.
Tape condition: Pre-record condition.
Tape speed: 2 m/s.
Read-track width: 410 μm.
Physical recording density: 972 ft/mm.
Write-gap length: 1.4 μm $\pm$ 0.2 μm.
Azimuth alignment: $\leq 6'$ between the mean write transitions and the read gap.
Write head saturation density: 0.34 T $\pm$ 0.03 T (3400 G $\pm$ 300 G).

Recording current: Standard measurement current.
Format: 18-track.
Tape tension: 2.2 N $\pm$ 0.2 N.

A.4.1. Dropouts

Definition. Pertains to a loss of the read-signal amplitude. A dropout exists when the base-to-peak read-signal amplitude is 25% or less than half of the average signal amplitude for the preceding 25 mm of tape, exclusive of dropouts. When a dropout is detected, a second dropout should not be counted until 64 consecutive dropout-free flux transitions are read. If a dropout persists for a distance of 1.0 mm, another dropout is counted.

Requirements. The average dropout rate will be less than one dropout for each 8×10^6 flux changes recorded. The average dropout rate is the total number of flux transitions recorded on tape divided by the number of dropouts counted.

A.4.2. Coincident Dropouts

Definition. A coincident dropout is simultaneous dropout condition on two or more tracks of a 9-track group. A coincident dropout is counted as a single event regardless of length. There are two 9-track groups in the 18-track format. One group consists of the odd-numbered tracks and the other group consists of the even-numbered tracks.

A coincident dropout occurring in both groups at the same time is counted as a single event. The length of a combined coincident dropout event is measured from the start of the earliest event of either group to the end of the latest event of either group. This is equivalent to the logical "OR" of the coincident-dropout-event indicators for the two groups.

Requirements. No 165-m length of tape will have more than 12 coincident dropouts. A coincident dropout will not be longer than 50 mm.

A.4.3. Tape Durability

The testing and measurements performed on the cartridge for tape durability are described below. These requirements must be met including cartridge exposure to the shipping environment (51.7° C/22% RH for 10 days) before test. A drive generally will experience no permanent or hard (uncorrectable) errors in its life. Most errors can be corrected using error correction codes (ECC) and methods used to boost signal amplitude-automatic gain control (AGC). Some tapes require a coating integrity failure test—adhesive or cohesive failure of the magnetic coating or the backcoat (Bhushan, 1990).

A.4.3.1 Short-Length Durability and Reliability Test

Definition. Pertains to the capability of the tape to withstand the wearing action encountered during repeated accesses to a short file of data.

Requirement. No permanent coincident dropouts are permitted for a minimum of 40,000 read-forward passes. In addition, no more than one permanent coincident dropout is permitted, on average, for each 80,000 read-forward passes. (A permanent coincident dropout is one that persists for ten consecutive read passes.)

Procedure. Ensure the tape drive is clean before starting this test. As a test sample, use a minimum of four cartridges, written in an area free of coincident dropouts. The area to be tested on each cartridge should be approximately 500 25K-byte records past the tape-load point. The test area should consist of 50 25K-byte records.

Each test cycle consists of a start operation at the beginning of the test area and accessing each record in the test area before returning, using the backspace file, to the beginning of the test area. For a complete test, 80,000 cycles should be made on each cartridge. Ten attempts to read forward should be made for each coincident dropout before a permanent coincident dropout is logged.

Tape-path cleaning between passes is not permitted for this test. The test must be performed in a normal operating environment for the tape and the tape drive.

A.4.3.2 Long-Length Durability and Reliability Test

Definition. Pertains to the capability of the tape to resist the wearing action encountered while cycling full length on a tape drive.

Requirement. The cartridges should meet the following criteria:

1. The coincident dropout rate (average per pass) for the first 200 full passes must be no more than 6 per 165 m of written tape.
2. There may be no more than 12 coincident dropouts for each 165 m of tape on any single pass.

Procedure. Clean the tape drive before starting this test. The data written should consist of at least 16K-byte records.

Tape-path cleaning between passes is not permitted for this test. The test must be performed in a normal operating environment for the tape and tape unit.

A.5. Mechanical Specifications of the Tape Cartridge

The IBM 3480 tape cartridge consists of a reel for the magnetic tape, cartridge case, brake mechanism, file-protect mechanism, leader block, and a latching mechanism for the leader block (for the figure, see Chapter 1). When fabricated, the final assembly must meet the mechanical specifications given. Particular attention should be given to the design of the reel and flange assembly to minimize tape offset. All dimensions have a tolerance of ± 0.25 mm unless otherwise noted.

A.5.1. Tape-Reel Hub

Straightness: 40 μm.
Perpendicularity to datum: 70 μm.
Taper (diametrical): 50 μm.
Rate-of-change (across the width): 25 μm per mm.
Total indicator runout (TIR) to datum: 200 μm.

Datum is a plane defined by the pitch line of the hub teeth at the 41-mm diameter (Anonymous, 1984a).

A.5.2. Metallic Insert

The reel will have a metallic insert made of 410 or 416 stainless steel. It will withstand a pull-out force of 300 N minimum. Its central opening will be concentric with the axis of rotation of the reel within 0.15 mm. The surface of the metallic insert will be parallel to datum (as defined earlier) to within 0.15 mm.

A.5.3. Cartridge Case

The cartridge case and reel assembly will be molded from polycarbonate material. The cartridge case top/bottom will be flexible.

A.5.4. File-Protect Mechanism

The write-protect mechanism will have a flat surface identified by a visual mark, e.g., a white spot, when in the position in which writing is inhibited. In the write-inhibit position the surface of the write-protect mechanism will be behind the front surface of the cartridge case, a minimum distance of 2.55 mm.

In the write-enable position, the write-protect mechanism will be within 0.25 mm of the front side of the case.

A.5.5. Brake Mechanism

Brake-Button Material. The brake-button material will be Nylon 6/6 with $2\% \pm 1\%$ molybdenum disulfide or its equivalent.

In-Hand Position. The brake mechanisms will inhibit the rotation of the reel when the cartridge is in the in-hand position. The brake mechanism angular resolution will not be more than $6°$ in the in-hand position. In the locked position, the reel will not rotate more than 10° when a torque of 0.32 N m is applied in the direction that will cause the tape to unwind.

In-Machine Position. The brake mechanism will be disengaged when the cartridge is in the in-machine position with the tip of the brake button displaced 8.1 mm ± 0.2 mm from the bottom of the cartridge case. The force required to move the brake button into this position will not exceed 12.25 N.

A.5.6. Leader-Block Pull-Out, and Insertion Forces

Definition. Pertains to the amount of force required to pull the leader block and the attached tape out of the cartridge. Insertion force is the amount of force required to push the leader block into its latched position in the cartridge.

Requirement. The leader-block pull-out force will be a minimum of 3 N and the insertion force will be a maximum of 12 N when measured at an angle of 48° ± 1° with respect to the cartridge edge (Anonymous, 1984a).

A.5.7. Bond Strength Between the Leader Block and Tape

Definition. Pertains to the amount of force required to pull the leader block free from the tape.

Requirement. The leader block will remain attached to the tape when a force of 10 N is applied to the leader block at an angle of 38° ± 2° with respect to its edge (Anonymous, 1984a).

A.5.8. Tape Winding

Tape will be wound around the tape-reel hub such that it forms a counterclockwise spiral when the cartridge is viewed from the top. The recording surface of the tape shall be toward the hub.

A.5.9. Winding Tension

Definition. Constant tension applied to the magnetic tape during cartridge spool winding.

Requirement. Wind, with a tension of 2.2 N ± 0.33 N.

A.5.10. Tape-Reel Circumference

Definition. The perimeter of the tape wound around the tape-reel hub.

Requirement. The tape will be wound to a circumference of between 280 mm and 307 mm.

A.5.11. Tape-Reel Moment of Inertia

Definition. The ratio of the torque applied to a tape reel (complete with tape, hub, and flanges) when it is free to rotate about a given axis to the angular acceleration thus produced about that axis.

Requirement. The moment of inertia of the reel and tape will be between 166×10^{-6} kg m^2 and 216×10^{-6} kg m^2. The empty reel moment of inertia will be 33.0×10^{-6} kg m^2 $\pm$ 3.63×10^{-6} kg m^2.

Procedure. Torsionally oscillate the reel assembly on an inertia dynamics unit. The oscillation period should be measured electronically with a universal counter. The oscillation time should then be converted to its rotational inertia value.

A.6. Cartridge Quality

A.6.1. Tape-Cleanliness Requirement

Definition. Contamination on the tape is defined as the presence of a visible particle combined with at least three (3) visible imperfections (dimples or pockmarks) caused by the particle on either surface of the magnetic tape. Other contaminations are oil, grease, or another substance which can cause data errors on the magnetic tape. Contamination on the reel hub or flanges is defined as the presence of a visible particle.

Requirement. When completed, the cartridge assembly will have no visible contamination (40 μm or larger) on the spool hub, on the internal surfaces of the spool flanges, or on either the top or bottom surface of the tape.

Procedure. The cartridge assembly can be visually inspected by unwinding the tape from the cartridge. Evidence of particle contamination on the surfaces of the tape is revealed by the presence of small dimples or pockmarks in the surface of the tape. The hub and internal flanges of the spool can be inspected through the leader block opening while rotating the spool through 360° of rotation. Inspection can be done with the unaided eye. Any visible particles should be considered to be 40 μm or larger.

A.6.2. Load–Unload

Definition. Pertains to the capability of the tape to withstand 5000 cycles each of load–unload within the drive and to be able to load to BOT after 5000 cycles of load–unload.

Requirement. One permanent load–unload failure in 50,000 load–unload cycles and one temporary failure in 600 load–unload cycles. A permanent failure is one that requires a field-engineer service call. A temporary failure is one that an operator can clear by pushing the unload button and reloading the cartridge.

A.6.3. Dropped Cartridges

Requirement. Customer data must be recoverable from a cartridge assembly after being dropped from a height of 1 m.

Procedure. Cartridge assemblies are to be tested by randomly dropping the sides, edges, or corners, from a height of 1 m, onto a tile-covered concrete floor. No single cartridge assembly will be dropped more than once.

A.7. Care and Handling of Tape Cartridges and Tape Drives

A.7.1. Cleaning a Tape Cartridge

Damaged or dirty cartridges can reduce subsystem reliability and cause the loss of recorded data. If dirt appears on the cartridge, wipe the outside surfaces with a lint-free cloth (part 2108930) that has been lightly moistened with IBM cleaning fluid (part 8493001) or its equivalent (Anonymous, 1984b).

Do not allow anything wet, including the cleaning fluid, to contact the tape. Make sure all cartridge surfaces are dry and the leader block is snapped into place before loading the cartridge.

A.7.2. Cleaning Cartridge Specifications

The drives have to be cleaned periodically to remove contaminant from the tape path (Bhushan, 1990). The cleaning cartridge consists of a cartridge case, a cleaning ribbon, a brake mechanism, a leader block, and miscellaneous associated parts. When fabricated, the final cleaning cartridge assembly must meet the mechanical specifications for the data cartridge presented earlier, and the additional specifications for the cleaning cartridge which follow (Anonymous, 1984b).

A.7.2.1. Cleaning Ribbon Properties Essential for IBM Tape Unit Use

The cleaning ribbon and cartridge properties that are required for the satisfactory operation on an IBM tape drive are described below. These properties have been established as the result of ribbon testing. Experience has demonstrated that a cleaning ribbon that meets these requirements performs satisfactorily if the tape drive is properly maintained.

(a) Ribbon Width

Definition. The distance measured from ribbon edge to ribbon edge when the ribbon is under no linear tension.

Requirement. The ribbon width must be 12.7 + 0.38−0.0 mm.

(b) Ribbon Length

Definition. The distance measured from one end to the other of the cleaning ribbon.

Requirement. The ribbon length must be 7.2 + 0.1−0.0 mm.

(c) Ribbon Thickness

Definition. The dimensions of the cleaning ribbon, as measured from the back to the front of the ribbon.

Requirement. The ribbon thickness must be 99 μm $\pm$ 8 μm.

(d) Ribbon Material

Definition. The composition of the cleaning ribbon.

Requirement. The ribbon material must meet the following specifications for thread count:

   Total: 114 $\pm$ 2 threads/cm^2.
   Minimum Warp: 66 ends/cm.
   Minimum Fill: 44 picks/cm.

A suitable ribbon material, Nylon fabric number 818, can be purchased from Bomont Inc. Equivalent ribbon material from other sources may also be used.

(e) Cleaning Cartridge Durability

Definition. Pertains to the ability of the cleaning ribbon to withstand the wearing action encountered during repeated usage.

Requirement. The cartridge must be capable of withstanding 500 load and unload cycles within the drive and to be able to load after 500 load and unload cycles.

(f) Cleaning Ribbon to Cartridge Reel Hub Bond Shear Strength

Definition. Pertains to the amount of force required to break the bond between the cleaning ribbon and the cartridge reel hub.

Requirement. The minimum tangential pull strength of the attached ribbon to the reel hub must be 35.5 N using two complete wraps of ribbon on the hub. Samples for testing must be taken from reels that have been fully wound for a minimum of 24 h.

(g) Leader Block Pullout and Insertion Forces

Definition. Leader block pullout pertains to the amount of force required to pull the leader block and attached cleaning ribbon out of the cartridge.

Insertion force pertains to the amount of force required to push the leader block into its latched position in the cartridge.

Requirement. The leader block pullout force must be a minimum of 3 N, and the insertion force must be a maximum of 12 N when measured at an angle of 48° ± 1° with respect to the cartridge edge.

(h) Leader Block to Ribbon Bond Strength

Definition. Pertains to the amount of force required to pull the leader block free from the ribbon.

Requirement. The leader block must remain attached to the ribbon when a force of 10 N is applied to the leader block at an angle of 38° ± 2° with respect to its edge.

A.7.2.2. Safety Data

The safety requirements for the cleaning cartridge are described below.

(a) Toxicity

No component of the cleaning cartridge assembly may contain any sensitizers, or known or suspected carcinogens. The components of the ribbon or cartridge assembly must not constitute any health hazard when used in the intended manner.

(b) Flammability

The materials used in the external cartridge covers must have a flammability rating of at least 94V-2 as described in the Underwriters Laboratory UL94 *Standard for Safety—Test for Flammability of Plastic Materials.*

A.8. Recording and Track: Format Requirements

A.8.1. Recording

The method of recording will be double-density NRZI (nonreturn-to-ZERO, change on ONEs):

A ONE is represented by a transition at the beginning of a bit-cell and no transition at the center of the bit-cell. A ZERO is represented by a transition at the beginning of the bit-cell followed by a transition at the center of the bit-cell.

A.8.1.1. Physical Recording Density

Physical recording density is defined as the number of recorded flux transitions per unit length of track. The physical recording density will be:

For all ZEROs: 1944 ft/mm.
For all ONEs: 972 ft/mm.

A.8.1.2. Bit-Cell Length

The resulting nominal bit-cell length is 1.029 μm. (A bit is a single digit in the binary number system; a bit-cell is the distance along the track between the initial flux transition of adjacent bits.)

A.8.1.3. Average Bit-Cell Length

The average bit-cell length is the sum of distances over n bit cells divided by n.

A.8.1.4. Long-Term Average Bit-Cell Length

The long-term average bit-cell length will be the average bit-cell length taken over a minimum of 972,000 bit-cells. It will be within $\pm 4\%$ of the nominal bit-cell length.

A.8.1.5. Short-Term Average Bit-Cell Length

The short-term average bit-cell length (STA) will be the average taken over 16 bit-cells. It will be within $\pm 7\%$ of the nominal bit-cell length.

A.8.1.6. Rate of Change

The rate of change of any two adjacent short-term average bit-cell length measurements will not exceed 1.6%.

A.8.1.7. Bit Shift

The maximum displacement of any ONEs zero crossing, exclusive of drop-outs, will not deviate by more than 28% from the expected nominal position as defined by the average bit-cell length.

A.8.1.8. Total Character Skew

No data-bit belonging to the same written transversal column will be displaced by more than 19 bit-cell lengths when measured in a direction parallel to the tape-reference edge.

A.8.1.9. Read Signal Amplitude

The average signal amplitude of an interchanged cartridge, averaged over 4000 flux transitions at 972 ft/mm, will be between 60% and 150% of the standard reference amplitude. Averaging for the interchange cartridge may be segmented into blocks. Traceability to the standard reference amplitude is provided by the calibration factors supplied with each secondary standard reference tape.

A.8.1.10. Coincident Dropouts

No coincident dropouts are permitted in a recorded block.

A.8.2. Track Format

A.8.2.1. Number of Tracks

There will be 18 tracks.

A.8.2.2. Tape-Reference Edge

The reference edge of the tape is the bottom edge, when viewing the recording side of the tape with the hub end (EOT) of the tape to the observer's right.

A.8.2.3. Track Positions

The distance from the centerline of the tracks to the tape reference edge will be:

Track 1: 11.68 mm.	Track 10: 6.01 mm.
Track 2: 11.05 mm.	Track 11: 5.38 mm.
Track 3: 10.42 mm.	Track 12: 4.75 mm.
Track 4: 9.79 mm.	Track 13: 4.12 mm.
Track 5: 9.16 mm.	Track 14: 3.49 mm.
Track 6: 8.53 mm.	Track 15: 2.86 mm.
Track 7: 7.90 mm.	Track 16: 2.23 mm.
Track 8: 7.27 mm.	Track 17: 1.60 mm.
Track 9: 6.64 mm.	Track 18: 0.97 mm.

The tolerance will be ± 0.040 mm for all tracks.

A.8.2.4. Track Width

The width of the written track will be 0.540 mm $\pm$ 0.017 mm.

A.8.2.5. Azimuth

The azimuth will be less than 3 minutes of arc. We note that at the time of tape writing, the azimuth should be less than 1 minute of arc. The remaining 2 minutes of arc are the allowance for tape distortion caused by the environmental conditions and aging.

A.8.2.6. Track Identification

Tracks are numbered consecutively, ending at the tape reference edge with track 18, and assigned as follows:

```
Physical track: 1  2  3  4  5  6  7  8  9  10 11 12 13 14 15 16 17 18
ECC track: 1a   2a   3a   4a   5a   6a   7a   0a   8a
           7b   6b   5b   4b   3b   2b   1b   0b   8b
```

References

Anonymous (1975). "Requirements Manual: IBM 3851 Mass Storage Facility—Data Cartridge Specification," 1st ed. IBM Corporation Order No. GA32-0032-0.

Anonymous (1977). "Standard for Unrecorded Magnetic Tape Cartridge for Information Interchange." ANSI X3.055, American National Standards Institute, New York.

Anonymous (1979). "Tape Requirements for IBM One-Half Inch Tape Units at: 556, 800, 1600 and 6250 BPI," 6th ed. IBM Corporation Order No. GA32-0006-5.

Anonymous (1983). "Standard for Information Systems—Unrecorded Magnetic Tape for Information Exchange." ANSI X3.40-1983, American National Standards Institute, New York.

Anonymous (1984a). "Tape and Cartridge Requirements for the IBM 3480 Magnetic Tape Drives," 1st ed. IBM Corporation Order No. GA32-0048-0.

Anonymous (1984b). "Care and Handling of the IBM Magnetic Tape Cartridge," 1st ed. IBM Corporation Order No. GA32-0047-0.

Anonymous (1991). "Standard for Magnetic Tape and Cartridge for Information Interchange." ANSI X3.180-1991, American National Standards Institute, New York.

Bhushan, B. (1990). "Tribology and Mechanics of Magnetic Storage Devices." Springer-Verlag, New York.

Bhushan, B. and Gupta, B. K. (1991). "Handbook of Tribology: Materials, Coatings, and Surface Treatments." McGraw-Hill, New York.

Wang, H. S. C. (1966). Gap loss function and determination of certain critical parameters in magnetic data recording instruments and storage systems. *Rev. Sci. Inst.* **37**, 1124–1130.

APPENDIX B

Analysis of Life Data

The rational handling of life data (or data of time-to-failure) and, particularly, the use of data to predict what is likely to happen in the future, or at some other operative condition, requires the recognition of a definable statistical failure distribution. The time of occurrence of a failure, or the duration of equipment operation between failures, is a random event. Methods of probability theory enable us to investigate the failure data as random large-number phenomena. There are many continuous and discrete distributions. Examples of continuous distributions are exponential, normal, lognormal, Weibull, extreme value, gamma and chi-square, and Student's t distributions; and examples of discrete distributions are geometric, Poisson, binomial, hypergeometric, and multinomial distributions (Bazovsky, 1961; Brunk, 1965; Johnson and Kotz, 1969, 1970; Polovko, 1968; Shooman, 1968; Dixon and Massey, 1969; Mendenhall, 1971; Meyer, 1973, 1975; Mann et al., 1974; Nelson, 1982).

The statistical distribution used should be rationally associated with the failure mechanisms involved, not simply that it fits the data best in a single experiment. There should be a proper distribution function reflecting the physical laws involved in the failure process. It does not make sense that one distribution is valid for results from one test, and that some other distribution is valid for a similar test (if the same failure mechanism occurs) simply because it provides a better "fit" to the data.

In this appendix, we describe and compare commonly used continuous and discrete distributions. We next describe the various types of data and methods to make probability plots, hazard plots, and analytical methods used to obtain distribution parameters. Finally, we describe in detail the Weibull analysis which is widely used for the analysis of the failure data of numerous machine components including that of magnetic head–medium interfaces.

B.1. Continuous Distributions

B.1.1. Basic Statistical Concepts

In this section, we present basic statistical concepts for continuous statistical distributions.

B.1.1.1. Probability Density Function (p.d.f.)

A continuous distribution with a probability density is a probability model, denoted by X, that consists of the possible numerical outcomes x that are a subset of the line $(-\infty, \infty)^1$ and a probability density function (p.d.f.) $f(x)$ with the properties

$$f(x) \geq 0 \qquad \text{for all } x, \tag{B.1a}$$

and

$$\int_{-\infty}^{\infty} f(x)\, dx = 1. \tag{B.1b}$$

The p.d.f. $f(x)$ is the mathematical model for the population histogram containing relative frequencies which (a) must be greater than or equal to zero and (b) must sum up to unity. The capital letter X denotes the distribution, which is also called a *random variable*. Of course, X is neither random nor a variable; it is a probability model consisting of outcomes and their corresponding probabilities. Lowercase letters x, x_1, x_2, ... denote particular outcomes. The $f(x)$ has units reciprocal to that x.

B.1.1.2. Cumulative Distribution Function (c.d.f.)

An *event* is any subset of all outcomes, and the probability of an event E is defined by

$$P\{E\} = \int_E f(x)\, dx, \tag{B.2}$$

where the integral runs over the outcomes in the event E. That is, the probability is the area under the probability density function for the points of the event. $P\{X \leq x_0\}$ denotes the probability of the event consisting of all outcomes of model X that satisfy the relationship, that is, are less than or equal to the value x_0.

The c.d.f. $F(x)$ with a p.d.f. $f(x)$ is

$$P\{X \leq x\} = F(x) = \int_{-\infty}^{x} f(x)\, dx, \qquad -\infty < x < \infty, \tag{B.3}$$

---

[1] A distribution that has a range from $-\infty$ to ∞ is not strictly appropriate as a life distribution, since lifetimes must be positive. The life distribution has a range from 0 to ∞; and such distributions are called *truncated distributions*.

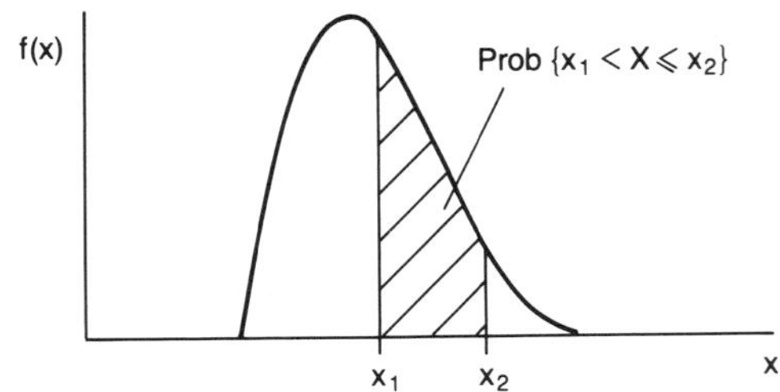

Fig. B.1. Probability calculation from probability density function.

where the integral runs over all outcomes less than or equal to x. (For a life data distribution, the lifetime x must be positive.) $F(x)$ and $f(x)$ are alternative ways of representing a distribution. They are related as shown in Eq. (B.3) and alternatively by

$$f(x) = dF(x)/dx. \tag{B.4}$$

Any such c.d.f. $F(x)$ is a continuous function for all x and has the following properties for all x:

$$F(-\infty) = 0 \quad \text{and} \quad F(\infty) = 1 \tag{B.5a}$$

and

$$F(x) \le F(x') \quad \text{for all} \quad x < x'. \tag{B.5b}$$

The $F(\infty) = 1.0$ expresses the fact that the probability is 1.0 and the endurance life is finite.

The probability $P\{x_1 < X \le x_2\}$ denotes the probability of the event consisting of outcomes which are between x_1 and x_2. The probability is given by the areas under the probability density curve between x_1 and x_2, as shown in Fig. B.1,

$$P\{x_1 < X \le x_2\} = \int_{x_1}^{x_2} f(x)\, dx$$

$$= F(x_2) - F(x_1). \tag{B.6}$$

B.1.1.3. Reliability Function

The reliability function, also called the survivorship function, $R(x)$, for a life distribution denotes the probability of (or population fraction) surviving beyond x.

$$P\{X > x\} = R(x) = \int_{x}^{\infty} f(y)\, dy = 1 - F(x) \tag{B.7a}$$

with the properties

$$R(0) = 1 \quad \text{and} \quad R(\infty) = 0 \tag{B.7b}$$

and

$$R(x) \ge R(x') \quad \text{for all} \quad x < x'. \tag{B.7c}$$

By definition, $R(x)$ is a decreasing function of x. It is also called the survivorship function. For a life data distribution, the lifetime x must be positive.

B.1.1.4. Hazard and Cumulative Hazard Functions

The hazard function $h(x)$ of a continuous distribution of life X with a probability density is defined for all possible outcomes x as

$$h(x) = P\{x < X < x + \Delta x\}/\{P\{X > x\}(\Delta x)\}$$

$$= \lim_{\Delta x \to 0} \frac{R(x) - R(x + \Delta x)}{\Delta x} \frac{1}{R(x)}$$

$$= -\frac{dR(x)}{dx} \frac{1}{R(x)} \tag{B.8a}$$

$$= \frac{f(x)}{R(x)}. \tag{B.8b}$$

Integrating $R(x)$ in Eq. (B.8a), we obtain

$$R(x) = \exp\left[-\int_{-\infty}^{x} h(y)\, dy \right] \tag{B.9a}$$

with the properties

$$h(x) \geq 0 \qquad \text{for} \quad -\infty < x < \infty. \tag{B.9b}$$

(We note that for a life data distribution, the lifetime x must be positive.) The $h(x)$ is the *instantaneous failure rate* (IFR) at age x. That is, in the short time Δx from x to $x + \Delta x$, a proportion $\Delta x h(x)$ of the population that reached the age x fails. Thus the hazard function is a measure of proneness to failure as a function of age. The hazard function is also called the hazard rate, the mortality rate, and the force of mortality. The units of $h(x)$ are the reciprocal to that of x.

A constant IFR means that choice of failure for an unfailed component over a specified short length of time is the same as that for an unfailed new component over the same length of time. A decreasing IFR is said to correspond to infant mortality. Such a failure rate indicates that the product is poorly designed or suffers from a manufacturing defect. An increasing IFR during the later life of a product is said to correspond to wear-out failure. Many products have an increasing failure rate over the entire range of life.

A typical failure rate curve, called the "bathtub" curve due to its shape, is shown in Fig. B.2. This curve is characteristic for the case of instantaneous failures of equipment with a wide usage of electronic components. It has three characteristic segments—the infant mortality segment, the segment of normal operation (with constant failure rate), and the segment of wear-out or aging, requiring replacement. The infant mortality or burn-in portion of the population may have an extremely high failure rate compared to that of the good population, but the proportion of these components may be very small, on

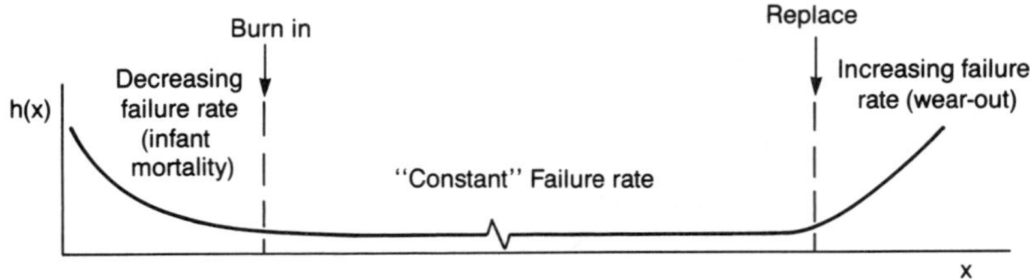

Fig. B.2. Hazard function (instantaneous failure rate) as a "bathtub curve."

the order of 0.1% upwards, so they may be removed by screening tests or are removed from the operating system in a relatively short time. In the wear-out region, the failure rate continuously increases with time.

The cumulative hazard function of a distribution is

$$H(x) = \int_{-\infty}^{x} h(x)\, dx, \qquad -\infty < x < \infty, \tag{B.10}$$

where the integral runs over all possible outcomes. Integrating Eq. (B.8a) yields the basic relationship

$$H(x) = -\ln R(x). \tag{B.11a}$$

Alternatively,

$$R(x) = \exp[-H(x)] \tag{B.11b}$$

with the properties

$$H(-\infty) = 0 \qquad \text{and} \qquad H(\infty) = \infty \tag{B.11c}$$

and

$$H(x) \le H(x') \qquad \text{for all} \quad x < x'. \tag{B.11d}$$

By definition, $H(x)$ is an increasing function. (We note that for a life data distribution, the lifetime x must be positive.)

B.1.1.5. The Percentile, Mode, Mean, and Variance

The $100P$th percentile, $100P$ percent point, Pth quantile, or P fractile is the life by which a proportion P of the population has failed. Figure B.3 shows a point on the life axis designated by the symbol x_P, and defined as the value of x for which the area under the probability density curve to its left is P,

$$P = F(x_P). \tag{B.12}$$

x_P is called the $100P$th percentile of the distribution.

The 0.632 (63.2%) point is called the "characteristic life" of a life distribution where 63.2% of the population is expected to fail. The 0.50 (50%) point is called the *median* ($x_{0.50}$ or $x_{\text{med.}}$) or the "typical life." It is the middle of the

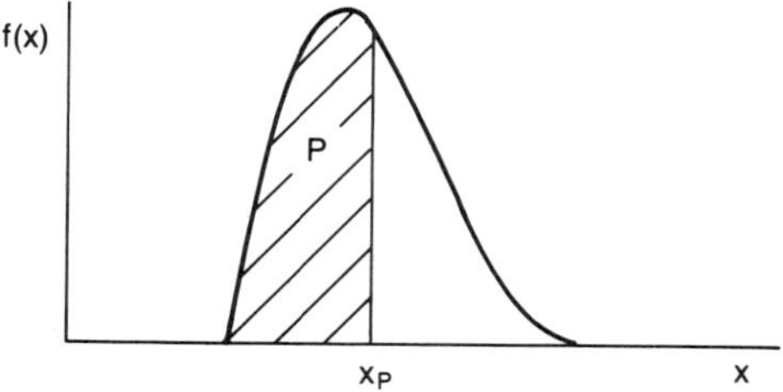

Fig. B.3. Definition of a 100Pth percentile (or Pth quantile).

distribution in the sense that 50% of the population fail before that age and 50% survive, i.e.,

$$\int_0^{x_{\text{med.}}} f(x)\,dx = \int_{x_{\text{med.}}}^{\infty} f(x)\,dx = F(x_{\text{med.}})$$

$$= 0.50. \tag{B.13}$$

The mode is the value where the probability density is a maximum. That is, the "most likely" time-to-failure, and may be regarded as a "typical" life for many distributions. The mode may usually be found by the usual calculus method for locating the maximum of the function; namely, by finding the life x where the derivative of the probability density with respect to life x equals zero and the second derivative is less than zero,

$$\left.\frac{df(x)}{dx}\right|_{x=x_{\text{mode}}} = 0,$$

$$\left.\frac{d^2 f(x)}{dx^2}\right|_{x=x_{\text{mode}}} < 0. \tag{B.14}$$

The mean $E(X)$ of a variable X with a probability density $f(x)$ is

$$E(X) = \int_{-\infty}^{\infty} xf(x)\,dx. \tag{B.15a}$$

The integral runs over all possible outcomes x. For a theoretical life distribution, the mean is also called the average, expected life, or "mean-time-between-failure" (MTBF), or "mean-time-to-failure" (MTTF). It corresponds to the arithmetic average of the lives of all units in a population, and is used as still another "typical" life. An alternative equation for the mean, if the range of X is positive (such as for a life data distribution), is

$$E(X) = \int_0^{\infty} R(x)\,dx. \tag{B.15b}$$

Finally, variance, the square of standard deviation (σ) of a variable X with a probability density function $f(x)$ is defined as the expected squared departure x from $E(X)$

$$\mathrm{Var}(X) = \sigma(X)^2 = \int_{-\infty}^{\infty} [x - E(X)]^2 f(x)\, dx$$

$$= \int_{-\infty}^{\infty} x^2 f(x)\, dx - [E(X)]^2$$

$$= \mathrm{rms}^2 - [E(X)]^2. \tag{B.16}$$

The integral runs over all possible outcomes x. (For a life distribution, the lifetime x must be positive.) This is a measure of the scatter or dispersion of the distribution. It describes the extent to which the distribution clusters around $E(X)$.

B.1.1.6. Graphical Representation of Probability Distributions

The probability density function (histogram) is constructed by plotting the number or proportion of outcomes (relative frequencies) lying between two periods as a function of a period x. The interval between two such periods is termed the class interval. It is generally recommended to use 15–50 class intervals for general random data. The cumulative distribution function is constructed by plotting the cumulative number or proportion of outcomes at a specific period as a function of that period. The reliability function is constructed by plotting the values obtained, by subtracting the cumulative distribution function from 1 at a specific period as a function of that period. The hazard function is constructed by plotting the instantaneous failure rate at a specific period as a function of that period.

The basic concepts are illustrated with an example of human mortality, Table B.1 (Nelson, 1982). The data are given in steps of 5 years. The probability density, cumulative distribution, reliability, hazard, and cumulative hazard functions for the mortality data are presented in Table B.1 and are plotted in Fig. B.4. For example, that 0.057 (5.7%) of the population died between the ages of 5 and 10. The probability of death by age 20 is the sum (corresponding to an integral) of the probabilities up to age 20, namely, $0.290 + 0.057 + 0.031 + 0.030 = 0.408$ (40.8%). The survivorship for 20 years is 0.592 (59.2%). The yearly mortality rate for 20-year-olds (who die before the age of 25) is calculated from the fraction 0.592 surviving to age 20 and the fraction 0.032 that die between the ages of 20 and 25 as $(0.032/0.592)/5 = 0.011\%$ or 1.1% per year.

B.1.2. Conditional Distribution

It is sometimes necessary to calculate the life distribution of units that have reached a specific age. This probability distribution is called the conditional distribution. The conditional probability density function of total age X, at

Table B.1. Human mortality data and associated distribution functions

x	$f(x)$	$F(x)$	$R(x)$	$h(x)$	$H(x)$
0	—	0	1.000		
0–5	0.290	0.290	0.710	0.058	0.058
5–10	0.057	0.347	0.653	0.016	0.074
10–15	0.031	0.378	0.622	0.010	0.084
15–20	0.030	0.408	0.592	0.010	0.094
20–25	0.032	0.440	0.560	0.011	0.105
25–30	0.037	0.477	0.523	0.013	0.118
30–35	0.042	0.519	0.481	0.016	0.134
35–40	0.045	0.564	0.436	0.019	0.153
40–45	0.049	0.613	0.387	0.022	0.175
45–50	0.052	0.665	0.335	0.027	0.202
50–55	0.053	0.718	0.282	0.032	0.234
55–60	0.050	0.768	0.232	0.035	0.269
60–65	0.050	0.818	0.182	0.043	0.312
65–70	0.051	0.869	0.131	0.056	0.368
70–75	0.053	0.922	0.078	0.081	0.449
75–80	0.044	0.966	0.034	0.113	0.562
80–85	0.034	1.000	0	0.200	0.762

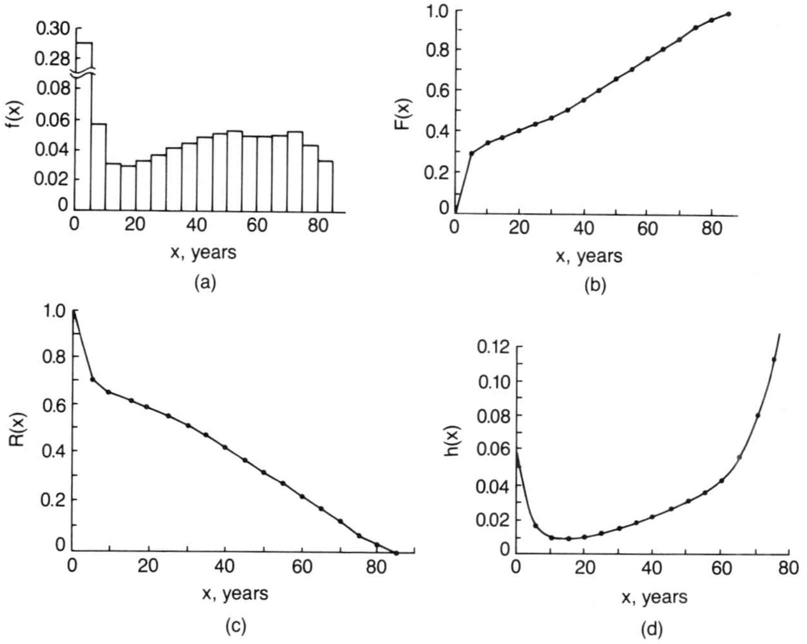

Fig. B.4. Plots of: (a) a probability density function (histograms of failure events); (b) a cumulative distribution function; (c) a reliability function; and (d) a hazard function for the data presented in Table B.1 (Nelson, 1982).

failure for units that have reached age x_0, is

$$f(x|x_0) = f(x_0)/R(x_0), \qquad x \geq x_0, \tag{B.17}$$

where $f(\ )$ is the probability density function and $R(\ )$ is the reliability function of the original distribution. The denominator $R(x_0)$ rescales the density so that the probability under it from x_0 to ∞ is 1, i.e.,

$$\int_x^\infty f(x|x_0)\, dx = \int_x^\infty f(x)[R(x_0)]^{-1}\, dx$$

$$= 1. \tag{B.18}$$

Figure B.5(a) depicts the rescaling. $f(x|x_0)$ has all the properties of a probability density function.

The conditional cumulative distribution function of age X at failure for units that reach age x_0 is

$$F(x|x_0) = \int_{x_0}^x f(x|x_0)\, dx = [F(x) - F(x_0)]/R(x_0), \qquad x \geq x_0, \tag{B.19}$$

where $F(\ )$ is the cumulative distribution of the original distribution and $R(x_0) = 1 - F(x_0)$. Figure B.5(b) shows the original and conditional cumulative distributions. $F(x|x_0)$ has all the properties of a cumulative distribution function.

The conditional reliability for failure at age X among units that reach age

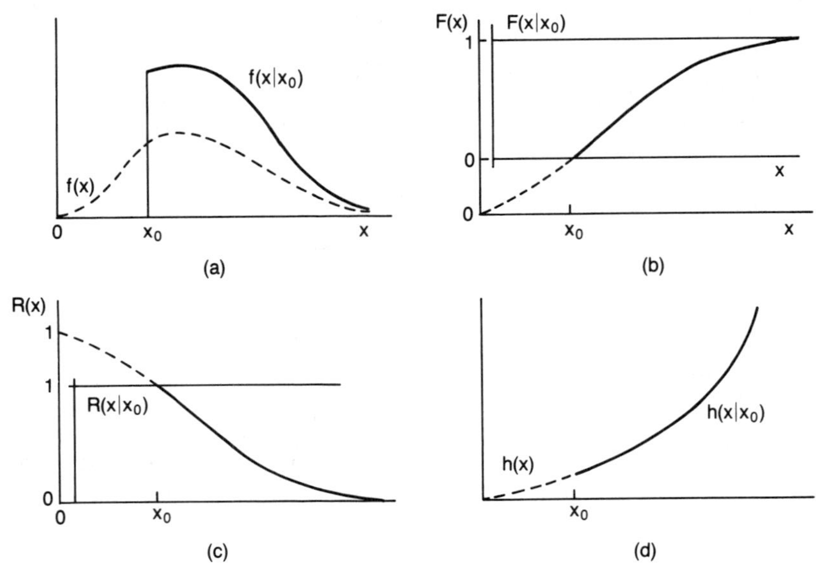

Fig. B.5. Original and conditional probability functions: (a) probability density; (b) cumulative distribution; (c) reliability; and (d) hazard functions.

x_0 is

$$R(x|x_0) = 1 - F(x|x_0) = R(x)/R(x_0), \qquad x \geq x_0. \tag{B.20}$$

Figure B.5(c) shows the original and conditional reliability functions.

The conditional hazard function for age X at failure among units that reach age x_0 is

$$h(x|x_0) = f(x|x_0)/[1 - F(x|x_0)] = f(x)/[1 - F(x)]$$

$$= h(x), \qquad x \geq x_0. \tag{B.21}$$

Thus the conditional failure rate at age x is the same as the unconditional failure rate, Fig. B.5(d). This fact indicates why burn-in and preventive replacement can reduce the failure rate in service. If units initially have a high failure rate, they can run in a factory burn-in; then survivors that are put into service have a lower failure rate. Also, such units can be replaced at an age when their failure rate begins to increase appreciably. Units are then used only over the low failure rate portion of their distribution.

The $100P$th percentile of the conditional distribution of age x at failure for units that reach age x_0 is the solution of

$$P = F(x_P^*|x_0)$$

$$= [F(x_P^*) - F(x_0)]/R(x_0). \tag{B.22}$$

x_P^* is the $100P_0$th percentile of the original distribution, where $P_0 = (1 - P)F(x_0) + P$.

The conditional mean (or expectation) of age X at failure among units that reach age x_0 is

$$E(X|x_0) = \int_{x_0}^{\infty} xf(x|x_0)\, dx. \tag{B.23a}$$

Alternatively,

$$E(X|x_0) = \left[\int_{x_0}^{\infty} xf(x)\, dx\right] \Big/ [1 - F(x_0)]. \tag{B.23b}$$

The conditional variance and standard deviation of age X at failure among units that reach age x_0 is

$$\mathrm{Var}(X|x_0) = \sigma(X|x_0)^2 = \int_{x_0}^{\infty} [x - E(X|x_0)]^2 f(x|x_0)\, dx. \tag{B.24a}$$

Alternatively,

$$\mathrm{Var}(X|x_0) = \int_{x_0}^{\infty} x^2 f(x|x_0)\, dx - [E(X|x_0)]^2. \tag{B.24b}$$

The remaining life for units reaching age x_0 is

$$Y = X - x_0. \tag{B.25}$$

This is equivalent to moving the origin of the x time axis to a value x_0, i.e., replace x by $y + x_0$. All quantities may be expressed in terms of Y.

Table B.2. Conditional human mortality data and associated distribution functions

Age x	$f(x\|40)$	$F(x\|40)$	$R(x\|40)$	$h(x\|40)$	$H(x\|40)$
40		0	1.000		
40–45	0.112	0.112	0.888	0.022	0.022
45–50	0.120	0.232	0.768	0.027	0.049
50–55	0.121	0.353	0.647	0.032	0.081
55–60	0.115	0.468	0.532	0.035	0.116
60–65	0.115	0.583	0.417	0.043	0.159
65–70	0.117	0.700	0.300	0.056	0.215
70–75	0.121	0.821	0.179	0.081	0.296
75–80	0.101	0.922	0.078	0.113	0.409
80–85	0.078	1.000	0	0.200	0.609

B.1.2.1. Graphical Representation

An example illustrates the basic concepts of the conditional distribution. We consider the life distribution of living 40-year-olds in the human mortality table (Table B.1). Table B.1 gives the population fractions dying in 5-year intervals beyond the age of 40. It is helpful to think of the reliability of 0.436 at age 40 as if 436 of 1000 people have reached age 40; other fractions can be similarly interpreted. Of the 436 people that reached 40, "49" died between ages 40 and 45. Thus, the fraction $f(45|40)$ of 40-year-olds dying between ages 40 and 45 is $f(45|40) = 0.049/0.436 = 0.112$ (11.2%). In contrast, 4.9% of the newly born die between the ages of 40 and 45. Similarly, the fraction $f(50|40)$ of 40-year-olds that die between ages 45 and 50 is $f(50|40) = 0.052/0.436 = 0.120$. Such calculations yield the conditional probability density of the distribution of age at death of those that reach age 40. This histogram or probability density function $f(x|40)$ and the associated distribution functions are tabulated in Table B.2 and are plotted in Fig. B.6 (Nelson, 1982).

B.1.3. Distribution of Products Which Fail Due to Competing Risks

A mix of competing risks occurs when units fail by different causes. For example, ball bearings can fail because of the failure of the inner ring, the outer ring, or a ball. In this section, we present the product rule for reliability and the addition law for failure rates, and apply these analyses to series systems to identical components and products which have failure rates proportional to size.

Suppose that a product has a potential time to failure from each of N causes (or competing risks or failure modes). Such a product is called a *series system* if its life is the smallest of those N potential times to failure; i.e., the system fails with the first "component" failure. In other words, if $X_1, X_2, \ldots, X_N$ are potential times to failure for the N causes (or components), then the

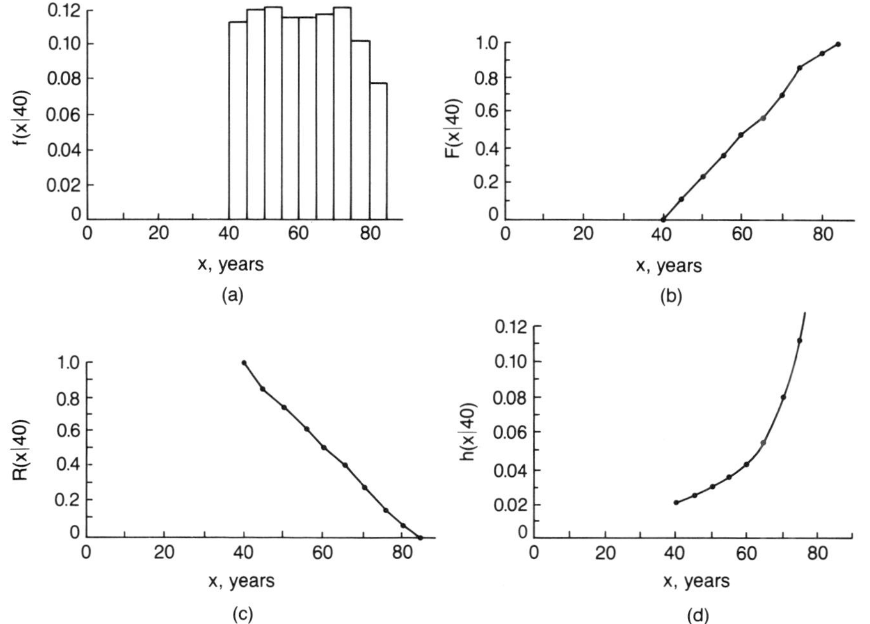

Fig. B.6. Plots of: (a) a probability density function; (b) a cumulative distribution function; (c) a reliability function; and (d) a hazard function for data presented in Table B.2 (Nelson, 1982).

system time X to failure is

$$X = \min(X_1, X_2, \ldots, X_N). \tag{B.26}$$

Let $R(x)$ denote the reliability function of the system, and let $R_1(x)$, $R_2(x)$, $\ldots$, $R_N(x)$ denote the reliability function of the N causes (each in the absence of all other causes). Suppose that the times to failure for different causes (components or risks) are statistically independent. Such products are said to have independent competing risks or to be series systems with independent components. For such systems

$$R(x) = P\{X > x\} = P\{X_1 > x \text{ and } X_2 > x, \ldots, \text{ and } X_N > x\}$$

$$= P\{X_1 > x\}P\{X_2 > x\}, \ldots, P\{X_N > x\}. \tag{B.27a}$$

Thus,

$$R(x) = R_1(x)R_2(x), \ldots, R_N(x) \tag{B.27b}$$

or

$$F(x) = 1 - [1 - F_1(x)][1 - F_2(x)], \ldots, [1 - F_N(x)]. \tag{B.27c}$$

This key result is the product rule for the reliability of series systems (with independent components). In general, for a system whose component distributions are of one type, the system distribution is not the same type. This is true of most distributions (e.g., normal, lognormal, and Weibull).

For series systems it is useful to look at the hazard function and the cumulative hazard function. The product rule [Eq. (B.27)], in terms of the cumulative hazard functions $H(x)$ of the system and $H_1(x)$, $H_2(x)$, ..., $H_N(x)$ of each of the N causes, is

$$\exp[-H(x)] = \exp[-H_1(x)]\exp[-H_2(x)], \ldots, \exp[-H_N(x)]$$

or

$$H(x) = H_1(x) + H_2(x) + \cdots + H_N(x). \tag{B.28}$$

Differentiating this to get the hazard function

$$h(x) = h_1(x) + h_2(x) + \cdots + h_N(x), \tag{B.29}$$

this is called the addition law for failure rates for independent failure modes (or competing risks). So for series systems, failure rates add.

As a special case, suppose that a series system consists of N statistically independent, identical components (with the same life distribution) with a component reliability $R_0(x)$, a cumulative distribution function $F_0(x)$, a cumulative hazard function $H_0(x)$, and a hazard function $h_0(x)$. For a series system of N such independent, identical components, let $R(x)$ denote the system reliability function, $F(x)$ the system cumulative distribution function, $H(x)$ the system cumulative hazard function, and $h(x)$ the system hazard function. Then

$$R(x) = [R_0(x)]^N, \tag{B.30a}$$

$$F(x) = 1 - R(x)$$

$$= 1 - [1 - F_0(x)]^N, \tag{B.30b}$$

$$H(x) = NH_0(x), \tag{B.30c}$$

and

$$h(x) = Nh_0(x). \tag{B.30d}$$

The life of such a system is the smallest of N (component) lives from the same distribution.

Some products that come in various sizes have failure rates that are proportional to the product size. For example, the failure rate of a capacitor dielectric is often assumed to be proportional to the area of the dielectric. In general, if $h_0(x)$ is the failure rate for an amount A_0 of such a product, then the failure rate $h(x)$ for an amount A of the product is

$$h(x) = (A/A_0)h_0(x). \tag{B.31a}$$

That is, such products are regarded as series systems of $N = A/A_0$ identical components from the same life distributions. A/A_0 need not be an integer. Equation (B.31) is based on the assumption that adjoining portions of the product have statistically independent lifetimes.

Formulas for an amount A of the product are

$$R(x) = [R_0(x)]^{A/A_0}, \qquad \text{(B.31b)}$$

$$F(x) = 1 - R(x) = 1 - [1 - F_0(x)]^{A/A_0}, \qquad \text{(B.31c)}$$

and

$$H(x) = (A/A_0)H_0(x), \qquad \text{(B.31d)}$$

where the zero subscript denotes the distribution for an amount A_0.

B.1.4. Probability Distributions

The commonly used continuous probability distributions for the representation of failure data are: the exponential, the normal (or Gaussian), the lognormal, the Weibull, extreme value, gamma, and chi-square. The mathematical expressions for these distributions are presented in Table B.3 and they are plotted in Fig. B.7 (Johnson and Kotz, 1970).

A distribution that has a range from $-\infty$ to $+\infty$ is not strictly appropriate as a life data distribution, since lifetimes must be positive. The lifetimes have a range from 0 to ∞; and they are called truncated distributions. The distributions can be shifted, and are sometimes used in life and failure data analysis. Such a distribution is obtained by shifting a distribution that is defined for outcomes between 0 and $+\infty$, so that the distribution starts at a value different from 0. The shift γ has the same units as used to measure life, and is referred to as the location parameter, the threshold parameter, or the guaranteed life. In Fig. B.7, only the Weibull distribution is shifted, other distributions (exponential, lognormal, and gamma distributions) can also be shifted by replacing the (x) by $(x - \gamma)$ in the definition of distribution functions, Table B.3.

The exponential distribution is the simplest, and is of interest where the failure rate is constant. It is used to describe product life subject to "chance failures." It is also used to describe the product life after units have gone through a burn-in, up to the time they start to wear out (see the bathtub failure rate curve). The normal distribution is symmetric and the mode and median (50% failure) are identical. The dispersion is normalized by the standard deviation, σ. The failure rate monotonically increases with time, giving the typical "wear-out" pattern. The normal distribution is often used to describe the dimensions of parts (natural, physical, and biological phenomena), and certain types of life data.

The curves shown for the lognormal distribution are for constant median life, $\exp(\mu)$. Although the normal distributions all have the same relative shapes (normalized by the value of σ), the lognormal distributions vary greatly because of the relative effects of the log scales of time interacting with the dispersion factor. The term σ refers literally to the standard deviation of the natural log of time. (Some authors have used σ with reference to $\log_{10}$ of the failure time, but this complicates the mathematical treatment.) The failure

Table B.3. Mathematical expressions for six continuous distributions

(a) Exponential Distribution

Distribution Functions
Probability density function: $f(x) = \lambda \exp(-\lambda x)$, $x \geq 0$.
Cumulative distribution function: $F(x) = 1 - \exp(-\lambda x)$, $x \geq 0$.
Hazard function (instantaneous failure rate): $h(x) = \lambda = $ constant.
Cumulative hazard function: $H(x) = \lambda x$.

Significant Distribution Properties
Mode $= 0$.
Mean $= 1/\lambda$, $\lambda > 0$.
Standard deviation $= 1/\lambda$.

Distribution Parameters
$\lambda = $ failure rate ($=1/$mean) with units reciprocal of x.

(b) Normal Distribution

Distribution Functions[a]
Probability density function:

$$f(x) = [1/\sigma(2\pi)^{1/2}] \exp[-(x - \mu)^2/(2\sigma^2)], \qquad -\infty < x < \infty.$$

Cumulative distribution function:

$$F(x) = [1/\sigma(2\pi)^{1/2}] \int_0^x \exp[-(y - \mu)^2/(2\sigma^2)]\, dy, \qquad -\infty < x < \infty.$$

Hazard function (instantaneous failure rate):

$$h(x) = \frac{\exp[-(x - \mu)^2/(2\sigma^2)]}{\int_{-\infty}^x \exp[-(y - \mu)^2/(2\sigma^2)]\, dy}, \qquad -\infty < x < \infty.$$

Cumulative hazard function:

$$H(x) = -\ln\left\{1 - [1/\sigma(2\pi)^{1/2}] \int_{-\infty}^x \exp[-(y - \mu)^2/(2\sigma^2)]\, dy\right\}, \qquad -\infty < x < \infty.$$

Significant Distribution Properties
Median $= \mu$.
Mode $= \mu$.
Mean $= \mu$.
Standard deviation $= \sigma$.

Distribution Parameters
Scale parameter $= \mu$.
   (units same as of x)
Shape parameter $= \sigma$, $\sigma > 0$.
   (units same as of x)

(c) Lognormal Distribution

Distribution Functions
Probability density function:

$$f(x) = [1/\sigma x(2\pi)^{1/2}] \exp\{-[\ln(x) - \mu]^2/(2\sigma^2)\}, \qquad x > 0.$$

Table B.3 (*continued*)

Cumulative distribution function:

$$F(x) = [1/\sigma(2\pi)^{1/2}] \int_0^x (1/y) \exp\{-[\ln(y) - \mu]^2/(2\sigma^2)\}\, dy, \qquad x > 0.$$

Hazard function (instantaneous failure rate):

$$h(x) = \frac{f(x)}{1 - F(x)}, \qquad x > 0.$$

Cumulative hazard function:

$$H(x) = -\ln[1 - F(x)], \qquad x > 0.$$

Significant Distribution Properties
  Median $= \exp(\mu)$.
  Mode $= \exp(\mu - \sigma^2)$.
  Mean $= \exp(\mu + \sigma^2/2)$.
  Standard deviation $= \exp(\mu + \sigma^2/2)[\exp(\sigma^2) - 1]^{1/2}$.

Distribution Parameters
  Scale parameter $= \exp(\mu)$.
  Shape parameter $= \sigma, \sigma > 0$.

(d) Weibull Distribution

Distribution Functions (*three parameter*)[b]
  Probability density function:

$$f(x) = (\beta/\eta)[(x - \gamma)/\eta]^{\beta-1} \exp\{-[(x - \gamma)/\eta]^\beta\}, \qquad x > \gamma.$$

Cumulative distribution function:

$$F(x) = 1 - \exp\{-[(x - \gamma)/\eta]^\beta\}, \qquad x > \gamma.$$

Hazard function (instantaneous failure rate):

$$h(x) = (\beta/\eta)[(x - \gamma)/\eta]^{\beta-1}, \qquad x > \gamma.$$

Cumulative hazard function:

$$H(x) = [(x - \gamma)/\eta]^\beta, \qquad x > \gamma.$$

Significant Distribution Properties
  Median $= \gamma + \eta(0.6931)^{1/\beta}$.

$$\text{Mode} = \begin{cases} \eta[1 - (1/\beta)]^{1/\beta} & \text{for } \beta > 1 \\ 0 & \text{for } 0 < \beta \le 1. \end{cases}$$

  Mean $= \gamma + \eta\Gamma[1 + (1/\beta)]$.
  Standard deviation $= \eta\{\Gamma[1 + (2/\beta)] - \{\Gamma[1 + (1/\beta)]\}^2\}^{1/2}$.

Distribution Parameters
  Scale parameter $= \eta, \eta > 0$.
    (units same as of x)
  Shape parameter $= \beta, \beta > 0$.
    (dimensionless)
  Location parameter $= \gamma$.
    (units same as of x)

Table B.3 (*continued*)

(e) Extreme Value Distribution

Distribution Functions
 Probability density function:

$$f(x) = (1/\eta) \exp[(x - \lambda)/\eta] \exp\{-\exp[(x - \lambda)/\eta]\}, \qquad -\infty < x < \infty.$$

 Cumulative distribution function:

$$F(x) = 1 - \exp\{-\exp[(x - \lambda)/\eta]\}, \qquad -\infty < x < \infty.$$

 Hazard function (instantaneous failure rate):

$$h(x) = (1/\eta) \exp[(x - \lambda)/\eta], \qquad -\infty < x < \infty.$$

 Cumulative hazard function:

$$H(x) = \exp[(x - \lambda)/\eta], \qquad -\infty < x < \infty.$$

Significant Distribution Properties
 Median $= \lambda - 0.367\eta$.
 Mode $= \lambda$ (63.2nd percentile).
 Mean $= \lambda - 0.5772\eta$.
 Standard deviation $= 1.283\eta$.

Distribution Parameters
 Scale parameter $= \eta, \eta > 0$.
  (units same as of x)
 Location parameter $= \lambda$.
  (units same as of x)

(f) Gamma and Chi-Square Distributions

Distribution Functions (gamma)[c]
 Probability density function:

$$f(x) = [1/\Gamma(\beta)](x^{\beta-1}/\eta^\beta) \exp(-x/\eta), \qquad x > 0,$$

where the gamma function, $\Gamma(\beta) = \int_0^\infty y^{\beta-1} \exp(-y)\, dy$ and for integer β, $\Gamma(\beta) = (\beta - 1)!$

 Cumulative distribution function:

$$F(x) = \Gamma^*(x/\eta, \beta), \qquad x > 0,$$

where the incomplete gamma function

$$\Gamma^*(x/\eta, \beta) = [1/\Gamma(\beta)] \int_0^{x/\eta} z^{\beta-1} \exp(-z)\, dz$$

and for integer β, $\Gamma(x/\eta, \beta) = 1 - P(\beta - 1, x/\eta)$ where $P(\beta - 1, x/\eta)$ is the cumulative
distribution function for $\beta - 1$ or fewer occurrences for a Poisson distribution with a mean
equal to x/η.

 Hazard function (instantaneous failure rate):

$$h(x) = [1/\Gamma(\beta)][x^{\beta-1}/\eta^\beta] \exp(x/\eta)/[1 - \Gamma(x/\eta, \beta)].$$

Significant Distribution Properties
 Median: $x_{0.50}$ is the solution of $0.50 = \Gamma(x_{0.50}/\eta, \beta)$

$$\text{Mode} = \begin{cases} \eta(\beta - 1) & \text{for } \beta > 1, \\ 0 & \text{for } 0 < \beta \leq 1. \end{cases}$$

 Mean $= \beta\eta$.
 Standard deviation $= \beta^{1/2}\eta$.

Table B.3 (*continued*)

Distribution Parameters
    Scale parameter $= \eta,\, \eta > 0$.
      (units same as of x)
    Location parameter $= \beta,\, \beta > 0$.
      (dimensionless)

---

[a] We can also write a *standard normal distribution* by replacing $(x - \mu)/\sigma$ in normal distribution by z. For this standard normal distribution, $\mu = 0$ and $\sigma = 1$.

[b] $\gamma = 0$ for a two-parameter Weibull distribution.

[c] The chi-square distribution is a special case of the gamma distribution with $\eta = 2$ and $\beta = v/2$, where v is an integer; v is the number of degrees of freedom. For example, the chi-square distribution with $v = 2$ is the exponential distribution with a mean equal to 2.

rate curve starts at zero, rises to a peak, then asymptotically approaches zero again, for all values of σ. For many products, the failure rate does not go to zero with increasing age. Even so, the lognormal distribution may adequately describe most of the range of life, particularly early life. For $\sigma \sim 0.5$, $h(x)$ is essentially constant over much of the distribution. For $\sigma \leq 0.2$, $h(x)$ increases over most of the distribution and is much like that of a normal distribution. For $\sigma \geq 0.8$, $h(x)$ decreases over most of the distribution. Thus, the lognormal can describe our increasing, decreasing, or relatively constant failure rate. This flexibility makes the lognormal distribution popular and suitable for many products. It is often used for economic data and life data such as metal fatigue and electrical insulation life.

The curves shown for the Weibull distribution are for a constant scale parameter, η, and for varying shape factors, β. These cumulative distributions look much like the lognormal in general shape. The $h(x)$ curve increases with life for $\beta > 1$ (a wear-out type of behavior) and decreases with time for $\beta < 1$ (infant mortality). For $\beta = 1$ (the exponential distribution), the failure rate is constant. The ability to describe increasing, decreasing, or constant failure rates contributed to making the Weibull distribution popular for life data analysis. The Weibull distribution is used for analysis of failure data of many tribological components such as rolling-element bearings. The relationship between the Weibull and extreme value distributions is similar to that between normal and lognormal distributions. The extreme value data are simpler to handle, as that distribution has a single shape and simple location and scale parameters, similar to the normal distribution. However, the $h(x)$ curve only increases with life (wearout). This distribution has been applied to human mortality data, and to a large number of tribological failures in industry. It is almost exclusively used to predict the life of rolling element bearings which generally fail by fatigue mechanisms.

In the case of the gamma distribution, for $\beta < 1$, $h(x)$ decreases to a constant value $1/\eta$, and for $\beta > 1$, $h(x)$ starts at zero and increases to a constant value $1/\eta$. Few products have such a failure rate behavior. This distribution has been applied to biomedical survival data.

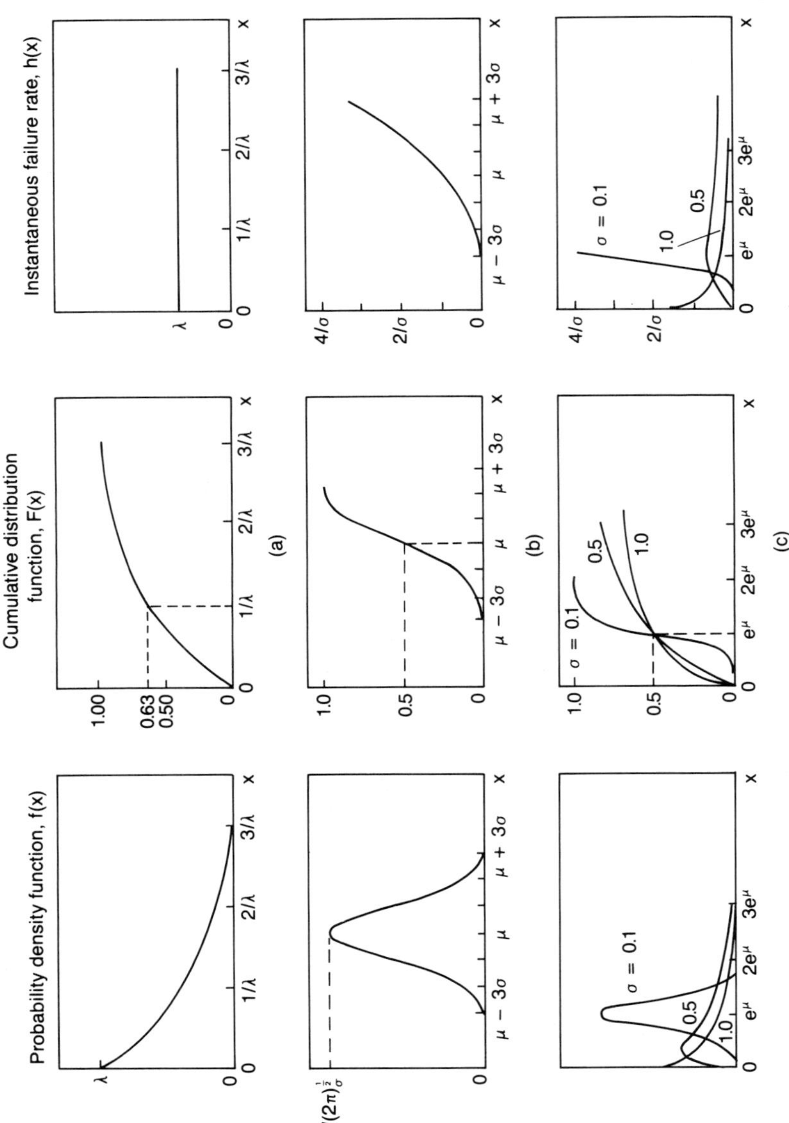

Probability density function, f(x)
Cumulative distribution function, F(x)
Instantaneous failure rate, h(x)
x
λ
0
1/λ
2/λ
3/λ
1.00
0.63
0.50
1/(2π)^(1/2) σ
μ − 3σ
μ
μ + 3σ
4/σ
2/σ
1.0
0.5
σ = 0.1
0.5
1.0
eμ
2eμ
3eμ
(a)
(b)
(c)

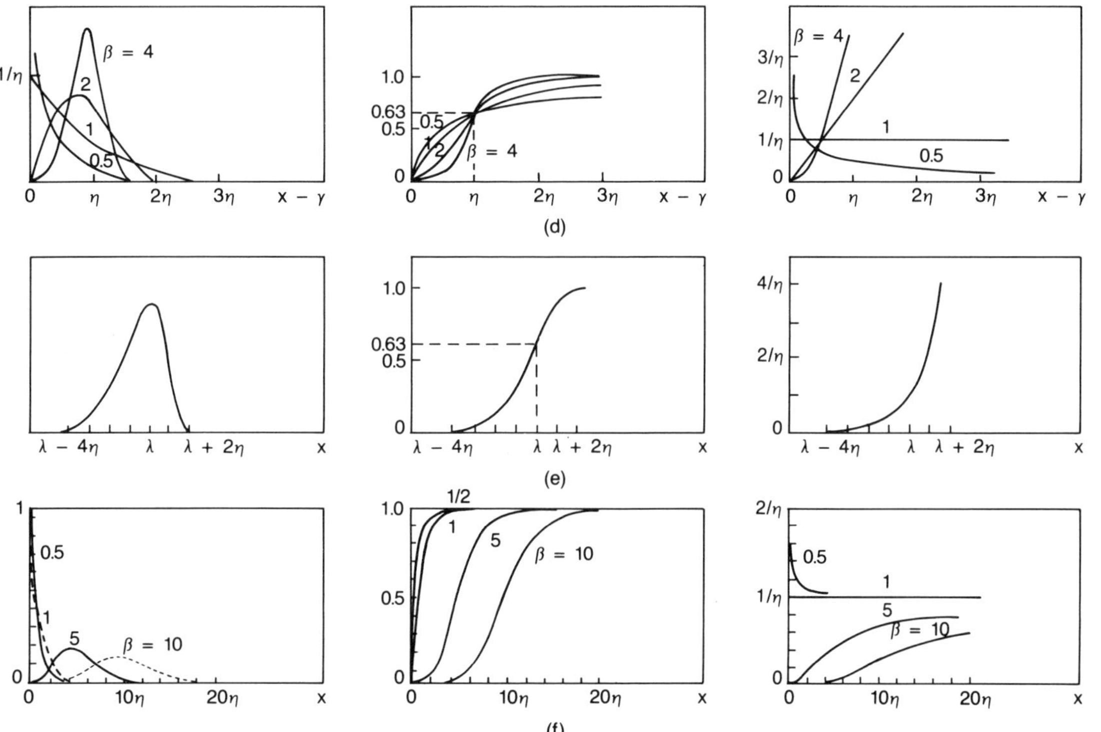

Fig. B.7. Schematics of six continuous probability distributions: (a) exponential; (b) normal; (c) lognormal [curves shown are for constant median life, $\exp(\mu)$]; (d) Weibull (curves shown are for a constant scale parameter η); (e) extreme value; and (f) gamma distributions.

B.2. Discrete Distributions

B.2.1. Basic Statistical Concepts

In this section, we present the basic concepts for discrete statistical distributions.

B.2.1.1. Probability Function

A discrete distribution is a probability life model X that consists of a list of possible outcomes (with integer values) x_1, x_2, x_3, etc., each with a corresponding probability $f(x_1)$, $f(x_2)$, $f(x_3)$, etc. Here $f(\ )$ is called the probability function. The probabilities $f(x_i)$ must be zero or greater and their sum must equal 1. The failure events may be *numerical* (e.g., 0, 1, 2, 3, etc.) *or categories* (e.g., dead, alive, undecided). For a population, $f(x_i)$ is the population fraction with the value x_i. It is also the chance that a randomly taken observation has the value x_i, i.e., $f(x_i)$ is the fraction of the time that x_i would be observed in an infinitely large number of observations.

B.2.1.2. Cumulative Distribution Function (c.d.f.)

An event is any set of outcomes of a probability model X. The probability of an event is the sum of the probabilities of the outcomes in the event. Notation for the probability of an event is $P\{\ \}$, where the relationship in the braces indicates that the event consists of the outcomes of X that satisfy the relationship, e.g., $P\{X = x\} = f(x)$ and $P\{X \le x\} = \sum_{x_i \le x} f(x_i)$. $P\{X \le x\}$ would be interpreted as the probability of failure by age x, and $P\{X > x\}$ is the probability of surviving beyond age x. An event is said to occur if the observed outcome is in the event.

The cumulative distribution function (c.d.f.) $F(x)$, of a discrete function with numerical outcomes, is

$$P\{X \le x\} = F(x) = \sum_{x_i \le x} f(x_i), \qquad -\infty < x < \infty, \qquad (\text{B.32})$$

where the sum runs over all outcomes x_i that are less than or equal to x. (For failure life data, the outcomes x cannot be less than zero.) For a population, $F(x)$ is the fraction of the population with a value x or less. $F(x)$ for a discrete distribution is a staircase function and the jump in the function at a failure event x_i equals the probability $f(x_i)$. Thus $F(x)$ and $f(x)$ may be obtained from each other. The function $F(x)$ has the following properties:

$$F(-\infty) = 0, \qquad F(\infty) = 1, \qquad (\text{B.33a})$$

and

$$F(x) \le F(x') \qquad \text{for all} \quad x < x'. \qquad (\text{B.33b})$$

B.2.1.3. Mean and Variance

The mean of a discrete random variable or the expected value of X is

$$E(X) = x_1 f(x_1) + x_2 f(x_2) + x_3 f(x_3) + \cdots. \tag{B.34}$$

The sum runs over all outcomes x_i. The mean corresponds to a population average, and the average of a large sample tends to be close to the distribution mean.

The variance (square of standard deviation σ) of a discrete random variable X is

$$\mathrm{Var}(X) = \sigma(X)^2 = [x_1 - E(X)]^2 f(x_1) + [x_2 - E(X)]^2 f(x_2)$$
$$+ [x_3 - E(X)]^2 f(x_3) + \cdots, \tag{B.35a}$$

where the sum runs over all possible outcomes x_i. Equivalently,

$$\mathrm{Var}(X) = [x_1^2 f(x_1) + x_2^2 f(x_2) + x_3^2 f(x_3) + \cdots] - [E(X)]^2. \tag{B.35b}$$

The variance is a measure of the spread of a distribution about its mean. It has dimensions of x^2.

B.2.2. Probability Distributions

The mathematical expressions for discrete geometric, Poisson, and binomal distributions are presented in Table B.4 and they are plotted in Fig. B.8 (Johnson and Kotz, 1969). The geometrical distribution is the simplest. The $f(x)$ in Table B.4 is the distribution of the number of x trials to "failure," where each trial is statistically independent of all other trials and each trial has same chance of failure. If a unit from a geometric life distribution has survived x trials, then its chance of failure on the next trial is p for any value of x. Thus each trial is statistically independent of all other trials and each trial has the same chance p of failure. Thus, such a unit does not become more or less prone to failure as it ages. This discrete distribution is analogous to the continuous exponential distribution. The geometric distribution is used for items that fail from a chance event.

The Poisson distribution describes situations where: (a) the occurrences occur independently of each other over time; (b) the chance of an occurrence is the same for each point in time; and (c) the potential number of occurrences are essentially unlimited. The Poisson distribution has been used to describe failure from a chance event, e.g., the number of defects in a length of wire or a sheet of material, the number of failures of a repairable product over a certain period, and the number of atomic particles emitted by a sample in a specified time.

Binomial distribution is widely used as a model for the number of sample units that are in a given category. The distribution is used if each unit is classified as in the category or not in the category, a dichotomy. For example,

Table B.4. Mathematical expressions for three discrete distributions

(a) Geometric Distribution

Distribution Functions
Probability density function:

$$f(x) = p(1 - p)^{x-1}, \qquad x = 1, 2, 3 \ldots.$$

Cumulative distribution function:

$$F(x) = p + p(1 - p) + p(1 - p)^2 + \cdots + p(1 - p)^{x-1}$$
$$= p - (1 - p)^x, \qquad x = 1, 2, 3 \ldots.$$

Significant Distribution Properties
Mean $= 1/p$.
Standard deviation $= (1 - p)^{1/2}/p$.

Distribution Parameters
$x =$ outcomes $(= 1, 2, 3, \ldots)$.
$p =$ chance (probability) of failure for each trial, $0 < p < 1$.

(b) Poisson Distribution

Distribution Functions
Probability density function:

$$f(x) = (1/x!)(\lambda t)^x \exp(-\lambda t), \qquad x = 0, 1, 2 \ldots.$$

Cumulative distribution function:

$$F(x) = \sum_{i=0}^{x} (1/i!)(\lambda t)^i \exp(-\lambda t), \qquad x = 0, 1, 2.$$

Significant Distribution Properties
Mean $= \lambda t$.
Standard deviation $= (\lambda t)^{1/2}$.

Distribution Parameters
$x =$ number of occurrences $(= 0, 1, 2, \ldots)$.
$\lambda =$ the failure rate or occurrence rate (occurrence per unit length), $\lambda > 0$.
$t =$ length or exposure of the observation, e.g., time, area. Sometimes λ and t are replaced
by a single parameter $\mu = \lambda t$.

(c) Binomial Distribution

Distribution Functions
Probability density function:

$$f(x) = \{n!/[x!(n - x)!]\}p^x(1 - p)^{n-x}, \qquad x = 0, 1, 2, \ldots, n.$$

Probability distribution function:

$$F(x) = \sum_{i=0}^{x} \{n!/[i!(n - i)!]\}p^i(1 - p)^{n-i}, \qquad x = 0, 1, 2, \ldots, n.$$

Significant Distribution Properties
Mean $= np$.
Standard deviation $= [np(1 - p)]^{1/2}$.

Distribution Parameters
$x =$ possible number of units in the category $(= 0, 1, 2, \ldots, n)$.
$n =$ number of units
$p =$ population proportion in the category or failure rate usually expressed as a fraction,
$0 \leq p \leq 1$.

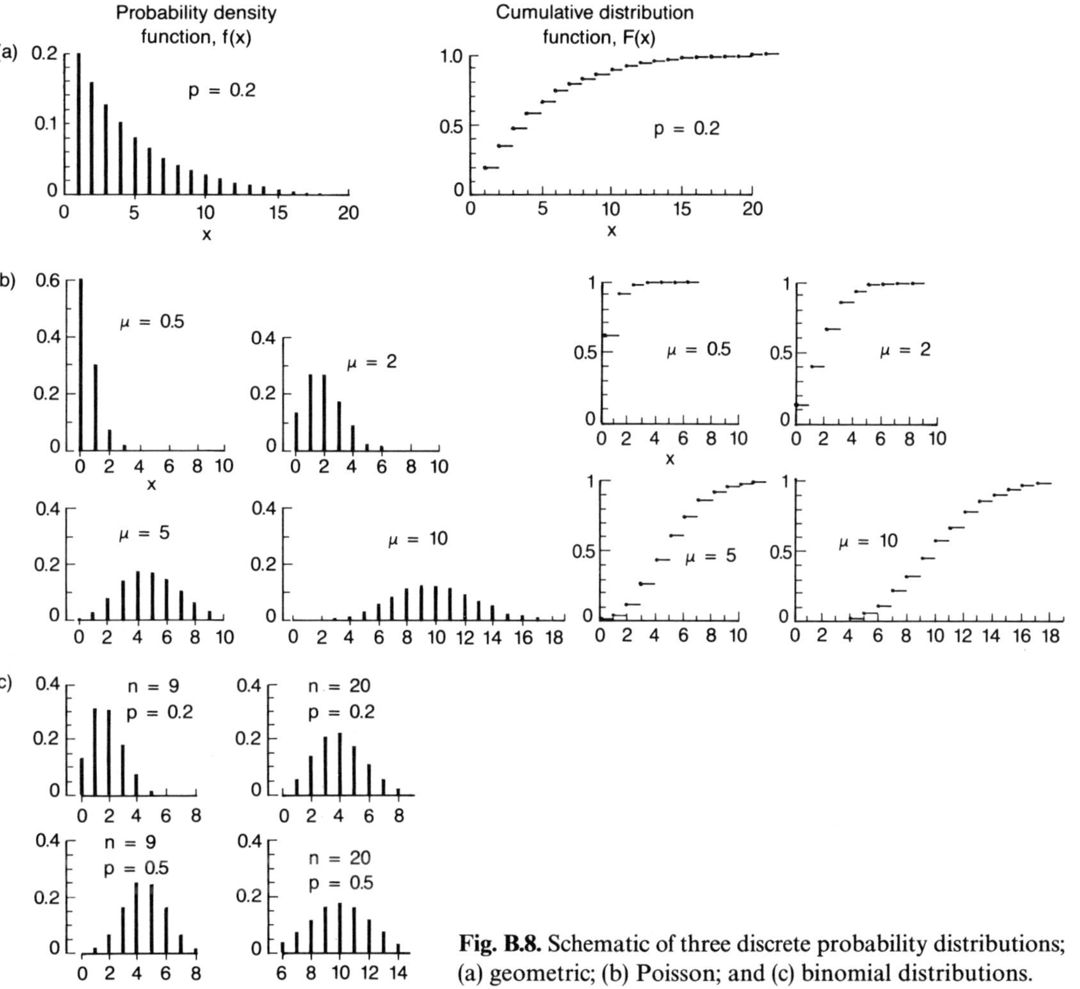

Fig. B.8. Schematic of three discrete probability distributions; (a) geometric; (b) Poisson; and (c) binomial distributions.

it is used for the defective units in samples from shipments and production, or the number of units that fail on warranty. Assumptions of the model are (a) each of n sample items has the same chance (probability) p of being in the category, and (b) the outcomes of the n sample items are statistically independent. Of course, the number of sample items in the category can range from zero to n. Binomial distribution is suitable for a small sample (about 1% or less) taken from a large population.

In summary, all the discrete distributions presented here are used for items that fail from a chance event.

B.3. Graphical Representation and Analysis of Data

In this section we describe the various types of data, and then describe the methods to make probability plots and hazard plots and the analytical methods that are useful for getting distribution parameters. Probability plots are often preferred over the analytical methods because they are fast and simple to use. They provide data in an easy-to-grasp form. They provide graphical estimates of the distribution parameters. However, they are not objective and they do not provide confidence intervals. Before using analytical methods, it is important to check with a probability plot that a chosen distribution fits the data. A combination of graphical and analytical methods is often most informative.

B.3.1. Types of Life Data

A sample of n units tested until all have failed is a complete sample. A sample of n items of which only r ($r < n$) units have failed is a censored sample. Censoring may be by design or random. Random censoring occurs when an arbitrary mechanism other than the failure mechanism under study acts to remove an unit from the test. A competing failure mode is an example. Planned censoring may be at a predetermined life (Type I or time censoring) or at a predetermined number of failures (Type II or failure censoring). Planned censoring may also be single or multiple, depending upon whether all units are removed when a prespecified life (or number of failures) is achieved or whether a portion of units is removed at each of several prespecified lives (or number of failures). Of course, a complete sample (all units to failure) yields more precise estimates than a censored sample of the same size. However, the reduced precision from a censored sample is often compensated for by the time saved from analyzing the data before all units fail.

B.3.1.1. Complete Data

The complete data consist of the failure times of all units in the sample.

B.3.1.2. Singly Censored Data

Singly censored data on the right consist of the earliest r failure times in a sample of n units. Such data arise when units start together and the data are analyzed before all units fail. Then the unfailed units all have the same running time, called the *censoring time*. The singly censored data have Type I censoring (or *time censoring*) if the censoring time is prespecified and the number of failures is random. The singly censored data have Type II censoring (or *failure censoring*) if the test is stopped at the time of the rth failure. Then the number of failures is fixed, and the length of the test is random. Such censoring is common since it is mathematically simpler than Type I censoring. If the failure time occurs too early to be recorded, such data are said to be *singly censored on the left*.

B.3.1.3. Multiply Censored Data

Multiply censored data consist of failure times intermixed with running times, called censoring times, as depicted in Fig. B.9. Such life data are common and can result from:

(1) units being put into service at different times;
(2) removal of units from use before failure;
(3) loss or failure of units due to extraneous causes; and
(4) collection of data while units are still running (common for field data and much test data).

The multiply censored data also have *Type I (time) or Type II (failure) censoring*. Multiply censored data are also called "progressively," "hyper-," and "arbitrarily censored."

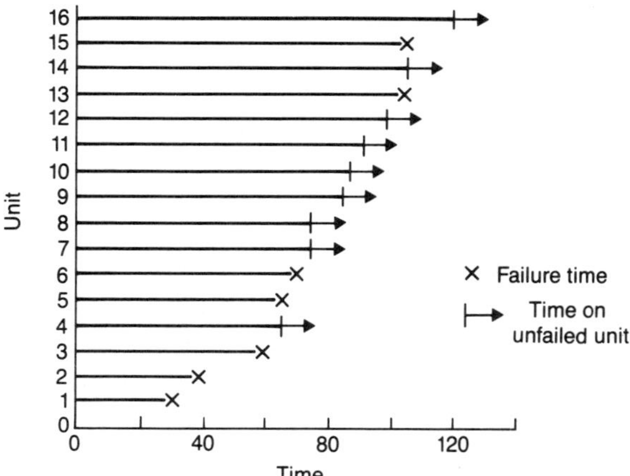

Fig. B.9. An example of multiply-censored life data.

B.3.1.4. Inspection Data (Quantal-Response and Interval Data)

For some products, a failure is found only on inspection, e.g., a cracked component inside a magnetic disk drive. Two types of such inspection data are commonly encountered: (1) quantal-response data and (2) interval data, also called grouped or coarse data.

In the quantal-response data, there is exactly one inspection on each unit to determine whether it has failed or not. If a unit has failed, we know that its failure time was before its inspection time. Similarly, if a unit does not fail, we know that its failure time is beyond its inspection time. Quantal-response data arise in many other applications besides product lives, e.g., particle velocity that cracks jet engine blades and the penetrating velocity of shells fired at a given thickness of armor. In such applications, each test unit is subjected to a value of some stress and is a failure or a success. A unit is not retested at another stress.

In the interval data, there is any number of inspections of a unit. If a unit has failed, we know that the failure occurred between that inspection and the previous one. Also, if a unit had not failed on its latest inspection, we know that its failure time is beyond the inspection time. The inspection intervals may differ.

Methods for the graphical representation and numerical analyses of inspection data are similar to that for the failure data discussed earlier. Methods for the graphical representation of inspection data will be demonstrated by an example in Section B.4.

B.3.2. Graphical Representation of Data

Data can be graphically represented as probability plots or hazard plots. Hazard plots give the same information as probability plots, but with less labor.

B.3.2.1. Probability Plotting of Data

In this section, we present methods of making the probability plots of various types of data. Probability plotting involves calculating the cumulative distribution function. This function is plotted against life (or age) on the probability paper. If the data fit with a straight line on the probability plot, the chosen distribution represents the data. Otherwise, a different probability plot should be used. In some cases, however, parts of the data may fit straight lines with different slopes which implies that more than one failure mode may be present, e.g., see the bathtub curve for failure rates, shown in Section B.1. In the Section B.4 we will show examples for the Weibull distribution.

Complete Data

To make a probability plot of a complete sample of life data, we follow the following steps:

1. Order the n failure times from smallest to largest.
2. Assign a rank to each failure. Give the earliest failure rank 1, the second failure rank 2, etc., and the last failure rank n.
3. Calculate the probability plotting positions F_i for rank i.

   Several methods have been proposed to calculate the probability plotting positions F_i (King, 1971). For the failure with rank i, the "midpoint" plotting position (corresponding sample cumulative failures) on the $F(x)$ axis for $(x_i - \gamma)$ (where γ is the location parameter) is

$$F_i = (i - 0.5)/n, \qquad i = 1, \ldots, n. \tag{B.36}$$

Many other choices have been advocated. In general, the ith plotting position is a typical population near to which the ith ordered observation falls. The expected value or "mean" plotting position is popular and is

$$F_i = i/(n + 1), \qquad i = 1, \ldots, n. \tag{B.37}$$

Johnson (1964) and Harter (1964) advocate and tabulate median plotting positions, often termed median ranks. An excellent approximation to the median ranks offered by Johnson is

$$F_i \sim (i - 0.3)/(n + 0.4), \qquad i = 1, \ldots, n. \tag{B.38}$$

In practice, plotting positions differ little, compared with the randomness of the data. We could use any of the formulas of the plotting positions depending on which is easier to use. Medium rank [Eq. (B.38)] is most commonly used (McCool, 1970).

4. Plot each failure against its time on the data scale and against its plotting position on the cumulative probability scale of a probability paper. There are probability papers for each distribution. The papers have the probability and data scales so that the theoretical cumulative distribution function (c.d.f.) is a straight line on paper. The distribution should be chosen from experience or an understanding of the physical phenomena. If the plotted points tend to follow a straight line, then the chosen distribution appears to be adequate.
5. Draw a straight line through the plotted data such that the deviations between the line and plotted points are minimum. The fitted line must pass through the origin. Depending on how the line will be used, it can be fitted to the whole data, the center of the data, the lower tail, or whatever is appropriate. Then obtain the distribution parameters from such plots.

Single-Censored Data

The method for plotting singly-censored life data is like that for complete data. In particular, plot the ith ordered observation against the plotting position F_i calculated by one of the methods discussed earlier [Eqs. (B.36)–(B.38)]. Here n is the total sample size *including the nonfailures*. Nonfailures are not plotted, since their failure times are unknown. Only the early failure times are observed, and they estimate the lower part of the life distribution, usually of greatest interest.

Sometimes we estimate the lower or upper tail of the distribution from a singly censored sample by extending a straight line beyond the plotted points. The accuracy of such extrapolation depends on how well the theoretical distribution describes the true one into the extrapolated tail. For left-censored data, we include the left-centered data in determining the sample size and the ranks of failures and their plotting positions.

Multiply-Censored Data

We present here three methods: the Herd–Johnson, the Kaplan–Meier (product-limit), and actuarial methods. Each provides a different nonparametric estimate of the reliability function. A basic assumption of each plotting method must be satisfied if it is to yield reliable results. Namely, units censored at a given age must have the same life distribution as units that run beyond that age. For example, this assumption fails to hold if units are removed from service when they look like they are about to fail.

(a) Herd–Johnson Method

Suppose that there are n times in a multiply censored sample. Order them from smallest to largest, and mark each running time "+." Number the times backwards with reverse ranks—the smallest time is labeled n, the second smallest is labeled $(n - 1)$, etc., and the largest is labeled 1. For the ith failure with reverse rank r_i, recursively calculate the reliability

$$R_i = [r_i/(r_i + 1)]R_{i-1}, \tag{B.39}$$

where $R_0 = 1$ is the reliability at time 0. The corresponding failure probability is the plotting position $F_i = (1 - R_i)$. The F_i are calculated only for failure times, but the running times determine the plotting positions of the failures. Johnson (1964) gives the plotting positions in terms of expected ranks, and he shows how to convert them to median plotting positions, a small refinement.

Plot each failure time against its plotting position F_i. Only failure times are plotted. Use the probability plot as described earlier to obtain information (Herd, 1960; Johnson, 1964).

(b) Kaplan–Meier Method

Kaplan and Meier (1958) give the *product-limit* estimate, much the same as the Herd–Johnson one. Kaplan and Meier use the recursion formula for reliability after failure i:

$$R_i = [(r_i - 1)/r_i]R_{i-1}, \tag{B.40}$$

where $R_0 = 1$ is the reliability at time 0. For a complete sample, $F_i = 1 - R_i = i/n$, the sample fraction failing; this is the usual nonparametric estimate. If the largest time in a sample is a failure, the corresponding F_i is unity and the failure cannot be plotted on most probability papers.

The Kaplan–Meier method is widely used in biomedical applications to obtain a nonparametric estimate of the survivorship function (Kalbfleisch and Prentice, 1980).

(c) Actuarial Method

Various actuarial methods for estimating life distributions have been used. They were developed and used to construct human life tables. Chiang (1968) describes such methods and the biomedical applications of them. The simplest such method is presented here. Such methods are suited to large samples; the Herd–Johnson and the Kaplan–Meier methods are better for small samples, since they plot individual failures times.

The actuarial method estimates the reliability or survivorship function from multiply-censored data. The method is usually used for large samples where the data are grouped into *time intervals*. For example, human mortality data are usually grouped into 1-year intervals. The method uses the number of units that enter, r_x, fail, f_x, are censored, c_x, and survive, r_{x+1}, in interval x. Of course, $r_{x+1} = r_x - f_x - c_x$. The intervals need not have the same length, and intervals with no failures need not be tabulated. In contrast, the preceding methods use and plot individual failure times.

Let $R(x)$ denote the product reliability (survival probability) at the end of interval x; this is the probability of a new unit surviving through interval x. The estimate of $R(x)$ uses the chain rule for conditional probabilities

$$R(x) = R(x|x - 1)R(x - 1|x - 2), \ldots, R(2|1)R(1), \qquad (B.41)$$

where $R(i|i - 1)$ denotes the conditional reliability that units that have survived interval $(i - 1)$ also survive interval i. It is convenient to work with the recursion relationship

$$R(x) = R(x - 1)R(x|x - 1). \qquad (B.42)$$

That is, the fraction $R(x)$ surviving interval x is the fraction $R(x - 1)$ surviving interval $(x - 1)$ times the fraction $R(x|x - 1)$ of those that reach interval x and survive it. The probability of a new unit failing by the end of interval x is $F(x) = 1 - R(x)$.

The actuarial estimate of $R(x)$ employs estimates of the conditional reliabilities $R(i|i - 1)$ and the preceding relationships. A simple estimate of $R(i|i - 1)$ is

$$R^*(i|i - 1) = 1 - [f_i/(r_i - 0.5c_i)]. \qquad (B.43)$$

The $-0.5c_i$ adjusts for the censored units, which run about half the interval. Other censoring adjustments appear in the literature (Chiang, 1968). Then the estimate of $R(x)$ is

$$R^*(x) = R^*(x - 1)R^*(x|x - 1), \qquad (B.44)$$

where $R^*(0) = 1$.

The recursive calculations of the $R^*(x)$ described above can readily be carried out in tabular form. This method provides an estimate $R^*(x)$ of the reliability at the end of each period. These $R^*(x)$ may be plotted against the interval upper end-points on square grid paper. Also, the sample (c.d.f.) fraction failing $F^*(x) = 1 - R^*(x)$ can be plotted against the interval upper end-points on probability paper. Such a plot is interpreted like any other probability plot described earlier.

B.3.2.2. Hazard Plotting of Data

Hazard plots involve calculating the cumulative hazard function. This function is plotted against life (or age) on the hazard paper. Hazard plots look like probability plots and are interpreted in the same way. They can be used for complete, singly, and multiply censored data (Nelson, 1972). Hazard plotting positions are slightly different from probability plotting positions.

We consider the data on n units consisting of the failure times and the running (censoring) times. For hazard plotting, the n times are ordered from smallest to largest without regard as to which are censoring or failure times. Times are labeled with reverse rank (k); i.e., label the first time with n, the second with $n - 1, \ldots$, and the nth with 1. Running times are marked with a "$+$," and failure times are unmarked. We calculate the hazard value for each failure as $1/k$, where k is its reverse rank. We then calculate the cumulative hazard value for each failure as the sum of its hazard value and the cumulative hazard value of the preceding failure. Finally, we plot the data on the hazard paper of a theoretical distribution (failure time on the vertical axis and the cumulative hazard value in percent on the horizontal axis). If the plot of failure times is roughly straight, we may conclude that the distribution adequately fits the data.

The fitted straight line estimates the cumulative percentage failing, read from the horizontal probability scale as a function of life (or age) read from the vertical axis.

B.3.3. Numerical Analysis of Data

In this section, we present the standard methods used for the analysis of complete data, linear methods used for the analyses of complete and singly-censored data, and the maximum likelihood (ML) methods for the analyses of all data (particularly for multiply-censored data) (Brunk, 1965; Johnson and Kotz, 1970; Mann et al., 1974; Bain, 1978; Nelson, 1982).

B.3.3.1. Standard Methods for Complete Data

We present the basic ideas for life data analysis. These ideas include a statistic and its sampling distribution, estimator, confidence intervals, sample size, and prediction (predictors and limits). These methods are used for the analysis of complete data (Nelson, 1982).

A Statistic and Its Sampling Distribution

A *statistic* is a numerical value determined from a sample by some procedure and denoted by $\theta^*(X_1, \ldots, X_n)$, a function of the data values. (The n observations in a sample are denoted by $X_1, \ldots, X_n$.) Sample statistics include the sample average, standard deviation, median, etc. The values of the statistic would differ from sample to sample and would have a distribution, which is called the *sampling distribution* of the statistic. The form of a sampling distribution depends on the statistic, the sample size, and the parent distribution.

Estimator

A statistic $\theta^*(X_1, \ldots, X_n)$ that approximates an unknown population value θ is called an *estimate* when we refer to its value for a particular sample; it is called an *estimator* when we refer to the procedure or the sampling distribution. In practice, we do not know how close a sample estimate is to the population value (called the true value). An estimator θ^* for θ is called *unbiased* if the mean $E\theta^*$ of its sampling distribution equals θ. If different from zero, $E\theta^* - \theta$ is called the bias of θ^*.

The spread in the sampling distribution of an estimator θ^* should be the same. The usual measure of spread is the distribution variance $\mathrm{Var}(\theta^*)$, or, equivalently, its positive square root, the standard deviation $\sigma(\theta^*)$, called the *standard error* of the estimator. We usually want estimators that are (almost) unbiased and have small standard errors. A useful biased estimator has a small bias compared to its standard error. If an estimator θ^* for θ is biased, then the *mean square error* (MSE) is a measure of uncertainty in the estimator

$$\mathrm{MSE}(\theta^*) \equiv E[(\theta^* - \theta)^2]$$

$$= \int_{-\infty}^{\infty} (\theta^* - \theta)^2 f(\theta^*)\, d\theta^*, \tag{B.45a}$$

where $f(\ )$ is the probability density of θ^*. The MSE, variance, and bias of an estimator are related by

$$\mathrm{MSE}(\theta^*) = \mathrm{Var}(\theta^*) + (E\theta^* - \theta)^2. \tag{B.45b}$$

Confidence Intervals (or Limits)

Estimation provides a single estimate of a population value. The uncertainty in such an estimate may be judged from its standard error. A *confidence interval* (or confidence limit) also indicates the uncertainty in an estimate. Calculated from sample data, it encloses the population value with a specified high probability. The length of such an interval indicates if the corresponding estimate is accurate enough for practical purposes. The following defines a confidence interval for a population value θ. Suppose $\underset{\sim}{\theta} = \underset{\sim}{\theta}(X_1, \ldots, X_n)$ and $\tilde{\theta} = \tilde{\theta}(X_1, \ldots, X_n)$ are functions of the sample data $X_1, \ldots, X_n$ such that

$$P_\theta\{\underset{\sim}{\theta} \leq \theta \leq \tilde{\theta}\} = \alpha, \tag{B.46}$$

no matter what the value of θ or any distribution parameter. Then the interval $[\underline{\theta}, \tilde{\theta}]$ is called a *two-sided* $100\alpha\%$ *confidence interval* for θ. $\underline{\theta}$ and $\tilde{\theta}$ are the lower and upper *confidence limits*, or bounds. The random limits $\underline{\theta}$ and $\tilde{\theta}$ enclose θ with probability α.

Similarly, suppose that $\tilde{\theta} = \tilde{\theta}(X_1, \ldots, X_n)$ satisfies

$$P_\theta\{\theta \leq \tilde{\theta}\} = \alpha, \tag{B.47}$$

no matter what the value of θ or any distribution parameter. Then the interval $(-\infty, \tilde{\theta})$ is called a "*one-sided* $100\alpha\%$ *upper confidence interval for* θ." One-sided lower confidence intervals and limits are defined similarly. In most applications, it is clear whether to use a one-sided or a two-sided interval. For example, we usually want a one-sided upper limit for a fraction failing on warranty or a one-sided lower limit for reliability, but two-sided limits for a Weibull shape parameter.

One- and two-sided confidence intervals are related. The upper (lower) limit of a two-sided $100\alpha\%$ confidence interval for a value θ is a one-sided upper (lower) $[100(1 + \alpha)/2]\%$ confidence limit for θ. For example, the upper limit of a two-sided 90% interval is the upper limit of a one-sided 95% confidence interval.

Most people would like an interval with high confidence. However, the width of a confidence interval increases with the confidence level, and a 99.9% confidence interval, for example, may be so wide that it has little value in an application. Most data analysts use 90%, 95%, and 99% confidence intervals. We can calculate and present a number of intervals for different confidence levels.

Approximate confidence limits for a population value θ can be obtained from an (almost) unbiased estimator θ^* that is approximately normally distributed. For example, then

$$P\{(\theta^* - \theta)/\sigma(\theta^*) \leq z_\alpha\} \simeq \alpha, \tag{B.48}$$

no matter what the value of θ, where z_α is the 100αth standard normal percentile.[2] Equation (B.48) becomes

$$P_\theta\{\theta \leq \theta^* + z_\alpha\sigma(\theta^*)\} \simeq \alpha. \tag{B.49}$$

This means that

$$\tilde{\theta} = \theta^* + z_\alpha\sigma(\theta^*) \tag{B.50}$$

is a one-sided approximate $100\alpha\%$ upper confidence limit for θ. Often $\sigma(\theta^*)$ is not known and must be estimated from the data. Similarly, a one-sided

---

[2] $z_\alpha = (x_\alpha - \mu)/\sigma$ for a *standard* normal probability density function (with mean $=0$, standard deviation $=1$). Some standard percentiles are:

$100P\%$	0.1	1	2.5	5	10	50	90	95	97.5	99	99.9
z_P	-3.090	-2.326	-1.960	-1.645	-1.282	0	1.282	1.645	1.960	2.326	3.090

approximate $100\alpha\%$ lower confidence limit for θ is

$$\underset{\sim}{\theta} = \theta^* - z_\alpha \sigma(\theta^*), \tag{B.51}$$

and two-sided approximate $100\alpha\%$ confidence limits for θ are

$$\underset{\sim}{\theta} = \theta^* - K_\alpha \sigma(\theta^*), \qquad \tilde{\theta} = \theta^* + K_\alpha \sigma(\theta^*), \tag{B.52}$$

where K_α is the $[100(1 + \alpha)/2]$th standard normal percentile. Such approximate intervals tend to be narrower than exact ones. Also, a two-sided interval tends to have a confidence closer to $100\alpha\%$ than does a one-sided interval if the sampling distribution of θ^* is unsymmetrical.

Sample Size

Often we must determine a suitable sample size. To do this, we specify how precise the chosen estimator θ^* for θ must be. Usually we want the estimate to be within a specified $\pm w$ of θ, with $100\alpha\%$ probability. The following is an approximate sample size n when θ^* is approximately normally distributed with variance V^*/n; V^* may depend on θ and other distribution parameters. Then

$$n \simeq V^*(K_\alpha/w)^2. \tag{B.53}$$

In practice, the value of V^* is usually unknown, and we must estimate it from experience, a preliminary sample, or similar data.

If θ is positive, we may specify that θ^* be within a factor f of θ with probability $100\alpha\%$. That is, θ^* is between θ/f and θf, with $100\alpha\%$ probability. Suppose that $\mathrm{Var}(\theta^*) = V/n$, where n is the sample size and V is a known factor. The appropriate sample size n is

$$n \simeq K_\alpha^2(V/\theta^2)/[\ln(f)]^2, \tag{B.54}$$

where K_α is the $[100(1 + \alpha)/2]$th standard normal percentile, V/θ^2 usually depends on unknown distribution parameters that must be estimated from experience, a preliminary sample, or similar data. The formula assumes that n is large enough that the distribution of $\ln(\theta^*)$ is approximately normal.

Prediction

The statistical methods above use data from a sample to get information on a distribution, usually its parameters, percentiles, and reliabilities. However, in business and engineering, we are often concerned about a future sample from the same distribution. In particular, we usually want to use a past sample to predict the future sample value of some statistic and to enclose it with prediction limits. The following presents such two-sample prediction theory.

(a) Predictor. Suppose that $X_1, \ldots, X_n$ and $Y_1, \ldots, Y_m$ are all independent observations in a past and future sample from the same distribution F_θ, where θ denotes one or more unknown distribution parameters. Suppose we wish to predict the random value $U = u(Y_1, \ldots, Y_m)$ of some statistic of the future sample. To do this, we use a *predictor* $U^* = u^*(X_1, \ldots, X_n)$ that is a function of the past sample.

The difference $U^* - U$ is the *prediction error*; its sampling distribution depends on one or more parameters θ. If its mean is 0 for all values of θ, U^* is called an *unbiased predictor* for U. The variance $\mathrm{Var}_\theta(U^* - U)$ of the prediction error or its square root, the standard prediction error, $\sigma_\theta(U^* - U)$ should be small.

(b) Prediction interval (or limits). Often we want an interval that encloses a single future statistic U with high probability. The interval width indicates the statistical uncertainty in the prediction. Statistics $\underset{\sim}{U} = \underset{\sim}{u}(X_1, \ldots, X_n)$ and $\tilde{U} = \tilde{u}(X_1, \ldots, X_n)$ of the past sample, that satisfy

$$P_\theta\{\underset{\sim}{U} < U < \tilde{U}\} = \alpha \tag{B.55}$$

for any value of θ, are called "two-sided $100\alpha\%$ *prediction limits* for U." One-sided prediction limits are defined in the obvious way. We can think about such an interval in terms of a large number of pairs of past and future samples. An expected proportion α of such intervals will enclose U in the corresponding future sample. Prediction limits for a random sample value (say, a future sample mean) are wider than confidence limits for the corresponding constant population value (say, a population mean).

Suppose that the sampling distribution of the prediction error is approximately normal. Then a two-sided approximate $100\alpha\%$ prediction interval for U has limits

$$\underset{\sim}{U} = U^* - K_\alpha\sigma_\theta(U^* - U), \qquad \tilde{U} = U^* + K_\alpha\sigma_\theta(U^* - U), \tag{B.56}$$

where K_α is the $[100(1 + \alpha)/2]$th standard normal percentile. One-sided approximate prediction limits use the 100αth percentile z_α in place of K_α. In practice, the standard prediction error $\sigma_\theta(U^* - U)$ is usually unknown and is estimated from the data.

B.3.3.2. Linear Methods for Complete and Singly Censored Data

Linear methods employ linear combinations of the ordered observations (Nelson, 1982).

Order Statistics

When n units start a test together, the shortest sample life is observed first, the second shortest is observed second, etc. The smallest sample value is the first-order statistic, the second smallest value is the second-order statistic, etc., and the largest value is the nth-order statistic; these are denoted by $X_{(1)}, X_{(2)}, \ldots, X_{(n)}$. For example, if n is odd, the sample median $X_{((n+1)/2)}$ is such an order statistic. If a sample is singly censored, only the first r order statistics $X_{(1)}, X_{(2)}, \ldots, X_{(r)}$ are observed.

Suppose that random samples of size n are taken from the same distribution. Then the ith-order statistic has different values in the different samples. That is, it has a sampling distribution, which depends on the population distribution, i, and n. Order statistics can be used to estimate the parameters and other characteristics of the population.

Linear Methods

A linear combination of K order statistics of a sample with Type II censoring is called a *systematic statistic*. It has the form

$$a_{i_1} X_{(i_1)} + a_{i_2} X_{(i_2)} + \cdots + a_{i_K} X_{(i_K)}, \qquad (B.57)$$

this contains any selected order statistics and known coefficients a_{i_K}. For example, if the sample is singly right censored, a linear combination of the r smallest order statistics has the form

$$a_1 X_{(1)} + a_2 X_{(2)} + \cdots + a_r X_{(r)}. \qquad (B.58)$$

The coefficients are chosen so that the linear combination is a good estimator for some parameter of the parent distribution. The coefficients depend on the parent distribution (exponential, normal, etc.), the quantity being estimated (mean, standard deviation, percentile, etc.), the size of the sample, the order statistics being used, and the properties of the estimator (unbiased, invariant, etc.).

A linear estimator θ^* is called a *best linear unbiased estimator* (BLUE) for a parameter θ if it is *unbiased* [its mean $E\theta^* = \theta$] and has minimum variance among linear unbiased estimators. A linear estimator θ^{**} is called a *best linear invariant estimator* (BLIE) for a parameter θ if it has a minimum mean squared error, $E(\theta^{**} - \theta)^2$, among linear estimators. There are tables of coefficients for both types of estimators, and one type may be obtained from the other. The two estimators differ little for practical purposes, compared to the scatter in the data. The choice of either estimator is mostly a matter of taste. The BLUEs are presented below. The BLIEs are used the same way, only the coefficients differ (Mann et al., 1974; Bain, 1978).

The linear methods below are exact only for data with Type II (failure) censoring. However, they are often used for data with Type I (time) censoring. They usually provide a satisfactory approximation for practical purposes.

BLUEs have good properties compared to other estimators, such as maximum likelihood ones. For small samples, their variances and mean squared errors are comparable to those of other estimators. For large samples, their asymptotic variances and mean squared errors usually equal the theoretical minimum variance (Cramer–Rao lower bound) for any estimators.

Linear estimators can readily be derived for two-parameter distributions that have a *scale parameter* σ and a *location parameter* μ. That is, the cumulative distribution function $F(x; \mu, \sigma)$ can be written as $F(x; \mu, \sigma) = G[(x - \mu)/\sigma]$, where $G(\ )$ is a function that does not depend on μ or σ. The normal and smallest extreme value distributions are such distributions. Linear estimators are used with the logs of lognormal and Weibull data, since the log data come from a normal and extreme value distribution, respectively. Then the true variances and covariance of the BLUEs μ^* and σ^* for the location and scale parameters have the form

$$\text{Var}(\mu^*) = A\sigma^2, \qquad \text{Var}(\sigma^*) = B\sigma^2, \qquad \text{Cov}(\mu^*, \sigma^*) = C\sigma^2, \qquad (B.59)$$

where A, B, and C depend on r and n and the distribution, but not on μ and σ. This is sometimes indicated with subscripts on the factors $A_{n,r}$, $B_{n,r}$, and $C_{n,r}$. Such factors for pooled estimates lack subscripts. Linear estimators can readily be derived for one-parameter distributions that have a scale parameter or a location parameter. The exponential distribution is such a distribution.

Distribution of an Order Statistic

A knowledge of how data (failed units) are distributed as a function of life is needed—more specifically, the probability of any unit failing within a given time interval.

Here we derive the distribution of an order statistic from a continuous parent distribution. Suppose random samples of size n are repeatedly taken from a continuous cumulative distribution $F(x)$. In each sample, the ith-order statistic X_i has a different value. So X_i has a distribution. This distribution depends on $F(x)$ and on n and i. We first derive $G_i(x) = P\{X_i \le x\}$, the cumulative distribution of X_i. The event $\{X_i \le x\}$ is equivalent to the event of that at least i observations fall below x. There are n independent observations and each has a probability $F(x)$ of falling below x. So $P(X_i \le x)$ is the binomial probability of i or more observations below x, namely,

$$G_i(x) = \sum_{k=i}^{n} \binom{n}{k} [F(x)]^k [1 - F(x)]^{n-k}$$

$$= 1 - \sum_{k=0}^{i-1} \binom{n}{k} [F(x)]^k [1 - F(x)]^{n-k}. \qquad (B.60)$$

Suppose the parent distribution has a probability density $f(x)$. Then the probability density of X_i is obtained from differentiative $G_i(x)$

$$g_i(x) = \frac{n!}{(i - 1)! \, 1! \, (n - i)!} [F(x)]^{i-1} f(x) [1 - F(x)]^{n-i}. \qquad (B.61)$$

The first term on the right is the reciprocal of the complete beta function $(n - i + 1, i)$ and the second term is the partial beta function evaluated at the median of the ith-order distribution. Figure B.10 motivates Eq. (B.61) as follows. The event that X_i is in the infinitesimal interval x to $x + dx$ corresponds to the event that $i - 1$ observations fall below x, one falls between x and $x + dx$, and $n - i$ fall above $x + dx$. Each of the n independent observations can fall into one of these three intervals with the appropriate probabilities $F(x)$, $f(x)\, dx$, and $1 - F(x)$. The trinomial probability of the above event

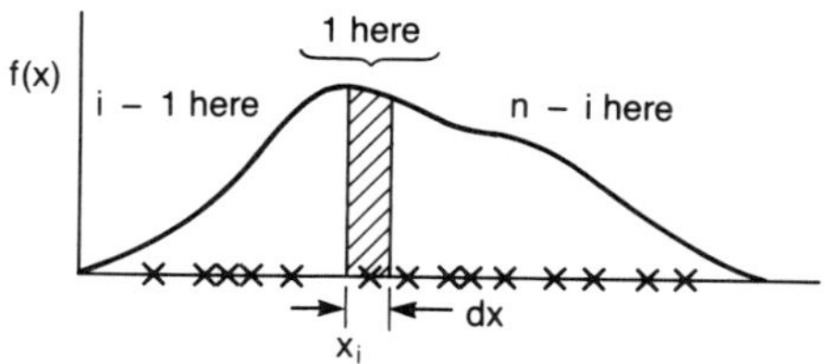

Fig. B.10. Schematic of the probability density function of order statistic X_i.

is approximately

$$g_i(x)\,dx \sim \frac{n!}{(i-1)!\,1!\,(n-i)!}[F(x)]^{i-1}[f(x)\,dx]^1[1-F(x)]^{n-i}. \quad \text{(B.62)}$$

Figure B.11 depicts the density of a parent distribution and those of the order statistic of a sample of size 4.

B.3.3.3. Maximum Likelihood Analyses for Complete and Censored Data

The maximum likelihood (ML) methods are very important in life data analysis and elsewhere because they are very versatile. That is, they apply to most theoretical distributions and kinds of censored data. The methods apply to multiple time-censored data (Type I), multiple-failure data (Type II), to Types I and II simply-censored data, and to complete data.

In principle, the ML method is simple. We first write the sample likelihood (or its logarithm, the log likelihood). It is a function of the assumed distribution, the distribution parameters, and the data (including the censoring or other form of the data). The ML estimates of the parameters are the parameter values that maximize the sample likelihood or, equivalently, the log likelihood. The exact distributions of many ML estimators and confidence limits are not known. However, they are given approximately by the asymptotic (large-sample) theory, which involves the asymptotic covariance and Fisher information matrices of the ML estimates. For the asymptotic theory to be a good approximation, the number of failures in the sample should be *large*. How large depends on the distribution, what is being estimated, the confi-

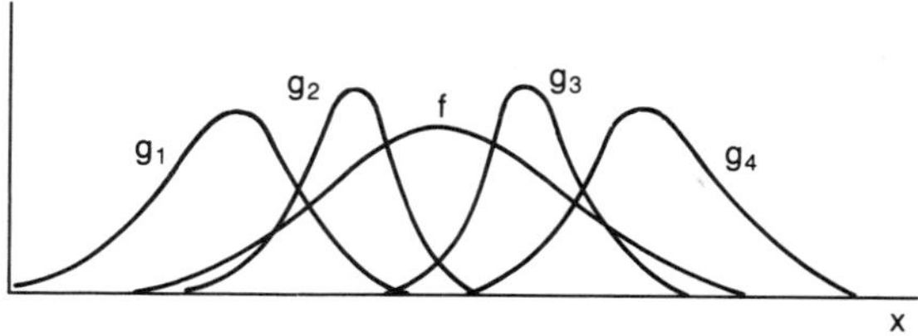

Fig. B.11. Schematics of the parent f and order static density functions g_i.

dence level of limits, etc. For practical purposes, the asymptotic methods are applied to small samples, since crude theory is better than no theory.

Like other methods for the analysis of multiply censored data, ML methods depend on a basic assumption. It is assumed that units censored at any specific time come from the same life distribution as the units that run beyond that time. This assumption does not hold, for example, if units are removed from service unfailed when they look like they are about to fail.

Computer programs are commercially available for the ML calculations.

B.3.4. Comparisons for Complete Data

Methods of comparing the parameters of distributions with specified values include hypothesis tests and confidence intervals for one or more samples. In real life, distribution parameters are not known, and we must take action on the basis of sample data. We then want convincing evidence in the data that a particular action is appropriate. For example, observed differences between a sample estimate for a parameter and a specified value or between sample estimates should be greater than normal random variations in such estimates; then it is convincing that the observed differences are due to real differences in true parameter values. Many comparisons can also be made with confidence intervals. Such intervals are usually equivalent to, but more informative than, a corresponding hypothesis test. We present here two practical examples (Nelson, 1982).

B.3.4.1. Analysis of Reliability Demonstration Test

In reliability demonstration testing, a product must demonstrate that its reliability, mean life, failure rate, or whatever is better than a specified value. A reliability demonstration test for equipment usually requires that a specified mean θ_0 (or failure rate h_0) be demonstrated with $100\alpha\%$ confidence. This means that the equipment passes the test if the observed lower $100\alpha\%$ confidence limit $\underset{\sim}{\theta}$ for the true θ is above θ_0. Otherwise the equipment fails. Equivalently, this means that the total running time T before the r_0th failure occurs (Type II censoring) must exceed T_0. The values of r_0 and T_0 are chosen so that the probability of passing the test when $\theta = \theta_0$ is

$$P_{\theta_0}\{T \geq T_0\} = 1 - \alpha. \tag{B.63}$$

This low probability $1 - \alpha$ is called the *consumer's risk*. So the true mean θ must be above θ_0 if the equipment is to have a high probability of passing the test. $100\alpha\%$ is the *confidence level* of the demonstration test.

Any number n of units may be run in such a test. Of course, the larger n is, the sooner the r_0 failures and total time T_0 are accumulated. For an exponential distribution, the total running time for a test with a multiple test unit is simply obtained by summing over all test units since the hazard rate is constant (see Section B.4 for more details).

B.3.4.2. Acceptance Sampling Plan

An acceptance sampling plan is used to decide whether a population, such as a manufactured lot or shipment, is acceptable. When units are categorized as good or bad, the plan usually states that the lot must have an acceptable quality level (AQL) of p_0, the fraction defective. Binomial data are usually called attribute data in acceptance sampling.

An *acceptance sampling plan* specifies the number n of sample units and the *acceptance number* x_0. If a consumer finds more than x_0 defectives in a sample, the shipment fails. The n and x_0 are chosen so the plan accepts most "good" shipments and rejects more "poor" ones. A plan had $n = 20$ and $x_0 = 1$. If a shipment (the binomial samples) has a defective proportion of $p = 0.01$, the chance of it passing inspection is $F(1) = 0.983$, from a binomial table. If a shipment has a defective proportion of $p = 0.10$, the chance of it passing inspection is $F(1) = 0.392$. When the true defective proportion is p, the probability of a lot passing the test with n and x_0 is

$$P_p\{X \le x_0\} = \sum_{X=0}^{x_0} \binom{n}{x} p^x (1 - p)^{n-x}. \tag{B.64}$$

In practice, n and x_0 are chosen so that the plan accepts lots with a low defective proportion p_0 with high probability $1 - \alpha$ and accepts lots with a high defective proportion p_1 with a low probability β. p_0 is called the *acceptable quality level* (AQL). The probability of such an acceptable lot failing is α and is called the *producer's risk*. The high proportion p_1 is called the *lot tolerance percent defective* (LTPD); it is an unacceptable quality level. The probability of such an unsatisfactory lot passing is β and is called the *consumer's risk*. For the plan above, if the AQL is 1% defective, the producer's risk is $1 - 0.980 = 0.020$, or 2%. Also, if the LTPD is 10%, the consumer's risk is 0.392, or 39%.

B.4. Weibull Analysis of Failure Data

Many failure data can be adequately represented by the Weibull distribution (Brunk, 1965; Meyer, 1973; Mann et al., 1974; Tsokos and Shimi, 1977; Nelson, 1982). The Weibull distribution was proposed by Waloddi Weibull (1951) and was originally associated with a fatigue failure mode of metallic rolling-element bearings (Lundberg and Palmgren, 1947, 1951). The Weibull distribution is now widely used to predict the failure lives of various machine components which fail by wear-out or aging, e.g., magnetic heads (Bhushan, 1990).

There are two forms of the Weibull distribution in current use, distinguished by the appearance of either two or three parameters in the mathematical expression for the distribution. We will first describe the two- and three-parameter forms of the Weibull distribution, and illustrate a number of computational procedures that will answer some questions that frequently arise in working with Weibull distributed random phenomena, when the parameters are assumed to be known. Second, we will examine methods for using the results of life tests to estimate the parameters or functions of the parameters when they are unknown.

B.4.1. General Properties of the Weibull Distribution

There are two forms of the Weibull distribution distinguished by the appearance of either two or three parameters in the mathematical expression for the distribution. Under the two-parameter Weibull failure model the probability that a component's life (X) is less than some specified value x is expressible as

$$P\{\text{life} \le x\} = F(x) = 1 - \exp\{-(x/\eta)^{\beta}\}, \qquad x > 0. \qquad \text{(B.65a)}$$

Weibull distribution parameters are sometimes expressed differently,

$$F(x) = 1 - \exp\{-\alpha x^{\beta}\}, \qquad \text{(B.65b)}$$

where $\alpha = 1/\eta^{\beta}$. $F(x)$ is called the *cumulative distribution function* or simply the *distribution function*. The $F(x)$ gives the summation of failure proportion to that life x. The quantities η and β are parameters. The parameter η has the same units as those in which life is measured (e.g., hours, cycles); β is dimensionless. For a given β value, plots of $F(x)$ versus x for two different η values can be drawn so as to differ only with respect to the scale of the life axis. η is therefore often called the *scale parameter* of the Weibull distribution. *Characteristic life* is another frequently applied name for the parameter η because when $x = \eta$, $F(x) = 1 - /e \sim 0.632$, irrespective of β. For any other value of x, $F(x)$ depends upon both η and β.

For the same η value, plots of $F(x)$ against x for two different values of β will differ in shape. β is therefore often called the Weibull *shape parameter*. Another frequent name for β is the *Weibull slope*, because, as discussed later, a plot of $F(x)$ against x is drawn on graph paper specially constructed so that the plot becomes a straight line, β is the slope of that line. Finally, β is sometimes called the dispersion parameter because it characterizes the degree of scatter or spread of the Weibull distribution.

The three-parameter Weibull distribution is obtained by shifting a distribution that is defined for outcomes between 0 and $+\infty$, so that the distribution starts at a value different from 0. Therefore, it is sometimes called the shifted Weibull distribution.

With the three-parameter Weibull distribution, $F(x)$ is written,

$$F(x) = 1 - \exp\{-[(x - \gamma)/\eta]^{\beta}\}; \qquad x > \gamma. \qquad \text{(B.66)}$$

γ has the same units as used to measure life, and represents the earliest life at which failure can occur. γ is known as the *location parameter*, the *threshold parameter*, or the *guarantee life*. Obviously, the two-parameter Weibull distribution is a special case of the three-parameter version, in which $\gamma = 0$.

The probability that life exceeds x is just the complement of the probability that life is less than x. The probability of no failure or reliability at life x, $R(x)$, is given by

$$R(x) = 1 - F(x) = \exp\{-[(x - \gamma)/\eta]^{\beta}\}. \tag{B.67}$$

$F(x)$ may be interpreted as the area under a curve $f(x)$ between the point γ and a general life x. This curve is called the Weibull probability density function. The $f(x)$ is the derivative of $F(x)$. For a three-parameter Weibull distribution, the expression for $f(x)$ is

$$f(x) = \frac{dF(x)}{dx} = (\beta/\eta)[(x - \gamma)/\eta]^{\beta-1} \exp\{-[(x - \gamma)/\eta]^{\beta}\}, \qquad x > \gamma, \tag{B.68a}$$

or

$$F(x) = \int_{\gamma}^{x} f(y)\, dy, \qquad x > \gamma. \tag{B.68b}$$

The $f(x)$ gives the relative frequency of failure at a life x. It is the mathematical model for the histogram of failure events.

The instantaneous failure rate or hazard function, $h(x)$, is the probability of failure per unit life at life x. The $h(x)$ is given by the proportion of failures in a life period Δx, of those components which were good at the beginning of Δx, divided by the period Δx. For a three-parameter Weibull distribution

$$h(x) = f(x)/R(x)$$

$$= (\beta/\eta)[(x - \gamma)/\eta]^{\beta-1}, \qquad x > \gamma. \tag{B.69}$$

We note that if $\beta = 1.0$ (exponential distribution) the hazard is constant, independent of x. Such a situation might occur where failure is caused by a random external event that is equally destructive to new and old units. When $\beta > 1.0$ the hazard increases uniformly with x and older items, i.e., items that have been in operation longer are more prone to failure than newer items ("wear-out"). When $\beta < 1.0$ the hazard decreases with life implying reduced vulnerability with age ("infant mortality"). Certain electronic components exhibit this type of failure pattern wherein components which have survived an "infant mortality" period have a lower failure rate than new items.

The cumulative hazard function for a three-parameter Weibull distribution is given as

$$H(x) = \int_{\gamma}^{x} h(y)\, dy, \qquad x > \gamma,$$

$$= [(x - \gamma)/\eta]^{\beta}, \qquad x > \gamma. \tag{B.70}$$

In Fig. B.7, plots of the cumulative distribution function, the probability density function, and the instantaneous failure rates for a constant scale parameter η and varying shape factors β were presented.

B.4.1.1. 100Pth Percentile

The 100Pth percentile, 100P percent point, Pth quantile, or P fractile is the life x_P by which a proportion of population P has failed. The 100Pth percentile of the three-parameter Weibull distribution is obtained using Eq. (B.65)

$$P = F(x_P) = 1 - \exp\{-[(x_P - \gamma)/\eta]^\beta\}$$

or

$$x_P = [-\ln(1 - P)]^{1/\beta}\eta + \gamma. \tag{B.71a}$$

A two-parameter Weibull distribution is completely determined if the value of the shape parameter and one percentile is known; because by definition x_P is the life that is exceeded with probability $100(1 - P)\%$. It is sometimes called the $100(1 - P)\%$ reliable life. In general, x_P is called the $100(1 - P)\%$ reliable life. Rolling bearings are rated by their tenth quantile life $x_{0.10}$. $x_{0.10}$ may be written as, from Eq. (B.71a),

$$x_{0.10} = (0.10536)^{1/\beta}\eta + \gamma. \tag{B.71b}$$

The relationship between two percentiles, say the $100P$th and $100Q$th, is expressed as

$$(x_P - \gamma)/(x_Q - \gamma) = \{[\ln(1 - P)]/[\ln(1 - Q)]\}^{1/\beta}. \tag{B.72}$$

B.4.1.2. Mode

The mode is the value where the probability density is a maximum. It is the "most likely" time-to-failure and it may be regarded as "typical" life. For a three-parameter Weibull distribution, the mode is given as

$$\left.\begin{aligned} x_{\mathrm{mode}} &= \eta[1 - (1/\beta)]^{1/\beta} \quad \text{for } \beta > 1, \\ &= 0 \qquad\qquad\qquad \text{for } 0 < \beta \le 1. \end{aligned}\right\} \tag{B.73}$$

B.4.1.3. Mean and Variance

The mean $E(X)$ [or expected life or mean-time-between-failure (MTBF) or mean-time-to-failure (MTTF)] of a variable X is the arithmetic average of the lives of all units in a population, and it is used as still another "typical" life. For a three-parameter Weibull distribution, the mean is given as

$$x_{\mathrm{mean}} = E(X) = \gamma + \eta\Gamma[1 + (1/\beta)], \tag{B.74}$$

where $\Gamma(\ )$ is the gamma function

$$\Gamma(y) = \int_0^\infty z^{y-1}\exp(-z)\,dz, \tag{B.75a}$$

and for integer y

$$\Gamma(y) = (y - 1)! \tag{B.75b}$$

A brief table of $\Gamma[1 + (1/\beta)]$ is given in Table B.5 (Abramowitz and Stegun, 1965; Harter and Dubey, 1967).

Another useful characteristic of probability distributions is the variance σ^2 (σ = standard deviation), defined as the expected squared departure of x from

Table B.5. Values of $\Gamma(1 + (1/\beta))$ and $\{\Gamma[1 + (2/\beta)] - \Gamma^2[1 + (1/\beta)]\}$

β	$\Gamma[1 + (1/\beta)]$	$\{\Gamma[1 + (2/\beta)] - \Gamma^2[1 + (1/\beta)]\}$
1.0	1.0	1.0
1.1	0.9649	0.7714
1.2	0.9407	0.6197
1.3	0.9236	0.5133
1.4	0.9114	0.4351
1.5	0.9027	0.3757
1.6	0.8966	0.3292
1.7	0.8922	0.2919
1.8	0.8893	0.2614
1.9	0.8874	0.2360
2.0	0.8862	0.2146
2.5	0.8873	0.1441
3.0	0.8930	0.1053
3.5	0.8997	0.0811
4.0	0.9064	0.0647
5.0	0.9182	0.0442

$E(X)$. The variance is a measure of scatter or dispersion. It describes the extent to which the distribution clusters around $E(X)$. For the Weibull distribution the variance is

$$\text{Var}(X) = \sigma^2 = \eta^2\{\Gamma[1 + (2/\beta)] - \Gamma^2[1 + (1/\beta)]\}. \tag{B.76}$$

Values of $\{\Gamma[1 + (2/\beta)] - \Gamma^2[1 + (1/\beta)]\}$ are listed in Table B.5. It is seen from Table B.5 that for a fixed value of η, σ^2 decreases uniformly with increasing β. β itself is therefore more customarily used than the variance to characterize the dispersion of the Weibull distribution.

B.4.1.4. Example

An item is drawn from a two-parameter Weibull population having a shape parameter $\beta = 1.3$ and a scale parameter $\eta = 50.0$ h.

(a) Find the probability that the item fails before $x = 25$ h.
From Eq. (B.65)

$$P\{\text{life} < 25.0\} = 1 - \exp[-(25/50)^{1\cdot3}]$$

$$= 0.334.$$

(b) Calculate the $x_{0.10}$ value for the Weibull population.
From Eq. (B.71b)

$$x_{0.10} = (0.10536)^{1/1\cdot3}50.0 = 8.9.$$

(c) Calculate the MTBF and the variance for the Weibull population.
From Table B.5 for $\beta = 1.3$, $\Gamma[1 + (1/\beta)] = 0.9236$ and $\{\Gamma[1 + (2/\beta)] -$

$\Gamma^2[1 + (1/\beta)]\} = 0.5133$. From Eq. (B.74)

$$MTBF = 50.0 \times 0.9236 = 46.2.$$

From Eq. (B.76)

$$\sigma^2 = (50)^2 \times 0.5133 = 1290.$$

B.4.2. Graphical Representation of a Weibull Distribution

B.4.2.1. Cumulative Distribution Function (c.d.f.)

Taking natural logarithms twice on both sides of Eq. (B.66) gives

$$\log_{10}\{-\ln[1 - F(x)]\} = \beta \log_{10}(x - \gamma) - \beta \log_{10} \eta. \qquad \text{(B.77)}$$

This may be written in the form of an equation for a straight line

$$y = \beta t + c, \qquad \text{(B.78a)}$$

where

$$y \equiv \log_{10}\{-\ln[1 - F(x)]\}, \qquad \text{(B.78b)}$$

$$t \equiv \log_{10}(x - \gamma), \qquad \text{(B.78c)}$$

and

$$c \equiv -\beta \log_{10} \eta. \qquad \text{(B.78d)}$$

Thus $\log_{10}\{-\ln[1 - F(x)]\}$ varies linearly with $\log_{10}(x - \gamma)$ with a slope equal to the shape parameter (β). Graph paper constructed with life in excess of the location parameter (i.e., $x - \gamma$) on a logarithmic scale on the abscissa and probability of failure $F(x)$ on the ordinate with the values of $F(x)$ located at $\log_{10}\{-\ln[1 - F(x)]\}$ will have the property that a Weibull population will plot thereon as a straight line, Fig. B.12. {For example, $F = 0.01$ is at $\log_{10}[-\ln(1 - 0.01)] \sim -2$; $F = 0.632$ at $\log_{10}[-\ln(1 - 0.632)] \sim 0$, and $F = 0.99$ at $\log_{10}[-\ln(1 - 0.99)] \sim 0.66$.} The scale parameter η is the 63.2th percentile. If the ordinate and abscissa logarithmic scales are chosen so that a cycle is the same length on both, then the Weibull shape parameter (β) will be the slope of the straight line, computed as the ratio of the vertical and horizontal distances measured between two points on the line. The shape parameter would indicate the type of failure mode. If the data fits into two or more straight lines, there may be more than one operative failure mode.

If the logarithmic scales do not have the same cycle lengths, an auxiliary scale is sometimes provided by which the shape parameter may be deduced from the slope of the straight line as plotted. An example of such paper is provided by Nelson (1967).

Example

The Weibull distribution having $\gamma = 0$, $x_{0.10} = 15 \times 10^6$ revolutions, and a shape parameter equal to 1.0 is plotted on Weibull paper in Fig. B.12. It is

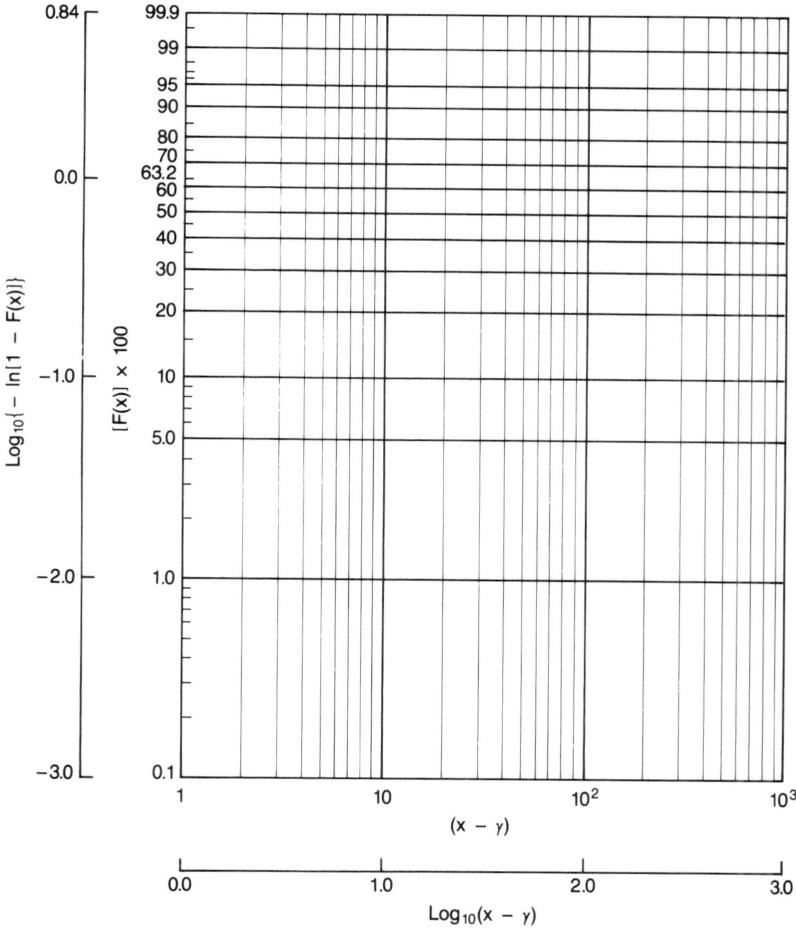

Fig. B.12. The Weibull paper.

constructed by locating the point associated with $F(x_{0.10}) = 0.10$ on the ordinate and $x_{0.10} = 15 \times 10^6$ revolutions on the abscissa and drawing a line of unit slope ($45°$ angle) through this point. Having drawn the line we may determine any other percentiles, or the probability of failing before any given life, at a glance, e.g., from Fig. B.13 the twentieth percentile, $x_{0.20} = 32.0 \times 10^6$ revolutions, and the probability of failing before 52×10^6 revolutions is 0.30.

B.4.2.2. Cumulative Hazard Function

Taking the logarithm of both sides of Eq. (B.70) gives

$$\log_{10}(x - \gamma) = (1/\beta) \log_{10}[H(x)] + \log_{10}(\eta). \tag{B.79}$$

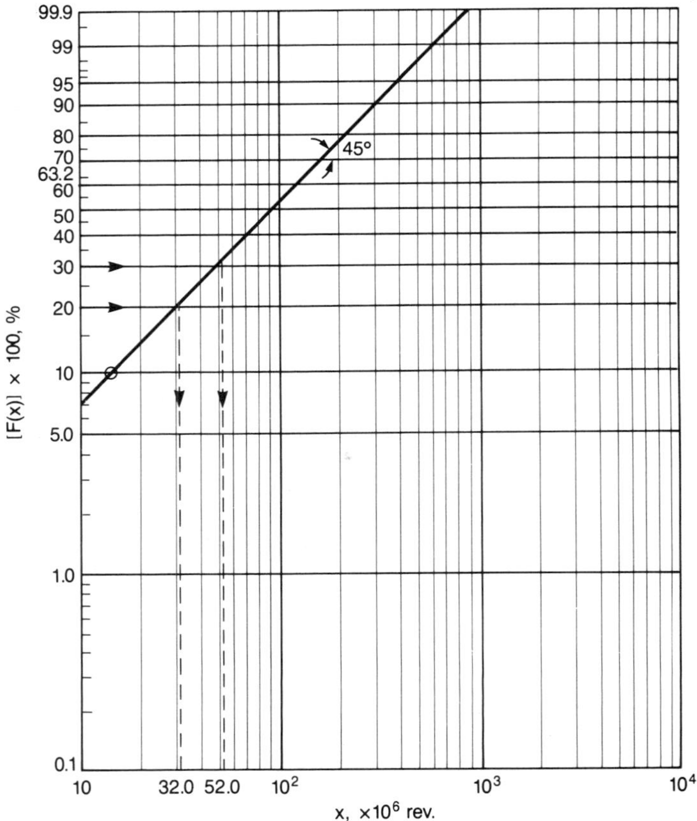

Fig. B.13. Graphical representation of the Weibull population (percent probability of failure as a function of life time) for uncensored random data, $\gamma = 0.0$, $\beta = 1.0$, $x_{0.10} = 15 \times 10^6$ rev.

This shows that $\log_{10}(x - \gamma)$ is a linear function of $\log_{10}[H(x)]$. Thus, Weibull hazard paper is a log-log graph paper with the cumulative hazard in percent plotted on the horizontal axis and life plotted on the vertical axis. The slope of the straight line equals $1/\beta$. For $H = 1$ (100%) (63.2% on the Weibull probability paper), the corrresponding life x equals η.

B.4.3. Conditional Weibull Distribution

It is sometimes necessary to calculate the probability that an item will fail within an additional time period x after it has already endured for some life $x_0 > 0$. This probability is expressed by the "conditional" distribution written $F(x|x_0)$, i.e., Prob[life $< x$ given that $x > x_0$] $= F(x|x_0)$.

When the underlying distribution is the three-parameter Weibull, the conditional distribution is

$$F(x|x_0) = \begin{cases} 0, & x < \gamma, \\ 1 - \exp\{-[(x - \gamma)/\eta]^\beta\}, & x_0 < \gamma, x > \gamma, \\ 1 - \exp\{[(x_0 - \gamma)/\eta]^\beta - [(x - \gamma)/\eta]^\beta\}, & x_0 > \gamma, \end{cases} \quad \text{(B.80)}$$

where β, η, and γ are the shape, scale, and location parameters of the original Weibull distribution. It is seen that when $x_0 = 0$ the conditional distribution reduces to the original unconditional form of Eq. (B.66).

When $0 \leq x_0 \leq \gamma$ the conditional distribution is still of Weibull form. For $x_0 > \gamma$ the conditional distribution is no longer of Weibull form. This means that groups of items formed from the survivors of an initial burn-in period $x_0 > \gamma$ will deviate from the Weibull model in their apparent failure behavior although the underlying population is truly Weibull.

Quantiles x_P^* of such previously run populations are found as before by setting $F(x_P^*|x_0) = P$, and solving for x_P^*,

$$x_P^* = [-\ln(1 - P_0)]^{1/\beta}\eta + \gamma, \qquad x_0 > \gamma, \quad \text{(B.81a)}$$

where

$$P_0 = 1 - (1 - P)\exp[-(x_0 - \gamma)/\eta]^\beta.$$

Alternatively,

$$x_P^* = (x_P - \gamma)\{1 + [(x_0 - \gamma)/(x_P - \gamma)]^\beta\}^{1/\beta} - (x_0 - \gamma). \quad \text{(B.81b)}$$

The conditional mean life X of units that reach an age x_0 is

$$E(X|x_0) = \eta\{\exp[(x_0 - \gamma)/\eta]^\beta\}\{\Gamma[1 + (1/\beta)]$$
$$- \Gamma^*\{[(x_0 - \gamma)/\eta]^\beta; 1 + (1/\beta)\}\}, \quad \text{(B.82)}$$

where $\Gamma^*(x, y) = [1/\Gamma(y)]\int_0^x z^{y-1}\exp(-z)\,dz$ is the incomplete gamma function. Note that $\Gamma^*(\infty, y) = \Gamma(y)$, the gamma function.

B.4.3.1. Example

A group of bearings is randomly selected from a Weibull population known (assumed) to have an $x_{0.10}$ life of 10.0×10^6 revolutions, a shape parameter β of 1.3, and a location parameter $\gamma = 0$. A new group consists of the members of this group which are still running after a burn-in period of $x_0 = 2.0 \times 10^6$ revolutions. Calculate the $x_{0.10}$ life of the population which this group represents.

Substituting in Eq. (B.81) with $x_P = x_{0.10}$ gives

$$x_{0.10}^* = \{10.0[1 + (2/10)^{1.3}]^{1/1.3} - 2.0\} \times 10^6 \text{ rev.} = 9.33 \times 10^6 \text{ rev.}$$

B.4.4. Distribution of Products Which Fail Due to Competing Failure Modes

Many products can fail for a number of reasons. Ball bearings, for example, can fail because of the failure of an inner ring, an outer ring, or a ball. Let $F_1(x)$, $F_2(x)$, and $F_3(x)$ denote for the inner, outer, and ball complement of a bearing, the probabilities of failure before a life x. The probability that the bearing as a whole survives x h is the probability that all of the three components survive. Assuming that the lives of the rings and ball set are independent of each other, the survival probabilities for each component may be multiplied to give the survival probability for the bearing.

$$P\{\text{Bearing survives a life } x\} = [1 - F_1(x)][1 - F_2(x)][1 - F_3(x)]. \quad (B.83)$$

$F(x)$, the probability that the bearing fails before life x, is just the complement of the probability that the bearing survives beyond x, hence

$$F(x) = 1 - [1 - F_1(x)][1 - F_2(x)][1 - F_3(x)]. \quad (B.84)$$

Suppose that the component failure distributions are each Weibull with shape parameters β_1, β_2, and β_3, scale parameters η_1, η_2, and η_3, and location parameters γ_1, γ_2, and γ_3, respectively. Then

$$F(x) = 1 - \exp\{-\{[(x - \gamma_1)/\eta_1]^{\beta_1} + [(x - \gamma_2)/\eta_2]^{\beta_2} + [(x - \gamma_3)/\eta_3]^{\beta_3}\}\}. \quad (B.85)$$

This, in general, is not a Weibull distribution. If the γ's and β's are all equal, it becomes the Weibull distribution with a scale parameter η_e. In this case,

$$F(x) = 1 - \exp\{-[(x - \gamma)/\eta_e]^{\beta}\}, \quad (B.86a)$$

where

$$\gamma = \gamma_1 = \gamma_2 = \gamma_3, \qquad \beta = \beta_1 = \beta_2 = \beta_3,$$

and

$$\eta_e = \left\{\frac{1}{\eta_1^{\beta}} + \frac{1}{\eta_2^{\beta}} + \frac{1}{\eta_3^{\beta}}\right\}^{-1/\beta}. \quad (B.86b)$$

The system hazard function is given by

$$h(x) = \beta(x - \gamma)^{\beta-1}/\eta_e^{\beta}. \quad (B.87)$$

Similarly, the percentile x_P for the distribution of bearing life is

$$x_P = \{1/x_{P_1}^{\beta} + 1/x_{P_2}^{\beta} + 1/x_{P_3}^{\beta}\}^{-1/\beta}. \quad (B.88)$$

These results generalize for any number of *independent* failure modes meeting the restrictions on equality of the shape and location parameters.

The independence restriction means that the probability distribution of the life at which an item will fail due to one mode is not influenced by the life at which the item will fail due to another mode. If, for example, bearing inner rings, outer rings, and balls were all made from the same physical piece of steel, the independence assumption would not hold, since a particularly good piece of steel would result in long lives for each of the components of a given bearing. Knowing that a given bearing has a long inner ring life would then lower the chances that the outer ring or ball complement would have a low life. As it happens, bearings are randomly assembled from separate production runs of components, so the independence assumption is reasonable.

B.4.4.1. Example

Calculate the $x_{0.10}$ life for a population of bearings assembled from inner ring, outer ring, and ball components, whose failure lives are Weibull distributed with a common shape parameter of $\beta = 1.3$, a common location parameter of $\gamma = 0.0$, and with respective $x_{0.10}$ lives of

$$(x_{0.10})_1 = 100.0,$$

$$(x_{0.10})_2 = 150.0,$$

$$(x_{0.10})_3 = 1000.0,$$

$$x_{0.10} = [(1/100.0)^{1.3} + (1/150.0)^{1.3} + (1/1000.0)^{1.3}]^{-1/1.3}$$

$$= 68.3.$$

The bearing $x_{0.10}$ life is lower than the $x_{0.10}$ life of any of the components. It is noted that improving the life of the ball component would have relatively little effect on the $x_{0.10}$ life of the bearing. The most improvement in bearing life is brought about by improving the life of the component with the smallest $x_{0.10}$.

B.4.5. Quantiles for Order Statistics of Weibull Samples

Order statistics are the individual values in a random sample, after they are sorted in ascending sequence, e.g., the first-order statistic is the smallest value in the sample and the second-order statistic is the second smallest value. Order statistics are random variables varying from sample to sample in accordance with a probability distribution that depends on the underlying parent distribution, the sample size, and the order number under consideration.

It is sometimes useful to be able to calculate an interval in which a given order statistic will fall with high probability. For example, if we wished to obtain five failed specimens for metallurgical investigation, and put ten specimens on test, the waiting time to completion of the test is the fifth-order statistic in the sample of size 10, designated $x_{5,10}$.

A 90% probability interval for the rth-order statistic in a sample of size n from a distribution is calculated as follows (Wilks, 1962)

$$F^{-1}[X(0.05, r, n - r + 1)] < x_{r,n} < F^{-1}[X(0.95, r, n - r + 1)], \quad \text{(B.89)}$$

where $F^{-1}(\ )$ is the inverse of the parent distribution function and $X(P, a, b)$

Table B.6. Values of
$X(0.05, r, n - r + 1)$ and
$X(0.95, r, n - r + 1)$ for $n = 5$

r	$X(0.05, r, 6 - r)$	$X(0.95, r, 6 - r)$
1	0.01021	0.45072
2	0.07644	0.65741
3	0.18926	0.81074
4	0.34259	0.92356
5	0.54928	0.98979

denotes the Pth quantile of the beta distribution having parameters a, b. Values of $X(P, a, b)$ are given in a table by Harter (1964). Values taken from Harter (1964), applicable for a sample size $n = 5$, are given in Table B.6.

Calculation of the above interval may be carried out graphically on a probability plot by entering the values of $X(P, a, b)$ on the probability ordinate and reading the interval values as the associated abscissa values (Mc-Cool, 1969).

It may also be performed analytically by using the mathematical expression for the inverse of the distribution function. For the Weibull distriubtion this yields

$$\gamma + \eta\{\ln\{1/[1 - X(0.05, r, n - r + 1)]\}\} < x_{r,n}$$

$$< \eta\{\ln\{1/[1 - X(0.95, r, n - r + 1)]\}\} + \gamma. \quad (B.90)$$

B.4.5.1. Example

Calculate a 90% interval for the third-order statistic in a sample of size 5 drawn from the two-parameter Weibull population having $\beta = 1.0$ and $\eta = 100$.

This population line is shown graphically in Fig. B.14. From Table B.6 for $r = 3$, we have

$$X(0.05, 3, 3) = 0.18926$$

and

$$X(0.95, 3, 3) = 0.81074.$$

Entering Fig. B.14 with these values on the ordinate we read the associated interval

$$21 < x_{3,5} < 166.$$

This is readily confirmed arithmetically. The upper limit of expression (B.90), for example, is calculated as

$$100\{\ln[1/(1 - 0.81074)]\} = 166.$$

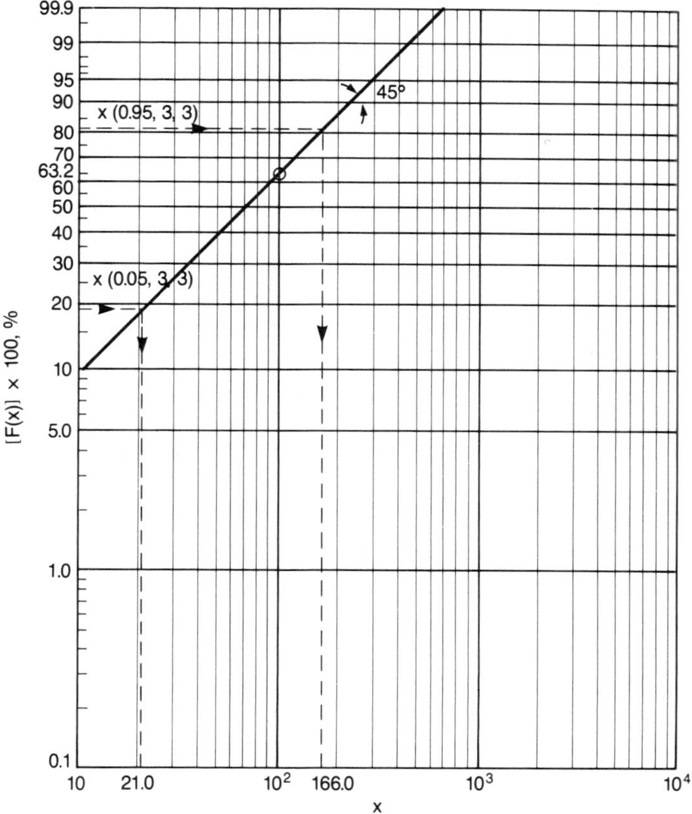

Fig. B.14. Graphical calculation of $x_{0.10}$ life for $x_{3,5}$ for a Weibull population, $\gamma = 0.0$, $\beta = 1.0$, and $\eta = 100$.

B.4.6. Evaluation of Weibull Parameters

All of the calculations described in the preceding sections are applicable where the Weibull parameters are known or assumed. Subsequent sections are concerned with the problem of evaluating one or more parameters or some functions of the parameters from a data sample, i.e., the results of an endurance test or field data.

A distinction is made between estimate and confidence limits (or intervals). A point estimate, or simply estimate, is a single numerical value computed from the observations in a data sample in such a way as to be a good guess at what the true but unknown parameter value is. Estimates, being functions of the values in a random sample, are themselves random variables and will vary in successive random samples drawn from the same distribution in accordance with a probability distribution known as the sampling distribution of the estimate. The confidence limit (or interval) or interval estimate consists of two numbers such that we are highly confident that the true parameter value is contained between them. The degree of confidence is expressed as the proportion of times that such statements would be true in an indefinitely large number of similar calculations performed on other independent random samples.

Many methods are available for evaluating the parameters of the Weibull distribution [Johnson and Kotz (1970); Mann et al. (1974)], each having its advantages and disadvantages. Standard methods are used for complete data which are simple but crude methods. The method of maximum likelihood (ML) is applicable to arbitrary censoring, and studies show it to be as precise as any other method in existence. Implementation of the method of maximum likelihood involves the solution of a nonlinear transcendental equation that would generally be regarded as too difficult for hand calculations when more than two failures have occurred. It is, however, readily programmed for implementation on a computer.

Graphical methods of parameter estimation are applicable to all kinds of censoring. They are not inherently as precise as many other methods and, moreover, are to a large extent subjective. There are no graphical techniques which yield confidence limits for the parameters. Graphical methods are, however, simple to employ and, unlike any other methods, provide valuable insight into data discrepancies (outliers) and possible non-Weibull behavior that no other method does. In view of the above, a reasonable approach to the analysis of Weibull data is to prepare a graphical estimate for a subjective appraisal of the validity of the data and the applicability of the Weibull model, and then use the method of maximum likelihood for the calculation of the estimate and confidence limits.

B.4.6.1. Graphical Estimation

Complete Data

We have seen that the Weibull distribution function with known parameters plots as a straight line against $x - \gamma$ on graph paper having a special set of grids. The essence of the graphical method of parameter estimation is to use the sample data to construct an estimate of $F(x)$ and hence its parameters using such graph paper.

We first order the n failure times from smallest to largest. Let $x_1 < x_2 < \cdots < x_n$ denote the ordered observations in a complete sample of size n. Then, even though $F(x)$ is not known, we may write, since $F(x)$ is a nondecreasing function of x, $F(x_1) < F(x_2) < \cdots < F(x_n)$. That is, the quantities $F(x_i)$ are an ordered set of random variables. If we took many samples, $F(x_i)$ would vary from sample to sample in accordance with a probability distribution. It happens that this probability distribution depends on i and n, but not on $F(x)$. Many choices have been advocated to estimate the plotting position on the $F(x)$ axis. In studies described by McCool (1970), it is shown that the median

Table B.7. Uncensored random sample of size $n = 10$ and calculated median ranks

Life, h	Failure rank (i)	Median rank = $(i - 0.3)/(n + 0.4)$
14.01	1	0.06731
15.38	2	0.16346
20.94	3	0.25962
29.44	4	0.35577
31.15	5	0.45192
36.72	6	0.54808
40.32	7	0.64423
48.61	8	0.74038
56.42	9	0.83654
56.97	10	0.93269

of $F(x_i)$, often termed the median rank, gave acceptably good average performance in 1000 simulated trials. Its use is therefore recommended on this basis. Tables of median ranks are contained in Johnson (1964). An excellent approximation to the median ranks offered by Johnson (1964) is $(i - 0.3)/(n + 0.4)$.

Table B.7 contains a sorted complete random sample of size $n = 10$. Also shown in Table B.7 are the median ranks correct to five decimal places which were calculated using the approximation $(i - 0.3)/(n + 0.4)$. Figure B.15 shows the values of the median ranks plotted against the ordered lives. The straight line fitted to the points in Fig. B.15 is the graphical estimate of the population line for $\gamma = 0$ (a two-parameter Weibull distribution). It yields an estimate of 1.55 for the shape parameter, 15.0 for $x_{0.10}$, and 39.0 h for the scale parameter.

In fitting a line to the data points we should focus on the departures in the horizontal direction, i.e., along the life axis rather than in the vertical direction (the probability axis) in subjectively determining the best fitting line. This is because the lives are random while the plotting positions are fixed for a given sample size. The points in another sample of size 10 would have the same vertical positions as those in Fig. B.15 but different horizontal positions along the life axis.

If the location parameter γ is not known, it is possible to estimate it graphically, by iteratively assuming a value of γ, say $\hat{\gamma}$, constructing a probability plot, and then revising $\hat{\gamma}$ and replotting until there is no obvious systematic curvature in the plot. When the value of $\hat{\gamma}$ is too small, the Weibull plot is concave downward, i.e., concave when viewed from below. When it is too large, it is concave upward.

Singly-Censored Data

When the censoring is Type I or Type II singly-censoring, probability plotting can proceed as above except that the lives associated with the unfailed elements are unknown and hence cannot be plotted. We can plot the failures that

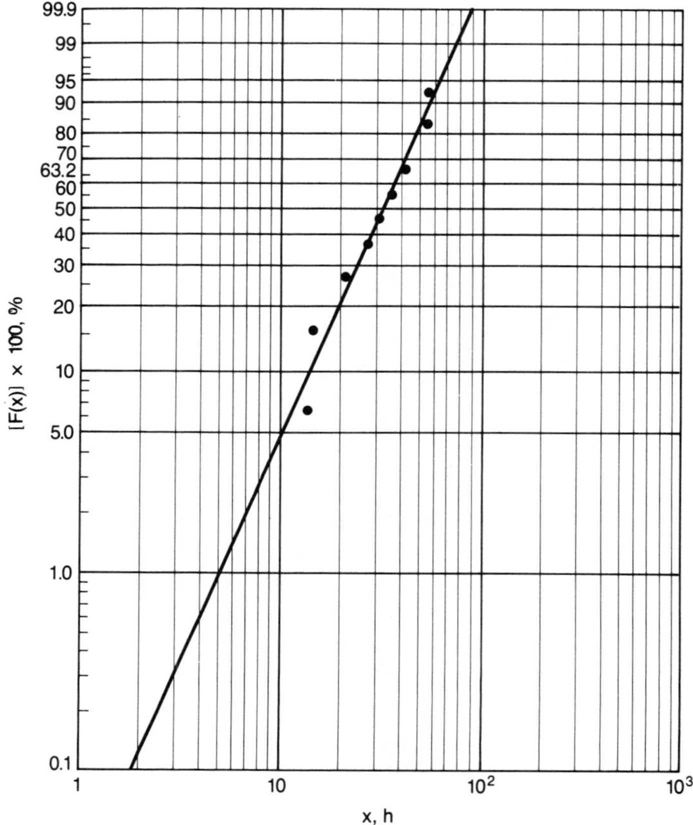

Fig. B.15. Weibull plot for an uncensored random sample of size $n = 10$. From Weibull population $\gamma = 0.0$, $\beta = 1.55$, and $\eta = 39$ h.

have occurred using their order numbers in the complete sample, and perform the fitting operation using just these points. Here, n represents the total sample size including the nonfailures.

Table B.8 shows the sorted lives obtained in a life test of seven units. A "+" marks the running time on each unfailed unit. Plotting positions appear in Table B.8, obtained using the approximation $(i - 0.3)/(n + 0.4)$. The data for failed units are plotted on a Weibull paper, Fig. B.16. The fitted straight line represents the graphical estimate of the two-parameter Weibull distribution of the life times. The estimate of the shape parameter is 0.96 and the estimate of the scale parameter is 74.0 h.

Multiply-Censored Data

We present here an example and make a probability plot using the Herd–Johnson method described earlier. Table B.9 presents the life data for 16

Table B.8. Singly-censored random sample
of size $n = 10$ and calculated median ranks

Life, h	Failure rank (i)	Median rank = $(i - 0.3)/(n + 0.4)$
8.9	1	0.1167
17.0	2	0.1757
39.3	3	0.3649
50.4	4	0.5000
55.0 +[a]	5	×
55.0 +	6	×
55.0 +	7	×

[a] (+)—Running.

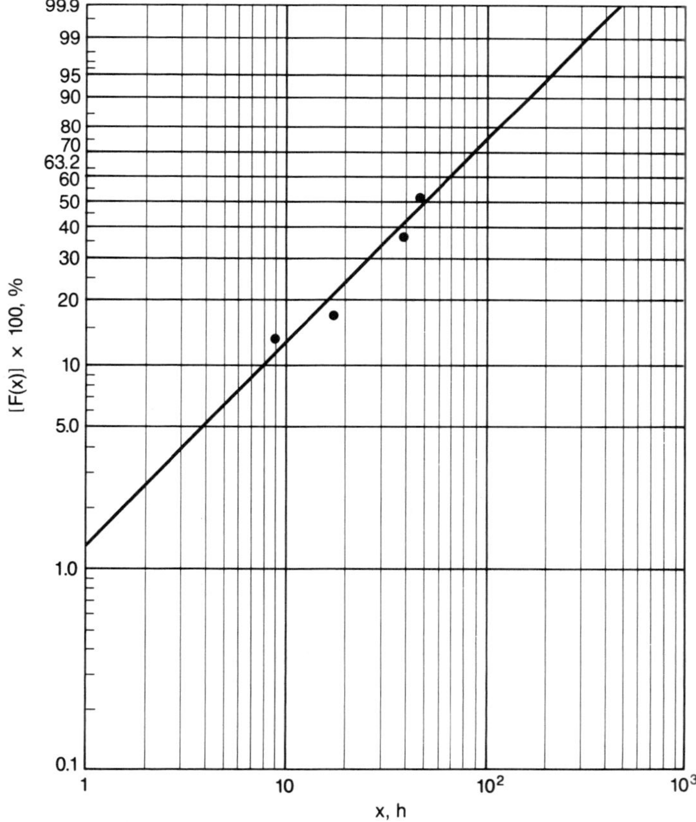

Fig. B.16. Weibull plot for a singly-censored random sample of size $n = 7$. From Weibull population $\gamma = 0.0$, $\beta = 0.96$, and $\eta = 74$ h.

Table B.9. Multiply-censored random sample of size $n = 16$ and the Herd–Johnson calculations

Life, h	Reverse rank (r_i)	conditional $r_i/(r_i + 1)$	$\times$	previous R_{i-1}	$=$	current R_i	Failure probability $(1 - R_i)$
31.7	16	(16/17)	$\times$	1.000	$=$	0.941	0.059
39.2	15	(15/16)	$\times$	0.941	$=$	0.883	0.117
57.5	14	(14/15)	$\times$	0.883	$=$	0.824	0.176
65.0 +[a]	13						
65.8	12	(12/13)	$\times$	0.824	$=$	0.761	0.239
70.0	11	(11/12)	$\times$	0.761	$=$	0.697	0.303
75.0 +	10						
75.0 +	9						
87.5 +	8						
88.3 +	7						
94.2 +	6						
101.7 +	5						
105.8	4	(4/5)	$\times$	0.697	$=$	0.557	0.443
109.2 +	3						
110.0	2	(2/3)	$\times$	0.557	$=$	0.372	0.628
130.0 +	1						

[a] (+)—Running.

bearings. Calculations for the probability distribution functions are shown in the table. A Weibull plot of the data is shown in Fig. B.17. From this figure, the estimate of the shape parameter is 1.85, and of the scale parameter 125.0 h.

Inspection Data

(a) Quantal-Response Data

We illustrate the graphical representation with data on failed bearings. Each bearing was inspected once to determine if it had failed or not. To make a probability plot, we first divide the range of the data into intervals that each contain inspections of at least 10 units. The plot will be crude unless the data set is moderately large, say, over 100 units. Table B.10 shows this for the entire set of bearing data. For the units inspected in an interval, calculate the percentage that are failed. For example, for the 30 bearings inspected between 2000 and 2400 h, the cumulative percentage failed is $100(5/30) = 16.7\%$. On probability paper for an assumed distribution, plot each percentage failed against the midpoint of its interval. For example, 16.7% is plotted against 2200 h in the Weibull probability plot in Fig. B.18. If such a percentage is 0% or 100%, the plotting position is off scale; the intervals might be chosen to avoid this.

The basis for the plotting method is that the units inspected in an interval are all regarded as having the interval midpoint as their age. Then the fraction failed among those units, estimates the population cumulative fraction failing

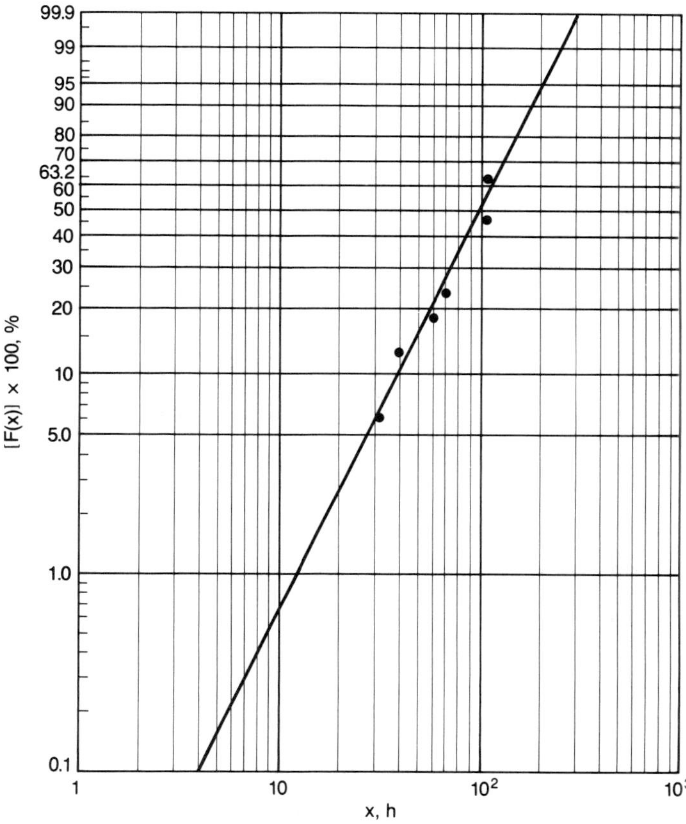

Fig. B.17. Weibull plot for a multiply-censored random sample of size $n = 16$. From Weibull population, $\gamma = 0.0$, $\beta = 1.85$, and $\eta = 125$ h.

Table B.10. Inspection (quantal response) data and their plotting positions and binomial confidence limits

Inspection time, h	Failed/ observed		% failed	Binomial 95% confidence limits	
4400+	21/36	=	58.4	40.8	74.5
4000–4400	21/40	=	52.5	36.1	68.5
3600–4000	22/34	=	64.8	46.5	80.3
3200–3600	6/13	=	46.2	15.9	74.9
2800–3200	9/42	=	21.4	10.3	36.8
2400–2800	9/39	=	23.1	11.1	39.3
2000–2400	5/30	=	16.7	5.6	34.7
1600–2000	7/73	=	9.59	4.9	18.8
1200–1600	2/33	=	6.06	0.7	20.2
800–1200	4/53	=	7.55	2.1	18.2
0– 800	0/39	=	0	0.0	9.0

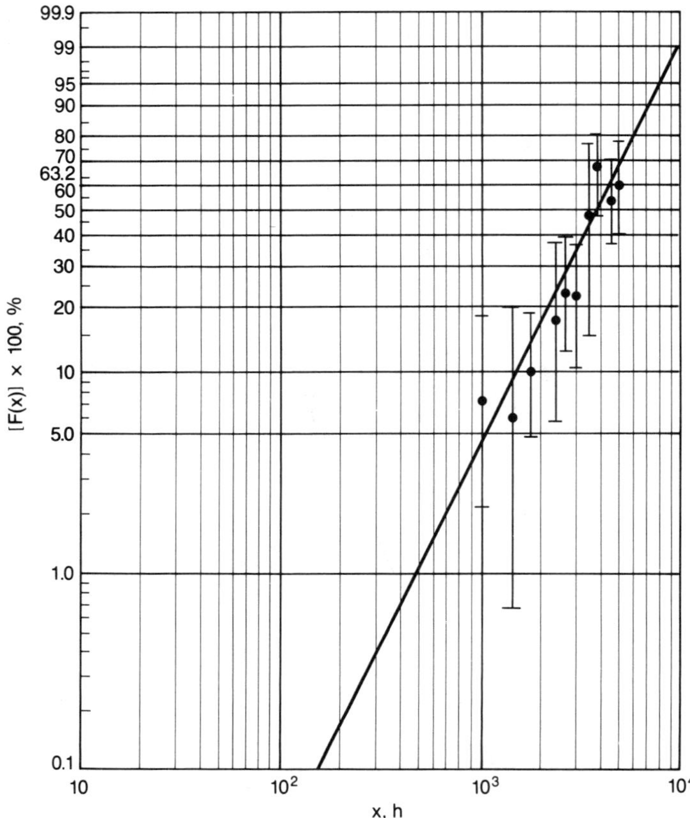

Fig. B.18. Weibull plot for inspection (quantal-response) data. From Weibull population, $\gamma = 0.0$, $\beta = 2.09$, and $\eta = 4800$ h (Nelson, 1982).

by that age. Such plots may be peculiar in that the observed cumulative fraction failing (an increasing function) may decrease from one interval to the next. When this happens we may wish to combine the data in two such intervals if that eliminates the decrease in the sample cumulative distribution. For example, in Fig. B.18 the points 7.55% at 1000 h and 6.06% at 1400 h could be replaced by the point $100(4 + 2)/(53 + 33) = 6.98\%$ at 1200 h. This simple subjective method gives an increasing cumulative distribution function.

If the plotted points follow a straight line reasonably well, the distribution adequately fits the data. Then draw a line through the plotted points to estimate the life distribution. In fitting the line, try to weight each point according to its number of units, or (better) try to weight each point according to its confidence interval described below. If the plotted points do not follow a straight line, plot the data on probability paper for another distribution.

Confidence limits for the sample cumulative distribution are obtained as follows. Each unit in an interval is a binomial trial that is either a failure or a success. The value of the cumulative distribution function for the interval midpoint is (roughly) the binomial probability of failure. So binomial confidence limits apply (Hahn and Shapiro, 1967; Nelson, 1982). For the bearing example, the interval of from 2000 h to 2400 h has five failures among 30 bearings. The binomial 95% confidence limits are 5.6% and 34.7%. Such two-sided 95% limits are given in Table B.10. Also, the bars in Fig. B.18 show these limits. These binomial limits involve no assumption about the form of the product life distribution; i.e., they are nonparametric.

The plot is interpreted like any other probability plot, as described earlier. The estimate of the shape parameter is 2.09 and the estimate of the scale parameter is 4800 h in Fig. B.18.

(b) Interval Data

If the inspection intervals differ, then each failure time can be approximated by the midpoint of its interval, and these times can be analyzed like complete or censored data, as described earlier. We illustrate the graphical representation with data on 167 identical parts in a machine. At certain ages the parts were inspected to determine which had cracked since the previous inspection. The data appear in Table B.11; it shows the months in service at the start and end of each inspection period and the number of cracked parts found in each period. For example, between 19.92 and 29.64 months, 12 parts cracked. Seventy-three parts survived the latest inspection at 63.48 months. The data are simple in that all parts were inspected at the same ages; this is not so for many sets of interval data.

To make a probability plot, for each inspection time, we calculate the cumulative number of failures by that time. For example, for the inspection at 29.64 months, the cumulative number of cracked parts is $5 + 16 + 12 = 33$. Table B.11 shows the cumulative number of cracked parts for each inspection time. Next, for each inspection time, we calculate the sample cumulative percentage failed. For example, for the inspection at 29.64 months, this is $100(33/167) = 19.8\%$. Table B.11 shows this percentage for each inspection time. We then plot each sample cumulative percentage against its inspection time. Figure B.11 shows this for the part data. A cumulative percentage of 100% cannot be plotted on the paper; for this reason, some people use the sample size plus one (168) to calculate the sample cumulative percentages. We then draw a straight line through the plotted points. This line is an estimate of the cumulative life distribution. The method above applies only when the units have a common inspection schedule and all unfailed units have run beyond all failed units.

The following method provides simple confidence limits for the cumulative distribution function at each inspection time. Each sample unit fails either before or after a particular inspection time. Each unit can be regarded as a

Table B.11. Inspection (interval) data and their plotting positions and binomial confidence limits

Inspection, months		Number		Cumulative percentage		
					Binominal 95% confidence limits	
Start	End	Cracked	Cumulative	Estimate		
0	6.12	5	5	2.99	0.98	6.85
6.12	19.92	16	21	12.6	7.95	18.6
19.92	29.64	12	33	19.8	14.0	26.6
29.64	35.40	18	51	30.5	23.7	38.1
35.40	39.72	18	69	41.3	33.8	49.2
39.72	45.24	2	71	42.5	34.9	50.4
45.24	52.32	6	77	46.1	38.4	54.0
52.32	63.48	17	94	56.3	48.4	63.9
63.48 +	Survived	73	167			
	Total	167				

binomial trial, and its binomial probability of failure is the value of the cumulative distribution function at the inspection time. So confidence limits for a binomial probability apply (Hahn and Shapiro, 1967; Nelson, 1982). For the part example, at 29.64 months, there are 33 cracked parts out of 167. Two-sided binomial 95% confidence limits at that age are 14.0% and 26.6%. Table B.11 gives such confidence limits. Also, the bars in Fig. B.11 show these binomial limits, which can be put on any plotting paper. The limits above apply only when the units have a common inspection schedule and all unfailed units have run beyond all failed units.

The plot is interpreted like any other probability plot, as described earlier. The estimate of the shape parameter is 1.49 and the estimate of the scale parameter is 72 months in Fig. B.19.

B.4.6.2. Numerical Analysis

We present standard analytical methods used for the complete data and maximum likelihood (ML) methods used for all data (particularly for multiply-censored data). Details on linear methods used for singly-censored and complete data, and ML methods used for inspection data, can be found in various textbooks, e.g., Mann et al. (1974), Bain (1978), and Nelson (1982).

Standard Methods for Complete Data

This section provides estimates and confidence limits for distribution parameters, percentiles, and reliabilities. We present an analysis for complete data for a two-parameter Weibull distribution of n observations, $X_1, \ldots, X_n$. The $Y_1, \ldots, Y_n$ (data from the extreme value distribution) are the natural logarithms of the Weibull data.

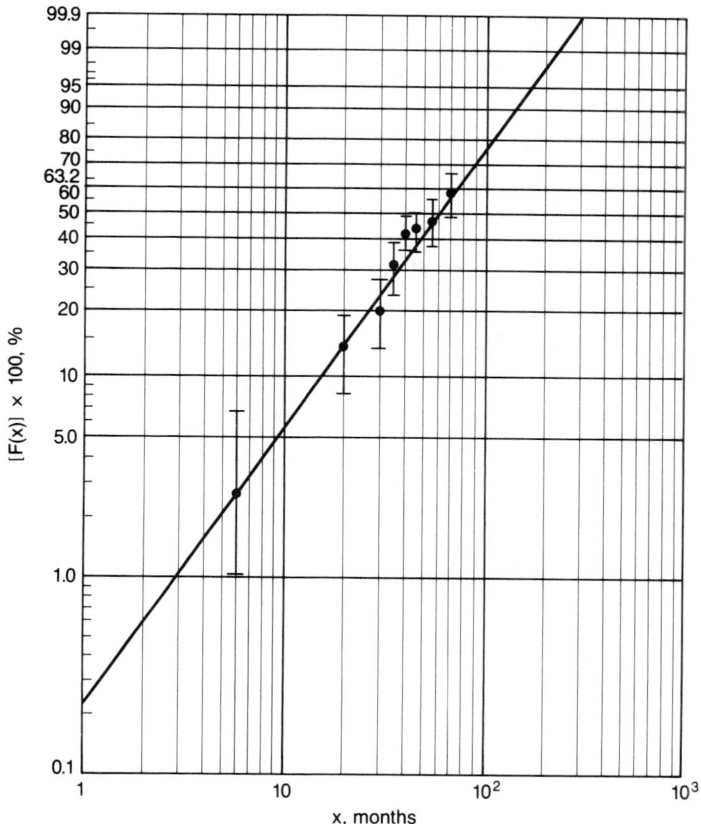

Fig. B.19. Weibull plot for inspection (interval) data. From Weibull population, $\gamma = 0.0$, $\beta = 1.49$, and $\eta = 72$ months (Nelson, 1982).

(a) Estimates and Confidence Limits for Distribution Parameters

An estimate of the Weibull shape parameter β is

$$B = 1.283/S, \tag{B.91}$$

where S is the sample standard deviation

$$S = \{[(Y_1 - \overline{Y})^2 + \cdots + (Y_n - \overline{Y})^2]/(n - 1)\}^{1/2}, \tag{B.92}$$

and

$$\overline{Y} = (Y_1 + \cdots + Y_n)/n \tag{B.93}$$

is the sample average.

An estimate of the Weibull scale parameter η is

$$A = \exp(\overline{Y} + 0.450S). \tag{B.94}$$

B and A are biased estimators.

Two-sided approximate $100\alpha\%$ confidence limits for β are

$$\underset{\sim}{\beta} = 1.283/S[\exp(K_\alpha 1.049/n^{1/2})] \tag{B.95a}$$

and

$$\tilde{\beta} = (1.283/S)[\exp(K_\alpha 1.049/n^{1/2})], \tag{B.95b}$$

where K_α is the $[100(1 + \alpha)/2]$th standard normal percentile. One-sided approximate $100\alpha\%$ confidence limits are obtained from Eq. (B.95), where K_α is replaced by z_α, the 100αth standard normal percentile.

Two-sided approximate $100\alpha\%$ confidence limits for η are

$$\underset{\sim}{\eta} \sim \exp[\bar{Y} + 0.450S - K_\alpha 1.081(0.7797S/n^{1/2})], \tag{B.96a}$$

and

$$\tilde{\eta} \sim \exp[\bar{Y} + 0.450S + K_\alpha 1.081(0.7797S/n^{1/2})]. \tag{B.96b}$$

(b) Sample Size to Estimate a Weibull Shape Parameter

The estimate B of a Weibull shape parameter β is between $f\beta$ and β/f, with about $100\alpha\%$ probability, if the sample size is

$$n \simeq 1.100[K_\alpha/\ln(f)]^2. \tag{B.97}$$

The accuracy of this formula increases with the sample size. It is based on the approximate normal distribution of $\ln(B)$ for large n.

Typically, a statistically significant sample size (n) of a minimum eight is chosen.

(c) Percentiles

The estimate of the $100P$th Weibull percentile is

$$X_P = \exp[\bar{Y} + (0.5772 + u_P)0.7797S], \tag{B.98}$$

where $u_P = \ln[-\ln(1 - P)]$ is the $100P$th percentile of the standard extreme value distribution.

Two-sided approximate $100\alpha\%$ confidence limits for the Weibull $100P$th percentile x_P are

$$\underset{\sim}{x_P} \sim \exp\{\ln(X_P) - K_\alpha 0.7797S[(1.1680 + u_P^2 1.1000 - u_P 0.1913)/n]\}^{1/2} \tag{B.99a}$$

and

$$\tilde{x}_P \sim \exp\{\ln(X_P) + K_\alpha 0.7797S[(1.1680 + u_P^2 1.1000 - u_P 0.1913)/n]\}^{1/2}. \tag{B.99b}$$

(d) Reliability

For a Weibull distribution, the estimate of reliability at age x is

$$R^*(x) = \exp[-(x/A)^B]. \tag{B.100}$$

Two-sided approximate $100\alpha\%$ confidence limits for the Weibull reliability

$R(x)$ at age x is

$$\underset{\sim}{R}(x) = \exp[-\exp(\tilde{u})] \qquad \text{(B.101a)}$$

and

$$\tilde{R}(x) = \exp[-\exp(\underset{\sim}{u})], \qquad \text{(B.101b)}$$

where

$$\underset{\sim}{u} = U - K_\alpha[(1.1680 + U^2 1.1000 - U0.1913)/n]^{1/2}, \qquad \text{(B.102a)}$$

$$\tilde{u} = U + K_\alpha[(1.1680 + U^2 1.1000 - U0.1913)/n]^{1/2}, \qquad \text{(B.102b)}$$

and

$$U = [\ln(x) - \overline{Y} - 0.450S]/0.7797S. \qquad \text{(B.103)}$$

One-sided approximate $100\alpha\%$ confidence limits are obtained where K_α is replaced by z_α.

(e) Example

Table B.12 shows $\ln$(times) Y_i $(i = 1, \ldots, n)$ bearing failures in an accelerated test of sample size $n = 19$. The mean of the 19 log observations is $\overline{Y} = [(-1.6608) + (-0.2485) + \cdots + 4.2889]/19 = 1.7863$, and their standard deviation is $S = \{[(-1.6608 - 1.7863)^2 + (4.2889 - 1.7863)^2]/(19 - 1)\}^{1/2} = 1.5252$. Therefore, the estimate of the Weibull shape parameter is $B = 1.283/S = 0.8409$. The estimate of the Weibull scale parameter is $A = \exp(1.7863 + 0.450 \times 1.5252) = 11.85$ h. The 90% confidence limits for the shape parameter

Table B.12. Uncensored random sample of size 19 and their failure ranks

Life, h	ln(Life, h)	Failure rank
0.19	−1.6608	1
0.78	−0.2485	2
0.96	−0.0409	3
1.31	0.2700	4
2.78	1.0224	5
3.16	1.1505	6
4.15	1.4231	7
4.67	1.5411	8
4.85	1.5789	9
6.50	1.8718	10
7.35	1.9947	11
8.10	2.0806	12
8.27	2.1126	13
12.06	2.4898	14
31.75	3.4578	15
32.52	3.4818	16
33.91	3.5237	17
36.71	3.6030	18
72.89	4.2889	19

are $\eta = \exp\{1.7863 + 0.450 \times 1.5252 - (1.645)1.081(0.7797 \times 1.5252/19^{1/2})\}$ $= 7.3$ h, $\tilde{\eta} = 19.3$ h. Each of these limits is a one-sided approximate 95% confidence limit. Suppose for the calculation of sample size, β is to be estimated within a factor of 1.25 [roughly within $100(1.25 - 1) = 25\%$] with 90% probability. The estimate of the needed sample size is $n \sim 1.100[1.645/\ln(1.25)]^2 \sim 60$.

The estimate of the tenth percentile of the Weibull distribution $x_{0.10}$ is $x_{0.10} = \exp\{1.7863 + (-2.2504)1.1892\} = 0.82$ h.

The 90% confidence limits for the Weibull $x_{0.10}$ are

$$\underset{\sim}{x}_{0.10} = \exp\{-0.2034 - 1.645(1.1892)\{[1.1680 + (-2.2504)^2$$
$$\times 1.1000 - (-2.2504)0.1913]/19\}^{1/2}\}$$
$$= 0.25 \text{ h},$$

and

$$\tilde{x}_{0.10} = 2.7 \text{ h}.$$

Each of these limits is a one-sided 95% confidence limit.

The bearing reliability at 2 h is

$$R^*(2.0) = \exp[-(2.0/11.85)^{0.8409}]$$
$$= 0.799.$$

The two-sided approximate 90% confidence limits for bearing reliability are obtained as

$$U = [\ln(2.0) - 2.4727]/1.1892 = -1.4964,$$
$$\underset{\sim}{u} = -1.4964 - 1.645\{[1.1680 + (-1.4964)^2 1.1000$$
$$- (-1.4964)0.1963]/19\}^{1/2} = -2.2433,$$
$$\tilde{u} = -0.7495,$$
$$\underset{\sim}{R}(2.0) = \exp[-\exp(-0.7495)] = 0.623,$$
$$\tilde{R}(2.0) = 0.899.$$

Figure B.20 shows a Weibull probability of bearing failure data. The fitted distribution with the parameter estimates is the straight line. The confidence limits for percentiles (and reliabilities) are curves and are narrowest near the center of the data.

Maximum Likelihood Method for All Data

We present maximum likelihood (ML) methods for fitting Weibull distributions to multiply-censored data. The methods also apply to singly-censored data and to complete data (Cohen, 1965; Harter and Moore, 1965; McCool, 1970; Nelson, 1982). We will limit discussion to the case of the two-parameter Weibull distribution. This includes those situations where the three-parame-

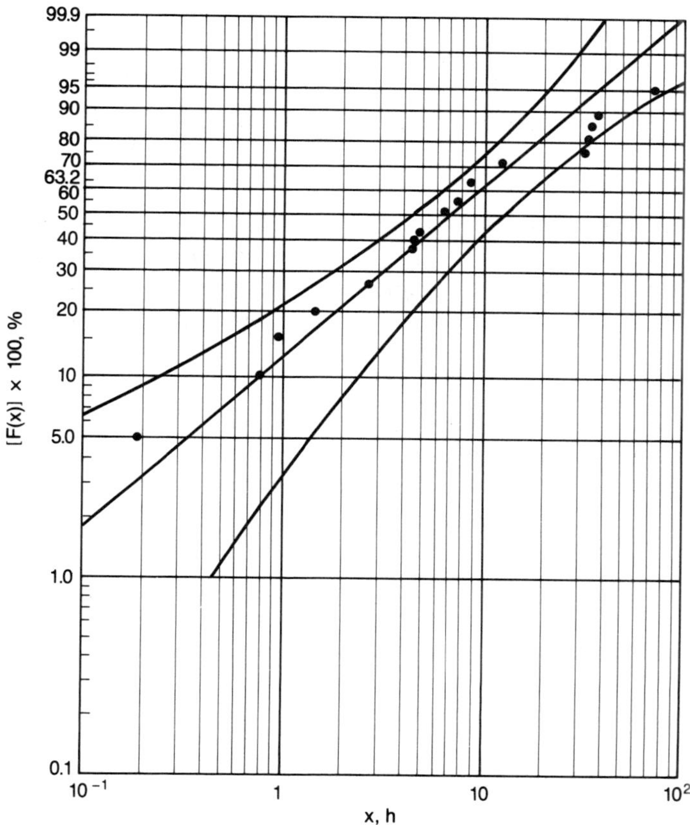

Fig. B.20. Weibull plot for an uncensored random sample of size $n = 19$. Also shown are Weibull fit and confidence limits (Nelson, 1982).

ter model applies but the location parameter γ is known, for then we need only transform the data by subtracting the known location parameter from each observation prior to performing the analysis. The theory distinguishes between three values of a parameter: the true population value θ_0, its ML estimate $\hat{\theta}$, and an arbitrary value θ. Computer programs that fit the Weibull distributions are commercially available.

(a) Estimates and Confidence Limits for Distribution Parameters

The Weibull log likelihood for a sample of n units with r failures is

$$L = \sum_i{}' \left[\ln(\beta) + (\beta - 1) \ln(x_i) - \beta \ln(\eta) - (x_i/\eta)^\beta \right] + \sum_i{}'' \left[-(x_i/\eta)^\beta \right]. \quad \text{(B.104)}$$

The sums $\sum_i, \sum_i{}'$, and $\sum_i{}''$, respectively, run over all, failed, and unfailed units.

The ML estimates $\hat{\eta}$ and $\hat{\beta}$ for η_0 and β_0 are the η and β values that maximize Eq. (B.104); $\hat{\eta}$ and $\hat{\beta}$ are unique as are also the solutions of the likelihood equations

$$0 = \partial L/\partial \eta = \sum_i' \left[-(\beta/\eta) + (\beta/\eta)(x_i/\eta)^\beta \right] + \sum_i'' (\beta/\eta)(x_i/\eta)^\beta, \tag{B.105a}$$

$$0 = \partial L/\partial \beta = \sum_i' \left[(1/\beta) + \ln(x_i/\eta) - (x_i/\eta)^\beta \ln(x_i/\eta) \right] - \sum_i'' (x_i/\eta)^\beta \ln(x_i/\eta). \tag{B.105b}$$

These nonlinear equations must be iteratively solved by computer to obtain $\hat{\eta}$ and $\hat{\beta}$. Otherwise, Eq. (B.104) must be numerically maximized to get them. Alternatively, the likelihood equations (B.105) can be combined to eliminate η. This yields a single equation in β that is easier to solve, namely,

$$\sum_i' \ln(x_i)/r = \left[\sum_i x_i^\beta \ln(x_i) \right] \left(\sum_i x_i^\beta \right)^{-1} - (1/\beta). \tag{B.106}$$

The left-hand sum runs over just the failure times. It is easy to solve iteratively to get $\hat{\beta}$, since the right-hand side is a monotone function of β. Then calculate

$$\hat{\eta} = \left(\sum_i x_i^{\hat{\beta}}/r \right)^{1/\hat{\beta}}. \tag{B.107}$$

For large samples, the joint cumulative distribution function of $\hat{\eta}$ and $\hat{\beta}$ is close to a joint normal one with means η_0 and β_0.

Two-sided approximate $100\alpha\%$ confidence limits for the true Weibull parameters η_0 and β_0 are

$$\underset{\sim}{\eta} = \hat{\eta}/\exp\{K_\alpha[\text{Var}(\hat{\eta})]^{1/2}/\hat{\eta}\}, \tag{B.108a}$$

$$\tilde{\eta} = \hat{\eta} \exp\{K_\alpha[\text{Var}(\hat{\eta})]^{1/2}/\hat{\eta}\}, \tag{B.108b}$$

$$\underset{\sim}{\beta} = \hat{\beta}/\exp\{K_\alpha[\text{Var}(\hat{\beta})]^{1/2}/\hat{\beta}\}, \tag{B.109a}$$

$$\tilde{\beta} = \hat{\beta} \exp\{K_\alpha[\text{Var}(\hat{\beta})]^{1/2}/\hat{\beta}\}. \tag{B.109b}$$

For one-sided approximate $100\alpha\%$ limits, replace K_α by z_α. The Var( ) is asymptotic variance and is calculated from the large-sample covariance matrix (Nelson, 1982).

(b) Percentiles

The $100P$th percentile of a Weibull distribution is $x_P = \eta_0[-\ln(1 - P)]^{1/\beta_0}$. Its ML estimate is

$$\hat{x}_P = \hat{\eta}[-\ln(1 - P)]^{1/\hat{\beta}}. \tag{B.110}$$

Two-sided approximate $100\alpha\%$ confidence limits for the Weibull percentile x_P are

$$\underset{\sim}{x}_P \cong \exp\{\ln(\hat{x}_P) - K_\alpha\{\text{Var}[\ln(\hat{x}_P)]\}^{1/2}\}, \tag{B.111a}$$

$$\tilde{x}_P \cong \exp\{\ln(\hat{x}_P) + K_\alpha\{\text{Var}[\ln(\hat{x}_P)]\}^{1/2}\}. \tag{B.111b}$$

For a one-sided approximate $100\alpha\%$ confidence limit, replace K_α by z_α.

(c) Reliability

The fraction of a Weibull distribution surviving age x is $R(x) = \exp[-(x/\eta_0)^{\beta_0}]$. Its ML estimate is

$$\hat{R}(x) = \exp[-(x/\hat{\eta})^{\hat{\beta}}]. \tag{B.112}$$

Two-sided approximate $100\alpha\%$ confidence limits for $R(x)$ are given by

$$\underset{\sim}{R}(x) = \exp[-\exp(\tilde{u})], \tag{B.112a}$$

$$\tilde{R}(x) = \exp[-\exp(\underset{\sim}{u})], \tag{B.112b}$$

where

$$\underset{\sim}{u} = \hat{u} - K_\alpha[\mathrm{Var}(\hat{u})]^{1/2}, \tag{B.113a}$$

$$\tilde{u} = \hat{u} + K_\alpha[\mathrm{Var}(\hat{u})]^{1/2}, \tag{B.113b}$$

and

$$\hat{u} = [\ln(x) - \ln(\hat{\eta})]/\ln(\hat{\beta}). \tag{B.113c}$$

For a one-sided approximate $100\alpha\%$ limit, replace K_α by z_α. The corresponding estimate and confidence limits for the fraction failing are

$$\hat{F}(x) = 1 - \hat{R}(x), \qquad \underset{\sim}{F}(x) = 1 - \tilde{R}(x), \qquad \tilde{F}(x) = 1 - \underset{\sim}{R}(x). \tag{B.114}$$

B.4.6.3. Analysis of Reliability Demonstration Test

A reliability demonstration test for equipment normally requires that a specified mean life and/or failure rate is demonstrated with $100\alpha\%$ confidence. We first determine the appropriate failure distribution for the failure data. We then calculate the sample size and the test life with a specified number of failures, necessary to verify that performance requirements are met.

The consumer risk (C.R.) for n test samples with no failure (first failure imminent) is equal to $1 -$ confidence level (C.L.). The consumer risk for a Weibull distribution is (see Section B.3.4)

$$
\begin{aligned}
\text{C.R.} &= 1 - \text{C.L.} \\
&= \{1 - F(x)\}^n \\
&= \{\exp[-(t_T/\eta)^\beta]\}^n \\
&= \exp[-(nt_T^\beta)/\eta^\beta], \tag{B.115a}
\end{aligned}
$$

or

$$nt_T^\beta = \eta^\beta[-\ln(1 - \text{C.L.})], \tag{B.115b}$$

where t_T is the minimum required test life with no failures for each sample. For accelerated tests, the predicted times must be divided by the acceleration factor. We note that t_T decreases with an increase in number of samples,

$$t_{T_1}/t_{T_2} = (n_2/n_1)^{1/\beta}. \tag{B.116}$$

[We note that for an exponential distribution ($\beta = 1$) the total test life can be predicted from Eq. (B.115) using $n = 1$, and this total test life is a summation of lives of individual tests. The tests can be run for different lengths in each unit as the hazard rate is constant.] Typically, a statistically significant sample size (n) of a minimum eight is chosen for reliability demonstration.

We demonstrate the method with two examples.

Example 1

A magnetic rigid-disk drive is expected to have the following flyability performance requirements: MTBF during flyability $= 25,000$ h or 2.85 years (over a product life of 8 years). Calculate the required test life with a confidence level of 90% for a given number of samples to satisfy the performance requirements. Assume that life times follow the two-parameter ($\gamma = 0$) Weibull distribution, based on the field data.

From the field failure data we estimate β, which generally ranges from 2 to 3 (a larger β implies a higher failure rate than that at lower β). We know that $\mathrm{MTBF} = \eta\Gamma[1 + (1/\beta)]$, therefore $\eta = 28210$ h for $\beta = 2$ and $\eta = 27996$ h for $\beta = 3$. (For an exponential distribution $\beta = 1$, and the total test duration is simply the summation of the test durations of individual tests.) Using Eq. (B.115), we can now calculate the test life t_T for various numbers of samples, Table B.13(a). For accelerated tests (e.g., reduced flying height, lower lubri-

Table B.13. Required minimum test lives with zero failures for an HDA test to meet the performance requirements

No. of samples	Test lives, h	
(a) Flyability Test Example		
	$\beta = 2$	$\beta = 3$
1	42,273	35,469
2	29,896	28,158
3	24,406	24,601
4	21,136	22,345
5	18,905	20,743
6	17,258	19,531
7	15,978	18,542
8	14,945	17,734
(b) Start/Stop Test Example		
	$\beta = 2$	$\beta = 3$
1	50,245	26,651
2	35,528	21,157
3	29,008	18,485
4	25,122	16,796
5	22,470	15,593
6	20,512	14,754
7	17,764	13,325
8		

cant level on the disk surface, increased disk defects, and/or high temperature and humidity), the required test life must be divided by the acceleration factor.

Example 2

A magnetic rigid disk drive is expected to have the following start-stop performance requirements: number of start/stops = 7500 (over a product life of 8 years) at 95% reliability (or fifth percentile). Calculate the test life for a given number of samples to satisfy the performance requirements with a 90% confidence level. Assume that start/stops follow two-parameters Weibull distribution.

From the field failure data, we estimate β which generally ranges from 2 to 3. From the allowable cumulative failures and β, we obtain η by using the following relationship:

$$R(t_P) = 1 - F(t_P) = \exp[-(t_P/\eta)^\beta]$$

or

$$\eta = t_P/[-\ln(1 - F(t_P))]^{1/\beta}$$

where t_P is the product life. We calculate $\eta = 33,115$ starts/stops for $\beta = 2$ and 48,734 starts/stops for $\beta = 3$. Using Eq. (B.115), we now calculate t_T as a function of n, Table 3.13(b). For accelerated tests (e.g., lower lubricant level on the disk surface, increased take off speed, and/or high temperature and humidity), the required number of starts/stops must be divided by the acceleration factor.

References

Abramowitz, M. and Stegun, I. A. (1965). "Handbook of Mathematical Functions." Dover, New York.

Bain, L. J. (1978). "Statistical Analysis of Reliability and Life-Testing Models: Theory and Methods." Marcel Dekker, New York.

Bazovsky, I. (1961). "Reliability Theory and Practice." Prentice-Hall, Englewood Cliffs, New Jersey.

Bhushan, B. (1990). "Tribology and Mechanics of Magnetic Storage Devices." Springer-Verlag, New York.

Brunk, H. D. (1965). "Introduction to Mathematical Statistics." Blaisdell, New York.

Chiang, C. L. (1968). "Introduction to Stochastic Processes in Biostatistics." Wiley, New York.

Cohen, A. C. (1965). Maximum likelihood estimation in the Weibull distribution based on complete and uncensored samples. *Technometrics* **7** (4).

Dixon, W. J. and Massey, F. J. (1969). "Introduction to Statistical Analysis," 3rd ed. McGraw-Hill, New York.

Hahn, G. J. and Shapiro, S. S. (1967). "Statistical Models in Engineering." Wiley, New York.

Harter, H. L. (1964). "New Tables of the Incomplete Gamma-Function Ratio and of Percentage Points of the Chi-Square and Beta Distributions." U.S. Government Printing Office, Washington, D.C.

Harter, H. L. and Dubey, S. D. (1967). "Theory and Tables for Tests of Hypothesis Concerning the Mean and Variance of a Weibull Population." Aerospace Research Laboratory Report ARL 67-0059.

Harter, H. L. and Moore, A. H. (1965). Maximum likelihood estimation of the parameters of gamma and Weibull populations from complete and from censored samples. *Technometrics* **7**, 639–643.

Herd, G. R. (1960). Estimation of reliability from incomplete data. In *Proc. 6th Nat. Symp. on Reliability and Quality Control*, pp. 202–217. IEEE, New York.

Johnson, L. G. (1964). "Theory and Technique of Variational Research." Elsevier, Amsterdam.

Johnson, N. L. and Kotz, S. (1969). "Distribution in Statistics—Discrete Distributions." Wiley, New York.

Johnson, N. L. and Kotz, S. (1970). "Continuous Univariate Distributions," Vols. 1 and 2. Houghton-Mifflin, Boston.

Kalbfleisch, J. D. and Prentice, R. L. (1980). "The Statistical Analysis of Failure Time Data." Wiley, New York.

Kaplan, E. L. and Meier, P. (1958). Nonparametric estimation from incomplete observation. *J. Amer. Stat. Assoc.* **53**, 457–481.

King, J. R. (1971). "Probability Charts for Decision Making." Industrial Press, New York.

Lundberg, G. and Palmgren, A. (1947). "Dynamic Capacity of Rolling Bearings." *Acta Polytechnica, Mech. Eng. Series* 1, No. 3, 7, R.S.A.E.E.

Lundberg, G. and Palmgren, A. (1951). "Dynamic Capacity of Roller Bearing." *Acta Polytechnica, Mech. Eng. Series* 2, No. 4, 96, R.S.A.E.E.

Mann, N. R., Schafer, R. E. and Singapurwalla, N. D. (1974). "Methods for Statistical Analysis of Reliability and Life Data." Wiley, New York.

McCool, J. I. (1969). "Graphical Determination of and Uses of Order Statistic Quantities." *J. Qual. Tech.* **1**, 237–241.

McCool, J. I. (1970). "Evaluating Weibull Endurance Data by the Method of Maximum Likelihood." *ASLE Trans.* **13**, 189–202.

Mendenhall, W. (1971). "Introduction to Probability and Statistics." Wadsworth, Belmont, California.

Meyer, P. L. (1973). "Introductory Probability and Statistical Analysis," 2nd ed. Addison-Wesley, Reading, Massachusetts.

Meyer, S. L. (1975). "Data Analysis for Scientists and Engineers." Wiley, New York.

Nelson, L. S. (1967). Weibull probability paper. *Industrial Quality Control* **23**, 452–453.

Nelson, W. (1972). "Theory and applications of hazard plotting for censored failure data." *Technometrics* **14**, 945–966.

Nelson, W. (1982). "Applied Life Data Analysis." Wiley, New York.

Polovko, A. M. (1968). "Fundamentals of Reliability Theory." Academic Press, New York.

Shooman, M. L. (1968). "Probabilistic Reliability: An Engineering Approach." McGraw-Hill, New York.

Tsokos, C. and Shimi, I. (eds.) (1977). "Theory and Applications of Reliability with Emphasis on Bayesian and Nonparametric Methods," Vol. 1—Theory, Vol. 2—Application. Academic Press, New York.

Weibull, W. (1951). "A statistical distribution of wide applicability." *J. Appl. Mech.* **18**, 293–297.

Wilks, S. S. (1962). "Mathematical Statistics." Wiley, New York.

Author Index

Abraham, R. C.   22, 79
Abramo, S. V.   163
Abramowitz, M.   518, 545
Ackerman, H.   286
Adams, W. W.   163
Altmann, H. C.   295, 298, 299, 300, 303, 304, 350
Amborski, L. E.   163
Andrade, E. N.   201, 282
Anonymous   22, 38, 73, 74, 79, 80, 295, 299, 350, 355, 363, 379, 423, 452, 468, 470, 475
Arai, T.   84
Argumedo, A. J.   424
Arwas, E. B.   425
Ashton, J. E.   150, 156, 160
Athey, S. W.   1, 80

Baasch, H. J.   46, 80
Baba, P. D.   82
Bahadur, S.   203, 282
Bain, L. J.   506, 536, 545
Bajorek, C. H.   21, 80
Baker, S.   162, 283, 351, 424
Barrall, E. M.   102, 103, 105, 128, 131, 132, 133, 134, 135, 136, 137, 138, 139, 160
Barrie, J. A.   140. 157, 160
Bashe, C. J.   21, 80
Bate, G.   81
Baumann, G. W.   363, 423
Bazovsky, I.   476, 545
Behren, R. A. von   38, 40, 84, 363, 425
Berghmans, H.   162, 284
Berghof, W.   57, 80
Berr, C. E.   163
Berry, B. S.   131, 160, 182, 254, 255,

257, 258, 259, 282, 285, 287, 291, 292
Bertram, N.   126, 160, 295, 298, 299, 302, 305, 350, 355, 358, 359, 360, 361, 379, 423
Bhushan, B.   21, 33, 45, 63, 67, 71, 80, 84, 88, 91, 93, 111, 118, 121, 123, 125, 126, 128, 150, 160, 161, 162, 163, 164, 189, 204, 206, 208, 210, 211, 212, 213, 218, 219, 224, 225, 226, 227, 228, 229, 230, 231, 232, 233, 236, 237, 238, 239, 240, 241, 242, 243, 244, 248, 249, 250, 251, 252, 253, 254, 276, 278, 279, 280, 281, 282, 283, 286, 295, 296, 305, 318, 325, 326, 330, 350, 351, 354, 355, 359, 362, 363, 369, 373, 377, 378, 387, 392, 393, 394, 395, 397, 400, 401, 402, 424, 438, 439, 450, 456, 457, 461, 465, 470, 475, 515, 545
Biangardi, H. J.   94, 95, 105, 118, 161
Blades, J. D.   56, 81
Bland, D. R.   182, 195, 283
Blok, H.   369, 381, 424
Blumentritt, B. F.   111, 116, 119, 120, 128, 129, 131, 132, 140, 141, 142, 145, 146, 147, 151, 152, 153, 161
Bogy, D. B.   117, 161, 210, 216, 217, 229, 283, 303, 332, 333, 335, 337, 339, 341, 342, 343, 344, 345, 349, 350, 438, 450
Bolli, P.   286
Boltzmann, L.   182, 283
Bowen, H. K.   82
Bradshaw, R. L.   161, 283
Brede, D. W.   81
Bretti, V.   80

548   Author Index

Breuer, H.   209, 212, 284
Brock, G. W.   10, 57, 80, 83
Brown, C.   162
Brunk, H. D.   476, 506, 515, 545
Bugdayci, N.   161, 283, 350
Bullock, J. G.   84
Bunn, C. W.   162
Burguette, M. D.   67, 80
Burton, R. L.   95, 105, 161
Buschman, A.   1, 81
Buttafava, P.   67, 78, 80

Camras, M.   1, 80
Cannon, D. M.   10, 34, 36, 80
Caporiccio, G.   80
Casey, D. C.   84
Cavallotti, T.   12, 63, 79, 80
Chan, A.   128, 129, 130, 145, 147, 161
Charap, S. H.   1, 3, 5, 80
Chen, M. M.   84
Chen, R.   82
Chen, T.   12, 16, 63, 79, 80
Chen, Y. T.   210, 283, 285
Chiang, C. L.   505, 545
Chiba, H.   81
Chiba, K.   65, 67, 80
Chikazumi, S.   1, 3, 5, 80
Chow, W. W.   84
Choy, C. L.   94, 128, 155, 161
Choy, I. C.   94, 161
Christensen, R. M.   182, 183, 184, 189,
    283, 284
Chu, W. H.   95, 101, 102, 105, 108, 111,
    119, 161, 207, 208, 283
Chubachi, R.   62, 80, 82
Ciardiello, G.   80
Cobbs, W. H.   95, 105, 161
Cohen, A. C.   540, 545
Cohen, M. S.   83
Connolly, D.   161, 204, 210, 212, 224,
    225, 226, 229, 238, 239, 240, 241,
    242, 243, 244, 248, 249, 250, 251,
    252, 254, 283, 295, 296, 299, 300,
    301, 304, 305, 306, 307, 308, 309,
    310, 313, 318, 350, 351, 354, 418,
    424, 431
Crandall, S. H.   363, 424
Croucher, T. G.   286
Cuddihy, E. F.   110, 117, 118, 119,
    120, 121, 122, 140, 143, 161, 244,
    283
Cullity, B. D.   6, 51, 80
Cutler, R. A.   80

Dahl, N. C.   363, 424
Daniel, E. D.   1, 9, 13, 21, 26, 51, 57, 76,
    82
Darlington, M. W.   332, 350
Daubeny, R.   95, 162
Dauer, F. W.   189, 283
Davis, L. P.   80
Dempwolf, W. R.   80
Dixon, W. J.   476, 545
Doetsch, G.   347, 348, 351
Doi, T.   81
Dubey, S. D.   518, 546
Dubin, R. R.   63, 80
Dulmage, W. J.   95, 108, 109, 110, 111,
    162
Dumbleton, J. H.   285

Ebine, Y.   80
Echigo, N.   81
Edwards, W. M.   163
Eiss, N. S.   84
Endrey, A. L.   163
Engel, P. A.   203, 283
Engh, J. T.   21, 43, 80
Enomoto, S.   84
Eshel, A.   123, 126, 150, 160, 162, 164,
    210, 283, 295, 296, 297, 298, 299,
    302, 305, 315, 326, 350, 351, 355,
    358, 359, 360, 361, 363, 379, 423,
    424

Fakirov, S.   105, 162
Farrell, W. R.   351
Fell, W.   363, 424
Ferry, J. D.   182, 200, 203, 204,
    221, 222, 240, 283, 286
Feuerstein, A.   62, 67, 76, 80, 81
Fischer, E. W.   162
Fisher, R. D.   56, 81
Flugge, W.   181, 182, 195, 201,
    283
Foss, G. D.   67, 80
Freeman, R. C.   22, 79
Froehlich, F. B.   84
Frye, K. F.   424
Frye, K. G.   295, 307, 351, 357,
    403
Fujii, Y.   285
Fujiki, M.   81
Fujiwara, H.   62, 81
Fujiwara, T.   12, 62, 81
Fursdon, J. W. M.   286

Gatzen, H. H.   57, 80
Geddes, A. L.   95, 108, 109, 110, 111,
    162
Geil, P. H.   101, 163
Geller, S. B.   358, 424
Gerow, C. W.   163
Gillen, K. T.   276, 283
Ginsburg, C. P.   25, 81
Goldstein, H.   170, 283
Goodier, J. N.   124, 163, 175, 286, 291,
    293, 302, 332, 351
Gorter, F. W.   84
Gottenberg, W. G.   189, 284
Goudreau, G. L.   286, 351
Graham, J. J.   162
Graham, S.   314, 351
Green, A. E.   332, 351
Greenberg, H. J.   141, 145, 147, 150,
    162, 164, 284, 438, 441, 442, 443,
    444, 445, 446, 447, 448, 449, 450
Groeninckx, G.   94, 95, 101, 147, 162,
    211, 212, 284
Gunning, J. R.   351
Gupta, A. K.   94, 95, 162, 284
Gupta, B. K.   67, 72, 80, 457, 475
Gupta, V. B.   94, 95, 162, 209, 265, 272,
    274, 284
Gurtin, M. E.   182, 284

Hahn, F. W.   37, 38, 81, 161, 283, 350,
    424
Hahn, G. J.   535, 536, 545
Halpin, J. C.   160, 332, 351
Hanafusa, H.   424
Hara, K.   83
Harker, J. M.   21, 81
Harris, J. P.   20, 21, 26, 81
Harter, H. L.   503, 518, 526, 540, 545,
    546
Hartman, A.   162, 283, 351
Hartman, S.   424
Hasegawa, T.   81
Hawthorne, J. M.   210, 212, 284
Hayaski, C.   82
Haynes, N. M.   1, 81
Heacock, J. F.   163
Hearmon, R. F. S.   167, 284
Heffelfinger, C. J.   85, 86, 87, 88, 90, 91,
    94, 95, 101, 108, 110, 111, 115, 151,
    162
Heinrich, J. C.   161, 283, 296, 315, 324,
    325, 326, 327, 328, 329, 330, 331,
    350, 351, 354, 424

Henderson, E.   201, 284
Herd, G. R.   504, 505, 530, 532,
    546
Hiratsuka, H.   370, 422, 424
Hirota, E.   56, 81, 83
Hoagland, A. S.   1, 81, 438, 450
Hoffman, R.   162
Hollman, W.   424
Honda, K.   67, 81
Hosokawa, S.   424
Howard, J. K.   57, 81
Hoyashida, H.   84
Hu, P. Y.   418, 424, 429, 430
Hussain, S. M.   305, 351

Ikeda, A.   81
Illers, K. H.   209, 212, 284
Imamura, M.   62, 81
Ishai, O.   93, 111, 162
Ishikawa, K.   285
Isshiki, M.   84
Ito, T.   84
Ito, Y.   81
Itoh, A.   82
Iwasaki, H.   82
Iwasaki, S.   10, 11, 12, 63, 78, 81
Izraelev, V.   441, 451

Johnson, L. G.   503, 504, 505, 529, 530,
    532, 546
Johnson, L. R.   80, 83
Johnson, N. L.   476, 489, 497, 506, 528,
    546
Jorgensen, F.   1, 6, 9, 51, 71, 72, 81
Judge, J. S.   63, 79, 81
Judge, J. T.   274, 284

Kadokura, S.   67, 78, 81
Kalbfleisch, J. D.   505, 546
Kalil, F.   1, 81
Kamei, K.   81
Kaplan, E. L.   504, 505, 546
Katoh, Y.   46, 62, 82
Kawai, H.   285
Kawana, T.   82
Kennedy, A. J.   201, 284
Kennedy, T. G.   363, 424
King, J. R.   503, 546
Kingery, W. D.   55, 82
Kirk, D.   25, 82
Kitamoto, T.   82

Knox, K. L.   85, 86, 87, 88, 95, 108, 110, 111, 162
Knudsen, J. K.   51, 82
Kohmoto, O.   54, 82
Kojima, M.   82
Kondo, H.   67, 82
Kotz, S.   476, 489, 497, 506, 528, 546
Kovacs, A. J.   275, 284
Krigbaum, W. R.   274, 284
Kryuchov, V. V.   163, 351
Kuhn, A. T.   63, 82
Kuijk, K. E.   84
Kunieda, T.   67, 77, 82
Kurokawa, H.   82

Lammermann, H.   81
Landel, R. F.   286
Lang, R. W.   286
Lasky, R. C.   203, 283
Lautner, W. K.   351
Lee, E. H.   203, 246, 247, 249, 250, 284
Lekhnitskii, S. G.   124, 162, 167, 170, 176, 284, 302, 332, 337, 338, 343, 344, 351, 440, 451
Lewis, A. F.   64, 66, 84
Logan, J. A.   102, 103, 105, 128, 131, 132, 133, 134, 135, 136, 137, 138, 139, 160
Love, A. E. H.   167, 285
Lowman, C. E.   1, 82
Luborsky, F. E.   53, 82
Ludema, K. C.   203, 282
Luecke, F. S.   46, 80
Lundberg, G.   515, 546

McCabe, J.   425
McCool, J. I.   503, 528, 540, 546
McCrum, N. G.   182, 209, 224, 237, 284
McMullen, J. F.   351
Magill, J. H.   275, 285
Majumdar, A.   162
Makoto, A.   285
Mallinson, J. C.   1, 6, 19, 82
Mallouk, R. S.   163
Mann, N. R.   476, 506, 515, 528, 536, 546
Manson, J. A.   286
Markovitch   285
Masayoshi, I.   161
Massey, F. J.   476, 545
Matsuda, H.   63, 79, 82
Matsumoto, M.   82

Matsuoka, M.   63, 82
Mayr, M.   62, 67, 76, 80
Mee, C. D.   1, 9, 13, 21, 26, 51, 57, 76, 82
Meier, P.   504, 505, 546
Mendenhall, W.   476, 546
Meyer, P. L.   424, 431, 515, 546
Meyer, S. L.   476, 546
Mihara, T.   81
Misra, A.   94, 163
Mitani, T.   67, 82
Miyagi, A.   102, 162
Miyake, A.   81
Miyasaka, K.   285
Mizushima, K.   81
Mizushima, M.   55, 56, 82
Moe, C. D.   67, 83
Monforte, F. R.   55, 82
Monk, D. W.   295, 351
Moore, A. H.   540, 546
Mori, T.   59, 82
Morisako, A.   63, 82
Morland, L. W.   203, 246, 247, 249, 250, 284
Morris, E. L.   224, 237, 284
Morrison, J. R.   81
Moyer, T.   425
Mukherjee, S.   305, 308, 311, 351, 355, 363, 424
Murai, M.   84
Murakami, Y.   81
Muraoka, T.   84
Murayama, T.   209, 212, 240, 285
Murray, S. F.   425

Naganawa, N.   81
Nagao, M.   67, 78, 82
Nahara, A.   82
Nakamura, K.   76, 78, 82, 424
Nakamura, Y.   10, 81
Nakayama, M.   82
Nakayasu, H.   203, 285
Naoe, M.   63, 81, 82
Nelson, L. S.   520, 546
Nelson, W.   424, 431, 476, 482, 483, 486, 487, 506, 510, 514, 515, 534, 535, 536, 537, 540, 541, 542, 546
Nielsen, L. E.   208, 285
Nishiguchi, T.   82
Nishikawa, R.   62, 78, 83, 84
Noda, M.   62, 83
Norton, D. G.   48, 83
Norwick, A. S.   182, 285

Oden, P. L.   92, 162
Oehme, W. F.   450
Ogisu, K.   83
Ohta, Y.   82
Ojima, T.   82
Okamoto, N.   82
Okazaki, Y.   83
Olivier, K. L.   163
Oono, R.   274, 285
Orcutt, F. K.   162, 283, 351, 424
Ostrowski, H. S.   64, 83
Oswin, C. R.   92, 163
Ouchi, K.   11, 63, 83
Owen, R. J.   295, 299, 351, 355, 393, 425

Padmanabhan, A.   162
Pagano, N. J.   332, 351
Pahud, V.   286
Palmer, J. H.   80, 83
Palmgren, A.   515, 546
Park, W. R. R.   150, 163
Pattison, R. E.   81
Pear, C. B.   1, 83
Peterson, M. L.   22, 83, 84
Petit, P. H.   160
Petrie, S. E. B.   139, 163
Pfeiffer, J. D.   295, 296, 305, 307, 315,
    351, 357, 367, 425
Pfleumer, F.   19
Phillips, W. B.   81
Phya, K.   82
Piano, M.   80
Pinnock, P. R.   209, 285
Pister, K. S.   286, 351
Plazek, D. F.   94, 145, 163
Plazek, D. J.   94, 161, 182, 201, 202,
    204, 260, 261, 262, 264, 266, 275,
    285, 286
Polovko, A. M.   476, 546
Porter, R. S.   161
Potter, R. I.   10, 16, 83
Poulsen, V.   19
Prager, R. H.   363, 425
Prentice, R. L.   505, 546
Pritchet, W. C.   131, 160, 254, 255, 257,
    258, 259, 282, 287, 291, 292
Protzman, T. F.   155, 163
Pugh, E. W.   21, 80, 83

Raj, R.   210, 283, 285
Rakovskii, V. A.   163, 351
Ramesh, C.   162, 284

Rand, W. M.   73, 83
Ravens, D. A. S.   140, 141, 143, 163
Read, B. E.   284
Reynaers, H.   162, 284
Richard, M.   314, 351
Richardson, M. J.   275, 285
Rienau, J. H.   73, 83
Roe, R. J.   274, 284

Saito, N.   84
Sakakima, H.   54, 83
Sakamoto, Y.   81
Sakemoto, A.   81
Samuels, R. J.   105, 163, 261, 274, 285
Sano, K.   82
Santana, G. R.   81
Sasaki, T.   80
Sato, K.   80
Satomi, M.   83
Saunders, D. W.   332, 350
Savill, N. G.   275, 285
Scarati, A. M.   80
Schafer, R. E.   546
Schmalhorst, J. M.   80
Schmidt, G. F.   162
Schmidt, P. G.   94, 95, 151, 162, 163
Schuele, W. J.   77, 83
Schwarzl, F.   202, 285
Senno, H.   83
Shapiro, S. S.   535, 536, 545
Sharma, B. S.   161, 283, 350, 424
Sharrock, M. P.   59, 61, 83
Shelledy, F. B.   10, 57, 80, 83
Shidlovskii, N. S.   351
Shimi, I.   515, 546
Shinohara, K.   82
Shooman, M. L.   476, 546
Sievers, J. A.   363, 425
Silkensen, R. D.   80
Simpson, D.   38, 83
Singapurwalla, N. D.   546
Singer, J.   162
Skorjanec, J.   67, 83
Skurdrzyk, E.   363, 425
Smets, G.   162, 284
Smith, D. O.   77, 83
Smith, D. P.   38, 40, 84, 363, 425
Smith, D. R.   121, 161, 276, 278, 279,
    280, 281, 282, 283
Smith, T. L.   95, 101, 102, 105, 108, 111,
    119, 128, 129, 130, 145, 147, 156,
    161, 163, 203, 207, 208, 240, 253,
    283, 285

552 Author Index

Snowdon, J. C. 363, 425
Sobajima, S. 81
Sokolnikoff, I. S. 167, 285
Sonoda, T. 83
Speliotis, D. E. 62, 81, 84
Srinivasan, R. 161
Sroog, C. E. 155, 163
Stannett, V. 141, 142, 143, 163
Staverman, A. J. 202, 285
Stegun, I. A. 518, 545
Stein, R. S. 94, 163, 274, 284
Stephens, R. L. 162, 284, 450
Sternberg, E. 182, 284
Sternstein, S. S. 182, 203, 204, 285
Stouffer, D. C. 246, 249, 250, 285
Stratton, R. A. 284
Struik, L. C. E. 260, 275, 286
Stuart, H. A. 139, 163
Sugaya, H. 21, 84
Sugita, R. 81
Suzuki, T. 62, 83, 84
Sweeting, O. J. 85, 163

Taft, L. G. 81
Tagami, K. 67, 78, 84
Tajiri, K. 209, 212, 285
Takahashi, K. 82
Takahashi, S. 46, 84
Takano, O. 82
Talke, F. E. 161, 162, 283, 284, 332,
    333, 335, 337, 339, 341, 342, 343,
    344, 345, 349, 350, 438, 450
Tamagawa, N. 62, 80
Tamai, H. 84
Tanaka, T. 62, 84
Taneka, Y. 82
Tang, S. 332, 351, 440, 451
Tant, M. R. 211, 212, 273, 274, 275,
    286
Tatum, W. E. 160, 163
Taylor, R. L. 223, 286, 322, 351
Teramura, N. 21, 84
Teranishi, K. 81
Thompson, A. B. 209, 286
Timoshenko, S. P. 124, 163, 175, 254,
    286, 291, 293, 302, 332, 351, 375,
    399, 425
Tjader, T. C. 155, 163
Tobolsky, A. V. 182, 286
Tochihara, S. 64, 84
Tomago, A. 65, 67, 82, 84
Tomie, T. 81

Tramposch, H. 295, 296, 298, 299, 303,
    304, 307, 310, 315, 351, 355, 425
Tripathi, K. 50, 84
Tsang, C. 10, 84
Tschoegl, N. W. 182, 185, 200, 286
Tse, M. K. 64, 66, 84
Tsokos, C. 515, 546
Tsukahara, N. 46, 84

Uematsu, I. 94, 129, 130, 131, 163
Uematsu, Y. 94, 129, 130, 131, 163
Uhlmann, D. R. 82
Umanskii, E. S. 124, 163, 295, 299, 303,
    305, 310, 351

Vallat, M. F. 94, 145, 163, 210, 211,
    212, 214, 215, 224, 225, 260, 261,
    262, 263, 264, 265, 266, 267, 268,
    269, 270, 271, 272, 273, 274, 275,
    286
Van Baren, F. 285
Van Gestel, W. J. 10, 84
Van Holde, K. 201, 286
Van Rossum, J. J. 369, 381, 424
Viswanathan, A. 95, 111, 163

Wahlstrom, E. E. 105, 163
Waid, D. D. 22, 84
Waites, J. B. 306, 352, 398, 425
Wallace, R. L. 16, 17, 84
Walowit, J. A. 371, 425
Walraven, A. 363, 425
Wang, H. S. C. 475
Ward, I. M. 108, 140, 141, 143, 163,
    182, 209, 285, 286
Weast, R. C. 241, 245, 286
Weber, S. M. 204, 286
Weibull, W. 515, 546
Weiss, G. P. 83
Weller, T. 162
Wells, J. F. 81
Wellwood, J. 63, 82
Wetton, R. E. 204, 205, 286
White, R. M. 1, 51, 84
Wiff, D. R. 163
Wilkes, G. L. 211, 212, 273, 274, 275,
    286
Wilks, S. S. 525, 546
Williams, G. 284
Williams, M. L. 202, 285, 286

Winarski, D. J.   30, 34, 84, 295, 296, 299, 300, 301, 304, 305, 306, 307, 308, 309, 310, 313, 314, 350, 351, 364, 365, 366, 381, 418, 424, 425
Wineman, A. S.   246, 285
Winger, W. D.   81
Winn, K. D.   80
Woinowsky-Krieger, S.   375, 399, 425
Woods, D. W.   209, 286
Wright, G. T.   48, 83
Wunderlich, B.   102, 162

Yagoda, H. P.   295, 299, 300, 310, 312, 352

Yamamori, K.   11, 62, 84
Yanaga, M.   46, 84
Yanaguchi, Y.   83
Yasuda, H.   141, 142, 143, 163
Yatagai, H.   82
Yeh, G. S. Y.   101, 163
Yonezawa, T.   82
Youngquist, R. J.   363, 425

Zachmann, H. G.   94, 95, 105, 118, 139, 161, 163
Zerna, W.   332, 351
Zienkiewicz, O. C.   320, 352
Zoller, P.   274, 286

Subject Index

Acceptance sampling plan 515
Active guidance control 414
Aging of PET film 268–270, 274–276
Air bearings 22, 29, 33–34, 91
Air-cushion shoes 71, 73
Alfenol alloys 7, 53
Amorphous magnetic alloys 53–54
Analog signals 12
Analog-to-digital conversion 12
Andrade creep 201–202, 267, 268, 275
Andrade plot 270–273
Anelasticity 182, 258
Anisotropy 11, 77, 149, 172, 263
Annealing 90, 119, 130, 160, 260–262,
    273–275, 449
  conditioning by 267–269
  effect on coefficient of hygroscopic
    expansion 143–145
  effect on thermal expansion 132–
    134, 137–139, 147, 149
Annular disk 335–341, 349
Areal expansion, coefficient of 130
Areal recording density 15
Arrhenius equation 240, 241, 244
Arrhenius temperature 203, 208, 238
Audio tape drives 22–24
Automatic gain control (AGC) 353

Backcoated tapes 125, 278, 373–374
Barium ferrite 8, 12, 47, 61–63
Barium titanate 58
Base film 59
Bernoulli disk drives 50
Bernoulli disks 48, 442
Bernoulli effect 50
Bernoulli plate 50
Bernoulli shoes 71, 73

Best linear invariant estimator (BLIE)
    511
Best linear unbiased estimator (BLUE)
    511, 512
Binary code 13
Binomial distribution 497, 498, 499
Birefringence 105, 108–111, 115–116,
    130
Bit length 16
Blocking property of film 90–91
Boltzmann linear superposition principle
    191, 194, 225, 226, 246
Brake button 27
Breaking stress 167
Bump geometries 374–376
Bump profile 426, 428
Burn-in 435

Calendering 72, 278–282
Cartesian coordinates 167–172
Censoring 500–501, 503–504, 513,
    528–531
Central clamping 341, 349
Centrifugal stresses 440
Chromium dioxide 8
Cinching 356, 357, 359
Circumferential stress. *See* Hoop stress
Coated tapes, viscoelastic behavior
    250–260, 331
Coating 70–76
Coating speeds 366, 390
Coating thickness 368, 369
Cobalt–iron oxide 8
Coercivity (Coercive force) 5, 8, 16, 37–
    38, 59, 150, 458–459
Composite tape 428
Compressible backcoat 41

Compressive creep data   276–282
   creep test apparatus   276–278
   experimental results   278–282
Computers   13
Conditional distributions   482–486
Continuous distributions   477–482
Continuous viscoelastic spectra   201
Creep   192–195, 212–218, 449–450
   test apparatus   212–218, 276–278
   see also Compressive creep data
Creep compliance   347, 348
Creep deformation   334
Creep strain   359
Creep tests   227–246, 359
   constitutive stress–strain relationships
         228–236
   time–humidity superposition   240–
         245
   time–temperature superposition
         189, 237–240
Cross-ply lamination   150, 151
Crowned rollers   412–414, 417
Crystalline materials   53–54, 211–212
Crystallinity, role in creep tests   273–
         274
Crystallization   99
Cumulative distribution function (c.d.f.)
         477–478, 496, 520
Cupping   370
Curie temperature   62
Curvilinear coordinates   172–174

Damping   363, 364
Data cartridges   37–42
   belt-driven   38–41
   high capacity   41–42
Data integrity   12
Data rates   21–22
Data storage systems   23
Data transfer rate   15
Data-processing tape drives   26–43
   instrumentation recorders   43
   mainframe computers   26–37
   mid-range computers   37–43
Decibels (dB)   18
Decoupler column   27, 30
Defect size distributions   431, 433, 436
Deformation kinetics   431
Demagnetization   11
Design considerations   15–19
   recording density   15–16
   reproduced signal amplitude   16–18
   signal-to-noise ratio   18–19

Deviatoric stress   178–179, 184
Differential scanning calorimetric (DSC)
         thermograms   265
Differential susceptibility   4
Digital signals   12
Discrete distributions   496–500
Disk deformation   439–450
   analysis   439–441
      centrifugal stresses and displace-
         ments   440–441
      hygroscopic expansion   439–440
      shrinkage   440
      thermal expansion   439
   measurements   441–450
      description of apparatuses   442–
         445
      experimental results   446–450
Disk drives   21
Diskettes. *See* Flexible disks
DMA (dynamic mechanical analysis)
         206–207, 208
DMTA (dynamic mechanical thermal
         analyzer)   204–205
Dynamic compliance   186–189
Dynamic modulus   186–189
Dynamic modulus data   204–209
   experimental results   207–209
   measurement techniques   204–207
Dynamic range   19

Elastic displacements   440
Elastic droop   395, 398–403
Elastic limit   166
Elastic modulus   121–128
Elastic–viscoelastic correspondence
         principle   345
Elasticity   166–181
   generalized Hooke's Law   167–174
   material constants for isotropic
         material   178–181
   material constants for orthotropic
         material   174–178
Electro/electroless plated media   63, 76,
         79
Electromagnetic induction   8
Electromagnets   6
Electroplated cobalt   12
Entrapped air pockets   368–382
Environmental requirements   67–70
Environmental stresses   310–314
Equilibrium recoverable shear com-
         pliance   193
Erase heads   15

Error-correction code   12
Estimator   507

Ferrites   7, 54–58
  polycrystalline   54–56
Fibers   156–157
Field lines   3
Filaments   156–157
Finite-element model   317–322
Flanges   422–423
Flexible disks (Floppy disks or
       Diskettes),   1, 15, 21, 43, 74
  clamping   46, 48
  diameters   46
  environmental requirements   69
  interchangeability   46
  jackets   64
  long-term reliability   438–450
    analysis of disk deformation   439–441
    measurements of disk deformation   441–450
  manufacture   74
  speed stability   45
  stress analysis   331–350
    elasticity solution   332–344
      annular disk: fixed inner boundary   335–339
      approximate scheme for annular disk: $a/b \ll 1$ 339–341
      estimates based on incomplete knowledge   341–344
      orthotropic solid disk   333–335
    viscoelasticity solution   347–350
      estimates for large times   347–349
      solution   345–347
Flexible media materials   58–67
  base film   59
  magnetic medium   59–63
  magnetic thin films   64–67
  particulate magnetic coatings   63–64
Flexible-disk drives   43–51
  dual-sided   46
  single-sided   46
Flexural fatigue   117–118
Floppy disks. *See* Flexible disks
Flux lines   4, 9, 11
Flying height   22, 34, 404, 407, 416–417
Force sensing resistors   305
Form factor disk drives   45, 49, 50
Form factor flexible disks   47
Frequency modulation (FM)   13, 20

Fringe field   13
Frontcoated tapes   373
Frozen-in stresses   139

Gamma–ferric oxide   8
Geometrical distribution   497, 498, 499
Glassy material   165, 166

Handling property of film   90
Hard magnetic materials   6–7, 9, 60
Head efficiency factor   17
Head materials   51–58
  Alfenol alloys   53
  amorphous magnetic alloys   53–54
  examples of head constructions,   56–58
  expected life   58
  ferrites   54–56
  Mu-metal and Hy-Mu 800B   53
  permalloys   53
  Sendust alloys   53
Head–medium separation (Flying height)   22, 34
Head–tape separation   353, 354, 382, 387, 395
Heat setting (Crystallization)   99
Helical scanning   24, 25, 37, 39, 43
Herd–Johnson method   504
Hexagonal-close-packed (HCP) structure   62
High temperature heat set   260, 272–273
History of magnetic recording   19–22
  storage hierarchy   21–22
Hooke's Law   166, 167–174, 179, 181–182
Hoop stress (Circumferential stress)   295, 296, 299–305, 307, 309, 313, 325–328, 353–358, 375–377
Horizontal magnetic recording   9
Hub   417–423
  compliance   304, 307, 417
  deformation   368
  elasticity   300
  geometry   305, 314
  taper   418, 419–422
Humidity, effects of in PET films   227–246
  time–humidity superposition   240–245
Hy-Mu 800B   53

Hydrodynamic air film   25
Hydrolytic stability   147
Hydrostatic stress   178
Hygroscopic deformations of disks
        446–448
Hygroscopic expansion   140–143, 439–
        441, 448
    coefficient of   140–143, 147–150, 152,
        157, 310, 355, 418, 439
Hysteresis loop   5, 6, 7
Hysteresis loss   6

Incremental susceptibility   4
Infrared dichroism   110–111, 115–116
Initial stress field   296–297
Initial susceptibility   4
Inspection data   502, 532–536
Instantaneous failure rate (IFR)   431,
        479
Instantaneous speed variations (ISV)
        360–366
    loose wraps   354, 362–363
    tape vibrations   363–366
Instrumentation recorders   43
Intensity of magnetization   3, 5, 8
Interlayer friction   369, 374
Interlayer pressure. *See* Radial stress
Interlayer slip (ILS)   314, 354–360, 362,
        398
    estimate of periods between rewinds
        359
    forms of tape distortion   357–358
    ILS containment methods   358–359
    reel slippage model   358
Interval data   502, 535–536
Iron   8
Irreversible susceptibility   5
Isothermal experimental data   219–
        224
Isotropic material   178–181, 184
Isotropic polymer films   154–155

Jumbos   366, 369, 379, 380

Kaplan–Meier method   504–505
Kapton film   88, 155, 157–160
Kronecker delta   178

Lame constants   178
Laminates   150, 151–154, 156

Laptop computers   46
Lay-on roller   415
Life data analysis   476–544
    continuous distributions   477–495
        basic statistical concepts   477–482
        conditional distribution   482–486
        distribution of products which fail
            due to competing risks   486–
            489
        probability distributions   489–495
    discrete distributions   496–500
        basic statistical concepts   496–497
        probability distributions   497–500
    graphical representation and analysis
            of data   500–515
        comparisons for complete data
            514
        graphical representation of data
            502–506
        numerical representation of data
            506–514
        types of life data   500–502
    Weibull analysis of failure data
            515–544
        conditional Weibull distribution
            522–523
        distribution of products which fail
            524–525
        evaluation of Weibull parameters
            527–544
        graphical representation   520–
            522
        properties of Weibull distribution
            516–520
        quantiles for order statistics
            525–527
Linear actuators   45
Linear density   15–16
Linear flux density   16
Linear hygroscopic expansion, coefficient
        of   157
Linear thermal expansion, coefficient of
        155–157
Linear variable displacement transducers
        (LVDTs)   206, 213–216
Linear voltage differential transducer
        (LVDT)   276
Loopers   6
Loose wraps   354, 361, 362–363
Loss tangent   187

Machine reels   366–369
Magnetic coatings   125

calendered versus uncalendered
278–282
cast films   278–279
thickness   278
viscoelastic behavior   250–260
Magnetic field   1–2
Magnetic flux reversal   13
Magnetic hysteresis   8
Magnetic medium   59–63
Magnetic moment per unit volume   3
Magnetic recording   8–10
Magnetic storage systems   19–70
  examples of modern storage systems
        using flexible media   22–51
    flexible-disk drives   43–51
    tape drives   22
  flexible media materials   58–67
    base film   59
    magnetic medium   59–63
    magnetic thin films   64–67
    particulate magnetic coatings   63–
      64
  functional requirements   67–70
  head materials   51–58
    Alfenol alloys   53
    amorphous magnetic alloys   53–54
    examples of head constructions
      56–58
    ferrites   54–56
    Mu-metal and Hy-Mu 800B   53
    permalloys   53
    Sendust alloys   53
  history of magnetic recording   19–22
    storage hierarchy   21–22
Magnetic substances   6
Magnetic susceptibility   3, 7
Magnetic tapes, long-term reliability
    353–437
  design of tape reels   417–423
  instantaneous speed variations (ISV)
    360–366
    loose wraps   362–363
    tape vibrations   363–366
  interlayer slip (ILS)   355–360
    estimate of periods between
        rewinds   359
    forms of tape distortion   357–358
    ILS containment methods   358–
      359
    reel slippage model   358
  mechanical print-through   390–393
  staggered wraps   393–416
    effect of winding parameters and
        storage conditions   398–406

errors caused by staggered tape
    406–411
    prevention of tape stagger   411–
      416
    sources of tape stagger   397–398
  uneven tape-stack profile,   366–390
    effect of uneven tape stack on data
        reliability   382–390
    sources and methods of preventing
        distortions   367–382
Magnetic thin films   62, 64–67
Magnetism   1–8
Magnetization curve   4–8
Magnetoresistive (MR) type read heads
    10, 57
Mainframe computers   20, 21, 26–37
Mandrel test   254–256, 291
Manufacturing processes of flexible
        magnetic media   70–79
  particulate media   70–74
    flexible disks   74
    tapes   70–74
  thin-film media   74–79
    electro/electroless plated media
      79
    metal evaporated media   77–78
    sputtered media   78–79
Marquardt algorithm   221, 223
Maximum likelihood (ML) methods
    506, 513–514, 528, 536, 540–
    543
Maximum susceptibility   5
Maxwell fluid   197
Maxwell models   196–197, 199–201,
    221, 222
Mean-bits-to-failure (MBTF)   384
Mean-time-between-failure (MTBF)
    481, 544
Mean-time-to-failure (MTTF)   481
Measurement techniques   204–207
  DMA   206–207, 208
  DMTA   204–205
Mechanical print-through   354, 390–
    393
Metal evaporated (ME) media   25, 76–
    78
Metal particles   62
Methyl ethyl ketone (MEK) solvent
    120–121, 122
Mid-range computers   37–43
Modified (delay) frequency modulation
    (MFM)   13
Moire grating   406
Mu-metal   7, 53

Multilayered tape
  flexural and tensile stress relaxation
      287–291
  thermal curling in elastic regime
      291–294
Multitrack heads   24
Mylar A film   85, 90, 100–102, 105–108
  annealing   120
  compressive creep compliance   278,
      279, 281, 282
  deformation   440, 441, 445, 446, 447
  effects of temperature   207, 209, 210,
      212, 260, 263, 265
  elastic modulus in thickness direction
      122, 123
  hydrolytic stability   148, 150
  hygroscopic expansion   140–143
  in-plane mechanical properties   111–
      117
  laminates   151
  mandrel test   258, 259
  relaxation moduli   319
  shrinkage   145, 146, 448, 449
  tensile creep compliance   226–234,
      239, 242, 243, 249, 251, 267, 269
  tensile relaxation modulus   235–239,
      252
  thermal expansion   129–133, 135–
      138
Mylar D film   85

Narrow-band–signal-to-noise ratio
      (NB–SNR)   463–464
Negative saturation magnetization   5
Newton's Law   181–182
Nickel–iron composites   7
Nonisothermal viscoelasticity   212, 224
Nonrecoverable (permanent) deforma-
      tion   193
Nonreturn-to-zero (NRZ) encoding   13
Nonreturn-to-zero-inverted (NRZI)
      encoding   13, 14
Notebook computers   46, 47

Offset wraps   403–406, 417
Optical disk drives   21
Oriented PET films   155–156
Orthotropic material   174–178, 184,
      309, 371
Orthotropic solid disk   333–335
Out-of-plane distortions (OPD)   354,
      456

Overcoatings   65, 67, 78
Oxidation treatments   65

Pack geometry   305, 306
Pancakes, 366–369, 376, 380, 381
Particulate magnetic coatings   59, 63–
      64
  size distributions   91
Particulate media   58, 70–74
  environmental requirements   69
  flexible disks   74
  tapes   69, 70–74
Permalloys   7, 53, 56, 58
Phase encoding (PE)   13
Physics of magnetic recording   1–19
  basic principle   1–8
      electromagnetic induction   8
      magnetic recording   8–10
      magnetism   1–8
  design considerations   15–19
      reproduced signal amplitude   16–
          18
      signal-to-noise ratio   18–19
  signal-processing methods   12–14
  vertical recording   10–12
Plane strain   175–176
Plane stress   175–177, 299
Poisson distribution   497, 498, 499
Poisson ratios   232
Poly(ethylene terephthalate) (PET) films
      59
  anisotropy   86, 88, 101, 149, 155
  commercial advantages   85
  crystallinity   86, 89–90, 94–95, 108,
      130, 148, 155, 211–212, 260,
      269–274
      tensilized film   265–266
  crystallization   94, 99, 139
  manufacturers   87
  manufacturing process   88–93
  outlook for improved substrates
      149–160
      coefficient of hygroscopic expansion
          157
      coefficient of thermal expansion
          154–157
          incorporation of fibers and fila-
              ments   156
          isotropic polymer films   154–155
          laminates   156
          oriented PET films   155–156
      high temperature polyimide sub-
          strate   158–160

long-term dimensional stability
    (shrinkage) 158
mechanical properties   150–154
physical and chemical properties
    104–149
  density 105, 108
  elastic modulus in thickness direc-
    tion 121–123
  hydrolytic stability 147–148
  hygroscopic expansion properties
    140–145
    effect of annealing 143–145
  in-plane mechanical properties
    111–118
    effect of annealing 119
    effect of solvents 119–121
    effect of temperature and strain
      rate 118
  long-term dimensional stability
    (shrinkage) 145–147
    effect of annealing 147
  radial elastic modulus of wound
    reels 123–128
    effect of magnetic coating 125
    effect of storage 125–126
    effect of winding parameters 125
    radial relaxation modulus 126–
    128
  refractive index, birefringence and
    infrared dichroism 105–111
  thermal expansion properties
    128–131
    effect of annealing 132–134
    heating rate effects 131–132
    mechanisms of thermal expansion
    134–140
shrinkage   131–139, 145–146, 149,
    152, 158, 260–263
stress–strain characteristics 85
structure 93–103
  commercial biaxially-oriented PET
    (Mylar A) 100–102
  heat setting (or crystallization) 99
  one-way stretching 95–97
  post-stretching 100
  strain relaxation (or annealing)
    100
  two-way stretching 97–99
thicknesses 93, 225–226
Polyimide film   88, 150, 158–160
Polymethyl methacrylate (PMMA)   155
Polypyromellitimide 154
Popping 361
Pressure sensor 429, 430

Principal optical axis 104
Probability distributions 489–495,
    497–500
Prony series   223, 228, 253, 317, 318,
    322, 331, 432
Pull tabs 305
Pure shear   185, 186, 191–194, 198

Quantal-response data   502, 532–535

Radial elastic modulus 123–128
Radial relaxation modulus 126–128
Radial stress (Interlayer pressure) 295–
    299, 305–309, 313, 325, 328,
    353–355, 376, 377
Random access memory accounting
    machine (RAMAC) 20
Read and write heads 9, 15
Reel geometry 305–310
Reel slippage 358
Refractive index   105, 109–111
Relative permeability 4
Relative susceptibility 3
Relaxation experiments 218–219
Relaxation matrix 317–318
Remanence 5, 8, 460
Remanent magnetization 5
Reproduced signal amplitude   16–18
Residual magnetization 5
Reversible susceptibility 4
Rewinds 359–360
Rigid disks 21
Ritz method 319
RLL code 13
Rotary actuators 45
Rotating head technology   20, 43
Rubbery polymers   165, 166

Safety 452, 454
Saturation magnetization 4
Scanning-laser-disk deformation
    measurement apparatus   443–
    445
Scratches 368, 370
Self-demagnetization 11
Semiconductor memory devices 21
Sendust alloys   7, 53, 54, 56
Separation loss 18
Series system 486
Servo techniques   15, 37, 41, 44, 48, 438,
    441

Shear cut slitting 73
Shear loss modulus 186, 187
Shear storage modulus 186, 187
Shear stress 298
Shear waves 365, 366
Shrinkage
  of disks 448–449
  of film 131–139, 145–146, 149, 152,
      158, 214, 260–263, 440, 441
Signal degradation 13, 57
Signal-processing methods 12–14
  design considerations 15
    recording density 15–16
Signal-to-noise ratio (SNR) 15, 18–
      19
Single-pole heads 10
Slip property of film 90–91
Soft magnetic materials 6–7
Solenoids 2–3
Solid-state pressure transducer 34
Solvents 117, 119–121, 122
Specific magnetization 8
Spoking 356, 357, 359
Sputtered media 78–79
Sputtering 56–57, 62, 75, 78
Squareness ratio (Orientation ratio)
      17, 71
Stack slip 398
Staggered wraps 354, 393–417
  effect of winding parameters and
      storage conditions, 398–406
  errors caused by staggered tape 406–
      411
  prevention of tape stagger 411–417
  sources of tape stagger 397–398
Stepping motors 45
Storage of flexible media 164, 398–406,
      417, 418
  humidity 353, 355, 453
  temperature 353, 355, 360, 453
Storage hierarchy 21–22
Strain history 182–184
Stress analysis of flexible media 295–
      350
  flexible disks 331–350
    elasticity solution 332–333
      annular disk: fixed inner bound-
        ary 335–339
      approximate scheme for annular
        disk: $a/b \ll 1$ 339–341
      estimates based on incomplete
        knowledge of elastic com-
        pliances 341–344
      orthotropic solid disk 333–335
      viscoelasticity solution 347–350

  estimates for large times 347–
      349
  solution 345–347
wound magnetic tape reels 295–331
  initial stress field 296–314
    analytical techniques 298–305
    environmental stresses 310–314
    measurement of radial stresses in
      wound tape reel 305
    role of reel geometry and winding
      parameters 305–310
  stress relaxation 314–331
    axisymmetric finite-element
      model 318–322
    constitutive relationships 315–
      317
    numerical results 323–328
    relaxation matrix from experi-
      mental measurements 317–
      318
Stress distribution 297, 302, 317, 332,
      357, 370
Stress field 295
Stress history 182–184
Stress relaxation 189–192, 314–315,
      354, 363, 370, 428
Stretched surface recording (SSR) 51,
      88
Stroboscopic disk deformation measure-
      ment apparatus 442–443
Surface roughness of film 91–92

Tape buffing 75
Tape cupping 398, 422, 456
Tape curvature 398, 417
Tape drives 21, 22–43
  audio tape drives 22–24
  data-processing tape drives 26–43
  video tape drives 24–25
Tape guiding 411, 414, 416
Tape reels 417–423
Tape stack 354, 356, 414
Tape stagger 397–398, 411–416
Tape substrate 427–428
Tape tension 27, 34, 41, 420
Tape velocity 34, 37, 41
Tape vibrations 363–366
Telescoping 357–358
Temperature, effects of in PET films
      227–246
  time–temperature superposition
      189, 237–240
Tensile relaxation and creep data 209–
      276

analysis of isothermal experimental
data 219–224
experimental results 224–276
effect of PET film thickness 225–226
effects of temperature and humidity
in PET films 227–246
effects of thermal treatment of PET
films 260–276
linearity of PET material response
224–225
thermoviscoelastic behavior of PET
films 246–250
viscoelastic behavior of coated tapes
and magnetic coatings 250–260
multilayered tape 289–291
test apparatuses 212–219
Tensilized film 90, 263–265
Tension gradient measurement 429–431
Tension ridges 307, 368–370, 374, 379, 382
Tension-to-flatten (TTF) 367, 384–387, 425–426
Tension transducer 33
Test methods for magnetic tapes and
tape reels 452–474
care and handling of tape cartridge
and tape drives 470–472
cartridge quality 469–470
environment and safety 452–454
mechanical specifications of the tape
cartridge 466–469
physical and magnetic characteristics
of tape 454–462
recording and track format requirements 472
tape magnetic-recording performance
462–464
tape quality 464–466
Thermal deformations of disks 446–448
Thermal expansion 439, 441
coefficient of 128–131, 134, 138, 147,
150–157, 314, 418, 439
and interlayer pressure 355
mechanisms 134–140
Thermal stress 310, 312
Thermal treatment of PET films 260–276
creep measurements 266–276
thermal effects 260–266
Thermomechanical analysis system
(TMS) 276, 277

Thermomechanical analyzer (TMA)
system 128, 132, 138
Thermoviscoelastic behavior of PET
films 246–250
Thin-film media 12, 27, 57, 59, 63, 74–79
advantages 59
coating methods 74–76
disadvantages 63
electro/electroless plated media 79
manufacture 74–77
metal evaporated media 77–78
sputtered media 78–79
Time-dependent deformation experiments 185–195
Toroidal geometry 5, 7
Total susceptibility 5
Track density 15
Track misregistration (TMR) 438
Transportation of tape cartridges 453
Transverse optical axis 104
Tunnel-erasing head 45, 46, 58

Ultimate tensile stress (UTS) 166
Undercoatings 65
Uneven tape-stack profile 354, 366–390
effect of uneven tape stack on data
reliability 382–390
examples 376–379
prevention 379–382
sources and methods of preventing
distortions 367–382

Vertical recording 10–12
Vibrating sample magnetometer (VSM)
6
Video recording 20
Video tape drives 24–25
Viscoelastic deformation 326–330, 354,
366, 384, 390, 403, 416, 432
nonuniform 370, 371
of offset region 410, 411
Viscoelasticity of PET substrate and
coated magnetic media 164–294
compressive creep data 276–282
creep test apparatus 276–278
experimental results 278–282
calendered versus uncalendered
coatings 279–281
magnetic coatings 278–279
recovery experiments 281–282

Viscoelasticity (*continued*)
  constitutive equation   182–185
  dynamic modulus data   204–209
    experimental results   207–209
    measurement techniques   204–207
  elasticity   166–181
    generalized Hooke's Law   167–174
    material constants for isotropic
      material   178–181
    material constants for orthotropic
      material   174–178
    mechanical model analogies of linear
      viscoelastic behavior   195–
      202
  multilayered tape   287–294
  tensile relaxation and creep data
      209–276
    analysis of isothermal experimental
      data   219–224
    experimental results   224–276
      effect of thickness   225–226
      linearity of response   224–225
      temperature and humidity   227–
      246
      thermal treatment   260–276
      thermoviscoelasticity   246–250
      viscoelastic behavior   250–260
    test apparatuses   212–219
  time (or frequency) and temperature
      effects   202–204
  time-dependent deformation experi-
      ments   185–195
  viscous liquids   181
Viscoplasticity   182
Viscous liquids   181
Voice-coil motor (VCM)   45
Voigt–Kelvin model   196, 197–201,
      221, 222
Voigt–Kelvin solid   199
Volumetric thermal expansion, coeffi-
      cient of   129, 130, 155

Wear resistant coatings   57
Weibull analysis of failure data   515–
      544
  conditional Weibull distribution
      522–523
  distribution of products which fail due
      to competing failure modes
      524–525
  evaluation of Weibull parameters
      527–544

general properties of Weibull distribu-
      tion   516–520
graphical representation of a Weibull
      distribution   520–522
quantiles for order statistics of Weibull
      samples   525–527
Weibull distribution   516–520, 522–
      523, 528–530, 542–543
Weibull paper   520–522, 530
Weibull parameters   433, 434, 437
Weibull population   520, 523, 526–527,
      533, 534
Winding parameters   305–310, 355, 370,
      398–406, 417
Winding speed   125
Winding stress   298
Winding tension   302, 307, 308, 404, 468
Windowing   356, 357
Wound magnetic tape reels   123–128,
      254
  initial stress field   296–314
    analytical techniques   298–305
    environmental stresses   310–314
    measurement of radial stresses in
      wound tape reel   305
    role of reel geometry and winding
      parameters   305–310
  stress relaxation   314–331
    axisymmetric finite-element model
      318–322
    constitutive relationships   315–
      317
    numerical results   323–328
    relaxation matrix from experimental
      measurements   317–318
Wound-in distortions   371–374
Writing   9

Yield-point stress   166
Young's modulus
  of amorphous films   95, 96
  anisotrophy ratio   116
  directional dependence   178
  of elasticity   86, 104, 105, 108, 109,
      148, 180
    in tension   111, 112, 114–119,
      124
  of laminated films   151–153
  major and minor axes   149
  of tensilized film   90
  in thickness direction   121, 123
  time-dependent   192

Dr. Bharat Bhushan is an internationally recognized expert in tribology, mechanics, and reliability of magnetic storage devices. He received an M.S. in mechanical engineering from the Massachusetts Institute of Technology, an M.S. in mechanics and a Ph.D. in mechanical engineering from the University of Colorado at Boulder, an M.B.A. from Rensselaer Polytechnic Institute at Troy, NY, and an D.Sc. (Tech.) from the University of Trondheim at Trondheim, Norway. He is a registered professional engineer (mechanical). He has authored or coauthored 3 technical books, more than 150 technical papers in referred journals, more than 60 technical reports, and 4 U.S. patents. He has delivered numerous lectures at national and international conferences and universities. He organized the first symposium on Tribology and Mechanics of Magnetic Storage Systems in 1984 and the first international symposium on Advances in Information Storage Systems in 1990 and continues to organize and chair the AISS symposia annually. He is editor-in-chief of the ASME series titled "Advances in Information Storage Systems" which he initiated. He has received more than a dozen awards for his contributions to science and technology from professional societies, industry, and U.S. government agencies. He is a fellow of ASME and the New York Academy of Sciences and is a senior member of IEEE.

Dr. Bhushan has worked for the Department of Mechanical Engineering at the Massachusetts Institute of Technology, Cambridge, MA; Automotive Specialists, Denver, CO; the R & D Division of Mechanical Technology Inc., Latham, NY; the Technology Services Division of SKF Industries Inc., King of Prussia, PA; the General Products Division Laboratory of the IBM Corporation, Tucson, AZ; the Almaden Research Center of the IBM Corporation, San Jose, CA, and the Department of Mechanical Engineering at the University of California, Berkeley, CA. He is presently an Ohio Eminent Scholar Professor in the Department of Mechanical Engineering at the Ohio State University, Columbus, OH.

He is married and has two children. His hobbies are music, photography, hiking, and traveling.